COMBUSTION MODELING IN RECIPROCATING ENGINES

PUBLISHED SYMPOSIA
Held at the
General Motors Research Laboratories
Warren, Michigan

Friction and Wear, 1959
Robert Davies, *Editor*

Internal Stresses and Fatigue in Metals, 1959
Gerald M. Rassweiler and William L. Grube, *Editors*

Theory of Traffic Flow, 1961
Robert Herman, *Editor*

Rolling Contact Phenomena, 1962
Joseph B. Bidwell, *Editor*

Adhesion and Cohesion, 1962
Philip Weiss, *Editor*

Cavitation in Real Liquids, 1964
Robert Davies, *Editor*

Liquids: Structure, Properties, Solid Interactions, 1965
Thomas J. Hughel, *Editor*

Approximation of Functions, 1965
Henry L. Garabedian, *Editor*

Fluid Mechanics of Internal Flow, 1967
Gino Sovran, *Editor*

Ferroelectricity, 1967
Edward F. Weller, *Editor*

Interface Conversion for Polymer Coatings, 1968
Philip Weiss and G. Dale Cheever, *Editors*

Associative Information Techniques, 1971
Edwin L. Jacks, *Editor*

Chemical Reactions in Urban Atmospheres, 1971
Charles S. Tuesday, *Editor*

The Physics of Opto-Electronic Materials, 1971
Walter A. Albers, Jr., *Editor*

Emissions From Continuous Combustion Systems, 1972
Walter Cornelius and William G. Agnew, *Editors*

Human Impact Response, Measurement and Simulation, 1973
William F. King and Harold J. Mertz, *Editors*

The Physics of Tire Traction, Theory and Experiment, 1974
Donald F. Hays and Alan L. Browne, *Editors*

The Catalytic Chemistry of Nitrogen Oxides, 1975
Richard L. Klimisch and John G. Larson, *Editors*

Future Automotive Fuels—Prospects, Performance, Perspective, 1977
Joseph M. Colucci and Nicholas E. Gallopoulos, *Editors*

Aerodynamic Drag Mechanisms of Bluff Bodies and Road Vehicles, 1978
Gino Sovran, Thomas Morel and William T. Mason, Jr., *Editors*

Mechanics of Sheet Metal Forming—Material Behavior and Deformation Analysis, 1978
Donald P. Koistinen and Neng-Ming Wang, *Editors*

Computer Vision and Sensor-Based Robots, 1979
George G. Dodd and Lothar Rossol, *Editors*

Combustion Modeling in Reciprocating Engines, 1980
James N. Mattavi and Charles A. Amann, *Editors*

COMBUSTION MODELING IN RECIPROCATING ENGINES

Edited by

JAMES N. MATTAVI and CHARLES A. AMANN

General Motors Research Laboratories

Springer Science+Business Media, LLC 1980

Library of Congress Cataloging in Publication Data

Symposium on Combustion Modeling in Reciprocating Engines, General Motors Research Laboratories, 1978.
Combustion modeling in reciprocating engines.

(General Motors symposia series)
Includes index.
1. Internal combustion engines – Combustion – Mathematical models – Congresses. I. Mattavi, James N. II. Amann, Charles A. III. General Motors Corporation. Research Laboratories. IV. Title
TJ756.S95 1978 621.43'4 80-10451
ISBN 978-1-4899-5300-1

Proceedings of the Symposium on Combustion Modeling in Reciprocating Engines, held at the General Motors Research Laboratories, Warren, Michigan, November 6 and 7, 1978

ISBN 978-1-4899-5300-1 ISBN 978-1-4899-5298-1 (eBook)
DOI 10.1007/978-1-4899-5298-1

Originally published by Plenum Press, New York in 1980
Softcover reprint of the hardcover 1st edition 1980

PREFACE

This book contains the Proceedings of the Symposium on COMBUSTION MODELING IN RECIPROCATING ENGINES held at the General Motors Research Laboratories on November 6 and 7, 1978. The Symposium was the twenty-third in an annual series presented by the Research Laboratories. Each Symposium has covered a different technical topic, selected as timely and of vital interest to General Motors as well as to the technical community at large.

In view of the importance of the internal combustion engine to our transportation sector and the pressing need to enhance our understanding of combustion in engines, the appropriateness of the 1978 Symposium topic was well established. Of ultimate interest is the development of combustion models which can provide direction for new and improved engine designs providing efficient energy utilization within environmental and driveability constraints. In this context the Symposium provided a forum for key individuals from the automotive industry, university research groups, independent research institutions, and government laboratories engaged in combustion modeling to exchange ideas and experiences and to identify directions for future research.

The Symposium was organized into four closely related technical sessions, preceded by an overview of combustion modeling. Subjects of the four sessions were (i) modeling of fluid motions in engines, (ii) modeling flame propagation and heat release in engines, (iii) modeling of engine exhaust emissions, and (iv) applications of engine combustion models.

Each session was chaired by a recognized authority in the subject area of that session whose background extended beyond modeling. To initiate each session and to set the stage for the modeling papers which followed, these session chairmen briefly reviewed the state of collective knowledge in their subject areas. Supplementing the presentations on the formulation of combustion models were two presentations which dealt with experimental work useful in testing the validity of the models. As a further step to avoid focusing too narrowly on the techniques of model *formulation,* a number of people active in engine research and development who are not modelers themselves, but rather potential *users* of models, were also invited to attend the Symposium. Regarding overall attendance, the Symposium was truly international, with participants coming from Australia, Canada, England, France, Germany, Italy, Japan, Sweden, Switzerland and the United States.

Time was allotted during the Symposium for informal discussions of the papers and related investigations by the attendees. These discussions were recorded on tape and subsequently transcribed and edited lightly for brevity and clarity. In addition to these recorded discussions, numerous discussions and interactions between attendees took place at the many social functions included in the total Symposium format. To conclude the Symposium, overall summary comments of the proceedings were presented by one of the Symposium cochairmen.

This Symposium could not have been held or these Proceedings published without the valuable assistance of many people to whom we are greatly indebted: Professors G. L. Borman, J. B. Heywood and W. C. Reynolds, Dr. R. B. Krieger and Dr. H. K. Newhall for assistance in organizing the technical program and acting as session chairmen; Dr. P. F. Chenea, Mr. J. B. Bidwell and Mr. J. D. Caplan for authorizing and supporting the Symposium; Dr. W. G. Agnew for his assistance in planning the Symposium; Mr. R. T. Beaman for efficiently handling the many details involving the physical arrangement of the Symposium; Mr. D. N. Havelock for assistance in editing the Symposium volume; Mses. M. D. Chowning, C. Dzedzie, D. A. Kenney, K. E. Russell and L. D. Sroda for participating variously in the correspondence, preparation of manuscripts, transcription of tapes, and reception of guests at the technical and social functions; Dr. E. G. Groff and Mr. C. W. Cooper for preparing the subject index of the Proceedings; and our many colleagues in the Engine Research Department for providing valuable assistance in conducting the Symposium and preparing the Proceedings.

James N. Mattavi and Charles A. Amann

CONTENTS

ENGINE COMBUSTION MODELING— AN OVERVIEW

J. B. HEYWOOD

Massachusetts Institute of Technology, Cambridge, Massachusetts

ABSTRACT

This paper presents a broad introduction to the field of engine combustion modeling. The past several years have seen a substantial growth in mathematical modeling activities whose interests are to describe the performance, efficiency and emissions characteristics of various types of internal combustion engines. The key element in these simulations of various aspects of engine operation is the model of the engine combustion process. Descriptions of the combustion process in different types of engine are presented to provide an experimental basis for the discussion. Photographs from high-speed movies of combustion are reviewed to illustrate the major features of the process being modeled. Current combustion modeling activities are then placed in the context of the historical development of engine cycle simulations. The relationship of engine performance and efficiency, engine emissions characteristics and the problem of detonation to the combustion process and the overall engine cycle are indicated.

Combustion models are then classified into three categories: zero-dimensional, quasi-dimensional and multidimensional models. Zero-dimensional models are built around the First Law of Thermodynamics, and time is the only independent variable. The rate of burning of the charge is obtained empirically. In quasi-dimensional models, the rate of burning is derived from a physical sub-model of the turbulent combustion process. In the multidimensional models mass-, momentum-, energy- and species-conservation equations are solved numerically in either one or two dimensions to follow the propagation of the flame through the engine combustion chamber. Key requirements in these multidimensional models are adequate sub-models of the turbulent transport coefficients and kinetic relations for the fuel-oxidation process.

Finally, examples of problems which have been examined with these three types of combustion model are described. It is shown that the zero-dimensional and quasi-dimensional models are already useful in parametric studies of the effects of changes in engine operating variables (and some design variables) on engine power, efficiency and emissions. The multidimensional models are proving most useful in examining problems characterized by the need for detailed spatial information and complex interactions of many phenomena simultaneously.

References pp. 33-35.

NOTATION

a	constant
$(A/F)_0$	overall air-fuel ratio
$(A/F)_{p,soc}$	air-fuel ratio in prechamber at start of combustion
A_f	flame front area
b	constant
B	cylinder bore
B_k	pre-exponential factor in Arrhenius rate constant for k^{th} reaction
c	constant
c_{p_i}	specific heat at constant pressure of ith species
D_{ij}	binary diffusion coefficient for species i and j
e	specific internal energy
E_k	activation energy in Arrhenius rate constant for k^{th} reaction
F_T/V_p	ratio of prechamber orifice area to prechamber volume
h_i	specific enthalpy of ith species
k	viscosity coefficient
K	constant; also number of chemical reactions
L	integral length scale of turbulence
m	constant in Wiebe function
m_b	mass burned
m_e	mass entrained in turbulent flame front
M_i	symbol for species i
N	number of species
NOx	oxides of nitrogen
Nu	Nusselt number
p	pressure

$\underline{p}$	pressure tensor
Pr	Prandlt number
$\underline{q}$	heat-flux vector
$\dot{Q}$	heat-transfer rate
R	gas constant
Re	Reynolds number
S_L	laminar flame speed
S_T	turbulent flame speed
t	time
T	temperature; subscripts a, b and u denote adiabatic core gas, burned gas and unburned gas temperature, respectively
u	velocity component
$\underline{u}$	velocity vector; the superscript T in Eq. 11, Table 1 denotes the transpose of the tensor
u_e	entrainment speed of unburned mixture into flame front
$\underline{u}_i$	diffusion velocity vector
v	velocity component
V	cylinder volume
V_p/V_c	ratio of prechamber volume to clearance volume
w_i	source term for the ith species
W_i	molecular weight of ith species
$\dot{W}$	work transfer rate
x	mass fraction burned
X_i	mole fraction of ith species
Y_i	mass fraction of ith species
α	constant
Δx	grid size

References pp. 33-35.

δ_T	thermal boundary-layer thickness
θ	crank angle
$\Delta\theta_{cm}$	combustion duration in main chamber
θ_s, θ_o, $\Delta\theta_{id}$, $\Delta\theta_b$	defined in Fig. 8
λ	thermal conductivity of gas mixture
μ	viscosity of gas mixture
μ_t	turbulent viscosity
ν_i	stoichiometric coefficient of ith species
ρ	gas density
ρ_u	unburned mixture density
$\underline{\underline{\hat{\sigma}}}$	viscous shear tensor
τ	characteristic reaction time
ϕ	fuel-air equivalence ratio

INTRODUCTION

Internal combustion engines—conventional spark-ignition, stratified-charge, diesel and rotary—are in the midst of a period of intensive research and development activity. Environmental and energy-conservation constraints are forcing rapid changes in long established design practices and market strategies. These new and changing requirements have greatly increased the complexity of engineering engine design to give optimum operating characteristics within the limitations of emissions and fuel-economy standards and fuel octane levels. One of the tools that is becoming increasingly important in meeting this challenge is the mathematical model. The level of effort devoted to internal combustion (IC) engine modeling and the capabilities of the resulting models have increased dramatically over the last ten years. This is a result of the simultaneous growth in the recognition of the power of mathematical models and the availability of ever-increasing computer capability. By mathematical model in the IC engine context we mean a set of equations governing the fluid mechanic and thermodynamic behavior of the engine working fluid as it passes through the cylinder of an operating engine, which are solved numerically on a computer.

The internal combustion engine is unique among heat engines in that the combustion process couples directly with the primary engine operating characteristics, power and efficiency, as well as emissions. Thus, the combustion model is one of the key components in any simulation of the complete IC engine operating cycle. At the same time, the details of all aspects of the operating cycle, the intake, compression, expansion and exhaust strokes, directly influence the combustion process.

The engine combustion process is exceedingly complex. Even in the conventional spark-ignition engine, where under many operating modes the fuel and air can be treated as premixed, the combustion process is initiated in a three-dimensional time-varying turbulent flow, concerns a fuel which is a blend of hundreds of different organic compounds whose combustion chemistry is poorly understood, and takes place in a space confined by the combustion chamber walls whose shape varies with time and whose walls directly influence aspects of the process. It is little wonder, given our inability to describe most of these processes in any exact sense, that combustion modeling is a controversial field!

Modeling activities of the type referred to above can make major contributions to engine developments in three general areas. These contributions come from different stages of any model-development process. They are: (i) the development of a more complete understanding of the important physical processes that emerges from the requirements of formulating the model; (ii) the identification of key controlling variables which provide guidelines for more rational and therefore less costly experimental programs, and for data reduction; and (iii) the ability to predict behavior over a wide range of design and operating variables which can be used to screen concepts prior to major hardware programs, to determine trends and trade-offs, and if the model is sufficiently accurate, to optimize design and control.

The three phases of model development which correspond to these contributions are: (i) model development through an analysis of the individual processes which are linked together in the engine operating cycle; (ii) the exploratory use of the model, its validation, and studies of sensitivity of model predictions to initial assumptions; and (iii) the use of the model in extensive parametric studies which examine the effects of changes in engine operating and design parameters on engine performance, efficiency and emissions.

My own view is that each of these three types of contribution is valuable and significant. Thus, even though the third and final goal for any particular problem may currently be unrealizeable, a modeling activity devoted to that problem which provides the gains described in (i) and (ii) above may still be very worthwhile.

The problems of interest in engine combustion modeling can be grouped under three headings: (i) engine performance and efficiency; (ii) engine emissions (oxides of nitrogen, carbon monoxide, unburned hydrocarbons, particulates, polycyclic aromatic hydrocarbons, aldehydes, odor and noise); and (iii) detonation. For a conventional spark-ignition engine, the relation of these problems to the combustion process is illustrated in Fig. 1. The flame-propagation process itself defines the thermodynamic state of the unburned and burned gas within the combustion chamber and thus determines the engine power and efficiency. Nitric oxide forms in the bulk burned-gas region, primarily behind the flame. Hydrocarbons originate in the quench regions adjacent to the wall. Carbon-monoxide

References pp. 33-35.

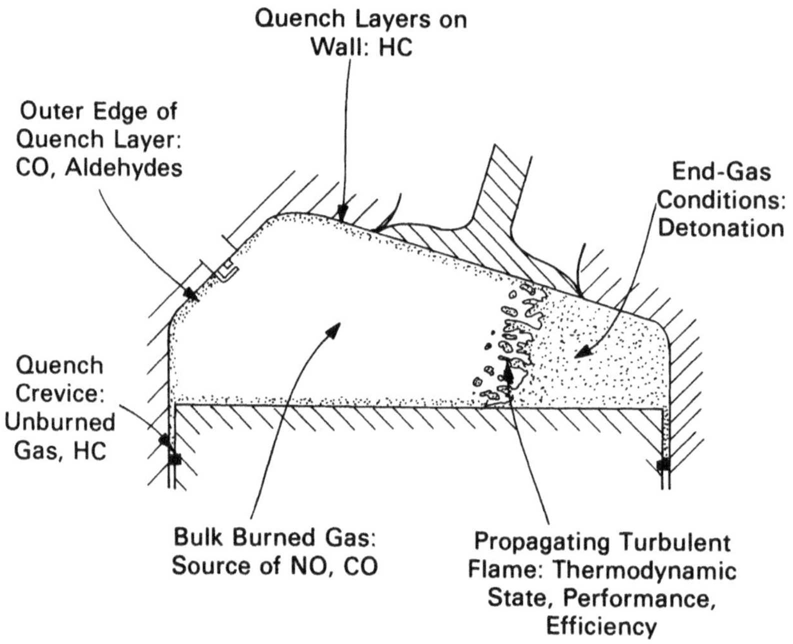

Fig. 1. Schematic of spark-ignition engine combustion process.

emissions are controlled by the bulk burned gas for fuel-rich mixtures, with the quench regions becoming increasingly important as the mixture is leaned out. Aldehydes originate at the outer edges of the quench regions. The end-gas conditions, along with the fuel characteristics, determine whether or not detonation occurs.

In fuel-injected engines, diesels and certain types of stratified-charge engines, the complexity of the combustion process is compounded by the non-uniform fuel distribution within the combustion chamber and the importance of the fuel-air mixing process. Fig. 2 illustrates how various parts of this type of combustion process affect the problems listed above. The development of the fuel jet as fuel evaporates and mixes with air controls the rate of heat release and therefore engine power and efficiency. Nitric oxide forms in the high-temperature burned-gas regions as before, but temperature and fuel-air ratio distributions within the burned-gas region are now nonuniform (a most important feature). Soot forms in the fuel-rich unburned zones within the flame region. Soot oxidizes in the flame zone, giving rise to the yellow luminous character of the flame. Hydrocarbons and aldehydes originate in regions where the flame quenches both on the walls and where excessive dilution with air prevents the combustion process from going to completion. Combustion-generated noise is controlled by the early part of the combustion process, the initial rapid heat release immediately following the ignition-delay period.

The events which take place during the engine cycle and the coupling between the combustion process and engine operating and emission characteristics are

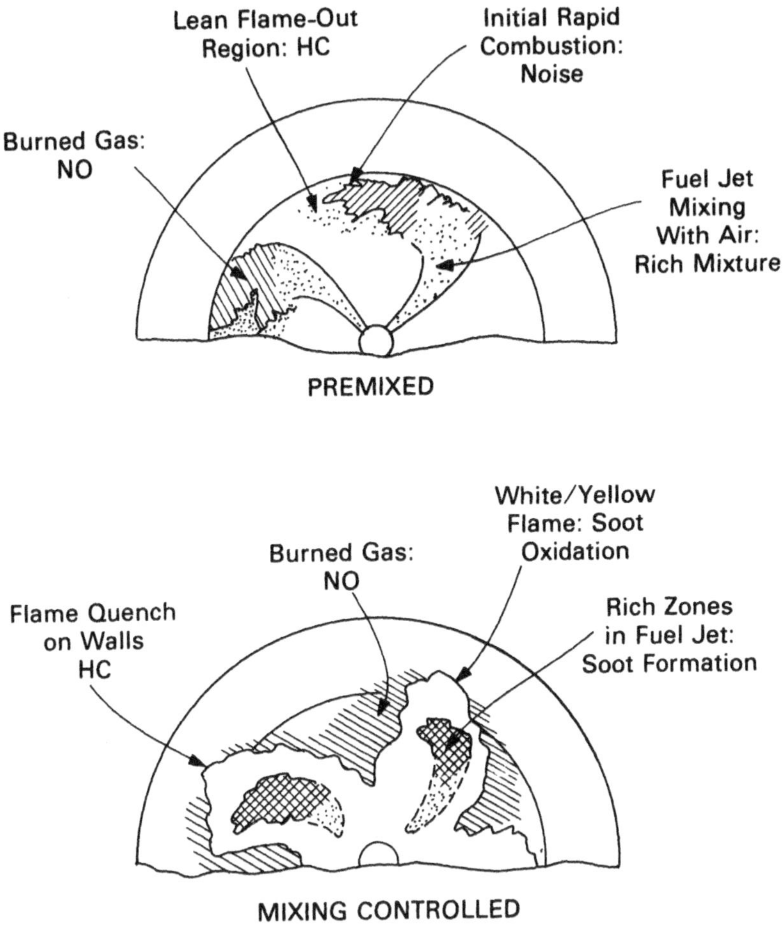

Fig. 2. Schematic of direct-injection compression-ignition engine combustion process during the initial "premixed" rapid combustion phase and subsequent "mixing controlled" slower combustion phases.

illustrated in Fig. 3, which shows the output of a spark-ignition engine cycle simulation at a mid-load and speed point. The top two graphs show details of the engine geometry—the cylinder volume and valve-lift profiles. Lower curves are mass flow rates through the inlet and exhaust valves, cylinder pressure, p, mass fraction burned, x, unburned-mixture temperature, T_u, mean burned-gas temperature, T_b, temperature of burned-gas adiabatic core, T_a, instantaneous heat-transfer rate, $\dot{Q}$ (normalized by the initial enthalpy of the fuel-air mixture within the cylinder), nitric-oxide concentration, NO, thermal boundary-layer thickness, δ_T (normalized by the cylinder bore B), all versus crank angle. Additional pertinent data are as follows: engine displacement, 5L; equivalence ratio, 1.0; speed, 1400 r/min; brake mean effective pressure (bmep), 328 kPa, and optimum combustion timing.

References pp. 33-35.

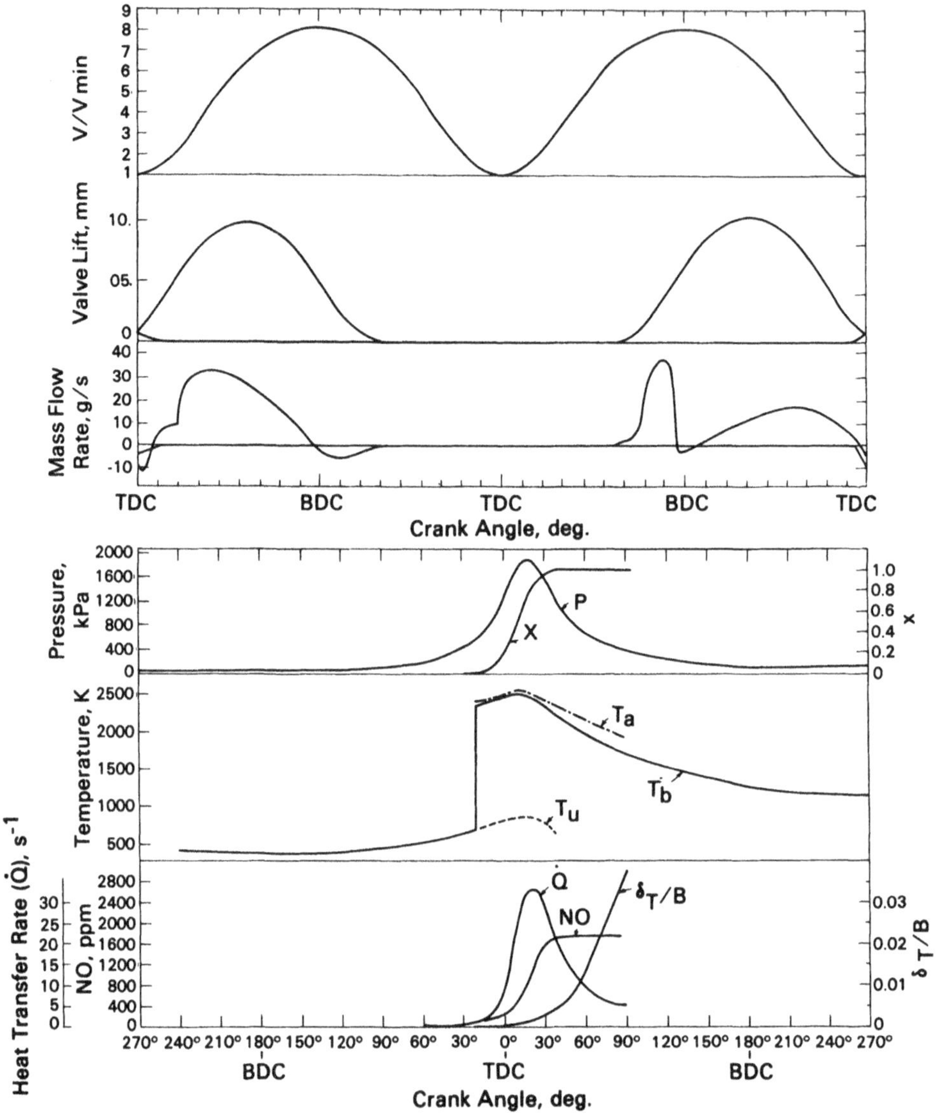

Fig. 3. Examples of output from a cycle-simulation calculation [1]. See text for details.

In this particular simulation the mass-fraction-burned curve—the combustion model—was an input to the calculation, and burned-gas temperatures were determined using a boundary-layer model for engine heat transfer. Because of heat transfer, the adiabatic burned-gas core shows a higher temperature than the mean burned-gas temperature. Burned-gas pressure and temperature were used as inputs to obtain the nitric oxide (*NO*) concentration through a parallel calculation based on the Zeldovich kinetic *NO* formation mechanism and equilibrium concentrations for the carbon-oxygen-hydrogen species in the burned gas [1]. The details of the combustion process control the pressure and temperature versus crank-angle curves, and thus the engine power, efficiency, fuel consumption, and emissions. For this example, these cycle average quantities were: indicated* mean effective pressure, 454 kPa; pumping mean effective pressure, 54 kPa; indicated specific fuel consumption, 232 g/(kW h); indicated efficiency, 35%; heat transfer, 29% of thermal input; indicated specific nitric oxide emissions, 7.8 g/(kW h).

ENGINE COMBUSTION PHOTOGRAPHS

Before we move on and focus exclusively on modeling, let us look at some photographs of the combustion process in various types of internal combustion engine. A good physical picture of the process to be modeled is most important. Fig. 4 shows photographs taken from high-speed color movies of the combustion process in various types of engine. In each case, the engine cylinder head has been modified to incorporate a transparent window, as shown.

Fig. 4a (Nakanishi et al. [2]) shows the shape and structure of the flame in an *L*-head spark-ignition engine shortly after ignition, at 6 crank-angle degrees before top dead center (6° BTDC), and close to the crank angle at which maximum cylinder pressure occurs (18° ATDC). The luminosity is radiated from the actual flame-reaction zone. The flame front is generally spherical in character and centered about the spark-plug location, contains smaller scale irregularities and possesses a smaller scale structure. Fig. 4b (Wong et al. [3]) shows two frames from a movie taken of combustion in a rapid compression machine set up to simulate the Texaco Controlled Combustion Process (TCCP) stratified-charge engine. The piston contains a deep cup of diameter about half the cylinder bore. The inlet air is given counter-clockwise swirl. The swirl rate is intensified as the air flows into the bowl. The first frame shows the fuel jet within the piston bowl prior to the start of combustion. Fuel is injected down into the rotating air flow within the piston cup. A non-uniform and rapidly varying fuel-air-ratio distribution is set up. The second frame shows the flame location about halfway through the heat-release process. The flame is stabilized in the wake of the spark plug. The burning gas region is being carried around by the swirling air flow and is entraining air as this transport process proceeds.

Fig. 4c (Alcock and Scott [4]) shows the flame in a small high-speed direct-injection diesel engine with counterclockwise swirl. In the first picture, at 3° BTDC, fuel injection is still taking place. Two of the fuel jets are visible. Combustion is also evident in the outer regions of the bowl in the piston as the earlier injected fuel which has "premixed" with air during the ignition delay period burns. The second frame, at 6° ATDC, shows the flame during the "mixing-controlled" combustion phase. An examination of a color version of this film frame revealed highly fuel-rich soot-formation zones, as indicated.

---

* *Indicated here means gross indicated quantities, i.e., evaluated over the compression and expansion strokes only.*

References pp. 33-35.

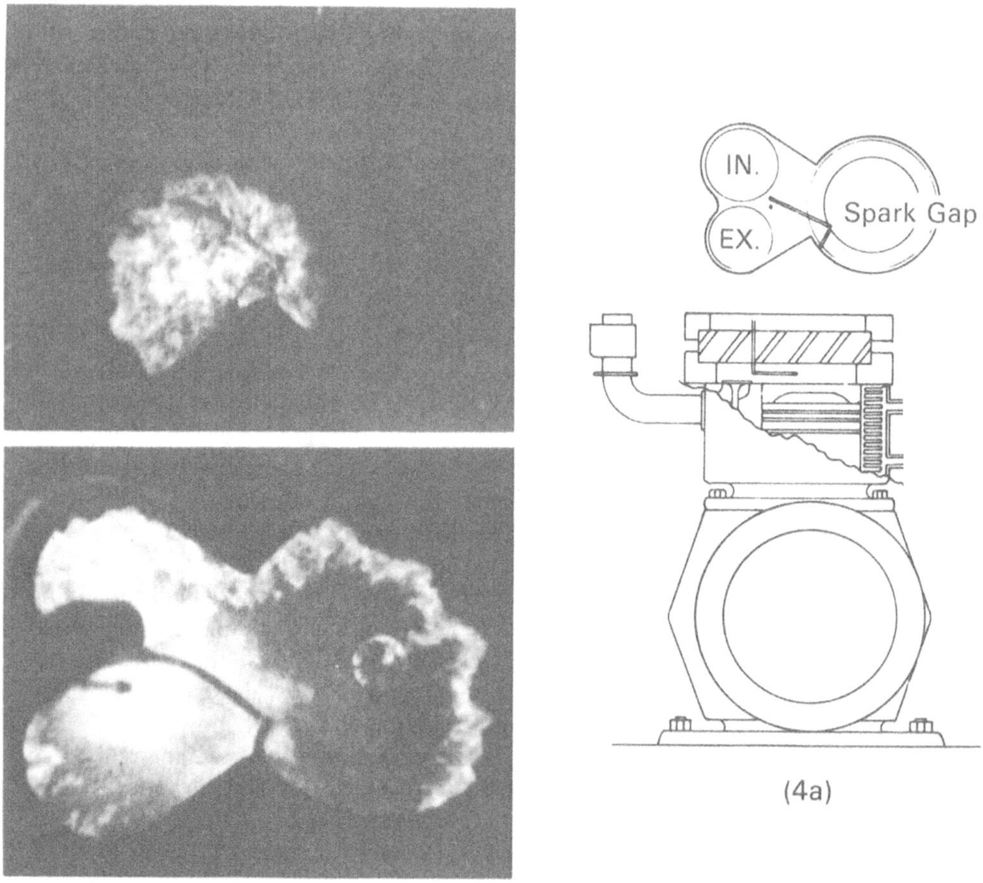

Fig. 4 (a-d). Photographs taken from high speed color movies of combustion process in different types of IC engine [2–4]. See text for details.

Fig. 4d (Alcock and Scott [4]) shows two frames from a movie of combustion in a Ricardo Comet V prechamber diesel. In the first of these, at 1° ATDC, combustion has started in the prechamber (here viewed along a line perpendicular to the cylinder axis), and a carbon-burning flame, surrounded by a leaner flame, follows the swirl-carried injected fuel. In the second frame, at 15° ATDC, the white carbon flame fills the prechamber. In the main chamber the burning jet issuing from the prechamber has impinged on the piston surface, interacted with the recesses in the piston crown and the cylinder wall opposite the nozzle, is entraining the excess air into the flame region, and shows (in the color version of this photograph) regions where soot is obviously present.

Even in the simplest case, the conventional spark-ignition engine, the detailed structure of the turbulent flame is not well understood. Some appreciation of the complexities can be obtained from schlieren photographs of the flame propagation across a spark-ignition engine combustion chamber. Fig. 5 (Iinuma and Iba [5])

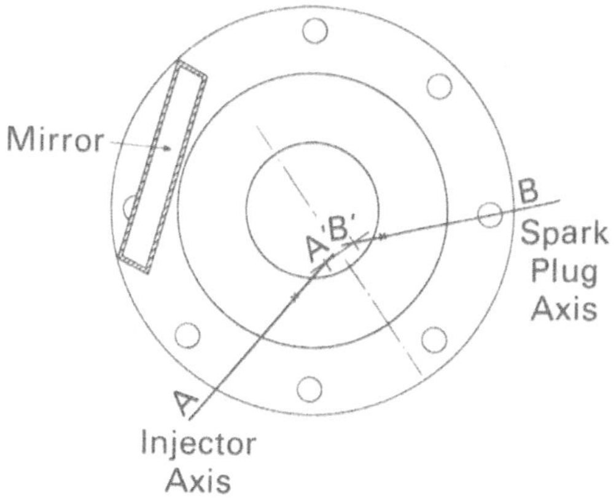

(4b)

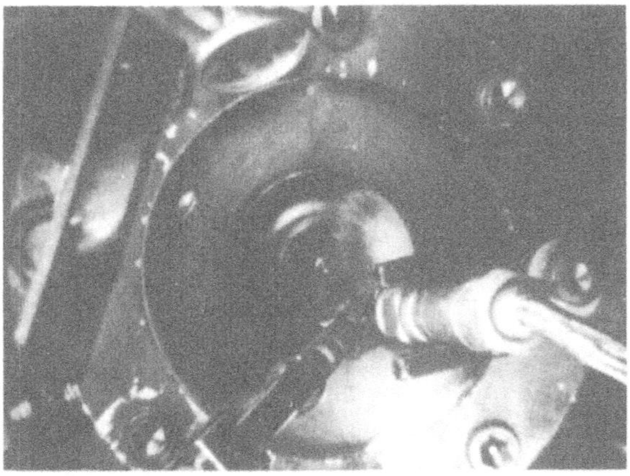

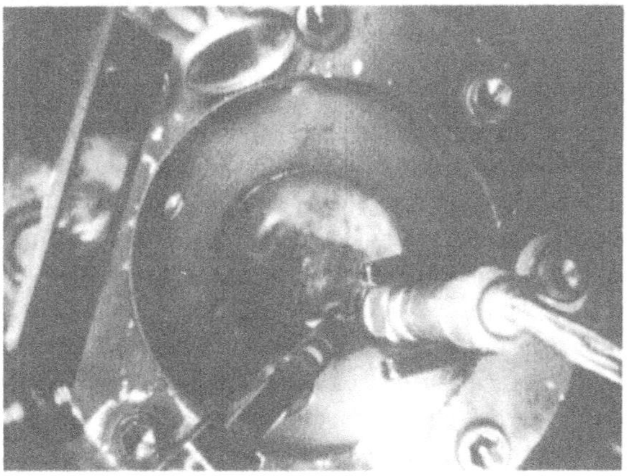

References pp. 33-35.

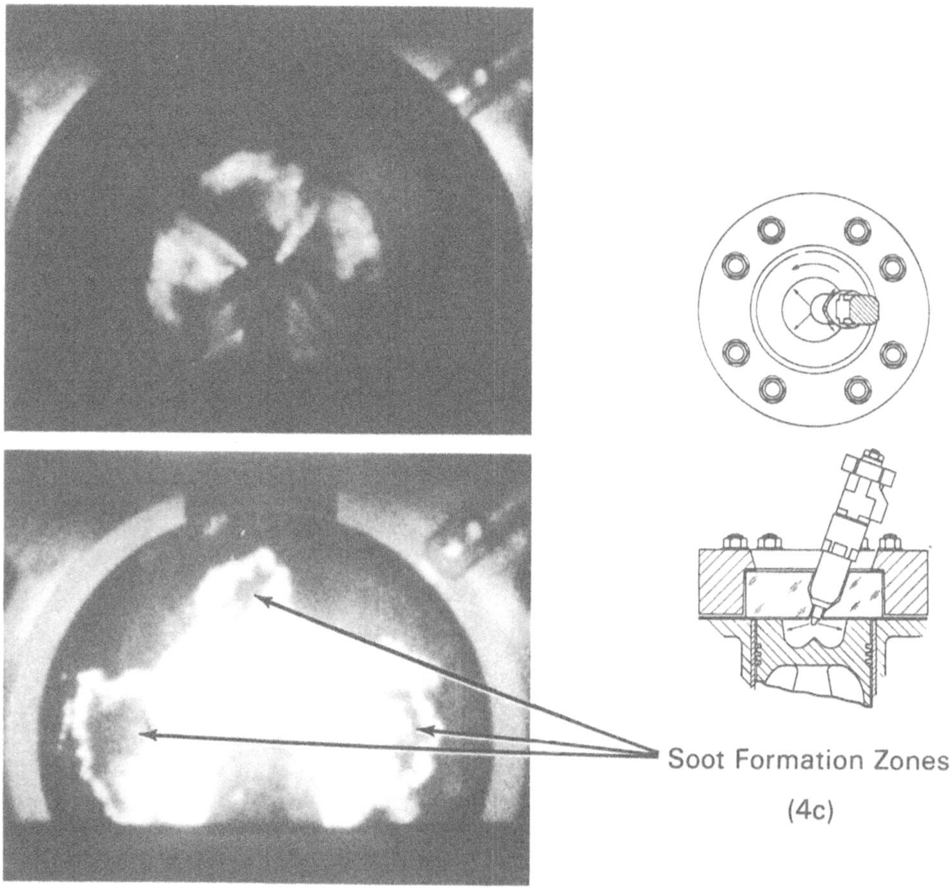

shows a sequence of three photographs, one early in the combustion process, one mid-way through the combustion process, and one at the end of the combustion process, and the details of the geometry of the engine used to obtain photographs. The engine, as shown, was designed with a long slot as part of the combustion chamber to provide an almost plain flame front. The middle photograph shows the fully developed propagating flame. It is "thick" overall but has a much finer scale internal structure which is believed to correspond to "thin" regions where the fuel oxidation process actually occurs. Once fully developed, the flame orientation is roughly normal to the direction of propagation. The initiation process is more difficult to describe, but again the small-scale structure within the flame kernel is apparent from these studies. When the leading edge of the flame reaches the combustion-chamber wall (the bottom photograph in Fig. 5), the front can no longer propagate forward, but clearly reaction is continuing within the flame itself.

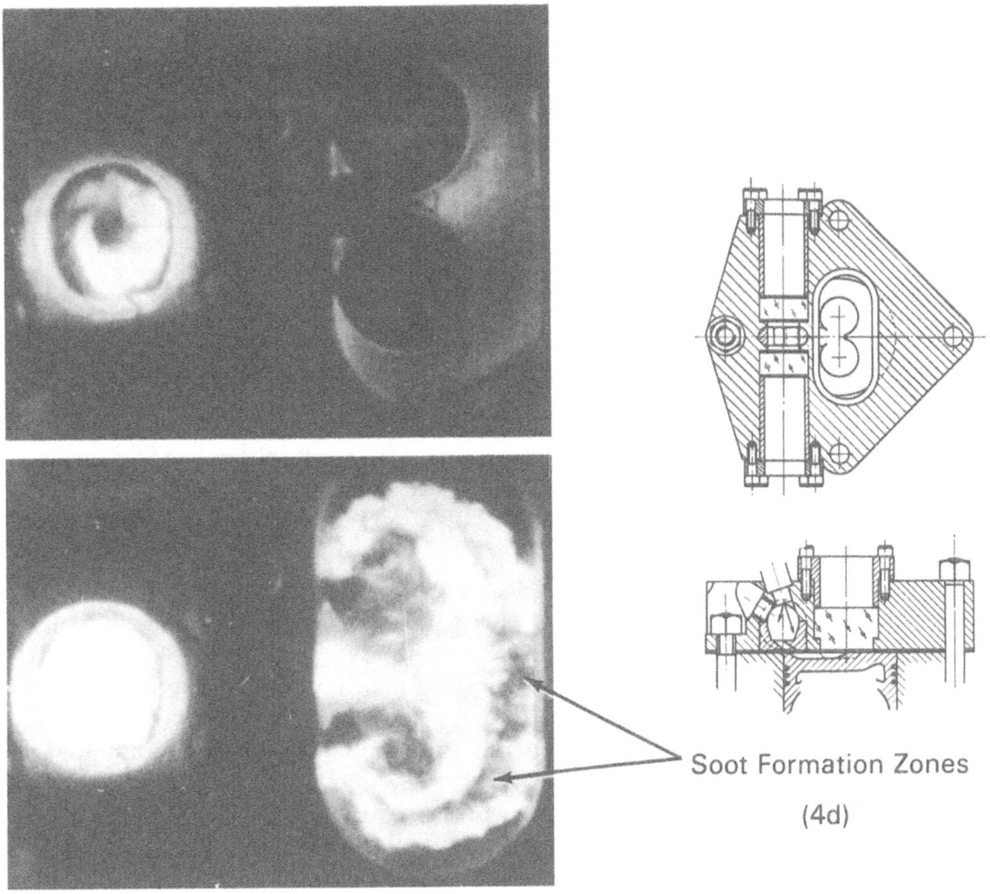

HISTORY OF ENGINE MODELING

While the word "modeling" is a relatively new addition to our technical vocabulary, the analysis of internal combustion engine processes and its use to guide engine development has a long and steady history. While the emissions aspects of the problem are new, the use of engine simulations for performance and efficiency estimates goes back to the early days of internal combustion engine development in the late 19th century. One can view this history as a series of activities aimed at developing both more realistic approximations to real engines processes (such as the intake, compression or expansion strokes) and more accurate methods for calculating the thermodynamic properties of the working fluids used in the engine (unburned mixture, i.e., fuel, air and residual gases, prior to combustion; burned mixture, i.e., the products of combustion of hydrocarbon fuel-air mixtures, after combustion has occurred).

The earliest use of IC engine cycle simulations was apparently by the English-

References pp. 33-35.

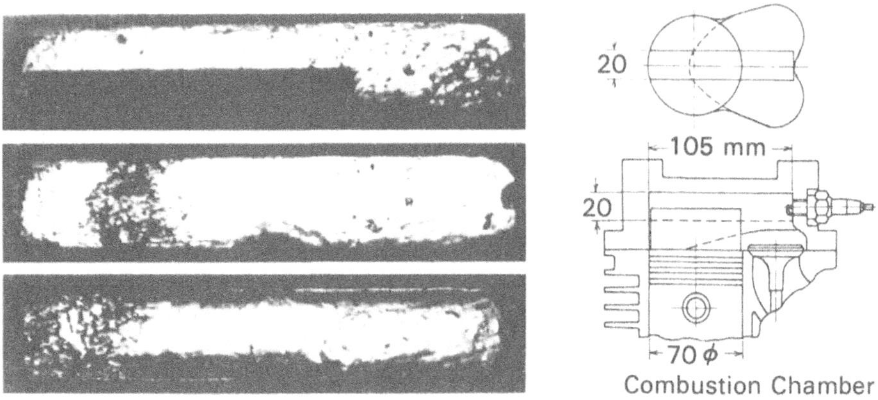

Fig. 5. Schlieren photographs indicating structure of turbulent flame in a spark-ignition engine during flame initiation (top picture), propagation (middle picture) and termination at the wall (bottom picture) [5].

man Sir Dugald Clerk (1854–1932). He was the first to use the technique of an air-standard cycle analysis to compare thermal efficiencies of IC engines. Rudolph Diesel (1858–1913) proposed a number of alternative engine cycle models with different modes of combustion (e.g., constant pressure, constant temperature). Though Nicolaus Otto's name is often associated with the constant-volume cycle, his pioneering IC-engine developments in the period 1860–1880 were not apparently guided by this type of cycle analysis [6].

Thus, the constant-volume, constant-pressure, and limited-pressure ideal cycles came into existence as models for IC-engine operation. For the next 80 years, developments concentrated on improving the accuracy with which the thermodynamic properties of the unburned- and burned-gas mixtures could be evaluated. A major advance came from the work of Hottel et al. [7], who developed charts for the thermodynamic properties of burned-gas mixture assuming the products of combustion were in thermodynamic equilibrium—a good approximation for performance calculations which need only be improved upon if the emission characteristics of the engine are to be evaluated also. Only with the advent of large high-speed computers did it become practical to improve the modeling of the engine operating cycle. The weakest link by far in the ideal-cycle models was the combustion-process model itself—constant volume, constant pressure or a combination of these idealizations. While it had long been known how to incorporate a finite burning time into the analysis (e.g., [8]), such calculations were too time consuming to be carried out on a large scale.

Much more recent additional developments which have expanded our engine combustion modeling capability are the following. Our understanding of the kinetics of nitric-oxide formation in engines—a non-equilibrium process in the bulk burned gas during the combustion process and early part of the expansion stroke—has been extensively developed. Accurate quantitative predictions are now feasible. The use of turbulent-flow heat-transfer correlations in the engine context has been explored and evaluated. Our understanding of processes which make up the hydrocarbon formation and oxidation mechanism has improved to the point where more extensive modeling activities are very worthwhile. Most importantly for this Symposium, our conceptual ideas of turbulent mixing and turbulent flame-propagation processes in the engine environment are rapidly developing. Thus, our ability to model the details of engine combustion processes, as influenced by engine design and operating parameters, has become more sophisticated, more realistic, and therefore more useful. A number of recent review papers summarize developments in many of these areas.*

COMBUSTION-MODEL CLASSIFICATION

A number of engine combustion-model classifications have been proposed. The most useful classification follows from one proposed by Bracco [12]. Its utility stems from the fact that different classes of combustion model, because of their formalism, are generally useful in examining different kinds of combustion-related engine problems. These categories of combustion model are:

(i) zero-dimensional (sometimes called thermodynamic) models,
(ii) quasi-dimensional (sometimes called entrainment) models,
(iii) multidimensional (sometimes called detailed) models.

Zero-dimensional and *quasi-dimensional* models are structured around a thermodynamic analysis of the contents of the engine cylinder during the engine operating cycle. Fig. 6 illustrates the general structure of these types of model [11] with the key sub-models and major assumptions indicated.

The flow into the engine cylinder during the induction process is handled using quasi-steady one-dimensional flow equations. Mass flows past valves and other restrictions are modeled by the equations for isentropic adiabatic flow through a nozzle. A discharge coefficient is used to relate the effective area for the particular constriction (an intake valve, high-aspect-ratio connecting passage, sharp-edge orifice, etc.) to the ideal area for the isentropic flow. System pressures upstream and downstream of the restriction are used for the pressure ratio in these calculations. Unless pipe dynamics are of interest, plenum assumptions are commonly used for intake and exhaust manifolds.

Where pipe dynamics are important, these are calculated using one-dimensional unsteady gas-flow equations, which are solved using the method of characteristics. These calculations generally include friction, heat transfer, and area change in the pipes. Empirical constants are used to determine boundary conditions at pipe junctions, valves, and open ends.

The unburned mixture is assumed to be a mixture of air, fuel vapor, and residual gas of frozen composition appropriate to the particular engine concept and operating conditions being modeled. The specific heat of each component

---

* *Spark ignition engines* [9–13]; *diesel engines* [14]; *heat transfer* [15], [16]; *induction and exhaust* [17].

References pp. 33-35.

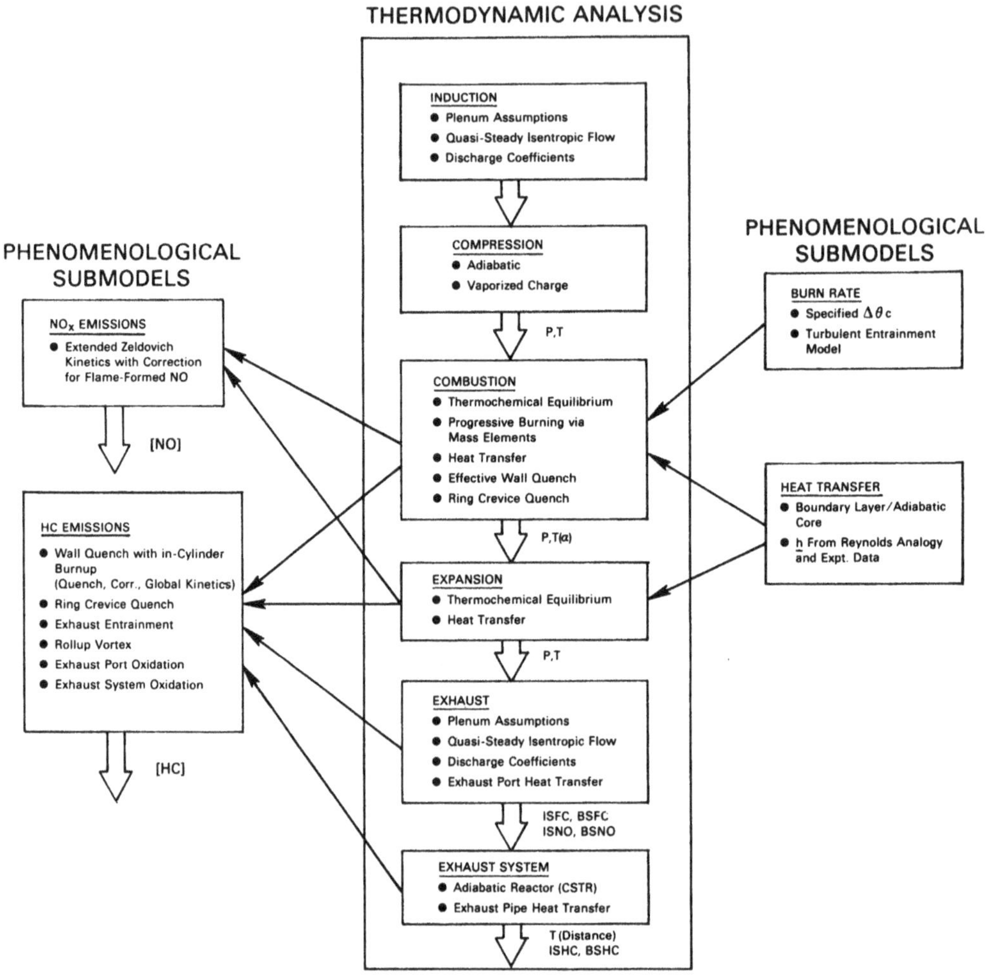

Fig. 6. Block diagram showing the structure of zero- and quasi-dimensional models.

gas in the gas mixture is modeled using polynomial functions of temperature. For most engine geometries under throttled-engine operation, burned gases will flow into the intake during the gas-exchange process. When this occurs, it is assumed that the burned gases that leave the cylinder do not mix with the fresh intake charge and are the first gases to be pulled into the cylinder during the intake process. This assumption has proved useful for calculating the residual-gas fraction in the chamber.

Heat transfer in IC engines is usually modeled using correlations between Nusselt, Reynolds and Prandtl numbers developed for heat transfer in steady turbulent flows in pipes and over flat plates, of the form

$$Nu = aRe^b Pr^c \tag{1}$$

The constants a, b and c are obtained by fitting experimental data from a particular engine system. The characteristic length and velocities used in the Reynolds number are the cylinder bore and mean piston velocity. Typical values for the constants are $a = 0.035$, $b = 0.8$ and $c = 0.333$. Models also exist which modify the velocity scale depending on the processes occurring during the cycle (e.g., gas-exchange, compression-expansion, and combustion) [16]. To improve the accuracy of predictions of burned-gas temperatures, boundary-layer heat-transfer models can be incorporated within the thermodynamic framework.

During combustion, burned-gas, unburned-gas and boundary-layer zones exist within the cylinder, as illustrated in Fig. 7. To complete the analysis of this phase of the cycle, the rate of mass burning must be determined. In the *zero-dimensional* class of models, the rate of mass burning is specified by some functional relationship obtained by matching with previous experimental data. In the quasi-dimensional models a turbulent combustion sub-model is used for this purpose, as explained below. Typical functional forms used for specifying the combustion rate are a cosine function:

$$x(\theta) = (\tfrac{1}{2})\{1 - \cos \pi[(\theta - \theta_0)/\Delta\theta_b]\} \tag{2}$$

or a Wiebe function:

$$x(\theta) = 1 - \exp\{-a[(\theta - \theta_0)/\Delta\theta_b]^{m+1}\} \tag{3}$$

where $x(\theta)$ is the mass fraction burned at crank angle θ, θ_0 is the crank angle at the start of combustion, and $\Delta\theta_b$ is the burn duration. a and m are parameters which can be varied. Typical values are 5 and 2, respectively. Fig. 8 illustrates the shapes of these curves and their relation to spark timing, ignition delay and experimental burn duration.* θ_s is spark timing and $\Delta\theta_{id}$ is the ignition

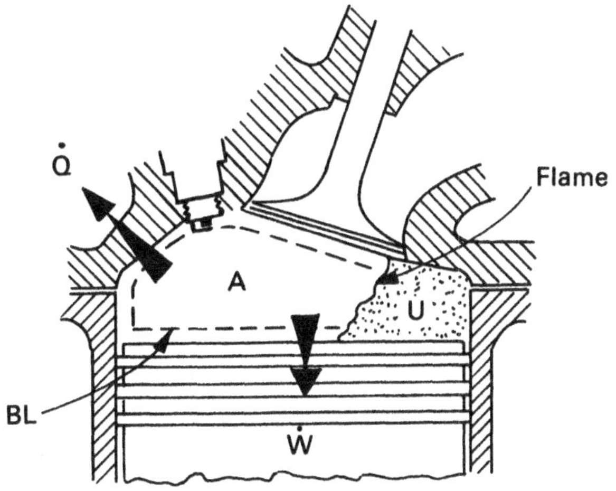

Fig. 7. Schematic of engine cylinder during combustion process.

* *In stratified-charge or diesel engines, more complicated functional forms are required to predict the rate at which unburned mass is consumed by the flame and the resulting rate of energy release within the cylinder.*

References pp. 33-35.

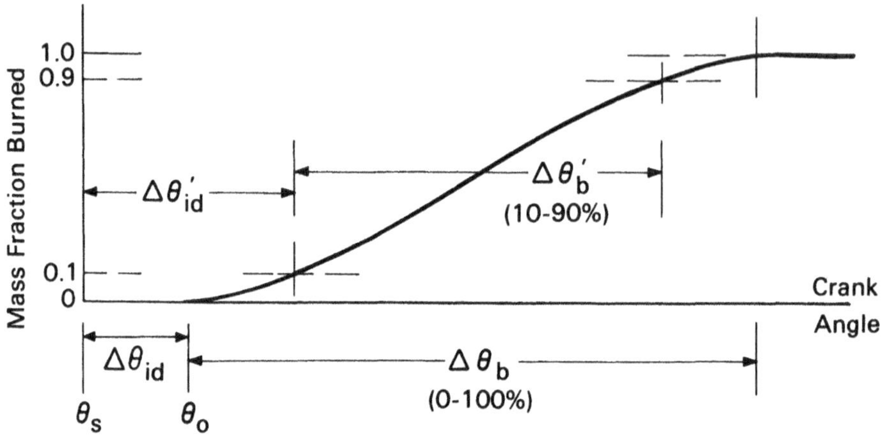

Fig. 8. Schematic of mass fraction burned profile.

delay. $\Delta\theta'_{\mathrm{id}}$ and $\Delta\theta'_{\mathrm{b}}$ are the usual empirical definitions of ignition delay and combustion duration (0–10% and 10–90% burned, respectively).

The burned gases in the cylinder following the fuel-oxidation processes are close to thermodynamic equilibrium. Burned-gas properties are usually specified through curve fits to thermodynamic-equilibrium values of burned-gas properties, or by approximate models of thermodynamic equilibrium, or by full thermodynamic-equilibrium calculations. Individual gases in both unburned- and burned-gas mixtures are modeled as ideal gases. Curve fits or other approximations to the full equilibrium calculations are used to reduce computation time.

One of the major drawbacks of zero-dimensional models is the need for "*a priori*" specification of the burn rate. For spark-ignition, stratified-charge, and diesel engines this has led to the development of *quasi-dimensional* models which attempt to predict the rate of burning from more fundamental physical quantities. These fundamental physical quantities are the turbulent intensity, the turbulent integral length scale, the turbulent micro-scale, the jet characteristics in any jet-mixing process, and the kinetics of the fuel-oxidation process. The intent here is to predict the ignition delay and combustion rate as a function of engine design and operating conditions.

For spark-ignition engines, one approach has been to model the burning process as a flame front of area A_f (usually assumed to be sections of the surface of a sphere) propagating through the unburned mixture at the turbulent flame speed S_T. Thus the rate of mass burning is

$$\frac{dm_b}{dt} = \rho_u A_f S_T \tag{4}$$

S_T is then related to the laminar flame speed (e.g., by $S_T = KS_L$, where K is a constant. See Andrews et al. [18] for a review of turbulent-flame-speed theories).

An alternative approach suggested by Blizard and Keck [19] and further developed by Tabaczynski et al. [20] is to model the flame-propagation process as first a turbulent entrainment of unburned mixture into the front, which is followed by a laminar burn-up process with a characteristic length scale. Thus the mass entrained into the front, m_e, is given by

$$\frac{dm_e}{dt} = \rho_u A_f u_e \tag{5}$$

where u_e is an entrainment speed (assumed proportional to the turbulent intensity). The rate at which mass is burned is given by

$$\frac{dm_b}{dt} = (m_e - m_b)/\tau \tag{6}$$

where τ is the characteristic reaction time to burn the mass of an eddy of size L.

Tabaczynski et al. [20] divided the propagation process into a developing phase corresponding to the ignition delay—the burning of an individual eddy, and a fully developed propagation phase—the entrainment and burning of many eddies. An eddy-burning model was developed, based on the following assumptions: Combustion on the Kolmogorov scale is instantaneous. Ignition sites in an individual eddy propagate at a rate governed by the sum of the turbulent intensity and laminar flame speed. Laminar burning takes place over the micro-scale. The integral scale is assumed proportional to the instantaneous chamber height at the time of spark. The turbulent intensity at the time of spark is proportional to engine speed. The integral scale and the turbulent intensity change with unburned mixture density due to compression of the unburned gas by the flame front, as defined by conservation of angular momentum.

The quasi-dimensionality introduced through the assumption of a spherical flame front and this turbulent combustion model provides reasonable agreement with flame-front shape and location as determined from high-speed combustion movies. Although this type of model of the turbulent combustion process is based on indirect experimental evidence, physical intuition, and to some extent mathematical convenience, the results indicate that it predicts the proper trends in ignition-delay period and combustion duration [21].

This type of quasi-dimensional model shows considerable promise of being able to predict the operating characteristics of conventional spark-ignition engines. Several questions remain, however, e.g., the character and uniformity of the turbulence field, especially in geometries involving squish and swirl; the use of spherical flame-front geometry; the effect of unburned-mixture compression due to combustion on the turbulence field; turbulence levels in the burned gases; a detailed physical model for the ignition process; heat transfer to various parts of the combustion chamber.

The third class of models—*multidimensional* models—are of a different character. In this type of model, the governing partial-differential conservation equations, along with appropriate sub-models which describe the turbulence processes, chemical processes, boundary-layer processes, etc., are solved numerically subject to the appropriate boundary conditions. The promise of this type of engine combustion model is that it should be able to provide detailed information on the spatial distribution of gas velocity, temperature and compo-

References pp. 33-35.

sition within the combustion chamber of the engine during the combustion process.

The set of conservation equations is made up of mass, momentum, energy and individual chemical-species conservation equations. Table 1 lists these conservation equations for multicomponent reacting gas mixtures in tensor notation [22]. The first four equations are the mass-, momentum-, energy- and species-conservation equations. The fluid density, ρ, velocity, $\underline{u}$, internal energy, e, heat-flux vector, $\underline{q}$, species mass fraction, Y_i, diffusion velocities, $\underline{u}_i$, species source terms, w_i, and pressure tensor, $\underline{\underline{p}}$, can be recognized. Eq. 11 gives the pressure tensor in terms of the (hydrostatic) pressure, p, and the viscosity shear tensor, $\underline{\underline{\hat{\sigma}}}$. Eq. 12 shows that the heat-flux vector includes both the flux due to temperature gradients and that transferred by diffusion velocities. These diffusion velocities are related to each other and to the concentration (X_i) gradients by Eq. 13. Eq. 15 gives a possible expression for the species source terms and represents chemical-kinetics phenomena; K elementary reaction steps, of the general form given by Eq. 14, are considered, each with its Arrhenius-type reaction rate constant. Eqs. 16 and 17 are the equations of state, Eq. 18 relates mole fractions to mass fractions and molecular weight [12].

To solve this set of equations one must introduce various approximations (both physical/chemical and numerical). These approximations can be categorized as [23]: (i) numerical algorithms/numerical schemes (e.g., explicit methods, implicit methods, differencing approximations); (ii) coordinate approximations (Eulerian, Lagrangian, mixed); (iii) subgrid scale models (e.g., for turbulence, chemistry, liquid fuel sprays, boundary layers, etc.); and (iv) dimensional approximations (one, two or three dimensions—steady or unsteady).

The solution of this set of partial differential equations in the unsteady multidimensional engine context, even with major simplifying assumptions, presents formidable difficulties. Because one has confined subsonic flow, often with flow reversals and swirl, the flame structure and propagation rate depend not only on the turbulent diffusivities and chemical kinetics but also on the entire time and space-changing flow field as well as the boundary conditions (which include heat transfer). One must contend with multiple time scales as well as multiple length scales. For example, the chemical rate processes themselves, such as the main oxidation reactions and the CO burn-out and NO chemistry, have substantially different characteristic time constants. This phenomena produces so-called "stiff" differential-equation sets. If radiation heat transfer is important as it is in diesels, an additional difficulty is introduced. A complicated coupled spatial-temporal integration needs to be performed to include radiation in a rigorous fashion. Phenomenological sub-models must also be employed to compensate for lack of process details or to compensate for lack of adequate spatial resolution imposed by computer storage limitations. Thus sub-grid scale models are required to describe the turbulence, the chemical processes, the boundary-layer processes and liquid fuel-spray processes which occur simultaneously within the engine combustion chamber. Other papers at this Symposium discuss these aspects of multidimensional modeling in more detail.

TABLE 1

Governing Equations for Multidimensional Models [12]

$$\partial\rho/\partial t + \nabla\cdot(\rho\underline{u}) = 0 \tag{7}$$

$$\partial(\rho\underline{u})/\partial t + \nabla\cdot(\rho\underline{u}\underline{u} + \underline{\underline{p}}) = 0 \tag{8}$$

$$\partial(\rho\underline{u}^2/2 + \rho e)/\partial t + \nabla\cdot(\rho\underline{u}^2\underline{u}/2 + \rho e\underline{u} + \underline{u}\cdot\underline{\underline{p}} + \underline{q}) = 0 \tag{9}$$

$$\partial(\rho Y_i)/\partial t + \nabla\cdot[\rho Y_i(\underline{u} + \underline{u}_i)] = w_i \tag{10}$$

where $i = 1 \ldots N$

$$\underline{\underline{p}} = p\underline{\underline{U}} - [(k - 2\mu/3)(\nabla\cdot\underline{u})\underline{\underline{U}} + \mu(\nabla\underline{u} + \nabla\underline{u}^T)] \equiv p\underline{\underline{U}} + \underline{\underline{\hat{\sigma}}} \tag{11}$$

$$\underline{q} = -\lambda\nabla T + \rho\sum_{i=1}^{N} h_i Y_i \underline{u}_i \tag{12}$$

$$\nabla X_i = \sum_{j=1}^{N}\left(\frac{X_i X_j}{D_{ij}}\right)(\underline{u}_j - \underline{u}_i) \tag{13}$$

where $i = 1 \ldots N$

$$\sum_{i=1}^{N} \nu'_{i,k} M_i \rightarrow \sum_{i=1}^{N} \nu''_{i,k} M_i \tag{14}$$

where $i = 1 \ldots N$ and $k = 1 \ldots K$

$$w_i = W_i \sum_{k=1}^{K} (\nu''_{i,k} - \nu'_{i,k}) \times [B_k T^{\alpha_k} \exp(-E_k/RT)] \prod_{j=1}^{N}\left(\frac{X_j p}{RT}\right)^{\nu'_{i,k}} \tag{15}$$

where $i = 1 \ldots N$

$$p = \rho RT \sum_{i=1}^{N} (Y_i/W_i) \tag{16}$$

$$e = \sum_{i=1}^{N} h_i Y_i - p/\rho = \sum_{i=1}^{N}\left(h_i^0 + \int_{T^\circ}^{T} c_{p_i}\,dT\right) Y_i - p/\rho \tag{17}$$

$$X_i = (Y_i/W_i) \sum_{j=1}^{N} (Y_j/W_j) \tag{18}$$

In the transition from the framework of zero- and quasi-dimensional engine models (where time is the only independent variable, and thus the governing equations are ordinary differential equations) to the multidimensional models (where one, two or three space dimensions and time are the independent variables, and the governing equations are partial differential equations), it is obvious from the above discussion that substantial numerical and computational and sub-model physical and chemical difficulties have been introduced. Currently, even with simple sub-models for the turbulence and chemistry, only the solution of one- and two-dimensional problems in the engine context is feasible.

So far I have discussed models for the conventional "premixed" spark ignition engine. In fuel-injected stratified-charge engines and diesel engines the heat-release rate is controlled primarily by the fuel-air mixing process. Thus attempts to predict the operating and emissions characteristics of these types of engine require models of this two-phase turbulent jet-mixing process and of ignition and flame-propagation in this developing non-uniform flow. Zero-dimensional models where the heat-release rate was empirically prescribed (e.g., Lyn [24]) have proved useful in developing our ideas of the basic processes involved. These have been replaced by quasi-geometric models of the liquid fuel-jet breakup, droplet deceleration, droplet evaporation, and air entrainment processes which, when coupled with an ignition model, can predict the combustion rate for these engine types. Mass- and momentum-conservation equations in one- or two-dimensional or axisymmetric form, in a simplified geometric configuration, are used as a basis for following the fuel-air mixing process (see [11] for a review). Because the geometric aspects of these types of engine are so important, this is a fruitful area for multidimensional models, and there are a number of active efforts in this area under U.S. Department of Energy sponsorship.

USE OF ENGINE COMBUSTION MODELS

The three classifications of engine combustion models described above are best suited to analyzing different kinds of engine-related problems. For example, the zero-dimensional and quasi-dimensional models can easily be incorporated in a complete simulation of the engine operating cycle. The computer cost associated with each calculation of the engine's operating characteristics and some of its emission characteristics is modest. Thus, these models are well suited to parametric studies of the effects of changes in design and operating variables on engine performance, efficiency and emissions. In the zero-dimensional models, the details of the combustion process are fixed as an input to the calculation. In the quasi-dimensional models, attempts are made to couple the combustion process with the conditions inside the engine cylinder while the combustion process is taking place.

However, there are no explicit links between the detailed geometry of the engine combustion chamber and the combustion process in these types of models. Hence, they are only able to predict approximately the effects of changes in combustion-chamber shape on the combustion process and thus on engine operating characteristics. Where details of engine geometry and flow are most important, the multidimensional models have a clear advantage. However these are at present limited to two-dimensional calculations; the computer costs associated with realistic three-dimensional calculations are currently prohibitive. While calculations with multidimensional models can provide valuable insights on aspects of combustion not treated in the zero- and quasi-dimensional models, the role of the multidimensional model is likely to remain an explorative and illustrative one for the forseeable future. They are not suited to the type of parametric studies associated with engine development.

To illustrate the current status and potential use of combustion models, I will

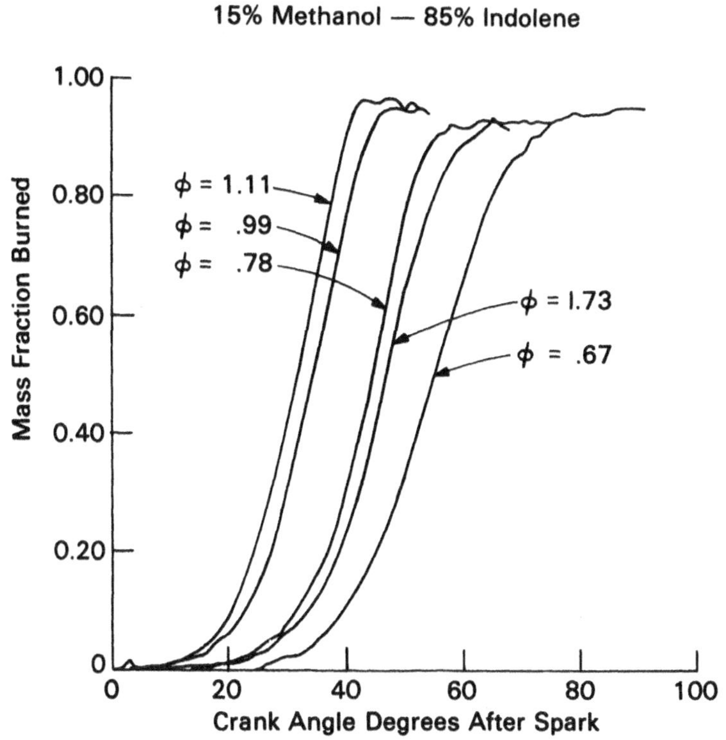

Fig. 9. Illustration of use of zero-dimensional model framework to obtain mass fraction burned curves from cylinder-pressure versus crank angle data [25].

now describe some examples of results of calculations which illustrate the characteristic features of the different types of model and the problems to which they are best suited. I have chosen to ignore questions regarding the validation of these particular models. In each of the cases described below, the authors of the model performed a model validation study where predictions were compared with appropriate experimental data. Details of these validation studies can be found in the references. That I have chosen several examples from my own laboratory's activities is primarily a matter of convenience. A great deal of work has been done elsewhere.

Zero-Dimensional Models—One important use of zero-dimensional models is in conjunction with engine experiments where cylinder pressure is recorded as a function of time to analyze details of the combustion process. With the zero-dimensional model structure, such data can be used to derive mass-fraction burned as a function of time. An example of mass-fraction-burned curves obtained for a series of equivalence ratios from engine cylinder pressure-time data is shown in Fig. 9 [25]. For this set of tests, the fuel was a methanol-gasoline blend.

References pp. 33-35.

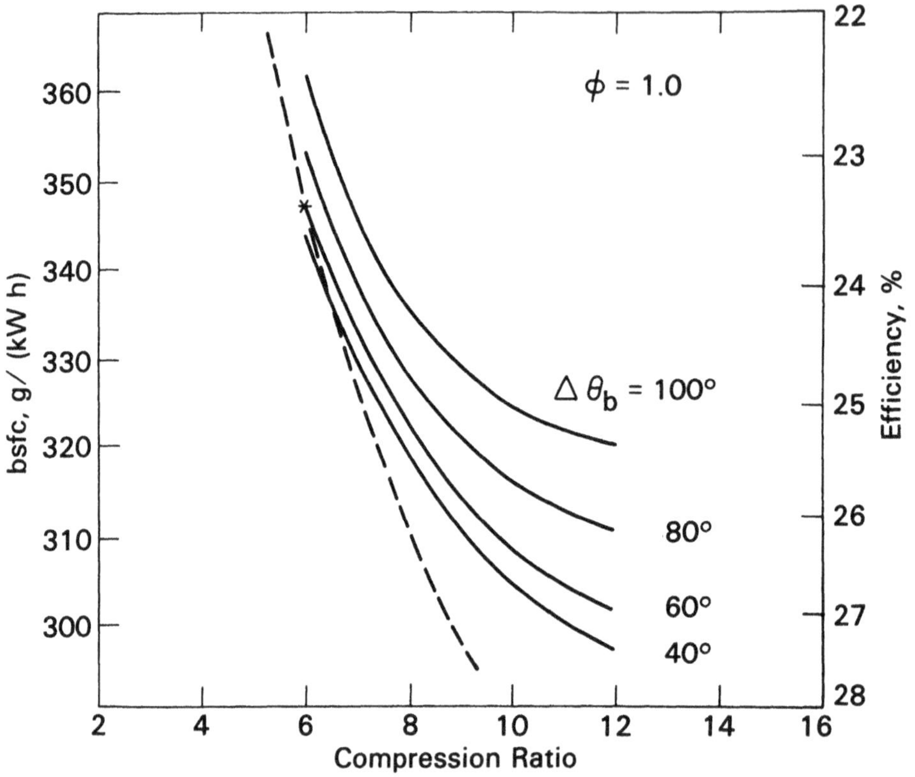

Fig. 10. Example of use of zero-dimensional model to examine the effect of variations in compression ratio on fuel consumption [1].

From the mass-fraction-burned curves, accurate measurements of ignition delay and combustion duration can be obtained. The increasing ignition delay and decreasing burn rate with increasingly lean mixtures are evident. Use of zero-dimensional models in this manner to extract additional information from engine experiments is now routine.

A second and growing use for zero-dimensional (and also quasi-dimensional) models is in parametric studies of the effects of changes in engine design and operating variables on engine power, efficiency, and *NOx* emissions. These parametric studies are usually done with all variables except the one being studied held constant. Such studies can be carried out at a number of engine speed and load points selected to represent those most typical of engine use in a particular vehicle application. As an example of this kind of study, consider the effect of compression-ratio changes on fuel consumption in a spark-ignition engine at a part-throttle operating point at constant brake load—an important yet still inadequately defined aspect of engine design.

Fig. 10 [1] shows brake specific tuel consumption and thermal efficiency for a

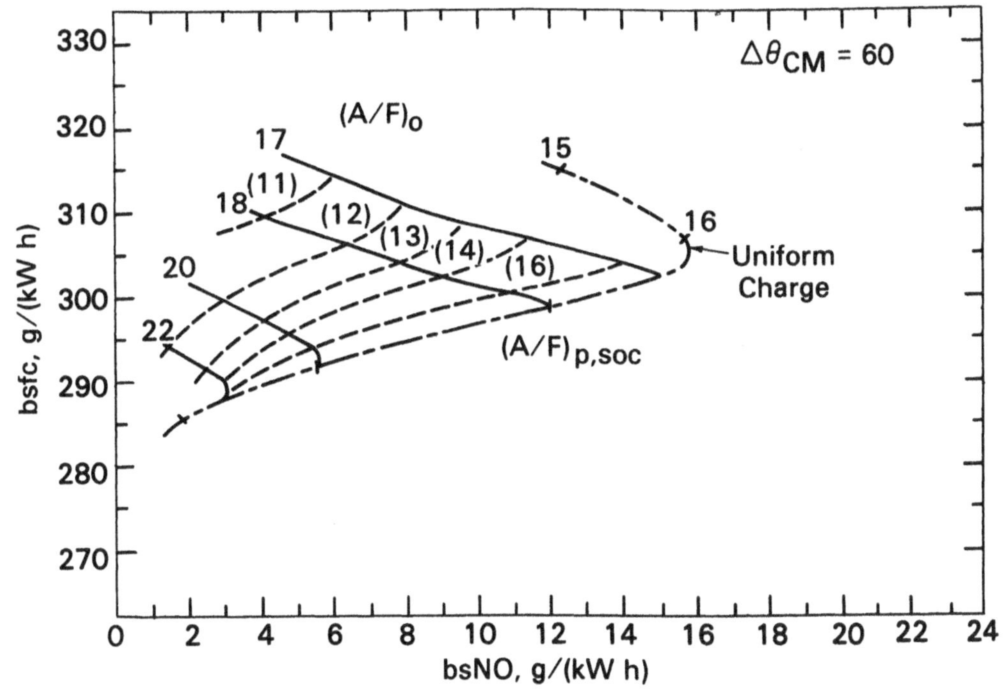

Fig. 11. Comparison of three-valve stratified charge engine with conventional premixed engine on brake specific fuel consumption versus brake specific *NO* plot [26].

four-stroke-cycle spark-ignition engine as a function of compression ratio at a mid-load and mid-speed operating point. In this particular study the combustion duration was held constant at the values indicated on the lines in the figure. For each calculation the start of combustion was adjusted to give maximum brake torque. The brake mean effective pressure was held constant at 328 kPa. Thus the mechanical friction and the pumping mean effective pressure, which of course increase as compression ratio increases, are included. Engine speed was 1400 r/min, equivalence ratio was 1.0 and spark timing was set at MBT. For comparison, the efficiency of the fuel-air cycle over this range of compression ratios has been scaled to pass through the point indicated by the asterisk. It is obvious that the ideal cycle calculations are not an accurate indicator of the effect of compression ratio on engine efficiency at constant load.

In the above study, the engine displacement was held constant at 5.7 L. The most useful framework for a design study of this problem would be to vary engine displacement as compression ratio is changed so that maximum engine power is held constant. Experimentally, such a study would be extremely difficult to carry out. With the model, the additional requirement of decreasing displacement as compression ratio increases can be readily incorporated.

References pp. 33-35.

Another area where zero-dimensional models have proved useful is in studying the effects of stratification on the efficiency and *NOx* emissions of the three-valve stratified-charge engine [26]. Fig. 11 shows the effect of stratification in this three-valve prechamber-engine concept on brake specific fuel consumption (*bsfc*) and brake specific nitric-oxide (*bsNO*) emissions. The dash-dot line corresponds to uniform-charge operation, where the mixture fed to the prechamber and main chamber have the same air-fuel ratio. Under these conditions the engine behaves like a conventional engine. The solid lines show the effect of increasing stratification as the prechamber is richened at constant overall air-fuel ratio $(A/F)_o$. (The numbers in parentheses are the air-fuel ratios in the prechamber at the start of combustion.) In this set of calculations, the combustion timing was optimum, the combustion duration in the main combustion chamber was constant at 60 CA°, the load was constant at a brake mean effective pressure (*bmep*) of 407 kPa, the speed was constant at 2100 r/min, and the engine displacement was 2.3 L.

The usefulness of these data can be illustrated as follows. The lean limit of

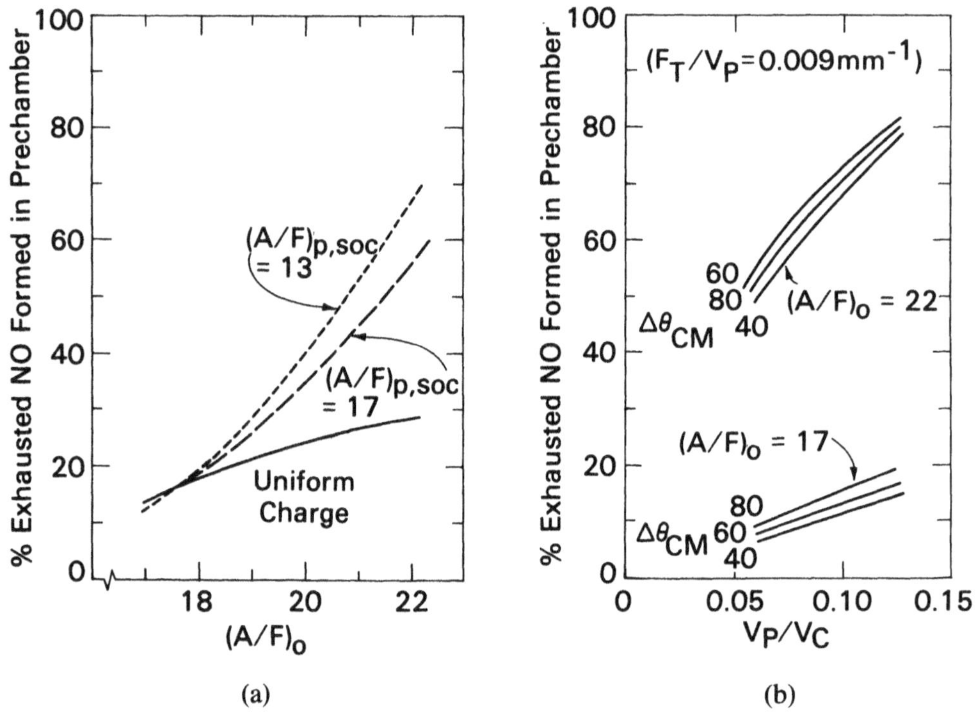

Fig. 12. Modeling study of the relative importance of nitric oxide formed in the prechamber of a three-valve stratified charge engine. Shown is percentage of exhaust *NO* which originated in the prechamber (a) as a function of overall air-fuel ratio at time of spark; (b) as a function of prechamber volume to clearance volume [26].

conventional engine uniform-charge operation is about 18 to 1 air-fuel ratio. At this point *bsNO* is 12 g/(kW h) and *bsfc* is 298 g/(kW h). This stratified-charge engine concept can operate up to 22:1 air-fuel ratio at least. At this point, with a prechamber air-fuel ratio of 13 (i.e., slightly rich of stoichiometric), *bsNO* has been reduced to below 2.7 g/(kW h) with a fuel consumption below 291 g/(kW h). Thus, substantial reductions in *NOx* emissions can be achieved in addition to slight gains in fuel consumption. However, the exhaust temperature is reduced from 1220 to 1067 K, a change likely to result in a substantial increase in hydrocarbon emissions. This type of simulation quantifies the effects of changes in engine operating conditions on fuel consumption and nitric-oxide emissions in a fashion extremely useful to the engine developer.

This same study also identified the fraction of the exhausted *NO* which originated in the prechamber. Such information would tell the developer whether changes in prechamber configuration are likely to affect the *NO* emissions significantly; for example, Fig. 12a shows the percentage of exhausted *NO* formed in the prechamber as a function of overall air-fuel ratio for various degrees of stratification. ($(A/F)_{p,soc}$ is the air-fuel ratio in the prechamber at the start of combustion.) We can see that at an overall air-fuel ratio of 18:1, the details of the prechamber are unlikely to be important. At 22:1 overall, however, where over 60% of the *NOx* originates in the prechamber, prechamber details are likely to be important. We see this in Fig. 12b, where the effect of prechamber volume on exhausted *NO* is shown. At 22:1 overall air-fuel ratio, the sensitivity to the prechamber volume fraction is high.

This particular figure also shows that changes in combustion duration ($\Delta\theta_{cm}$ in Fig. 12b) have little effect on this result. Thus the fact that the combustion process is specified as an input and has been held constant for these particular calculations is not important.

Quasi-Dimensional Models—In the quasi-dimensional class of models, an *ad hoc* geometric model for the fuel-air mixing process (if relevant) and the combustion process, based on an appropriate set of physical and chemical assumptions, is added to the purely thermodynamic framework of the zero-dimensional model. The intent is to introduce the coupling which exists between engine design and operating conditions and the combustion process without the full complexity of the multidimensional models. Because our knowledge of this coupling is still quite limited, this is a rapidly developing research area. The models currently available must still be regarded as speculative.

A model of the combustion process in a conventional spark-ignition engine of this type, based on a turbulent eddy-entrainment process which brings fresh mixture into the flame and subsequent mixture burn-up at the appropriate scale through a laminar burning process, has already been briefly described [19, 20]. This model attempts the above mentioned coupling through the turbulence characteristics of the unburned mixture—the turbulent intensity and the turbulent length scales—and the mixture composition and state. The turbulence characteristics and mixture state ahead of the flame change during the combustion process

References pp. 33-35.

as the unburned mixture is compressed. Approximate [20] as well as more fundamental methods [27] for dealing with this aspect of the problem have been proposed. Results from the study by Hires et al. [21], who used the combustion model of Tabaczynski et al. [20] to predict ignition delay and burn duration in a *SI* engine, indicate why this part of the problem is important.

Fig. 13 shows the effect of variations in exhaust gas recirculation (*EGR*) and timing on ignition delay and combustion duration as predicted by this model. Increasing *EGR* always increases the ignition delay and the combustion duration, as is well known. However, these predictions indicate that a strong interaction exists between the timing of the start of combustion (crank angle at 1 percent mass burned) and this increase in combustion duration as *EGR* increases. If optimum timing is always used—that is, the start of combustion is advanced appropriately as combustion duration increases—the deceleration effect of *EGR* on combustion duration is minimized. But these results show that if a combination of *EGR* and timing retard from the optimum is used for emission control, the effect of *EGR* on combustion duration will be substantially greater.

The reason for this interaction is the dependence of the local burning rate on the turbulence intensity ahead of the flame, the turbulence scale, and the laminar flame speed. The turbulence characteristics change due both to changes in piston position and to compression of the unburned mixture as combustion proceeds. The laminar flame speed is affected by the local unburned mixture temperature

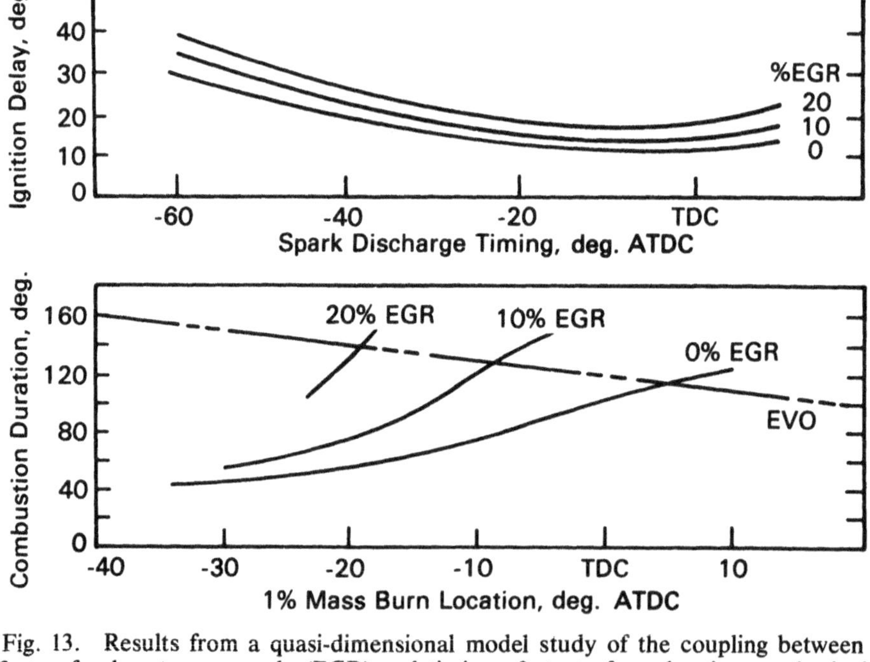

Fig. 13. Results from a quasi-dimensional model study of the coupling between the effects of exhaust gas recycle (EGR) and timing of start of combustion on the ignition delay and the combustion duration [21].

and pressure (which, of course, vary during the combustion process) as well as by the dilution with *EGR*. The implications of such an increased understanding of practical problems of this type are substantial.

Consider now an example of a quasi-dimensional model where fuel-air mixing processes are important. A model for combustion and *NOx* formation in the Texaco stratified-charge engine has been developed by Hiraki [28]. The mode of operation of this engine concept—high-pressure injection of a single fuel jet directly into the swirling airflow in the cup-in-piston combustion chamber just before the crank angle at the start of combustion, with ignition effected by a long-duration spark discharge—has already been illustrated in the combustion photographs in Fig. 4b.

In this analysis, the major geometric features of the mixing of air into the fuel jet and of the jet-dominated combustion process have been incorporated into the thermodynamic framework of Fig. 6 as follows. The basis for the combustion model is a turbulent jet-entrainment analysis in a cylindrical coordinate system appropriate to the high air swirl of this engine concept. In addition, because the time histories of the mixing processes of fuel elements in the fuel jet with air, both prior to and after combustion, are most important for performance and *NOx* emissions predictions, the jet is divided into a large number of elements which are individually followed through the calculation. Combustion is initiated at the spark plug, and the growth of the flame kernel within the mixing fuel-air jet is computed using concepts from the turbulent-entrainment combustion models proposed for conventional *SI* engines [19, 20], suitably modified to fit the stratified-charge engine environment. In this fashion, the rate of burning is related to the local turbulent intensity, turbulent scale, and laminar flame speed.

Once the burned-gas pressure and temperature in each element in the jet after combustion have been determined, performance and *NOx* emissions predictions can be made. *NOx* emissions levels and trends provide a good check on the model's accuracy. *NOx* emissions predictions are especially sensitive to gas temperature and fuel-air equivalence-ratio distribution. Fig. 14 compares predictions of *NOx* emissions as functions of injection timing and equivalence ratio with experimental data. *NOx* emissions vary substantially with injection timing and overall equivalence ratio. Good agreement exists between predictions and data.

This model has been constructed for one specific fuel-injected engine concept and a particular combustion-chamber geometry. Its usefulness lies primarily in evaluating effects of changes in engine operating conditions and fuel characteristics, and modest changes in engine geometry, on engine performance, efficiency, and *NOx* emissions.

Multidimensional Models—The multidimensional class of models have the capability, in principle, for predicting the details of the fluid flow within the engine cylinder and the propagation rate and geometrical shape of the flame. The complete simulation of the combustion process with this type of model is still in an exploratory stage. The turbulence and overall reaction-rate models have not yet been developed to an adequate level of sophistication. Sample results from

References pp. 33-35.

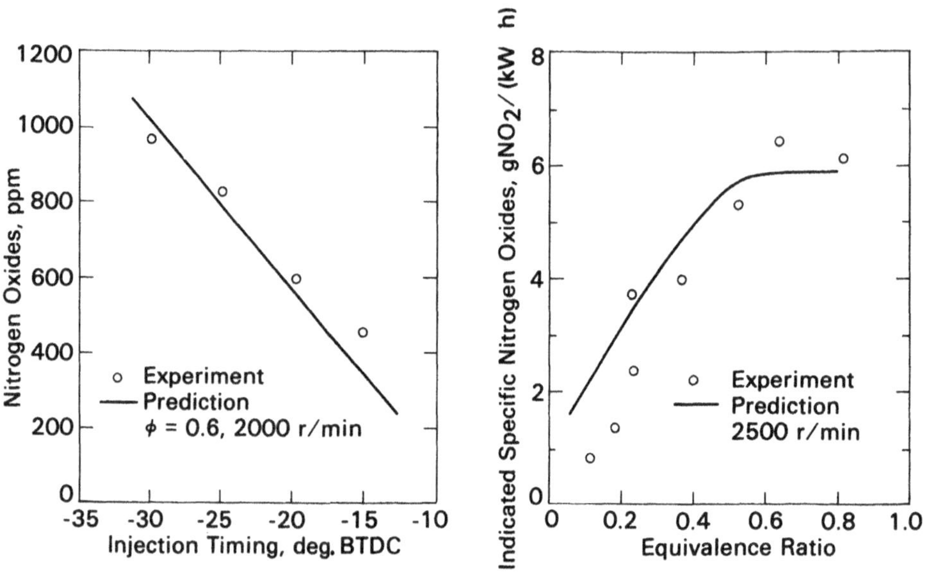

Fig. 14. Comparison of predictions of *NOx* emissions from a quasi-dimensional model study with experimental data for Texaco stratified charge engine [28].

this type of calculation (with one- or two-dimensional geometries) which indicate its potential have already been published, e.g., [12, 13].

While calculations of the complete engine combustion process are still highly speculative (though nonetheless very worthwhile as a research activity), studies of the flow pattern inside the engine cylinder with these techniques are reaching the stage where much physical insight can be gained. For example, Gosman et al. [29] have examined the effect of the shape and size of the bowl-in-piston combustion chamber used in direct-injection diesel engines on the flow during compression. The airflow pattern at the time of fuel injection is one of the important factors in determining fuel-air mixing and hence heat-release rates. There is no other way of examining the details of this type of geometry-induced fluid flow.

The example I have chosen to illustrate the utility of multidimensional models is the study by Haselman and Westbrook [30] of a gaseous fuel jet mixing with air in an engine cylinder. The full set of fluid-mechanic conservation equations in Table 1 can be simplified in the absence of any combustion. In this study, a two-dimensional configuration was assumed. The jet is injected into a thin disc-shaped combustion chamber, and all variables are assumed uniform over the height of the chamber. Piston motion is neglected; thus only the flow near top dead center is examined.

The model equations are solved using a finite difference approach. The physical domain is subdivided into computational cells, and each conservation equation is solved for every cell. Since the computational grid is much too coarse to define the complete turbulent motions of the gas, a turbulence model is used to represent those turbulent motions which are below the resolution of the grid. This was done by using a turbulent viscosity, μ_t, in place of the molecular viscosity. It is assumed that the turbulence is locally isotropic and generated by the large-scale fluid motions. The turbulent viscosity is calculated from

$$\mu_t = \alpha\rho \left| \frac{\partial u}{\partial r} + \frac{\partial v}{\partial z} \right| (\Delta x)^2 \tag{19}$$

where α is an input constant, ρ the gas density, u and v are velocity components and Δx is the grid size. This formulation is equivalent to Prandtl's mixing-length hypothesis with the mixing length equal to the grid spacing. The gas-jet injection process is simulated by defining a constant mass source rate into a selected computational cell from start of injection to end of injection. Associated with this mass source is an injector velocity which gives the injector momentum source rate.

Fig. 15 shows contours of constant equivalence ratio for one particular injector location, with no air swirl, at selected times after injection [30]. Contour values are for ϕ = 2.0, 1.5, 1.0 and 0.5, with the richest values nearest the jet center. The overall equivalence ratio is 0.8. As the gas jet moves, air is continually entrained, both at the front of the jet and along the side boundaries. Shear stresses are large at the interface between the jet and the surrounding air, resulting in

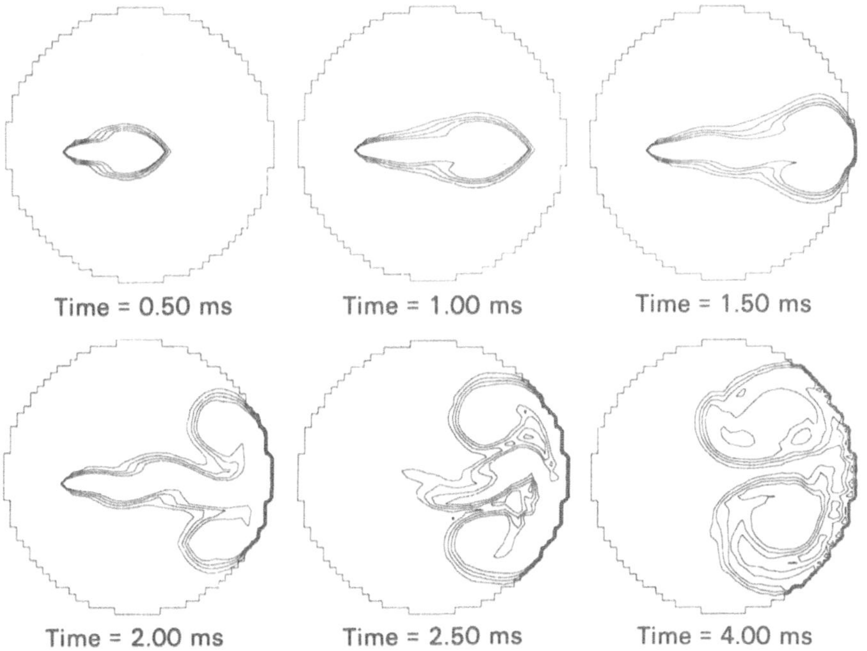

Fig. 15. Results from a multidimensional model study of a gaseous fuel jet mixing with quiescent air in an engine cylinder [30]. Contours of constant equivalence ratio (ϕ = 2.0, 1.5, 1.0 and 0.5 with the richest values nearest the jet center) are shown.

References pp. 33-35.

large values for the viscosity in the individual species transport equations. This produces rapid mixing of fuel and air near the interface. The core of the jet remains very rich, however. As the jet interacts with the wall, it splits and rolls up. Large fuel-rich regions persist for some considerable time. Haselman and Westbrook used this model to examine the effects of injector location, injection velocity and air-swirl on fuel-air mixing rates.

The similarity between the computed gaseous-fuel jet shape during this mixing process and the jet-mixing process which occurs in the swirl-chamber diesel-engine combustion process shown in the photograph in Fig. 4d is striking. In the photograph (one of a sequence of photographs in the original reference [4]), the persistence of fuel-rich soot-laden regions in the core of the jet for long periods of time is also evident, both before and after the jet interacts with the wall. Similar modeling studies of this gaseous jet-mixing process might suggest how modifications to the prechamber nozzle geometry and location would improve rates of mixing of the burning fuel-rich jet with excess air and thus enhance the soot burn-up process.

SUMMARY COMMENTS

In reviewing the history, current status, and some examples of the use of various types of *IC* engine combustion model, I have tried to develop these points:

(i) Several different types of engine combustion model have already been and are being developed to predict certain aspects of engine operation. These were classified as zero-dimensional, quasi-dimensional and multidimensional models according to whether and how the governing equations are organized to include the geometry-induced fluid-flow aspects of the problem.

(ii) These different types of model, because of their different underlying assumptions, organization, stages of development, and computational cost, are best suited to analyzing different aspects of how the engine "combustion process" influences engine operation and emissions.

(iii) Zero-dimensional models can provide information on the effects that changes in engine design and operating variables exert on engine operation and emissions when the corresponding changes in the combustion process are not a dominant factor in the problem.

(iv) Quasi-dimensional models include the effects of these changes in engine design and operation on the details of the combustion process through a phenomenological sub-model where the geometric details are only approximated. They thus extend the range of problems to which the thermodynamic framework of the zero-dimensional model can be applied to problems where changes in the combustion process may be one of the dominant factors. Since computational costs are modest, in the long-term as better intuitive physical models for the combustion process are developed, quasi-dimensional models are likely to be the type of model used for extensive parametric studies of engine operating and emissions characteristics. However, quasi-dimensional models cannot examine in a truly predictive fashion the details of the fluid-flow engine-geometry interaction.

(v) Multidimensional models do have the potential for examining the details of this fluid-flow engine-geometry interaction. Currently they are limited by the inadequacy of sub-models for turbulence, combustion chemistry, etc., and by computer size and cost of operation, to crude approximations to the real flow and combustion problem. Though the impact of these limitations will decrease as these models develop, even in the longer term they are likely to be most useful in developing our conceptual understanding of the details of processes occurring inside the engine which are important in determining engine operating characteristics, rather than in extensive quantitative parametric studies.

Finally, it should be obvious that the complexity of the real combustion process is so overwhelming that any modeling activity that attempts to quantify the major features will always have a speculative character. Substantial simplifying assumptions must always be made to obtain solutions. Yet combustion modeling can provide an extremely useful framework—sometimes qualitative and conceptual, sometimes quantitative—for engine problems strongly influenced by the combustion process. But given this speculative aspect of combustion modeling, the activity will flourish best in an environment where its successes are appreciated, and its inadequacies and occasional inevitable failures are forgiven. In other words, don't expect a home run every time you send a batter to the plate.

ACKNOWLEDGMENT

My own activities in this area have been supported recently by the U.S. Department of Energy under Contract E(11-1)-2881. I am indebted to Dr. K. Nakanishi of Toyota Motor Co., Ltd., Dr. V. W. Wong of M.I.T. and Mr. W. M. Scott of Ricardo Consulting Engineers, Ltd. for kindly supplying me with slides or prints from the excellent combustion movies which they had taken in different types of internal combustion engines.

REFERENCES

1. *J. B. Heywood, R. J. Tabaczynski, J. M. Higgins and P. Watts, "Development and Use of a Cycle Simulation to Predict SI Engine Efficiency and NOx Emissions," SAE Paper No. 790291, 1979.*
2. *K. Nakanishi, T. Hirano and T. Inoue, "The Effects of Charge Dilution on Combustion and Its Improvement—Flame Photograph Study," SAE Trans., Vol. 84, Paper No. 750054, pp. 352–364, 1975.*
3. *V. W. Wong, J. M. Rife and M. K. Martin, "Experiments in Stratified Combustion in a Rapid Compression Machine," SAE Paper No. 780638, 1978.*
4. *J. F. Alcock and W. M. Scott, "Some More Light on Diesel Combustion," Proc. I. Mech. E. (Auto Division), No. 5, pp. 179–200, 1962–63.*
5. *K. Iinuma and Y. Iba, "Studies of Flame Propagation (Structure of Flame Zone and Burning Velocity)," JARI Technical Memorandum No. 15, Combustion and Emission Research Committee, Japan Automobile Research Institute, Inc., December 1973.*

6. C. L. Cummins, Jr., "Internal Fire," Carnot Press, Lake Oswego, Oregon, 1976.
7. A. Hershey, J. Eberhardt and H. C. Hottel, "Thermodynamic Properties of the Working Fluid in Internal-Combustion Engines," SAE J., Vol. 31, pp. 409–424, 1936.
8. H. C. Hottel and J. E. Eberhardt, "A Mollier Diagram for the Internal-Combustion Engine," Proceedings of the Second Symposium on Combustion (held at the Ninety-Fourth Meeting of the American Chemical Society at Rochester, New York, September 9–10, 1937), First and Second Symposia (International) on Combustion, The Combustion Institute, Pittsburgh, Pennsylvania, pp. 234–248, 1965.
9. J. B. Heywood, "Pollutant Formation and Control in Spark-Ignition Engines," Progress in Energy and Combustion Science, Vol. 1, pp. 135–164, 1976.
10. R. J. Tabaczynski, "Turbulence and Turbulent Combustion in Spark-Ignition Engines," Progress in Energy and Combustion Sciences, Vol. 2, pp. 143–165, 1976.
11. P. N. Blumberg, G. A. Lavoie and R. J. Tabaczynski, "Phenomenological Models for Reciprocating Internal Combustion Engines," Paper presented at the U.S. Department of Energy Division of Power Systems sponsored workshop on Modeling of Combustion in Practical Systems, Los Angeles, January, 1978; to be published in Progress in Energy and Combustion Science.
12. F. V. Bracco, "Introducing a New Generation of More Detailed and Informative Combustion Models," SAE Paper No. 741174, November, 1974.
13. F. V. Bracco, "Modeling of Two-Phase, Two-Dimensional, Unsteady Combustion for Internal Combustion Engines," I. Mech. E. International Conference on Stratified-Charge Engines, London, England, November 1976.
14. N. A. Henein, "Analysis of Pollutant Formation and Control and Fuel Economy in Diesel Engines," Progress in Energy and Combustion Science, Vol. 1, pp. 165–207, 1976.
15. W. J. D. Annand, "Heat Transfer in the Cylinders of Reciprocating Internal Combustion Engines," Proc. I. Mech. E., Vol. 177, No. 36, pp. 973–990, 1963.
16. G. Woschni, "A Universally Applicable Equation for the Instantaneous Heat Transfer Coefficient in the Internal Combustion Engine," SAE Trans., Vol. 76, Paper No. 670931, pp. 3065–3083, 1968.
17. R. H. Sherman and P. N. Blumberg, "The Influence of Induction and Exhaust Processes on Emissions and Fuel Consumption in the Spark-Ignited Engine," SAE Trans., Vol. 86, Paper No. 770880, pp. 3025–3040, 1977.
18. G. E. Andrews, D. Bradley and S. B. Lwakabamba, "Turbulence and Turbulent Flame Propagation—A Critical Appraisal," Comb. and Flame, Vol. 24, pp. 285–304, 1975.
19. N. C. Blizard and J. C. Keck, "Experimental and Theoretical Investigation of Turbulent Burning Model for Internal Combustion Engines," SAE Trans., Vol. 83, Paper No. 740191, pp. 846–864, 1974.
20. R. J. Tabaczynski, C. R. Ferguson and K. Radhakrishnan, "A Turbulent Entrainment Model for Spark-Ignition Engine Combustion," SAE Trans., Vol. 86, Paper No. 770647, pp. 2414–2433, 1977.
21. S. D. Hires, R. J. Tabaczynski and J. M. Novak, "The Prediction of Ignition Delay and Combustion Intervals for a Homogeneous Charge, Spark Ignition Engine," SAE Paper No. 780232, 1978.
22. F. A. Williams, "Combustion Theory," Addison-Wesley Publishing Co., Inc., Reading, Massachusetts, 1965.
23. A. A. Boni, "An Overview of Numerical Simulation of Combustion for Automotive Applications," National Science Foundation Workshop on the Numerical Simulation of Combustion for Application to Spark and Compression Ignition Engines, La Jolla, California, April 1975.
24. W. T. Lyn, "Study of Burning Rate and Nature of Combustion in Diesel Engines," Ninth Symposium (International) on Combustion, The Combustion Institute, Pittsburgh, Pennsylvania, pp. 1069–1082, 1962.
25. J. A. LoRusso and R. J. Tabaczynski, "Combustion and Emissions Characteristics of Methanol, Methanol-Water and Gasoline-Methanol Blends in a Spark-Ignition Engine," Paper No. 769019, Proceedings of 11th Intersociety Energy Conversion Engineering Conference, pp. 122–132, Lake Tahoe, Nevada, September 1976.
26. J. C. Wall, J. B. Heywood and W. A. Woods, "Parametric Studies Using a Cycle-Simulation

Model on Performance and NOx Emissions of the Prechamber Three-Valve Stratified Charge Engine,'' SAE Paper No. 780320, 1978.

27. *D. P. Hoult and V. W. Wong, "Generation of Turbulence in an Internal Combustion Engine," General Motors Research Laboratories Symposium on Combustion Modeling in Reciprocating Engines, Warren, Michigan, November 1978.*
28. *S. Hiraki, "Performance and NOx Modeling in a Direct Injection Stratified Charge Engine," M.I.T. S. M. Thesis, 1978.*
29. *A. D. Gosman and R. J. R. Johns, "Development of a Predictive Tool for In-Cylinder Gas Motion in Engines," SAE Paper No. 780315, 1978.*
30. *L. C. Haselman and C. K. Westbrook, "A Theoretical Model for Two-Phase Fuel Injection in Stratified Charge Engines," SAE Paper No. 780318, 1978.*

DISCUSSION

F. C. Gouldin *(Cornell University)*

In your presentation you seemed to be very much in favor of quasi-dimensional models, but I wonder if, in practice, there is a significant difference between a quasi-dimensional model and a zero-dimensional model. My reasoning is the following: In the zero-dimensional model the specification for heat-release rate depends on a few parameters that generally are a function of crank angle. In the quasi-dimensional model that specification is replaced with a flame-speed model. The flame-speed model is based on intuition and allows one to introduce a number of other parameters or variables. But the problem with that type of quasi-dimensional model is in the verification of the flame-speed model. It seems to me that what one has to do is compare model calculations for different engine operating conditions with the actual engine operating conditions, and then back out the various empirical parameters that go into the model. That's the best one can do in verifying that particular turbulent flame-speed model, because I don't think that model would be applicable to steady-state flames in a turbulent premixed-flame burner. The model would be specific to that particular engine because the turbulent combustion process is specific to that engine. So there is always the difficulty of verifying any turbulent-flame model beyond a particular engine.

J. B. Heywood

The difficulty that you are describing is very real, but I think it's there in all types of models. There has to be this calibration to experiments at some point in developing the model. But I feel there is a very important distinction between the zero-dimensional and quasi-dimensional models. The quasi-dimensional models have the potential for incorporating a sufficiently accurate description of the turbulent flame-propagation process, so that one can achieve the coupling between engine details and the combustion process of the type I've shown by example, into these calculations. It's very important to try to achieve that coupling, because the extensive use of these models in parametric studies (for example, to look at changes in design and operating variables) is going to require

J. B. Heywood

a very large number of calculations. A quasi-dimensional model is well suited to this kind of use, and if we can get the physics in there in a sufficiently appropriate fashion, I believe that the parametric study will be its major application. The dimensional models, I feel, are likely to be used in a much more conceptual, but nonetheless important, fashion because I cannot visualize using that kind of calculation procedure to do hundreds of calculations for slightly different engine details. That's my own view. Maybe some of you disagree and would want to comment on that.

W. G. Agnew *(General Motors Research Laboratories)*

John, you didn't comment on the differences between one-dimensional, two-dimensional and three-dimensional models in your multidimensional classification. I'm wondering what you think about the likely usefulness of higher-dimensional models.

J. B. Heywood

I have some qualitative comments about that kind of modeling. Numerical modeling has progressed to the point where two-dimensional simulations under a restrictive set of assumptions can be done, but people are not quite sure about the feasibility of three-dimensional models. I'm sure there will be lots of discussion about that later because several people actively involved in that area are in the audience and giving talks. I don't think one-dimensional calculations are very useful, other than in a purely conceptual sense, and I think that's well recognized. The two-dimensional calculations, I feel, are already useful in a conceptual sense, but they're not going to be precise because of the two-dimensional limitations and some other limitations currently required to solve the required sets of equations. I won't comment in more detail on the potential for three-dimensional calculations because there are people here who can do that much better; I'd like to ask them to comment on whether they see these calculations being used in the not-too-distant future.

C. A. Amann *(General Motors Research Laboratories)*

I would observe that while three-dimensional techniques are being developed, how much they are used has a lot to do with how much computer time one can afford. Perhaps some of the people in that business can tell us what the outlook is for decreasing computer requirements with three-dimensional models.

A. D. Gosman *(Imperial College, England)*

I have two comments about three-dimensional calculations. First, I think that the two-dimensional area needs to be further explored in order to allow us to assess the accuracy of the computational methods being employed. This assessment is, of course, much easier to make for two-dimensional than three-dimensional calculations. Secondly, I think that when we get to three-dimensions we're going to have to have methods which are cost competitive with the alternative—a long series of experiments in real engines. I think one has to accept the fact that these kinds of calculations are going to be more expensive than cycle models regardless of how efficient we make them. They have to be able to produce something, in return, cheaper than the alternative.

J. B. Heywood

What they have to offer in return for this greater expense is that the geometry is built into the problem in an appropriate fashion. There is no other way of including the geometry from a fundamental point of view.

F. F. Pischinger *(Institute of Applied Thermodynamics, Germany)*

As to the question of usefulness, it seems to me it's very important to link to reality, as you mentioned. That means to link to measurements. When one uses a three-dimensional engine model and its associated turbulence model, then one must also be able to measure turbulence. I think the final success of the three-dimensional model will therefore be linked with the possibility of measuring turbulence in a real engine, and also of evaluating the influence, for instance, of the inflow into the engine on the turbulence exchange process. Hand-in-hand with the modeling progress there must be progress in measuring turbulence parameters within the engine. Success will depend on this.

W. A. Sirignano *(Princeton University)*

I'd like to take issue with your model classification and some of your points of view. Your quasi-dimensional characterization is at best a Madison Avenue change of words from zero-dimensional, and at worst its a step backwards in that you are now inserting into the model phenomenological statements which you would find difficult to support. You showed some diagrams, for example, from a diesel engine and made some detailed statements about the way this engine operates, which I think you would be a bit hard put to justify. Beyond that, I think the one-dimensional model should be separated out into a category between zero-dimensional and higher dimensional models (for example, two- and three-dimensional), because the one-dimensional approach is a relatively inexpensive way to make some calculations, and it allows one to talk about things like flame structure with reverence for things like conservation laws, details of heat and mass transfer, etc. I believe we have to go to multidimensional models, and in fact my own research is now concentrating on two-dimensional models. If anything, the place which you reserve for quasi-dimensional models should be taken by one-dimensional models. Higher dimensional models at this point require a good deal of research with regards to accuracy and cost. They are on the very frontier.

J. B. Heywood

Let me say that in answering Fred Gouldin's point earlier, I tried to show where I thought the value of the quasi-dimensional model lies. In talking about three classifications of models I did not intend to make any value judgments about which class of model I believed was better. I really believe one cannot say which is better because each class of model is useful for different kinds of problems. I think the points you raised, however, will be intensely debated during the next two days.

SESSION I

MODELING OF FLUID MOTIONS IN ENGINES

Session Chairman
W. C. REYNOLDS

Stanford University
Stanford, California

MODELING OF FLUID MOTIONS IN ENGINES—AN INTRODUCTORY OVERVIEW

W. C. REYNOLDS

Stanford University, Stanford, California

ABSTRACT

This paper begins with a broad overview of the general nature of fluid motion in engine cylinders, from the intake through the exhaust processes. Particular attention is given to the processes by which turbulence is formed, convected and distorted during these processes. The general magnitudes of the pertinent turbulence length and velocity scales are reviewed, and problems in determining physically meaningful scales in the presence of large cycle-to-cycle variations are discussed. It is shown that cycle times are short compared to relaxation times for important flow-adjustment phenomena, and hence that prediction methods based on statistically steady turbulent flows may not work well in engine analysis.

The paper then reviews the essential ingredients of three types of flow-prediction methods, emphasizing the problems in extending these methods to engine flows. Zonal models require knowledge of the location of important shear layers and simple correlations for their behavior. Full-field models require complex turbulence-closure models, and present models are based almost entirely on incompressible, steady-flow data. Large-eddy simulation, an approach currently in the very early stages of development, may offer the best hope for engine flow modeling. The method is described, and the extensions that must be made before it can be applied in engines are outlined.

NOTATION

C, C_0, C_1, C_2	model constants
e	internal energy per unit mass
f, g, h	flow variables
$\tilde{f}$	phase-averaged flow variable

References pp. 64-65.

$\bar{f}$	mass-weighted phase-averaged variable
f'	fluctuating component of flow variable
$G(\underline{\mathbf{u}}, \underline{\mathbf{u}}')$	filter function
H_i	diffusive flux of $\bar{\rho}\epsilon$
J_ϕ	diffusive transport of quantity ϕ
k	turbulence kinetic energy per unit mass
l	large eddy turbulence scale
l_I	integral scale
l_T	Taylor microscale
l_K	Kolmogorov microscale $\equiv (\nu^3/\epsilon)^{1/4}$
P	rate of turbulence production per unit volume
Q	thermal energy source term
q_j	heat flux
r_{ij}	(negative) subgrid-scale turbulent stress
S	rms strain rate of the large-scale field
$\bar{S}_{ij}$	strain rate of the $\overline{u_i}$ field $\equiv \frac{1}{2}\left(\frac{\partial \overline{u_i}}{\partial x_j} + \frac{\partial \overline{u_j}}{\partial x_i}\right)$
S_α	species source term
U	mean velocity of the flow
ΔU	maximum velocity difference in a shear layer or boundary layer
$\underline{\mathbf{u}}$	velocity vector
u_i	velocity component in the i direction
W	source term of $\bar{\rho}\epsilon$
Y_α	species concentration per unit mass
Δ	filter width
δ	thickness of a shear layer or boundary layer
ϵ	rate of turbulence energy dissipation per unit mass

μ_T	turbulent viscosity
ν	kinematic viscosity
ρ	density
σ_ϕ	turbulent Prandtl number for ϕ
τ_d	development time
τ_e	characteristic time scale of a large eddy turnover time
τ_f	flow-past time
τ_I	integral time scale
τ_{ij}	stress tensor
D/Dt	mass convective operator
∇_0	divergence operator
$\{\ \}$	phase averaging
$\langle\ \rangle$	mass-weighted phase-averaging

INTRODUCTION

The fluid motion within the cylinder of a piston engine has a major influence on the performance of the engine [1]. The general motion within the cylinder and the associated turbulence affect the charge stratification, combustion, heat transfer and purging processes. In order to predict the charge stratification, one must understand and be able to predict the turbulent mixing processes. Combustion analysis requires knowledge of the turbulence structure and turbulence scales. The flow in wall boundary layers controls the heat transfer and quenching. Since pollutants are formed in various amounts in different regions, analysis of the exhaust emissions requires the ability to predict which fluid elements will be purged during the exhaust and which will remain. Hence, a complete predictive capability requires a great deal of knowledge about the general circulation in the cylinder, the turbulence and the boundary layers. The purpose of this paper is to provide an overview of what is known about these phenomena, together with an overview of current turbulence models and suggestions for key experiments needed for model development and validation.

A QUALITATIVE PICTURE

Let's begin by developing a qualitative picture of the important flow features in a typical engine cycle. This picture is based on measurements in motored

References pp. 64-65.

engines, photographs in operating engines and, to a great extent, on the behavior of other flows of similar character.

During the intake process the flow through the valve produces sharp shear layers off the valve lip and seat (Fig. 1a) [2]. These shear layers are dynamically unstable and break down initially into ring-like vortices which merge to form larger-scale vortices. These larger-scale vortices in turn break down into three-dimensional turbulent motions [3, 4]. The resulting conical turbulent jet contains a broad range of turbulence scales ranging from large eddies of the order of the thickness of the jet to small-scale eddies that dissipate the turbulent motion [5]. The jet flow induces a general circulation in the cylinder, which might be compounded by inlet swirl produced in the manifold. The separation of the shear layers off the valve lip and seat sets up recirculation regions [2], and the downward motion of the piston induces a boundary layer on the cylinder wall. The large-scale circulations probably induce smaller reverse circulation in the corners.

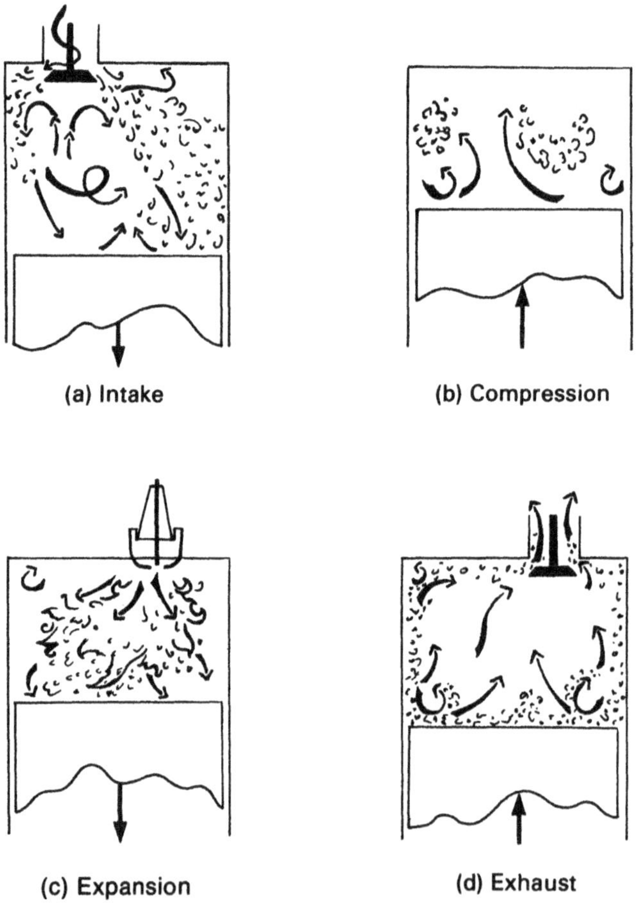

Fig. 1. Flow structure in a spark-ignition engine.

Recirculating flows of this type are typically very sensitive to minor variations in the flow, and hence there probably are substantial cycle-to-cycle variations in the locations and sizes of the recirculating regions.

If the engine has a separately aspirated prechamber, the situation is even more complex. Within the prechamber the flow will have many of the features mentioned above, and in addition there will be flow from the prechamber to the main chamber through the connecting passages. The resulting jet or jets will further complicate the motion within the main chamber.

The details of the recirculating patterns will depend heavily on the system geometry. For example, there are additional recirculations in deep-dished pistons [2].

As the valve closes, the sharp shear layers disappear, but the turbulence generated by them remains. The large-scale circulations convect the small-scale turbulence, which will exist in local patches until diffused through the cylinder by the large-scale motions (Fig. 1b). The structure of the turbulence is altered during the compression process. (This phenomenon is treated in detail by the paper of Hoult and Wong in these proceedings.) In the absence of viscous effects, particles of fluid try to maintain their angular momentum, and therefore the compression process tends to increase the angular velocity. These changes in the small-scale turbulence are produced both by the piston-induced compression and by the strain induced by the general circulation in the cylinder. Near TDC the flow may be complicated by ejection of material from squish regions. In prechamber engines there will be flow into the prechamber during compression, and the prechamber turbulence will be enhanced by these jet flows. In direct-injection engines the general circulation could be enhanced or reduced by the fuel injection, depending on the injector location and direction, and small-scale turbulence will be formed by the injection process.

Ignition and combustion depend very significantly on the flow structure, particularly in stratified-charge engines. Of particular importance at this stage will be the small-scale structure of the turbulence, which influences flame propagation [6, 7]. The rapid rise in temperature associated with combustion suddenly increases the fluid viscosity, thereby increasing the rate of decay of turbulence and increasing the scales of motion that are effective for molecular mixing. In prechamber engines the flow of reacting gases from the prechamber will alter the turbulence structure in the main chamber, and the subsequent combustion in the main chamber can lead to back-flow into the prechamber [8].

During the expansion stroke (Fig. 1c) the flow is again altered by the straining and continued combustion. A boundary layer is formed on the cylinder wall. Large-scale circulations, the residues of earlier processes, certainly exist during this period. Their general nature probably varies significantly with the engine design [9], and perhaps even from cycle to cycle in a given engine.

During the exhaust process the fluid must be peeled off the cylinder wall, and there is some evidence [10] that this gives rise to the formation of a large central vortex (Fig. 1d). Quench layers, formed at the walls during combustion, are expelled at some point in the process [11, 12]. In prechamber engines the exhaust from the prechamber will produce additional turbulence and mixing in the main

References pp. 64-65.

chamber. These complex events determine which fluid elements (and hence which pollutants) remain in the cylinder at the conclusion of the exhaust process.

In summary, the flow in a piston engine involves a complicated system of turbulent shear layers, boundary layers, and recirculating regions. It is very unsteady and may exhibit substantial cycle-to-cycle variations. Large changes in fluid transport properties accompany the temperature rise during combustion. Both the large-scale and small-scale turbulent motions are important factors in regulating the overall processes. The detailed modeling of this flow presents a severe challenge to fluid mechanicians and should not be undertaken with undue optimism.

LENGTH AND TIME SCALES

A number of length scales characterize the flow in engine cylinders. The general circulations are of the scale of the cylinder, and any turbulence of this scale will be very anisotropic. These large-scale, nearly two-dimensional, general circulations are fed by the continual merging of smaller-scale motions of nearly two-dimensional character [3]. In essence, the nearly two-dimensional vorticity of the shear layers diffuses into the recirculating regions, where it coalesces to form the large-scale recirculations.

The eddies responsible for most of the turbulence production during the intake process are the large eddies in the conical inlet jet flow. These are of a size roughly equal to the local thickness of the jet [5]. Hence, these eddies are approximately the size of the valve lift and become larger in size as they move away from the valve in the widening jet flow. The character of energy-producing eddies depends strongly on the nature of the flow. In the conical valve jet they may be nearly two-dimensional, ring-like vortices [3] very near the valve; but as the jet expands, these vortices become unstable and the jet's large eddies become very three-dimensional. The largest of these structures are quite anisotropic; their nature depends critically on the system geometry and flow history [4].

Superposed on the general recirculation and large-scale turbulence is a range of eddies of smaller and smaller size, fed by the continual breakdown of larger three-dimensional eddies. Since the smaller eddies respond more quickly to local changes [5], they tend to be more isotropic in structure and more universal in character. Hence, the small-scale structure of turbulence in an engine is believed to be very much like that found in other turbulent flows, and this has important implications for turbulence modeling. The smallest structures appear to be very thin shear layers distributed throughout the flow like a tangle of ribbons [13]. The dissipation of turbulence energy into molecular energy takes place in these thin layers.

Quantitative measurement of turbulence scales requires measurement of the flow field as a function of space and time. In a flow that is statistically homogeneous in space, the correlation between velocity fluctuations at two separated points is used to define two important scales. A typical correlation has the form of Fig. 2. The *integral scale*, l_I, is defined [5] in terms of the area under the (normalized) correlation curve. The integral scale is determined principally by

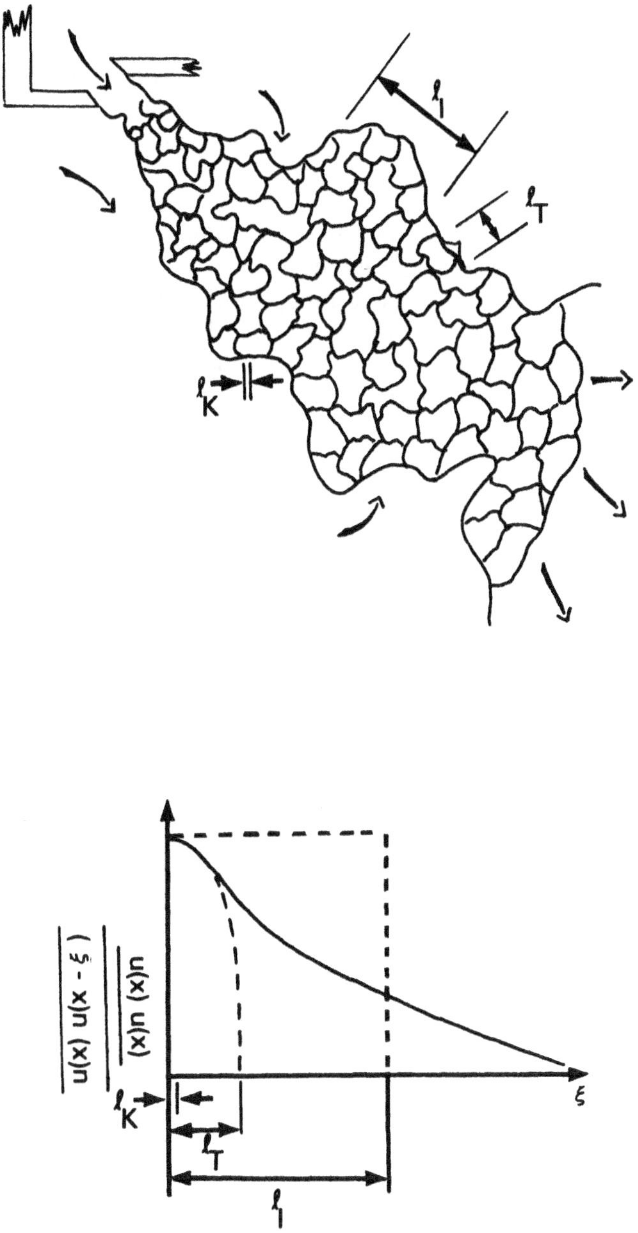

Fig. 2. Length scales in turbulent flow.

References pp. 64-65.

the extent of coherent motions and hence is a measure of the large eddies. The *Taylor microscale*, l_T, is defined [5] by the osculating parabola (Fig. 2) and is believed by some to be a rough measure of the spacing of the very thin shear layers in which viscous dissipation occurs.

A measure of the size of the smallest turbulence structures is determined by the rate at which energy must be dissipated per unit mass, ϵ, and the fluid kinematic viscosity, ν. This is the *Kolmogorov microscale*, l_K [5].

$$l_K = (\nu^3/\epsilon)^{1/4} \tag{1}$$

Several time scales are also important. The *Kolmogorov time scale* [5], $\tau_K = (\nu/\epsilon)^{1/2}$, characterizes the momentum-diffusion time of the smallest structures. In flows that are statistically steady in time, the correlation between the motion at two points in time is used to define an *integral time scale*, τ_I. If it is reasonable to assume that the turbulence pattern is convected past the observing probe without significant distortion and that the turbulence itself is rather weak, then the integral time and length scales are related by

$$l_I = U\tau_I \tag{2}$$

where U is the mean velocity of the flow. In flows without mean motion, the integral time scale provides a rough indication of the lifetime of a large eddy. However, in flows where the large eddies are convected, the integral time scale is instead a rough measure of the time that it takes for a coherent structure to pass a point, and therefore it is not necessarily a measure of the lifetime or mixing time of large eddies.

In shear layers and boundary layers, a time-scale characteristic of the large-eddy turnover time is (Fig. 3)

$$\tau_e = \delta/\Delta U \tag{3}$$

where δ is the thickness of the layer and ΔU is the maximum velocity difference in the layer. In stratified flows, the large-eddy turnover time provides a rough measure of the time scale on which large eddies mix the flow *macroscopically*, which must be accomplished before *molecular* mixing can occur.

The conditions imposed on boundary layers and shear layers in engines change during the engine cycle, and the response of these layers to external changes is not instantaneous. The *flow-past time*

$$\tau_f = L/U \tag{4}$$

measures the time required for a fast-moving fluid particle to travel the length of the boundary layer. The *development time*, τ_d, is the time required for a boundary layer to reach equilibrium with imposed external conditions. Typically of the order of ten flow-past times are required [14], so $\tau_d \approx 10\tau_f$. This has very important implications for engine-flow modeling, since the piston moves from TDC to BDC in only *one* flow-past time for the cylinder wall boundary layer.

Separated and recirculating flow regions are known to have extremely long development times compared to the flow-past times of the shear layers that drive the recirculation [15]. The primary recirculations in the engine survive at most

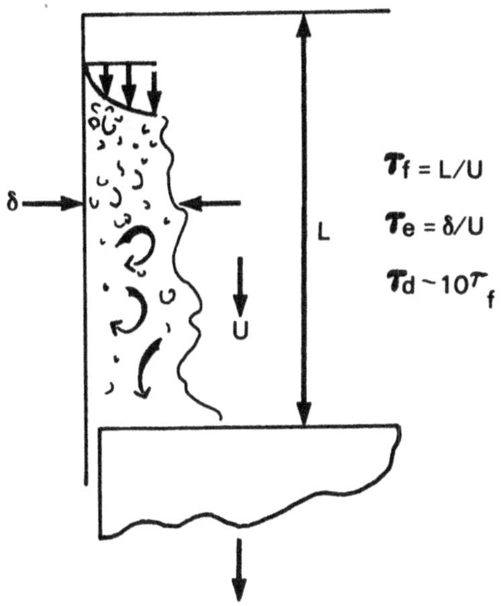

Fig. 3. Time scales in turbulent flow.

one cycle, and this would not appear sufficient for full development. Hence, the rates of entrainment from the recirculating regions may be different from those that would exist in a similar steady flow.

In summary, there is a range of important length and time scales in engine flows. The engine cycle time is short compared to the time required for development of the circulating regions and adjustment of the boundary layers, and models of these features must be constructed with this in mind.

TURBULENCE DEFINITIONS AND MEASUREMENTS

In the engine system there is a serious problem with the basic definitions of turbulence. Since the flow is not steady, the turbulent component of a variable cannot be defined as the departure from its time-averaged value, as one does in statistically steady flows. In periodic flows, the concept of the *phase average* has proven useful [16, 17, 18, 19]. The phase average is defined as the average of values at a given phase in the basic cycle. Fig. 4 displays the phase-average concept for a two-stroke cycle with large cycle-to-cycle variation. The phase-average velocity would be the average over a large number of measurements taken at the same crank angle, indicated by the dots. By repeating this sampling at a number of crank angles, the phase average can be found over a full cycle.

The difference between the instantaneous velocity and the phase-average velocity is defined as the "turbulent" component of the flow field. Under this definition the "turbulence" is the departure from the average over many realizations, but this will not necessarily relate to the random fluctuations in space or time that exist in any single realization. Hence, a laminar flow with large cycle-

References pp. 64-65.

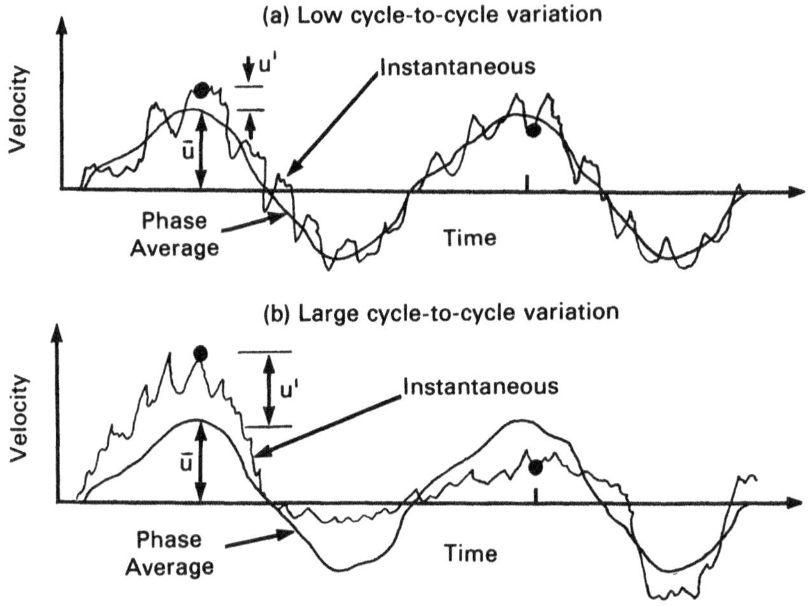

Fig. 4. Phase averaging.

to-cycle variations will appear to have a high degree of "turbulence". In such a case "turbulence" defined in this way can have no relation to the mixing processes occurring in each single realization.

Measurements of the "turbulence" in motored engines, defined in this way, shows rms fluctuation velocities that at times are as great or greater than the phase averages [2, 20, 21]. In boundary layers, jets, and wakes in steady flow, the rms turbulence velocity is typically of the order of 20% of the mean velocity. This striking difference between engine turbulence and steady-flow turbulence suggests that what has been interpreted as turbulence in the engine may instead be cycle-to-cycle variations in the general flow pattern. Therefore, one should be rather cautious in interpreting the reported data on engine turbulence. The implication for flow modelers is that reported values of the length and time scales of the turbulence, determined from one-point histories, may not properly characterize the phenomena to be modeled during a single cycle. The implication for experiments is that two-point measurements may be necessary to determine the physically significant scales during a single cycle.

The nonhomogeneity of engine turbulence further complicates engine turbulence measurements. The large-scale motions are quite anisotropic in structure and very dependent upon system geometry, and their strength varies across the cylinder [2]. The smallest scales certainly are locally isotropic, but their intensity and scale may vary with position. These spatial variations are especially pronounced during the intake process, and it is meaningless to talk about *the* value of the rms turbulence velocity or *the* turbulent length scale during the intake process.

Once the intake valve has closed, mixing tends to reduce the anisotropies. Experiments indicate that the dissipative processes remove energy from the smaller eddies quickly, and hence the large-scale structures become more dominant as time progresses [5]. Hence, the anisotropy of the turbulence responsible for mixing (the largest eddies) probably *increases* during the compression processes.

Those measurements of turbulence scale that have been made [22, 23] suggest that the integral scales are of the order of 10 mm and the Taylor microscales of the order of 0.5 mm in typical engines. Some measurements also suggest that the integral scales do not change appreciably during the compression process. However, turbulence spectra measurements [2, 21] suggest that low-frequency (large-scale) motions become more pronounced during the relaxation process, which suggests that the integral scales should increase after intake-valve closure.

In summary, there is need to develop new ways to define engine turbulence in a more meaningful way, and a great need for good experimental data from which the physically important characteristics of engine turbulence can be determined as functions of position and phase in the engine cycle. These data would be very useful in guiding the development of flow models and are essential to their validation.

BASIC APPROACHES TO FLOW MODELING

We shall consider three quite different approaches to modeling the flow in piston engines: *zonal modeling, full-field modeling,* and *large-eddy simulation.*

In zonal modeling one seeks to compute a complex flow by treating various zones or "flow modules" in different ways. Different models are used for boundary layers, jets, and the large-scale, nearly irrotational recirculating regions. The various analyses must be coupled in appropriate ways so that their interactions are properly represented. A successful example of this approach is the recent diffuser flow model of Ghose and Kline [24]. They used a very simple integral boundary-layer model, properly coupled to a potential-flow solution, and were able to predict quite accurately the peak in diffuser pressure recovery, which occurs when there is a small region of transitory separation in the boundary layer.

The advantage of zonal modeling is that one does not have to have a general model but, instead, can patch together good models of different types of flow. The disadvantage is that one must know what zones to include and where to put them. Thus, the approach is not well suited to flows in which one does not have a clear and accurate picture of the flow structure. Therefore, it seems likely that other approaches will be more successful in engine flows.

In full-field modeling (FFM), one works with the partial differential equations describing suitably averaged quantities, using the same equations everywhere in the flow. In steady incompressible flows, the averages are time averages. The variables include at least the velocity field and may also include various mean turbulence parameters, such as the turbulence kinetic energy, the turbulent stress

References pp. 64-65.

tensor, etc. For compressible flows, other thermodynamic state variables also must be calculated. For periodic flows, such as in piston engines, the time averaging must be replaced by phase averaging.

In FFM, models are needed for various averages of turbulence quantities. These models must reflect the contributions of *all* scales of turbulent motion. The paper by Gosman, Johns, and Watkins in these proceedings [25] is an example of this approach, referred to by those authors as "statistical flux modeling".

Large-eddy simulation (LES) is an approach in which one actually calculates the large-scale three-dimensional time-dependent turbulence structure in a single realization (or set of realizations) of the flow. Thus, only the small-scale turbulence need be modeled. As we have noted, the small-scale structure of turbulence is much more isotropic than the large-scale turbulence, is quite universal in character, and responds rapidly to changes in the large-scale field. This makes the modeling of the statistical fluxes associated with the small-scale motions a simple task compared to that faced in FFM, where the effects of large-scale turbulence must be included in the models.

An important difference between FFM and LES is in the definitions of the "turbulence". In FFM the "turbulence" is the deviation of the flow at any point at any instant from the average over many cycles of the flow at the same point in space and oscillation phase. Thus, FFM "turbulence" contains some contribution from cycle-to-cycle variation. LES, on the other hand, defines turbulence in terms of variations about a local average, and hence in LES the "turbulence" really is related to events in the current cycle. LES should therefore be useful in predicting cycle-to-cycle variations, while FFM cannot be used for this purpose.

Although with projected computer development it does appear that it will be a routine matter to do LES calculations in engine-like systems within a decade, at present the computer time requirements of LES are such that LES calculations are beyond the capabilities of most groups. If one is limited by computer capability to two-dimensional time-dependent calculations, then LES is not possible, and FFM is the only choice. However, if one has the computer capability to do a three-dimensional time-dependent calculation, then it might as well be LES as FFM, and LES is likely to give more accurate results.

Much can be learned now from two-dimensional FFM calculations, and so FFM is expected to remain an important approach for some time. It is by no means a perfected technique, and it really remains to be shown that it can be used successfully in piston-engine flows. The paper by Gosman et al. [25] is a start in this direction.

In the following section the fundamentals of FFM are outlined as applied to the piston-engine problem, and the areas where future development is needed are mentioned. The next section provides an overview of the foundations of LES and its current state of development. Following that section is a very brief discussion of another new flow simulation technique, vortex chasing, which shows promise for use in piston-engine flows, but for which the basic foundations need to be tightened before the method can be applied to compressible turbulent flows with much confidence.

FULL-FIELD MODELING

In this section we develop the foundations of full-field modeling as applied to piston engines. The equations governing the statistical properties of the flow are derived from the basic equations governing the density, ρ, the velocity, u_i, the internal energy per unit mass, e, and the species concentration per unit mass, Y_α. Reduced to their essence, these equations may be written as

$$\frac{D}{Dt}\begin{pmatrix}1\\ u_i\\ e\\ Y_\alpha\end{pmatrix}=\begin{pmatrix}0\\ 0\\ Q\\ S_\alpha\end{pmatrix}-\frac{\partial}{\partial x_j}\begin{pmatrix}0\\ \tau_{ij}\\ q_j\\ J_{\alpha j}\end{pmatrix} \tag{5}$$

where D/Dt is a mass-convective operator

$$\frac{Df}{Dt}\equiv\frac{\partial(\rho f)}{\partial t}+\frac{\partial}{\partial x_j}(\rho u_j f) \tag{6}$$

The first equation expresses conservation of mass, the second (set of three) are the momentum equations, the third is the thermal-energy equation (derived by combination of the true energy equation, the momentum equations, and the species equations), and the fourth is derived from a species balance. The D/Dt operator provides the convective transport terms, the first vector on the right gives the source terms, and the second the diffusive transports.

The thermal-energy source term, Q, involves a viscous term and source terms arising from the chemical reactions. Both Q and the species source, S_α, will depend upon the chemical rate equations, which must be known to close the problem. Thermodynamic state equations are also required. In addition, constitutive equations for the diffusive fluxes must be specified.

A careful development shows that the diffusion of the various species contributes to the diffusive flow of internal energy, q_j, in addition to conductive heat diffusion. Radiative transport, neglected in this formulation, may be important during part of the engine combustion process.

In the FFM approach, one forms equations for the *averaged* variables from Eq. 5. Most turbulence-model development has been done with time-averaging. However, in periodic flows it is more appropriate to use the phase-averaging process described earlier. For engine-flow analysis one needs to consider compressibility. One possibility is just to use the simple phase average throughout, as done (implicitly) by Gosman and co-workers in their paper in this proceedings [25]. Another approach, which has proven useful in other types of compressible flow, is mass-weighted (or Favre) averaging [26]. Favre averaging makes the averaged compressible flow equations look almost exactly the same as the averaged equations for incompressible flow. Thus, the Favre-averaged equations do not have extra terms for which new models must be posed (see for example the footnote to Table 1 in the Gosman paper). Hence, models found successful

References pp. 64-65.

for incompressible flows can be extended directly to the Favre compressible-flow equations*.

The combined phase-Favre averaging approach is fundamental to the analysis of engine flows, and hence we shall go through it here in some detail. We shall indicate the phase-averaging process by *braces*

$$\{\rho(\underline{\mathbf{x}}, t)\} = \lim_{N\to\infty} \frac{1}{N} \sum_{n=1}^{N} \rho(\underline{\mathbf{x}}, t + n\tau) \tag{7}$$

where τ is the cycle period. We will also denote $\{\rho\} = \tilde{\rho}$. Then we decompose ρ as $\rho = \tilde{\rho} + \rho'$. Also, we will indicate the mass-weighted phase-averaging process by *brackets,* and mass-weighted phase-averaged quantities by an overbar,

$$\tilde{\rho}(x, t)\langle f(\underline{\mathbf{x}}, t)\rangle = \lim_{N\to\infty} \left[\frac{1}{N} \sum_{n=1}^{N} \rho(\underline{\mathbf{x}}, t + n\tau)\, \bar{f}(\underline{\mathbf{x}}, t + n\tau)\right] = \tilde{\rho}\bar{f} \tag{8}$$

We decompose all flow variables (except density and pressure) as $f = \bar{f} + f'$. Note that, under these definitions and operations,

$$\begin{gathered}
\langle f'\rangle = 0 \quad \{\rho'\} = 0 \quad \text{but} \quad \{f'\} \neq 0 \\
\{\rho f\} = \tilde{\rho}\bar{f} \quad \{\rho f'\} = 0 \quad \{\bar{f}\} = \bar{f} \\
\{\rho f g\} = \tilde{\rho}(\bar{f}\bar{g} + \overline{f'g'}) \\
\{\rho f g h\} = \tilde{\rho}[\bar{f}\bar{g}\bar{h} + \bar{f}\overline{g'h'} + \bar{g}\overline{f'h'} + \overline{f'g'h'}]
\end{gathered} \tag{9}$$

Phase-averaging Eq. 5, one obtains

$$\frac{\bar{D}}{Dt}\begin{pmatrix} 1 \\ \bar{u}_i \\ \bar{e} \\ \bar{Y}_\alpha \end{pmatrix} = \begin{pmatrix} 0 \\ 0 \\ \{Q\} \\ \{S_\alpha\} \end{pmatrix} - \frac{\partial}{\partial x_j}\begin{pmatrix} 0 \\ \{\tau_{ij}\} \\ \{q_j\} \\ \{J_{\alpha j}\} \end{pmatrix} - \frac{\partial}{\partial x_j}\begin{pmatrix} 0 \\ \tilde{\rho}\overline{u_i'u_j'} \\ \tilde{\rho}\overline{e'u_j'} \\ \tilde{\rho}\overline{Y_\alpha'u_j'} \end{pmatrix} \tag{10}$$

where

$$\frac{\bar{D}\bar{f}}{Dt} \equiv \frac{\partial}{\partial t}(\tilde{\rho}\bar{f}) + \frac{\partial}{\partial x_j}(\tilde{\rho}\bar{u}_j\bar{f})$$

The terms on the left in Eq. 10 involve only the solution variables $\tilde{\rho}$, $\bar{u}_i$, $\bar{e}$ and $\bar{Y}_\alpha$, and hence require no modeling. However, all of the terms on the right, particularly the last terms that represent turbulence transport, involve turbulence fluctuation quantities and must be modeled in terms of the solution variables.

The source terms, $\{Q\}$ and $\{S_\alpha\}$, present special difficulties to the modeler. For example, if the kinetic equation for the heat source term Q is

$$Q = \sum_{\alpha,\beta} \rho A_{\alpha\beta} Y_\alpha Y_\beta \exp[-E_{\alpha\beta}/(RT)] \tag{11}$$

then $\{Q\}$ will contain terms like

* *The extension is merely mathematical; the physical validity of the extension requires experimental validation.*

$$\bar{\rho} A_{\alpha\beta} \overline{Y_\alpha' Y_\beta' \{\exp[-E_{\alpha\beta}/(RT)]\}'} \tag{12}$$

Because of the extreme nonlinear temperature dependence, this term will be very strongly influenced by the positive temperature fluctuations, and this effect must be reflected in the modeling. The use of probability-density functions can be quite helpful in this modeling [27].

The momentum equations contain terms $-\overline{\bar{\rho} u_i' u_j'}$ that represent turbulent stresses. In "stress-equation modeling", equations for these stresses are developed and closed by modeling. A less complex approach which seems worth exploring for engine studies is two-equation turbulence modeling [28]. Here one adds to Eq. 10 two partial differential equations describing velocity and length scales of the turbulence, and then expresses all other turbulence quantities in terms of these scales.

Two-equation turbulence models make use of the equation governing the turbulence kinetic energy per unit mass, $k \equiv \overline{u_i' u_i'}/2$. An equation governing k can be developed by multiplying the u_i equation in Eq. 5 by u_i, subtracting from this the equation formed by multiplying the $\bar{u}_i$ equation in Eq. 10 by $\bar{u}_i$, and phase averaging the result. One finds the k equation as

$$\frac{\bar{D}\bar{k}}{Dt} = \bar{\rho}(P - \epsilon) - \frac{\partial}{\partial x_j} J_k \tag{13a}$$

where P is the rate of turbulence production per unit mass

$$P = -\overline{u_i' u_j'} \frac{\partial \bar{u}_i}{\partial x_j} \tag{13b}$$

ϵ is the rate of turbulence energy dissipation per unit mass and J_k represents diffusive transport.

Practically all two-equation models use as the second turbulence equation a model equation for the dissipation, ϵ. The rationale for this is the assumption that the rate of energy dissipation is controlled not by viscosity but by the rate at which the *large* eddies feed energy to the small dissipative scales, which in turn adjust in size to handle this energy. Thus, if l is a large-eddy turbulence length scale, the dissipation should scale on k and l.

$$\epsilon \propto k^{3/2}/l \tag{14}$$

In a two-equation turbulence model all of the unknown turbulence quantities are modeled in terms of the turbulent velocity scale, $k^{1/2}$, and length scale, $k^{3/2}/\epsilon$, or, alternatively, in terms of k and ϵ. Hence, these models are called k-ϵ models. The usual approach involves the definition of a turbulent viscosity,

$$\mu_T \equiv C_0 \bar{\rho} k^2/\epsilon \tag{15}$$

where C_0 is a model constant. The turbulent stress terms appearing in Eq. 10 and Eq. 13 are then modeled in a quasi-Newtonian manner,

$$\overline{\bar{\rho} u_i' u_j'} = \bar{\rho} \frac{2}{3} k \delta_{ij} + \frac{2}{3} \mu_T \nabla \cdot \underline{\bar{\mathbf{u}}} \delta_{ij} - 2\mu_T \bar{S}_{ij} \tag{16a}$$

References pp. 64-65.

where $\bar{S}_{ij}$ is the strain rate of the $\bar{u}_i$ field,

$$\bar{S}_{ij} = \frac{1}{2}\left(\frac{\partial \bar{u}_i}{\partial x_j} + \frac{\partial \bar{u}_j}{\partial x_i}\right) \tag{16b}$$

The quasi-Newtonian model means that the principal axes of the turbulent stress tensor align with those of the strain-rate tensor. It is well known that these axes are *not* aligned in real turbulent flows, and this is often used as the basis for criticism of the quasi-Newtonian model.

The viscous-stress terms in the momentum equations are evaluated using a Newtonian constitutive equation. They include additional turbulence terms involving fluctuations in the molecular viscosity. These terms may be important in combustion fronts, where large viscosity variations occur, but are neglected in current models.

The turbulent-diffusion terms in the various transport equations are modeled using the turbulent diffusivity. The diffusing flux of a quantity ϕ is modeled by

$$J_{\phi_i} = -\frac{\mu_T}{\sigma_\phi}\frac{\partial \phi}{\partial x_i} \tag{17}$$

where σ_ϕ is a turbulent Prandtl number for ϕ.

Finally, the model is completed with a transport equation for ϵ. An exact equation for ϵ can be developed by suitable manipulation with the Navier-Stokes equations, and this is useful as a guide in formulating the ϵ equation. All ϵ equation models are of the form

$$\frac{\bar{D}\epsilon}{Dt} = W - \frac{\partial H_i}{\partial x_i} \tag{18}$$

where W is a source term and H_i is the diffusive flux of $\tilde{\rho}\epsilon$, which is modeled as are the other diffusion terms. The form of W has been the subject of controversy [28]. While the exact equation for ϵ does provide guidance, in the end one really is reduced to constructing a W that leads to the proper physical behavior in special cases. For incompressible flow one can make good arguments for the model [28]

$$W = [-C_2 + C_1 P/\epsilon]\tilde{\rho}\epsilon^2/k \tag{19}$$

The model of Gosman et al. [25] in these proceedings is of this form, and the constants C_1 and C_2 are as denoted there. The C_2 term produces the proper behavior of homogeneous isotropic turbulence, and the C_1 term modifies the behavior reasonably well for homogeneous shear.

An additional term in W is needed to account for the changes in ϵ produced by dilation of the flow. To illustrate a line of attack used in turbulence modeling, we will now make an argument about this term. Consider the rapid spherical expansion of homogeneous isotropic turbulence, at the start of which

$$\overline{\tilde{\rho} u_i' u_j'} = \tilde{\rho}\frac{2}{3}k\delta_{ij} \tag{20}$$

For a spherical expansion with $\tilde{u} = \Gamma r$,

$$\nabla \cdot \underline{\tilde{u}} = \frac{1}{r^2} \frac{\partial(\tilde{u} r^2)}{\partial r} = 3\Gamma \tag{21}$$

Then, from the continuity equation, treating $\tilde{\rho} = \tilde{\rho}(t)$,

$$\frac{d\tilde{\rho}}{dt} = -\tilde{\rho} \nabla \cdot \underline{\tilde{u}} = -3\Gamma\tilde{\rho} \tag{22}$$

Considering only the production term linear in the mean strain, the k equation is (see Eqs. 10 and 13a)

$$\frac{dk}{dt} = -\frac{2}{3} k \nabla \cdot \underline{\tilde{u}} = -2k\Gamma \tag{23}$$

In rapid distortion the angular momentum of the turbulence should be conserved [29]. Hence, the product of the turbulent length and velocity scales should remain fixed during the expansion. Thus,

$$k^2/\epsilon = \text{constant} \tag{24}$$

$$\frac{d\epsilon}{dt} = 2\frac{\epsilon}{k}\frac{dk}{dt} = -4\epsilon\Gamma \tag{25}$$

Thus, the ϵ equation for this case should be

$$\frac{d(\tilde{\rho}\epsilon)}{dt} = -4\tilde{\rho}\epsilon\Gamma \tag{26}$$

This suggests that, under rapid distortion, W should be $-\frac{4}{3}\tilde{\rho}\epsilon\nabla \cdot \underline{\tilde{u}}$. For this case the W model of Eq. 19 is (neglecting the C_2 term) $-\frac{2}{3}C_1\tilde{\rho}\epsilon\nabla \cdot \underline{\tilde{u}}$. Hence, in order to obtain the proper behavior in rapid dilation, we should add to Eq. 19 a term $(-\frac{4}{3} + \frac{2}{3}C_1)\tilde{\rho}\epsilon\nabla \cdot \underline{\tilde{u}}$. An analysis of the value of C_1 required to produce the correct ϵ behavior for homogeneous straining and shearing of homogeneous incompressible flow [28] suggests $C_1 = 1$. The proper prediction of the decay of isotropic turbulence [28] requires $C_2 = 11/6$. Using these constants, the W model for compressible flows would be

$$W = [-11/6 + P/\epsilon]\tilde{\rho}\epsilon^2/k - \frac{2}{3}\tilde{\rho}\epsilon\nabla \cdot \underline{\tilde{u}} \tag{27}$$

For comparison, the model used by Gosman et al. [25] in these proceedings is

$$W = [-1.92 + 1.44P/\epsilon]\tilde{\rho}\epsilon^2/k + \tilde{\rho}\epsilon\nabla \cdot \underline{\tilde{u}} \tag{28}$$

The constants in their model were determined by fits to selected turbulent shear flows, not by reference to the simpler experiments described above. Their model will not predict the rapid dilation behavior outlined above. Good experimental data for some simple compressible turbulence experiments are needed before choices of this sort can be properly resolved.

The models described above must be modified at low turbulence Reynolds numbers, for example in the region very near a wall or (perhaps) in a flame front [28].

Two-equation models have been used successfully in both free and wall-bound shear layers. With modifications mentioned above, the models work reasonably well right to the wall. However, if one tries to carry the calculations to the wall, a very fine mesh is needed near the wall, and so the usual procedure is to fit the solution near the wall to an assumed "universal" wall layer form. An example of this is detailed in the paper by Gosman et al. [25] in these proceedings. However, in an engine the wall boundary layers do not have time to relax to local equilibrium, and thus this wall treatment is a potential source of error.

In summary, the most complex form of full-field turbulence modeling that seems reasonable to use for engine studies is a two-equation turbulence model. Current models are based almost entirely on statistically steady incompressible flow data. There is considerable uncertainty as to how best to extend these models to time-dependent compressible flows. Proper formulation requires phase averaging, and the use of mass-weighted averaging appears to offer desirable simplifications.

LARGE-EDDY SIMULATION

A large-eddy simulation is a three-dimensional time-dependent calculation of the large-scale turbulence field. Modeling is necessary only for turbulence of scales smaller than the computational grid spacing. Thus, one does not have to concoct a universal model for large-scale turbulence, as one must with full-field modeling. The small-scale turbulence that must be modeled is much more universal in character and hence can be modeled easily in terms of the local large-scale field.

An engine flow is a time-dependent, three-dimensional flow, so even if one elects to use a full-field model the calculation will have to be time-dependent and three-dimensional. This calculation will require a fine grid to resolve the regions of rapid variations. Given the complexity of this full-field calculation, a large-eddy simulation might actually be simpler, shorter in execution and more accurate.

It might be thought that a time-dependent three-dimensional full-field calculation is itself a large-eddy simulation, but it is not. In full-field simulations the length scales of the turbulence usually are much larger than the grid spacing. FFM will only reveal unsteady motions of scales larger than the model's turbulence scale. Motions with scales of the order of the turbulence length scale and smaller are considered as part of the modeled turbulence, and these motions will not appear as part of the computed time-dependent flow field in FFM.

LES is like FFM except that the length scales of the subgrid-scale turbulence are set by the grid and not by a model equation. Thus, in an LES the computed time-dependent three-dimensional field captures *all* of the turbulence down to the grid size.

In regions of the flow where the grid size is larger than the turbulence scales in the flow, the length scale of the subgrid turbulence must be smaller than that of the grid. Thus, the length scales appropriate for FFM are useful upper limits on the length scales of subgrid motions in LES. In other words, coarse-grid LES should reduce to FFM.

In the engine problem there is another important difference between LES and FFM. FFM is intended to calculate phase-average properties and requires models for such quantities as phase-average turbulent stresses. We have commented on the way in which substantial cycle-to-cycle variations can make these phase averages be rather unrelated to the events during a single realization. In contrast, turbulence in LES is defined as the departure from a local average at a single instant in time. Hence, LES carried out over several cycles should reveal the magnitudes of cycle-to-cycle variations, which cannot be found by FFM. These considerations suggest that the best way to attack piston engine flow computation is through LES.

The first step in LES is the precise definition of the large-scale velocity, pressure, density, temperature and concentration fields. These should represent a smoothing of the actual fields, with the smoothing removing the small-scale fluctuations. Hence, the large-scale field of a variable at a point, $\underline{\mathbf{x}}$, should be defined as some sort of a local average of the actual field at points near $\underline{\mathbf{x}}$. The deviation of the actual field from the local average field is then the small-scale "turbulence". This idea leads to the mathematical definition of *filtering,* shown schematically in Fig. 5. The large-scale field of a variable $f(\underline{\mathbf{x}}, t)$ is denoted by $\bar{f}(\underline{\mathbf{x}}, t)$ and defined by

$$\bar{f}(\underline{\mathbf{x}}, t) = \int G(\underline{\mathbf{x}}, \underline{\mathbf{x}}')\, f(\underline{\mathbf{x}}', t)\, d\underline{\mathbf{x}}' \tag{29}$$

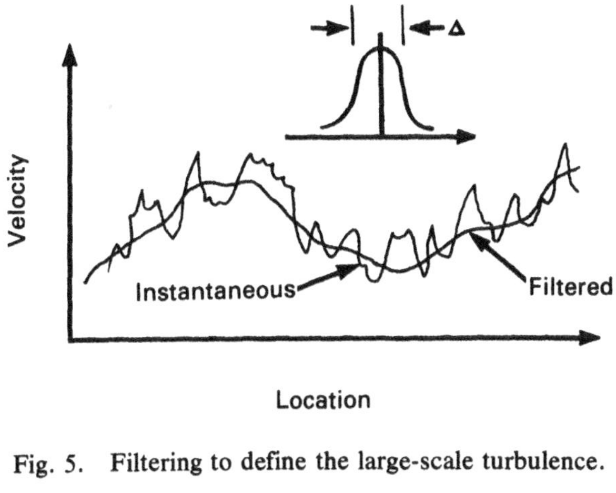

Fig. 5. Filtering to define the large-scale turbulence.

References pp. 64-65.

where the integration is over the entire flow volume. The function $G(\underline{x}, \underline{x}')$ is the filter function; it determines the weighting to be given to points near and away from $\underline{x}$. Every filter function will have a width parameter that indicates the region over which significant contributions are made to the filtered field at any point. This width parameter defines the *largest* scale of *subgrid* turbulence, and hence is the relevant length scale for the subgrid scale turbulence modeling. For a constant field one requires $\bar{f} = f$; hence G must satisfy

$$\int G(\underline{x}, \underline{x}')d\underline{x}' = 1 \tag{30}$$

A variety of filters have been explored. For homogeneous flows a Gaussian filter has advantages [30]. For flows with solid boundaries, in which a variable mesh is required, the box filter is more convenient. The box-filtered value of a variable at point $\underline{x}$ is the average over a box surrounding point $\underline{x}$. If the box extends from $x_i - \Delta_i^-(\underline{x})$ to $x_i + \Delta_i^+(\underline{x})$ in each direction, the box filter is

$$G = \begin{cases} \dfrac{1}{\prod_{i=1}^{3} (\Delta_i^- + \Delta_i^+)} & x_i - \Delta_i^-(\underline{x}) \leq x' \leq x_i + \Delta_i^+(x) \quad \text{for } i = 1, 2, 3 \\ 0 & \text{otherwise} \end{cases} \tag{31}$$

The equations for the large-scale fields are obtained by filtering the governing Eq. 5. For compressible flows the concept of mass-weighted filtering is again useful. We define

$$\bar{\rho}(\underline{x}, t) = \int \rho(\underline{x}', t)G(\underline{x}, \underline{x}')d\underline{x}' \tag{32a}$$

$$\bar{\rho}(\underline{x}, t)\tilde{f}(\underline{x}, t) = \int \rho(\underline{x}', t) f(\underline{x}', t)G(\underline{x}, \underline{x}')d\underline{x}' \tag{32b}$$

Then, to filter the continuity equation, Eq. 5, we multiply it by the filter function and integrate, obtaining

$$\frac{\partial \bar{\rho}}{\partial t} + \frac{\partial(\bar{\rho}\tilde{u}_j)}{\partial x_j} = 0 \tag{33}$$

Filtering the momentum equation yields

$$\frac{\partial(\bar{\rho}\tilde{u}_i)}{\partial t} + \overline{\frac{\partial}{\partial x_j}(\rho u_i u_j)} = -\frac{\partial \bar{\tau}_{ij}}{\partial x_j} \tag{34}$$

Now we decompose each variable into large- and small-scale components

$$\rho = \bar{\rho} + \rho' \qquad u_i = \tilde{u}_i + u_i' \tag{35}$$

Then Eq. 34 becomes

$$\frac{\partial(\tilde{\rho}\bar{u}_i)}{\partial t} + \frac{\overline{\partial(\tilde{\rho}\bar{u}_i\bar{u}_j)}}{\partial x_j} + \frac{\partial r_{ij}}{\partial x_j} = -\frac{\partial \tau_{ij}}{\partial x_j} \tag{36}$$

Here the term r_{ij} containing the small-scale quantities u_i' and ρ' is the (negative) subgrid-scale turbulent stress, which must be modeled.

The second term in Eq. 36 must be reduced to an expression in terms of $\bar{u}_i$ and ρ. In general,

$$\frac{\overline{\partial(\tilde{\rho}\bar{u}_i\bar{u}_j)}}{\partial x_j} = \frac{\partial}{\partial x_j}(\tilde{\rho}\bar{u}_i\bar{u}_j) + \text{correction terms} \tag{37}$$

The correction terms, which do *not* appear in the conventional time-averaged or phase-averaged equations of full-field modeling, occur here because of the local averaging (filtering) used in defining $\tilde{\rho}$ and $\bar{u}_i$. They give rise to "Leonard stresses", which make moderately important contributions to the solution [31].

A model for the subgrid-scale turbulence stresses, r_{ij}, that is simple but remarkably effective is the Smagorinsky model [32]. It is based on the notion that the length scale of the *largest* subgrid scale motions is the filter width parameter, Δ. (Experience shows that Δ should be approximately twice the computational grid spacing.) If one assumes that the subgrid-scale turbulence is in equilibrium with the local large-scale field, then the velocity scale of the subgrid turbulence should be proportional to ΔS, where $S = \sqrt{r\bar{S}_{nm}\bar{S}_{nm}}$ is the rms strain rate of the large-scale field. So the turbulent viscosity of the subgrid field should be given by

$$\mu_T = \tilde{\rho} C \Delta^2 S \tag{38}$$

Then, if one makes a quasi-Newtonian turbulence stress assumption,

$$r_{ij} = \tilde{\rho}\frac{2}{3}k\delta_{ij} + \frac{2}{3}\mu_T \nabla\cdot\underline{\bar{u}} - 2\mu_T\bar{S}_{ij} \tag{39}$$

where k is now the kinetic energy per unit mass of the subgrid-scale turbulence. The k term can be absorbed with the pressure, and need not be calculated explicitly. The constant C is the *only* empirical constant in this LES approach.

An alternative approach, which has not yet been examined in any detail, is to use a model equation for the subgrid-scale kinetic energy, k, similar to the approach of FFM. However, in LES one does not need a length scale (or ϵ) equation because the size of the largest and most important subgrid-scale eddies is determined by the filter width, or grid mesh size. There is some indication that a dynamic model for k is necessary in transition flows. This would allow the strength of subgrid motions to be zero in a laminar flow and to rise as energy is pumped into them from the large eddies.

This is essentially all that is needed for large-eddy modeling. The calculations

References pp. 64-65.

require initial conditions for the field variables, constructed in some way with appropriate randomness. Boundary conditions also are required, and ways to introduce appropriate randomness in the boundary conditions are now being explored [33].

To date, LES research has concentrated on the development of the foundations of the approach and evaluation of the method in simple incompressible flows. Clark et al. [34], and more recently McMillan and Ferziger [35], have performed some important subgrid scale model evaluations. They began by calculating a turbulence field representative of homogeneous isotropic turbulence, using a 64 × 64 × 64 grid sufficiently fine to resolve *all* important scales of motion. No subgrid scale modeling was required for this calculation. Then they overlaid an 8 × 8 × 8 grid and treated the field computed on the fine grid as actual subgrid-data for the coarse grid. This enabled them to calculate the subgrid stress terms exactly and compare these with the predictions of various subgrid models. By this process they actually computed the value of the subgrid model constant C, obtaining essentially the same value found empirically by others.

Kwak et al. [31] and Shaanan et al. [36] and later others [30] studied LES in homogeneous isotropic turbulence. They found that good predictions of the turbulence spectrum evolution could be obtained with grids as coarse as 16 × 16 × 16, for which a LES calculation requires only a few minutes of CDC 7600 time.

Mansour et al. [37] and more recently Cain et al. [38] have studied LES applications in free-shear flows. They were able to replicate the observed features of vortex pairing and three-dimensional disturbance growth in such flows.

Moin et al. [39] applied LES to turbulent channel flow. They used a variable mesh to resolve the motion very near to the walls. Their simulation displayed many observed features of turbulent channel flows—wall layer streaks, bursting, and large-scale motion coherence. The predicted mean and turbulence characteristics of this flow were in very good agreement with experiment.

Feiereisen et al. [40] are currently doing the first LES extensions to compressible flows. Bardina et al. [41] are investigating problems associated with boundary conditions in LES. These and other basic studies need to be completed before a LES study of piston engine flow is attempted. Such a calculation probably will be feasible in the early 1980's.

In summary, LES is a three-dimensional time-dependent calculation procedure in which the large-scale motions are calculated directly and the subgrid scale turbulence is modeled. Each calculation deals with a single realization of the flow. LES is in an early stage of development and is not yet ready for use in engine flows, but it may offer the best hope for soundly based, accurate flow calculations in practical engine systems.

VORTEX CHASING, AN ALTERNATIVE LES

A completely different approach to turbulent flow simulation is being explored by W. Ashurst at Sandia-Livermore, A. Leonard at NASA-Ames, and O. Buneman at Stanford. The flow field is represented by a large number of discrete vortices which induce motions in each other, producing the flow. The method is attractive in that one needs vortices only where vorticity is present. Ashurst has produced some two-dimensional shear-layer simulations that look remarkably like experiments. Leonard has recently calculated many of the features of a three-dimensional turbulent spot in a flat-plate boundary layer. Ashurst has also produced some striking simulations of compressible flow in two-dimensional and axisymmetric "motored engines", and these mirror much of the flow behavior suggested by Fig. 1.

The models used in this "vortex chasing" approach need to be placed on stronger theoretical grounds. Ways to incorporate "subgrid"-scale turbulence and density variations need to be developed. The potential of this method is such that continued effort in formulating and experimenting with the method is strongly encouraged.

CONCLUDING REMARKS

What can one really expect to get from flow models? Models that are able to predict the flow structure up to the point of combustion should be able to give values for turbulence parameters used in simple combustion models. In stratified-charge systems a good model should be able to predict the charge distribution at the moment of ignition. If the models can be made to handle the combustion processes, then at the very least it should be possible to predict the pressure-time history and estimate cycle-to-cycle variations. It may even be possible to predict the history of pollutant exhausting if the models can handle the quench layers and flow circulation properly.

If the models are even roughly correct in their quantitative predictions of these factors, they should be reasonably successful in predicting the trends with design parameters. When they have reached this state and when engine designers are confident that this has indeed been achieved, then the models will begin to become design tools.

In order to reach this point there is a great need for good experiments. Experiments of two types are needed. First, simple well-controlled thoroughly documented experiments are needed to guide the development of turbulence models. Second, well-documented engine experiments are needed as a basis for model validation.

The type of experiment most useful to the modeler is one in which most effects are absent except one. For example, a good experiment on the compression of homogeneous isotropic turbulence would be very useful. The experimenter should measure at least $k(t)$, $\epsilon(t)$, and $\tilde{\rho}(t)$, plus the histories of the integral scale, Taylor microscale, and Kolmogorov scales. Other basic experiments that would be useful include an experiment on unsteady, compressible boundary layers, an experiment on unsteady entrainment in a compressible flow, and an experiment involving only combustion and expansion, starting with a well-documented turbulence field.

The experiments most useful for model validation are well-documented experiments in simple but realistic engines. Good combustion experiments in rapid-compression devices would be useful for model validation. A fired axisymmetric

References pp. 64-65.

engine experiment would be useful, particularly if the cycle-to-cycle variations were documented.

Ultimately, validation should rest on real engine experiments in which the flow-field, cylinder-pressure, and exhaust-composition histories are documented. These should be conducted with systematic parametric variations, so it can be seen if the models are able to predict the observed trends with design parameters.

It may be useful to note that computational fluid dynamics now *is* a basic design tool in the aircraft industry. It has already led to significant improvements in wing design that have been implemented *without* wind tunnel testing. Fluid-flow computations could become an important part of the design process for advanced piston engines. I believe that a serious commitment to this goal on the part of industry, academic researchers, and government research agencies would enable this goal to be achieved within a decade.

REFERENCES

1. R. J. Tabaczynski, "Turbulence and Turbulent Combustion in Spark-Ignition Engines," Prog. Energy Combust. Sci., Vol. 2, pp. 143–165, 1976.
2. A. P. Morse, J. H. Whitelaw and M. Yianneskis, "Turbulent Flow Measurements by Laser-Doppler Anemometry in a Motored Reciprocating Engine," Imperial College, London, Paper FS/78/24. To appear in J. Fluids. Engrg.
3. E. Bouchard and W. C. Reynolds, private communication, Stanford Univ., 1978.
4. C. Chandrsuda, R. P. Mehta, A. D. Wier and P. Bradshaw, "Effect of Free-stream Turbulence on Large Structures in Turbulent Mixing Layers," J. Fluid Mech., Vol. 85, pp. 693–704, 1978.
5. H. Tennekes and J. L. Lumley, "A First Course in Turbulence," MIT Press, Cambridge, Mass., 1972.
6. R. J. Tabaczynski, C. R. Ferguson and K. Radhakrishnan, "A Turbulent Entrainment Model for Spark-Ignition Engine Combustion," SAE Trans., Vol. 86, Paper No. 770647, pp. 2414–2433, 1977.
7. D. R. Lancaster, R. B. Krieger, S. C. Sorenson and W. L. Hull, "Effects of Turbulence on Spark-Ignition Engine Combustion," SAE Trans., Vol. 85, Paper No. 760160, pp. 689–710, 1976.
8. M. W. Richman and W. C. Reynolds, private communication, Stanford Univ., 1978.
9. M. Tsuge, M. Kido, K. Kato and Y. Nomiyama, "Decay of Turbulence in a Closed Vessel," Bull. JSME, Vol. 16, 1973.
10. R. J. Tabaczynski, D. P. Hoult and J. C. Keck, "High Reynolds Number Flow in a Moving Center," J. Fluid Mech., Vol. 42, pp. 249–256, 1970.
11. R. J. Tabaczynski, J. B. Heywood and J. C. Keck, "Time-Resolved Measurements of Hydrocarbon Mass Flow Rate in the Exhaust of a Spark-Ignition Engine," SAE Trans., Vol. 81, Paper No. 720112, pp. 379–391, 1972.
12. A. Ekchian, J. B. Heywood and J. M. Rife, "Time-Resolved Measurements of the Exhaust from a Jet Ignition Prechamber Stratified Charge Engine," SAE Trans., Vol. 86, Paper No. 770043, pp. 153–173, 1977.
13. H. Tennekes, "Simple Model for the Small-scale Structure of Turbulence," Phys. Fluids, Vol. 11, pp. 669–671, 1968.
14. W. J. McCroskey, "Some Current Research in Unsteady Fluid Dynamics," ASME Trans., J. Fluids Engrg., Vol. 99, pp. 8–39, 1976.
15. L. Smith and S. J. Kline, "An Experimental Investigation of the Transitory Stall Regime in Two-dimensional Diffusers Including the Effects of Periodically Disturbed Inlet Conditions," J. Fluids Engrg., ASME Trans., Vol. 96 (I), pp. 11–15, 1974.
16. A. K. M. F. Hussain and W. C. Reynolds, "The Mechanics of an Organized Wave in Turbulent Shear Flow," J. Fluid Mech., Vol. 41, pp. 241–258, 1970.

17. A. K. M. F. Hussain and W. C. Reynolds, "The Mechanics of an Organized Wave in Turbulent Shear Flow," Part 2, J. Fluid Mech., Vol. 54, pp. 241–262, 1972.
18. M. Acharya and W. C. Reynolds, "Measurements and Predictions of a Fully Developed Channel Flow with Imposed Controlled Oscillations," Report TF-8, Dept. of Mech. Engrg., Stanford Univ., May 1975.
19. H. L. Norris and W. C. Reynolds, "Turbulent Channel Flow with a Moving Wavy Boundary," Report TF-7, Dept. of Mech. Engrg., Stanford Univ., May 1975.
20. P. O. Witze, "Hot-Wire Turbulence Measurements in a Motored Internal Combustion Engine," Second European Symposium on Combustion, Orleans, France, September 1975.
21. J. C. Dent and N. S. Salama, "The Measurement of the Turbulence Characteristics in an Internal Combustion Engine Cylinder," Paper No. C83/75, I. Mech. E. Conference on Combustion in Engines, Cranfield, England, pp. 23–32, 1975.
22. J. C. Dent and N. J. Salama, "The Measurement of the Turbulence Characteristics in an Internal Combustion Engine Cylinder," SAE Paper No. 750886, 1975.
23. D. R. Lancaster, "Effects of Turbulence on Spark-Ignition Engine Combustion," Ph.D. Thesis, University of Illinois, 1975.
24. S. Ghose and S. J. Kline, "A Numerical Method for Calculating the Performance of Two-Dimensional Diffusers Operating in the Transitory Stall Regime, Including Predictions of Optimum Recovery," Proc. Joint Symposium on Design and Operation of Fluid Machinery, ASME, June 1978.
25. A. D. Gosman, R. J. R. Johns and A. P. Watkins, "Assessment of a Prediction Method for In-cylinder Processes in Reciprocating Engines," General Motors Research Laboratories Symposium on Combustion Modeling in Reciprocating Engines, Warren, Michigan, November 1978.
26. M. W. Rubesin and W. C. Rose, "The Turbulent Mean-flow, Reynolds-stress and Heat-flux Equations in Mass-averaged Dependent Variables," NASA TMX 62,248, March 1973.
27. S. B. Pope, "The Probability Approach to the Modeling of Turbulent Reacting Flows," Combustion and Flame, Vol. 27, pp. 299–312, 1976.
28. W. C. Reynolds, "Computation of Turbulent Flows," Ann. Rev. Fluid Mech., Vol. 8, pp. 183–208, 1976.
29. D. P. Hoult and V. W. Wong, "The Generation of Turbulence in an Internal Combustion Engine," General Motors Research Laboratories Symposium on Combustion Modeling in Reciprocating Engines, Warren, Michigan, November 1978.
30. N. N. Mansour, P. Moin, W. C. Reynolds and J. H. Ferziger, "Improved Methods for Large-Eddy Simulation of Turbulence," Proc. Symp. on Turbulent Shear Flows, Pennsylvania State Univ., 1977.
31. D. Kwak, W. C. Reynolds and J. H. Ferziger, "Three-Dimensional, Time-Dependent Computation of Turbulent Flows," Report TF-5, Dept. of Mech. Engrg., Stanford Univ., 1975.
32. J. Smagorinski, "General Circulation Experiments with the Primitive Equations," Monthly Weather Rev., Vol. 91, p 99, 1963.
33. W. C. Reynolds, Stanford University, work in progress, 1978.
34. R. A. Clark, J. H. Ferziger and W. C. Reynolds, "Evaluation of Subgrid-Scale Turbulence Models Using a Fully Simulated Turbulent Flow," Jour. of Fluid Mech., Vol. 91, pp. 1–16, 1979.
35. D. J. McMillan and J. H. Ferziger, private communication, 1978.
36. S. Shaanan, J. H. Ferziger and W. C. Reynolds, "Numerical Simulation of Turbulence in the Presence of Shear," Report TF-6, Dept. of Mech. Engrg., Stanford Univ., 1975.
37. N. N. Mansour, J. H. Ferziger and W. C. Reynolds, "Large Eddy Simulation of a Turbulent Mixing Layer," Report TF-11, Dept. of Mech. Engrg., Stanford Univ., 1978.
38. A. Cain, W. C. Reynolds and J. H. Ferziger, Stanford University, private communication, 1978.
39. P. Moin, W. C. Reynolds and J. H. Ferziger, "Large Eddy Simulation of Incompressible Turbulent Channel Flow," Report TF-12, Dept. of Mech. Engrg., Stanford Univ., 1978.
40. W. I. Feiereisen, W. C. Reynolds and J. H. Ferziger, Stanford University, private communication, 1978.
41. J. Bardina, J. H. Ferziger and W. C. Reynolds, Stanford University, private communication, 1978. J. Fluid Mech. (1979) Vol. 91, Part 1 pp.1–16 Evaluation of Subgrid-Scale Models using an accurately simulated turbulent flow.

DISCUSSION

Editors' Note

In his oral presentation Professor Reynolds showed a film by W. T. Ashurst* in which the "vortex chasing" method was used to simulate flow in motored two-dimensional and axially symmetric engines. This film was used to display the ideas covered in the written paper.

J. W. Daily *(University of California)*

First of all I'd like to say that Bill Ashurst's film is beautiful. In fact, it's far more beautiful than anything we can actually duplicate in the laboratory. Furthermore, in many ways it is better than reality because in reality we have to look at the three-dimensional flow field with, so far, two-dimensional imaging techniques, and we can't see the flow field in as great a detail as you can see it in his movies.

I have a couple of questions. First of all you mentioned that it's hard to deal with the filtering problem experimentally, and I don't quite understand that statement. Assuming one can make experimental measurements of, say, a velocity vector with sufficient spatial and time resolution, why couldn't one go back and filter the data numerically?

W. C. Reynolds

Filtering requires a *spatial* average, which requires simultaneous measurements at more than one point. How are you going to make measurements that are going to resolve the field spatially? It can be done, but it's difficult.

W. C. Reynolds

* *The referenced film is based on work reported in the following publication: W. T. Ashurst, "Vortex Dynamic Calculation of Fluid Motion in a Four-Stroke Piston 'Cylinder'—Planar and Axisymmetric Geometry," Report SAND78-8229, Sandia Laboratories, 1978.*

T. Morel *(General Motors Research Laboratories)*

I have two questions. One question is about Bill Ashurst's movie, which was very nice. Because it is a two-dimensional configuration, all the vortices we see are infinite lines perpendicular to the screen. That would correspond to something, perhaps, like two-dimensional turbulence. Now, you made quite a point about the fact that if you compress this Ashurst flow, you see a spin-up, an increase in energy. However, a real flow has three components of vorticity which interact to some degree during compression, because one can never do it quite fast enough to have no interaction (so-called rapid distortion). Would you still expect an increase in turbulence level due to compression?

W. C. Reynolds

I certainly expect to see some of that. It depends on the cycle three-dimensionality. Vortex chasing has been used in three dimensions. Tony Leonard at Ames has a beautiful simulation* of a turbulent spot. It starts out with vortices that are across the flow. After a kink is put in one of them that kink lifts up, moves downstream, then goes out and develops a kink in the vortices alongside of it. The turbulent spot then spreads and grows and hangs out in front, just like it does in pictures. While it's qualitatively correct, it's not quantitatively correct at this point. Qualitatively it is a three-dimensional turbulent situation, analyzed by that type of approach. So I think there's potential for analyzing it in three dimensions. Now there's some problems with that technique. If Bill Ashurst will excuse me for saying so, I think that the conceptual foundations need to be strengthened. It's not at all clear how one should really handle turbulence and compressibility.

T. Morel

You showed us a very simple sub-grid-scale model that really is impressive in its simplicity. It doesn't take much to incorporate it into present codes. However, what I saw missing was perhaps the term that you normally use in sub-grid-scale modeling to take account of the cross-spectrum transfer of energy, for example, the Leonard term. How important is that term? Can you really drop it and use a simple model like this one?

W. C. Reynolds

I didn't elaborate on that point. When you do filtering, you introduce some important differences between the filtered equations and the phase-averaged equations. There is something called the Leonard stress that results from that

---

* *A. Leonard, "Vortex Simulation of Three-Dimensional, Spot-Like Disturbances in a Laminar Boundary Layer," Proc. of Second Symposium on Turbulent Shear Flows, Imperial College, London, England, July, 1979.*

approach. It's like a Reynolds stress. It turns out that if you have the large-eddy field, the $\bar{\underline{u}}$ field, you can calculate that term directly. In all of our simulations we're doing that directly, and we no longer have to evaluate the Leonard stress explicitly. It's just a computational detail that exists. I don't think there's any really essential physics missing except that the model assumes that the small scales are in equilibrium with the large scales. Now if you have a very rapid transient, that might not be the case. Then you might want to put some dynamics in for the small-scale energy, with the simple "k" equation. And Deardorff* has done that in some simulations of the atmosphere.

* *J. W. Deardorff, "Use of Subgrid Transport Equations in a Three-Dimensional Model of Atmospheric Turbulence," J. Fluids Engr., pp. 429–438, 1973.*

DEVELOPMENT OF PREDICTION METHODS FOR IN-CYLINDER PROCESSES IN RECIPROCATING ENGINES

A. D. GOSMAN, R. J. R. JOHNS and A. P. WATKINS

Imperial College, London, England

ABSTRACT

A new generation of computer-based methods is emerging for calculating the detailed patterns of gas flow, heat transfer and combustion in reciprocating engines by solving numerically the governing partial-differential conservation laws of physics. The present paper describes a program of research aimed at quantitatively assessing the capabilities of a method of this kind, code-named "*RPM*", for the prediction of gas flow in motored engines. An outline is provided of the essential features of the method and the principal sources of uncertainty which enter into its development. Then comparisons are shown between *RPM* predictions and experimental measurements of laminar and turbulent gas flow in axisymmetric model "engines" of both compressing and non-compressing varieties. The comparisons reveal areas of both good and poor agreement, although the global behavior is adequately reproduced in nearly all cases. Attempts to identify the reasons for disagreement are, however, impeded by inadequate knowledge of the distributions of the velocity components and other quantities in the inlet aperture during the intake stroke.

NOTATION

a	acceleration
C_μ, C_1, C_2	constants of the turbulence model, assigned values 0.09, 1.44 and 1.92 respectively
C_p, C_v	specific heats at constant pressure and constant volume respectively
d	diameter
D	cylinder bore
E	turbulence model constant, assigned the value 9.793

References pp. 123-124.

G	generation of turbulent energy
h	specific stagnation enthalpy
k	turbulent energy
L	length
N	number of time-interval subdivisions per cycle
p	pressure
P	laminar-sublayer resistance factor
r	radial coordinate
R	gas constant
s	source term of differential equation
S	stroke
t	time
T	temperature
u	axial velocity
v	radial velocity
V	velocity tangential to a wall
y	normal distance from wall
z	axial coordinate
Γ	effective diffusivity
ϵ	dissipation rate of turbulent energy
κ	turbulence model constant, assigned the value 0.4187
ϕ	a dependent variable of the differential equations
μ	viscosity
ρ	density
∇	divergence operator
σ_h, σ_k, σ_ϵ	turbulent Prandtl/Schmidt numbers assigned values 0.9, 1.0 and 1.2 respectively

ξ	dimensionless axial coordinate ($\equiv z/z_p$)
θ	crank angle

Subscripts

eff	effective
p	piston
R	resultant
t	turbulent
o	reference value (spatial)
ϕ	relating to ϕ

Superscripts

$+$	dimensionless quantity
$\sim$	relative value
$-$	mean value
*	analytic solution
$\circ$	reference value (temporal)
$'$	fluctuating quantity

INTRODUCTION

Background—The pressures which have arisen during the past decade for the development of reciprocating engines which, on the one hand, meet the increasingly stringent pollutant emission standards and, on the other hand, consume less fuel have stretched traditional design techniques to their limit, and perhaps beyond. The reasons for this state of affairs are not hard to identify; thus the traditional approach was perforce limited, because of inadequate knowledge of the detailed in-cylinder processes, to a combination of modest extrapolation based on past experience, trial and error experimentation, and semi-empirical analysis of the "lumped parameter" variety. Although this approach proved to be remarkably successful when allowed to proceed at its natural pace, when pressed beyond this by the need for radical changes to meet the new requirements, the extrapolations have become too large, the required experiments too numerous, and the analytical techniques too empirically based to allow extrapolation, with the result

References pp. 123-124.

that the expense of development has increased substantially. More fundamentally, it has been recognized that these inadequacies stem primarily from the fact that, despite the long history of development of the reciprocating internal combustion engine, very little quantitative information is available about structure of the flow, mixing and combustion processes within the combustion chambers, and the important governing parameters and their effects; for it is these, after all, which determine the performance of an engine. Of course, this state of affairs is not one of choice, for the hostile and complex environment of the combustion chamber has long posed formidable problems for both experimenters and theoreticians alike, and it is only comparatively recently that techniques have become available to them for probing this environment.

The present paper is concerned with one of the major new theoretical techniques for in-cylinder analysis which has emerged during the past five years, namely computer-based prediction methods which operate by solving numerically the partial-differential conservation equations of physics governing the forementioned processes. These methods are capable, in principle, of providing the kind of detail mentioned above; however, as discussed later, there are in practice numerous obstacles which remain to be surmounted before the new methods can gain full acceptance as reliable design tools.

The emergence of the new class of method appears to date from about 1973, when two of the present authors developed a finite-difference procedure, which was code-named "*RPM*" (standing for Reciprocating Piston Motion), for calculating laminar, axisymmetric, non-combusting flow in a closed cylindrical chamber. Following on from this initial work, which is described in Refs. 1 and 2, the *RPM* procedure was subsequently extended to allow turbulent flow simulation by incorporation of a "turbulence model" (about which more will be said later) and simulation of the opening and closing of a valve by means of an orifice representation. These extensions allowed computer explorations to be made of both closed- and valve-equipped cylinders and produced what were perhaps the first predictions, some of which are reported in Ref. 3 and in the present paper, of turbulent flow in circumstances representative of reciprocating engines. These same explorations also yielded predictions of the transient gas and wall temperatures and heat fluxes, which are described and compared with the "accepted" heat-transfer correlations in Refs. 4 and 5. More recently, the *RPM* procedure has been further extended to allow simulation of the swirling motion imparted by certain designs of inlet tract and also representation of non-cylindrical combustion chambers such as are found in "cup-in-piston" and prechamber designs. Refs. 5 and 6 contain examples of such applications. Ref. 5 contains, as well, an outline of how the procedure is currently being developed to enable calculations of a firing "engine" of idealized form, using contemporary mathematical models of the turbulent combustion process.

Research on the new generation of computer methods was also initiated elsewhere by groups headed by Anderson [7, 8] and Boni [9, 10]. The former have based their methods on the numerical technique of MacCormack [11] and have produced the first three-dimensional engine simulations, albeit on a very coarse computational grid and with relatively crude representations of turbulence and combustion. The method of Boni and colleagues is based on a numerical technique developed by Hirt et al. [12] for two-dimensional configurations of arbitrary shape and has been applied to an idealized divided-chamber spark-ignition engine, using a multi-step combustion representation and a simple uniform eddy-viscosity simulation of turbulence. Mention should also be made of the related research by Bracco and coworkers on methods for constant-volume calculations employing relatively sophisticated turbulence and combustion models [13, 14].

As will be evident from the above comments, the obvious potential of the new methods has spurred their developers to strive towards the goal of a "complete" computer model of real engines, with all the attendant complexities. Unfortunately it is important to inject a note of realism into this process by observing that the predictions of these methods are subject to various sorts of errors, none of which has yet been quantified to any significant degree by the developers of the methods. The errors in question result, as will be more fully discussed later, from approximations inherent in the numerical techniques and mathematical models of the physical processes which the methods embody, and also from uncertainties about the boundary conditions appropriate to the particular engine being simulated. It is also important to note that it is currently not possible to make *a priori* estimates of the magnitudes of these effects.

It is clear that until some definitive statements can be made about the accuracy of prediction of the new methods, they will be (quite rightly) viewed with some skepticism by the engine community. The present authors have accordingly subjected the *RPM* method to a program of assessment, involving comparisons with both experimental in-cylinder measurements and analytical solutions to the governing equations (where available), and it is this study which is the central theme of the present paper. We have chosen to focus at this stage on the situations involving air motion in the absence of combustion and in relatively simple axisymmetric "model" engines because we believe there are decisive arguments for this approach, as follows. Firstly, from the practical point of view such systems are far more amenable to in-cylinder measurements than are real engines, for which the necessary techniques are only just beginning to emerge. Also, the simplification of axial symmetry substantially reduces the costs of both the experiments and the computations. Secondly, on the theoretical side it is clear that the ability to predict the air motion correctly (and by this is also meant the important facets of the turbulence structure) is a prerequisite to predicting combustion correctly. It is equally clear that successful airflow prediction in simple configurations is a necessary condition for success in the real situation. Finally we would also suggest, but admittedly without much justification as yet, that there may well turn out to be strong similarities between the air motions in axisymmetric and real engines.

Contents of Paper—As stated above, the purpose of this paper is to provide, as far as the available evidence will allow, an assessment of the current ability of the *RPM* method to predict air motion in non-firing axisymmetric model engines of various kinds.

The method itself is presented in the following section, commencing with a

References pp. 123-124.

statement of the differential equations, including those of the particular turbulence model which it currently embodies. This is followed by an outline of the numerical aspects. The presentation is necessarily brief, but further details are available in other publications which are cited.

The third major section contains the main body of the paper, where the various test cases to which the method has been applied are first described (see Fig. 2), and then details of each application and the results obtained are presented. Of the seven cases which are discussed, two are concerned with laminar motion and five with turbulent flow. The former set includes one situation possessing an analytical solution (one-dimensional isentropic flow) and one for which measurements are available (cylindrical chamber with a central orifice in the head). The turbulent-flow cases include four situations for which measurements are available, all being variations on the theme of a cylindrical chamber equipped with a central opening or valve. The fifth case is one of a closed chamber, for which neither data nor analytical solutions are available. It has been included because the results are, in addition to being of interest in their own right, helpful in the interpretation of the remaining cases.

In the spirit of the last remark, the third major section also describes some calculations which were performed in order to shed light on certain aspects of the turbulence behavior and its modeling, namely the structure produced by the action of normal stresses alone as produced by pure compression and expansion, and the influence of the wall on the nearby structure in an oscillating fully developed flow.

The overall performance of the method is assessed in the fourth major section, which also highlights the major areas of uncertainty revealed by the exercise and makes recommendations in this regard concerning the numerical technique, turbulence modeling, and experiments. Finally, the main conclusions of the study are summarized in the fifth major section.

DESCRIPTION OF THE *RPM* METHOD

Preliminary Remarks—Strictly speaking, the *RPM* method as such consists of a single general numerical procedure for solving sets of simultaneous non-linear partial differential equations which describe engine processes. The equations of the sets may vary in both form and number according to the nature of the processes involved and the "mathematical models" formulated to represent them. In the present context, however, the "method" will be taken to refer to the particular version of the procedure embodying the equation set used for the test cases.

For those cases involving laminar flow the assembly of the governing equations is straightforward; they simply consist of the familiar set describing the conservation of mass, momentum, and total energy. However, the more practically interesting turbulent flows are by no means as straightforward to deal with since, although they are strictly governed by precisely the same equations as the laminar cases, it is not feasible to solve them at the Reynolds numbers in question for the well-known reason that the small-scale energy-containing turbulent eddies cannot be economically resolved by any known numerical scheme. The activity of "turbulence modeling" has its foundations in this problem and attempts to circumvent it by providing alternative, more economical means of representing the effects of the small-scale turbulent motion. The considerable importance of this activity to engine prediction is reflected by the fact that one of the papers in this Symposium is specifically devoted to it [15]. In the light of the existence of this paper, and the fact that the present authors are users of turbulence models rather than originators, we shall not attempt to do more here on this topic than to describe briefly the alternatives as we saw them, state the reasons for our particular choice, and then present the equations of the selected model.

Assembly of Equations

Selection of Turbulence Model—There are two main classes of model which seem suitable for engine applications, namely:

(i) Statistical Flux Models (*SFM*):—This name has been coined here to denote the approach whereby the basic conservation equations governing the instantaneous flow are ensemble-averaged to yield a new set containing additional unknowns which have the significance of turbulent fluxes of the entity in question (e.g., the "Reynolds stresses" in the case of the momentum equations). Closure of the equation set is then effected by deriving additional equations, of either algebraic or differential form, for the new unknowns, in which process certain approximations and assumptions necessarily enter which render the results inexact. The success or failure of such methods therefore resides in the validity of the approximations which they embody.

(ii) Large-Eddy Simulation (*LES*):—This name refers to an alternative approach in which the averaging process is limited to eddies below a certain scale which can (hopefully) be made of the same order as the mesh size of computational grids for which numerical calculations are feasible. Motions larger than this are calculated in the normal way, i.e., as for laminar flow. The new unknowns arising from the averaging process (termed "subgrid-scale Reynolds stresses" in the case of the momentum equations) are approximated by a *SFM*, but the small-scale turbulence is believed to possess certain regularities (e.g., isotropy) which should allow them to be modeled accurately in particularly simple ways.

More detailed expositions of these methods can be found in Refs. 16 and 17, among others.

Of the two approaches, the *LES* appears to be the more direct and potentially accurate, since it only approximates that part of the eddy-scale spectrum which is not accessible to direct calculation, whereas the *SFM* attempts to cover the whole range. We nevertheless chose the *SFM* route for our turbulent-flow calculations for two main reasons. Firstly, the *LES* was, and still is, comparatively new and untested, although it should be said that even for the *SFM* we could not find evidence of previous applications to flows possessing engine-like character-

References pp. 123-124.

istics, particularly periodicity of the mean motion and compressibility. Secondly, and perhaps more importantly, the *LES* does not allow the full economies of the restriction to axisymmetric flow to be realized for the simple reason that even in such flows the turbulence structure remains three-dimensional and must be treated as such in the calculations (at, of course, considerably greater expense). As time proceeds and more is learned about the capabilities of the various models, the arguments may turn in favor of the *LES*: indeed it is already clear that it may have decisive advantages when the axisymmetric restriction is released and the full three-dimensional problem is considered.

Considerations of capabilities and costs have also guided our selection from the many models which are available under the *SFM* heading. Since whatever evidence of capability exists is almost completely confined to steady flows, and since even for these it is extremely limited for flows exhibiting the complexities which may occur in engine cylinders (e.g., strong recirculation), we decided to select the simplest model which, in our view, had any prospect of performing adequately.

Our choice fell on the so-called "energy-dissipation" model of turbulence, in the form proposed by Jones and Launder [18] for fully turbulent flows. In this version the Reynolds stresses are assumed to be linked to the mean strain field via a Newtonian-type law which contains a new unknown in the form of a scalar eddy viscosity. The latter, which may vary in space and time, is calculated from the local ensemble-averaged values of two parameters of the turbulence structure, namely the kinetic energy of turbulence, k, and its dissipation rate, ϵ. These are in turn obtained from their own differential transport equations. The forms of these and the other members of the equation set embodied in the *RPM* method are given below.

The Equations Solved—For the axisymmetric test cases considered herein, the governing differential equations are those describing the conservation of mass, momentum (axial and radial components*), thermal energy (required for cases involving compression), and turbulent energy and its dissipation rate (turbulent flows only). The ensemble-averaged dependent variables of these equations are: the velocity components u and v in the axial (z) and radial (r) directions; the pressure p; the stagnation enthalpy $h \equiv C_vT + p/\rho + (u^2 + v^2)/2 + k/2$, where C_v, T and ρ are respectively the constant-volume specific heat, temperature and density of the fluid; and of course k and ϵ. The entire set may be compactly represented by the following general transport equation:

$$\frac{\partial \phi}{\partial t} + \frac{\partial}{\partial z}(\rho u \phi) + \frac{1}{r}\frac{\partial}{\partial r}(r\rho v \phi) - \frac{\partial}{\partial z}\left(\Gamma_\phi \frac{\partial \phi}{\partial z}\right) - \frac{1}{r}\frac{\partial}{\partial r}\left(r\Gamma_\phi \frac{\partial \phi}{\partial r}\right) + s_\phi = 0 \quad (1)$$

in which t stands for time, ϕ stands for the dependent variable of the equation in question, and Γ_ϕ and s_ϕ are the corresponding effective diffusivity and source term respectively, whose definitions are given in Table 1.

* *Swirl is not present in these cases, but such problems may be handled by the RPM method, as is demonstrated in Ref. 6.*

TABLE 1

Definitions of Coefficients Appearing in the General Transport Eq. 1

Equation	Dependent Variable	Γ_ϕ	s_ϕ
u momentum	u	μ_{eff}	$\frac{\partial}{\partial z}\left(\mu_{eff}\frac{\partial u}{\partial z}\right)+\frac{1}{r}\frac{\partial}{\partial r}\left(r\mu_{eff}\frac{\partial v}{\partial z}\right)-\frac{\partial p}{\partial z}-\frac{2}{3}\frac{\partial}{\partial z}(\mu_{eff}\nabla\cdot\underline{u}+\rho k)+s_u'$
v momentum	v	μ_{eff}	$\frac{\partial}{\partial z}\left(\mu_{eff}\frac{\partial u}{\partial r}\right)+\frac{1}{r}\frac{\partial}{\partial r}\left(r\mu_{eff}\frac{\partial v}{\partial r}\right)-2\mu_{eff}\frac{v}{r^2}-\frac{\partial p}{\partial r}-\frac{2}{3}\frac{\partial}{\partial r}(\mu_{eff}\nabla\cdot\underline{u}+\rho k)+s_v'$
Continuity	1	0	s_ρ'
Total energy	h	μ_{eff}/σ_h	$\frac{\partial p}{\partial t}+\frac{\partial}{\partial z}\left\{\mu_{eff}\left(1-\frac{1}{\sigma_h}\right)\frac{\partial}{\partial z}\left(\frac{u^2+v^2}{2}\right)\right\}+\frac{1}{r}\frac{\partial}{\partial r}\left\{r\mu_{eff}\left(1-\frac{1}{\sigma_h}\right)\frac{\partial}{\partial r}\left(\frac{u^2+v^2}{2}\right)\right\}$
			$+\frac{\partial}{\partial z}\left\{\mu_{eff}\left(\frac{1}{\sigma_k}-\frac{1}{\sigma_h}\right)\frac{\partial k}{\partial z}\right\}+\frac{1}{r}\frac{\partial}{\partial r}\left\{r\mu_{eff}\left(\frac{1}{\sigma_k}-\frac{1}{\sigma_h}\right)\frac{\partial k}{\partial r}\right\}+s_h'$
Turbulence energy	k	μ_{eff}/σ_k	$G-\rho\epsilon+s_k'$
Dissipation rate	ϵ	$\mu_{eff}/\sigma_\epsilon$	$\frac{\epsilon}{k}(C_1G-C_2\rho\epsilon)+\rho\epsilon\nabla\cdot\underline{u}+s_\epsilon'$

Notes 1. $\nabla\cdot\underline{u}\equiv\frac{\partial u}{\partial z}+\frac{1}{r}\frac{\partial}{\partial r}(rv);\ G\equiv\mu_{eff}\left[2\left\{\left(\frac{\partial u}{\partial z}\right)^2+\left(\frac{\partial v}{\partial r}\right)^2\right\}+\left(\frac{\partial u}{\partial r}+\frac{\partial v}{\partial z}\right)^2+2\left(\frac{v}{r}\right)^2\right]-\frac{2}{3}\nabla\cdot\underline{u}(\mu_{eff}\nabla\cdot\underline{u}+\rho k)$

2. The s' terms contain correlations involving density fluctuations which have been ignored, in the absence of guidance on how they should be modeled.

In addition to the differential equations, auxiliary relations are required connecting the "effective viscosity", μ_{eff}, and density to the other variables. The density is evaluated here from the Ideal Gas Law, while μ_{eff} is given by

$$\mu_{eff} \equiv \mu + \mu_t = \mu + C_\mu \rho k^2/\epsilon \tag{2}$$

where μ is the molecular viscosity and C_μ is an empirical coefficient. The values assigned to this and the other coefficients C_1, C_2, σ_h, σ_k and σ_ϵ of the energy/dissipation model appearing in Table 1 were taken from Ref. 18 and are quoted in the Notation section.

Specification of Boundary Conditions—It should be clear from both the nature of the flows considered and the form of the governing equations that the complete specification of a particular problem requires statements of the initial and boundary conditions applying to all equations. If, however, "start-up" effects are not of interest (and it seems sensible to focus first on the steady operating state, in which the flow becomes periodic), then the initial conditions will have no influence on the solution obtained, and it is only the boundary conditions which are required.

For practical reasons it will probably be necessary for the present to confine the domain of solution to the combustion chamber itself, for to attempt to extend it outside into the ports would entail additional, probably prohibitive, expense, with little prospect of a return on the investment. This at least is the view of the present authors. The boundaries to be considered are therefore of two kinds, namely, the confining walls and, where they exist, the planes of the inlet and exhaust apertures.

The wall boundary conditions are straightforward to specify and implement when the flow is laminar; thus the usual no-slip conditions pertain to the velocities and the temperatures are either known or calculable. The *specifications* for turbulent flow are of the same kind, but *implementation* is more difficult, because it is not feasible in the present circumstances to solve the equations in near-wall regions, where molecular and turbulence effects are of the same order of magnitude, due to the fine computational grids required. (An impression of the additional effort involved can be gained from Ref. 18.) This problem is here circumvented by the usual practice, described in Ref. 19, of "bridging" the near-wall region by use of special formulas which are intended to describe the flow there and whose implications are used to modify the main equations appropriately in this region. Unfortunately such formulas do not exist for the circumstances of engine flows, with their inherent periodicity and other complexities, so recourse has had to be made to those which have been used for steady-flow calculations.

The particular formulas used are those cited in Ref. 20 as being appropriate to a one-dimensional turbulent Couette flow which obeys the "logarithmic law of the wall". They are

$$V^+ = [\ln E y^+] \tag{3}$$

$$\partial k/\partial y = 0 \tag{4}$$

$$\epsilon = C_\mu^{3/4} k^{3/2} / \kappa y \tag{5}$$

$$T^+ = \sigma_h (V^+ + P) \tag{6}$$

where V^+ and y^+ are respectively the tangential velocity and normal distance in dimensionless law-of-the-wall form, κ and E are constants given in the Notation section, T^+ is the dimensionless temperature, and P is a laminar-sublayer resistance factor. Further details about the origins and implementation of these formulas may be found in the references cited.

The aperture inlet-plane boundary conditions also present difficulties, but of a different kind. Here the problem is that because the required distributions of *all* the main dependent variables (i.e., the velocity components, pressure, temperature and the turbulence parameters) depend on upstream conditions; they cannot be deduced from knowledge of the chamber configuration alone. Reliance must therefore be placed on experimenters to provide this information, but unfortunately the need for this has not yet been fully recognized.

In the absence of details of the inflow, our practice has been to connect the instantaneous mass flow with the instantaneous pressure difference across the aperture via an orifice model in the traditional way, and then attempt to make plausible estimates of the distributions of the variables. For the calculations reported herein, we have used the orifice equations of Woods and Khan [21] except in those instances where it was justifiable to ignore the pressure drop and simply base the flow on the piston displacement. Details of the assumed distributions will be given when the individual cases are described.

The Numerical Solution Procedure

Coordinate Transformation and Grid—The *RPM* procedure is of the finite-difference variety and employs, in its most general form, a flexible curvilinear computational grid which is fitted to the shape of the combustion chamber and expands and contracts with the piston motion, as described in Ref. 6. For the present test cases, however, a simpler variant was adequate, in which the grids are of the cylindrical-polar form shown in Fig. 1. The expanding/contracting feature is introduced through a coordinate transformation* in which the axial coordinate z of Eq. 1 is replaced by $\xi \equiv z/z_p$, z_p being the instantaneous piston displacement. Eq. 1 then becomes

$$\frac{1}{z_p}\frac{\partial}{\partial t}(\rho z_p \phi) + \frac{1}{z_p}\frac{\partial}{\partial \xi}(\rho \tilde{u} \phi) + \frac{1}{r}\frac{\partial}{\partial r}(r \rho v \phi) - \frac{1}{z_p}\frac{\partial}{\partial \xi}\left(\frac{\Gamma_\phi}{z_p}\frac{\partial \phi}{\partial \xi}\right) - \frac{1}{r}\frac{\partial}{\partial r}\left(r\Gamma_\phi \frac{\partial \phi}{\partial r}\right) + s_\phi = 0 \tag{7}$$

where $\tilde{u} \equiv u - \xi u_p$ is the local relative velocity between the fluid and the coordinate plane, u_p being the instantaneous piston velocity. Solutions to this

* *The more flexible transformation of Ref. 6 was used for one application (Test Case 6) involving a 'cup-in-head' chamber, but this is not so different as to warrant elaboration here.*

References pp. 123-124.

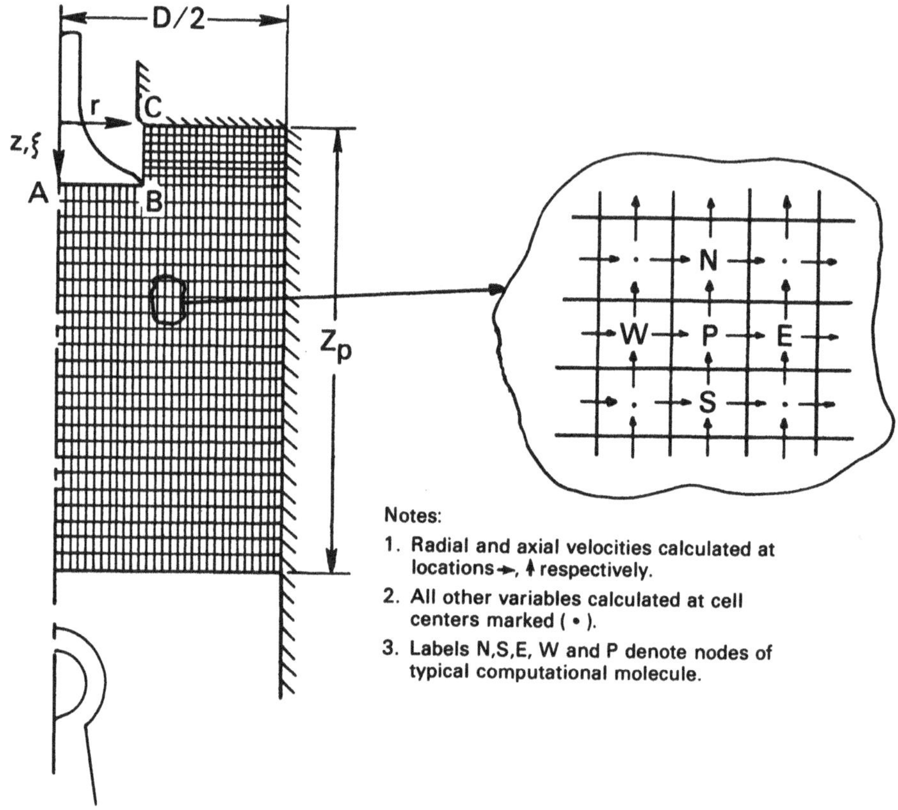

Fig. 1. Illustration of grid arrangement employed for finite-difference calculation and location of variables.

equation are obtained in the ξ, r, t space and then transformed back to physical space when required. As indicated in the inset of Fig. 1, the majority of the dependent variables are calculated by the finite-difference procedure at the control volume centers, the exceptions being the velocities, which are displaced in the now conventional fashion so as to lie between the pressures which drive them.

Finite-Difference Transformation—Conventional practices are also employed in deriving the finite-difference counterparts of the equation set represented by Eq. 7. A five-point "computational molecule" is employed in the $\xi \sim r$ plane, as shown in Fig. 1. With one exception to be mentioned later, forward differencing in time is used throughout, giving rise to fully implicit difference forms. Considerable effort has also been put into formulating alternative time-centered schemes, but the improved accuracy which these are observed to yield for somewhat simpler problems (see, e.g., Ref. 22) has yet to be realized for the more complex circumstances of the engine calculations.

A self-selecting hybrid central/upwind spatial differencing scheme is used, as described in Ref. 19, to represent the combined advection and diffusion terms of Eq. 7, although in most practical applications the "mesh Reynolds number" formed from the local velocity and mesh spacing is almost invariably in the range where the upwind option is invoked, on grounds of superior numerical stability and accuracy in these circumstances. The components of the "source term", s_ϕ, of Eq. 7 are linearized where necessary and approximated by central differences. Care is taken throughout the derivation to preserve the conservation properties of the parent differential equations.

The continuity equation is accorded special treatment, being first forward-differenced in time and central-differenced in space and then transformed, with the aid of the momentum equations, into the finite-difference counterpart of a Poisson equation for a "pressure correction", which acts to produce a pressure field that drives the velocities in the direction of satisfying local continuity, in the manner described in Ref. 23.

Incorporation of Boundary Conditions—The boundary information required by the finite-difference equations effectively takes the form of relations connecting the boundary fluxes of the entity in question to its values at nearby grid nodes, e.g., for the tangential momentum equation a "drag law" type of relation is required. In the case of walls, these relations emerge quite naturally for laminar flow from the finite-difference analysis. For turbulent flow, they are deduced from the formulas given by Eqs. 3–6.

As remarked earlier, the boundary conditions at the aperture should, for circumstances of inflow, ideally take the form of *prescriptions* of the fluxes crossing the aperture plane, which may be imposed straight-forwardly. However there is equally no difficulty in incorporating orifice formulas, when necessary, for these may simply be viewed as another kind of relation between a boundary flux and interior conditions.

The opening and closing of a valve results in an annular aperture whose time-varying area is determined by the valve geometry and lift characteristics. In our first calculations of valve-equipped "engines", such as are reported in Ref. 3, we chose to simplify the representation by ignoring the fact that the valve protrudes into the cylinder space and treating the aperture as lying in the plane of the head. This simplification, which must result in some error, especially at high valve lifts, has since been removed. Instead the protrusion of the valve is now simulated by imposing the appropriate boundary conditions on surfaces *AB* and *BC* of Fig. 1, which are caused to move through the interior space in a suitable fashion. Details cannot be given here, but they will appear in a subsequent publication.

Solution Algorithm—Given details of the chamber configuration, piston and valve motion, and boundary conditions, the *RPM* method then solves the difference equations on a prescribed grid by a time-marching process, in which solutions are obtained at small intervals of time (or, equivalently, crank angle) until the desired period has been covered. Unless start-up effects are of especial interest, the calculations are continued for as many simulated engine cycles as

References pp. 123-124.

are required for the solution to become periodic. All results presented herein are of this kind.

For each new time interval the method updates the fields of variables from the preceding interval in the following sequence of steps: (i) first approximations to the mean pressure and temperature levels are obtained by solving global continuity and energy equations formed from the finite-difference equations, but otherwise similar to those used in conventional engine "cycle" methods (e.g., Ref. 24); (ii) a preliminary estimate of the velocity field is then obtained by solving the momentum equations; (iii) a field of pressure corrections is obtained from its governing equations and used both to adjust the spatial distribution of pressure and to bring the velocities into local continuity balance; (iv) the fields of the remaining variables (i.e., h, k and ϵ) are then updated in turn from their respective equations. Since during this process the equations are linearized by taking the coefficients to be constant, and the calculation of each variable is "lagged" with respect to its predecessor, it is necessary to repeat steps (i) through (iv), using newly-assembled coefficients on each occasion, until the prevailing solution satisfies the full non-linear set to the required level, which is stringently monitored. Solution of the linearized equations is effected by an *ADI* method.

As is well known, numerical solution procedures are subject to errors resulting from the approximations which they embody concerning the spatial and temporal behavior of the variables. Provided due care is taken in the formulation of the difference equations, these errors diminish as the grid and time intervals are decreased. Formal analysis of the magnitude of the error and its dependence on the forementioned intervals is only possible for especially simple problems (an example of which is given in the next section) and not for the general case. Thus it is necessary to rely for the latter on so-called "grid-refinement tests" in which the intervals are ideally systematically reduced until the solution shows no significant change. Tests of this kind have been performed in the present exercise, but for reasons to be explained later, the ideal situation was only partially realized.

APPLICATION TO TEST CASES

Description of Cases—The seven test cases for which predictions obtained with the *RPM* method are presented and assessed in this section are summarized in Table 2 and Fig. 2. The former contains brief descriptions, including the important dimensions and operating conditions, together with an indication of the nature and source of the available data; the latter shows the geometrical configurations of the model engines which comprise the majority of the cases.

As can be seen from Table 2, two of the cases relate to laminar flows, occurring in one instance in a hypothetical one-dimensional "engine" for which analytical solutions to the governing equations are available for comparison, and in the other in a real, experimental non-compressing engine in which measurements have been made. The intention in selecting these cases was to allow assessment of the accuracy of the method in circumstances in which the governing equations are exact.

TABLE 2

Summary of Test Cases

Case No.	Description	Chamber Configuration/ Fig. No.	Inlet/Exhaust Arrangement	Engine Dimensions: Bore D (m)	Engine Dimensions: Stroke (m)	Engine Dimensions: Comp. Ratio	Speed (r/min)	Nature of Data Available	Source of Data
1	One-dimensional laminar isentropic motion	—	None		arbitrary		Low	Analytical solution: see third major section	—
2	Non-compressing engine with orifice inlet (laminar)	Disc/2b	Central circular orifice of d/D = 0.25	0.075	0.060	3	10	*LDA** meas. of axial velocity distribution	[25]
3	Compressing engine with closed chamber (turbulent)	Disc/2a	None	0.130	0.130	14	2000	None	—
4	Non-compressing engine with pipe inlet (turbulent)	Disc/2b	Central pipe of d/D = 0.25 and L/d = 97	0.075	0.060	3	200	*LDA** meas. of axial velocity and turbulent intensity distributions	[26]
5	Non-compressing engine with annular orifice inlet (turbulent)	Disc/2b	Central annular orifice with $d_{inner}/D = 0.459$, $d_{outer}/D = 0.56$, seat angle = 30°	0.075	0.060	3	200	*LDA** meas. of axial velocity and turbulent intensity distributions	[26]
6	Compressing single-valve engine with disc chamber (turbulent)	Disc/2c	Central valve operated in 4-stroke cycle; seat angle = 45°	0.076	0.084	5.8	1500	*HWA*** meas. of resultant velocity and turbulent intensity at single location position in Fig. 2c	[27]
7	Compressing single-valve engine with cup-in-head chamber (turbulent)	Cylindrical cup-in-head/2c	As Case 6	0.076	0.084	5.8	1500	As Case 6	[28]

* *LDA* ≡ Laser Doppler Anemometer; ** *HWA* ≡ Hot Wire Anemometer.

References pp. 123-124.

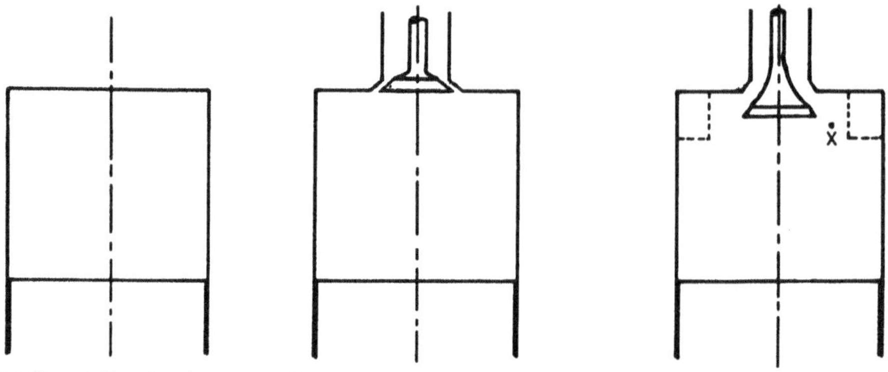

(a) Closed disc chamber (b) Non-compressing chamber with various head arrangements (c) Single-valve 4-stroke engine of Witze [27, 28] with optional bowl-in-head arrangement.

Fig. 2. Illustration of model engine configurations examined.

The remaining five cases all involve turbulent flows. Here, of course, the predictions are subject to the additional uncertainties inherent in the turbulence modeling. One of these (Case 3) relates to the closed-cylinder configuration of Fig. 2a, for which predictions are shown purely for their intrinsic interest and relevance to the other cases, since no data could be found for this situation. Measurements are, however, available for the two non-compressing model engines (Cases 4 and 5). These differ in the single respect that one has an inlet/exhaust arrangement consisting of a long pipe opening into the center of the cylinder head and the other has a valve-like arrangement which gives a fixed annular opening. There are likewise data available for the two compressing single-valve engines (Cases 6 and 7) depicted in Fig. 2c, which share an identical inlet/exhaust arrangement consisting of a single centrally located poppet valve opened and closed in a simulated four-stroke cycle, but differ in that one has a disc-shaped chamber while the other has an annular insert that produces a cup-in-head configuration.

Further information about the individual cases will be given as they are dealt with in turn in the following sub-sections. Reference will also be made in these to Table 3, which summarizes the boundary conditions which were imposed in the computer simulations.

Laminar Flow Cases

One-Dimensional Isentropic Motion (Case 1)—An important limiting situation is obtained when conditions are such that the motion in the cylinder is essentially isentropic and axially one-dimensional, for this is the simplest form of compressible engine flow which possesses any realism. These conditions will strictly prevail if the piston motion is slow, the chamber walls are adiabatic, and radial motion is negligible (i.e., as would be the case in the central region of a cylinder of large bore/stroke ratio). For such conditions, the variations about their absolute

TABLE 3

Summary of Boundary Conditions Assigned in Computer Predictions

Case No.	Inlet/Exhaust Aperture: Axial velocity (m/s)	Radial velocity (m/s)	Temperature (K)	Turbulent energy (m^2/s^2)	Energy dissipation rate	Chamber Surface Temperatures (K): Head	Wall	Piston
1	—	—	—	—	—	Adiabatic	—	Adiabatic
2	(a) Extrapolated measurements.	0	Isothermal	—	—	—	—	—
	(b) As above.	Estimated from calculations.	As above	—	—	—	—	—
3	—	—	—	—	—	350	350	Adiabatic
4	Extrapolated measurements.	0	Isothermal	$0.0135u_c^2(1 + r/R^2)$*	$C_\mu^{0.75}k^{1.5}/0.03R$*	—	—	—
5	(a) Uniform, to give inflow angle of 30°* (b) Uniform, to give inflow angle of 60°*	Uniform,* from piston displacement	Isothermal	$0.0135u_c^2[1 + 2(r - \bar{R}/2)/\Delta R]^2$* as above	$C_\mu^{0.75}k^{1.5}/0.03R$*	—	—	—
6	Uniform, to give inflow angle of 45°*	Uniform,* from orifice analysis	310	10^{-6}*	10^{-6}*	350*	350*	350*
7	As above*	As above	310	10^{-6}*	10^{-6}*	400*	350*	325*

Notes: * denotes guessed behavior; subscript c stands for centerline, inlet; R ≡ inlet pipe radius; $\Delta\bar{R}$ ≡ mean annulus radius; R ≡ annulus gap; L ≡ valve gap.

levels of the pressure and temperature will be negligibly small, so the density will be essentially uniform, as will be the viscosity and thermal conductivity, if the flow is laminar. In these circumstances the governing continuity, axial momentum and thermal energy equations simplify, as described in Ref. 29, to such a state that analytical solution is possible, with the following result:

$$u^* \equiv \frac{u}{u_p} = \xi \tag{8}$$

$$p^* \equiv \frac{p - p_0}{\rho^o S |a_{p,\max}|} = \frac{-a_p}{|a_{p,\max}|} \frac{\xi^2}{2} \tag{9}$$

$$T^* \equiv \frac{T}{T^o} = \left(\frac{S + z_{p,\min}}{z_p}\right)^{\frac{R}{C_v - R}} \tag{10}$$

where $p - p_0$ is the pressure difference between the locations ξ and $\xi = 0$ (the cylinder head), a_p and $a_{p,\max}$ are respectively the instantaneous and maximum values of the piston acceleration, S is the stroke, R is the gas constant, and the 'o' superscript denotes conditions at the reference state, here taken as BDC (i.e., $z_p = S + z_{p,\min}$).

The above results are of interest in their own right, for apart from confirming that the temperature obeys the usual reversible thermodynamic law, they also show that: the velocity increases linearly from the cylinder head to the piston; the normal strain du/dz acting on the fluid is consequently uniform at any instant and equal to u_p/z_p (a fact of some interest in connection with later observations about Case 3); and the pressure varies quadratically with ξ, with sign opposite to that of the prevailing acceleration so that, for example, as TDC is approached the fluid experiences an adverse pressure gradient (also relevant to later observations).

The main object of the present inquiry is, however, to assess the accuracy of the *RPM* method in these circumstances, reasonable performance for which could be regarded as one of the minimum requirements of any method of its kind. As demonstrated in Ref. 29, it is possible to perform the assessment analytically, with the following conclusions for a piston oscillating in simple harmonic motion:

(i) The continuity requirement, and hence Eq. 8, are satisfied exactly.

(ii) The dimensionless pressure given by Eq. 9 is estimated to within a fractional error given by π/N, where N is the number of time intervals in the finite-difference subdivision of the complete cycle. Thus, for example, a subdivision of 3° crank angle, which is a typical value used in more general situations, yields an error of about 2.6%.

(iii) The dimensionless temperature given by Eq. 10 is likewise reproduced to within a small fractional error, equal in this instance to $8\pi^3/N^3$, which amounts to about 0.5% for a 3° crank angle interval. Moreover, the finite-difference equations reproduce the reversibility displayed by the analytical solution.

The performance for this case is therefore generally satisfactory, but of course this is not accidental for it was heavily relied upon for guidance in formulating the finite-difference equations of the method. In this connection, it is worth noting that the several plausible difference approximations to the thermal energy equation that were initially tried were rejected on the grounds that they did not yield reversible behavior and could therefore produce gross errors. The particular formulation which finally evolved employed time-centering of the energy source term associated with pressure work, as described in Refs. 3 and 29. Also noteworthy was the finding that time-centering of the momentum equation will reduce the error in pressure for this case by a factor of 5. Unfortunately, as mentioned earlier, no such improvement has been realized in other, more general circumstances.

Non-compressing Model Engine with Orifice Inlet (Case 2)—For this second laminar-flow example, measurements and predictions will be compared of the velocity field in the model engine of Fig. 2b, equipped in this instance with a head fitted with a central circular orifice. Some of this information has been reported elsewhere [25], but is felt to be sufficiently important to the overall assessment to merit repetition here.

The engine was operated at low speed (10 r/min) and measurements were made, by Laser-Doppler Anemometry (LDA), of the radial variation of the axial velocity component at various axial locations. Regrettably, for reasons of inadequate optical access, these measurements did not include the plane of the orifice. Accordingly, in making the predictions it was necessary to resort to the practices summarized in Table 3 to estimate the inlet conditions, with the consequences described below.

An impression of the evolution of the flow structure throughout the full cycle, as suggested by the computer predictions, can be gained from the series of vector diagrams* shown in Fig. 3 for various values of the crank angle, θ, measured from TDC.

The diagrams reveal that during inlet a jet-like flow eminates from the orifice with a velocity in excess of that of the piston (whose velocity is indicated by the overlying vectors) on which the jet fluid therefore impinges. This fluid then flows radially outward along the piston face, and backwards along the cylinder wall, forming a toroidal vortex rotating in the anticlockwise sense in the view shown. Shortly after the stroke commences, a second smaller vortex of opposite sense appears in cylinder head/wall corner, caused by the adverse axial pressure gradient acting on the cylinder-wall boundary layer. Both vortices persist until BDC, but the smaller one has disappeared at the 216° stage of exhaust, and the larger one is no longer visible at 288°, by which time the overall pattern is one of a simple sink flow. However, towards the end of this phase the corner vortex is recreated, presumably by the same mechanism as before, and at the end of the stroke the inertia of the fluid provokes a slow-moving double-vortex structure.

---

* *The vector length is generally made proportional to the magnitude of the velocity. However, in order to preserve clarity, vectors which would either be excessively large or excessively small have been drawn truncated or extended and marked, thus: (-•»)*

References pp. 123-124.

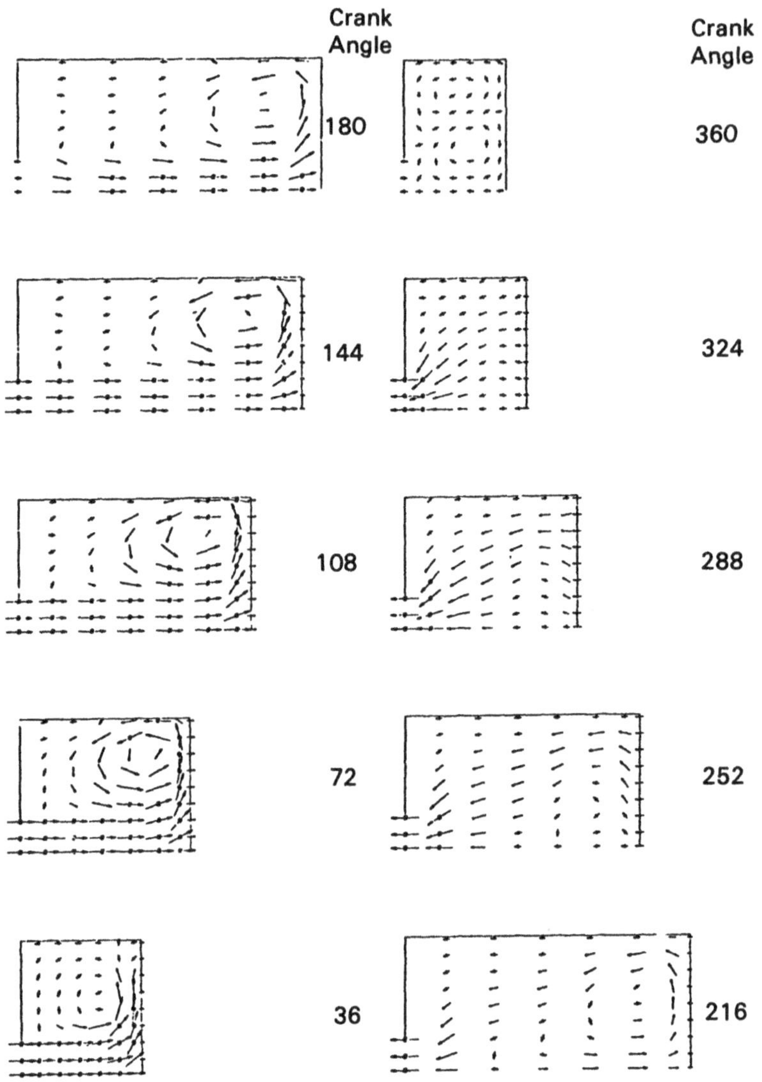

Fig. 3. Predicted velocity fields for the non-compressing engine with orifice inlet, operating at 10 r/min (Case 2).

Before a full set of comparisons is made between these predictions and the data, some explorations will be described into the effects of the assumed inlet conditions and the fineness of the computational grid. Concerning the first matter, Fig. 4 contains comparisons between the data for $\theta = 72°$, shown as points, and two sets of predictions corresponding to inlet conditions (a) and (b) of Table 3. The former, which are based on extrapolations of the measured interior axial-velocity distributions to the orifice plane (the extrapolated profile appears in the diagram), and the assumption that flow enters parallel to the axis, clearly overestimate the rate of spread of the jet and therefore underpredict the jet velocities. The refinement of set (b), wherein the inward radial component, which evidently must exist in the incoming flow due to the vena-contracta effect, is estimated in the absence of better information by using the *predicted* radial velocities for the orifice plane during the exhaust phase, clearly produces better agreement, especially within the jet. However, both predictions underestimate the size of the second vortex.

An indication as to whether numerical inaccuracies are responsible for the remaining errors can be gained from Fig. 5, which shows predictions corresponding to inlet condition (b) for two grids of different fineness. In both instances non-uniform spacing was used in order to concentrate the lines in regions of anticipated steep variations near the walls and in the shear layer emanating from the orifice tip. Evidently changes do occur with refinement, but not of the desired kind. Thus the marginal effects in the outer region are in the right direction in some respects, but there is no sign of appreciable enlargement of the vortex. The best agreement in the jet region would actually be obtained with the predictions of the *coarser* grid. Further refinement (e.g., explorations of the effect of crank angle interval showed no appreciable change below 6°) therefore seems unwarranted until the uncertainties about the inlet conditions can be resolved.

Fig. 6 contains a set of comparisons at various crank angles spanning the entire cycle. The calculations are for inlet conditions (b) and the finer grid, and required about 4 minutes of IBM360/195 time to produce. These results are far too complex and numerous to allow detailed comment on all aspects; however the following are believed to be the main points:

(i) The jet behavior is still incorrectly predicted in important respects, even

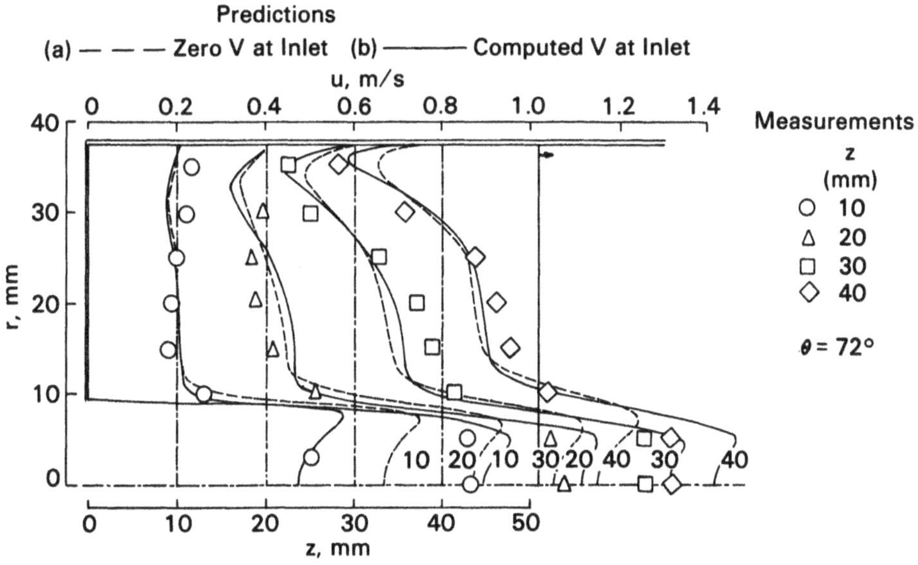

Fig. 4. Effect of inlet conditions on the predictions for Case 2.

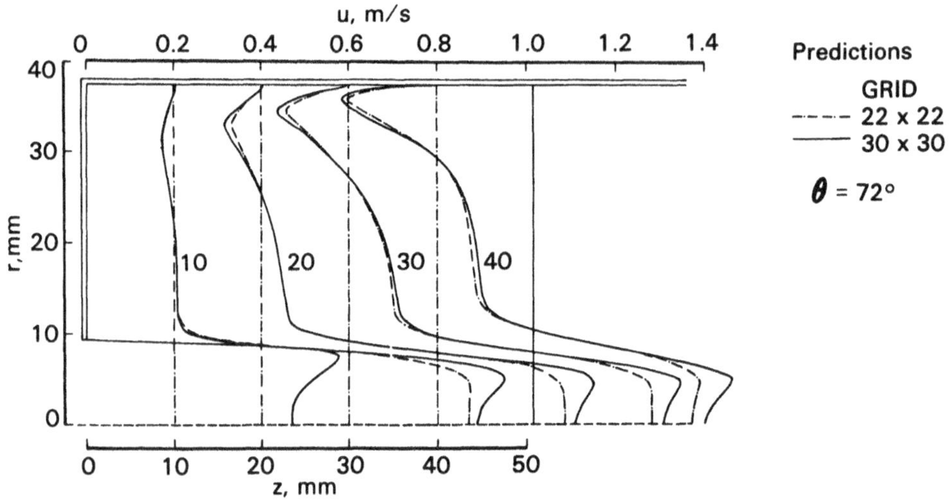

Fig. 5. Effect of grid refinement on the predictions for Case 2.

though the errors within the jet itself are in the range 7 to 25%, apart from the vicinity of the piston, where they are larger. With the exception of the initial phase of the inlet stroke, where the rate of spread is overpredicted and the calculated velocities are consequently too low (see results for $\theta = 36°$), the general tendency is for the predicted jet to be too narrow and velocities too large. The explanation for this probably lies in unsatisfactory specification of the inlet radial-velocity distribution. It is also noticeable that the measured velocity distributions are generally flatter than the predicted ones, suggesting that the inlet axial-velocity specification which, it will be recalled, was deduced by extrapolation, may also be incorrect.

(ii) The initial inaccuracies in jet prediction are immediately felt elsewhere, which is not surprising. For example, at $\theta = 36°$ and 54° the measured maxima in the near-piston profiles which lie about midway between the axis and the cylinder wall are not reproduced, nor are the magnitude and extent of the reverse flow region. Although better agreement for these features is observed in some later phases (see, e.g., the results for $\theta = 72°$ and 90°), the accuracy thereafter deteriorates again.

(iii) The earlier observation of the failure to predict the relatively slow-moving structures in the remainder of the cylinder space correctly appears to apply over the whole cycle. For example, the measurements indicate that the corner vortex is formed by $\theta = 72°$, whereas the predicted one emerges later and remains smaller in both size and strength. This behavior can also be plausibly attributed to the incorrect prediction of the jet, whose excessively large velocities cause the surrounding fluid to be impelled forward by the action of the shear stresses at an excessive rate, thereby shifting the location of the initial flow reversal closer to the cylinder wall. The main eddy is consequently enlarged, and this in turn constrains the size of the smaller one.

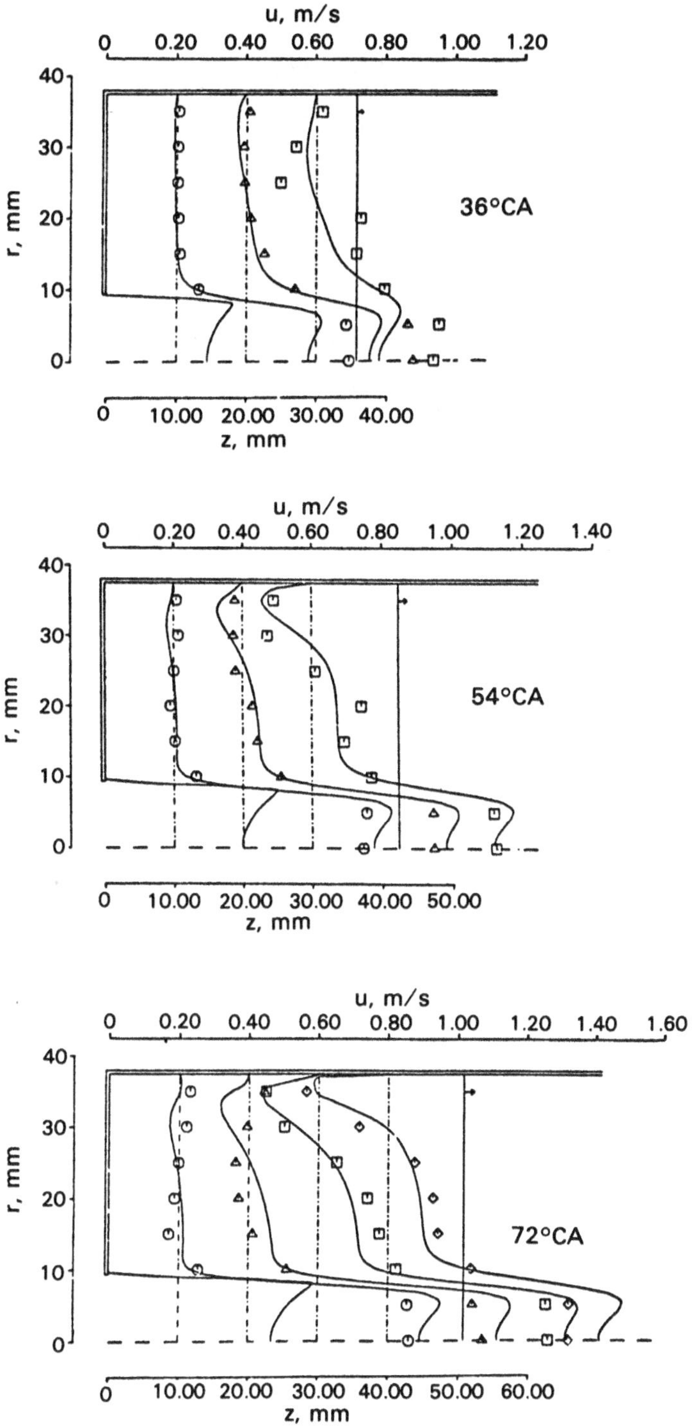

Fig. 6. Comparison between measured and predicted axial velocities for Case 2.

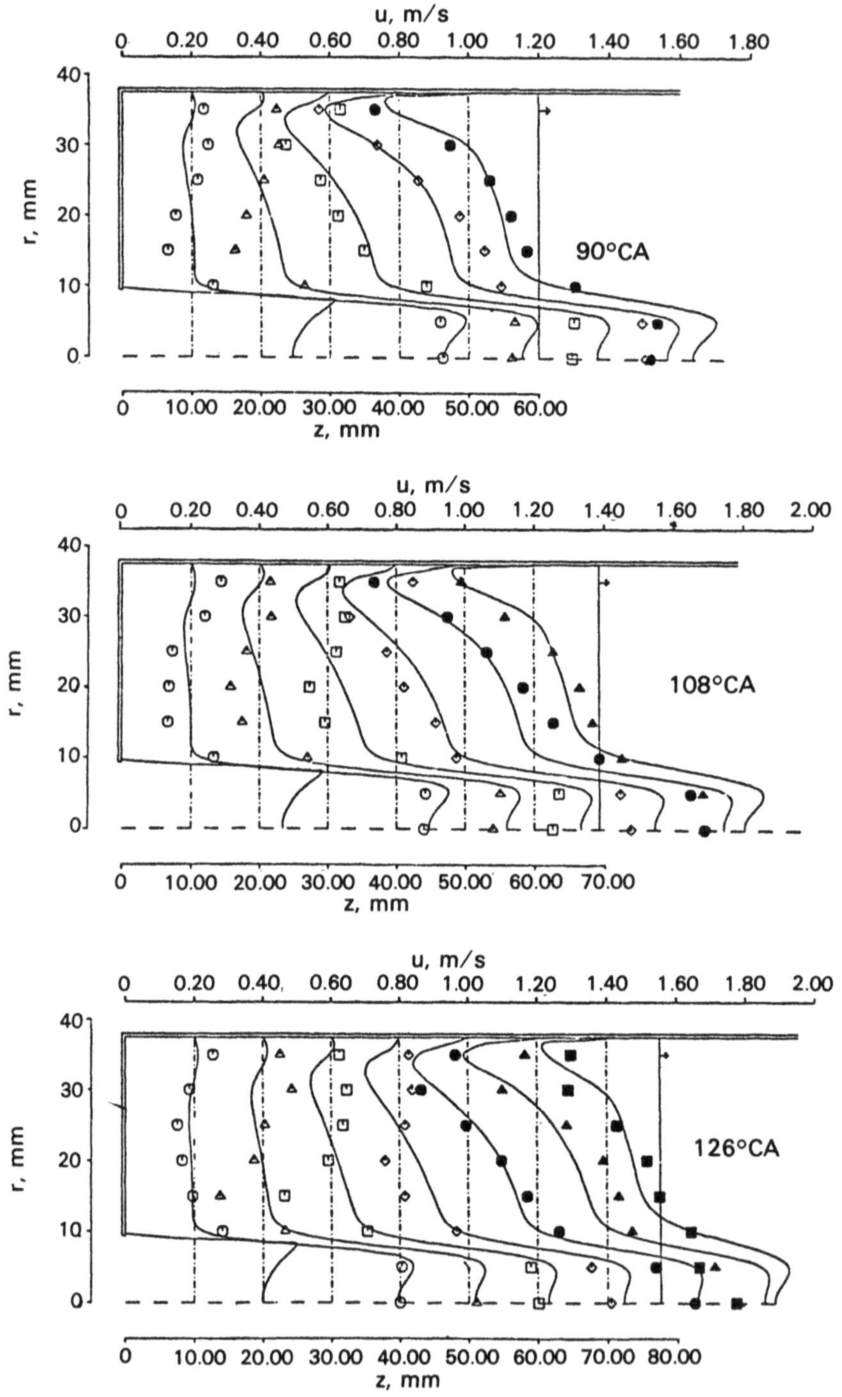

Fig. 6. (Continued)

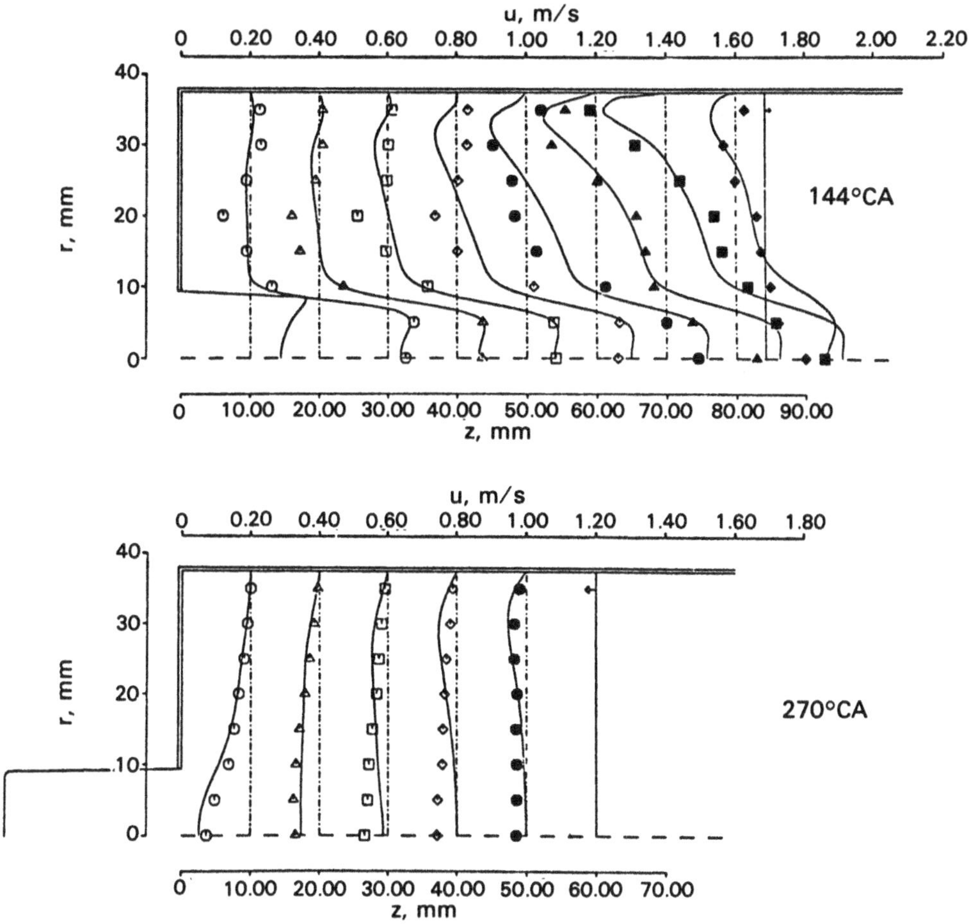

Fig. 6. (Continued)

In summary, it can be said that although the calculations correctly mirror the qualitative behavior of the flow and quantitatively predict some aspects, notably the jet behavior, with moderate accuracy, the slower motions are not well predicted. Uncertainties in the inlet conditions are probably the major source of error, but the question must remain open until more precise inlet information is available.

Turbulent Flow Test Cases.

Compressing 'Engine' with Closed Chamber (Case 3)—In this section predictions are described of the situation of a piston oscillating at high speed in a closed disc-shaped cylinder. Somewhat surprisingly, no experimental data could be found for this circumstance, despite its obvious relevance to engines, compressors and the like. It was nevertheless selected for study by us because of the unique opportunity it affords to explore the fluid motions produced by an oscillating piston in the absence of other complicating effects due to geometrical protuberances, inlet processes, etc.

The results reported here are but one sample of an extensive parametric exploration which is described in more detail in Ref. 4 and other publications in preparation. The simulated operating conditions are summarized in Table 2 and correspond to an engine of unity bore/stroke ratio and 14:1 compression ratio operating at 2000 r/min. The predictions entailed solving the full set of governing equations for mass, momentum, the two turbulence parameters, and thermal energy, with boundary conditions on the latter arbitrarily specified as adiabatic on the piston surface and uniform temperature (350 K) on the remaining surfaces. Non-uniformly spaced grids were used, with concentration near the walls. Refinement tests led to the selection of a 21 × 21 arrangement and a crank-angle increment of 6°, with corresponding computing times of 5 minutes per cycle.

Fig. 7 shows two characteristics of the predicted flow behavior at various crank angles, namely the ensemble-mean velocity fields (illustrated by means of the vector diagrams) on the left and the turbulence energy distributions (in the form of contours of "total intensity", defined as $\sqrt{2k/3}/\bar{u}_p$, where $\bar{u}_p$ is the mean piston velocity) on the right.

The vectors reveal that over the greater part of the cycle the interior flow is essentially one-dimensional. More detailed examination shows that the velocities seldom depart from the values given by Eq. 8 by more than 10%. Exceptions occur near TDC and BDC, where the combined effects of fluid inertia, wall shear and adverse pressure gradient give rise to the formation of single, weak toroidal vortices*. The velocities near the wall exhibit the expected maxima characteristic of high-frequency oscillating flows, but this feature is barely discernible in the plots.

The turbulent intensity contours indicate small spatial variations in the central region and somewhat steeper gradients near the walls. Not surprisingly, the peak levels are generated in the high-stress region near the corner formed by the sliding piston and the cylinder wall. A rather unique property of this class of flow is signaled by two features. First, the intensity increases during compression and subsequently decreases during expansion. Secondly, although the shear stresses must be small everywhere except in the near-wall region, the turbulence levels in the interior are of the same order as or greater than the near-wall values (with the exception of the forementioned corner). The property in question is the generation of turbulence by the action of normal stresses associated with the compression/expansion processes. In view of the earlier observations about the near linearity of the axial velocities, and the observation discussed earlier that this gives rise to a uniform strain proportional to u_p/z_p, it may be expected that since the generation of turbulence by normal stresses is directly connected with

* *It is of interest to note that no evidence exists in these calculations of the corner vortex which has been claimed to be formed by the action of the piston scraping the boundary layer off the cylinder wall [30], nor does it appear in any of the other predictions or detailed measurements which are described in this paper.*

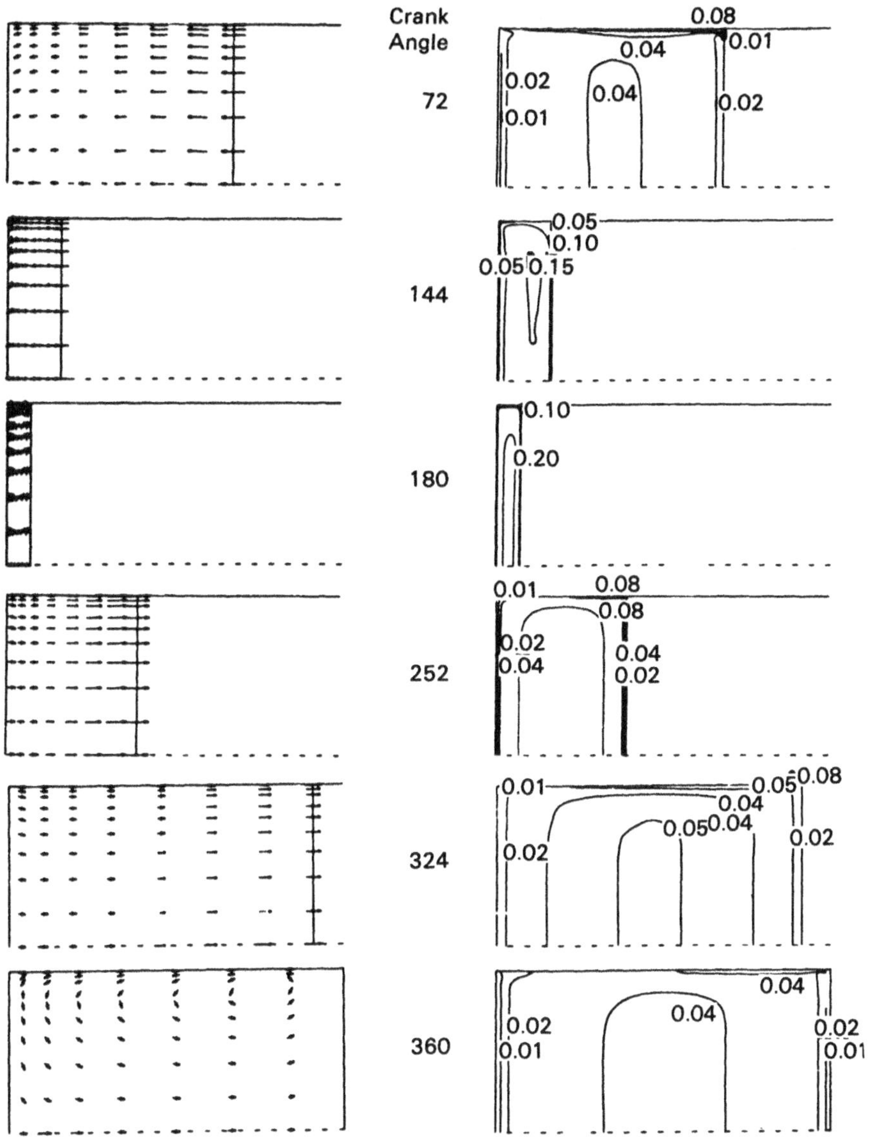

Fig. 7. Predicted velocity and turbulent intensity fields for closed disc chamber, operating at 2000 r/min (Case 3).

the strain, the intensity would indeed be uniform (were it not for the influence of the walls) and vary with crank angle as observed. This point will be examined later in greater detail.

As a final comment on these results, note should be taken of the general level of turbulent intensity produced in these circumstances, so that it may be later compared with the levels observed in situations more akin to real engines.

Non-compressing Engine with Pipe Inlet (Case 4)—What appear to be the first extensive measurements of the turbulent flow behavior in a model engine were performed by Morse et al. [26] on the apparatus of Case 2, but fitted in this instance with an inlet/exhaust tract consisting of a long pipe and operated at the higher speed of 200 r/min. In these circumstances the Reynolds number based on the mean piston velocity and cylinder diameter is about 3200 and the Reynolds number in the pipe is a factor of 4 larger, so the flow is turbulent, but perhaps not in the fully developed sense.

In the experiments, profiles of the ensemble-mean axial velocity component and its fluctuation intensity were determined by LDA, but again it proved not possible to measure the inlet conditions. As Table 3 indicates, it was therefore necessary for the calculation to rely on a combination of guesswork and a few measurements by Melling [31] of the axial velocity profiles in the pipe several diameters from the cylinder inlet plane. The latter profiles were assumed to prevail at the inlet and crudely interpolated in space and time (with due care to preserve the correct instantaneous volumetric flow), while the radial velocity was taken to be zero, and the turbulence parameters were calculated from rough approximations to the observed profiles in *steady* pipe flow. The results reported here were obtained with a 30 × 30 grid non-uniformly spaced as for Case 2. Then with the crank angle interval set at 3°, a cycle simulation required 15 minutes computing time. Predictions for a coarser 21 × 21 grid differed at most by about 15% from the present results.

In this instance the overall behavior of the flow field will be illustrated by means of the *measured* streamline patterns of Fig. 8. These show the structure to be very similar to that of Case 2, with a jet-like flow during intake impinging on the piston face and generating first one, and then two, toroidal vortices. These grow steadily until BDC and then disappear soon thereafter, to be replaced by a straightforward sink flow.

The measured and predicted normalized (as $u/\bar{u}_p$) axial-velocity profiles are compared in the sequence of plots shown in Fig. 9, which also shows similar comparisons at corresponding crank angles for the axial turbulence intensity, $u'/\bar{u}_p$. The latter was deduced from the predicted turbulence energies on the assumption of isotropy and is given by $\sqrt{2k/3}/\bar{u}_p$.

Before the predictions are assessed it is of interest to examine the behavior of the measured turbulent intensity. Throughout the entire intake stroke the profiles of this intensity exhibit maxima which roughly correspond with the edge of the jet, and are the consequence of turbulence generation by the intense shear which occurs between this and the surrounding fluid. As the jet approaches the piston and is decelerated and turned, the strong, mainly normal stresses which are produced cause further augmentation of the turbulence. Other regions of high stress occur between vortices and near the walls and are again responsible for turbulence augmentation there. Turbulence is transported away from these regions by the combined actions of convection and diffusion. This is why, for example, there are appreciable levels on the jet axis and in the central cores of the vortices. During the exhaust stroke, when the interior shear layers decay away, the transport mechanisms tend to smooth the profiles, while viscous

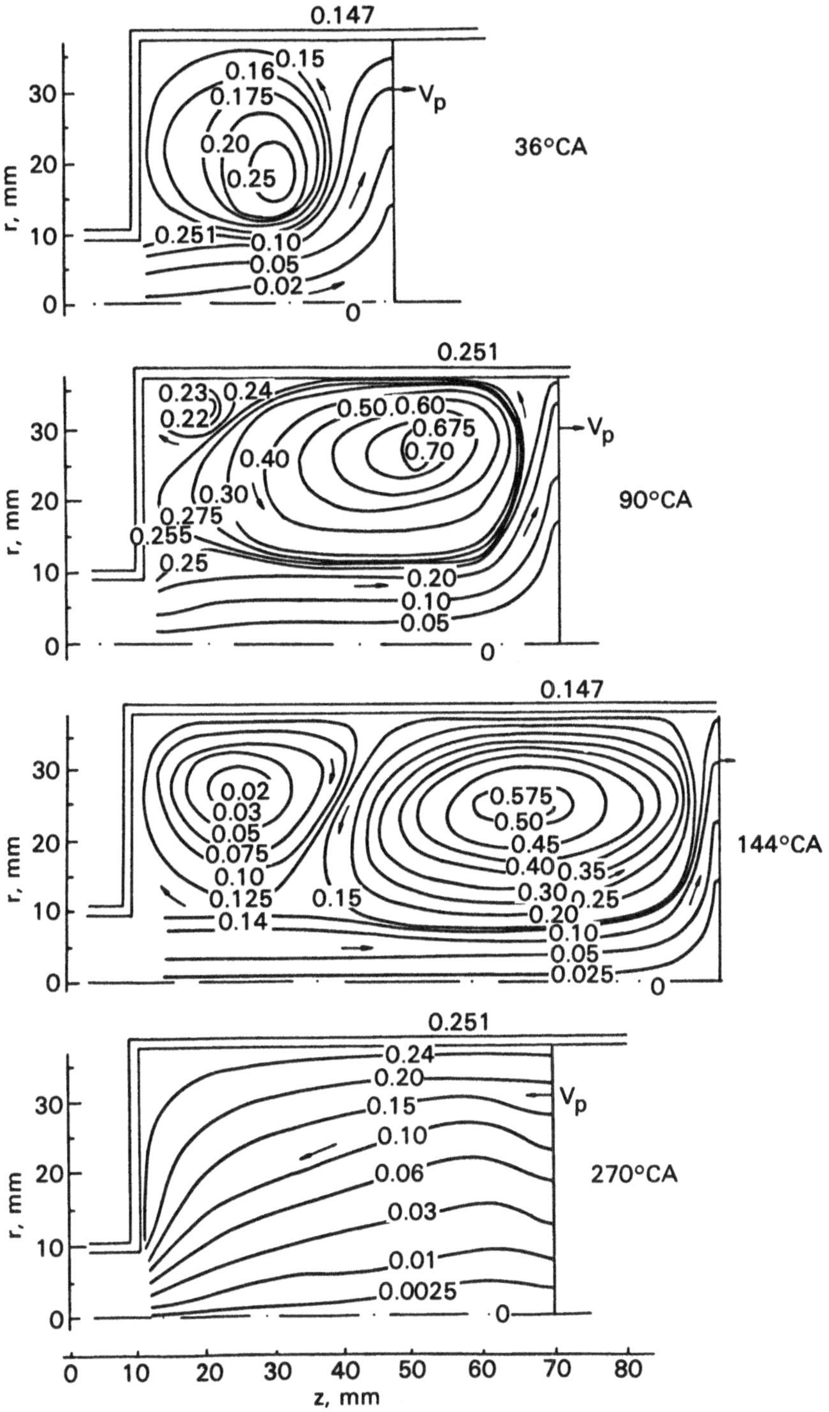

Fig. 8. Measured velocity fields for non-compressing engine with pipe inlet, operating at 200 r/min (Case 4).

References pp. 123-124.

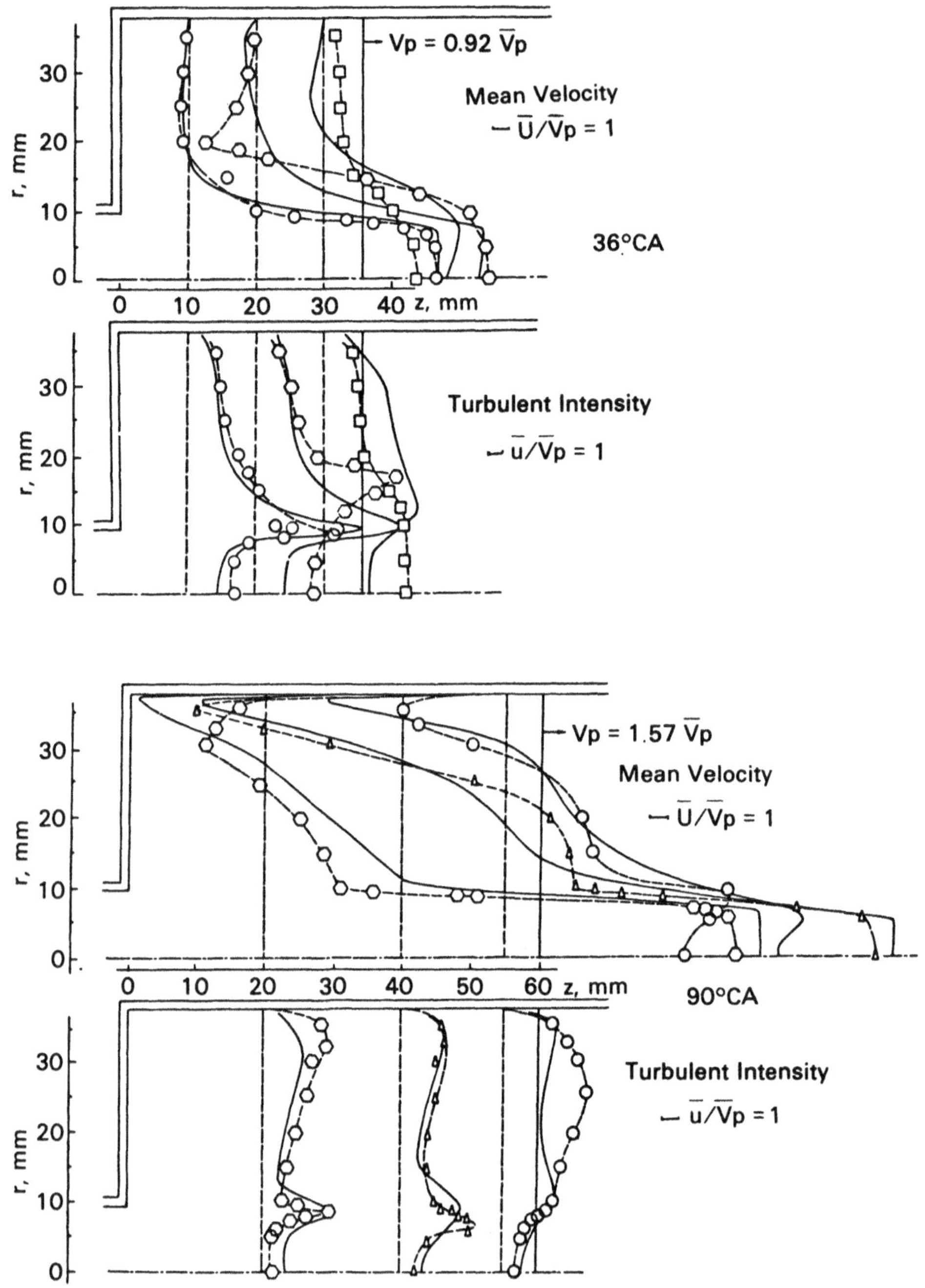

Fig. 9. Comparisons between measured and predicted axial velocity and turbulent intensity profiles for Case 4.

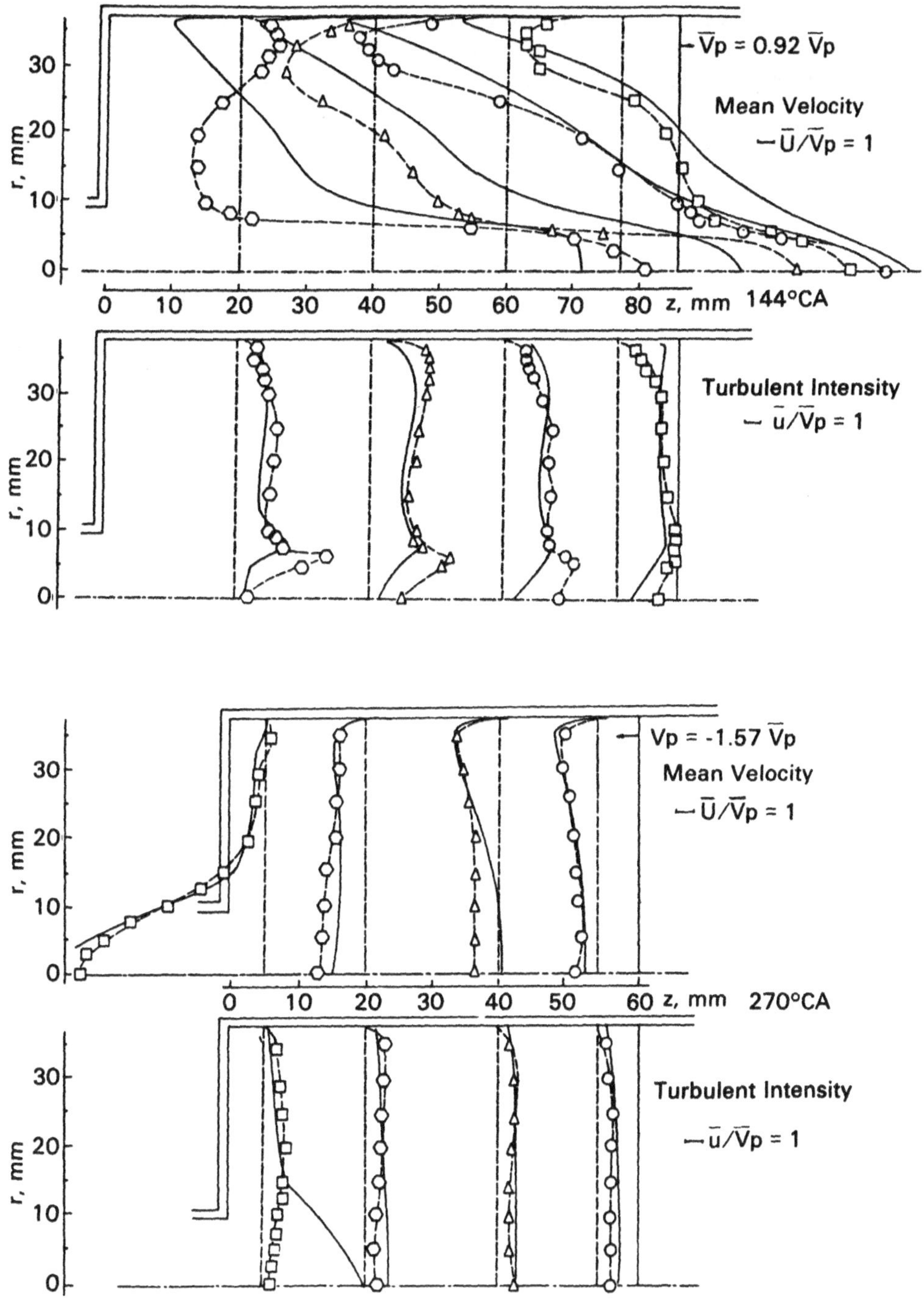

Fig. 9. (Continued)

dissipation causes the magnitude to decay. Finally, it is noteworthy that, especially during the inlet stage, the general level of turbulent intensity is high, with values in excess of unity not uncommon. (Recall that the normalizing factor is, however, the mean piston velocity.) These are an order of magnitude greater than the peak levels observed for the closed cylinder of Case 3 and suggest that, as will be further supported by the evidence of the remaining cases, inlet-generated turbulence far outweighs that produced by the action of compression-generated normal stresses alone.

Turning now to the assessment, the overall picture for the velocity field predictions is one of reasonable reproduction of the jet behavior, apart from delayed decay and spread near the piston; poor initial agreement at $\theta = 36°$ for the peripheral flow as it spreads radially outwards and backwards, although the level of error diminishes at the later crank angles of 90° and 144°; underprediction of the size of the secondary vortex and its influence on the return flow throughout the entire inlet stroke, as signaled by the predicted velocity maxima being too large and lying too close to the cylinder wall, as well as the absence of positive velocities at the closest station to the corner; and finally some measure of agreement at 270° in the exhaust stroke, but indications in the central region that the predicted vortex has decayed away more slowly than the measured one.

In view of the intimate connection between the mean flow and turbulence fields described earlier, it should come as no surprise that the turbulent intensity predictions also exhibit errors. However, if allowance is made for the observed differences in the locations of measured and predicted shear layers at, for example, $\theta = 36°$, which must inevitably displace the turbulence profiles, then the agreement is surprisingly good. The locations of greatest error appear to be near the piston face during intake and near the outlet during exhaust. It may or may not be significant that these are regions dominated by normal stresses for which, as will be shown later, the energy/dissipation turbulence model appears to be prone to error.

In view of the uncertainties associated with the inlet conditions and turbulence modeling, it would probably not be fruitful to speculate on the cause of the discrepancies at this stage, although some general comments will be made later. It is interesting to observe, however, that in addition to the various problems with inlet conditions already alluded to, there is the distinct possibility that the fluid entering the pipe during the exhaust phase separates, forming a vortex structure within it which is subsequently drawn into the cylinder during intake. This illustrates, among other things, the difficulty of designing an inlet system which will produce a well ordered flow.

Non-compressing Engine with Annular Orifice Inlet (Case 5)—Morse et al. [26] also report a set of measurements in which the model engine was fitted with the alternative inlet/exhaust arrangement shown in Fig. 2b. This consisted of a mock "valve" inserted in an oversize seat, so that with the valve face flush with the cylinder head, an annular opening remained for the entry and exit of air. This arrangement has the obvious advantage of simulating the flow past a real valve in some respects, without the complications introduced when it protrudes into the cylinder space.

In the absence of any information about the inlet flow structure whatsoever in this instance, two alternative sets of boundary conditions were explored. They have in common the presumptions that the profiles of both velocity components are uniform and those of the turbulence parameters are roughly characteristic of flows in annuli. (The precise specifications are given in Table 3.) The sole difference was that for conditions (a) the flow was presumed to enter at the valve seat angle of 30°, while a steeper angle of 60° was prescribed in set (b), for reasons which will shortly become apparent. The specifications of grid and crank angle interval were similar to those of the previous case, as were the computing times.

The measured flow behavior is shown by way of streamline patterns in the sequence of plots of Fig. 10, where it is immediately obvious, particularly in the earlier phases of intake, that the entering annular jet assumes a steeper angle than that of the seat, at least within the region near the opening accessible to measurement. Separation occurs at the inner and outer edges of the annulus and results in two vortices of unequal size and strength being formed at $\theta = 36°$, with the larger and weaker of the pair located in the lee of the valve and the smaller one centered near the cylinder head corner. The jet meanwhile flows between these two and impinges almost vertically onto the piston. Further on at $\theta = 90°$ the central eddy has enlarged and strengthened at the expense of the corner one, and the jet now follows a shallower angle, albeit still greater than that of the seat. At the same time, the adverse pressure gradient brought about by impingement on the piston has been sufficient to cause the cylinder-wall boundary layer to separate and form a third vortex near the piston. This behavior persists at 144°, where the main vortex has increased in strength and the near-piston vortex is larger, but weaker. As now seems to be the pattern, this relatively complex structure is very quickly suppressed as soon as the piston reverses direction, so that hardly any trace remains by $\theta = 270°$, corresponding to the last diagram in the series.

The profile comparisons to be examined shortly demonstrate quite conclusively that the predictions based on inlet conditions (b) are in best agreement with reality, so the associated vector* fields are displayed in Fig. 11 to allow an overall impression to be gained first. It is indeed possible to see, among other similarities, initial formation of the twin-vortex structure and a suggestion of the emergence of the third vortex, although its appearance appears to have been delayed until somewhere between 90° and 144°. It is equally obvious that even better correspondence might have been obtained by causing the effective inlet angle to reduce gradually from 60° to 30° during intake, but the temptation to do so has been resisted on the grounds that there is insufficient information to justify such a refinement.

---

* *Drawn in this instance to scale, irrespective of the magnitude of the velocity.*

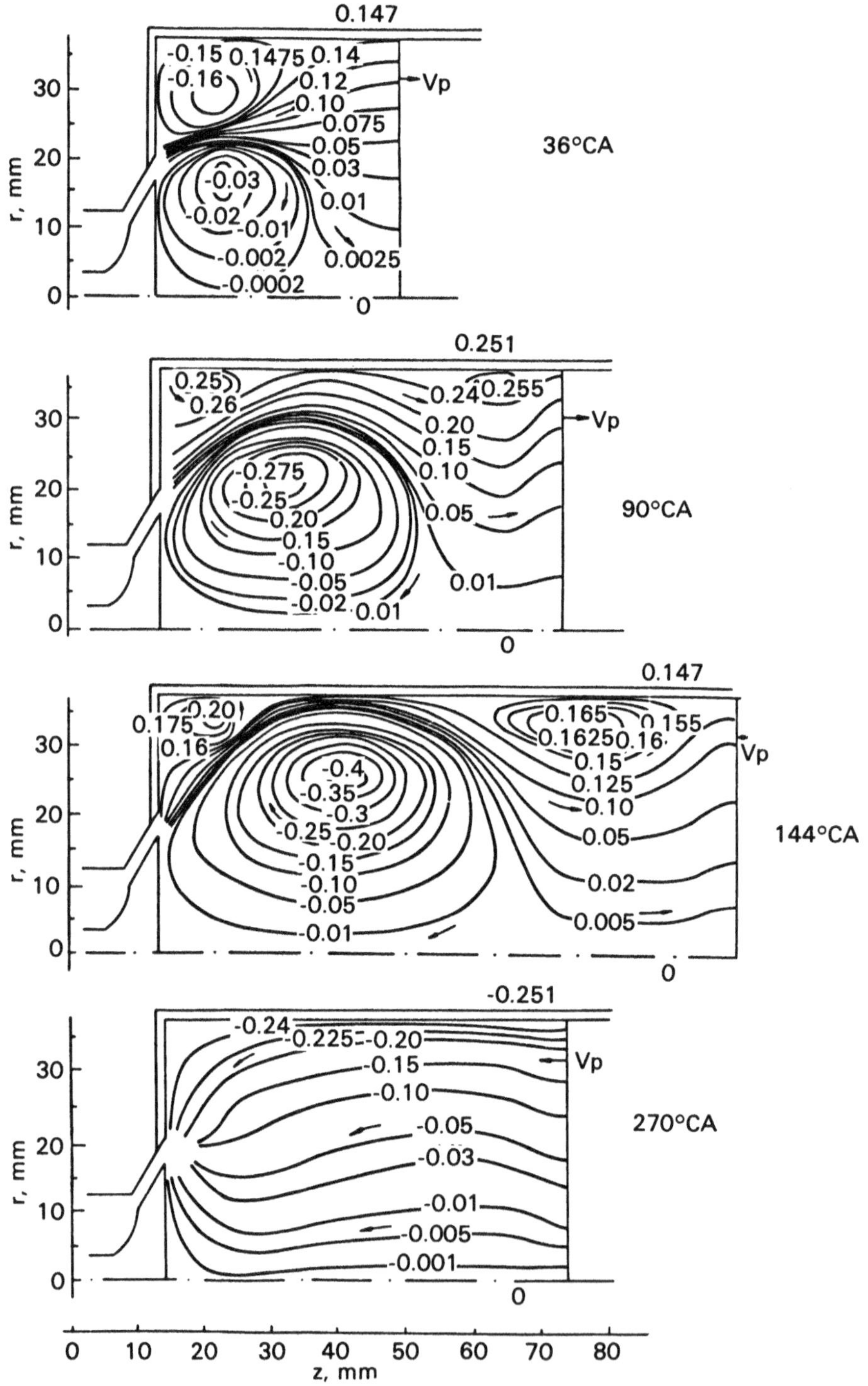

Fig. 10. Measured velocity fields for non-compressing model engine with annular orifice inlet, operating at 200 r/min (Case 5).

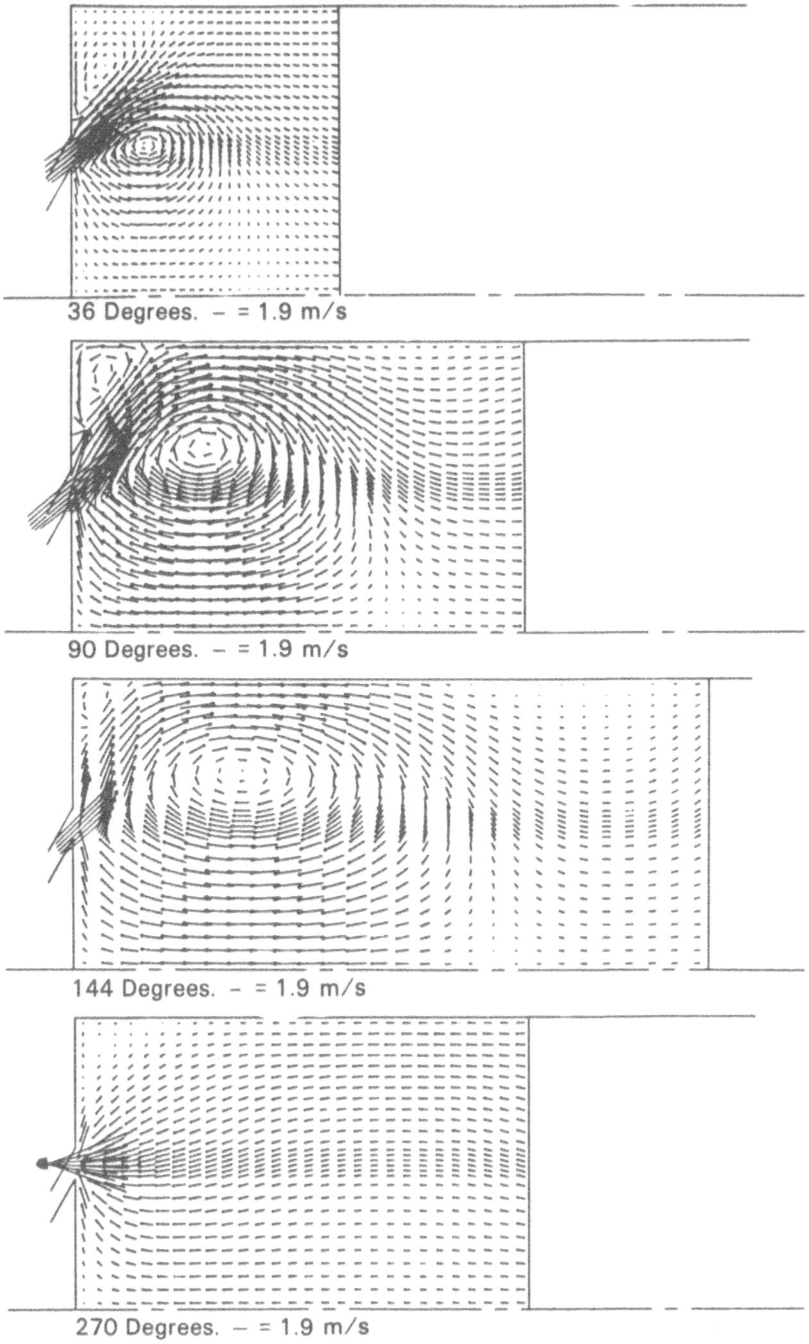

Fig. 11. Predicted velocity fields for Case 5 with inlet angle specified as 60°.

References pp. 123-124.

The plots of Fig. 12 showing comparisons of measured and predicted axial velocities and turbulent intensities demonstrate clearly how the calculations based on inlet conditions (a)—represented by the solid curves—produce the wrong jet trajectory and other consequential errors throughout the entire intake stroke. Attention will therefore be focused on the chain-dashed lines representing the results for conditions (b), which are evidently much better, albeit still not satisfactory. The plots for 36° crank angle show the predicted jet to have about the right trajectory but to spread and decay too rapidly, with the result that the associated shear-generated turbulence maxima are hardly visible. Later, at θ = 90° and 144°, the agreement is better for the velocity profiles near the inlet but not, somewhat surprisingly, for the turbulence structure there. The prediction of the mid-plane profiles is very poor indeed, although some improvement occurs adjacent to the piston. Finally, the comparisons for θ = 270° in the exhaust stroke are generally satisfactory for velocity, but some errors still persist in turbulent intensity.

The overall picture which emerges seems therefore to be one of strong sensitivity to the inlet flow angle (and perhaps other as-yet-unexplored inlet characteristics). There is some suggestion that if the inlet flow angle were specified to vary in the manner suggested by the data, the gross features might be better predicted, but there might well still remain appreciable discrepancies of indeterminate cause. The message from this is again quite clear; detailed inlet information is urgently required.

Compressing Single-Valve Engine with Disc Chamber (Case 6)—The pioneering hot-wire anemometer (HWA) measurements obtained by Witze [27, 28] in his specially constructed axisymmetric model engine form the testing ground in this and the following case. The engine, which is depicted in Fig. 2c, was sufficiently robust to be operated at typical engine speeds (i.e., 1500–2000 r/min) and was unique in having a single centrally-located conventional poppet valve which was cam-actuated in a sequence simulating a four-stroke engine cycle, i.e., induction, compression, expansion and exhaust. Of interest here are the measurements obtained in an early version of the engine having a disc-shaped combustion chamber. The data used for comparison were obtained at the location labeled X in Fig. 2c. The orientation of the single-wire probe was such that, in the absence of measurement errors, it would detect the resultant of the axial and radial velocity components. The HWA signal was processed to yield the variation over the four-stroke cycle of the apparent ensemble-averaged resultant velocity, v_R, and associated turbulence intensity, v_R'. No measurements of inlet conditions are reported.

The computer simulations are made for the conditions given in Table 2 using the single set of hypothetical inlet conditions of Table 3. Thus the flow was assumed to enter at the valve seat angle of 45° with uniform distributions of the velocity components, temperature and the turbulence parameters. The instantaneous mass flow through the valve was calculated as described earlier with the exterior pressure taken as atmospheric, the orifice discharge coefficient set to 0.6 (which value resulted in peak cylinder pressures identical to those measured),

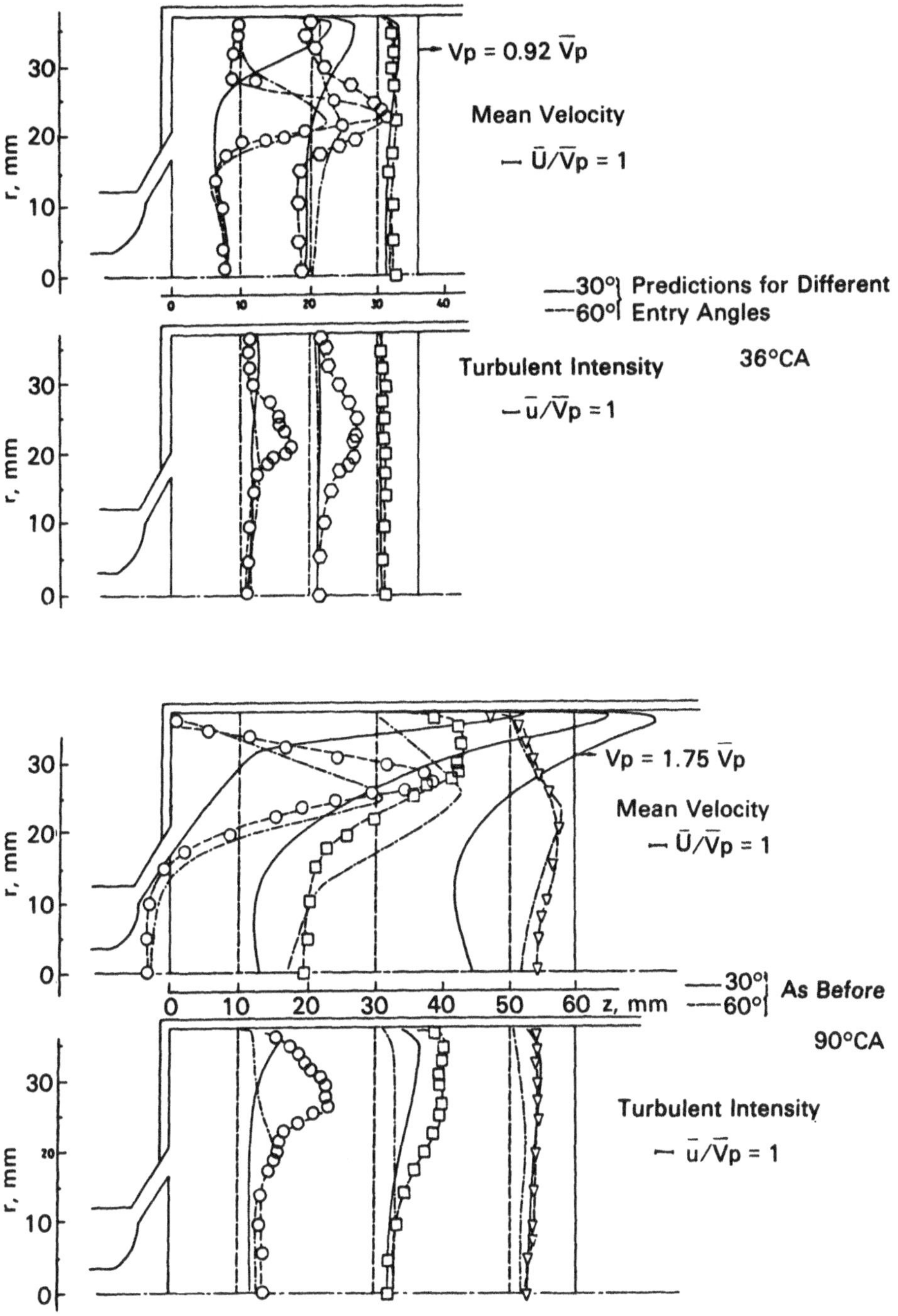

Fig. 12. Comparisons between measured and predicted axial velocity and turbulent intensity profiles for Case 5.

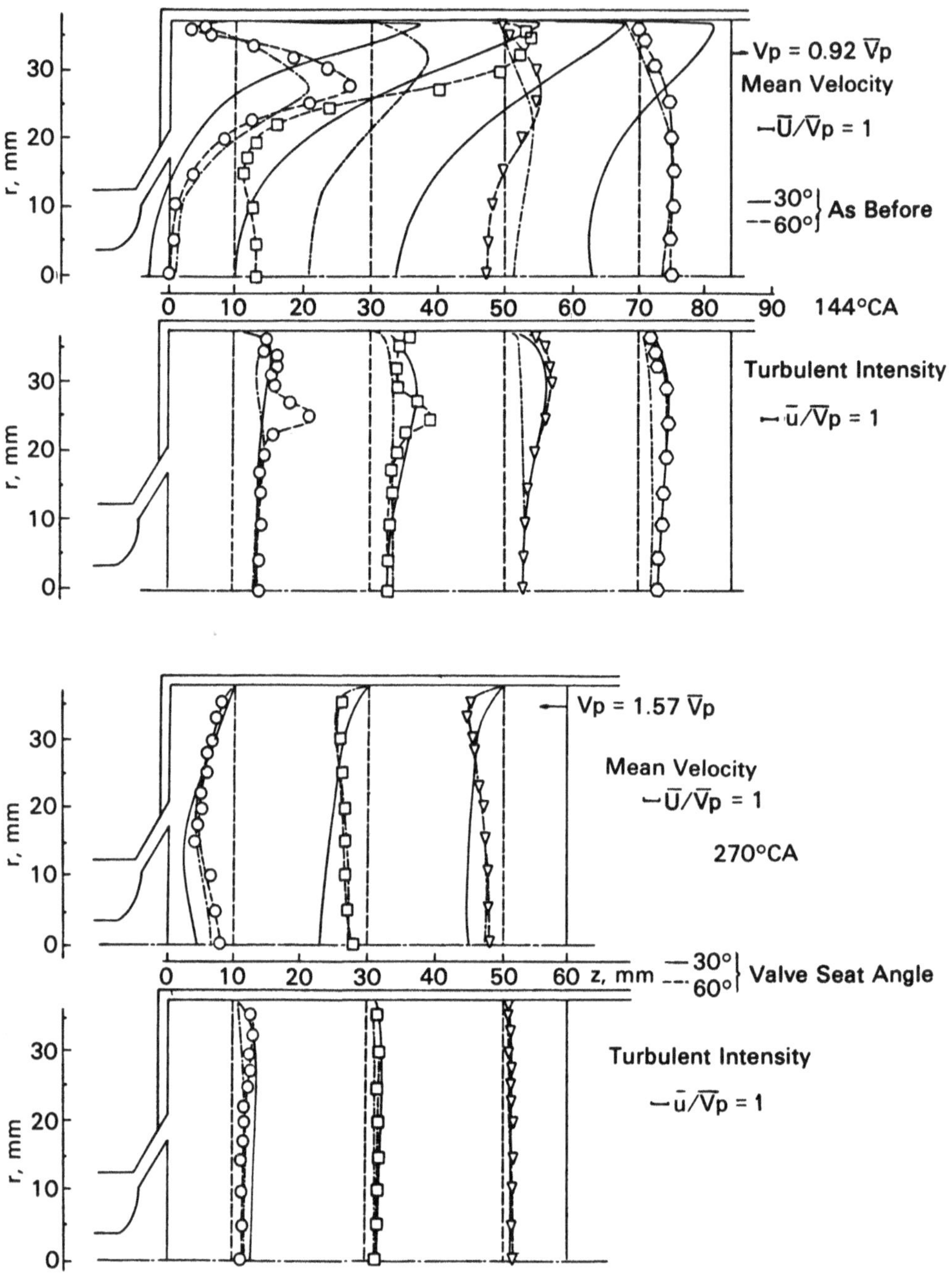

Fig. 12. (Continued)

and the instantaneous orifice opening prescribed from the dimensions and valve-lift data provided by Witze. The present results were obtained with a 30 × 30 grid which was concentrated near the valve area and the walls. With a time step equivalent to 3° crank angle, a four-stroke cycle simulation required about 40 minutes computing time.

The calculated velocity and turbulence fields are shown by way of vectors and contours respectively in Fig. 13 for a number of crank angles spanning the entire cycle. The flow structure which evolves during the inlet stroke is similar to that of the previous case with the entering annular flow separating and provoking the two annular eddies centered midway between the piston and valve and in the cylinder head/wall corner, while the remainder of the flow continues as a jet and impinges on the cylinder wall and piston. During this process high levels of turbulence are produced near the valve opening due to the intense shear there and in other areas of high stress, such as where the flow impinges on the cylinder wall and valve face. The transport of this locally generated turbulence to other regions in the cylinder is signaled by the general shape of the contours, which resemble the expected streamlines in several respects, and also by the appreciable intensities prevailing in regions of low stress.

In the compression and expansion phases the mean motion and turbulence gradually die away so that, for example, by TDC of compression the remaining eddy is very weak and disappears shortly thereafter to be replaced by an essentially one-dimensional structure. There is no sign of the compression-generated turbulence augmentation observed in Case 3, but this is hardly surprising since here, as in other situations where there is intake-generated turbulence, the latter mechanism tends to swamp the former.

Finally, the exhaust stroke produces the usual sink-flow structure, with turbulence being generated in increasing amounts as the valve opening is approached and the flow distorts to reach it.

The predictions are compared with Witze's single-point measurements in Fig. 14, with the left diagram showing the ensemble-average resultant velocity (deduced in the calculations from the axial and radial components) and the right one the turbulent intensity (with the calculated value based as usual on the isotropy assumption, giving $v_R' \equiv \sqrt{2k/3}$). Two additional curves are shown in the left diagram which represent the predicted maximum and minimum resultant velocities that would be obtained from probes located at ±1.29 mm in the axial and radial directions of the nominal probe position. The peak inlet velocity is overestimated by about 25%, but as the results at the surrounding locations indicate, this region is experiencing particularly steep velocity gradients. A small error in the predicted jet trajectory, produced by the wrong inlet specification, could easily produce errors of the observed magnitude. The turbulent intensity at this time is lower than the measured value, which is somewhat surprising in view of the high velocity gradients and the overestimation of the velocity. However, this seems to be an emerging trend, observed in the previous Case 5. As will be seen later, it is also true for Case 7. During the compression and expansion phases the measurements indicate a maximum in both mean and turbulent velocities at TDC,

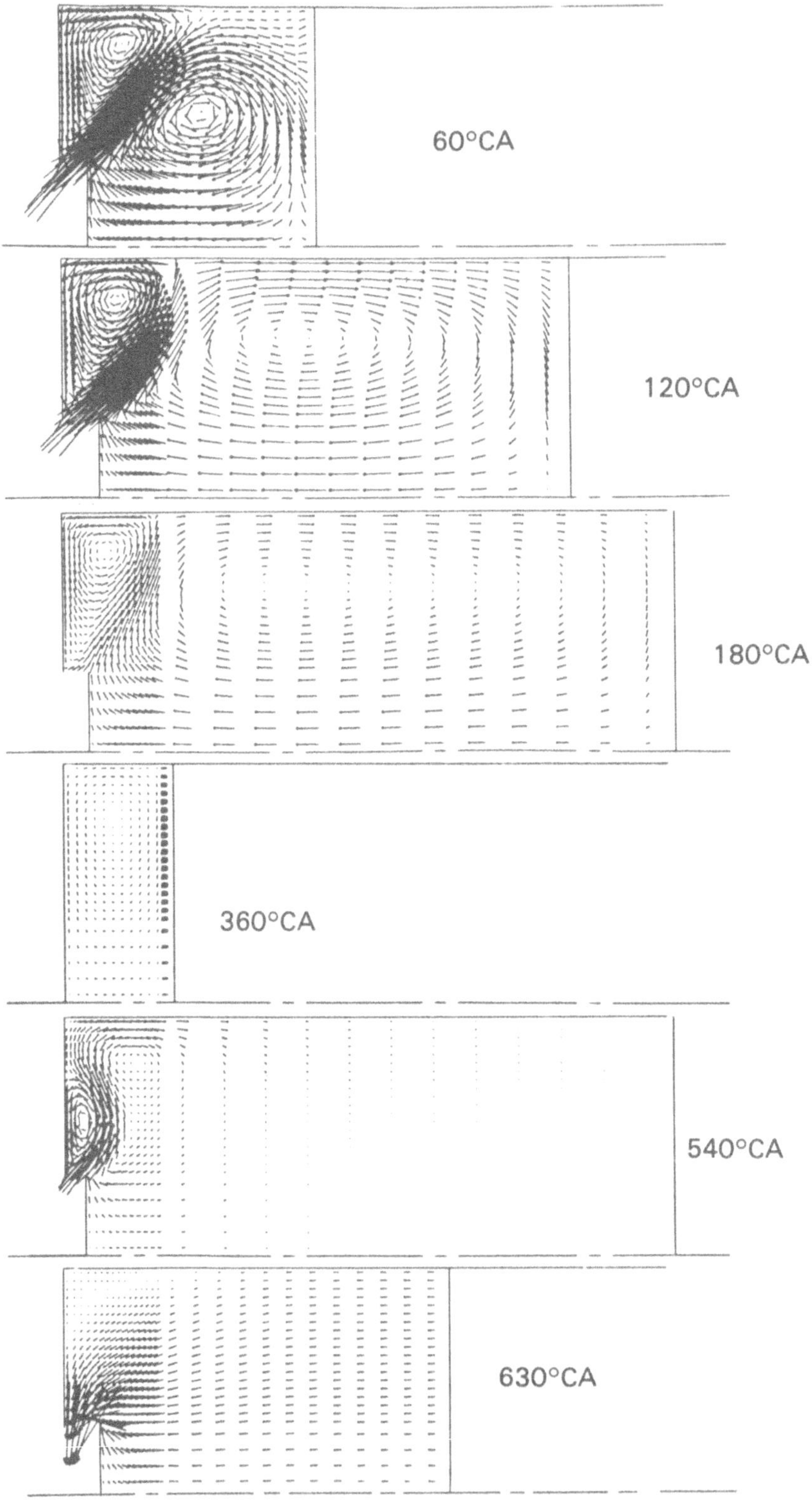

Fig. 13. Predicted velocity and turbulent fields for central-valve engine with disc chamber of Witze [27] (Case 6).

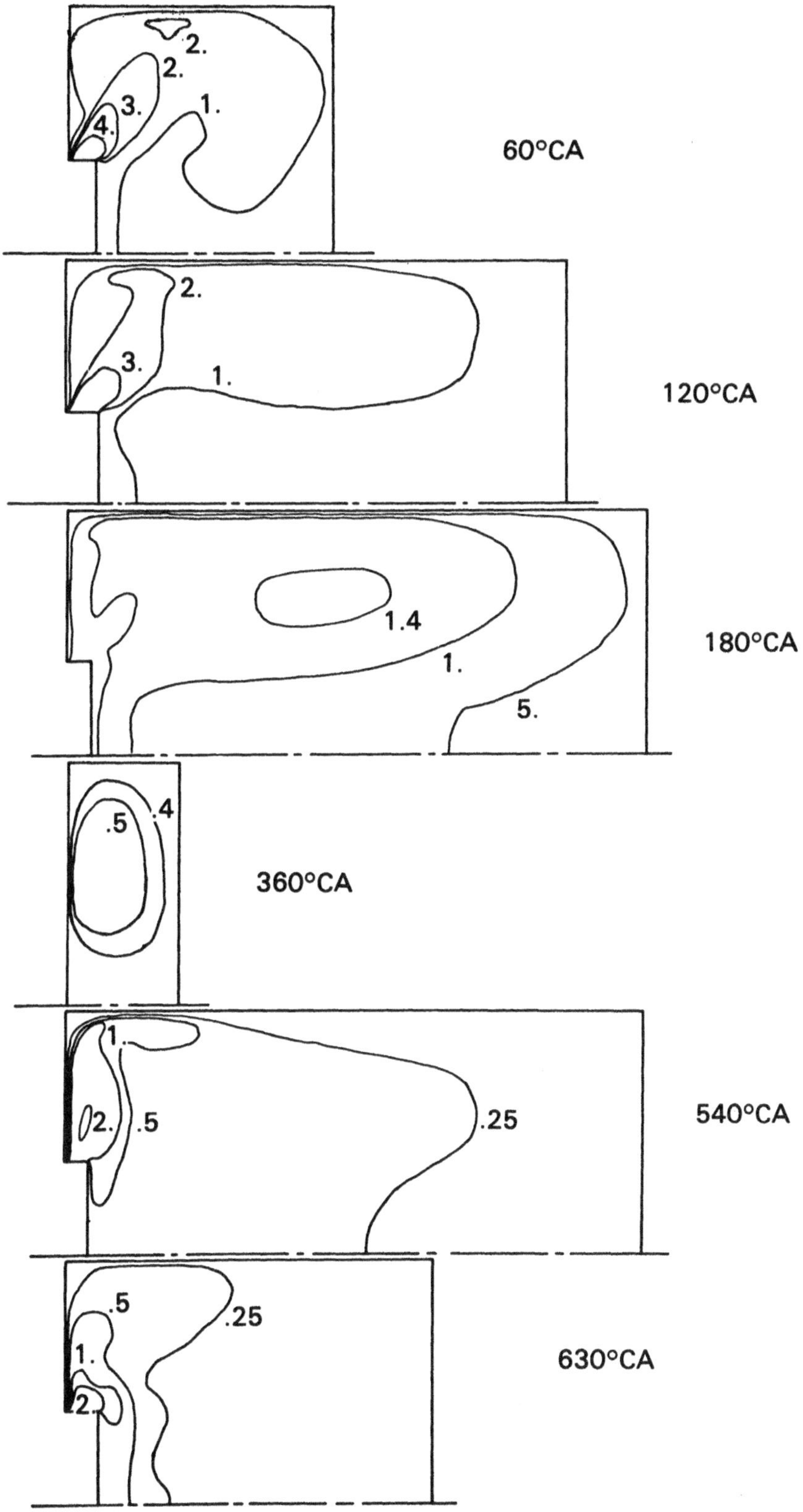

Fig. 13. (Continued)

References pp. 123-124.

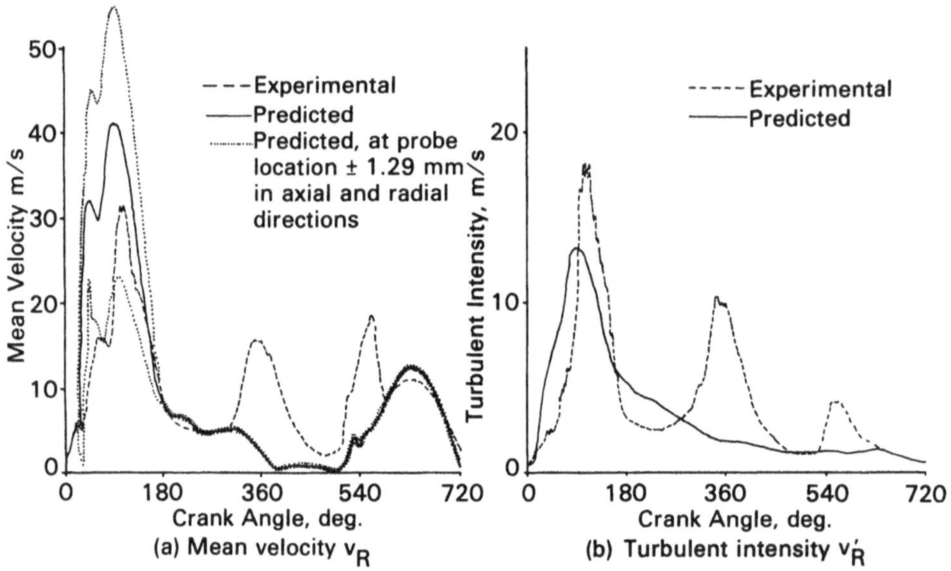

Fig. 14. Comparisons between measured and predicted variations of resultant velocity and turbulent intensity for Case 6.

while the predictions show the steady decay observed earlier. In the final exhaust phase the initial velocity and intensity peaks are not predicted, but agreement improves in the latter part of the stroke.

In order to put the above results into proper perspective it is important to note that Witze has subsequently endeavored to assess independently the accuracy of the HWA method he used, for in common with other investigators, he was unable to calibrate the instrument for engine conditions. In a recent report of his [32] showing HWA and LDA measurements at the same location in a motored L-head engine, it is clear that the HWA method has overestimated the mean velocity by about 10% during inlet and by much larger amounts during compression and expansion, with similar comments applying to the turbulent intensities during the latter phases. It is likely, therefore, that the present data were subject to errors of the same direction, if not necessarily magnitude. It further follows that agreement with the predictions is probably better than is indicated.

Comparisons of profiles, where available, provide a more balanced picture than do point measurements. However, although some heart can be taken from these observations, the fact remains that once again experimental uncertainties have prevented any firm assessment of the accuracy of the prediction method.

Compressing Single-Valve Engine with Cup-in-head Chamber (Case 7)—For this last case the situation examined corresponds to Witze's later experiments [28] in which annular inserts were used to modify the "combustion" chamber of the single-valve engine to a cup-in-head configuration, as indicated by the dashed lines in Fig. 2c. The particular experiment simulated here involved an insert of the proportions shown in the Fig. 2c. The HWA measurements were obtained at location X of that figure. The engine was operated at 1500 r/min. Other relevant information is given in Table 2.

The details of the calculations were very much the same as for the previous case (as evidenced by the inlet specifications of Table 3), although in this application the more flexible coordinate transformation of Ref. 6 was used in order to accommodate the cup region.

The velocity and turbulent intensity fields which emerged from the simulation are shown in Fig. 15. Like the previous case, the entering flow initially separates at the valve exit to form two vortices, but here a further separation occurs at the protruding edge of the cup to produce a third major vortex in the outer cylinder volume. This grows to a large size as the piston descends because the jet eminating from the valve at 45° strokes the side of the cup and is deflected axially. There is the suggestion of a fourth, weak vortex emerging near the center of the piston from about 90° onwards. The build-up of turbulence during the inlet phase is in line with the flow behavior. Maximum levels coincide, as expected, with regions of high stress, the most pronounced of which is near the valve exit.

Following the usual pattern, the vortex motions tend to die out rather rapidly during compression, while the turbulent intensity also becomes more uniform and decays. Due to the rather large TDC clearance gap between the head and the piston, there is relatively little squish-induced motion in the cup as compared with more practical circumstances (see, e.g., Ref. 6) and the locally high zone of turbulence which it contains was probably convected from elsewhere. There is likewise no appreciable "reverse squish" during expansion, so that very little happens until the valve opens, as shown by the sequence up to 510°. The rather complex behavior which thereafter occurs, with vortices initially appearing in the cup and then dying away, is caused by the initial rapid "blowdown" effect, followed by the slower displacement of the gas by the piston. These motions produce rather high turbulence levels near the valve and some augmentation in the surrounding region.

The comparisons with the measurements (which are subject to the same kinds of error as in the previous case) are shown in Fig. 16, where it can be seen that, with the exclusion of the compression and expansion phases, the agreement is probably better on the average than for the disc chamber. Significant errors still exist, however. For example, the peak velocity during inlet is now overpredicted by about 35%. The overall picture is otherwise much the same as for the disc chamber.

Further Explorations

Comparison of Turbulence Models for One-dimensional Compression/Expansion—The observed importance of compression-induced turbulence in the closed-cylinder predictions of Case 3 raises questions about the accuracy of the energy/dissipation turbulence model in these circumstances, for which it has never been

References pp. 123-124.

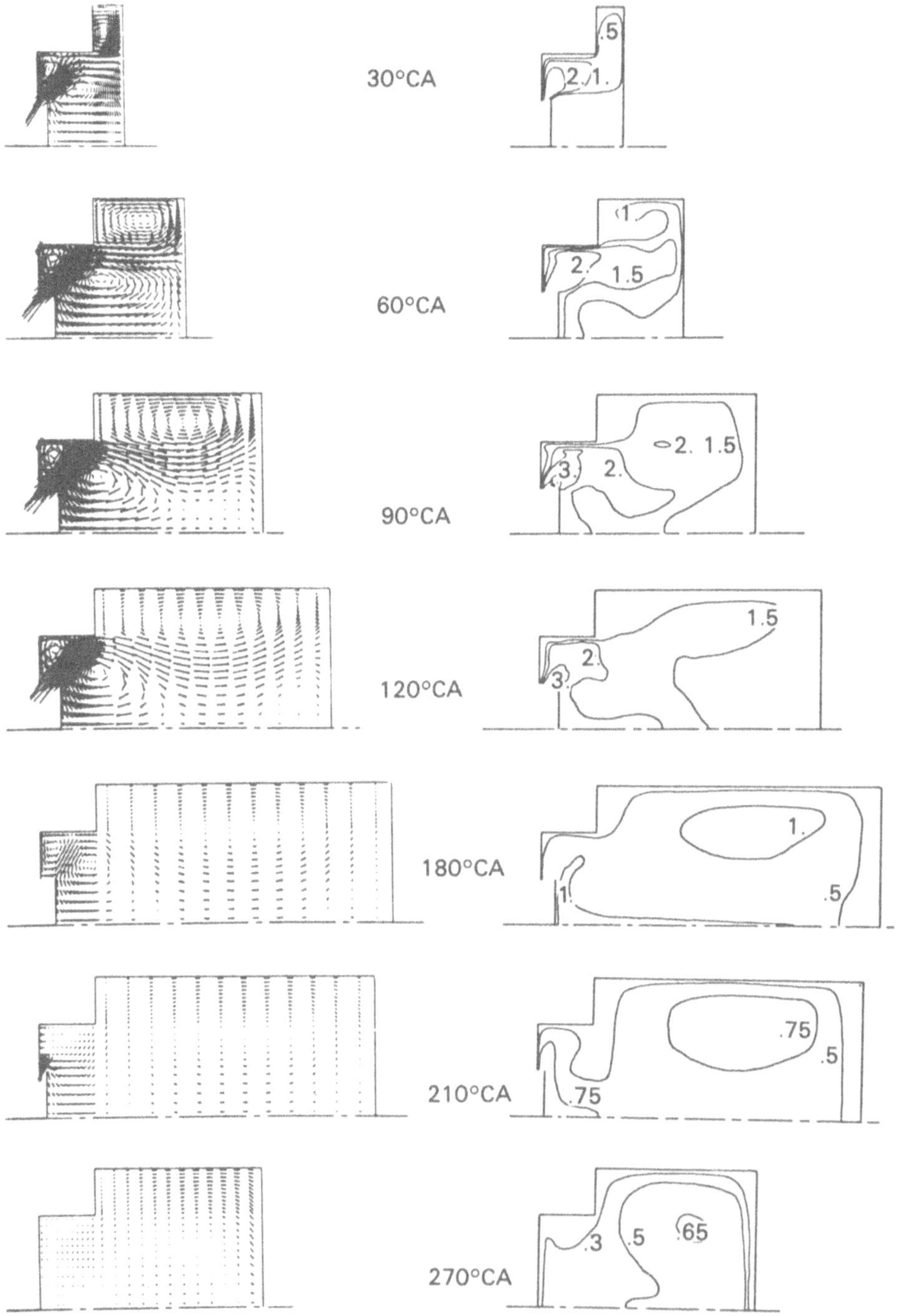

Fig. 15. Predicted velocity and turbulent fields for central-valve engine with bowl-in-head chamber of Witze [28] (Case 7).

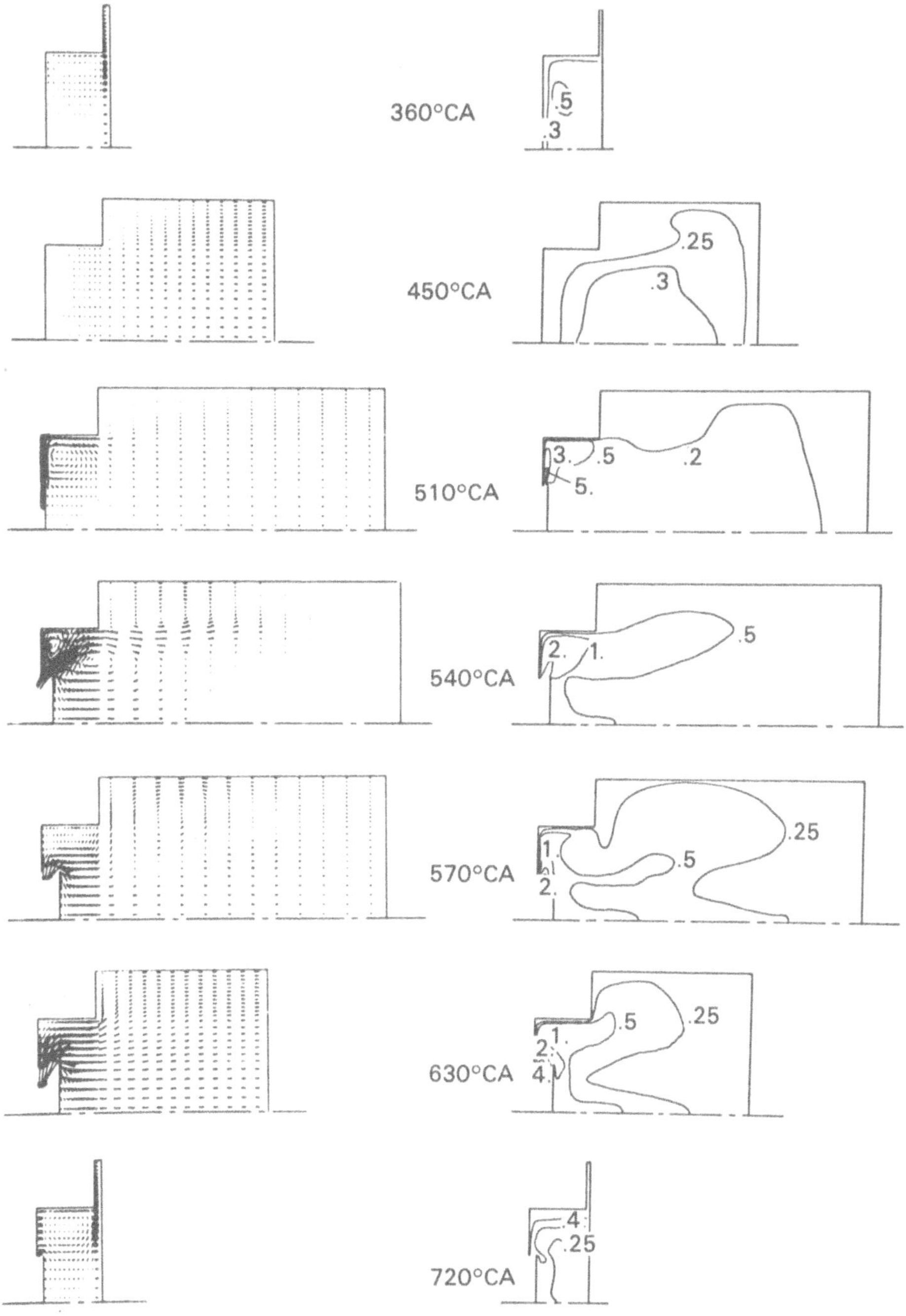

Fig. 15. (Continued)

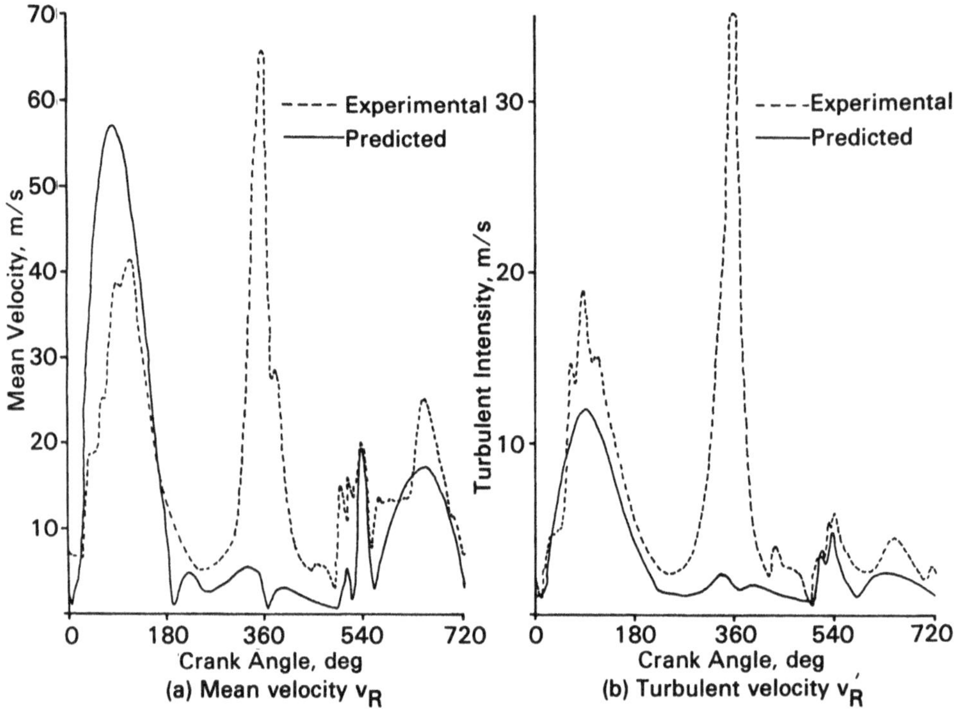

Fig. 16. Comparisons between measured and predicted variations of resultant velocity and turbulent intensity for Case 7.

tested. Since no experimental data could be found for this situation, it was decided to construct an idealized version of it which would be particularly amenable to analysis, and then to compare the predictions of the forementioned model with those of the more general, and therefore potentially more accurate, "Reynolds-stress" model of Launder et al. [33], in which the individual stresses are calculated from their own equations. Of particular interest was the degree of anisotropy suggested by the latter model, since the energy/dissipation model does not allow for this.

The axial velocity of the idealized flow was presumed to vary linearly in the fashion given by Eq. 8. It was further presumed that the resultant uniform strain field produces homogeneous turbulence, as indeed is suggested by the calculations of Case 3 for the region remote from the walls. These simplifications reduce the mathematical problem to an initial-value one, involving for each model a set of ordinary differential equations for the turbulence parameters with the specified strain variation as a forcing function. Solution of the equations is still non-trivial, but is easily accomplished numerically.

The connection with Case 3 was further strengthened by performing the present calculations for analogous conditions and using as initial values for the compression stroke the (averaged) predictions of Case 3 at BDC. (In the case of the Reynolds stress model, initial isotropy of the stress was arbitrarily assumed because there was no obvious alternative.) The energy/dissipation calculations were re-initiatized at TDC, again using the mean two-dimensional values, because these had been clearly affected by the effects of wall proximity and recirculation. No such measure was taken with the Reynolds-stress predictions, however.

Sample results from the study, which will be reported more extensively elsewhere, are shown in Figs. 17 and 18. The former shows the predicted variations in k and ϵ over the cycle obtained from the energy/dissipation model, which evidently correspond reasonably well with the Case 3 predictions shown for comparison. Fig. 18 displays the predictions of the Reynolds-stress model, for which in the present circumstances the only non-zero stress components are the three normal stresses. Of these, the two acting parallel to the wall may be taken to be equal. Plotted, therefore, are the axial component $\overline{u'^2}$, the tangential component $\overline{v'^2}$, the dissipation rate ϵ, and for convenience, k, which is of course equal to one-half the sum of the normal stresses.

A feature immediately obvious from Fig. 18 is that despite the isotropic starting field, the Reynolds-stress model predicts highly anisotropic behavior, with the axial stress component being considerably higher than the tangential one. On reflection this is hardly surprising, since the only non-zero mean strain component du/dz will clearly have its most direct effect on $\overline{u'^2}$. A second feature of interest is that the turbulence energy levels predicted by this model are almost an order of magnitude lower than those produced by the energy/dissipation calculations. Detailed analysis shows that this discrepancy is intimately connected with the departure from anisotropy, and that it is possible to devise a correction to the energy/dissipation model which will bring the two results into closer agreement. It is questionable, however, whether this would be worthwhile, in view of the apparent unimportance of compression-generated turbulence in more practical circumstances and the general uncertainty at present about the performance of the model in these much more complex strain fields.

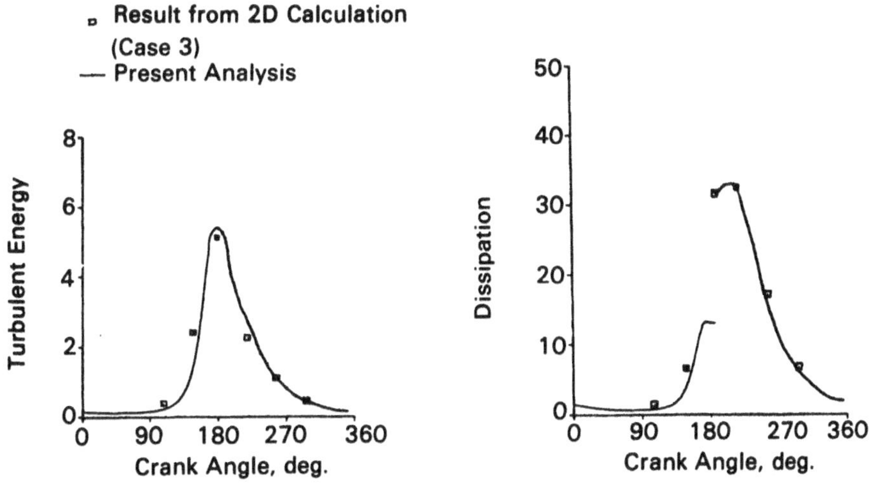

Fig. 17. Predictions of turbulent behavior for one- and two-dimensional compression/expansion based upon k-ϵ model.

References pp. 123-124.

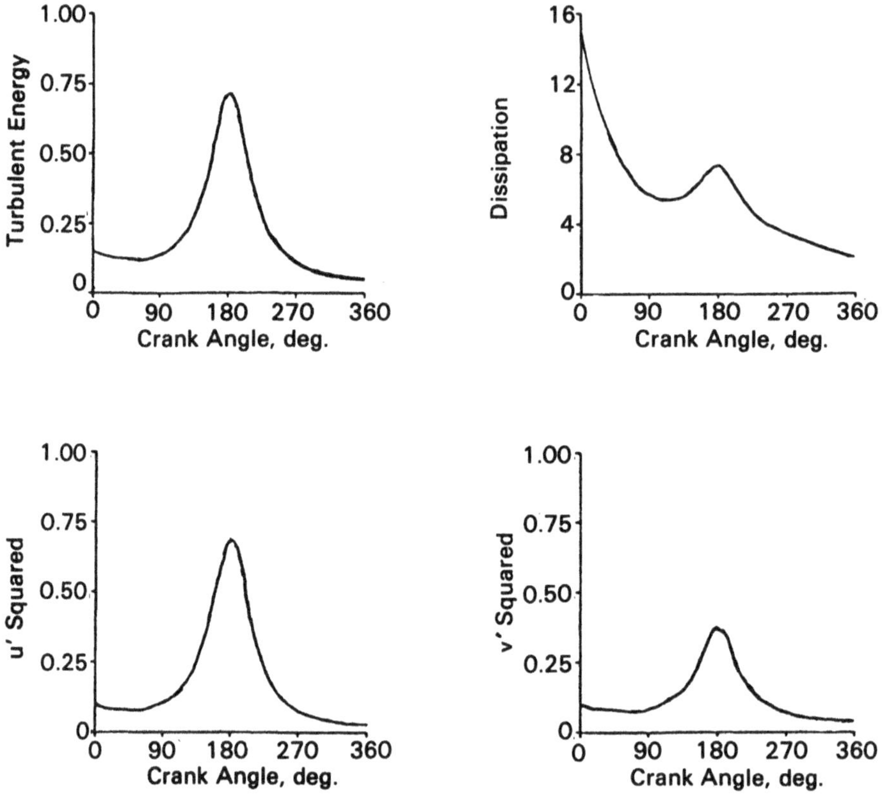

Fig. 18. Predictions of turbulent behavior for one-dimensional compression/expansion based on Reynolds-stress model.

Comparison of Turbulence Models for the Near-Wall region of an Oscillating Flow—The uncertainties attached to the energy/dissipation turbulence model are at least equaled, if not overshadowed, by those associated with the (enforced) use in the calculation method of Eqs. 3 through 6 to bridge the near-wall region, for these, it will be recalled, were originally formulated for steady flows, and there is no real justification for assuming that they are applicable to the high-frequency oscillating flows occurring in engines.

Since experimental data are lacking here, too, recourse has again been made to the combination of a more sophisticated turbulence model—in this instance the "low Reynolds number" version of the energy/dissipation model developed by Jones and Launder [18] and subsequently refined by Launder et al. [34]—and a simple prototype problem, namely that of fully developed incompressible flow in a round tube under the influence of a sinusoidally oscillating axial pressure gradient. The particular merit of this combination is that the low Reynolds number model is claimed to apply right down to the wall and therefore does not require bridging formulas, while the one-dimensional (in physical space) nature of the flow to which it is applied compensates for the much finer computational grid which is required to resolve the near-wall region.

The objectives of this particular investigation, which is still in progress, are to determine the near-wall flow structure and its dependence on the frequency and amplitude of the oscillations. A sample of the type of information which is emerging* is contained in Fig. 19, which shows the predicted profiles of the ensemble-average axial velocity at various stages in the cycle plotted in "law-of-the-wall" coordinates, with $\theta = 0$ defined as the instant at which the pressure gradient is zero. The conditions of the calculations roughly correspond to those of an automotive engine of conventional size operating at 2000 r/min. Also plotted in this diagram, with its log-linear scales, are a straight line representing the logarithmic law of the wall as given by Eq. 3 and, for interest's sake, a curve representing the usual linear $u^+ = y^+$ relation of the laminar sublayer.

The fact that the predictions clearly do not obey Eq. 3 in what is usually termed the fully turbulent region beyond about $y^+ \approx 40$, where it is usually assumed to apply, suggests that the initial doubts about its applicability may be justified. It is also interesting to note that a linear sublayer relation does appear to hold on an instantaneous basis. However it is still too early to draw any firm conclusions, which must await a more searching analysis of the results.

DISCUSSION

Assessment of Progress so Far

Performance of Prediction Method—Although the original intention of obtaining a definitive assessment of the capabilities of the *RPM* method for air-flow prediction has been to a considerable extent thwarted by the problems of unknown boundary conditions, there are nonetheless some useful lessons to be derived from the results. Thus, on the positive side, it is encouraging to note that the method was able to produce results for all the test cases, with their varied circumstances, including that of an operating valve. This is no mean accomplishment in itself. The fact that the predictions were found to be sensitive to the inlet conditions is also heartening, for despite the associated problems which this posed, it is precisely because this same behavior occurs in real engines that so much effort is focused on optimization through inlet-port design.

Less encouraging have been the indications that adequate numerical resolution may entail finer grids and therefore greater computing costs than were originally anticipated, although refinements are in hand. Certainly, although it is still quite feasible to perform axisymmetric calculations with the method, substantial improvements will be necessary before realistic computations of three-dimensional flows will be feasible. (It should be added that similar conclusions will probably also apply to other methods of the present kind, all of which are based on differencing practices of comparable accuracy.) Another less encouraging aspect

* *The authors are grateful to the investigator in question, Mr. B. Younis, for providing early access to his results.*

References pp. 123-124.

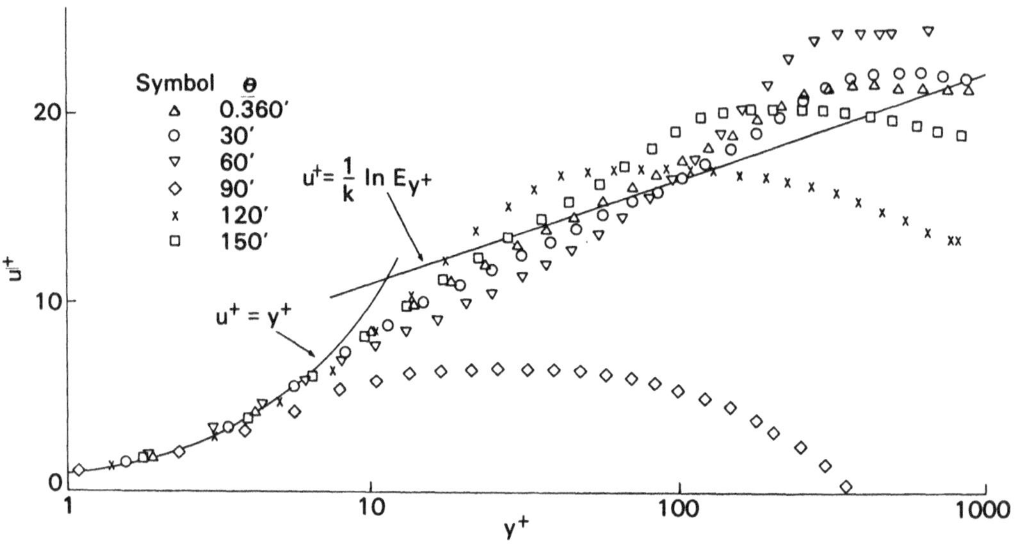

Fig. 19. Predictions of axial velocity profiles for oscillating, fully developed, turbulent pipe flow.

has been the not entirely unexpected indications that the particular combination of turbulence model and associated wall treatment employed may not be able to cope with all the complexities of the engine airflow. Perhaps, however, it would be wise not to draw final conclusions about this until more reliable evidence is available, ideally in the form of direct comparisons with experimental data for well defined circumstances.

Practical Implications of Study—It is right that developers of prediction methods should be strongly critical of their performance, for it is only in this way that deficiencies will be swiftly identified and improvements will be wrought. It is equally important, however, that perspective should be maintained about the potential usefulness of the methods as they stand to the practical engine analyst, for it would be wrong to discourage the implementation of a new but imperfect tool if it could already be of value.

The present authors believe that the *RPM* method does merit consideration as it stands as an analysis/design tool for two main reasons. Firstly, there is reason to suppose, on the basis of the evidence so far, that given correct boundary conditions the method will produce airflow predictions of at least qualitative correctness and will possibly also correctly predict the trends produced by changes in the major parameters, especially the chamber configuration. This is more than earlier methods are capable of. Secondly, there is currently no other way of obtaining the extensive and detailed in-cylinder flow information which the method provides, apart from performing expensive experiments, and even for these the necessary techniques are still at an incomplete state of development.

In connection with the practical side, it should also be noted that there is a great deal of useful information contained in the predictions already obtained that we have not attempted to extract in this paper. To cite one obvious example, the calculations clearly show that the structure produced during inlet is strongly influenced by the presence of the piston, due to jet impingement and other effects. This finding has important implications with regard to the conventional practice of characterizing an inlet design by performing flow-field measurements in a cylinder with the piston removed and a steady flow introduced through the inlet port. It is now clear that although such experiments may have some significance, they are certainly not fully representative of the real situation.

Areas For Further Research

Finite-Difference Analysis—In this and the following sub-sections the specific deficiencies revealed by the present research will be discussed under the appropriate headings, and suggestions will be made as to how they may be remedied.

Errors associated with the numerical-analysis aspects of the present method and others of its kind arise from two sources, namely in the "discretization" of the differential equations, i.e., the process of reducing them to algebraic form, and in the solution of the simultaneous non-linear sets of discretized equations. The former source is by far the most important and occurs, in simple terms, when the assumed distributions of the dependent variables between the nodal points of the computational mesh embodied in the discretized equations do not correspond with reality.

Of course as intuition suggests, and formal truncation-error analysis (TEA) confirms, any reasonable assumptions about the distributions, however simple, can be made to fit the real solution if the mesh is made sufficiently dense. Some assumptions are undoubtedly better than others, and TEA has been traditionally used to distinguish the good from the bad. As the mesh density increases, so however does the cost of the calculations, as well as the size of computer required. These factors combine to place upper limits on mesh densities for practical calculations. There are two important consequences of such limits. Firstly, it may not be possible to make the mesh sufficiently fine to reduce the discretization errors to tolerable levels. Secondly, the estimation of these errors by the TEA may become unreliable, since the TEA is itself only valid in the limit of small mesh spacing (or more precisely, in the limit of the "mesh Reynolds number", formed from the local spacing, fluid velocity and viscosity, becoming less than unity).

The possibilities of discretization errors are greatest in regions of steep gradients of the calculated variables and where the flow is appreciably inclined to the computing mesh. The first mentioned regions occur near walls, in the shear layers associated with the jet-like structures produced during intake, and in certain geometries, by the phenomena of "squish" and "reverse squish" (see, e.g., Ref. 6). Gradients of particular severity also occur in the immediate vicinity of sharp protruding corners, such as the edge of a valve or of a piston cup. As for the second cause, it is clear that the process of induction and the formation

References pp. 123-124.

of vortices from this and other mechanisms may result in inclination of the flow to the mesh.

If the flow were steady and the structure could be anticipated (which is infrequently the case), the errors could be alleviated somewhat while still operating under an overall constraint on mesh numbers by employing non-uniform spacing, as in the *RPM* method, so that the density may be selectively increased where required with compensating reductions elsewhere. However the unsteady character of engine flows entails that no single mesh distribution will be optimum during the entire cycle, so this measure cannot be wholly effective.

What can be done about these difficulties? One improvement which suggest itself is to increase the flexibility of the computing mesh to the extent that it can be dynamically re-optimized as the structure of the flow evolves. Computational procedures like, for example, the ALE method of Hirt et al. [12] use mesh systems which approach the degree of flexibility required, but this extension is relatively straightforward by comparison with the problems of evolving a workable re-optimization strategy, of which none appear to exist at present.

Mesh optimization cannot in any case entirely compensate, in the present authors' view, for the generally low level of accuracy of current finite-difference schemes. These can be traced to two causes. One is excessive reliance on the TEA for guidance in circumstances where it is not applicable, and the other is insufficient inquiry into the physical implications of the discretized equations. Thus, for example, there is now increasing recognition that the equations should exhibit the correct limiting behavior as the mesh Reynolds number becomes *large*. Many of the so-called higher-order schemes, as classified under the TEA, do not. Various promising differencing schemes have been developed (e.g., Refs. 35 and 36) with this guideline in mind. Mention should also be made in this regard of an idea which can be traced back to Allen and Southwell [37], and perhaps earlier, that the discrete approximations may be constructed from local, approximate analytical solutions to the parent differential equations under appropriate conditions. This approach, if further developed, is likely to lead to substantial improvements in accuracy.

A further idea which holds promise is that of what might be termed "sub-grid analysis" of certain well defined small-scale flow structures which are difficult to resolve, such as the initial jet-like intake flow at small valve lift or the initial phase of the direct injection of a gaseous fuel from a nozzle. The point about such structures is that, to a good approximation, they may be regarded as relatively insensitive to surrounding conditions and therefore capable of separate, and often simplified, analysis. The results may then be used *in place of* the regular finite-difference equations in the regions in question. The advantage of this approach, which has been tried with some success in the steady-flow calculations reported in Refs. 38 and 39, is that the regular equations need only "take over" at the stage when the required degree of resolution is comparable with that elsewhere in the chamber. This idea may also be extended to situations where the small-scale structure interacts with its surroundings, in which case the separate calculations would be performed repeatedly in an iterative fashion, with appropriate alterations being made to the boundary conditions.

In summary, although current numerical techniques appear to be stretched to the limits of their capabilities, the prospects for improvement seem good. The ultimate goal is not, of course, sophistication for its own sake but rather acceptable accuracy at lowest cost.

Turbulence Modeling—The main point to be made about turbulence models is that so much empiricism enters into their derivation that their range of applicability cannot be accurately delineated without appeal to experiment. As implied by earlier remarks, there appear to be so little data available for engines or related unsteady flows that proper assessment is not possible for these applications. It is clear, therefore, that the first priority in this area should be to perform a series of "benchmark" experiments which should in the first instance focus on relatively simple situations, like the fully developed oscillating pipe flow described earlier. Until this has been done and the required data base is established, there seems little point in conjecturing about the performance of this or that model.

Engine Experiments—The case for further engine airflow measurements with well quantified inlet conditions has already been justified by earlier comments and will not be pursued here. Something should however be said about the kinds of experiments which are required, however. The following are some suggestions:

(i) Axisymmetric model engines are clearly capable of providing, at minimum effort, all the information currently required for assessment and should therefore continue to be used for this purpose. Measurements in practical three-dimensional configurations would, however, also be of interest, both in their own right and also as indicators of the degree of similarity between the real and axisymmetric flows.

(ii) There is merit in performing further measurements in non-compressing engines, especially of the Case 4 variety, since these appear to produce a realistic intake-stroke flow structure while being much easier to construct and operate than compressing engines.

(iii) Although at least one complete set of data for laminar flow would be useful for assessment of numerical accuracy, the greatest need is for turbulent-flow experiments, preferably at engine speeds high enough for the motion to be fully turbulent.

(iv) In view of the marked inhomogeneity of the flow already indicated by the available predictions and measurements, future efforts should be directed at obtaining the spatial, as well as temporal, variations of the important variables. These include the velocity components, the turbulence intensities, and ideally, the remaining components of the Reynolds stress tensor and the dissipation length scale.

(v) Ideally the distributions of all the fore-mentioned quantities across the inlet-valve orifice should be determined as functions of crank angle in the operating engine. In view of the difficulties of making such measurements, however, it would be valuable to ascertain the degree to which the incoming flow can be simulated by out-of-cylinder tests on the head alone, in

References pp. 123-124.

which an oscillating pressure gradient is imposed and the valve is opened and closed in the conventional manner. Such tests could also be used to examine critically the validity of using orifice formulas as described earlier to calculate the instantaneous mass flow. It would similarly be useful to establish the relation between the transient experiment and like measurements of the conventional steady-flow kind in which the valve is held in various fixed positions.

SUMMARY AND CONCLUSIONS

The following are the main points to emerge from the present assessment exercise:

(i) Calculations with the *RPM* method have been reported for a series of seven test cases, embracing both laminar and turbulent, non-compressing and compressing flows in axisymmetric model engines of various inlet configurations.

(ii) For the single case for which an exact analytical solution is available for comparison, the method gives acceptably accurate predictions of the one-dimensional isentropic flow involved.

(iii) Agreement with the measurements available for one laminar-flow case and four turbulent-flow cases is reasonable in some respects but not satisfactory overall. However uncertainties about inlet conditions, especially of flow angle, prevent the causes of the discrepancies from being pinpointed.

(iv) Indications are that improvements in computational efficiency and turbulence modeling will be necessary, especially for three-dimensional calculations, but in the meantime it is believed that the *RPM* method can already provide useful information to the engine analyst.

(v) Among the more interesting findings of the calculations are that the inlet process produces mean and turbulent motions which far outshadow those induced by the piston motion (for disc-shaped chambers) and compressive normal stresses, that the general inlet-flow structure is strongly nonhomogeneous and clearly influenced by the presence of the piston, and that in the absence of squish effects, the tendency is for both mean and turbulent motions to decay monotonically during the compression and expansion strokes.

(vi) The energy/dissipation turbulence model currently employed in the *RPM* method probably overestimates the level of compression-generated turbulence, due in part to the neglect of the anisotropy suggested by a higher-level model. The use of wall formulas based on steady-flow theory is likewise probably also in error, but the net consequences of these defects on the overall airflow prediction has yet to be quantified.

(vii) The need for further experiments of both fundamental and applied kinds has been identified, in which priority should be given to quantifying boundary conditions.

ACKNOWLEDGMENT

The authors wish to acknowledge the encouragement and assistance which their work has received from the U.K. Science Research Council and the Perkins Engines Company, as well as many individuals, including Mr. W. Tipler, Prof. J. H. Whitelaw and the late Prof. R. S. Benson.

REFERENCES

1. A. P. Watkins, "Calculation of Flow and Heat Transfer in the Combustion Chamber of a Reciprocating Engine," M.Sc. Thesis, University of London, 1973.
2. A. D. Gosman and A. P. Watkins, "A Calculation Procedure for Flow and Heat Transfer in Piston Cylinder Assemblies," Imperial College Mech. Eng. Dept. Report, December 1975.
3. A. D. Gosman and A. P. Watkins, "A Computer Prediction Method for Turbulent Flow and Heat Transfer in Piston/Cylinder Assemblies," Presented at Symposium on Turbulent Shear Flows, Pennsylvania State University, April 18-20, 1977. Also Imperial College Mech. Eng. Dept. Reports FS/77/10, 1977.
4. A. D. Gosman and A. P. Watkins, "Predictions of Local Instantaneous Heat Transfer in Motored Engines," Imperial College Mech. Eng. Dept., Report FS/78/37, 1978.
5. A. D. Gosman, R. J. R. Johns, W. Tipler and A. P. Watkins, "Computer Simulation of In-Cylinder Flow, Heat Transfer and Combustion: A Progress Report," Imperial College Mech. Eng. Dept. Report FS/78/38, 1978. Submitted for presentation at Thirteenth International Congress on Combustion Engines, Vienna.
6. A. D. Gosman and R. J. R. Johns. "Development of a Predictive Tool for In-Cylinder Gas Motion in Engines," SAE Paper No. 780315, 1978.
7. R. Diwakar, J. D. Anderson, M. D. Griffin and E. Jones, "Inviscid Solutions of the Flow Field in an Internal Combustion Engine," Dept. of Aero-Space Engr. Report, University of Maryland, 1976.
8. J. D. Anderson, R. Diwakar, M. D. Griffin, and E. Jones, Jr., "Computational Fluid Dynamics Applied to Flows in an Internal Combustion Engine," AIAA 16th Aerospace Sciences Meeting, Huntsville, Alabama, 1978.
9. A. A. Boni, M. Chapman, J. L. Cook and G. P. Schneyer, "Computer Simulation of Combustion in a Stratified Charge Engine," Sixteenth (International) Symposium on Combustion, The Combustion Institute, Pittsburgh, Pennsylvania, pp. 1527–1541, 1976.
10. A. A. Boni, "Numerical Simulation of Flame Propagation in Internal Combustion Engines: A Status Report," SAE Paper No. 780316, 1978.
11. R. W. MacCormack, "The Effect of Viscosity in Hypervelocity Impact Cratering," AIAA, Paper No. 69-354, 1969.
12. C. W. Hirt, A. A. Amsden and J. L. Cook, "An Arbitrary Lagrangian-Eulerian Computing Method for All Flow Speeds," J. Comp. Phys., Vol. 14, pp. 227-253, 1974.
13. F. V. Bracco, H. C. Gupta and R. L. Steinberger, "Divided-Chamber, Stratified-Charge Engine Combustion: A Comparison of Calculated and Measured Flame Propagation," SAE Paper No. 780317, 1978.
14. F. V. Bracco, "Computed and Measured Two-Dimensional Unsteady Flames in Reciprocating Engines," General Motors Research Laboratories Symposium on Combustion Modeling in Reciprocating Engines, Warren, Michigan, November 1978.
15. W. C. Reynolds, "Modeling of Fluid Motions in Engines—An Introductory Overview," General Motors Research Laboratories Symposium on Combustion Modeling in Reciprocating Engines, Warren, Michigan, November 1978.
16. W. C. Reynolds, "Computation of Turbulent Flows," Ann. Rev. Fluid Mech., Vol. 8, pp. 183–208, 1976.

17. P. Bradshaw, ed., "Turbulence," Springer-Verlag, 1976.
18. W. P. Jones and B. E. Launder, "The Calculation of Low Reynolds Number Phenomena with a Two-Equation Model of Turbulence," Int. J. Heat and Mass Transfer, Vol. 16, pp. 1119–1130, 1973.
19. A. D. Gosman and F. J. K. Ideriah, "User Manual for the TEACH-T Computer Program," Imperial College Mech. Eng. Dept. Report, June 1976.
20. B. E. Launder and D. B. Spalding, "Mathematical Models of Turbulence," Academic Press, 1972.
21. W. A. Woods and S. R. Khan, "An Experimental Study of Flow through Poppet Valves". Proc. I. Mech. E., Vol. 180, pt. 3N, 1965–66.
22. A. K. Runchal, "Comparative Criteria for Finite-Difference Formulations for Problems of Fluid Flow," Int. J. Num. Methods in Eng., Vol. 11, pp. 1667–1697, 1977.
23. L. S. Caretto, A. D. Gosman, S. V. Patankar and D. B. Spalding, "Two Calculation Procedures for Steady, Three-Dimensional Flows with Recirculation," Proc. Third International Conf. on Numerical Methods in Fluid Mech., Vol. 2, Springer-Verlag, pp. 60–68, 1972.
24. N. Watson and M. Marzouk, "A Non-Linear Digital Simulation of Turbocharged Diesel Engines under Transient Conditions," SAE Trans., Vol. 86, Paper No. 770123, pp. 491–508, 1977.
25. A. D. Gosman, A. Melling, J. H. Whitelaw and A. P. Watkins, "Axisymmetric Flow in a Motored Reciprocating Engine," Proc. I. Mech. E., Vol. 192, No. 11, pp. 213-223, 1978.
26. A. P. Morse, J. H. Whitelaw, and M. Yianneskis, "Turbulent Flow Measurements by Laser-Doppler Anemometry in a Motored Reciprocating Engine," Imperial College Mech. Eng. Dept. Report FS/78/24, 1978.
27. P. O. Witze, "Preliminary Hot-Wire Measurements in a Motored Engine with an Axisymmetric Cylinder Geometry," private communication, July 1976.
28. P. O. Witze, Private Communication, 1976.
29. A. D. Gosman and A. P. Watkins, "A General Calculation Method for Flow and Heat Transfer in Reciprocating Engines," Report in preparation.
30. R. J. Tabaczynski, D. P. Hoult and J. C. Keck, "High Reynolds Number Flow in a Moving Corner," J. Fluid Mech., Vol. 42, pp. 249–256, 1970.
31. A. Melling, "Axisymmetric, Turbulent Flow in a Motored Reciprocating Engine," Imperial College, Mech. Eng. Dept. Report CHT/77/4, 1977.
32. P. O. Witze, "Application of Laser Velocimetry to a Motored Internal Combustion Engine," Sandia Laboratories Report SAND 78-8722. Proc. Third Int'l Workshop on Laser Velocimetry, Purdue University, July 1978.
33. B. E. Launder, G. J. Reece and W. Rodi, "Progress in the Development of a Reynolds-Stress Turbulence Closure," J. Fluid Mech., Vol. 68, pp. 537–566, 1973.
34. B. E. Launder, C. H. Pridden and B. I. Sharma, "The Calculation of Turbulent Boundary Layers on Spinning and Curved Surfaces," ASME Trans., Vol. 99, J. Fluids Eng., pp. 231–239, 1977.
35. G. D. Raithby, "Skew Upstream Differencing Schemes for Problems Involving Fluid Flow," Comp. Meth. Appl. Mech. Engng., Vol. 9, pp. 153–164, 1976.
36. B. P. Leonard, M. A. Leschziner and J. J. McGuirk, "Third-Order Finite Difference Method for Steady Two-Dimensional Convection," Proc. 1st Int'l Conf. on Num. Meth. in Laminar and Turbulent Flow, University College Swansea, 1978.
37. D. N. de G. Allen and R. V. Southwell, "Relaxation Methods Applied to Determine the Motion in Two Dimensions of a Viscous Fluid past a Fixed Cylinder," Quart. J. Appl. Math., Vol. 8, No. 2, pp. 129–145, 1955.
38. M. M. M. Abou Ellail, A. D. Gosman, F. C. Lockwood and I. E. A. Megahed, "The Prediction of Reaction and Heat Transfer in Three-Dimensional Combustion Chambers," Proc. AIAA/ASME Thermophysics and Heat Transfer Conference, Palo Alto, 1978.
39. A. D. Gosman, P. Nielson, A. Restivo and J. H. Whitelaw, "The Flow Properties of Rooms with Small, Rectangular Ventilation Openings," Imperial College Mech. Eng. Dept. Report FS/78/14, 1978.

DISCUSSION

W. A. Sirignano *(Princeton University)*

First, let me congratulate you on quite an effort. What you have accomplished is remarkable. I would like to ask several questions. The first concerns the exercise of the energy equation. I have not seen the energy equation exercised in your published works, except in the trivial closed-chamber case. Now, from our own experience with a similar code, which was also derived from the *TEACH* code, we found in determining the temperatures that the basic algorithm would not work for the two-dimensional unsteady case. We had to modify it before we could get meaningful results for the temperature. Did you have the same problem?

The second question concerns the inlet conditions. Presumably it's obvious to you that the flow is turning in the inlet before it's coming in, and that's why you have the problem. So you must have thought of the possibility of taking some grid points away from the main chamber and putting them in the inlet port. I wonder what your philosophy is concerning that approach.

The third point has to do with the question of compression in relation to Case 5. Now, even with the port annulus open, if the port is small enough in area compared to the piston cross-sectional area and if the engine speed is high enough, one will still get compression. Are you allowing for that possibility? Or are you in a low enough speed range and large enough port area so that the pressure and density aren't changing?

A. D. Gosman

First, on the matter of the solution of the energy equation, we have taken quite a lot of care to obtain realistic solutions*: thus we have insured that our method in the limit of isentropic compression and expansion agrees very closely with the standard thermodynamic laws. And yes, we initially had difficulty in getting that agreement. It didn't happen accidently. But we reckon we have it.

The second question concerns inlet conditions. I've seen calculations where people have taken the computational grid back up into the inlet, but in all due respect to the people who have done them, I don't believe them. That is simply because in our experience of calculating flow around sharp-edged obstructions,

---

* *A. D. Gosman and A. P. Watkins, "A Computer Prediction Method for Turbulent Flow and Heat Transfer in Piston/Cylinder Assemblies," Presented at Symposium on Turbulent Shear Flows, Pennsylvania State University, April 18-20, 1977. Also Imperial College Mech. Eng. Dept. Report FS/77/10, 1977.*

A. D. Gosman and A. P. Watkins, "Predictions of Local Instantaneous Heat Transfer in Motored Engines," Imperial College Mech. Eng. Dept., Report FS/78/37, 1978.

A. D. Gosman, R. J. R. Johns, W. Tipler and A. P. Watkins, "Computer Simulation of In-Cylinder Flow, Heat Transfer and Combustion: A Progress Report," Imperial College Mech. Eng. Dept. Report FS/78/38, 1978. Submitted for presentation at Thirteenth International Congress on Combustion Engines, Vienna, Austria.

the degree of resolution required to obtain realistic predictions of that separation is far in excess of what we could afford to do in the engine. We'd need as many grid points around the valve as we currently have in the whole combustion chamber. I don't think that is a feasible proposition. We can do it; that is, we can do the exercise, but what use the results will be I'm not sure.

A. D. Gosman

Third, you raised the point about compression versus noncompression in relation to Case 5. In that particular example we did both estimates and computer predictions using the orifice equations and energy equations, and so on. The estimates suggested that the pressure drop would be negligible for that geometry at the low engine speed and relatively large flow area. That is one of the reasons why the flow is turning. The gap is fairly large.

F. C. Gouldin *(Cornell University)*

I'd like to come back again to the question of the inlet specification. It's not surprising that the flow should be so sensitive at the inlet because, as Bill Reynolds pointed out, the major source of vorticity in the chamber comes from separation around the valve. So if you're going to have a predictive code, it seems to me you have to have some technique or model to determine *a priori* the nature of the separation and the strength of the vorticity source.

A. D. Gosman

I think there are two possibilities here. I think that it's probably quite feasible to do a separate calculation of the flow past the valve by just taking the prescribed cylinder pressure variation at the valve orifice as the outflow boundary condition. In other words, one could calculate out into the port and around the valve, but stop in the plane of the orifice. But that's not trivial, and most inlet ducts are, after all, three-dimensional. So, that's a fairly major problem. The alternative is simply to make measurements. We're now accustomed to doing steady-flow measurements in out-of-cylinder rigs to characterize ports, but normally the measurements are done downstream of the orifice. I suggest that in the future, as we get more proficient in using these computational methods, people will be characterizing inlet ports by measuring in the plane of the valve orifice. That information will then provide the input to the in-cylinder calculations, and also will provide the kind of information that one can use to infer what's going to happen in the cylinder.

J. C. Keck *(Massachusetts Institute of Technology)*

Those are very nice experiments you've done, but I think it is important to make contact with real experiments. There are two things which are relatively easy to measure in an engine with which I think you might be able to compare your results. One is the propagation velocity of the turbulent flame after it's fully developed as it crosses the chamber. I think there's ample evidence which suggests that this velocity is controlled primarily by the turbulence intensity. So if you can predict the way turbulence intensity varies with compression, you should be able to make reverse comparisons fairly directly. It should be a critical test of the theory. Second, of course, is the heat-transfer rate. It can be measured at various points on the wall of the cylinder and is again dependent upon the turbulence intensity. I, for one, have been very dissatisfied with the Woschni* correlation because it depends on pipe-flow analogy and clearly what we have to deal with here is jet turbulence. I never felt that it represented the physics of a very good model. How far are you from being able to make these two types of comparisons?

A. D. Gosman

While I take your points, let me first say I really think that both combustion and heat transfer are one step removed from the fluid mechanics, and therefore we still need to ask the question, "How well do we predict the fluid mechanics?" I think that the only way one can answer that is to compare with the fluid mechanics data, but we haven't been able to resist the temptation to go further. We certainly have done heat-transfer studies** and compared the results with existing correlations, but those predictions depend crucially on what one uses for a wall boundary-layer treatment. We get results which fall within the accepted

---

* *G. Woschni, "A Universally Applicable Equation for the Instantaneous Heat Transfer Coefficient in the Internal Combustion Engine," SAE Trans., Vol. 76, Paper No. 670931, pp. 3065–3083, 1968.*

** *A. D. Gosman and A. P. Watkins, "Predictions of Local Instantaneous Heat Transfer in Motored Engines," Imperial College Mech. Eng. Dept.. Report FS/78/37, 1978.*

A. D. Gosman, R. J. R. Johns, W. Tipler and A. P. Watkins, "Computer Simulation of In-Cylinder Flow, Heat Transfer and Combustion: A Progress Report," Imperial College Mech. Eng. Dept. Report FS/78/38, 1978. Submitted for presentation at Thirteenth International Congress on Combustion Engines, Vienna, Austria.

correlations, but this is not really a very stringent test. Unless we can find some very precise heat-transfer data, I don't think that we can draw very quantitative conclusions. As to the combustion side of things, it's very appropriate that you should ask that question because one of the things we are doing is trying one of the models that you proposed, and has subsequently been developed by Tabaczynski*, to predict turbulent flame speeds. That calculation is just, I think, emerging from the computer now, so I'm not going to say anything more about it.

T. Morel *(General Motors Research Laboratories)*

I have two questions, one of which is very short. What kind of boundary condition did you impose in the inlet plane? Was it related to the velocity itself or perhaps the pressure?

A. D. Gosman

For the cases where there was appreciable pressure drop, we couldn't make the perfect displacement assumption. We used instead a conventional orifice-type equation to get the instantaneous mass flow rate. We then made assumptions about the distributions of the velocity components and the turbulence energy and its dissipation rate. The assumptions are specified in the paper. You could argue that these assumptions weren't very good ones, but at least they're given there.

T. Morel

So you fixed the velocity vector and its direction?

A. D. Gosman

That's right.

T. Morel

The other question is this. We've talked a little bit up to now about the problem of vortex roll-up in the corner. There are a number of people who are trying to chase this vortex and have found it. Yet, you haven't seen it in your calculations. What do you think it would take to get to the point where you could conclusively say, "All right—I gave it an honest try and it is there "or" It isn't there." How much more do you think it would take? What is the required grid resolution for such modeling to resolve this problem?

---

* *R. J. Tabacynski, C. R. Ferguson and K. Radhakrishnan, "A Turbulent Entrainment Model for Spark-Ignition Engine Combustion," SAE Trans., Vol. 86, Paper No. 770647, pp. 2414-2433, 1977.*

A. D. Gosman

That is an open-ended question that I don't think I could fully answer. We have made attempts to pick up the vortex in laminar flow where in principle—if our method is correct and it's there—we should be able to pick it up, unless it's a three-dimensional phenomenon and we're suppressing it by making the axisymmetric assumption. In turbulent flow the vortex is quite small at the typical speeds that we're making calculations at, according to accepted correlations. Sometimes the size of the vortex is even within the dimensions of the region covered by the wall boundary-layer formula that we're using, so it could be lost. We do get a kind of vortex produced during our calculations, but it appears in the corner very late in the compression stroke and is due to separation of the wall boundary layer. Now we may be talking about the same thing, but what we don't see is a vortex emerging immediately as the piston starts going up, and then peeling off and getting bigger and bigger with time.

C. R. Ferguson *(Purdue University)*

What do you think of using your calculation technique to tell us what the hot-wire anemometer is measuring? We know experimentally it's one technique that we can use, but we don't really understand what the hot wire is measuring. Perhaps your theory can address the hot-wire anemometer installation and tell us what we're measuring—or things like that.

A. D. Gosman

I think I could get myself into hot water here by even claiming that we could use our method to predict hot-wire behavior. The hot wire, of course, is sensing the resultant velocity perpendicular to the wire axis. Therefore, when we're comparing with hot-wire measurements, and we do have some data kindly provided to us by Pete Witze and others, we calculate the local resultant velocity and make comparisons on that basis. So in that sense, I suppose we are in a way simulating the hot wire. But are you suggesting a more detailed calculation where one attempts to compute the boundary layer on the wire?

C. R. Ferguson

It is very difficult, for example, to get two hot wires into an engine. Therefore, to obtain the physical length scale one has to use the Taylor hypothesis. Could you examine the validity of that approach with your techniques?

A. D. Gosman

At this stage, I would rather rely on measurement techniques to tell us what the fluid flow is rather than rely on our predictions.

THE GENERATION OF TURBULENCE IN AN INTERNAL-COMBUSTION ENGINE

D. P. HOULT

Massachusetts Institute of Technology, Cambridge, Massachusetts

V. W. WONG

Cummins Engine Company, Columbus, Indiana

ABSTRACT

This paper reviews three aspects of the generation of turbulence in engines. First, flow visualization using a water model for the intake shows two distinct ring vortices. The stability of the vortices depends on the parameters of valve location, bore-to-stroke ratio and engine speed.

Second, a rapid distortion theory is applied to the evolution of both the ring vortex and turbulence during the compression stroke. Turbulence computations for the compression and expansion phases are compared with available hot-wire measurements. The theory works best during the combustion phase, when the gases are compressed rapidly. A simple strain field due to the piston-induced mean motion is assumed.

During combustion the strain field is established by the expanding flame front. Rapid distortion theory is used to calculate the turbulent properties of this flow. Flame-front position and gas densities are related by thermodynamics. For a cylindrical flame, the turbulent intensity just ahead of the flame front initially increases rapidly, then stays at an approximately constant value for the remainder of the combustion duration. Turbulent-intensity amplification depends very weakly on initial pressure at the time of spark. The dependence is stronger at lean fuel-air ratios than at rich fuel-air ratios.

NOTATION

A_{in}	parameter defined by Eq. 4a
B/S	bore-to-stroke ratio
c	$\frac{1}{\rho}\frac{\partial \rho}{\partial t}$

References p. 155.

c_1, c_2	constants
D	operator $\frac{D}{Dt} = \frac{\partial}{\partial t} + U_i \frac{\partial}{\partial x_i}$
D_m	major diameter of vortex
D_v	valve diameter at seat
e	specific internal energy
$E(t)$	turbulent energy spectra at time t
$E_0(k)$	von Karman spectrum (Eq. 11)
f	frequency
h	chamber height
I	turbulence intensity amplification $I = \int_0^1 \frac{u'}{u_0'} dx$ (x = mass fraction burned)
k_1, k_2, k_3	wave numbers in x, y, z space
$M_{i,n}$	transfer function
N	shaft angular velocity
p, P	pressure
Q	circulation parameter $= \Omega R_d^2$
r	radial coordinate position of a particle which was released at r_0, t_0
r, θ, z	cylindrical coordinates
r_c	distance from center of vortex
r_f	position of flame in radial direction
$R(t, t + \tau)$	correlation function
R	radius of bore
R_d	dimensionless distance $= r_c/R$
S_{oi}	initial Fourier amplitude of the random velocity field

t	time
T	temperature
T_I	integral time scale
U	base velocity in z direction
U_p	piston velocity
U_r	radial velocity
U_T	turbulent flame speed
U_L	laminar flame speed
$\langle u' \rangle$	rms turbulent intensity
x	mass fraction burned
$\underline{\mathbf{x}}$	general spatial coordinate
$\underline{\mathbf{X}}$	local wave number
X_p	local wave number component
z	distance of a particle above piston crown moving parallel to axis of cylinder with base velocity
$Z(t)$	distance between piston and cylinder head
z_c	vortex displacement from top dead center
Z_P	piston position from top dead center
α	r_o/r
β	$Z_o/Z(t)$
Γ	vortex strength
γ	circulation
γ_{ij}	Cauchy tensor
ϵ_{ikl}	cyclic tensor
ρ	density
σ	angle in θ direction
τ	time increment

References p. 155.

Φ	equivalence ratio
Φ_{onm}	three-dimensional spectrum tensor
ω	vorticity
ω_i	i^{th} component of vorticity
Ω	angular speed in vortex

Subscripts

o	value of variable at $t = 0$
i, j, k	tensor indices or components in cartesian coordinates
r, θ, z	components in cylindrical coordinates
b	burned part
ub	unburned part

Other Symbols

$\langle\ \rangle$	ensemble averaged
$(\underline{\ \ })$	vector representation
$(\ )'$	turbulent part of variable

INTRODUCTION

Turbulence generated in an internal combustion engine is quite complex. During the intake process, flow passing the valve separates and results in a highly unsteady motion. This flow contains both large- and small-scale turbulence. Its intensity is determined by the detailed geometry of the port and the valve, the geometry of the chamber, and the speed of the piston. After bottom dead center and after closure of the intake valve, the existing turbulence field is compressed, first by the motion of the piston during the compression stroke and subsequently by the expanding burned gases during combustion. Since turbulence has a major effect on the combustion process and on the heat transfer near top dead center of the power stroke, it is important to achieve a better understanding of turbulence generation during the intake process and of the evolution of turbulence during compression and combustion.

There are a variety of methods to study these phenomena. We can categorize these methods as follows. First, measurements can be made at single points. Generally there are two techniques that can be used to measure velocities at a single point. One technique is hot-wire anemometry, the virtue of which is its excellent frequency response. But there are calibration problems with hot-wires

[1, 2]. The second technique for single-point measurement is laser doppler anemometry. With the laser doppler method the absolute velocity is measured, so the calibration problems are avoided [3].

The second method of study comprises flow visualization. Since the secondary flow in the reciprocating engine, particularly during intake, is complex, it is difficult to deduce the characteristics of the overall flow from measurements taken at points in the flow. Flow-visualization studies are normally done with gas or water, where one scales the Reynolds number of the flow but not the physical speed. In the following section we discuss such a flow-visualization study of the intake process [4]. The motion of the working fluid is tagged with polystyrene beads which are neutrally buoyant, and the Reynolds number of this process is matched with the values that occur in an engine during induction.

A third broad method of study of turbulence in engines is by theory. There are a variety of semi-empirical turbulence theories [5]. The theory which this paper presents is a rapid distortion theory (RDT). We apply it to two problems: first to compression [6] and second to combustion [7]. Although the rapid distortion theory has some very extreme approximations, especially when it is applied to the compression process, its virtue is that it has no adjustable parameters and that it can be solved analytically, at least to quadrature. This theory is a rational theory in the sense that specific terms in the equations of motion, the sizes of which can be evaluated, can be ignored. In summary, these three methods comprise the tools that can be applied to determine the characteristics of the flow in the combustion chamber of an engine.

FLOW VISUALIZATION IN THE INTAKE PROCESS

In this study, water is used as a working fluid. The apparatus consists of a plexiglass cylinder and a piston which is driven by a hydraulic system so as to match the intake process of a reciprocating engine. That is, the motion of the piston approximates one-quarter of a sinusoid. Fig. 1 shows a schematic of the experimental setup.

The water is seeded with polystyrene beads which are a fraction of a millimeter in diameter and nearly neutrally buoyant. The valve, as shown in Fig. 1, is axisymmetrically located in the cylinder for most of the experimental conditions. The flow is illuminated by a slit of light, the plane of which coincides with the axis of the cylinder. Photographs of the flow field were taken perpendicular to this plane.

Results were obtained in the following way. First the fluid motion was photographed using a known exposure time. A typical photograph is shown in Fig. 2. From the exposure time and the streak length of an individual bead, the velocity of the bead was obtained. In this manner the velocity distribution in the flow field was determined. From these photographs it can be observed that the flow during induction is comprised of basically two ring-vortex structures, the cross sections of which can be seen in Fig. 2. The dashed curve shown in Fig. 3 is the velocity distribution in the vortex that would be obtained if all the vorticity in the flow was concentrated at a point and if the motion outside the

References p. 155.

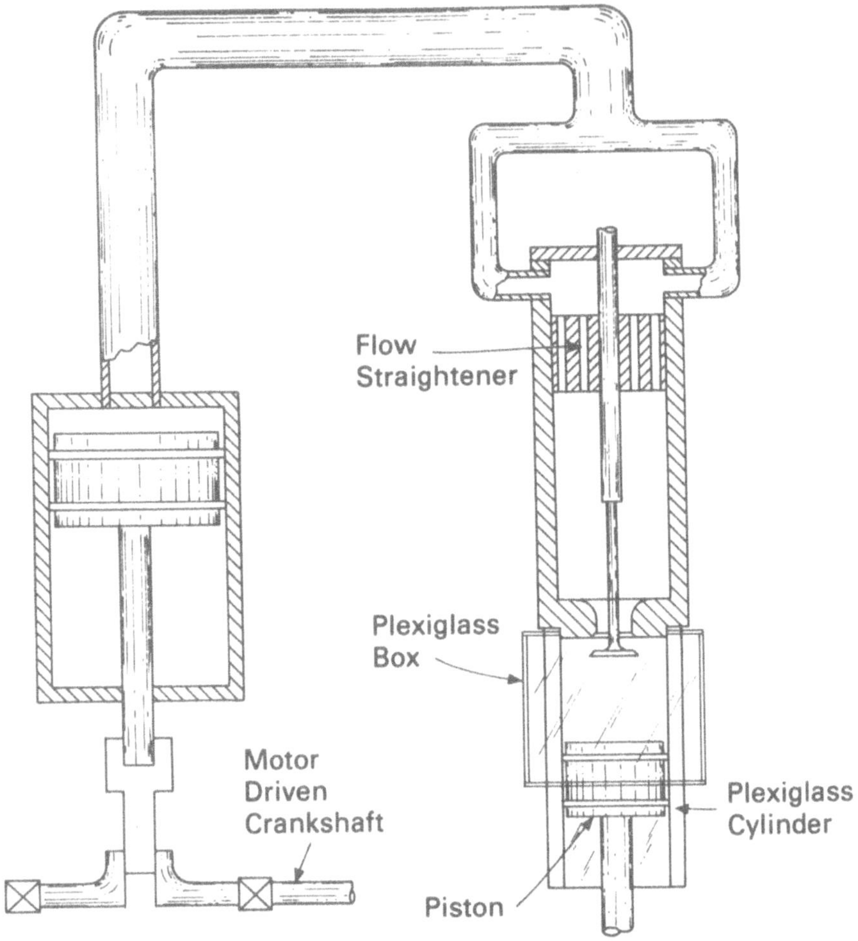

Fig. 1. Schematic of experimental apparatus.

core of the vortex was inviscid. It is important to note that within the scatter of the data the motion does appear to be inviscid. Since Γ is constant, the amount of vorticity in the vortex scales with engine speed.

The result suggests that the motion of the vortex could be described neglecting viscosity. That is, its locations and its strength can be determined from a set of inviscid scaling rules. Dimensional analysis then shows that if viscosity is not important, the location of the vortex and its diameter must depend only on geometric properties of the bore and stroke and valve diameter. Figs. 4 and 5 show how axial location and diameter of the vortex, respectively, scale with piston position.

In summarizing these results on the flow-visualization process, we wish to point out an aspect of the problem which we do not yet thoroughly understand, but which we know to be important. We find that when the bore-to-stroke ratio is greater than one, the vortex does not last until bottom dead center, but it breaks up due to a well known instability. On the other hand, when the bore-to-stroke ratio is one or less, the vortex does last to bottom dead center and then does maintain itself as the compression stroke starts. The sign of the vortex is the same as the sign of the vortex found by Tabaczynski et al. [8] on the compression stroke. So the vortex generated by the intake valve may be maintained, and even perhaps amplified, during the compression process, provided the bore-to-stroke ratio is one or less.

We have repeated the experiments with nonaxisymmetric head geometries and found that the vortex motion is still present, although its geometry is more complex.

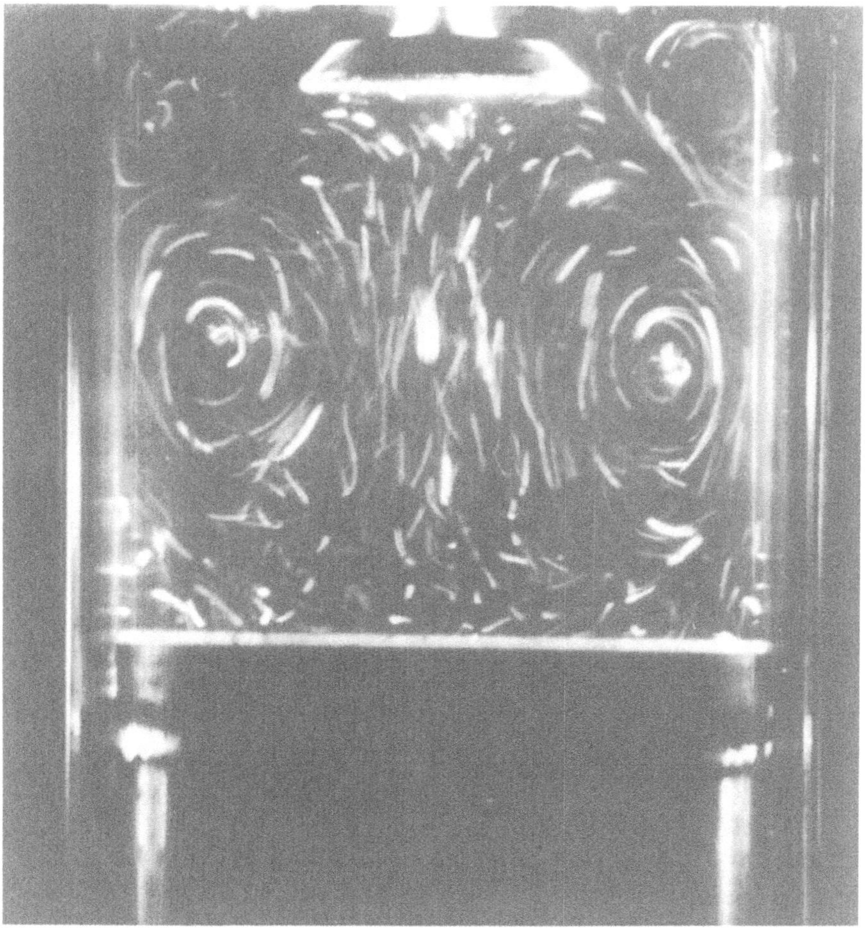

Fig. 2. Typical photograph of the structure of the flow during the intake process.

References p. 155.

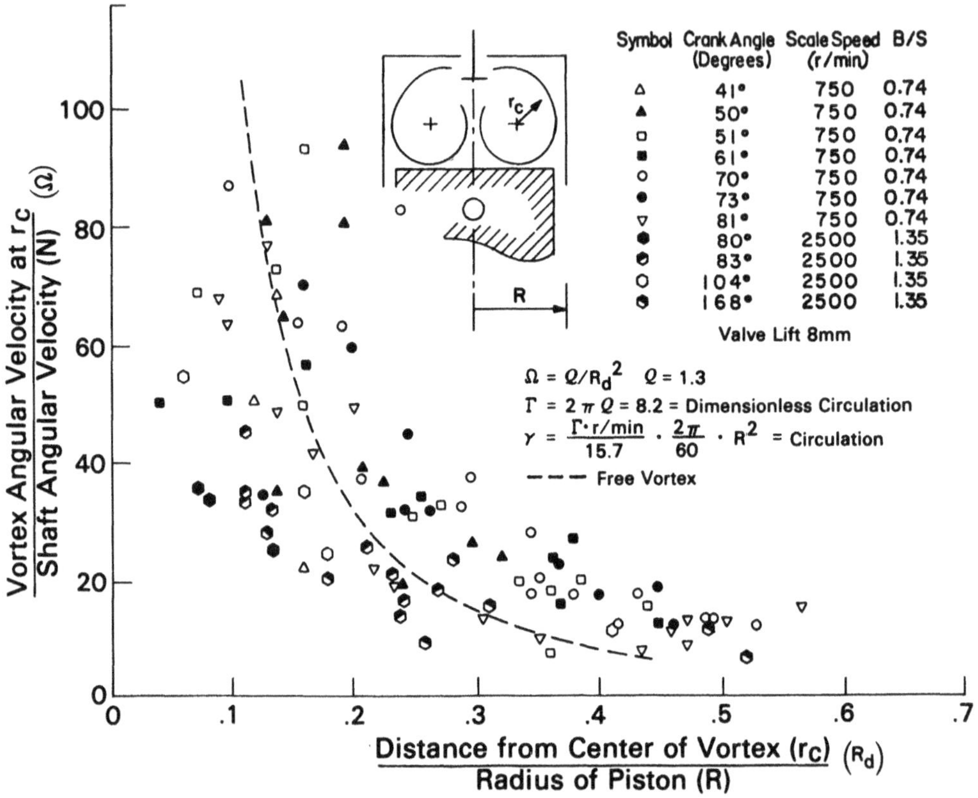

Fig. 3. Distribution of angular velocity in the lower vortex as a function of radius.

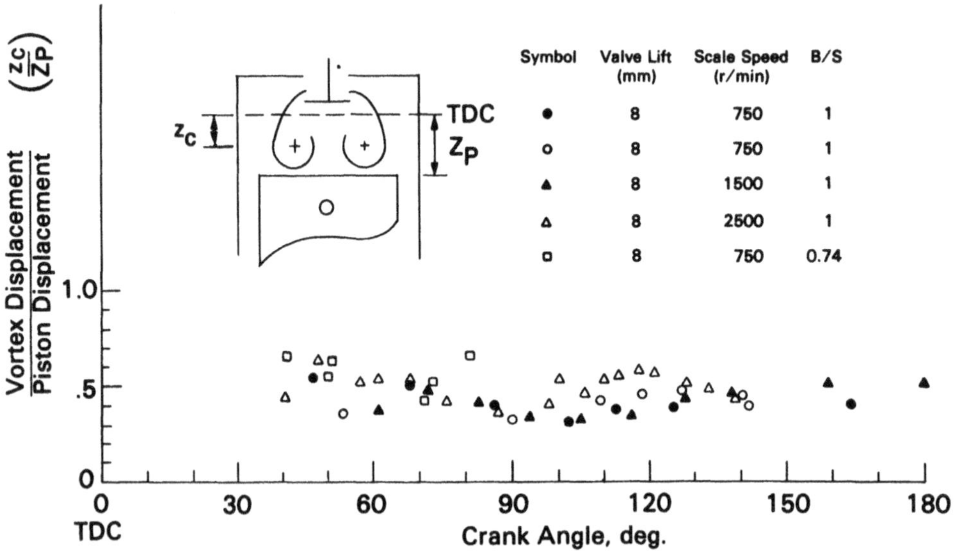

Fig. 4. Variation of non-dimensional lower vortex displacement with crank angle.

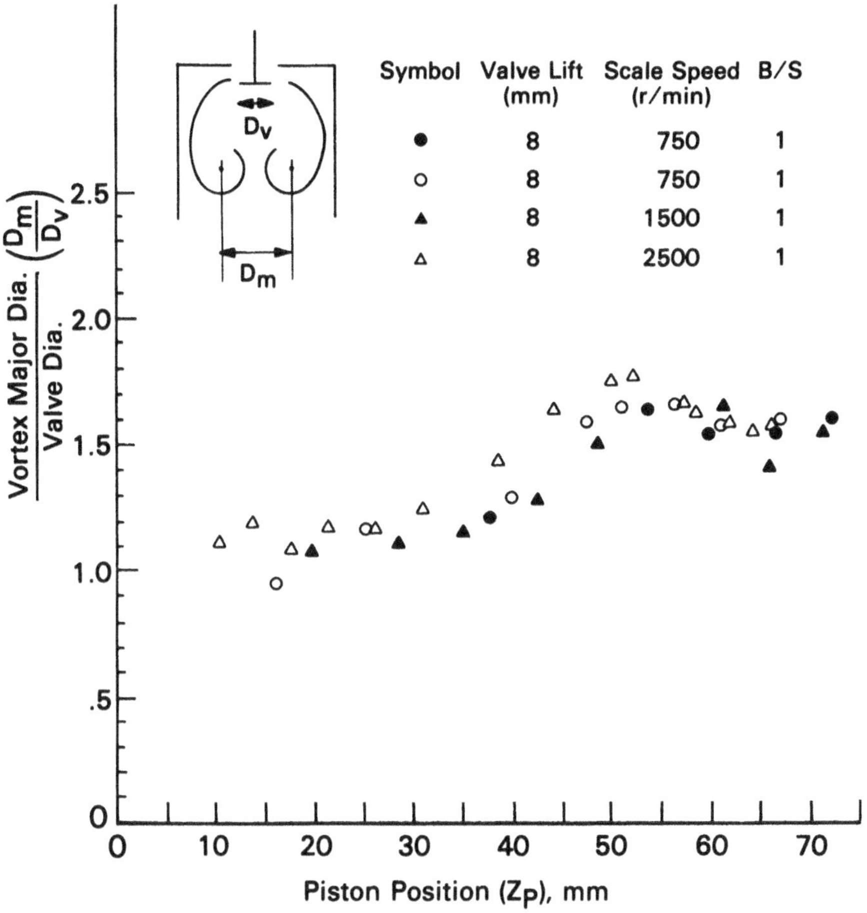

Fig. 5. Variation of non-dimensional lower vortex major diameter with piston displacement.

DISTORTION OF TURBULENCE AFTER THE INTAKE PROCESS

Rapid Distortion Theory—It is known experimentally that the turbulent flow properties are changed drastically by large imposed distortions. Flows around obstacles, contracting streams in wind tunnels, gases in rapid compression and expansion are examples.

Turbulent motions are characterized by high levels of fluctuating vorticity. Rapid distortion theory accounts for (i) the transport of the main turbulent motion by bulk convection of the fluctuating vorticity and (ii) the distortion of the main turbulent motion due to the stretching of the fluctuating vorticity by the base flow. The theory does not consider:

(i) viscous effects
(ii) nonlinear interactions of fluctuating quantities.

The validity of applying rapid distortion theory to a given flow situation depends directly on the degree to which the above two conditions are satisfied. The

References p. 155.

mathematical method of the theory applies when the initial state of turbulence is homogeneous, at least locally, so that superposition of the Fourier components of the fluctuating quantities is possible.

The interaction between the base flow and the main turbulent motion governs the evolution of the fluctuating vorticity. For irrotational base flows and subsonic fluctuating speeds, it can easily be shown that the linearized vorticity transport equation is

$$\frac{D(\underline{\omega}'/\rho)}{Dt} = \left(\frac{\underline{\omega}'}{\rho}\cdot\nabla\right)\underline{U} \tag{1}$$

the solution to which can be written in the form

$$\omega_i(\underline{x}, t) = \left(\frac{\rho}{\rho_o}\right)\gamma_{ij}\,\omega_{oj} \tag{2}$$

where

$$\gamma_{ij} = \frac{\partial x_i}{\partial x_{oj}}(\underline{x}_o, t) \tag{2a}$$

is the distortion tensor. Now it is clear that given the Lagrangian description of the fluid particles, $\underline{x}_i(t)$ as a function of $\underline{x}_o$ and t, and the initial state of turbulence, one can calculate the vorticity field at any later time. The Lagrangian path of a fluid particle can be determined by assuming that the particle convects essentially with the base flow, i.e.,

$$\frac{dx_i(\underline{x}_o, t)}{dt} = U_i(\underline{x}, t) \tag{3}$$

The general mathematical procedure of obtaining the solution from Eqs. 2 and 3 is given in Ref. 7. The solution for the space-time-velocity auto-correlation is

$$\langle u_i(\underline{x}, t)u_j(\underline{x}', t')\rangle = \iiint_{-\infty}^{\infty} A_{in}(\underline{x}, t)A^*_{jm}(\underline{x}', t')\,\Phi_{onm}(\underline{k})d\underline{k} \tag{4}$$

where * denotes complex conjugate, and $\Phi_{onm}(k)$ is the initial three-dimensional spectrum tensor. The variance of the fluctuation is given when $t = t'$ and $i = j$. Hence the turbulent intensities, spectra, and length scales are derived directly from Eq. 4. In Eq. 4,

$$A_{in} = \frac{-\epsilon_{ipq}\epsilon_{jln}X_p k_l\left(\frac{\rho}{\rho_o}\right)\gamma_{qj}}{X^2} \tag{4a}$$

where ϵ_{ijk} is the permutation tensor, and $\underline{X}$ is the local wave number related to $\underline{k}$ and γ_{ij}. Hence from Eq. 4 it can be seen that the turbulence parameters at time t are known once the distortion tensor γ_{ij} of the flow is determined.

Compression as Described by the Rapid Distortion Theory—Although we expect the motions inside the engine cylinder to be quite complex as the crank angle changes, it is a basic approximation of our approach that the strain field is comprised only of the z component of the motion.

As the mean flow, U, must be zero on the cylinder head and equal to the piston speed at the piston, it is clear from continuity that U varies linearly with the distance from the piston:

$$U = U_p(t)z/Z(t) \tag{5}$$

Here z is the distance in the direction parallel to the axis of the cylinder, with $z = 0$ being the position of the cylinder head. U_p is the piston speed. $Z(t)$ is the distance between the piston and cylinder head at time t.

In this simple model of the motion, fluid particles move up and down, in a cyclic manner, parallel to the z axis. Hence the Lagrangian calculation for a particle located at z yields

$$z/Z = \text{constant} \tag{6}$$

As the motion proceeds, the particle position relative to the cylinder head and piston remains the same. For this compression problem, we find easily from Eq. 6 that

$$\gamma_{ij} = \frac{1}{\beta}\begin{pmatrix} \beta & 0 & 0 \\ 0 & \beta & 0 \\ 0 & 0 & 1 \end{pmatrix} \tag{7}$$

where $\beta = Z_0/Z(t)$.

In the applications which follow we will specify the turbulence spectrum at a given crank angle and compare the predictions of the theory with measured spectra at later crank angle degrees during the compression process.

To compare with experiment we do not apply a Taylor hypothesis in this development to relate the time fluctuations to space fluctuations. The theory itself can be written either in terms of time fluctuations or space fluctuations. Since the data are obtained at a fixed point and with time varying, we apply the theory in the same way. One investigator Witze [1] has shown spectra which have the form

$$E(t) = 4\int_{-\infty}^{0} R(t, t+\tau)\cos(2\pi\tau)d\tau \tag{8}$$

where

$$R(t, t+\tau) = \frac{\langle U'(t)U'(t+\tau)\rangle}{\langle U'^2(t)\rangle^{1/2}\langle U'^2(t+\tau)\rangle^{1/2}} \tag{9}$$

These spectra are normalized. For Witze's data we prefer this normalization because the uncertainties in the hot-wire corrections are minimized.

Given this form of the observation, one can work out the corresponding theoretical statement of how the theory predicts the evolution of spectra:

$$\langle U_z'(z,t)U_z'(z,t')\rangle = \frac{\beta\beta'}{2}\int_{\sigma_1=0}^{\pi} d\sigma \cdot \int_0^{\infty} \frac{dk\cos((\beta-\beta')k_3 z)\sin^3\sigma_1 E_0(k)}{(\beta^2\cos^2\sigma_1+\sin^2\sigma_1)(\beta'^2\cos\sigma_1+\sin^2\sigma_1)} \tag{10}$$

References p. 155.

where $E_o(k)$ is the von Karman spectrum, as given by

$$E_o(k) = \frac{c_1[k]^4}{(c_2 + [k]^2)^{17/6}} \tag{11}$$

where the constants c_1 and c_2 are adjusted to match the total energy and integral length scale of Witze's observations at an initial crank angle.

There are no adjustable parameters in this theory. The theoretical calculations require only the parameters in the von Karman spectrum which were fitted to Witze's data at 45 crank angle degrees after bottom dead center. Figs. 6, 7 and 8 then show the evolution of the spectra with crank angle degrees and compare the theory, which is the solid line, with the measured data points. We can see that this theory appears to be as good as the data. Similar comparisons with Lancaster's data give similar results, as shown in Figs. 9 and 10.

It is worthwhile here to point out that life is not really as simple as described

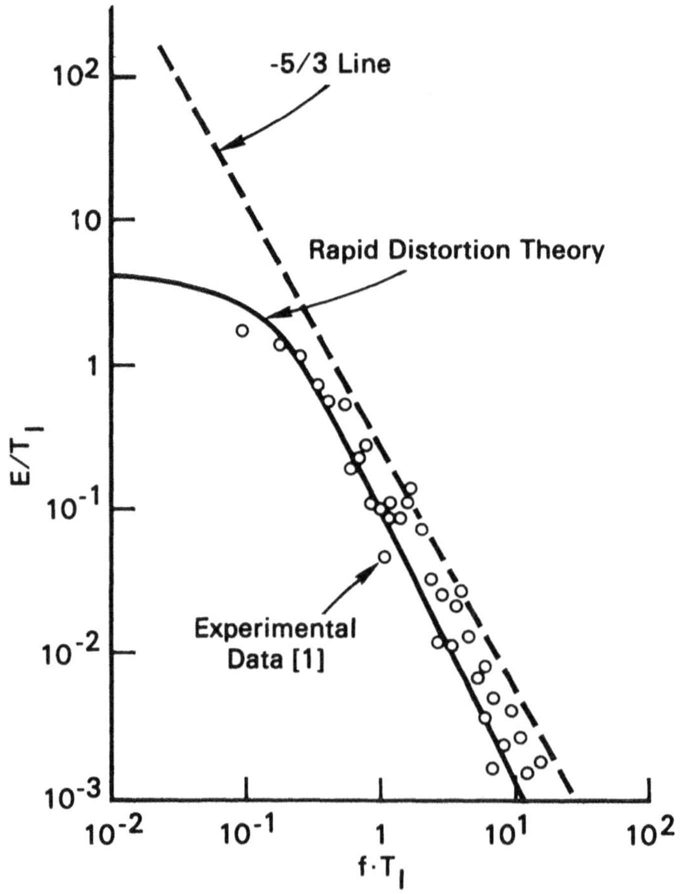

Fig. 6. Measured and computed spectra. Piston position at 90° BTDC during compression.

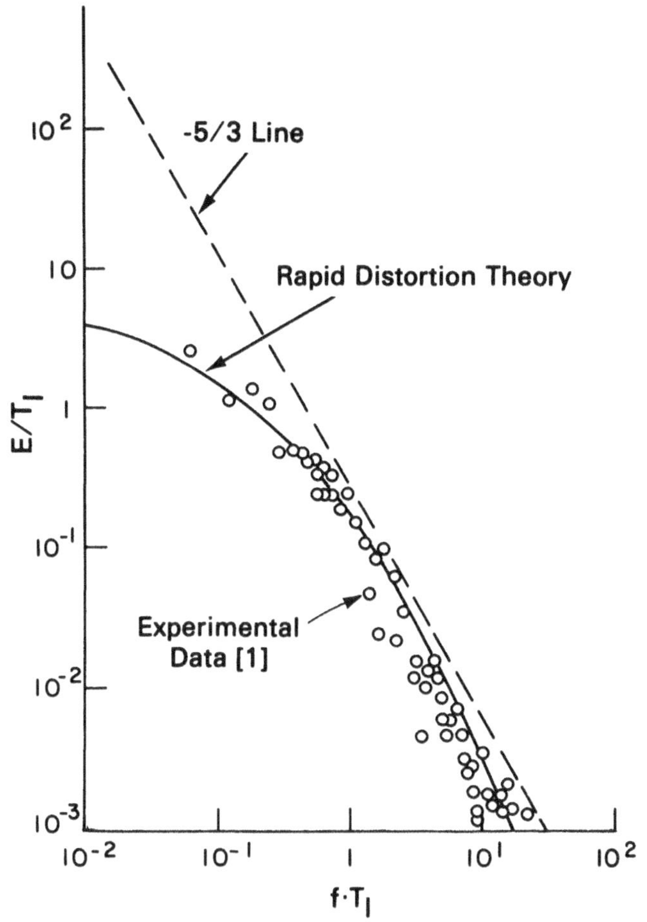

Fig. 7. Measured and computed spectra. Piston position at 45° BTDC during compression.

by this comparison between theory and experiment. There is a disagreement between investigators on how to correct the hot-wire measurements in terms of a mean motion and a fluctuating motion. For high frequencies we think this ambiguity is small, but for low frequencies we do not think it is small. The location of the hot wire is important because if the probe is located in or near a vortex ring, as described earlier, the measured velocity might be very large and fluctuating but might not correspond to small scale turbulence. Reported hot-wire measurements of turbulent intensity during compression are compared with the theory [1, 2, 9] in Fig. 11.

Combustion as Described by the Rapid Distortion Theory—The reader will find it convenient to think of the results outlined in this section as consisting of rather idealized statements. The first statement describes the assumption that the combustion chamber is divided into two parts which are separated by a thin flame

References p. 155.

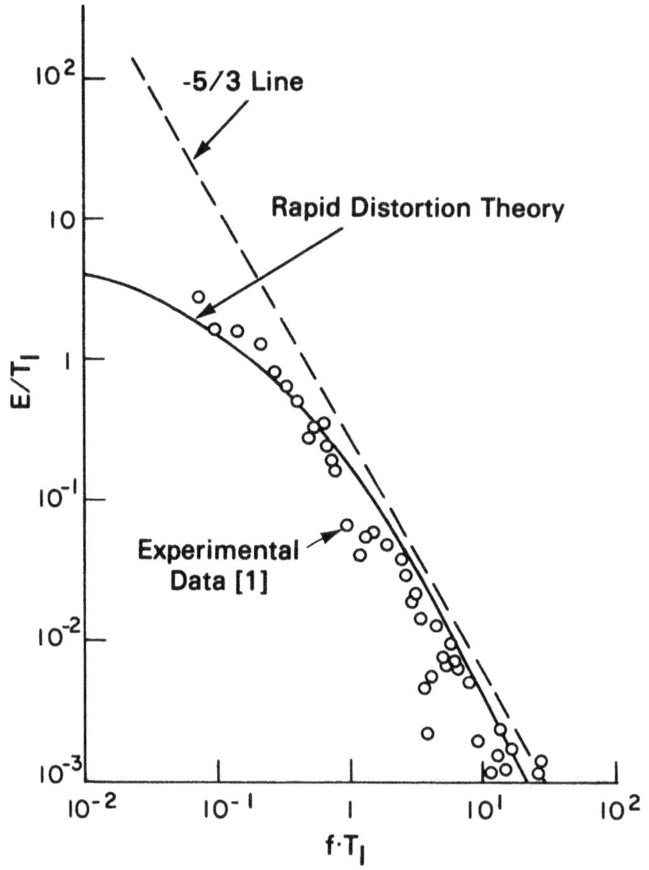

Fig. 8. Measured and computed spectra. Piston position at TDC during compression.

front. This allows the density ahead of the front to be determined as a function of the position of that front.

The second statement shows that the strain field ahead of the gas is determined solely by the density ahead of the gas and the initial conditions. This remarkable result implies that the turbulence properties of the flow ahead of the flame front are independent of the specific relationship (if it exists) between the turbulent flame speed and the turbulent intensity. The method of solution follows that presented in the previous section. Several computations are presented.

The third statement assumes the existence of a relationship between laminar flame speed, turbulent flame speed, and turbulent intensity. With this assumption, one can make a detailed comparison between observation and theory. One such comparison is presented.

The geometry we chose was a cylindrical pancake-shaped combustion chamber in which the piston does not move and in which a spark is fired along the axis of the cylinder. The chamber was conceived of as being very thin, i.e., height $h \ll R$. The motion was presumed to be axisymmetric, consisting of a radial flame front which moves outward and compresses the gas ahead of it. The density in the burned region, ρ_b, and the density in the unburned region, ρ_{ub}, therefore depend only on time.

We divided the flow into two parts, a burned gas region and an unburned gas region, as shown in Fig. 12. One specifies the driving force by writing down the thermodynamics of pressure, temperature and density in the two regions.

The statement of mass conservation is

$$\frac{x}{\rho_b} + \frac{1-x}{\rho_{ub}} = \frac{1}{\rho_o}, \tag{12a}$$

and

$$e = xe_b + (1-x)e_{ub} \tag{12b}$$

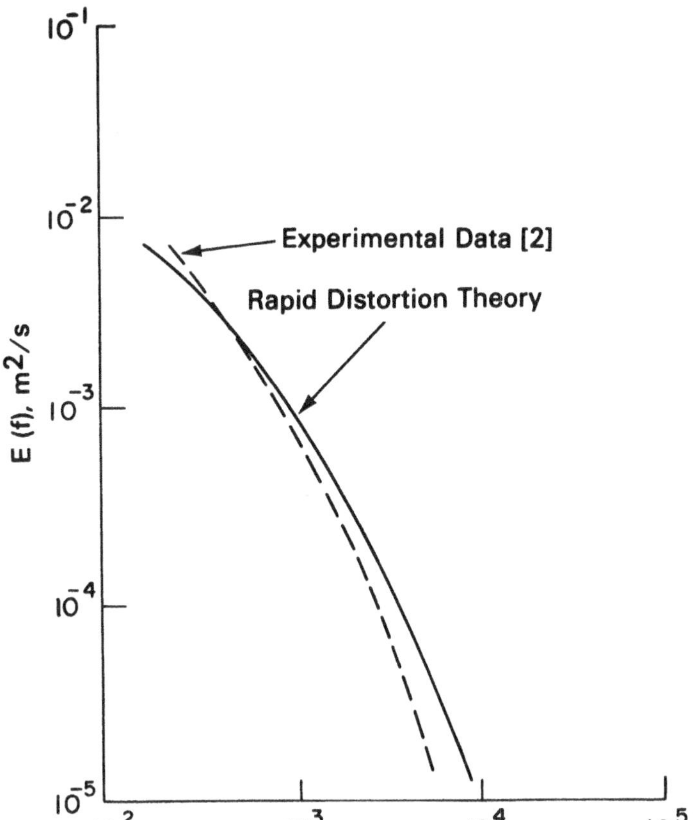

Fig. 9. Comparison of computed spectrum with experimental results from Lancaster [2] for non-shrouded valve. Piston position at 45° ABDC during compression.

References p. 155.

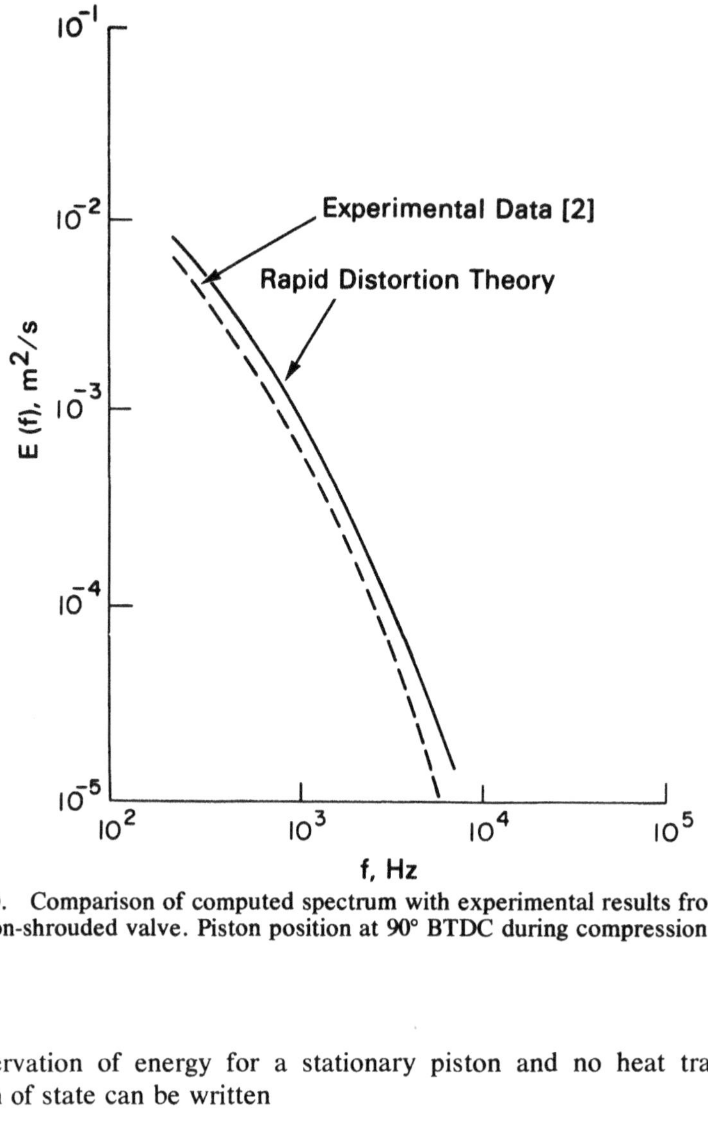

Fig. 10. Comparison of computed spectrum with experimental results from Lancaster [2] for non-shrouded valve. Piston position at 90° BTDC during compression.

is conservation of energy for a stationary piston and no heat transfer. The equation of state can be written

$$\begin{aligned} e_{ub} &= e_{ub}(p, T_{ub}) \\ e_b &= e_b(p, T_b) \end{aligned} \tag{12c}$$

We assumed the unburned charge was compressed adiabatically, and that the combustion process was adiabatic. These assumptions allowed T_{ub} and T_b to be calculated from the initial values (ρ_o, T_o).

The unburned gas moves because, as the burning proceeds, the pressure rises, causing the density in the unburned gas to increase. The changing density causes the unburned gas to move outward radially.

The continuity equation for the unburned gas, in cylindrical coordinates (see

Fig. 12), is

$$\frac{1}{r}\frac{d}{dr}(rU_r) = -\frac{1\partial\rho}{\rho\partial t} \tag{13a}$$

where U_r is the local base motion due to density change. Let $c(t) = \frac{1}{\rho}\frac{\partial\rho}{\partial t}$, where $c(t)$ is related to the time rate of change of density, which is prescribed by the combustion process. The solution to Eq. 13a, subject to the boundary condition

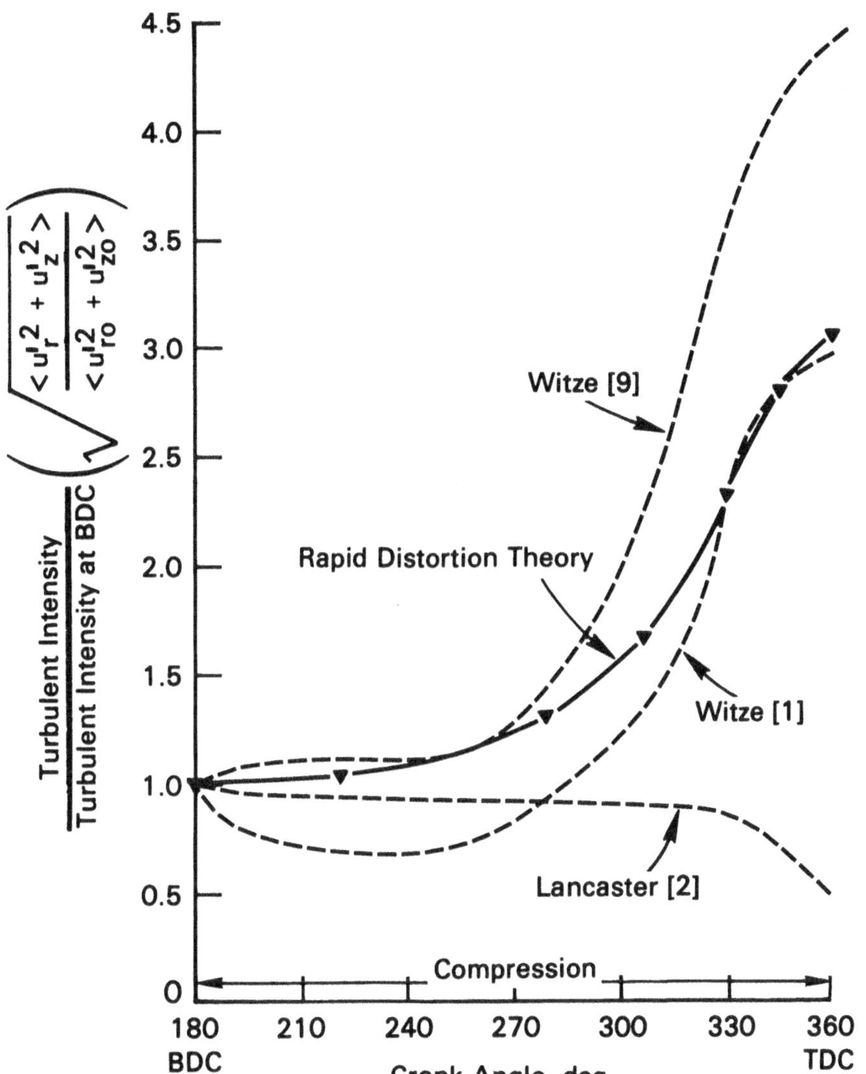

Fig. 11. Comparison of calculated turbulent intensity with measurements during compression.

References p. 155.

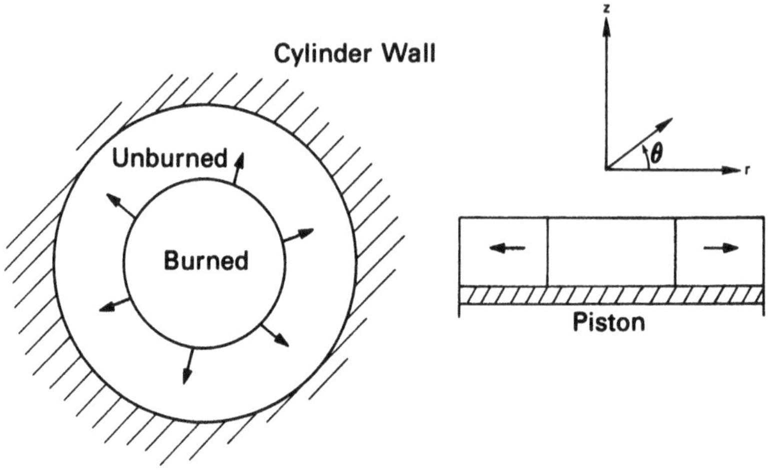

Fig. 12. Schematic of cylinder flame propagation.

that $U_r = 0$ at the wall ($r = R$), for the mean flow, U_r, can be written as

$$U_r = \frac{c(t)R}{2}\left(\frac{R}{r} - \frac{r}{R}\right) \tag{13b}$$

It is a basic assumption in the rapid distortion theory that the trajectory of a fluid particle is determined uniquely by the distortion history of the mean flow and not by the turbulent fluctuations. The distortion tensor can therefore be calculated from a knowledge of the Lagrangian path of a fluid particle. The Euler-Lagrange equation can thus be written as

$$\frac{dr}{dt} = U_r = \frac{c(t)R}{2}\left(\frac{R}{r} - \frac{r}{R}\right) \tag{13c}$$

which has the solution

$$\frac{R^2 - r_0^2}{R^2 - r^2} = \frac{\rho}{\rho_o} \tag{13d}$$

Now the strain field associated with the combustion process can be worked out for the gas ahead of the flame front. The Lagrangian position of a particle which was initially at r_o and now is at r is related to the density by

$$r_o = \sqrt{\frac{\rho}{\rho_o} r^2 + R^2\left(1 - \frac{\rho}{\rho_o}\right)} \tag{14a}$$

Defining

$$\alpha(\rho/\rho_o, r/R) = \frac{r_o}{r} = \sqrt{\frac{\rho}{\rho_o} + \frac{(1 - \rho/\rho_o)}{(r/R)^2}} \tag{14b}$$

Eq. 2a can be written as

$$\gamma_{rr} = \frac{\alpha}{(\rho/\rho_o)} \tag{14c}$$

$$\gamma_{\theta\theta} = \frac{1}{\alpha} \tag{14d}$$

$$\gamma_{zz} = 1 \tag{14e}$$

The distortion tensor is only a function of the density ratio, ρ/ρ_o, and the radial position, r/R. In cylindrical coordinates the off-diagonal terms of the tensor are zero. It turns out that this problem also can be solved by quadrature in a manner quite analogous to the compression problem previously discussed.

The changes in turbulent intensity just ahead of the flame front are shown as a function of mass fraction burned and flame position in Figs. 13 and 14. The intensities initially rise sharply and increase only slightly for the rest of the combustion duration. The initial state in Figs. 13 and 14 is defined as the state at which r_f/R = 0.015 and $r_f = r_{fo}$ (as shown implicitly in Fig. 14).

In reality, the flame starts at a finite distance from the ignition point. This effect on the turbulent intensity is a decrease in the initial amplification rate. The initial flame kernel may be formed with little change in the unburned gas density due to heat transfer. Figs. 15 and 16 show the change in turbulent intensity for various values of the initial flame radius.

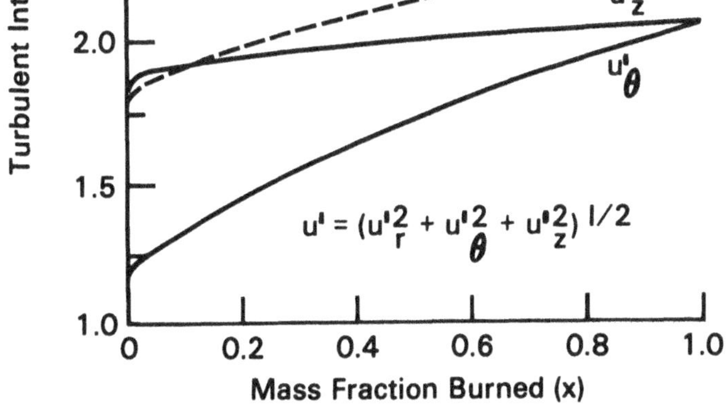

Fig. 13. Turbulent intensities in the unburned gas just ahead of the flame front versus mass fraction burned. P_o = 8 atm., T_o = 800 K, Φ = 1.0, isooctane-air mixture.

References p. 155.

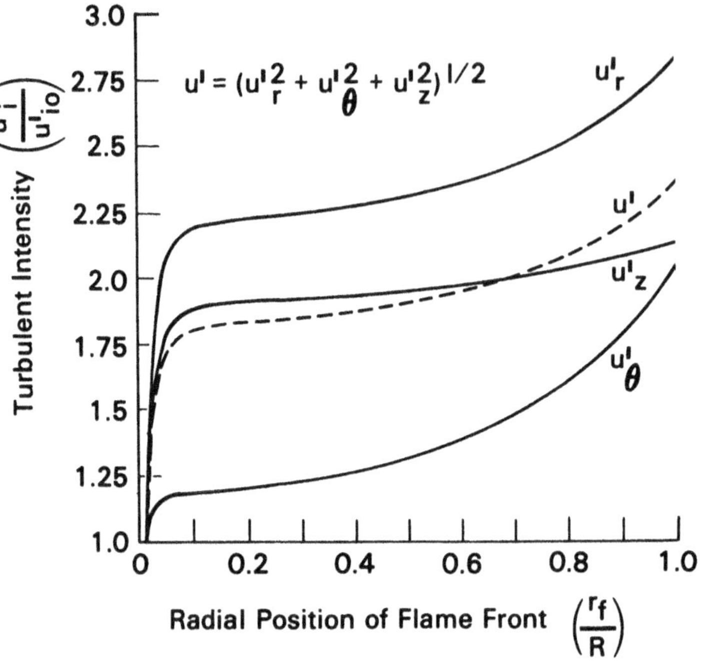

Fig. 14. Turbulent intensities versus flame position. P_o = 8 atm, T_o = 800 K, Φ = 1.0, isooctane-air mixture.

The spatial distribution of turbulent intensity amplification at a given instant is shown in Fig. 17. Since the turbulent intensity decreases rapidly with distance from the ignition site, the sharp increase of turbulent intensity is not observed at locations far from the ignition point. In Fig. 17 the dashed line shows the relationship between the flame front position and the unburned gas density ratio for

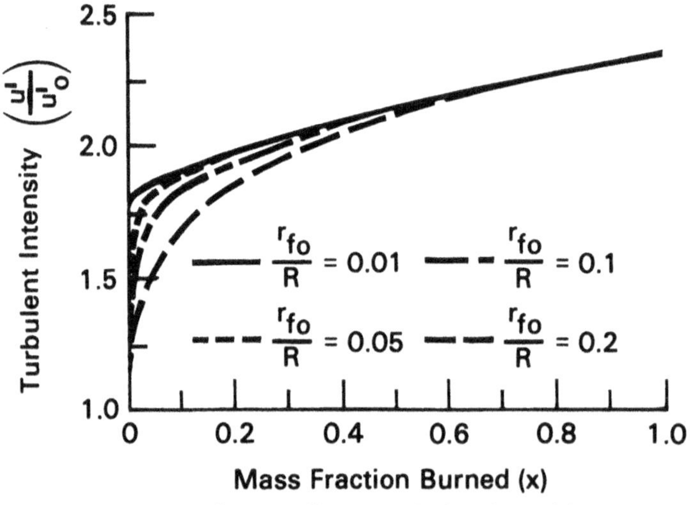

Fig. 15. Effect of the initial flame radius on turbulent intensities versus mass burned fraction.

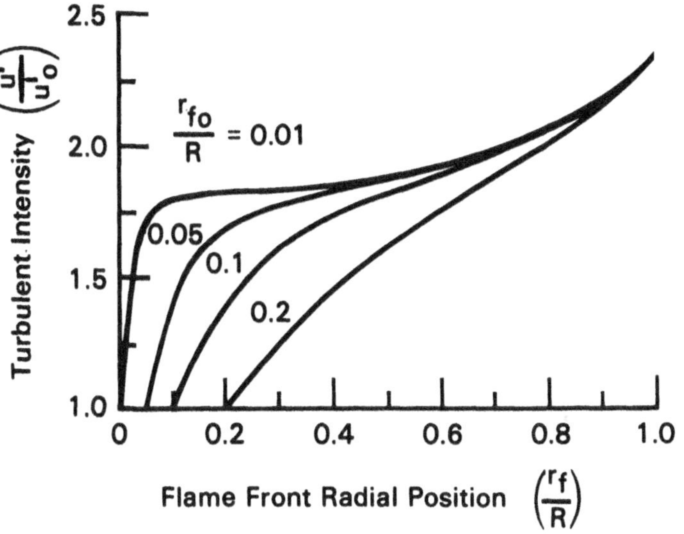

Fig. 16. Effect of the initial flame radius on turbulent intensity versus radial position of the flame front.

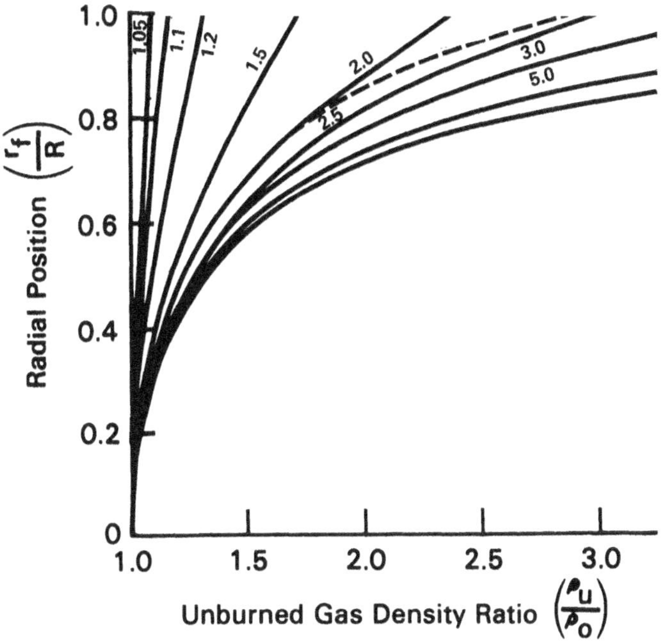

Fig. 17. Lines of constant turbulent intensity showing its dependence on location and compression. Numerical values shown are for u'/u_0'. - - - - Path of the flame front, i.e., relationship between the radial position of the flame front and the unburned gas density ahead of the flame. P_0 = 8 atm., T_0 = 800 K, Φ = 1.0, isooctane-air mixture.

References p. 155.

one typical set of initial conditions. It can be seen that during combustion the flame path follows closely the lines of constant turbulent intensity. The actual rate of burning determines how fast the flame moves along the flame path.

Fig. 18 shows the overall turbulence amplification versus equivalence ratio for a case which corresponds to an intercooled turbocharged engine. The initial pressure "just before the spark fires" varies from 6 to 10 atmospheres while the initial temperature "just before the spark fires" is fixed at 800 K, due to the action of the intercooler. Of course, since the piston does not move in this elementary model, one cannot ascribe crank angle positions to specific events, except in a rough sense. Hence the quotation marks. Note the change in amplification with equivalence ratio on the lean side but not on the rich side of stoichiometric.

Fig. 19 shows similar results at a fixed pressure and varying initial temperature. It can be seen that the turbulent intensity amplification is virtually independent of initial pressure but varies with initial temperature.

Fig. 20 is a cross plot of Figs. 18 and 19 summarizing the results.

At this point we emphasize, again, that the results to date are independent of the rate at which combustion occurs, as well as any detailed scheme by which combustion might occur. This corollary follows from Eqs. 14 which showed that the distortion tensor can be written without explicit reference to time.

Now we suppose that there exists a relationship between laminar flame speed, turbulent flame speed, and turbulent intensity. On dimensional grounds, the relationship must have the following form

$$\frac{U_T}{U_L} = F\left(\frac{U'}{U_L}\right) \tag{15}$$

Physically, it is clear that $F(0) = 1$, and that F is monotonically increasing. The

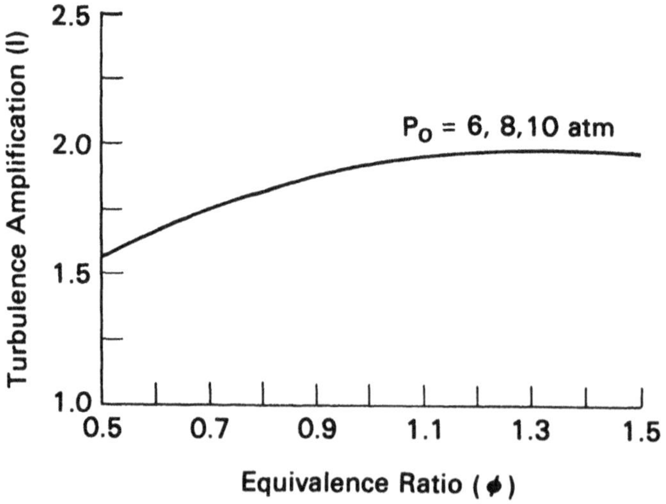

Fig. 18. Overall turbulence amplification versus equivalence ratio for different initial pressures. T_o = 800 K.

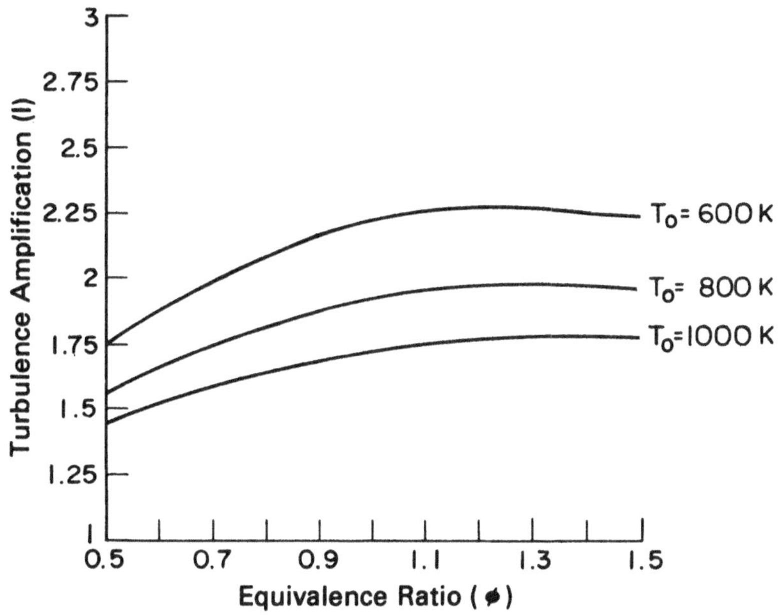

Fig. 19. Overall turbulence amplification versus equivalence ratio for different initial temperatures. P_0 = 8 atm.

literature suggests some thirty specific forms for F, based on a variety of *ad hoc* arguments. We prefer to use the relationships proposed by Shchelkin [10] modified by Lancaster et al. [11] and Leason [12], and that by Damköhler [13] and Karlovitz et al. [14], which according to our interpretation relate the turbulent flame speed, U_T, to the laminar flame speed, U_L, and the turbulent intensity, U', just ahead of the flame front, as calculated, for example, in Fig. 14.

Comparison with Experiments—We can now compare this theory with some measurements of Lancaster et al. [11], who measured the turbulent intensity in a motored engine and the burning rates in the same engine while firing it. By applying the rapid distortion theory to the turbulent intensity data, turbulent flame speeds were calculated using the different combustion theories which relate the flame speed with the instantaneous turbulence parameters. The calculated flame speeds were then compared with experimental measurements, as shown in Fig. 21. The experimental results lie in the range of values predicted by the different correlations, although the scatter is wide.

It should be stated that the engine used in Lancaster's experiments did not have the exact ideal geometry in this problem. More direct comparison between the theory and experiment has yet to be made. Disagreements in the comparison arise from approximations in the turbulence theory, from uncertainties in the combustion model, and to an unknown extent from the differences in the combustion chamber contours employed for theory and experiment. It is believed that when more experimental data on turbulent velocities during combustion, or

References p. 155.

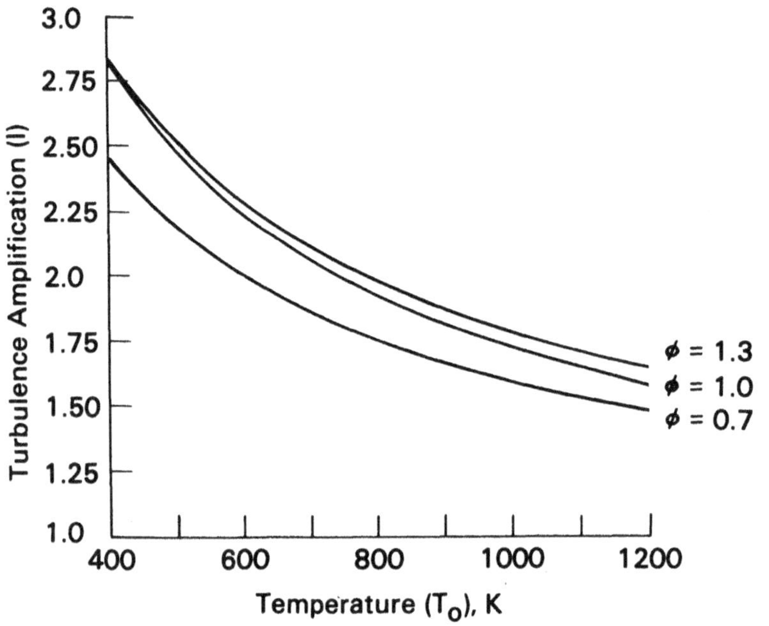

Fig. 20. Overall turbulence amplification versus initial temperature for different equivalence ratios. $P_0 = 8$ atm.

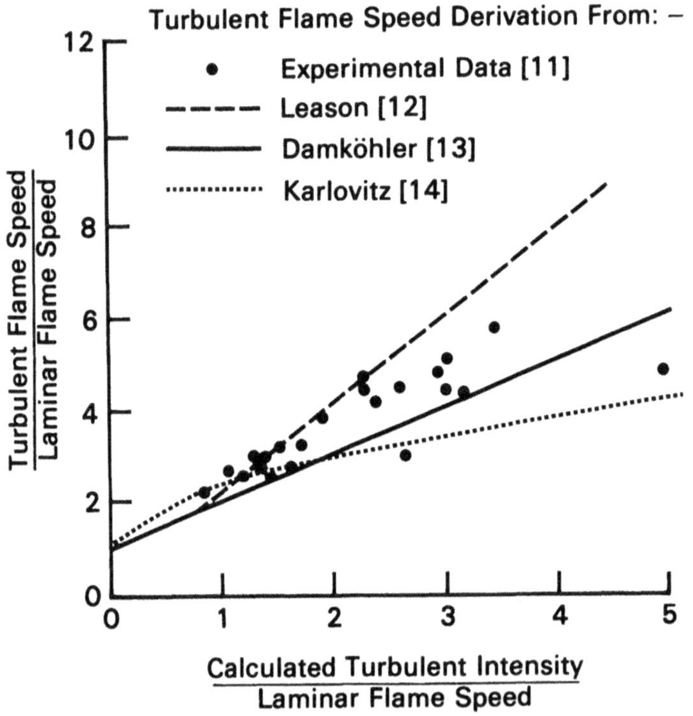

Fig. 21. Comparison of computed flame speed ratio using different combustion theories with experimental results.

at least under conditions simulating the combustion phase, become available, the virtues of the rapid distortion theory can be demonstrated more readily by more direct comparisons between turbulence calculations and experimental measurements.

REFERENCES

1. P. O. Witze, "Measurements of the Spatial Distribution and Engine Speed Dependence of Turbulent Air Motion in an I.C. Engine," SAE Trans., Vol. 86, Paper No. 770220, pp. 1012-1023, 1977.
2. D. R. Lancaster, "Effects of Engine Variables on Turbulence in a Spark-Ignition Engine," SAE Trans., Vol. 85, Paper No. 760159, pp. 671-688, 1976.
3. L. D. Nystrom, "Hot Film and Hot Wire Anemometry, (theory and applications)," TSI Technical Bulletin TB5, TSI Inc., St. Paul, Minnesota, October 1970.
4. A. Ekchian and D. P. Hoult, "Flow Visualization Studies in an Internal Combustion Engine," SAE Paper No. 790095, 1979.
5. N. C. Blizard and J. C. Keck, "Experimental and Theoretical Investigation of Turbulent Burning Model for Internal Combustion Engines," SAE Trans., Vol. 83, Paper No. 740191, pp. 846-864, 1974.
6. V. W. Wong, D. P. Hoult and J. Hunt, "Turbulence Distortion and Vortex Motion in a Reciprocating Engine Cylinder," submitted to J. Fluid Mech. for publication.
7. V. W. Wong and D. P. Hoult, "Rapid Distortion Theory Applied to Turbulent Combustion," SAE Paper No. 790357, 1979.
8. R. J. Tabaczynski, D. P. Hoult and J. C. Keck, "High Reynolds Number Flow in a Moving Corner," J. Fluid Mech., Vol. 42, pp. 249-256, 1970.
9. P. O. Witze, "Hot-Wire Measurements of the Turbulence Structure in a Motored Spark Ignition Engine," Sandia Report SAND 778233, Sandia Laboratories, Livermore, California, May 1977.
10. K. I. Shchelkin, "On Combustion in a Turbulent Flow," NACA TM 1110, 1947.
11. D. R. Lancaster, R. B. Krieger, S. C. Sorenson and W. L. Hull, "Effects of Turbulence on Spark-Ignition Engine Combustion," SAE Trans., Vol. 85, Paper No. 760160, pp. 689-710, 1976.
12. D. B. Leason, "Turbulence and Flame Propagation in Premixed Gases," Fuel, Vol. 30, pp. 233-238, 1951.
13. G. Damköhler, "A Theoretical Discussion of the Influence of Turbulence on the Propagation of Flames in Gas Mixtures and a Description of a Method of Measuring the Velocity of Flames in Mixtures of C_3H_8 and O_2," Electrochemistry and Applied Physical Chemistry, Vol. 46, pp. 601-626, 1940.
14. B. V. Karlovitz, D. W. Denniston, Jr. and F. E. Wells, "Investigation of Turbulent Flames," J. Chem. Phy., Vol. 19, pp. 541-547, 1951.

DISCUSSION

J. W. Daily *(University of California)*

I have a facetious comment. I have had the advantage of reading the paper in advance, and I see now why some people do numerical analyses—because the theory requires real knowledge of mathematics. Sometimes the numerics are easier to do than the experiments, but the simple kinds of insights that you get from this kind of analysis are intriguing. I think the independence theorem is an especially intriguing concept. Let me back up and ask some questions that will lead up to a question about that. It seems to me the rapid distortion theory, because of the fact that the Cauchy tensor, or the distortion tensor, is diagonal, neglects interactions between coordinate directions. As a result you're forced into a situation where you can't transfer energy between various coordinate directions. If that's the case, then that might have a bearing on the independence theorem that comes later for the combustion case, and the question I have is, "What is the physical assumption of the rapid distortion theory for both the non-burning and burning cases, and how do those physical assumptions play back on the independence theorem that comes out of the theory?"

D. P. Hoult

Let me answer the first part. The fact that the tensor is diagonal in these particular cases does not prevent the flow from conducting energy from one component to another and does not change the independence theorem, but it would be a much more laborious task to calculate. The two basic physical assumptions are mentioned in the first or second slide. Those are, the eddy-to-eddy interactions are ignored, and the theory is an inviscid one. The independence theorem is not correct if those two assumptions are not met. Also, there may be

D. P. Hoult

small corrections if there is very large squisn in the chamber geometry or if the burn time is so slow that there is appreciable piston displacement during the course of the burn process. But for engines operating the way engines do today in practice, the independence theorem is presumably a valid theorem. What is important is that it suggests to other modelers, who deal with turbulent combustion, that there is a natural theoretical way to split apart certain problems. One does not need to deal with turbulent combustion problems necessarily with the same full set of equations that one deals with in analyzing the turbulent state ahead of the burning gas. We think that's conceptually a great simplicity.

F. V. Bracco *(Princeton University)*

This is not a question, but an effort to understand your work a little bit better. Essentially you say that for the high Reynolds number case you can neglect the viscous effects. You are using some kind of a conservation of vorticity perturbation law. You also say that vorticity perturbation is maintained up to top dead center and that when the flame propagates, it does not alter the field ahead of itself, which is the only field that the flame itself really feels in the first place. Then the vorticity perturbation is conserved and the field ahead of the flame is not. This is my understanding. As a result, the field is not perturbed by the flame itself as it propagates. Therefore you can relate in a direct way the turbulent flame speed to something which could be called the thermodynamic flame speed, that is, the laminar flame speed, plus the opposite of the perturbation of the velocity. You can relate that *a priori* certainly.

D. P. Hoult

Yes, that seems a fair summary.

F. V. Bracco

If, however, there were squish in the chamber, or if the chamber were divided into a prechamber and main chamber, then there would be other means of generating turbulence during compression, and one would not necessarily expect your conclusion to be applicable.

D. P. Hoult

Certainly, modifications may be required for a prechamber and main chamber geometry. But for any turbulent state existing before the spark is fired, regardless of how the turbulence is generated, that is, through squish or anything else, the statement is that the flow ahead of the flame front can only respond to certain properties of the burning process. In particular, it only responds to the numerical value of the density field ahead of the flame. Of course, the combustion process squeezes the flow ahead of the flame, but the turbulence cannot respond, according to this theory, to any time-dependent process relating, say, to the reaction rates or to the detailed combustion properties. There's no doubt, of course, that the turbulence is affected by the burning process because the theory predicts a substantial change in intensities as the burning process develops. In the curve I was showing, there's a rapid rise and then a relative plateau in turbulent intensity. But the important feature about the theorem is that it says the detailed kinetic rates associated with this turbulent combustion process are not important regarding how the turbulence evolves.

F. A. Matekunas *(General Motors Research Laboratories)*

In the development of your theory for the burning case, you had three equations, the third one being a solution that basically looked like the ratio of the volumes of the unburned is equal to the ratio of the density of the unburned to the initial density of the unburned. That is, there is an "$R^2 - r_0^2$" term.

D. P. Hoult

Yes, everybody that reads the paper for the first time tries to relate it that way. That's a solution, but not the equation you think. It's a representation of the Lagrangian solution of a fluid particle which is released at r_0 and presently is at a position r.

F. A. Matekunas

But in calculating the strain field don't you really depend on knowing the velocity of the flame relative to the unburned gases, which is just the burning velocity?

D. P. Hoult

No, it's independent of the burning and the propagation velocity. It depends only on the density, that is, the compression resulting from burning. The turbulence, it is claimed, cannot care how you did the compression. That's some private affair with you and the spark plug.

F. A. Matekunas

I agree with that, but I just wondered if you were calculating the rate of strain, assuming that the distortion theory holds. I just wondered how you calculated that strain field when it appeared that there was a discrepancy in that equation in that it basically gave you the propagation velocity.

D. P. Hoult

No, it doesn't give you the propagation velocity. Let me try to say it one more time and then I'll quit. Ignore combustion! Suppose that inside a roughly cylindrical pancake-shaped chamber you have a sudden machine that expands the gas ahead of it, and you ask a different question. You ask, "How does the turbulent flow ahead respond?" Now clearly the turbulent flow gets squeezed, due to the changing density field, but it is not *a priori* clear that the time rate of change of the density field or the position of the squeezer don't make a difference. What I claim is that those latter two things don't count. The turbulence only feels what its density is right there and how far it is from the wall. That's it.

P. O. Witze *(Sandia Livermore Laboratories)*

You mentioned in the paper that you were going to talk about the water analog, I want to raise a question about that.

One of your conclusive results in the water test was that the large ring vortex that forms during intake always positions itself basically midway between the piston surface and the top of the head, irrespective of valve lift or what have you. Then you go on to show some Imperial College measurements showing that that vortex stays at the top of the cylinder. Can you give any explanation for why that might have happened?

D. P. Hoult

I don't know. I think there are a couple of points different between those measurements and ours, and I don't understand the difference. It is true that we were running at much higher rotational speeds, that is scaled speeds. The Reynolds numbers in our experiments were 1500 to 2500 or so, and that may be the difference, but that's the only thing I can think of. The measurement methods were different, but I don't think they should make that big a difference.

A. D. Gosman *(Imperial College, England)*

I would like to make this comment. I haven't read your paper so I don't know how your experiments were organized, but the one at Imperial College, which Jim Whitelaw conducted, was just steady flow out of the operating engine. Then you have the residual motion of the preceding cycle.

D. P. Hoult

Ours was an intake process, and then the motion stopped. That may be an important difference. I would expect it is.

P. N. Blumberg *(Ford Motor Company)*

Can you comment on the limitations of the rapid distortion theory or the theorem of independence that arise either from compressing the unburned gases into a region with irregular geometry, where there might be substantial wall influences and perhaps the generation of bulk flow, or from squish action that induces bulk flow motion into the unburned gas ahead of the flame front? It seems that those two situations might have had some impact on the validity of saying that the only thing that matters is the density field.

D. P. Hoult

Let me be very frank. We don't know the answers for each of those questions. The calculations I've been able to work out in a rough form suggest that if the chamber is not too irregular in geometry, the theorem holds. But I'm not in a position to make more general statements. That's a very fair question, though, but I don't know the answer to it.

D. R. Lancaster *(General Motors Research Laboratories)*

My question deals with your last comparison, in which you used some of my experimental results. It's not clear as to what you're really comparing because you have a different geometry in your calculations than that used in my experiments.

D. P. Hoult

I've wrongly glossed over that in trying to save time. We took the simple disc-shaped geometry. We took your measured turbulent intensity in a motored engine, and your measured turbulent flame speed and calculated laminar flame speed to make the comparison. We did not try to correct the theory for the specific details of the geometry which you were studying, so one should not conclude from that comparison that the theory has been proven to be correct.

A. K. Oppenheim *(University of California)*

At the risk of saying something very obvious, I would just like to clarify the situation. In order to generate turbulence one has to deposit energy, unless you have the intensity already. You did not have any conservation of the differential equation of energy, so whenever you don't deposit energy you haven't generated the turbulence.

D. P. Hoult

No, you solve the full Navier-Stokes equations. That is, you can write them down in terms of an energy equation. All this calculation says is that if you start with a turbulent flow and you follow the rules of evolution according to the approximations that I came up with, you get this kind of independence. From the work on the intake process, we believe that the vortex ring breaks up before top dead center and deposits turbulent energy in the cylinder.

SESSION I WRAP-UP

W. C. REYNOLDS

Stanford University, Stanford, California

In wrapping up the session I'd just like to say one thing. It appears to me that rapid distortion analysis is an exact analysis in the case for which it applies, that is, when the distortion is very quick. This case should be a very useful place to test models. The phenomenological models, like the k-ϵ models, probably should be tested in order to make sure that they fit rapid distortion analysis.

SESSION II

MODELING FLAME PROPAGATION AND HEAT RELEASE IN ENGINES

Session Chairman
G. L. BORMAN

University of Wisconsin
Madison, Wisconsin

MODELING FLAME PROPAGATION AND HEAT RELEASE IN ENGINES—AN INTRODUCTORY OVERVIEW

G. L. BORMAN

University of Wisconsin, Madison, Wisconsin

ABSTRACT

An overview of experimental information and models available for the analysis of flame propagation and heat release in internal combustion engines is presented.

The charge preparation period for homogeneous engines is reasonably modeled by current thermodynamic cycle analysis, but such factors as liquid-phase fuel distribution and residual mixing have not been adequately modeled. Fuel spray mixing and vaporization in diesel engines are currently modeled practically by gas-jet mixing theory, but this is inadequate. Promising numerical techniques are being developed but are limited by cost of computation and inadequate data for initial conditions. The modeling tools currently available could, however, be profitably exercised to a greater extent to explore qualitative phenomena.

A major experimental need for use in understanding both charge preparation and ignition is a method of mapping fuel-air ratio in the cylinder. Sampling methods are slow and spatially restricted, but less restrictive optical methods are applicable only to specially prepared engines.

Ignition modeling for spark-ignition (SI) engines has concentrated on thermal theories, but cycle-by-cycle variations are believed to be caused by turbulent structure variations in the region of the spark plug. Adequate data on the effects of turbulent intensity and scale on kernel growth is lacking. Analysis of various types of jet ignition by use of computational fluid mechanics would be timely, but may be hampered by lack of chemical kinetics data. Analysis of ignition delay in compression-ignition (CI) engines depends on adequate charge preparation models, and thus is currently empirical. Detailed experimental study of the ignition region in CI engines is thus needed.

Flame propagation in homogeneous engines is currently reasonably well modeled by assuming spherical flame-front propagation. Turbulent flame-speed models based only on turbulent intensity and laminar flame speed, and other models based on turbulent entrainment of eddies, both give promising results. The effects of turbulent scale need to be experimentally documented.

Models for stratified-charge engines with strong fluid motion effects are best handled by numerical solution of the equations of change. Such

References pp. 185–190

models are currently restricted to two-dimensional flows and are thus very limited in application. Methods to couple empirical models with the equations of change are needed for the immediate future.

Combustion models in which droplets play an active role are limited by lack of flame theory for such two-phase cases. Available thick-spray models are mechanistic and require considerable empirical input. Much more time- and space-resolved experimental data are thus needed in order to make progress in diesel modeling.

NOTATION

B	engine bore
C	empirical constant
FSR	flame speed ratio, S_T/S_L
FSR_m	fully developed flame speed ratio
h	piston-to-head distance
$imep$	indicated mean effective pressure
L	integral scale of turbulence
m,n	exponents
N	engine speed
r	radius of spherical flame
r_c	flame radius for critical ignition kernel size
r_m	flame radius at start of fully developed flame period
S_L	laminar flame speed
S_P	average piston speed
S_T	turbulent flame speed
t	time
u'	turbulent intensity
η	Kolmogorov scale
ν	kinematic viscosity
ρ_i	end-gas density at start of flame propagation period
ρ_u	end-gas density

δ_L laminar flame thickness

ϵ volumetric efficiency

INTRODUCTION

The successful application of modeling techniques to engine design requires a strong interaction between modelers, research experimentalists and development engineers. Without such interaction, modeling can become sterile and pontifical. The purpose of this paper is thus to provide an overview of the experimental information and techniques available and their potential interaction with modeling needs and methods. Because of the organization of the Symposium Agenda, the present discussion will be restricted to problems involving engine charge preparation, combustion and heat release. Problems concerning emissions and design applications will be taken up in later overviews by Drs. Newhall and Krieger. Even with the restriction to combustion phenomena, the task represents an ambitious undertaking, and thus much detail has been omitted due to limitations of length and the writer's knowledge. In this same vein, the references cited are meant to be a representative sampling rather than a complete bibliography.

Before starting on the major tasks of this paper, it may be helpful to make a few preliminary comments concerning the purpose, classification and philosophy of engine modeling. With regard to purpose, it hardly seems necessary to dwell upon the usefulness of models, nor to point out that modeling is an aid to, but not a replacement for, ingenuity, intelligence and experience. It may be appropriate, however, to note that the most detailed and least empirical model is not necessarily "best". Rather, the simplest model that can accomplish the objective in the most cost-effective manner is usually "best" in engineering practice. The more sophisticated our demands, however, the more likely it is that models will require greater detail and generality.

The particular approach used in modeling should be carefully considered for each problem. The idea of a universally applicable approach such as starting from the partial differential equations of change and solving them numerically is much too restrictive to be practical, given our current technology. A much better idea is to balance the various assumptions used in the model and to incorporate empirical correlations whenever they are available and warranted by the level of accuracy expected. Nevertheless, long-range research on currently impractical but potentially powerful methods must continue so that new modeling tools will become available as technology advances. It is unfortunate, however, that research of long-range potentiality and pressures to satisfy immediate needs have sometimes been mixed to produce results which lack credibility.

In the following overview, it will be helpful to refer to models in terms of a classification based on the extent to which the models can predict results without need of specific test data. The following classification will be used for the purposes of this discussion.

References pp. 185–190

Empirical Simulation:
Fitting of data by use of regression techniques such that interpolation allows simulation of the system. An example is the simulation of diesel engine heat release based on analysis of test results [1].

Analysis Models:
Models which require specific and usually extensive experimental data input from the system process being modeled. An example is the calculation of a mass burning rate history based on pressure-time data and estimated trapped mass for a given engine and given test. [2].

Incomplete Models:
Models which require that certain parameters are empirically adjusted for the specific system being modeled. An example is the calculation of burning rate in a homogeneous charge spark-ignition engine based on an assumed (spherical) flame shape and a turbulent flame speed which contains a number of adjustable parameters [3, 4, 5, 6, 7].

Complete Models:
Models which require only fundamental data or correlations of a general nature and not data specific to the system being modeled. The author knows of no engine combustion model in this category. The incomplete model given by example above might be complete if the turbulent flame speed could be correlated to engine parameters and if the other assumptions of the model did not lead to an unacceptable level of inaccuracy for some engines. Models based on the three-dimensional unsteady equations of change offer the possibility of being complete if general formulations for turbulence and chemical kinetics become available [8].

One should note that a model for a subsystem could be complete even if it requires initial values or boundary conditions empirically obtained from the system being modeled. Also, given such factors as ease of use, cost and accuracy, it is clear that a complete model (even if available) is not necessarily always the best model for a given situation.

The major sections of the paper, which follow, deal with phenomena as they take place chronologically in the engine: the charge-preparation period, the ignition period, and the heat-release period. The discussion is then summarized with some general comments on future needs and the blending of the modeling methods.

THE CHARGE PREPARATION

Engine charge preparation may be divided into two categories: those in which the charge is mixed primarily outside the cylinder (premixed charge) and those in which the charge is mixed in the cylinder (direct injection).

In all cases, it is necessary to know the trapped mass and the residual fraction. In addition, for the premixed case, the uniformity of the charge and residual mixture temperature and composition and the charge quality may enter into determining the subsequent combustion, but to date, models have incorporated the assumption of a homogeneous system. If a cycle analysis program is not used, the concentration of residual may be measured by taking a large sample directly from the cylinder. This is probably necessary for two-cycle engines because models for scavenging are rather primitive. Even in four-cycle engines, computation of trapped mass can only be estimated by using the experimental intake mass flow rate because of valve overlap, cycle-to-cycle variations and cylinder-to-cylinder variations. Calculation of the trapped mass based on detailed thermodynamic cycle analysis can give the proper variation with engine parameters, but normally predicts mass flow to only 3–4% accuracy. Similarly, initial absolute pressure must be estimated because the pressure transducer gives pressures relative to some selected point in the cycle. Use of a balanced diaphragm pickup to determine absolute pressure during the essentially constant-pressure portion of the intake stroke is one means of obtaining this absolute reference pressure. Methods for measuring turbulence parameters and/or charge uniformity are only now in development for a fired engine. One might use a variety of optical methods [9, 10, 11], including the one to be discussed by Professor Dent in this session. Another method currently being developed at the University of Wisconsin–Madison uses a pulsed tracer-gas technique. Qualitative measures of uniformity can be obtained from schlieren movies, but otherwise data are only available for motored engines, where hot-wire transducers can be used [12].

In addition to questions of gas-phase uniformity and turbulence, there is the question of charge quality. In many cases, the fuel-air mixture contains liquid fuel as it enters the cylinder. Although a uniform vapor-air mixture might seem ideal, the effects of the liquid phase on charge stratification and combustion can sometimes have a salutary effect [13]. Modeling of vaporization in such vapor-air-liquid droplet mixtures can be carried out if the initial droplet size distribution is known, although the appropriate vaporization model for multicomponent (practical) fuels is not certain [14]. Droplet size data might be obtained by various optical methods [15], but such data are not yet available for engines.

In summary, the premixed-charge engine-combustion model starts with a charge of uncertain mass and homogeneity but known pressure. Turbulence and mixture motion parameters generally available at this time are from motored engine data and thus do not contain the effects of hot-residual mixing and backflow found in the fired engine. It may be that the best way to investigate some of these effects is by computational fluid mechanics [16]. Thus far, however, such computational methods have been mostly limited to two-dimensional flows with very imprecise modeling of the boundary layer.

For calculations where engine cycle performance is desired, but not the details of combustion, a detailed homogeneous ideal-gas thermodynamic-cycle analysis is quite satisfactory. Such cycle programs are particularly useful in predicting trends of breathing, throttling losses and indicated mean effective pressure (*imep*). They currently provide a practical means of estimating the thermodynamic state of the charge prior to ignition.

Estimation of the charge state prior to ignition in a premixed engine, despite the many uncertainties, is a simple matter compared to making the same estimates for a direct-injected engine. While considerable data and a number of incomplete models are available for steady-state sprays, very little detailed information is available for the transient injection of liquid into an engine atmosphere.

Starting with the mechanism of transient high-pressure spray formation, one finds no theory and very little data, although recently, ultrahigh-speed movies have begun to shed some light on possible breakup mechanisms [17]. The general outline of the spray has been observed by numerous investigators, dating back to the pioneering works of Lee [18] and Schweitzer [19]. Such observations, plus the utilization of gas-jet theory in some cases, have led to a large number of formulas for spray-tip penetration [20]. But none of the formulas given in Ref. 20 addresses the problem of multiple sprays in an enclosure with air swirl, and only limited data are available for jets with rapid fuel vaporization. More recently, Morris and Dent [21] have obtained experimental data for injection into a cross flow, but again the experimental conditions were different from those found in an engine. Rife and Heywood [22] have studied injection in a rapid compression machine, thus obtaining data at temperatures and pressures similar to those in engines. They suggest that the dense gas-jet theory of Abramovich [23] can be used to predict the spray behavior for nonswirl cases.

Information on the fuel mass, droplet size and velocity distributions within high-pressure sprays is limited primarily to early investigations which, understandably, used rather crude instrumentation [24]. This information is further limited to droplet size distribution measured at a large distance from the orifice with injection into an ambient-temperature environment [25]. The reason for the obvious lack of information is the very short duration of the spray, coupled with its high density. Optical methods traditionally used for continuous-burner spray analysis can only give information about the fringes of such thick sprays. If one wishes to model the spray approximately using single-droplet ballistic theory [26, 27], or more accurately using distribution theory [28], the required initial droplet spatial, size and velocity distributions are not known. If one assumes the temporal size distribution at the break-up point (very close to the orifice) is the same as farther downstream, then either the larger droplets found downstream have a much lower than average velocity when formed, or they manage to survive because they do not have time to break up before they slow down [29]. It is probably reasonable to assume that vaporization plays a minor role in the region close to the nozzle [26]. Drag coefficients from less-dense spray studies [30, 31] or single-droplet studies [32] can probably be applied except for the very dense spray core. Parenthetically, one should note that except for Stokes drag (droplet Reynolds numbers less than 0.1), the drag force in a given direction is a nonlinear function of all three components of the droplet relative velocity. Thus using the Stokes drag formula outside of its range gives both improper coupling of the gas flow field and droplet motion and much too low a value for the drag coefficient. Recent models for diesel sprays have tended to use gas-jet theory [22] or empirically based gas-jet-like equations [33, 34] or a combination of gas-jet theory and droplet vaporization [35, 36]. Such models seem to provide a useful approach even though the physical mechanisms of transient liquid and gas jets are quite different in detail [28]. Use of the steady gas-jet theory allows rapid calculations of the fuel-air concentration profiles as well, but inclusion of the cooling caused by vaporization [37] cannot be predicted without finding an expression for the vaporization rate of the droplets. If, in fact, all the droplets tended to vaporize in a very short distance from the nozzle, the temperature depression would be quite large, and adiabatic saturation would be reached, thus preventing further vaporization until more air became available through entrainment. At chamber pressures and temperatures above the critical point of the liquid, the liquid will tend to heat up until its critical point is reached. For such conditions, if the droplet surface is in equilibrium with the gas molecules in contact with it, we would find that the surface reaches a mixture critical point temperature, and the energy required to change the phase becomes zero. The droplet temperature thus continues to rise without reaching a steady-state temperature even if the ambient temperature and pressure are constant. Although modeling of the transient history with variable properties and spherical symmetry has been carried out for such supercritical cases [38], the problem of similar modeling for real fuels is much more complex and requires a number of questionable assumptions. Because of the computational speeds required for spray models, it is necessary to use quasi-empirical heat- and mass-transfer correlations, although these correlations are inaccurate in the critical region [39, 40]. Much more additional data are thus needed for engine conditions despite the huge amount of droplet research results already available [41, 42]. Additionally, it appears that droplets may sometimes reach a critical state where they suddenly disintegrate [43], giving a very rapid transition to the vapor state. However, in a recent experimental study of combustion of a high-pressure steady spray in stagnant air, Shearer and Faeth [44] found no unusual behavior under conditions where droplets were calculated to have gone through their critical point.

Modelers also need to know the effects of internal droplet liquid circulation to estimate the composition of the vaporizing species [14]. One should note that, even if vaporization is not rate-controlling, the rate various fuel components are vaporized by different size drops in different spray regions could result in a stratification of the vapor components. This problem is thus additionally related to chemical changes taking place in both the liquid and vapor phase.

Even without much additional data, it would seem that models for unsteady multicomponent vaporization in a limited supply of air could be used now to obtain a qualitative understanding of the coupling between the physical and chemical parameters. It is clear from both experiment and theory that gas-jet models of liquid sprays in engines are not adequate. Such models, however, provide the opportunity for the rapid calculation required when the model is to be coupled to a combustion and emissions model. The detailed two-dimensional calculations of Haselman and Westbrook [28], even though they incorporate a simplified vaporization model for a pure fuel, are extremely lengthy. It would thus seem that two paths need be followed. First, a well coordinated effort is needed to provide the experimental inputs needed for the detailed models and to evaluate such modeling results. Second, the detailed-model research should be used to provide the insight necessary to evolve a simplified model which might require empirical adjustment for each combustion system design application. It should be clear, however, that only the very detailed calculations using the equations of change can predict the interaction between sprays and the fluid motion in the cylinder enclosure. To be complete, detailed models must incorporate multicomponent fuels and three-dimensional flow. It is unlikely that this can be done without considerable improvement in computer speed and capacity. It is thus necessary that designers pose very carefully considered, limited questions that can be answered given current computing capabilities.

Although the above discussion has been devoted to thick sprays, one should note that some stratified-charge engines such as the PROCO engine use less dense sprays. The PROCO in fact, uses an injector with a vibrating pintle designed, presumably, to reduce the penetration while not compromising atomization. The modeling of such a spray should be well suited to the computational approaches of Ref. 28. Given the fluid motion, one might then treat the spray three-dimensionally, provided the spray to fluid coupling is neglected [45], but the validity of this assumption even for lower-pressure sprays in engines has not been established.

THE IGNITION PERIOD

For the homogeneous, spark-ignited engine the conditions conducive to good ignition are well established: low flame temperature, high initial temperature, high heat of combustion, high mean reaction rate, low volumetric heat capacity, low thermal conductivity, high total pressure, nearly stoichiometric mixture, low gas velocity, low intensity of turbulence, and electrode separation distance close to the quench distance. The lean ignition limit can be extended by increasing mixture homogeneity, decreasing charge dilution, increasing compression ratio, decreasing engine speed, locating the spark more centrally in the chamber and using multiple sparks [46]. A good deal of practical information is also available on the design of the plug for good ignition and performance [47, 48, 49]. Coupled with this mass of information is the advent of improved high-energy ignition systems. Despite all of this technology, however, most of the details concerning the chemistry and fluid mechanics of the initial flame-kernel growth in an engine are unknown.

Experimental analysis of the chemistry of species in a spark can be carried out with the time and space resolution required [50], but interpretation of the very complex case of a spark in a reactive mixture has yet to be done. At this time it is thus impossible to appraise the importance of spark-related chemistry in the ignition process.

The effects of charge motion on ignition have been investigated both in engines and bombs. Although the flow near the plug affects the required ignition energy by changing the surface-to-volume ratio of the kernel and the heat transfer from it, a serious design problem centers around cycle-to-cycle variations. These variations can be caused by variations in local composition, particularly poor residual mixing, but the more common cause is probably the random variations in the turbulent field near the spark plug [51]. Comparison of the slow- and fast-burn cycles (low and high peak pressures respectively) shows that slower-burning cycles also have a longer induction period. That is, the mass-burned-versus-time curve is "S" shaped, and the initial low burning rate portion is extended for low-peak-pressure cycles [52]. The induction time is sometimes misnamed "ignition delay" but is actually the time required for the kernel to grow to a critical size where it will be self-sustaining. In Ref. 7, the authors assume that this time is the same as the time necessary to burn a turbulent eddy.

It is not surprising that the small flame kernel can be greatly affected by local variations in velocity, and that these variations are random in time at a particular location in the flow field. It is interesting however that Evers [53] found cyclic pressure variations were essentially eliminated when a strong jet of compressed air was directed into the cylinder just prior to ignition. The added turbulence caused the interval between spark and peak pressure to decrease from 40 to 15 crank angles (CA) and MBT spark timing to change from 25 CA° BTDC to 6 CA° BTDC. This increase in flame speed with decreased variability corresponds to the observations of Cole and Mirsky [54], obtained from use of a propane-air jet directed at the initial flame kernel in a combustion bomb. Similar behavior has also been observed in engine experiments [55]. Thus we may conclude that although increased velocity or turbulence could cause misfire by increasing the required minimum ignition energy, the effect of increased motion at a fixed engine speed is more likely to be a decrease in cycle-to-cycle variation. In the case cited (Evers), the local values of small scale (random) turbulence which could cause cycle-to-cycle variations may have been overcome by a consistent large-scale motion induced by the jet and by improved mixing of residual and charge. A similar effect may be responsible for the decrease in dispersion sometimes experienced with increasing engine speed. In addition to time-averaged turbulence values, the turbulence variation with time (crank angle) must be considered. In some cases with squish and/or swirl, the turbulent intensity and mean velocity may peak near TDC [56]. Thus shifts in the induction time will cause the propagating flame to encounter more or less turbulence depending upon the phasing between the flame growth and the intensity variation. Shifting the combustion to give the effect of retard will also cause a change in the laminar flame speed. The result will generally be a decrease in the slope of the mass-burned-fraction curve at its inflection point. In chambers of simple geometry (CFR engines), the turbulent intensity may be essentially constant, with only a small decay, so that increases in induction time at fixed spark timing will always also cause a decrease in rate of combustion. Such phasing could also play a role when initial kernel growth is changed by decreasing the chemical energy released per unit volume through leaner mixtures or exhaust gas recirculation.

Cyclic variation in the velocity at the plug as measured under motoring conditions correlates very well with the standard deviation in peak pressure observed during firing [57]. It must be noted, however, that to obtain the experimental data really needed for ignition models one should measure the velocity at the plug, the shape and volume of the growing kernel, and the cylinder pressure, all during the same cycle. Such data could be obtained by combining use of laser velocimetry and direct or schlieren high-speed photography [58]. In theory at least, one could also obtain the mixture composition at the plug at the time of spark from Raman spectroscopy. Such data could help to clarify the role of turbulent scale and of composition variation in the production of cyclic variations, as well as to give a better basis for evaluation of previous models [7, 59, 60].

Although modelers and experimentalists alike have mostly ignored chemical effects on ignition, recent work on jet ignition [61] has caused some renewed interest in the idea that chemically active species might play a larger role than has been appreciated [62].

Modeling of the ignition process, even based on simplified thermal theories, can give some insight regarding the role of lean limit, turbulence and heat transfer in homogeneous spark ignition. Effects due to chemical kinetics, heterogeneous mixtures and droplets require more understanding gained by experimental inputs before modeling is likely to have much impact. In this regard, data for steady sprays [63] and mists [64] show an important influence of droplet size on minimum ignition energy. Although no similar data are available for engine conditions, Peters and Quader [13] have shown an improved lean misfire limit for fuel injection in the manifold during the intake-valve-open period as compared to the homogeneous case. The effect was dependent upon the use of a shrouded intake valve. Their indirect evidence indicates that there may be an optimum droplet size and spacing for lean ignition. We note that the combustion rate also increased with fuel injection so that an increase in initial rate of kernel growth may be the underlying reason for the improved lean limit. This trend agrees with both bomb [65] and engine [66] work, which show flame velocity can be increased by addition of droplets to a lean fuel-vapor and air mixture.

Ignition in divided-chamber three-valve engines is similar to the homogeneous engine except that residual mixing may play a more important role. But most other stratified-charge spark-ignition engines present the additional problems of possible large variations in mixture fuel-air ratio and the presence of liquid droplets. For direct-injection engines, the placement of the spark is crucial. Both the Texaco and Ford stratified engines, for example, use special plug geometry and electronics to try to overcome this problem [67, 68]. Given the fluctuations in composition and velocity inevitable in a turbulent mixing situation it would seem that some form of multiple-point ignition would be desirable in stratified engines. Such ignition could be obtained by use of an ignition-cell jet or, more simply, a small cavity in the head with the spark plug mounted at the orifice. The disadvantages of such designs are an increase in energy loss due to heat transfer and possible loss in stratification. Modeling of such jet-ignition systems should be an excellent area of application for two-dimensional unsteady computational fluid mechanics. This should be particularly helpful because experience has shown that jet-ignition systems require rather exacting design.

No matter what ignition design is used, the mapping of fuel-air ratio in the cylinder is most important. Experimental methods of mapping used thus far have been both expensive and time consuming [67, 68]. With great care, both cycle-averaged and individual-cycle samples can be obtained [69]. The probe position is generally limited, however, and the sample times of 1–2 ms (12–24 CA° at 2000 r/min) give a blurred image of the map. On the other hand, trial and error location of the ignition source and injector, while direct, gives very few clues about possible general design improvements. Given the experimental difficulties, modeling would seem to offer a viable alternative. Unfortunately, models thus far have been restricted to two spatial dimensions because of computational speed and accuracy. Although the fuel spray is inherently axisymmetric, neither the chamber nor the flow share this axis. What is thus badly needed is a three-dimensional (non-reactive) analysis which can handle the case of an unsteady (gas) jet in a practical chamber geometry. Until such programs are possible, however, the two-dimensional programs can be used to generate concepts about the concentration field. Such concepts will, of course, need to be adapted to the true engine geometry and evaluated experimentally. For such purposes, the axisymmetric jet model should do reasonably well for quiescent chambers as long as one is interested in early portions of the injection where wall effects might be neglected.

Thus far the discussion of spark ignition has been concerned with the positive aspects of igniting the charge. In some engines, deposit formation leads to preignition from hot spots. Such uncontrolled ignition, which results in engine rumble, could be modeled if the deposit formation processes, deposit properties, and the mechanism by which deposits come off the wall were known. A fair amount is known about deposit technology [70], but past efforts relating to rumble were often concerned with the effect of tetraethyl lead.

As in the case of charge preparation, the compression-ignition process is much more complex than spark ignition. The physical processes of droplet formation, vaporization and mixing control the preparation for combustion, but the chemical reactions leading to ignition dominate the length of ignition delay. Many experimental studies of ignition delay in engines have, of course, been conducted over the years, and theories for ignition delay of gas-phase mixtures have been developed based on an energy balance for the mixture and simple global chemical schemes.

For engines the delay has usually been modeled by prediction of a physical portion of the delay based on droplet vaporization [71–76] combined with a chemical delay period based on some simplified kinetics scheme. Although ignition in the engine takes place at the edge of the spray, it is unlikely that single-droplet vaporization theory can give the proper formulation for the engine. Ignition normally takes place during the transient portion of the spray development. During this time, fluid flow between the spray tip and entrained airflow closer to the injector may well be an important mechanism for determining the fuel-air ratio near the injector end of the spray. Thus outflow from the spray tip may contribute to determining the fuel-air ratio near the tip end of the spray. At this juncture, it is therefore necessary to depend on empirical formulations such as those of Wolfer [77, 78] or Henein and Bolt [79]. Such formulas are very limited by their use of bulk temperature, with no incorporation of local temperature or composition values. However, movies of diesel combustion show that for a given engine the location of ignition is relatively consistent for each cycle. It should thus be possible to determine and then study this small ignition region by means of sampling or perhaps by use of absorption or Raman spectroscopy.

Although models may be developed to predict spray behavior in a rough, quantitative way, it is unlikely that such models will be able to predict accurately what is happening at a particular location in a given engine. One may thus speculate that future models could predict the correct trends by use of, say, a transient gas-jet model combined with an empirical kinetics model, but that they

References pp. 185–190

will not be very good for quantitative prediction of ignition delay without some empirical adjustment for each engine. Such a future model could possibly combine the physical model of Ref. 80 with a chemical model similar to that of Ref. 81. If indeed small pockets of richer fuel-air mixture resulting from droplet vaporization are important, a much more complex model will be necessary [82]. Droplet mixing effects may be particularly important in high-swirl engines where droplets can be swept sideways out of the spray core.

THE COMBUSTION PERIOD

As in the cases of charge preparation and ignition, the homogeneous spark-ignition engine offers the least complexity during combustion. Nevertheless, the unique unsteady character of this combustion causes many problems in both measurement and modeling. Thus the homogeneous engine will be discussed in some detail before going on to the more complex cases of stratified-charge engines.

When discussing modeling of the combustion in homogeneous SI engines we must first recall that the initial status of the gas with regard to inhomogeneity in temperature and composition is not well known, and detailed initial values can only be supplied by as yet unevaluated two-dimensional models. The same situation holds for the turbulent velocity field. We must also remember that the ignition varies from cycle to cycle, so that the combustion is not the same for each cycle, even if the spatially averaged values of turbulent intensity, composition, temperature, etc., do not vary much from cycle to cycle. Fortunately, these cyclic variations do not result in correspondingly as large variations in *imep* [52]. Thus models which predict average cycle data are useful both for predicting performance and helping to explain phenomena. It should be noted that even data-analysis models are best applied to ensemble-averaged data because of the noise in individual pressure records and the uncertainty in trapped mass for a given cycle.

Gathering ensemble-averaged data by using modern digital electronics is possible for such data as pressure and velocity. Measuring and averaging the flame-front position and shape is much more difficult [83]. Enhancement and automated methods used for analysis of satellite pictures might be adapted to engine studies but would require use of very large computers. Ion-gap data could be used [84] and more easily averaged, but the placement of the intrusive electrodes and interpretation of the data remains a problem. Another method would be the use of surface temperature transducers which, although nonintrusive, offer data limited to the surfaces of the chamber and restricted by the same problems of placement as the ion gaps. The use of cineholography [9] would appear to be the most attractive method, but, even if developed, could only be applied to special visualization engines. Location of the flame front on the reconstructed holographic images by automatic scanning and averaging would be most desirable and thus presents a stimulating challenge for optical-electronic specialists. Such a specialized and undoubtedly very expensive method would obviously be limited to use by only a few laboratories. If complete models for homogeneous engines can be formulated with the help of such detailed data, this limitation is not important. If, however, the models remain incomplete, then we must ask what easily applied methods can be used in engines where only the pressure pick-up hole or equivalent is available for transducer access. Obviously, the data obtained could only give values at an extremely limited number of spatial locations. The parameters measured can thus only provide a few critical points for evaluation of model constants or must be measures of homogeneous quantities. Although various hot-wire measurements in motored engines indicate that for chambers with no squish the turbulence is isotropic, it is doubtful that it is also homogeneous except in simple pancake-shaped chambers [85]. Thus for practical fitting of incomplete models we are left primarily with the pressure-time diagram. It is therefore worthwhile to consider carefully data-analysis models which use the pressure-time data as input before going on to discuss the various incomplete models now available.

Data-analysis models for homogeneous engines all have many assumptions in common. They assume: ideal gas, a set of thermodynamic-equilibrium control volumes, uniform pressure, and an infinitesimally thin flame [2]. In the simplest models only two control volumes are used—the products and the reactants. More complex models have divided the products control volume into a number of systems, mainly in an attempt to model nitric-oxide formation [86]. For strictly thermodynamic calculations, neglecting the temperature gradient in the product system has a negligible effect on the computed mass burning rate. The heat transfer from the control volumes to the chamber surfaces is calculated by use of one of the many empirical heat-transfer-coefficient formulas [87, 88, 89]. These formulas are based on very sparse experimental data, predict only average values at a given instant, use bulk properties for correlation and can easily be in error by 200% or more. Given such inaccuracy, it is generally advisable to calibrate the formula for the given engine by computing the total heat transfer between the time of spark and the end of combustion. This can be done by computing the mechanical work using the pressure data, computing the internal energy of the reactants at time of spark and the equilibrium products at the end of combustion, and substituting these quantities into an energy balance to find the heat transfer. The instantaneous heat-transfer-coefficient formula can then be multiplied by an appropriate constant so that the total heat transfer in the model for the combustion period will agree with the overall energy balance. This procedure does not correct the shape of the instantaneous heat-transfer curve, but comparisons with experimental data indicate that most formulas predict the shape reasonably well. Another serious problem concerning heat transfer is the phasing between the instantaneous rate at the wall surface and the rate at which the energy is removed from the bulk gas. One can assume a wall boundary-layer system which transfers heat to the wall by conduction and exchanges energy with the bulk gas by turbulent mixing. The wall heat-transfer rate can be computed using the empirical formulas discussed above, but the mixing rate between the bulk gas and boundary layer system can only be estimated. If the mixing rate is assumed zero, so that the bulk gas is adiabatic, the results for *NO* calculations indicate improved accuracy [90]. Obtaining better models for the wall heat-transfer rate and the

References pp. 185–190

phasing with bulk temperature depends on new (probably optical) methods of probing the very thin, unsteady boundary layer. Fortunately, accuracy in prediction of heat transfer is not vital to burning-rate models. Emission models are, of course, another matter.

None of the homogeneous models known to the author include energy release in the end-gas due to prereactions. Such energy release can be negligible for high-octane fuels but could be important for lower-octane fuels or near knocking situations, where 10% or more of the total heat release could be released in the end-gas [91]. Because the data analysis models use the experimental pressure, they attribute any chemical energy released in the end-gas to the burning rate.

Despite the many approximations incorporated in the data-analysis models, they can be used to evaluate the effects of design and operating parameters on burning rate. Empirical simulations of the burning rate based on such analyses can also be used to good advantage in total-cycle analysis programs.

The major use of the burning-rate analysis programs from a combustion modeling viewpoint is to determine flame speed given a model for the flame geometry. The analysis-model mass burning rate divided by the end-gas density and assumed flame area gives the apparent one-dimensional flame velocity. The area is a function of the assumed flame geometry and the product volume is computed by the analysis model. Errors associated with the usual assumption of a spherical flame front could be sizable. To take an extreme example, if the assumed flame is hemispherical while the actual flame is torn away from the plug and is spherical [58], the area would be in error by almost 60%. Such deviations could be caused by large-scale fluid motion, turbulence inhomogeneity, and wall interaction effects.

In order to progress from a data analysis to an incomplete burning-rate model, the turbulent flame speed must be predicted in addition to the flame area, or at least the product of flame speed and area must be predicted. The correlation of turbulent flame speed is, of course, at the heart of almost all practical combustion problems. Correlations for open flames and flames in ducts have been based on the wrinkled-flame concept and use the intensity and laminar flame speed as correlating variables. For higher flow rates in ducts, the inlet flow velocity has been added as a correlating variable [92, 93, 94]. Because these correlations are largely empirical, they are not proof of the validity of the wrinkled-flame concept. The role of theory in these cases is primarily to act as a guide to the appropriate mathematical form, but the wrinkled-flame hypothesis evidently leaves some ambiguity in this regard. The Damkohler [95] formula sets the sum of the laminar velocity and turbulent intensity equal to the turbulent velocity, while many other formulas set the ratio of turbulent to laminar velocity equal to a function of the intensity.

Engine combustion is quite different from the steady-flow combustion experiments. First, the flow is unsteady, so that the variation with time of both mean velocity and turbulence may be important during the lifetime of the flame. Second, the end-gas is contained, so that its compression can have an effect on its turbulent structure and composition. Third, the piston motion induces fluid motion and changes the geometry of the chamber during combustion. Considering all of these factors, it is difficult to believe that any accurate universal correlation for turbulent flame speed is likely to be found without considerably more knowledge than is currently available. It may be, however, that a correlation can be found for a given combustion chamber shape or family of chambers with similar shapes. Certainly such correlations must include the geometric factors which determine the turbulent intensity and scale. Lancaster et al. [5] found that the ratio of turbulent to laminar velocity is proportional to intensity, which, for the simple CFR chamber, was proportional to engine speed and independent of compression ratio. Their correlation, like most of those for open steady flames, indicates the flame speed is independent of turbulent scale. One would expect that this simple relationship might not always hold for more practical chamber shapes.

Another view of the turbulent flame process which has received some attention in the past [94] is that the flame is a thick zone in which lumps of reactant are entrained and burned. Ballal and Lefebvre [96] and Ballal [97] have observed three regimes of turbulent behavior in steady-flow combustors. For the most turbulent regimes, where the intensity is more than twice the laminar flame speed and the isotropic Kolmogorov microscale is less than the laminar flame thickness, the flame becomes, " . . . a fairly thick matrix of burned gas interspersed with eddies of unburned mixture" [97]. Such a description of a flame is consistent with the mechanistic models for engine flames formulated by Blizard and Keck [98] and Tabaczynski et al. [99]. In the formulation of Ref. 99 the combustion period is divided into an "ignition-delay period" followed by a turbulent flame-propagation period. The argument for this is that the initial burning takes place over a small spatial region, so that the average turbulent parameters are not at work. Rather, the model postulates the burning of an individual eddy having a size proportional to the instantaneous chamber height. The burnup time of the eddy is obtained by assuming that each microcell burns at the laminar flame speed. The Taylor microscale is correlated in terms of the turbulent Reynolds number [100], assuming isotropic turbulence. These assumptions lead to a formula of the form

$$(\Delta t \cdot S_p/h) = C(\nu/hS_p)^m(S_p/S_L)^n \tag{1}$$

where

Δt = ignition delay time
S_p = average piston speed
h = average chamber height during the time Δt
C = an empirical constant adjusted to fit each engine
ν = kinematic viscosity
S_L = laminar flame speed at the average conditions prevailing during time Δt.

The theory of Ref. 99 suggests $m = 1/3$ and $n = 2/3$. This formula is formally applied by the authors to obtain the time to burn the first one percent of mass. The formula should be compared to other theories for spark-ignition kernel growth. We note that the other theories place their emphasis on heat transfer and energy input rather than on turbulence parameters. In Ref. 5, Lancaster et al. divided the combustion period into four regimes: establishment of the critical kernel, flame development, fully developed flame travel and the termination regime. The flame-termination regime is not treated extensively in the literature except to note that wall heat-transfer effects begin to dominate, and that the flame thickness may become an important parameter. Flame quenching in engines has been modeled by several authors, mostly for purposes of predicting unburned hydrocarbon. This subject is thus left to the next session of this Symposium. It is only necessary to comment that the heat transfer during the termination period has not been established experimentally and thus requires additional work [101].

In Ref. 99 the developing and fully developed regimes are lumped together. The model used a spherical, effective flame area and a turbulent flame speed derived from the entrainment hypothesis. The ratio of integral scale to the time calculated to burn up an eddy the size of the integral scale is set proportional to the turbulent flame speed. Rapid distortion theory is used to predict the variation of intensity, u', and integral scale, L, caused by the change in charge density during burning. One form of the equation for the turbulent to laminar flame speed ratio (FSR) given in Ref. 99 is

$$\mathrm{FSR} = S_T/S_L = C(u'/S_L)^{1/3}(\rho_u/\rho_i)^{1/9}(u'L/\nu)^{1/3} \tag{2}$$

where the integral scale, L, is used to form a turbulent Reynolds number, $u'L/\nu$. The hypotheses which gave Eq. 2 may be contrasted with the approach of Lancaster et al. [5]. They argue that only turbulent eddies of diameter less than the spherical flame radius, r, can distort the flame shape and thus influence the flame velocity during its development following the establishment of the initial flame-kernel critical radius, r_c. Using the spectral distribution of turbulent energy, they speculate that the flame speed ratio as a function of flame radius, r, the fully developed flame speed ratio, FSR_m, and the flame radius at the start of the fully developed period, r_m, is given by

$$\mathrm{FSR} = \mathrm{FSR}_m(r/r_m)^{(m-1)/2} \tag{3}$$

where

$$r_c \leq r \leq r_m$$

The number m lies between $^5/_3$ and 2, so that the exponent in Eq. 3 lies between $^1/_3$ and $\frac{1}{2}$. Ref. 5 gives FSR_m proportional to u', which in turn is proportional to the average intake-volume flow rate. Thus we may take

$$\mathrm{FSR} = C\,\epsilon\, B^2 S_p(r/r_m)^{(m-1)/2} \tag{4}$$

The value of r_m was found to be about twice the clearance volume height, so that $r_m \sim 2h$ near TDC. The development regime thus accounts for about 10% of the mass burned and about one third of the combustion duration. Direct comparison of Eqs. 2 and 4 is best done by application to a given engine where extensive data are available. One can see, however, that Ref. 5 gives a linear dependence of FSR on engine speed while Ref. 99 gives FSR proportional to speed to the $^2/_3$ power. Interestingly, both papers support their respective correlations with engine data. In the comparison of Ref. 99 the speed was varied from 1000 to 3000 r/min with fixed spark timing, giving a crank angle duration which went from approximately 40° to 60°, thus giving 60/40 = 1.5 as compared to $3^{1/3}$ = 1.44. Such an increase in crank angle duration with speed is not commonly found, however [102]. Rather, the crank angle duration for fixed spark timing is generally found to be essentially constant, with only a slight increase with speed that is proportional to the engine speed, N, raised to about the 0.15 power.

In Ref. 99 the model results are compared to engine data for three different engines. The comparisons are impressively encouraging. Nevertheless one must question the general hypothesis which was used in the model. Using the CFR engine data of Refs. 5 and 12, one obtains the approximate parameter values shown in Table 1. The flame thickness is unknown, but the clearance height in the cylinder was 14.8 mm, so that the flame might be expected to be of similar thickness, perhaps slightly less than 10 mm. Estimating the Kolmogorov scale, η, using the data of Table 1 and $\nu = 5 \cdot 10^{-6} m^2/s$: $\eta \sim 0.02$ mm for all of the cases shown.

Ballal [97] studied a steady-flow stoichiometric premixed propane-air flame at 0.2 atm. pressure. Turbulence was created by a grid. The values of mean velocity, intensity, integral scale and microscale were all within the ranges shown in Table 1. His values of η were much larger, however, because of the much lower pressure. The ratio of intensity to laminar flame speed for the engine case is relatively low because of the high value of S_L for engine conditions. Ballal indicates that if $u'/S_L < 2$ and $\delta_L < \eta$, the flame behaves as a wrinkled flame (his region 1). It is difficult to determine the laminar flame thickness corresponding to engine conditions, but taking the engine value equal to one tenth the laboratory value should be representative. This gives $\delta_L \sim 0.02$ mm, or $\delta_L \sim \eta$, for the engine case.

Complicating the picture even further is the increase in turbulent intensity caused by flame-generated turbulence. Considering the proportionality between FSR and flame-generated intensity and the large values of FSR found in engines,

TABLE 1

Combustion Parameters (Refs. 5 and 12)
CFR Engine, 8.72 Compression Ratio

MBT Spark Timing, Values at 50% Mass Burned
for Shrouded (S) and Nonshrouded (NS) Intake Valves

Valve	N, r/min	S_T,m/s	FSR_m	S_L,m/s	u',m/s	u'/S_L	λ,mm
NS	1000	6.0	2.25	2.7	0.91	0.34	0.6
NS	1500	9.0	2.75	3.3	1.70	0.52	0.8
NS	2000	12.0	3.25	3.7	2.11	0.57	0.7
S	1000	12.5	3.8	3.3	2.73	0.83	1.0
S	1500	16.0	4.6	3.5	3.76	1.07	1.2
S	2000	20.0	5.8	3.4	4.96	1.46	1.7

References pp. 185–190

one would expect a rather significant flame-induced increase in u'. Tabaczynski et al. [99] propose that based on rapid distortion theory

$$u' = u_i'(\rho_u/\rho_i)^{1/3} \tag{5}$$

For a typical engine condition $\rho_u/\rho_i \sim 3$ at the end of combustion, so that $u'/u_i' \sim 1.44$. This increase is in the same range as that found for steady-flow combustors.

The major problem with both the wrinkled flame and the entrainment mechanistic models of the flame is that they are based on average values of scale and intensity. The real flame responds to the spectrum of turbulent eddy sizes, perhaps by engulfing some and by becoming wrinkled because of others. Promising new methods for modeling such effects are just evolving [103].

In both types of models the one-dimensional laminar flame speed is applied to either the flame front or the eddy burnup. Laminar flame speeds are not well established for engine conditions [104], but even if they were, they would be suspect in this application.

In summary, a great deal more data of a detailed nature is needed in order to sort out the various mechanistic turbulent flame models. Until such data become available, it would seem that empirical dimensionless-group models which incorporate easily measured quantities (as opposed to turbulent parameters) should be developed. Such formulas will certainly require the laminar flame speed as a useful input. Thus work on obtaining values of S_L at engine conditions should be encouraged. At the same time, experimental work on the influence of scale and intensity is much needed. One type of useful experiment might be to change the turbulence parameters artificially by use of a jet as in the work of Evers [105]. Such variation over a wide range of measured turbulence parameters, combined with accurate mass burning rate analysis, could give a better understanding of homogeneous flames in engines.

Before leaving the subject of homogeneous spark-ignition engines it should be noted that even if the model for normal combustion were complete, the model for end-gas reactions leading to knock remains to be resolved. The problem of end-gas uncontrolled autoignition severely limits both engine design and fuel composition. As already noted, end-gas reaction can be appreciable even in non-knocking situations. Although an enormous effort has been expended on the study of cool flames, induction kinetics and knocking phenomena, the elementary reaction scheme for the end-gas is unknown. Recent spectroscopic measurement by Abata [106] of OH radical concentration in the end-gas suggests that the reaction scheme proposed by Trumpy [107] is quantitatively inaccurate. The work of Kirsch and Quinn points to a rational method of approaching an incomplete model for knocking [81]. Because an additional paper by these authors follows this overview, their work will not be discussed further here, except to point out that such analysis, while practical, cannot lead to the complete model needed for understanding detailed fuel effects.

The increase in required octane number with engine age is closely connected to changes in heat transfer caused by deposits. The deposit heat-transfer problem is virtually untouched despite the ubiquity of heat-transfer modelers. The problem is particularly intriguing when one considers porous deposits. Prediction of deposit formation rate and properties and their effect on octane-number increase for various chamber designs is thus still an engineering art.

The mechanistic flame models discussed above have the appeal of relatively easy and fast calculation. An alternative method is the direct application of the equations of change with appropriate empirical or incomplete models for turbulence and chemical kinetics [8, 108]. The power of these models comes in their ability to calculate the flow field. The current status of chemical kinetics for practical fuel combustion greatly limits the representation of the reaction zone, so that detailed calculation results can only be viewed as heuristic. If indeed the effects of large scale (bulk) motion on the flame are small and if the spherical flame area is good to within say 10% of the true value, then the potential power of these methods is lost on the homogeneous SI problem. This is not to say, however, that all engine problems are of this sort. If jet ignition, stratification and/or fuel injection are brought into the design, the fluid bulk motion becomes important or even dominates. In those cases thermodynamic models are inadequate, so that we must either be content with empirical or very incomplete models or turn to numerical solutions of the equation of change.

An excellent example of an engine design for which bulk fluid motion is very important is the three-valve engine as exemplified by the Honda CVCC. For some designs of this engine, thermodynamic models based on two zones for each chamber may suffice [109, 110, 111]. Such models then provide a very useful way to sort out the complex interactions among the parameters. For many other possible designs, however, the jet action of the fluid flowing out of the prechamber orifice is important [112]. For extreme cases, the jet can become essentially a line source of ignition. Models which attempt to include such fluid motions can, of course, be based on quasi-steady gas-jet equations, but the ability to model the leading portion of the jet, its interaction with the chamber fluid, and the reverse flow back into the prechamber is severely restricted. Models based on computational fluid mechanics are much more promising [113, 114, 115] but still primitive. Evaluations of the models, even with bomb data, are scanty.

As previously discussed, detailed models for evaporation and mixing of fuel jets in engines are just emerging. Extension of such modeling methods to include combustion is beyond current computer capability without the adoption of very severe simplifications. As a result, models for diesel and other fuel-injected engines have been developed by dividing the chamber into a number of thermodynamic control volumes. The connection between the fuel-air mixture in each volume and the spray has been made by use of gas-jet mixing equations [33, 80, 116, 117] or by a combination of the gas-jet equations and droplet-vaporization equations [35, 36]. In a recent limited comparison [118] between data and the models of Refs. 35, 36, 80, 116, the Cummins model [80] gave the best overall result, but the models of Hiroyasu and Kadota and Kau et al. also predicted general trends reasonably well. The Cummins model incorporated empirically adjusted jet equations while the later two models use mechanistic equations with undetermined parameters. It is of course possible that proper adjustment of these parameters would have produced considerably better agreement with experiment.

One may thus conclude that at the current state of modeling for diesels, considerable empirical input is required, and that one should not expect models developed for one engine family necessarily to work for other engines. In particular, the nature of the combustion may vary widely in fuel-air ratio. Quiescent chambers may develop very rich zones with mixing-controlled combustion, while engines with considerable swirl may show nearly stoichiometric combustion [119]. The importance of droplet-promoted mixing and droplet (or vapor) centers imbedded in flame zones also probably differs considerably between engines. The reason for the success of models which ignore droplet phenomena [80] is obscured by the fact that the equation parameters were obtained empirically.

Because detailed models of direct-injection combustion leave so many open questions, the use of data-analysis, empirical and highly simplified incomplete models has been the route most often followed by designers [1, 120, 121]. Such heat-release models are useful in cycle-analysis programs [122, 123]. They suffer inaccuracy in providing an accurate energy balance, however, because of the effects of inhomogeneity on the internal energy and the poor accuracy of existing heat-transfer correlations. Prominent among the problems of calculating the heat transfer is the very large radiation component caused by carbon-particle radiation during burning [89, 124, 125]. Again, an energy balance using pressure-time data can give the total heat transferred during the combustion period and thus provide a means of empirical adjustment.

SUMMARY

In looking at the many unanswered questions, our ignorance of detailed chemical kinetics, the complex nature of unsteady-flow turbulence, and the limitations placed on even two-dimensional models by computer speed, it would be easy to overlook the tremendous growth and use of engine modeling during the past twenty years. Detailed cycle analysis has grown in use to the point where it can and is used routinely to predict trends in engine parameters, match components, and evaluate new engine-system concepts. Data-analysis models have, through the advent of improved transducers and data processing, allowed consistent calculation of burning rates in homogeneous engines. Such burning-rate data coupled with one-dimensional models have produced models capable of predicting effects of spark plug location and chamber geometry variation. Although as yet not predictive, the use of two-dimensional computational fluid mechanics offers the ability to analyze the complex fluid motions in engines. In short, modeling has come a long way despite our ignorant state.

In this overview I have tried to point out the many areas where both theory and experiments are needed. The question still remains if there is any new thrust in modeling which is currently lacking. Study of the modeling techniques discussed throughout this overview show that a large number of models use the equations of mass and energy conservation but neglect the momentum equation. These models usually also incorporate some mechanistic model for burning rate and/or mixing. Because such models work with a limited number of adjustable parameters, they can easily utilize experimental data to improve the model. Whole sections of these models may, in fact, be empirical submodels. The weaknesses of these models are their inability to produce spatial resolution and to resolve the coupling between the flow field and the energy release. The foregoing weaknesses are in direct contrast to those models which incorporate the momentum equation. Computational fluid-mechanics models, however, need fundamental data or detailed models for kinetics, turbulence and transport properties. These differences are very familiar. They are very similar to the difference between using a convective heat-transfer coefficient and solving the boundary-layer equations.

What is lacking, and it would appear is needed, are hybrid models which can couple the fluid motion to empirical models for reaction zones or jets. Some of this sort of thing can be easily done with existent computational fluid-mechanics models. For example, if one applies a two-dimensional model to the x and y directions, the heat transfer and friction at the $z = 0$ and $z = l$ ends of the computational elements can be included by use of empirical formulas. Similarly, empirical or integral boundary-layer methods could be applied at the containing surfaces. The extention of this idea to coupling with a mechanistic reaction-zone model is more difficult. The problems may be similar to those of adaptive grid programming, but with the additional concern of providing compatible conditions at the interface between the models.

Whatever the modeling technique, it is important that more-detailed models be evaluated by equally detailed experiments. Because the reciprocating engine environment is unique, this means data taken directly from engines. Modelers must thus be ready to say what measurements will provide critical tests and to propose experiments required to answer specific questions. With such cooperation the modeling community may yet produce a complete model of a combustion engine before the last drop of petroleum fuel is burned.

REFERENCES

1. *J. Shipinski, O. A. Uyehara and P. S. Myers, "Experimental Correlation Between Rate-of-Injection and Rate-of-Heat Release in a Diesel Engine," ASME Paper No. 68-DGP-11, 1968.*
2. *R. B. Krieger and G. L. Borman, "The Computation of Apparent Heat Release for Internal Combustion Engines," ASME Paper No. 66-WA/DGP-4, 1966.*
3. *G. G. Lucas and E. H. James, "A Computer Simulation of a Spark Ignition Engine," SAE Paper No. 730053, 1973.*
4. *B. S. Samaga and B. S. Murthy, "Investigation of a Turbulent Flame Propagation Model for Application for Combustion Prediction in the S. I. Engine," SAE Paper No. 760758, 1976.*
5. *D. R. Lancaster, R. B. Krieger, S. C. Sorenson and W. L. Hull, "Effects of Turbulence on Spark-Ignition Engine Combustion," SAE Trans., Vol. 85, Paper No. 760160, pp. 689–710, 1976.*
6. *R. S. Benson and P. C. Baruah, "Performance and Emission Predictions for a Multi-cylinder Spark Ignition Engine," Proc. I. Mech. E., Vol. 191, 32/77, pp. 339–354, 1977.*
7. *S. D. Hires, R. J. Tabaczynski and J. M. Novak, "The Prediction of Ignition Delay and Combustion Intervals for a Homogeneous Charge, Spark Ignition Engine," SAE Paper No. 780232, 1978.*
8. *A. A. Boni, "Numerical Simulation of Flame Propagation in IC Engines—A Status Report," SAE Paper No. 780316, 1978.*
9. *J. D. Trolinger, H. T. Bentley, A. E. Lennert and R. E. Sowls, "Application of Electro-Optical Techniques in Diesel Engine Research," SAE Trans., Vol. 83, Paper No. 740125, pp. 633–643, 1974.*

10. P. Hutchinson, A. Morse and J. H. Whitelaw, "Velocity Measurements in Motored Engines: Experience and Prognosis," SAE Paper No. 780061, 1978.
11. S. Lederman, "Laser Based Diagnostic Techniques for Combustion Research," Prog. in Astronautics and Aeronautics, AIAA, Vol. 58, "Turbulent Combustion," Ed. L. A. Kennedy, Chapter III, 1978.
12. D. R. Lancaster, "Effects of Engine Variables on Turbulence in Spark-Ignition Engines," SAE Trans., Vol. 85, Paper No. 760159, pp. 671–688, 1976.
13. B. D. Peters and A. A. Quader, "Wetting the Appetite of Spark-Ignition Engines for Lean Combustion," SAE Paper No. 780234, 1978.
14. S. Prakash and W. A. Sirignano, "Liquid Fuel Droplet Heating with Internal Circulation," Int. J. of Heat and Mass Transfer, Vol. 21, pp. 885–895, 1978.
15. J. B. McVey, J. B. Kennedy, F. K. Owen and C. T. Bowman, "Diagnostic Techniques for Measurements in Burning Sprays," Presented at 1976 Fall Meeting of Western States Section, The Combustion Institute, LaJolla, Calif., October 1976.
16. A. D. Gosman and R. J. R. Johns, "Development of a Predictive Tool for In-Cylinder Gas Motion in Engines," SAE Paper No. 780315, 1978.
17. R. D. Reitz, "Atomization and Other Breakup Regimes of a Liquid Jet," Princeton University, Ph.D. Thesis, Dept. of Aerospace and Mechanical Sciences, 1978.
18. D. W. Lee, "Experiments on the Distribution of Fuel in Fuel Sprays," NACA Report No. 438, 1932.
19. K. J. DeJuhasz, O. F. Zahn and P. H. Schweitzer, "On Formation and Dispersion of Oil Sprays," Penna. State Univ., Engr. Expt. Station Bull. No. 40, August 1932.
20. N. Hay and P. L. Jones, "Comparisons of the Various Correlations for Spray Penetration," SAE Paper No. 720776, 1972.
21. C. J. Morris and J. C. Dent, "The Simulation of Air-Fuel Mixing in High Swirl Open Chamber Diesel Engines," Proc. I. Mech. E., Vol. 190, 47/76, pp. 503–513, 1976.
22. J. Rife and J. B. Heywood, "Photographic and Performance Studies of Diesel Combustion with a Rapid Compression Machine," SAE Trans., Vol. 83, Paper No. 740948, pp. 2942–2961, 1974.
23. G. N. Abramovich, "The Theory of Turbulent Jets," MIT Press, 1963.
24. D. W. Lee, "The Effect of Nozzle Design and Operating Conditions on the Atomization and Distribution of Fuel Sprays, NACA Report No. 425, pp. 505–521, 1932.
25. H. Hiroyasu and T. Kadota, "Fuel Droplet Size Distribution in Diesel Combustion Chamber," SAE Trans., Vol. 83, Paper No. 740715, pp. 2615–2624, 1974.
26. G. L. Borman and J. H. Johnson, "Unsteady Vaporization Histories and Trajectories of Fuel Drops Injected into Swirling Air," SAE Paper 598C, 1962. (Also in, "Burning a Wide Range of Fuels in Diesel Engines," Progress in Technology, Vol. 11, pp. 13–29, 1961).
27. V. Pirouz-Panah and T. J. Williams, "The Influence of Droplets on the Properties of Liquid Fuel Jets," Proc. I. Mech. E., Vol. 191, 28/77, pp. 299–306, 1977.
28. L. C. Haselman and C. K. Westbrook, "A Theoretical Model for Two-Phase Fuel Injection in Stratified Charge Engines," SAE Paper No. 780318, 1978.
29. A. A. Ranger and J. A. Nicholls, "Aerodynamic Shattering of Liquid Drops," AIAA J., Vol. 7, No. 2, pp. 285–290, February 1969.
30. R. D. Ingebo, "Drag Coefficients for Droplets and Solid Spheres in Clouds Accelerating in Airstreams," NACA TN 3762, 1956.
31. E. Rabin, A. R. Schallenmuller and R. B. Lawhead, "Displacement and Shattering of Propellant Drops," Rocketdyne AFO SR TR 60-75, 1960.
32. M. C. Yuen and L. W. Chen, "On Drag of Evaporating Liquid Droplets," Comb. Sci. and Tech., Vol. 14, pp. 147–154, 1976.
33. S. M. Shahed, W. S. Chiu and W. T. Lyn, "A Mathematical Model of Diesel Combustion," I. Mech. E., Conf. on Combustion in Engines, C94/75, Cranfield, England, pp. 119–128, 1975.
34. C. M. Vara Prasad and Subir Kar, "An Investigation on the Diffusion of Momentum and Mass of Fuel in a Diesel Fuel Spray," ASME Paper No. 76-DGP-1, 1976.
35. H. Hiroyasu and T. Kadota, "Models for Combustion and Formation of Nitric Oxide and Soot in Direct Injection Diesel Engines," SAE Trans., Vol. 85, Paper No. 760129, pp. 513–526, February 1976.

36. C. J. Kau, M. P. Heap, T. J. Tyson and R. P. Wilson, "The Prediction of Nitric Oxide Formation in a Direct Injection Diesel Engine," Sixteenth Symposium (International) on Combustion, The Combustion Institute, Pittsburgh, Pennsylvania, pp. 337-350, 1976.
37. M. M. El-Wakil, P. S. Myers and O. A. Uyehara, "Fuel Vaporization and Ignition Lag in Diesel Combustion," SAE Transactions, Vol. 64, pp. 712-729, 1956.
38. J. A. Manrique, "Theory of Droplet Vaporization in the Region of the Thermodynamic Critical Point," NASA CR-72558, June 1969.
39. C. W. Savery, D. L. Juedes and G. L. Borman, "n-Heptane, Carbon Dioxide, and Chlorotrifluoromethane Droplet Vaporization Measurements at Supercritical Pressures," Ind. Eng. Chem. Fundam., Vol. 10, No. 4, pp. 543-553, 1971.
40. T. Kadota and H. Hiroyasu, "Evaporization of a Single Droplet at Elevated Pressures and Temperatures," Trans. JSME, Vol. 42, No. 356, pp. 1216-1223, 1976.
41. G. M. Faeth, "Current Status of Droplet and Liquid Combustion," Prog. Energy Combust. Sci., Vol. 3, pp. 191-224, 1977.
42. A. Williams, "Fundamentals of Oil Combustion," Prog. Energy Combust. Sci., Vol. 2, pp. 167-179, 1976.
43. R. E. Sowls, "An Experimental Study of Carbon Dioxide Droplets Falling Through an Inert High Pressure High Temperature Environment," Ph.D. Thesis, Dept. of Mech. Engr., University of Wisconsin, Madison, 1972.
44. A. J. Shearer and G. M. Faeth, "Combustion of Liquid Sprays at High Pressures," NASA CR-135210, 1977.
45. C. K. Westbrook, "Three Dimensional Numerical Modeling of Liquid Fuel Sprays," Sixteenth Symposium (International) on Combustion, The Combustion Institute, Pittsburgh, Pennsylvania, pp. 1517-1526, 1976.
46. A. A. Quader, "Lean Combustion and the Misfire Limit in Spark Ignition Engines," SAE Trans., Vol. 83, Paper No. 741055, pp. 1517-1526, 1976.
47. R. J. Craver, R. S. Podiak and R. D. Miller, "Spark Plug Design Factors and Their Effect on Engine Performance," SAE Trans., Vol. 79, Paper No. 700081, pp. 229-239, 1970.
48. R. R. Burgett, J. M. Leptich and K. V. S. Sangwan, "Measuring the Effect of Spark Plug and Ignition System Design on Engine Performance," SAE Trans., Vol. 81, Paper No. 720007, pp. 48-66, 1972.
49. T. W. Ryan, III, S. S. Lestz and W. E. Meyer, "Extension of the Lean Misfire Limit and Reduction of Exhaust Emissions of an SI Engine by Modification of the Ignition and Intake Systems," SAE Paper No. 740105, 1974.
50. J. P. Walters, "Spark Discharge: Application to Multielement Spectrochemical Analysis," Science, Vol. 198, pp. 787-797, November 1977.
51. D. J. Patterson, "Cylinder Pressure Variations, A Fundamental Combustion Problem," SAE Trans., Vol. 75, Paper No. 660129, pp. 621-632, 1966.
52. B. D. Peters and G. L. Borman, "Cyclic Variations and Average Burning Rates in a SI Engine," SAE Paper No. 700064, 1970.
53. L. W. Evers, "Nitrogen Oxide Control with the Delayed Mixing Stratified Charge Engine Concept," Ph.D. Thesis, University of Wisconsin, Madison, 1976.
54. D. E. Cole and W. Mirsky, "Mixture Motion—Its Effect on Pressure Rise in a Combustion Bomb: A New Look at Cyclic Variation," SAE Trans., Vol. 77, Paper No. 680766, pp. 2993-3007, 1968.
55. G. A. Harrow and P. L. Orman, "A Study of Flame Propagation and Cyclic Dispersion in a Spark-Ignition Engine," Adv. in Auto. Eng., Part IV, Pergamon Press, Oxford, 1966.
56. P. O. Witze, "Hot Wire Turbulence Measurements in a Motored Internal Combustion Engine," Second European Symposium on Combustion, Orleans, France, September 1975.
57. R. K. Barton, S. S. Lestz and W. E. Meyer, "An Empirical Model for Correlation of Cycle-by-Cycle Cylinder Gas Motion and Combustion Variations of an SI Engine," SAE Trans., Vol. 80, Paper No. 710163, pp. 695-707, 1971.
58. K. Iinuma and Y. Iba, "Studies of Flame Propagation Process," JARI Tech. Memo, No. 10, 1972.
59. G. G. DeSoete, "The Influence of Isotropic Turbulence on the Critical Ignition Energy,"

Thirteenth Symposium (International) on Combustion, The Combustion Institute, Pittsburgh, Pennsylvania, pp. 735–743, 1973.
60. *R. E. Winsor and D. J. Patterson, "Mixture Turbulence—A Key to Cyclic Combustion Variation," SAE Trans., Vol. 82, Paper No. 730086, pp. 368–383, 1973.*
61. *A. K. Oppenheim, K. Teichman, K. Holm and H. E. Stewart, "Jet Ignition of an Ultra-Lean Mixture," SAE Paper No. 780637, 1978.*
62. *L. A. Gussak, "High Chemical Activity of Incomplete Combustion Products and a Method of Prechamber Torch Ignition for Avalanche Activation of Combustion in Internal Combustion Engines," SAE Trans., Vol. 84, Paper No. 750890, pp. 2421–2445, 1975.*
63. *K. V. L. Rao and A. H. Lefebvre, "Minimum Ignition Energies in Flowing Kerosene-Air Mixture," Comb. and Flame, Vol. 27, pp. 1–20, 1976.*
64. *C. E. Polymeropoulos and V. Sernas, "Ignition and Propagation Rates for Flames in a Fuel Mist," Report No. FAA-RD-76-31, U.S. Dept. of Trans., November 1976.*
65. *Y. Mizutani and A. Nakajima, "Combustion of Fuel Vapor-Drop-Air Systems: Part II—Spherical Flames in a Vessel," Comb. and Flame, Vol. 20, pp. 351–357, 1973.*
66. *Y. Mizutani and S. Matsushita, "Fuel Vapor-Spray-Air Mixture Operation of a Spark-Ignition Engine," Comb. Sci. and Tech., Vol. 8, pp. 85–94, 1973.*
67. *R. E. Canup, "The Texaco Ignition System—A New Concept for Automotive Engines," SAE Paper No. 750345, 1975.*
68. *A. Simko, M. A. Choma and L. L. Repko, "Exhaust Emissions Control by the Ford Programmed Combustion Process—PROCO," SAE Trans., Vol. 81, Paper No. 720052, pp. 249–264, 1972.*
69. *T. Ayusawa, S. H. Jo, T. Nemoto and Y. Koo, "Relationship Between Local Air- Fuel Ratio and Combustion Character in a Spark Ignition Engine," SAE Paper No. 780147, 1978.*
70. *J. C. Guibet and A. Duval, "New Aspects of Preignition in European Automotive Engines," SAE Trans., Vol. 81, Paper No. 720114, pp. 399–417, 1972.*
71. *M. M. El-Wakil and M. I. Abdou, "The Self Ignition of Fuel Drops in Heated Air Streams," Fuel, Vol. XLV, pp. 177–205, May 1966.*
72. *G. M. Faeth and D. R. Olson, "Ignition of Hydrocarbon Fuel Droplets in Air," SAE Trans., Vol. 77, Paper No. 680465, pp. 1793–1802, 1968.*
73. *N. A. Henein, "A Mathematical Model for the Mass Transfer and Combustible Mixture Formation Around Fuel Droplets," SAE Paper No. 710221, 1971.*
74. *P. Sunn Pedersen and B. Ovale, "A Model for the Physical Part of the Ignition Delay in a Diesel Engine," SAE Trans., Vol. 83, Paper No. 740716, pp. 2625–2638, 1974.*
75. *J. J. Sangiovanni and A. S. Kesten, "A Theoretical and Experimental Investigation of the Ignition of Fuel Droplets, Comb. Sci. and Tech., Vol. 16, pp. 59–70, 1977.*
76. *J. J. Sangiovanni and A. S. Kesten, "Effect of Droplet Interaction on Ignition in Monodispersed Droplet Streams," Sixteenth Symposium (International) on Combustion, The Combustion Institute, Pittsburgh, Pennsylvania, pp. 577–592, 1976.*
77. *H. H. Wolfer, "Ignition Lag in the Diesel Engine," VDI Forschaft, No. 392, Vol. 15, 1938.*
78. *J. H. Shipinski, P. S. Myers and O. A. Uyehara, "A Spray Droplet Model for Diesel Combustion," I. Mech. E., Symposium on Diesel Engine Combustion, 1970.*
79. *N. A. Henein and J. A. Bolt, "Correlation of Air Charge Temperature and Ignition Delay for Several Fuels in a Diesel Engine," SAE Paper No. 690252, 1969.*
80. *W. S. Chiu, S. M. Shahed and W. T. Lyn, "A Transient Spray Mixing Model for Diesel Combustion," SAE Trans., Vol. 85, Paper No. 760128, pp. 502–512, 1976.*
81. *L. J. Kirsch and C. P. Quinn, "A Fundamentally Based Model of Knock in the Gasoline Engine," Sixteenth Symposium (International) on Combustion, The Combustion Institute, Pittsburgh, Pennsylvania, pp. 233–244, 1976.*
82. *M. Labowsky and D. E. Rosner, "Conditions for Group Combustion of Droplets in Fuel Clouds; I, Quasi-steady Predictions," Symposium on Evaporation and Combustion of Fuel Droplets, San Francisco, ACS, Div. Petro-Chem., pp. 663–675, 1976.*
83. *G. M. Rassweiler and L. Withrow, "Motion Pictures of Engine Flames Correlated with Pressure Cards," SAE J., Vol. 33, pp. 185–204, May 1938.*
84. *S. Curry, "A Three-Dimensional Study of Flame Propagation in an SI Engine," SAE Trans., Vol. 71, pp. 628–650, 1963.*

85. E. S. Seminov, "Studies of Turbulent Gas Flow in Piston Engines," NASA Technical Translation F-97, 1963.
86. G. A. Lavoie, J. B. Heywood and J. C. Keck, "Experimental and Theoretical Study of Nitric Oxide Formation in Internal Combustion Engines," Comb. Sci. and Tech., Vol 1, pp. 313–326, 1970.
87. W. J. D. Annand and T. A. Ma., "Instantaneous Heat Transfer Rates to the Cylinder Head Surface of a Small Compression Ignition Engine," Proc. I. Mech. E., Vol. 185, 72/71, pp. 976–987, 1970-71.
88. G. Woschni, "A Universally Applicable Equation for the Instantaneous Heat Transfer Coefficient in the Internal Combustion Engine," SAE Trans., Vol. 76, Paper No. 670931, pp. 3065–3083, 1967.
89. J. C. Dent, "Convective and Radiative Heat Transfer in a High Swirl Open Chamber Diesel Engine," SAE Trans., Vol. 86, Paper No. 770407, pp. 1758–1783, 1977.
90. G. A. Lavoie and P. N. Blumberg, "A Fundamental Model for Predicting Emissions and Fuel Consumption for the Conventional Spark-Ignition Engine," Eastern States Section, Combustion Institute, Hartford, Conn., November 1977.
91. J. H. Johnson, P. S. Myers and O. A. Uyehara, "End-Gas Temperatures, Pressures, Reaction Rates, and Knock," SAE Paper No. 650505, 1965.
92. A. H. Lefebvre and R. Reid, "The Influence of Turbulence on the Structure and Propagation of Enclosed Flames," Comb. and Flame, Vol. 10, pp. 355–366, 1966.
93. J. M. Beer and N. A. Chigier, "Combustion Aerodynamics," Halsted Press, N.Y., 1972.
94. G. E. Andrews, D. Bradley and S. B. Lwakabamba, "Turbulence and Turbulent Flame Progagation—A Critical Appraisal," Comb. and Flame, Vol. 24, pp. 285–304, 1975.
95. G. Damkohler, "The Effect of Turbulence on the Flame Velocity in Gas Mixtures," Z. Elektrochem, Vol. 46; NACA Tech. Memo 1112, 1947.
96. D. R. Ballal and A. H. Lefebvre, "The Structure and Propagation of Turbulent Flames," Proc. Roy. Soc. (London), A344, pp. 217–234, 1975.
97. D. R. Ballal, "An Experimental Study of Flame Turbulence," Sixth (International) Colloquium on Gas Dynamics of Explosions and Reactive Systems, Stockholm, Sweden, 1977.
98. N. C. Blizard and J. C. Keck, "Experimental and Theoretical Investigation of Turbulent Burning Model for Internal Combustion Engines," SAE Trans., Vol. 83, Paper No. 740191, pp. 846–864, 1974.
99. R. J. Tabaczynski, C. R. Ferguson and K. Radhakrishnan, "A Turbulent Entrainment Model for Spark-Ignition Engine Combustion," SAE Trans., Vol. 86, Paper No. 770647, pp. 2414–2433, 1977.
100. H. Tennekes and J. L. Lumley, "A First Course in Turbulence," MIT Press, Cambridge, Mass., 1972.
101. J. K. Kilham and M. R. I. Purvis, "Heat Transfer From Normally Impinging Flames," Comb. Sci. and Tech., Vol. 18, pp. 81–90, 1978.
102. E. F. Obert, "Internal Combustion Engines and Air Pollution," Intext Educational Publishers, N.Y., 1973.
103. D. B. Spalding, "Development of the Eddy-Breakup Model of Turbulent Combustion," Sixteenth Symposium (International) on Combustion, The Combustion Institute, Pittsburgh, Pennsylvania, pp. 1657–1663, 1976.
104. G. A. Lavoie, "Correlations of Combustion Data for S.I. Engine Calculations-Laminar Flame Speed, Quench Distance and Global Reaction Rates," SAE Paper No. 780229, 1978.
105. L. W. Evers, P. S. Myers and O. A. Uyehara, "A Search for a Low Nitric Oxide Engine," SAE Paper No. 741172, 1974.
106. D. L. Abata, "Spectroscopic Investigation of Hydroxyl Radical Formation in the End Gases of a Spark Ignited Engine Utilizing a Dye Laser," Ph.D. Thesis, University of Wisconsin, Madison, 1977.
107. D. R. Trumpy, O. A. Uyehara and P. S. Myers, "Kinetics of Ethane in a Spark Ignition Engine," SAE Trans., Vol. 78, Paper No. 690518, pp. 1849–1874, 1969.
108. F. V. Bracco, "Introducing a New Generation of More Detailed and Informative Combustion Models," SAE Paper No. 741174, 1974.

109. G. C. Davis, R. B. Krieger and R. J. Tabaczynski, "Analysis of the Flow and Combustion Processes of a Three-Valve Stratified Charge Engine with a Small Prechamber," SAE Trans., Vol. 83, Paper No. 741170, pp. 3534–3550, 1974.
110. T. Asanuma, M. K. G. Babu and S. Yagi, "Simulation of the Thermodynamic Cycle of a Three-Valve Stratified Charge Engine," SAE Paper No. 780319, 1978.
111. J. C. Wall, J. B. Heywood and W. A. Woods, "Parametric Studies Using a Cycle-Simulation Model on Performance and NOx Emission of the Prechamber Three-Valve Stratified Charge Engine," SAE Paper No. 780320, 1978.
112. R. B. Krieger and G. C. Davis, "The Influence of the Degree of Stratification on Jet-Ignition Engine Emission and Fuel Consumption," I. Mech. E., Conf. on Stratified Engines, London, pp. 109–119, 1976.
113. F. V. Bracco, H. C. Gupta, L. Krishnamurthy, D. A. Santavicca, R. L. Steinberger and V. Warshaw, "Two-Phase, Two Dimensional, Unsteady Combustion in Internal Combustion Engines, Preliminary Theoretical-Experimental Results, SAE Paper No. 760114, 1976.
114. A. A. Boni, M. Chapman, J. L. Cook and G. P. Schneyer, "Computer Simulation of Combustion in a Stratified Charge Engine," Sixteenth Symposium (International) on Combustion, The Combustion Institute, Pittsburgh, Pennsylvania, pp. 1527–1541, 1976.
115. H. C. Gupta, R. L. Steinberger and F. V. Bracco, "Combustion in a Divided Chamber, Stratified-Charge Reciprocating Engine: Comparisons of Calculated and Measured Flame Propagation," Seventeenth Symposium (International) on Combustion, The Combustion Institute, Pittsburgh, Pennsylvania, 1978.
116. I. M. Khan, G. Greeves and C. H. T. Wang, "Factors Affecting Smoke and Gaseous Emissions from Direct Injection Engines and a Method of Calculation," SAE Trans., Vol. 82, Paper No. 730169, pp. 687–709, 1973.
117. M. Meguerdichian and N. Watson, "Prediction of Mixture Formation and Heat Release in a Diesel Engine," SAE Paper No. 780225, 1978.
118. I. A. Voiculescu and G. L. Borman, "An Experimental Study of Diesel Engine Cylinder-Averaged NOx Histories," SAE Paper No. 780228, 1978.
119. K. T. Rhee, P. S. Myers and O. A. Uyehara, "Time and Space-Resolved Species Determination in Diesel Combustion Using Continuous-Flow Gas-Sampling," SAE Paper No. 780226, 1978.
120. N. D. Whitehouse and B. K. Sareen, "Prediction of Heat Release in a Quiescent Chamber Diesel Engine Allowing for Fuel/Air Mixing," SAE Paper No. 740084, 1974.
121. N. D. Whitehouse and N. Baluswamy, "Calculations of Gaseous Products During Combustion in a Diesel Engine Using a Four Zone Model," SAE Paper No. 770410, 1977.
122. R. S. Benson, "A Comprehensive Digital Computer Program to Simulate a Compression Ignition Engine Including Intake and Exhaust Systems," SAE Paper No. 710173, 1971.
123. E. E. Streit and G. L. Borman, "Mathematical Simulation of a Large Turbocharged Two-Stroke Diesel Engine," SAE Trans., Vol. 80, Paper No. 710176, pp. 733–768, 1971.
124. T. K. Kunimoto, K. Matsuoka and T. Oguri, "Prediction of Radiative Heat Flux in a Diesel Engine," SAE Trans., Vol. 94, Paper No. 750786, pp. 1908–1917, 1975.
125. P. Flynn, M. Mizusawa, O. A. Uyehara and P. S. Myers, "An Experimental Determination of the Instantaneous Potential Radiant Heat Transfer Within an Operating Diesel Engine," SAE Trans., Vol. 81, Paper No. 720022, pp. 95–126, 1972.

DISCUSSION

A. K. Oppenheim *(University of California)*

I wanted to make a comment earlier, after John Heywood's remarks, and now I feel that it is very appropriate to make it. We should realize where we are. On the one hand, we have the modelers, who are essentially using data and would like to get results. On the other hand we have engines with cylinders and all the complications associated with them. I would like to make a suggestion regarding the integration of models and experiments. That is, let us not forget about the possibility of making model experiments. These experiments have been referred to by Bill Reynolds. For example, he noted that the cylindrical flame front process analyzed by Wong and Hoult using rapid distortion theory might be examined experimentally. I think such intermediate experiments would play a key role in joining the modelers with the experimentalists.

G. L. Borman

I would agree except that I would say that there is a great tendency, once having done those modeling experiments, to think that those experimental models are engines. In the spirit of your comment, I believe it necessary to test the computations with experimental models or even with sub-models. I see many computations that could be applied to experimental sub-models. I think what

G. L. Borman

you're saying is that we could do the latter. But one can think of many other problems where it's clear that only the engine can provide the atmosphere which is appropriate for testing the mathematical model. This is so, of course, because the engine is so complicated. I think the modelers are over here [motions left], though, and the experimentalists are over there [motions right].

THE APPLICATION OF A HYDROCARBON AUTOIGNITION MODEL IN SIMULATING KNOCK AND OTHER ENGINE COMBUSTION PHENOMENA

S. L. HIRST and L. J. KIRSCH

Shell Research Limited, Chester, England

ABSTRACT

The spontaneous oxidation of hydrocarbon fuels is a process that is very complex, both chemically and phenomenologically. Even for relatively simple fuels, vast numbers of chemical intermediates are involved, and the physical manifestations of the process include multiple cool flames, negative temperature coefficients, and two-stage ignitions. These phenomena can be explained in terms of a relatively simple thermokinetic mechanism that constitutes the basic kinetic framework for the oxidation of most hydrocarbon fuels. We have developed a generalized mathematical model that incorporates this mechanism and, therefore, can simulate the autoignition behavior of hydrocarbons in practical situations.

In this paper, we review the development of the chemical model and its optimization to give a quantitative prediction of the behavior of selected fuels, using data obtained in a rapid compression machine. We also review some of the areas where the model can be of direct use in modeling combustion processes of practical significance. These include knock, run-on, and the action of antiknock agents in the gasoline engine, and the ignition of fuel sprays and the action of ignition promoters in the diesel engine.

The most important application of the autoignition model to date has been the development of a working model of knock in the gasoline engine. This model has given an encouraging qualitative simulation and rationalization of the phenomenon of engine severity. In the present paper, we describe more detailed studies designed to evaluate the potential of the knock-modeling approach as a tool for the quantitative prediction of the fuel quality requirement of engines while in the design stage. An experimental program was carried out to characterize knock in terms of in-cylinder pressure diagrams under octane rating conditions in a CFR engine. The results were compared with computer predictions, using an optimized form of the knock model. Good agreement between experiment and simulation was found, and the model correctly ranked the various fuels incorporated in the study. However, the accuracy of

References pp. 222-223.

the computer predictions was very sensitive to the accuracy of the basic engine cycle simulation, particularly the description of the trapped-charge conditions, heat transfer, and flame propagation. Considerable experimental data were necessary to optimize these aspects of the model, and the inclusion of a more sophisticated description of them is an important step towards a wider application of the knock model.

NOTATION

A/F	air-fuel ratio
A_p	piston area
A_T	combustion chamber surface area
B	total degenerate branching agent (intermediate in chemical model)
$BRFAC$	burning rate coefficient
$BIFAC$	initial burning increment
C_P	specific heat
H/C	hydrogen/carbon ratio by weight (in fuel molecule)
M	air entrainment coefficient
n_R, n_B etc.	total number of moles (species defined by subscript)
O_2	oxygen (reagent in chemical model)
P	ignition promoter (in diesel simulations)
Q	total labile intermediates (intermediate in chemical model)
QK	heat release rate from chemical reaction
Q_L	heat loss rate
R	total radicals (intermediate in chemical model)
RH	total fuel (reagent in chemical model)
r_p	piston radius
T	system temperature
T_{ch}	cylinder head temperature
T_{cw}	cylinder wall temperature

T_{ph}	piston crown temperature
V	system volume
W	engine speed
α	heat-loss coefficient
ψ	molar expansion ratio on combustion
τ	ignition delay ($\tau = \tau_1 + \tau_2$ for two-stage ignition)
f_1, f_2, f_3, f_4, k_B, k_p, k_q, k_t, m, p, q	parameters of autoignition model—for further explanation see Ref. 9
τ_1, τ_2	induction periods for two-stage ignition

Subscripts

b	burned gas in knock model
tot	total system in knock model
u	unburned gas in knock model
w	wall in knock model

References pp. 222-223.

INTRODUCTION

The problem of knock has been with us as long as the gasoline engine itself, and today, as much as ever, octane quality represents the single most important property of a gasoline. Up to the 1960's the primary emphasis on engine design was towards higher specific output, and oil companies had successfully coped with the problem of continually improving the quality of their products to meet more stringent market demands. This was achieved through an ever-increasing appreciation of the way in which engine design and fuel composition influenced the tendency of an engine to knock, and the solution of the problem was much facilitated by the widespread use of upgrading-facilities in the refinery and the availability of a cheap, highly-effective class of antiknock additives—the lead alkyls. The approach was basically an empirical one, and although there was some concensus that knock was derived from autoignition of part of the fuel-air charge, a fundamental analysis of the physico-chemical processes involved was not considered practical or necessary.

More recently, with the pressures for lower emissions and better fuel economy, the problem of matching fuels to engines has taken on a new dimension. This has necessitated substantial changes in engine design, taking the range of operation beyond the realm of established experience. The rapid solution of the many problems posed has meant that a fundamental understanding of engine combustion has become a necessity rather than a luxury, and knock has proved to be one of the many aspects of the overall process where our understanding has been shown to be less than adequate.

In this situation, computers have come to the fore as a means of processing and developing, in the form of models, the knowledge gained through intensive experimental studies. The scope of this Symposium illustrates the range of topics that has been approached in this way in recent years. The modeling of complex combustion processes in engines is essentially an interdisciplinary problem, requiring input from mathematicians, engineers, physicists and chemists. The engine designer, who must employ the end-product from such a consortium, might be forgiven for considering himself less than adequately served by combustion chemists. The majority of engine combustion models contain only a rudimentary description of the chemical processes involved, generally limited to the inclusion of a single Arrhenius rate term. Now, it is fair to concede that in many instances this approach is entirely justified—either detailed chemical knowledge is not available or its application is unnecessary or impractical. However, in one aspect of hydrocarbon oxidation—the so-called pre-flame reactions which can lead to autoignition—we feel the extent of chemical understanding has progressed sufficiently for the chemist to make a much more substantial contribution to engine modeling problems. This has been the aim of our work at Thornton in recent years. The work described in this paper originated as a fundamental study of hydrocarbon oxidation processes—in their own right, a most interesting and unusual class of chemical reactions. It has been our aim to take the knowledge gained and apply it to a number of areas of practical significance, of which the modeling of engine knock is the most important. The present paper is devoted mainly to describing our evaluation of the knock-modeling concept, but we shall also take the opportunity of reviewing the various applications of autoignition modeling in engines and elsewhere. We shall conclude by describing a preliminary approach to one such application, namely spray ignition in a diesel engine.

THE PRACTICAL SIGNIFICANCE OF HYDROCARBON AUTOIGNITION

Engine Knock—Sir Harry Ricardo [1] was the first to develop the autoignition theory of engine knock, and his basic interpretation of the phenomenon has been generally accepted since that time. Other hypotheses, including in-cylinder detonations or some form of flame acceleration, have been proposed from time to time [2]. However, the weight of evidence favors the former theory, and we believe the quantitative success of our engine knock model, which was developed from autoignition measurements obtained *outside* an engine, provides confirmation of Ricardo's interpretation. Indeed, physical observations highly characteristic of spontaneous oxidation phenomena, i.e., cool flames and two-stage ignitions, were made in motored and fired engines by Barber et al. [3] as early as 1947. Thus, chemists working in the field of hydrocarbon autoxidation have been well aware of the practical significance of their work with regard to engine knock, and many studies have been directed accordingly.

The essential approximation to knock that we have built into our model is, therefore, that of a uniform, homogeneous and gaseous fuel-air mixture that undergoes spontaneous oxidation as it is compressed and heated by the motion of the piston and the advancing flame front. This relatively simply physical picture is a fair approximation to reality under most engine operating conditions and enables a quite complicated description of the chemical kinetics to be incorporated.

Run-On—The problem of run-on, when the ignition of a gasoline engine is switched off, is a familiar one and has possibly become rather more severe with the use of various methods of emission control. A very straightforward simulation of run-on, via partial or complete autoxidation of the fuel-air charge, can be achieved by operation of the knock model without a spark. In reality, run-on may be affected by hot-spots and local heat transfer. Thus, a more detailed treatment would include spatial analysis to take account of their effects, but we believe the chemical autoignition model offers the possibility of a quite realistic simulation of this phenomenon.

Antiknock Action—Although lead alkyls were discovered early in the development of the gasoline engine, their highly effective inhibition of the knock process has never been fully clarified. More seriously, totally acceptable alternatives to lead alkyls, from the standpoint of cost, toxicity and particulate emissions, have never been discovered. An effective model of the chemical processes that cause autoignition provides an ideal vehicle for testing hypotheses and exploring the practical limits on antiknock action. Some work that we have

References pp. 222-223.

carried out at Thornton in this context has been outlined in a previous publication [4].

Diesel Ignition—The diesel engine relies on spontaneous autoignition to initiate combustion of the injected fuel charge. As with gasolines, ignition quality represents the over-riding determinant of quality for a diesel fuel—in this case, readiness to autoignite being a favorable property. Difficulty in achieving prompt and reliable ignition under all operating conditions is responsible for many of the undesirable features of diesel engines, notably cold-starting problems, combustion noise, and engineering problems associated with very high compression ratios.

Although diesel fuels are refined with different chemical and physical priorities in mind, the essential chemical reactions that lead to autoignition of a diesel spray may be expected to relate very closely to those that cause knock in the gasoline engine. The same chemical model, with suitable parameter re-optimization, is therefore suitable for describing both processes. The physical processes of evaporation and mixing that occur in a diesel spray between injection and ignition are complex and play a critical role in determining the length of ignition delays. To describe these processes realistically in the form of a mathematical model is itself a major undertaking and, in contrast to the gasoline combustion, some spatial analysis of the combustion volume is essential. To incorporate a chemical model of any complexity, i.e., one requiring stepwise integration of a series of differential equations, into a spatial grid model is clearly a daunting task. Later in this paper we shall illustrate some preliminary studies we have made towards the goal of modeling diesel ignition.

Diesel ignition improvers have received some renewed interest with the growth of the passenger-car diesel market. These additives probably owe their action to their releasing of radicals into the reacting spray through thermal decomposition. As with the case of antiknock agents, the chemical autoignition model provides a suitable vehicle for investigating the mode of action and potential effectiveness of ignition improvers.

Autoignition Hazards—Although not directly relevant to the present Symposium, this important application of autoignition modeling is worth mentioning. Many industrial accidents, for example in tankers, pipelines and static installations, are attributed to the autoignition of a combustible fuel-air mixture that is subject to suitably high temperatures and/or pressures. At Thornton we have an ongoing research effort to evaluate potential autoignition hazards, and our autoignition model has proved valuable in understanding experimental and practical situations. Typical applications of the model incorporate spatial treatments of conduction and convection in a gaseous chemically reacting medium [5].

HYDROCARBON OXIDATION—DEVELOPMENT OF THE CHEMICAL MODEL

A number of excellent reviews have recently appeared that bring the current understanding of hydrocarbon oxidation mechanisms up to date [6–8]. Anyone reading these works in the hope of applying the knowledge gained might be excused for concluding that our understanding of the chemistry is by no means sufficiently advanced. Even the processes involved in the oxidation of a pure hydrocarbon fuel are very complicated and have not been fully clarified, either quantitatively or qualitatively. What hope, therefore, do we have for describing the oxidation of the complex mixture of hydrocarbons in a typical gasoline?

To answer this question one must look more closely at the phenomenological nature of hydrocarbon oxidation, remembering that the engineer is interested in the physical manifestation of the process rather than in the chemistry for its own sake. This phenomenology is in fact very complicated—hydrocarbon oxidation systems exhibit oscillatory effects in the form of cool flames at low temperatures and pressures; ignition diagrams are very complex, and at high pressures and temperatures appropriate to engine combustion, two-stage ignitions are generally observed. The reactions also show well-defined regions of negative temperature coefficient. The saving grace from the modeler's standpoint is that this pattern of behavior is highly characteristic of most hydrocarbons, regardless of their complexity. The behavior of a gasoline is essentially no more complicated than that of propane—differences are of degree rather than of kind. This leads us to believe that the phenomenological complexity in these systems is not related to their undoubted chemical complexity and that at the root of it all could lie a relatively simple common mechanism. It is this basic kinetic framework that we have tried to distill out of the enormous body of literature information—in other words, the simplest kinetics scheme capable of generating all the observed patterns of behavior.

The chemical mechanism that we have used in our models is represented in the series of coupled differential equations shown below:

$$\frac{1}{V}\frac{dn_{\mathrm{R}}}{dt} = 2\{k_q[RH][O_2] + k_B[B] - k_t[R]^2\} - f_3k_p[R] \qquad (1)$$

$$\frac{1}{V}\frac{dn_{\mathrm{B}}}{dt} = f_1k_p[R] + f_2k_p[Q][R] - k_B[B] \qquad (2)$$

$$\frac{1}{V}\frac{dn_{\mathrm{Q}}}{dt} = f_4k_p[R] - f_2k_p[Q][R] \qquad (3)$$

$$\frac{1}{V}\frac{dn_{\mathrm{O_2}}}{dt} = -pk_p[R] \qquad (4)$$

with

$$n_{RH} = \frac{(n_{O_2} - n_{O_2}(t=0))}{pm} + n_{RH}(t=0) \qquad (5)$$

The heat-release rate, QK, is given by

$$QK = k_p q V[R] \qquad (6)$$

where q is the exothermicity of the chain-propagation cycle. Further details of the model and the symbolism employed may be found in Ref. 9.

References pp. 222-223.

This scheme incorporates the two basic assumptions concerning the mechanism that are now generally accepted. The first is that the reaction proceeds via a degenerate branched-chain mechanism. The second is that the various phenomena are of thermokinetic origin, arising from coupling between the heat release and the reactions responsible for self-acceleration. The description of the chemical system is reduced to one containing three generalized intermediates—total radicals, R, total degenerate branching agent, B, and a third class of labile intermediates, designated Q. Although these equations demonstrate the phenomenon known as 'stiffness', they may be integrated with a suitable modern technique [10]. The model simulates all the essential features of spontaneous hydrocarbon oxidation systems under the appropriate physical conditions.

A rapid compression machine may be used to generate gas conditions representative of those experienced by the end-gas in a gasoline engine [11]. We have used data obtained in our machine at Thornton both to develop the basic structure of the model [12] and to carry out the essential parameter optimization to enable the model to predict quantitatively the autoignition behavior of certain selected fuels. This work has been reported in full detail elsewhere [9]. The performance of the model in simulating the characteristic two-stage ignition behavior of a paraffinic fuel is shown in Fig. 1. Some examples from the parameter optimization study are shown in Figs. 2 and 3. These illustrate the prediction of the temperature and pressure dependence of ignition delays for a Primary Reference Fuel (PRF) with a Research Octane Number (RON) of 90 (Fig. 2) and the ability of the model to distinguish the performance of a sensitive Toluene Reference Fuel (TRF) from that of the PRF of the same RON (Fig. 3). (The TRF has a lower Motor Octane Number (MON).)

We have characterized and developed model parameters to simulate the autoignition behavior of five fuels as follows:

Non-sensitive paraffinic fuels:
- 100 RON, Primary Reference Fuel (100 PRF)
- 90 RON, Primary Reference Fuel (90 PRF)
- 70 RON, Primary Reference Fuel (70 PRF)

Sensitive aromatic-containing fuels:
- 99.6 RON, 89.5 MON, Toluene Reference Fuel (99.6 TRF)
- 89.5 RON, 77.9 MON, Toluene Reference Fuel (89.5 TRF)

To obtain a suitable fit for these fuels, experimental data were compared with model predictions over a range of temperatures, pressures, and stoichiometries. The quality of the fit obtained was highly satisfactory considering the rather complex nature of the data and the relative simplicity of the model [9].

EVALUATION OF THE COMPUTER MODEL OF KNOCK

Background—The chemical autoignition model contains the basic properties necessary for a computer simulation of engine knock. Unlike more simplified approaches, it is capable of taking into account the coupling between the chemical and physical systems that is an essential feature of thermokinetic reactions, and

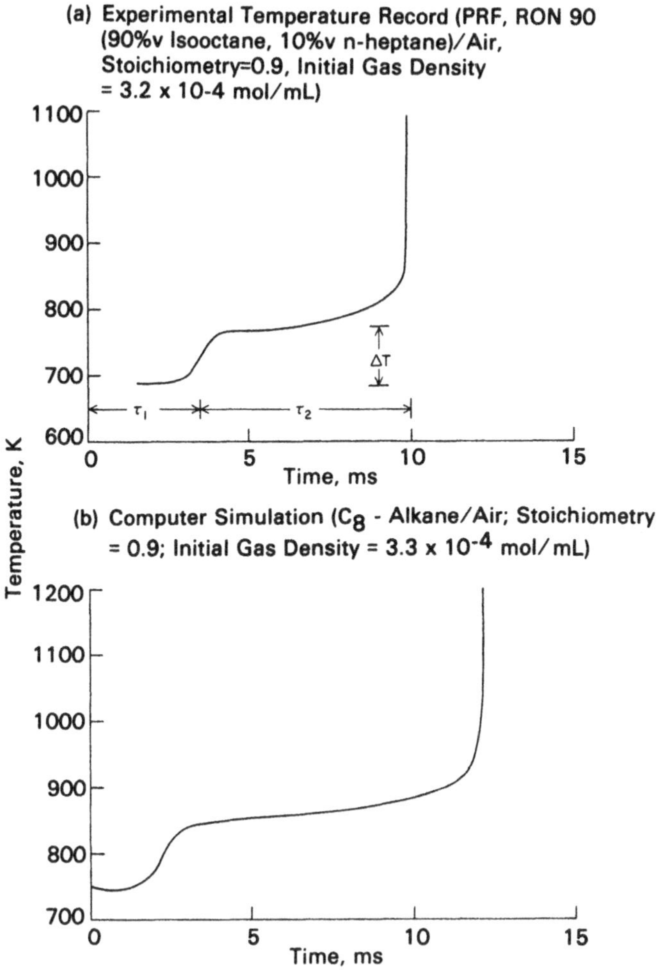

Fig. 1. Two-stage ignition in the rapid compression machine.

it embodies a realistic, albeit generalized, description of the chemical interactions involved in the end-gas. To achieve the knock simulation, it is necessary to integrate, as a coupled set of differential equations, the chemical rate expressions and the thermodynamic relationships that determine the physical state of the end-gas.

We have published the results of our initial studies, using a model of this kind [13]. Figs. 4 and 5, taken from this work, show the nature of the simulations obtained for the variation of end-gas conditions (Fig. 4) and of the end-gas intermediates (Fig. 5) during knocking combustion. We were also able to demonstrate the ability of the model to reproduce and rationalize the phenomenon of

References pp. 222-223.

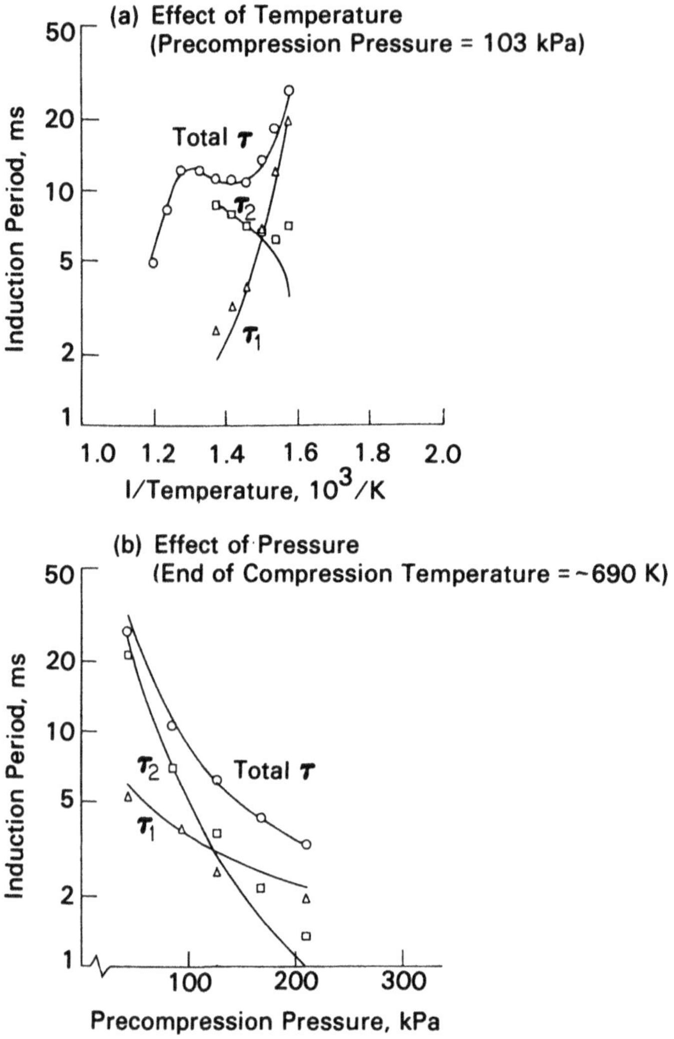

Fig. 2. Comparison of experiment (points) and model prediction (curves) for the autoignition of 0.9 stoichiometric mixtures of a 90 RON PRF in a rapid compression machine. Experiments were at a compression ratio of 9.6:1, with a wall temperature of 373 K.

engine severity, whereby an engine derates so-called 'sensitive' fuels, depending on the operating conditions. This study was carried out with a minimum of input from the actual engine operating data that are necessary for an accurate fixing of the engine-cycle-simulation element of the overall computer program. While the results of this work were highly encouraging to our overall approach, it became clear that a more quantitative evaluation of the model would require quite exten-

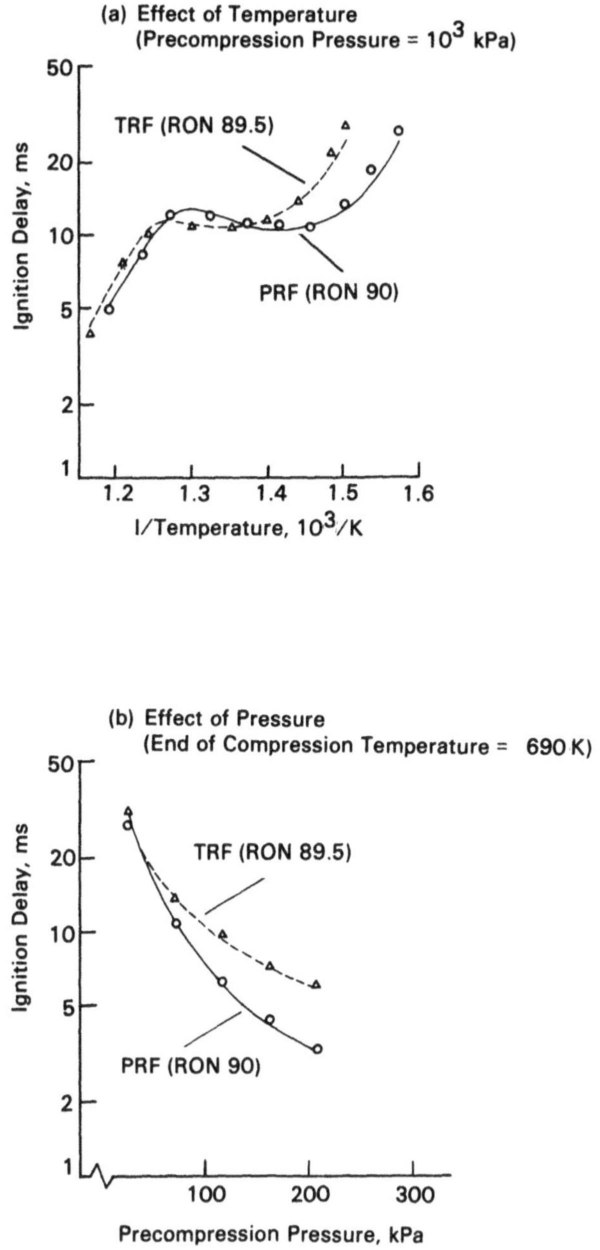

Fig. 3. Comparison of experiment (points) and model prediction (curves) for the total ignition delays in the autoignition of 0.9 stoichiometric mixtures of a 90 RON PRF and an 89.5 RON TRF in a rapid compression machine. Experiments were at a compression ratio of 9.6:1, with a wall temperature of 373 K.

References pp. 222-223.

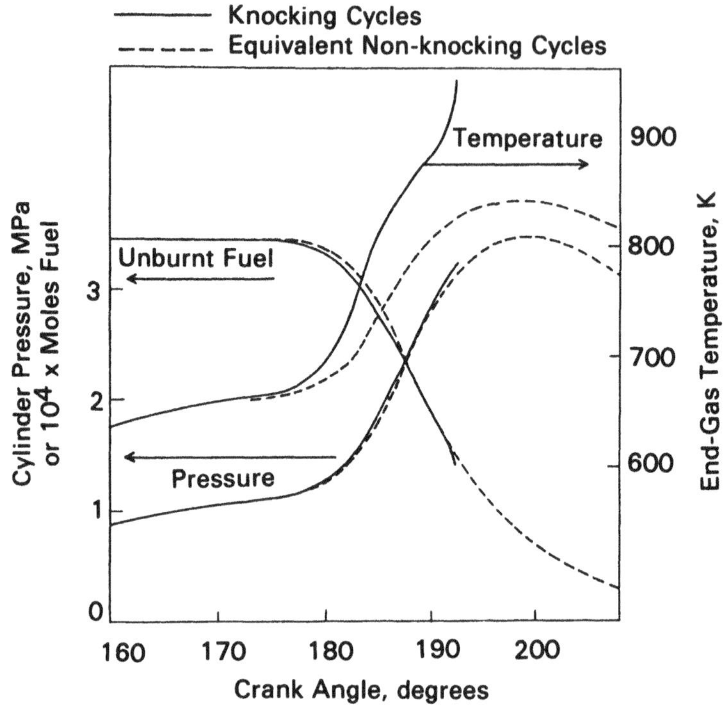

Fig. 4. Variation of end-gas conditions during a simulation of engine knock.

sive dedicated engine measurements. It is this phase of the work that we describe in this section of the present paper.

The purpose of this work was to take the model beyond the stage of providing insight and rationalization of basic observations and towards being a working tool for the quantitative predictions of engine performance. Our demands in this respect are dictated by our status as a supplier of fuels rather than of engines. Thus, detailed optimization of a specific engine design is of rather less importance than an across-the-board identification of future market requirements and an early identification of design trends that can lead to unusual or novel demands on fuel quality. As in the earlier study [13], the CFR engine under RON and MON rating conditions, which shows a good correlation in terms of fuel quality requirement with the overall car population [14], was chosen as the basic evaluation tool. However, the long-term goal is a very demanding one, namely the quantitative prediction of the octane requirement of engines while still in the design stage.

The Engine Cycle Simulation Model—The autoignition model is incorporated into an engine cycle simulation capable of predicting separately the states of the burned and unburned gases in the engine cylinder. Both systems are assumed homogeneous, gaseous, and separated by a flame front of infinitesimal thickness. Eight independent conservation equations are required to define fully the states

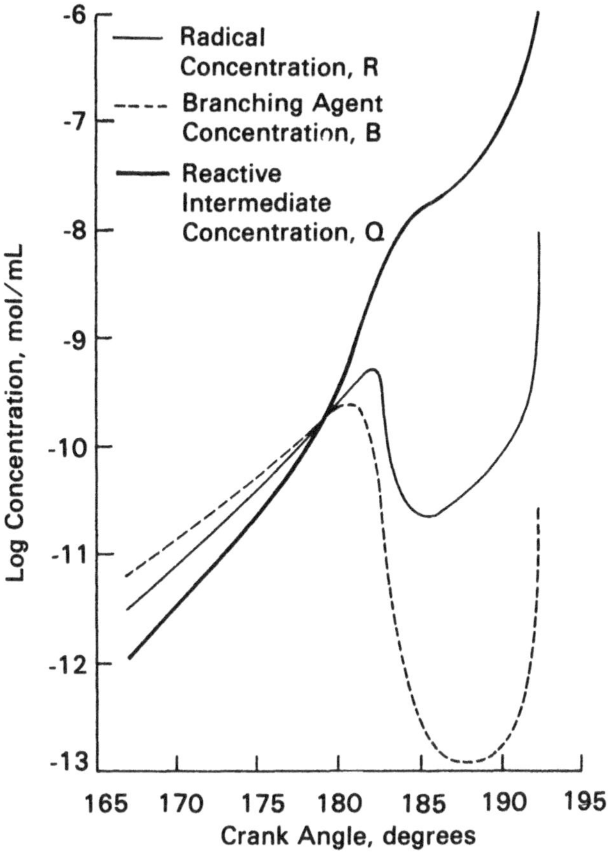

Fig. 5. Variation of concentration of chemical intermediates in the end-gas during a simulation of engine knock.

of both systems. These were derived according to the following assumptions:

(i), (ii) perfect gas law obeyed by both systems.

(iii) volume equation, as defined by piston motion—a right cylindrical shape was assumed for the combustion chamber.

(iv) conservation of matter—a fixed molar expansion ratio, ψ, was assumed upon combustion, with products consisting of CO_2, H_2O and N_2, with the balance made up by CO or O_2 for combustion of non-stoichiometric mixtures.

(v) conservation of total system energy.

(vi) pressure differences across the flame front neglected.

(vii) heat transfer across the flame front neglected.

(viii) flame propagation, i.e., mass transfer across the flame front, was described empirically, thus

$$\frac{dn_b}{dt} = \left[\psi \, BRFAC \left(\frac{r_p}{A_p} \right) V_{tot}(\sin \pi x) \frac{n_u}{V_u} \right] \tag{7}$$

References pp. 222-223.

where

$$x = \frac{V_b}{V_{tot}} \tag{8}$$

The empirical coefficient *BRFAC* will, in a complete model, depend on the engine operating conditions and will be a function of such factors as engine speed, stoichiometry, pressure, and exhaust dilution. To develop a suitable parametric form for *BRFAC* would require extensive experimental data which we did not attempt to obtain in the present work. Instead, *BRFAC* was determined for each relevant operating condition by optimization of predicted pressure diagrams against experimental measurements. An additional parameter, *BIFAC*, describing the initial burned increment generated by the spark, was treated similarly.

In addition to the above constraints, an empirical model, describing heat transfer from burned and unburned systems separately, was incorporated, using a development of Annand's approach [15]. Thus, a corrective term was applied to the unburned gases, with an empirical coefficient, α_u, describing the rate of loss to the cylinder head, piston head, and cylinder wall. Estimates of these surface temperatures were incorporated in the calculation. Thus, the rate of heat loss was written as

$$Q_{L,u} = \alpha_u C_{p,u}\left(W\frac{n_u}{V_u}\right)^{0.7}\frac{V_u}{V_{tot}}A_T[T_u - \bar{T}_w] \tag{9}$$

with

$$\bar{T}_w = \frac{\pi(r_p)^2(T_{ch} + T_{ph}) + 2\dfrac{V_{tot}}{r_p}T_{cw}}{A_T} \tag{10}$$

where

$$A_T = 2\pi(r_p)^2 + 2\frac{V_{tot}}{r_p} \tag{11}$$

T_{ch} = cylinder head temperature

T_{ph} = piston crown temperature

T_{cw} = cylinder wall temperature

It was taken that $T_{ch} \simeq T_{ph} = 525$ K and $T_{cw} = 425$ K. Radiative heat loss from the burned gases was written as

$$Q_{L,b} = \alpha_b \frac{V_b}{V_{tot}} A_T T_b^4 \tag{12}$$

Exhaust gas dilution was also treated empirically. From measurements of cylinder pressure at the point of opening of the exhaust valve, approximate relationships were derived to describe, in terms of the engine compression ratio, the amount of gas remaining to dilute the incoming charge.

The Evaluation Study—Outline—The computer program integrates an individual engine cycle from the point of closure of the inlet valve to the point of autoignition of the end-gas or, for non-knocking cycles, to the opening of the exhaust valve. The program requires as input, for any given fuel, determination of the boundary conditions, i.e., the physical state of the trapped charge and of the various empirical coefficients outlined above. The present study was therefore conducted as follows:

(i) Using an instrumented engine, a thorough study was made of the variation of cylinder pressure during knocking cycles in the CFR engine. The study was concentrated on the standard RON and MON rating conditions.
(ii) From these and additional measurements in non-knocking cycles, the input conditions for the computer simulations were determined.
(iii) Computer predictions of knocking cycles were then carried out and compared with the actual measurements recorded in (i) above.

Experimental Procedure—The CFR engine is a single-cylinder engine of 114.4 mm stroke and 82.7 mm bore, with a continuously variable compression ratio. The inlet system is designed such that both inlet air temperature and mixture temperature can be controlled. The engine and its ancillary equipment are fully described in Ref. 16.

Engine knock was characterized by knock amplitude and maximum cylinder pressure derived from cylinder pressure-time diagrams (see Fig. 6). The crank angle at which autoignition occurred was also identified to give a direct compar-

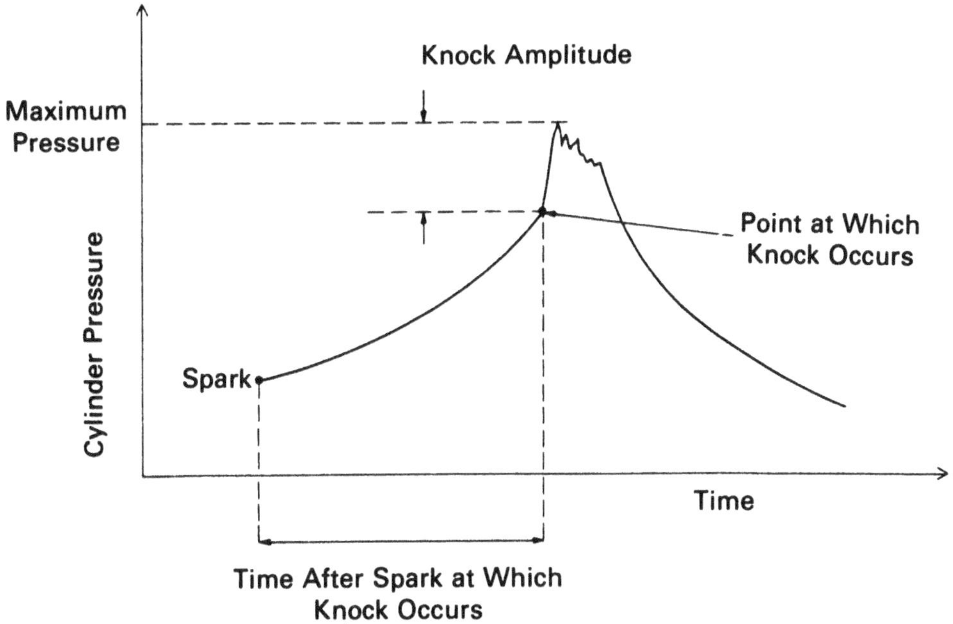

Fig. 6. Characterization of knocking combustion from pressure-time diagram.

ison with the output from the model. In addition, measurements of pressure-time diagrams were needed to derive some input parameters for the computer program. To make these measurements, a piezoelectric quartz-crystal transducer (Type SLM PZ14) was mounted flush with the cylinder wall. By using a transient recorder (Datalab DL905), individual cycles could be recorded and displayed on an oscilloscope or recorded on a paper chart. An inductive-type pick-up on the spark-plug lead provided a trigger for the recording system.

Measurement of cylinder pressure at the closure of the inlet valve (for input to the computer program) was achieved using the in-cylinder transducer at higher amplification. Further measurements were made with a Bell and Howell strain-gauge transducer (model BHL 4003-00-03M0) mounted in the inlet manifold. Good agreement was found between the two methods.

Fuel consumption was determined by measuring, with a stopwatch, the time taken for the engine to consume 10 mL of fuel, as measured with a burette. Fuel temperature was also recorded so that the specific gravity of the gasoline could be calculated.

Air-fuel ratio was measured using a modified Gerrish and Meem method [17]. A sample of the exhaust gases was oxidized in a copper-oxide furnace, maintained at a temperature of 700°C to give complete combustion. After the water was removed with a water-cooled condenser and a drying agent, the percentage by volume of carbon dioxide in the gas flow was measured by infrared gas analysis. From the measured carbon-dioxide concentration and known fuel composition, the air-fuel ratio was calculated [17].

In addition, inlet air temperature, mixture temperature, ignition timing and engine speed could be measured using the standard instrumentation of the CFR engine.

Experimental Results—The engine operating conditions for the tests were the same as those used in RON and MON determinations, thus defining the engine speeds, ignition timing, air inlet temperatures and mixture temperatures for this program. The operating conditions are summarized in Table 1.

As discussed earlier in the paper, autoignition model parameters are available for five fuels—100 PRF, 90 PRF, 70 PRF, 99.6 TRF(RON = 99.6, MON = 89.5) and 89.5 TRF(RON = 89.5, MON = 77.9). All five fuels were used in the test program.

In both RON and MON ratings, the octane number of a gasoline is determined at the air-fuel ratio that gives maximum knock intensity [16]. This air-fuel ratio was determined for each of the five fuels under both RON and MON rating conditions, using the standard knock detector, the Waukesha transducer and knockmeter [16].

The ASTM test procedure defines that the octane number of a gasoline be measured at a standard knock intensity. Guide curves, relating the compression ratio needed to give this standard knock intensity with octane number, have been established for RON and MON ratings, and these are shown in Fig. 7. For example, under RON rating conditions, a gasoline of RON 90 will give the standard knock intensity at a compression ratio of 6.65:1.

TABLE 1

Engine Operating Conditions for RON and MON Determinations

Conditions	Research Test	Motor Test
Engine speed, r/min	600 ± 6	900 ± 9
Spark timing	13° BTDC	Automatically varied from 17.5° to 26° BTDC depending on compression ratio
Coolant temperature, °C	100	100
Humidity, grains H_2O/454 g dry air	25–50	25–50
Air-intake temperature, °C	Varied with barometric pressure	38
Mixture temperature, °C	Variable*	149

For further details see Ref. 16.

* *For the* RON *test, the mixture temperature depends on the volatility of the fuel and the air-intake temperature.*

The following tests were carried out for each fuel under both RON and MON conditions:

The compression ratio was adjusted to the appropriate guide-curve compression ratio, and the fuel height in the carburetor float bowl was adjusted to give the air-fuel ratio for maximum knock intensity. When the engine reached equilibrium, a number of cylinder pressure-time diagrams were recorded. For each cycle recorded, the crank angle at which knock occurred, the knock amplitude, and the peak pressure were determined. The average values of these parameters over the cycles gave the knock characteristics for this test condition. The

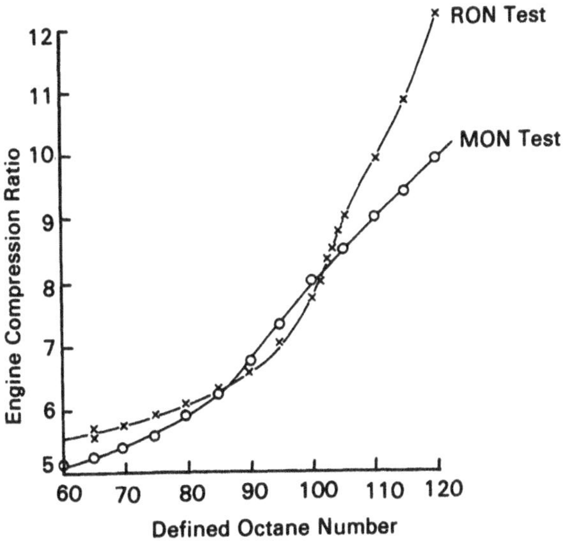

Fig. 7. Relationship between compression ratio and octane number for the standard knock intensity.

References pp. 222-223.

compression ratio was then reduced until the knock amplitude fell to zero (i.e., no visible discontinuity on the pressure-time diagram), and at this low compression ratio, pressure-time diagrams of individual cycles were recorded. Measurements of pressure-time diagrams were also made for a further two compression ratios which were lower than the guide-curve value but which gave visible knocking combustion.

The results of the test program are summarized in Tables 2 and 3.

In Fig. 8, the effect of compression ratio on knock amplitude is shown for each fuel at both RON and MON rating conditions. A small decrease in compression ratio causes a large decrease in knock amplitude. The graphs also show the wide variation in knock amplitude between the fuels at their guide-curve compression ratios, which are defined by the ASTM manual to give a 'standard knock intensity'.

The knock characteristics of each fuel at its guide-curve compression ratio may be compared from Tables 2 and 3. The expression "standard knock intensity" has no meaning in terms of the crank angle and pressure at which knock occurs, or of knock amplitude, i.e., these parameters are not the same for each fuel at the guide-curve compression ratios. However, a relationship does exist between peak pressure in the cylinder and octane number at these compression

TABLE 2

Measured Knock Parameters for RON Rating Conditions

Fuel	Air-fuel ratio	Compression ratio	Crank angle at which knock occurs, (deg. ATDC)	Pressure at which knock occurs (MPa)	Knock amplitude (MPa)	Peak pressure (MPa)
		7.82*	192.1	3.62	0.80	4.42
100 PRF	13.6	7.30	196.6	3.50	0.34	3.85
		6.80	197.8	3.36	0.08	3.44
		6.65*	191.6	2.78	1.01	3.79
90 PRF	13.2	6.10	195.1	2.70	0.55	3.25
		5.60	198.5	2.68	0.15	2.84
		5.78*	188.8	1.94	1.40	3.34
70 PRF	13.1	5.30	193.5	2.17	0.78	2.95
		4.80	196.9	2.25	0.25	2.50
		7.74*	191.3	3.34	0.95	4.30
99.6 TRF	13.6	7.20	193.3	3.19	0.57	3.76
		6.70	195.6	3.17	0.20	3.37
		6.61*	189.6	2.56	1.21	3.78
89.5 TRF	13.2	6.00	194.1	2.63	0.61	3.24
		5.40	197.4	2.58	0.14	2.72

* *Guide-curve compression ratios*

The above results are the average of two sets of data recorded on separate days.
The reproducibility of the results is defined as follows for 95% confidence level:

(i) crank angle at which knock occurs ± 1°
(ii) knock amplitude ± 0.03 MPa
(iii) peak pressure ± 0.03 MPa

TABLE 3

Measured Knock Parameters for MON Rating Conditions

Fuel	Air-fuel ratio	Compression ratio	Crank angle at which knock occurs, (deg ATDC)	Pressure at which knock occurs, (MPa)	Knock amplitude, (MPa)	Peak pressure, (MPa)
		8.03*	192.8	2.44	0.74	3.18
100 PRF	15.1	7.51	194.7	2.39	0.42	2.82
		7.00	195.6	3.36	0.24	2.59
		6.76*	193.4	2.26	0.48	3.75
90 PRF	14.4	6.30	193.5	2.23	0.30	2.53
		5.80	195.8	2.12	0.14	2.26
		5.43*	186.8	1.83	0.60	2.43
70 PRF	13.1	5.00	188.6	1.83	0.32	2.16
		4.63	190.9	1.79	0.11	1.90
		6.61*	191.1	2.05	0.62	2.66
99.6 TRF (89.5 MON)	14.1	6.10	191.9	2.03	0.39	2.42
		5.61	194.4	1.92	0.20	2.13
		5.79*	188.3	1.75	0.70	2.45
89.5 TRF (77.9 MON)	14.0	5.30	190.1	1.73	0.44	2.18
		4.79	191.5	1.68	0.14	1.82

* *Guide-curve compression ratios*

The above results are the average of two sets of data recorded on separate days.
The reproducibility of the results is defined as follows for 95% confidence level:

(i) crank angle at which knock occurs ± 1°
(ii) knock amplitude ± 0.03 MPa
(iii) peak pressure ± 0.03 MPa

ratios. Fig. 9 is a plot of peak cylinder pressure against octane number for both RON and MON rating conditions. The graph shows that, for the specific test condition, fuels with the same octane number have the same peak cylinder pressure. Further tests were carried out to measure the peak pressures at the same compression ratios, but with higher octane number fuels that did not knock, so that the increase in peak pressure due to the knock could be calculated for each fuel. The findings are shown in Fig. 10. The increase in peak pressure could be considered a constant for both RON and MON rating conditions, within the error estimation. It appears that the term "standard knock intensity," as applied to the guide-curve compression ratio, refers to a "standard" increase in peak cylinder pressure due to the knocking combustion.

Derivation of Model Input Parameters—

(i) Initial gas conditions: Measurement of the cylinder pressure at the closure of the inlet valve, together with the mass of the inducted charge, is sufficient to define the state of the charge gases at the start of the integrations,

References pp. 222-223.

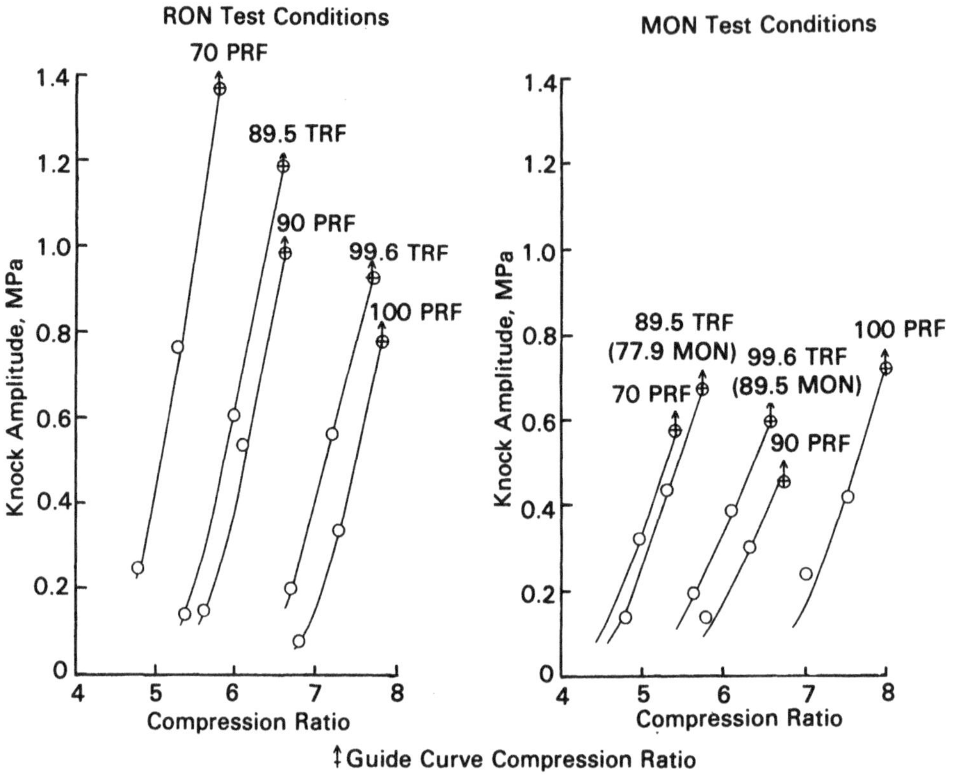

Fig. 8. The effect of compression ratio on knock amplitude.

since the volume of the cylinder at this point is known. The mass and composition of the charge were calculated from the measured air-fuel ratios and rates of fuel consumption, taking into account the empirical correction for exhaust-gas dilution.

(ii) Heat-transfer coefficients: The coefficient for the unburned gases, α_u, was derived by optimizing the prediction of cylinder pressure at the point of the spark. Separate values of α_u were obtained for RON and MON test operating conditions, the latter value being about 50% higher than the former. This is indicative of some shortcomings in the modified heat-transfer model adopted in this work. The radiative coefficient, α_b, was derived by fitting the prediction of pressure drop after the combustion peak. A single value was found to be suitable for both RON and MON test conditions.

(iii) Burning rate coefficients: These were derived by optimizing the predictions of the overall cylinder pressure diagram for non-knocking simulations. The initial burning increment, *BIFAC*, was found to be independent of operating condition. The propagation rate coefficient, *BRFAC*, was determined separately for each and every operating condition where simulations were required. Emphasis was placed on achieving an accurate prediction of peak cylinder pressure.

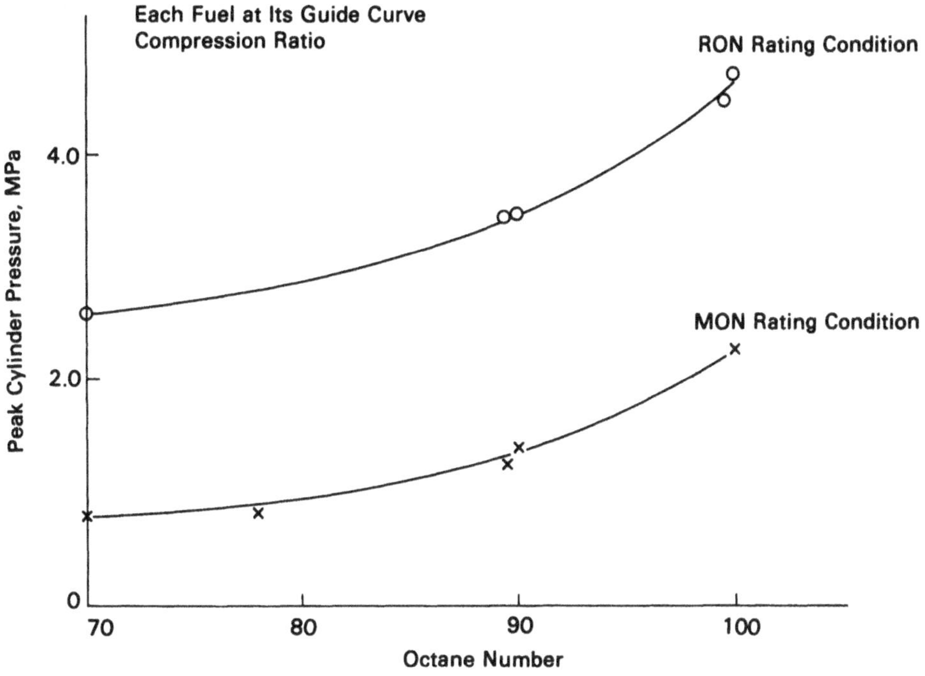

Fig. 9. Relationships of octane number to peak cylinder pressure.

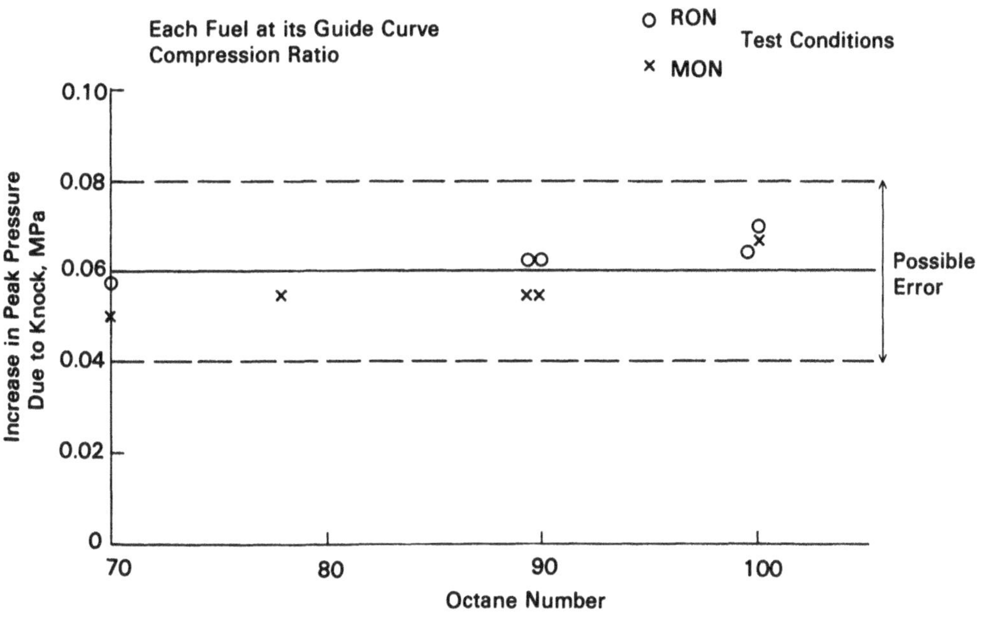

Fig. 10. Increase in peak pressure in relation to the octane number of the fuel.

References pp. 222-223.

TABLE 4

Comparison of the Predicted and Measured Crank Angles at which Knock Occurs for the Guide Curve Compression Ratios—RON Test Conditions

Fuel	Crank angle at which knock occurs, degrees		Difference, degrees
	Measured	Predicted	
100 PRF	192.1	192.4	+0.3
90 PRF	191.6	192.1	+0.5
70 PRF	188.8	190.5	+1.7
99.6 TRF	191.3	191.0	−0.3
89.5 TRF	189.6	191.8	+2.2

Comparison of Experimental Results and Mathematical Simulations—The comparison could now be made between the experimental results and the model simulations of the engine operation. In Figs. 11 and 12, measured and predicted pressure-time diagrams are compared for the engine running at the guide-curve compression ratios. For all the test fuels, the model gave a good simulation of engine knock, especially in predicting the crank angle at which the end-gas autoignites. In Tables 4 and 5, a direct comparison is made of the measured and predicted crank angles at which knock occurs for the guide-curve compression ratios. Similar results were found for the tests at lower compression ratios.

The model, in its present form, makes no attempt to simulate the processes in the combustion chamber after autoignition occurs and thus cannot predict knock amplitude or peak cylinder pressure. However, the model does predict the number of moles of fuel in the end-gas at the point of autoignition. Assuming that all the fuel ignites, the energy released in the autoignition per unit volume of combustion chamber can be calculated, and this parameter can be considered an indicator of knock intensity. The effect of compression ratio on the energy release per unit volume is shown in Fig. 13. The relationship of compression ratio to octane requirement is correctly described by the model—increasing compression

TABLE 5

Comparison of the Predicted and Measured Crank Angles at which Knock Occurs for the Guide-Curve Compression Ratios—MON Test Conditions

Fuel	Crank angle at which knock occurs, degrees		Difference, degrees
	Measured	Predicted	
100 PRF	192.8	192.4	−0.4
90 PRF	193.4	192.6	−0.8
70 PRF	186.8	188.5	+1.7
99.6 TRF (89.5 MON)	191.1	192.4	+1.3
89.5 TRF (77.9 MON)	188.3	191.3	+3.0

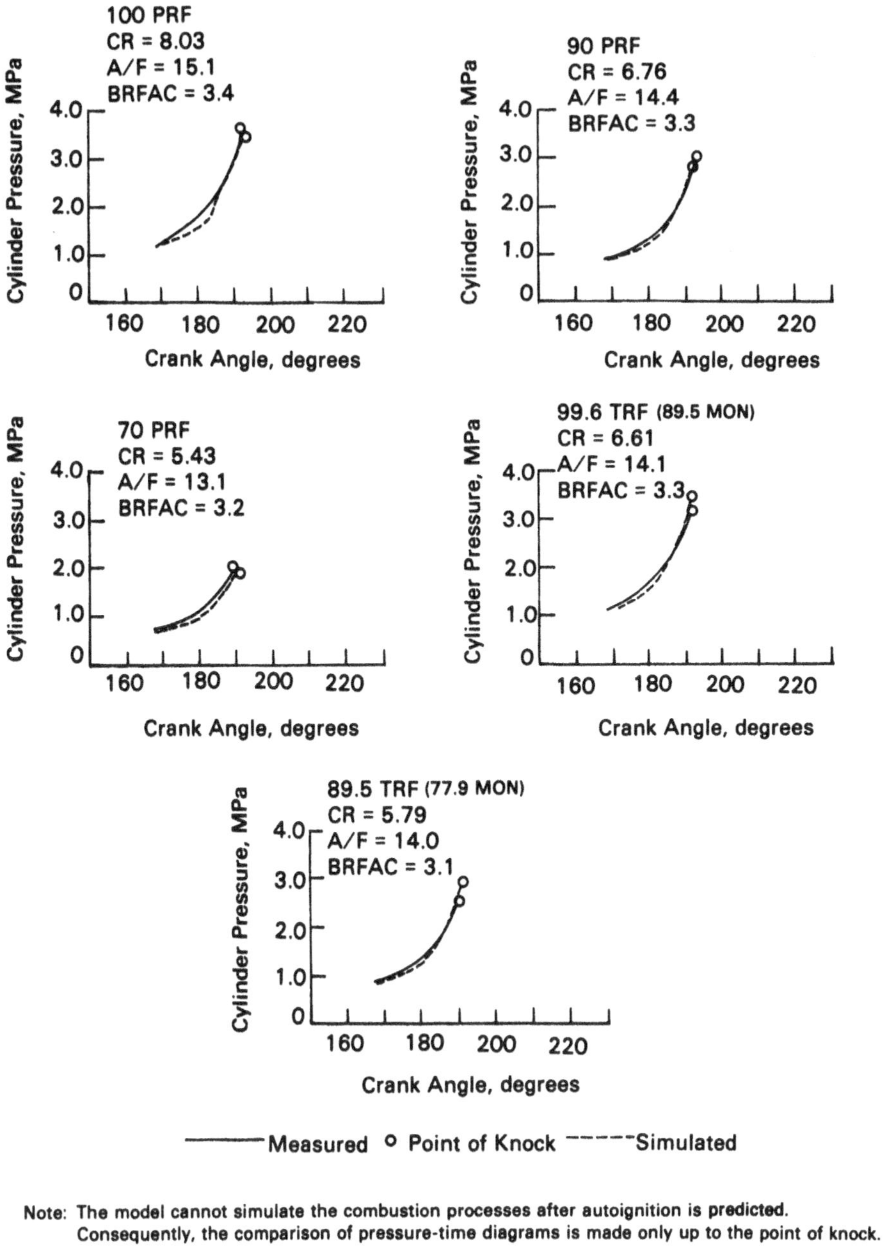

Fig. 11. Comparison of the model simulations of knock with experimental measurements for each fuel at its guide-curve compression ratio under RON test conditions.

References pp. 222-223.

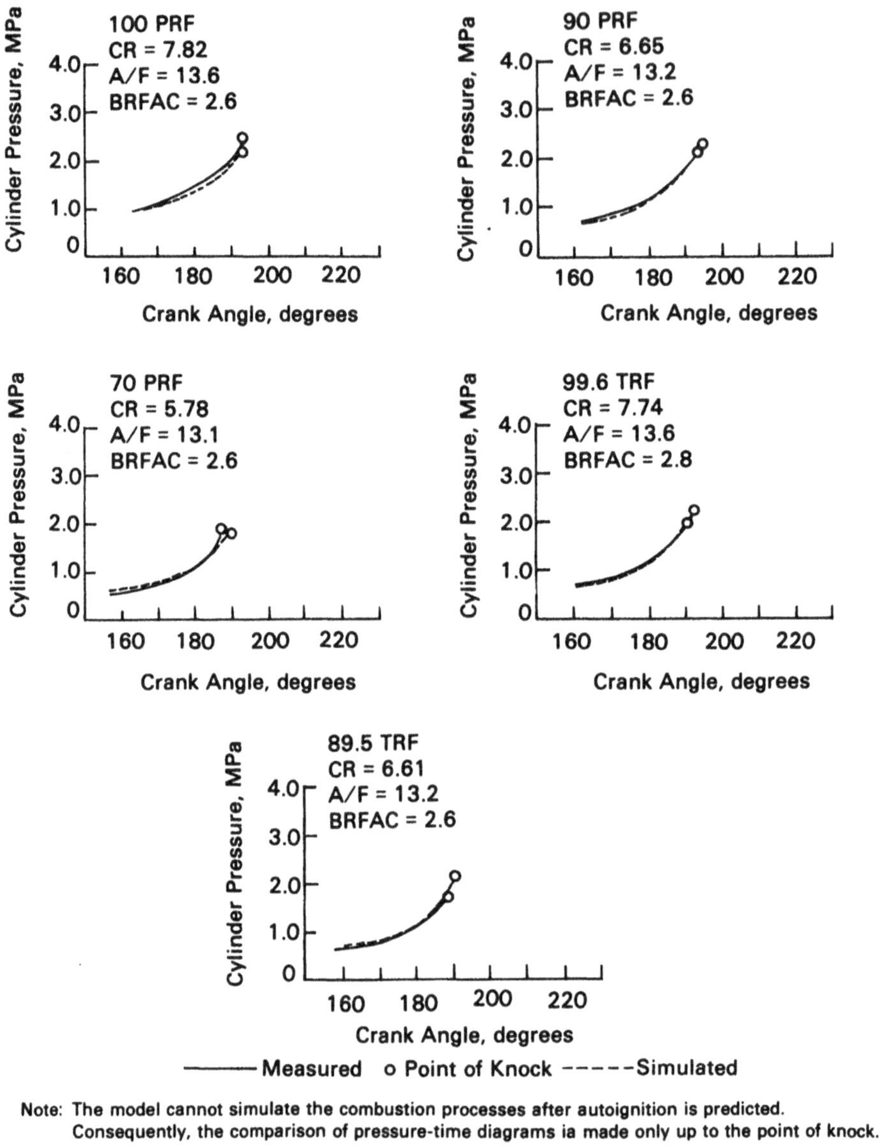

Fig. 12. Comparison of the model simulations of knock with experimental measurements for each fuel at its guide-curve compression ratio under MON test conditions.

ratio increases knock intensity. The model also correctly ranks the fuels in order of their octane number under both RON and MON test conditions.

A direct correlation should exist between knock amplitude and this simulated knock intensity. In Fig. 14, measured values of knock amplitude are plotted against the predicted energy release, and this shows that there is a general correlation between these parameters.

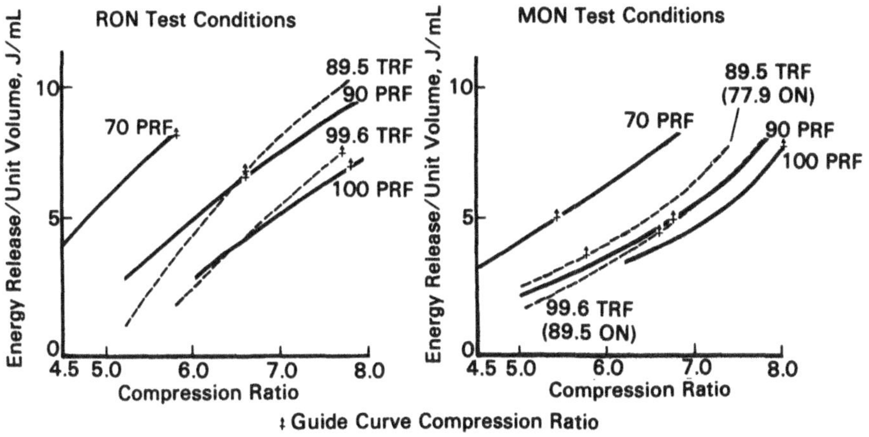

Fig. 13. Relationship of the energy release per unit volume by the autoignition with compression ratio.

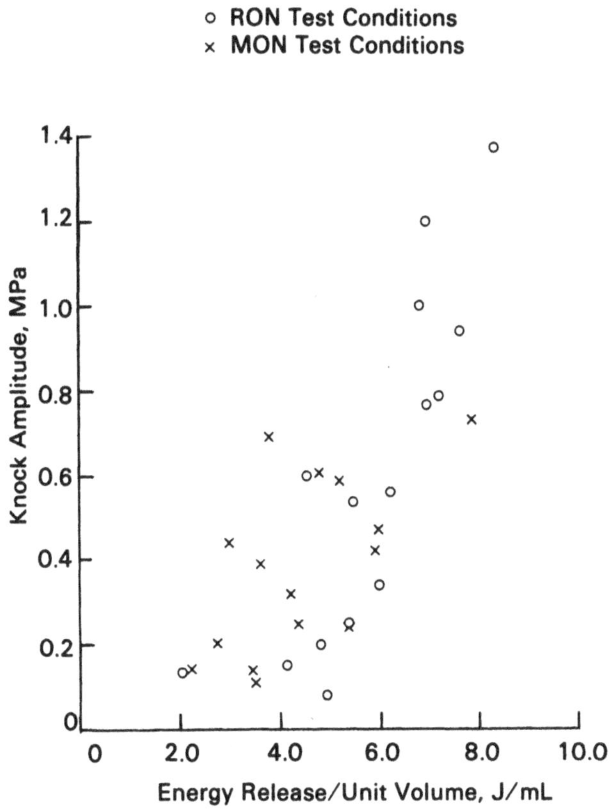

Fig. 14. Correlation of knock amplitude to the predicted energy release per unit volume by the autoignition.

References pp. 222-223.

Discussion—The validity of the model has been established by comparing simulations of engine operation to experimental results. The model gave a very good simulation of knock in the CFR engine under RON and MON rating conditions, especially in predicting the crank angle at which the end-gas ignites. From the model output, the energy release by the autoignition per unit volume of the combustion chamber could be derived (assuming that all the fuel in the end-gas ignites), and this can be considered a measure of knock intensity. In terms of this simulated knock intensity, the model correctly ranked the fuels in order of their octane numbers under both RON and MON rating conditions. By using this simulated knock intensity, the model could be used to look at the effect of engine design and operating conditions on octane requirement.

However, extensive experimental measurements were required to establish some of the input parameters of the model. This mainly involved measurements of cylinder pressure-time diagrams to determine the heat-transfer coefficients and the burning rate coefficients. To see how important it is to have accurate values for the input parameters in terms of their effect on the knock predictions, deviations of 5–10% were applied to calculated values, and the effect on the simulation was evaluated. The following parameters were considered:

(i) Concentrations of oxygen and nitrogen in the inlet charge
(ii) Exhaust gas dilution
(iii) Pressure on closing of the inlet valve
(iv) Temperature on closing of the inlet valve
(v) Heat-transfer coefficient for unburned gases, α_u
(vi) Heat-transfer coefficient for burned gases, α_b
(vii) Burning rate coefficient, *BRFAC*

These were not simple calculations in that there are a number of interactions between the parameters. However, it was found that the critical parameters were the initial pressure and temperature, the heat-transfer coefficient for unburned gases and the burning rate coefficient. Errors of 5–10% in these parameters gave errors of between 1° and 3° in the prediction of the crank angle at which knock occurs. However, a small change in the predicted crank angle has a large effect on the simulated knock intensity—e.g., a change of 1° in the predicted crank angle will give a 10–25% change in the knock intensity. Thus, the deviations of 5–10% in the critical input parameters gave large errors in the simulated knock intensity. Consequently, some input parameters must be accurately known if the model is to give a good simulation of knock intensity, and this requires a significant amount of experimental testing.

To summarize, the work has established the basic validity of the mathematical model of knocking combustion. However, to determine some of the input parameters with the necessary degree of accuracy, a significant amount of experimental testing of the engine has been needed. This need for experimental data would be much alleviated by the development and incorporation of a more sophisticated model for the normal combustion of the charge. This is the most important step that must now be taken towards the goal of predicting the behavior of an engine in the development stage.

MODELING OF DIESEL IGNITION

In the final section of this paper we present some preliminary modeling studies that illustrate a future application of the autoignition model, namely to simulate the process of fuel ignition in a diesel engine. The magnitude of the problems involved in achieving such a model has been referred to briefly earlier in the paper. The present study makes no attempt to address the considerable task of combining the autoignition model with a spatial description of a fuel spray. Instead, to gain some initial feeling for the potential of such an approach, we have examined the behavior of a reacting fuel-air mixture under conditions of changing stoichiometry and temperature.

We take as our frame of reference an element of fuel after injection into the hot gas in a diesel engine and follow the progress of this element as it proceeds through the combustion chamber. As this occurs, the relatively cold fuel will entrain hot air, so that the local stoichiometry decreases and the temperature rises with time. The rate of the autoignition reactions will maximize somewhere near stoichiometric, a factor which is 'built-in' to the autoignition model, and under appropriate conditions, ignition will occur. We assume a purely gaseous, homogeneous system and neglect heat and mass transfer from adjacent reacting fuel elements. An empirical mixing model is adopted such that the rate at which air is entrained per mole of fuel is given by

$$\frac{dn_{\text{air}}}{dt} = \frac{M}{t^{0.5} + 0.001} \tag{13}$$

where M is an arbitrary coefficient. The $t^{0.5}$ dependence was incorporated to allow for slowing down of the spray with time, and is based upon the empirical correlation given by Dent [18]. (The rate of entrainment is taken as being proportional to the rate of penetration of the spray.) Of course, different elements of fuel will experience different mixing rates, and we can take this into account by exploring a range of values for the mixing coefficient, M, within the context of the mixing model.

The parameters of our autoignition model used in this work describe the behavior of the lowest octane PRF examined, 70 PRF, which has a cetane number of 25, but were modified from those used in the knock model. The modifications were designed to give a more realistic prediction of total radical concentrations during the reactions, following independent studies of this factor. Simulations were carried out at constant pressure (3 MPa) and fuel temperature (373 K).

Results obtained using this very simple model are shown in Figs. 15 and 16. In Fig. 15, variation of simulated ignition delays with mixing rate parameter is shown at two air temperatures. The explanation of the shape of these curves is as follows. At low mixing rates, the delay is mixing-rate limited, i.e., determined by the time taken for the fuel to be diluted to a "reactive" stoichiometry and temperature. At much higher mixing rates, ignition delays increase because the reactions can be quenched through overly rapid dilution with air. Thus, there is an optimum mixing rate, yielding a minimum ignition delay. Considering now the

References pp. 222-223.

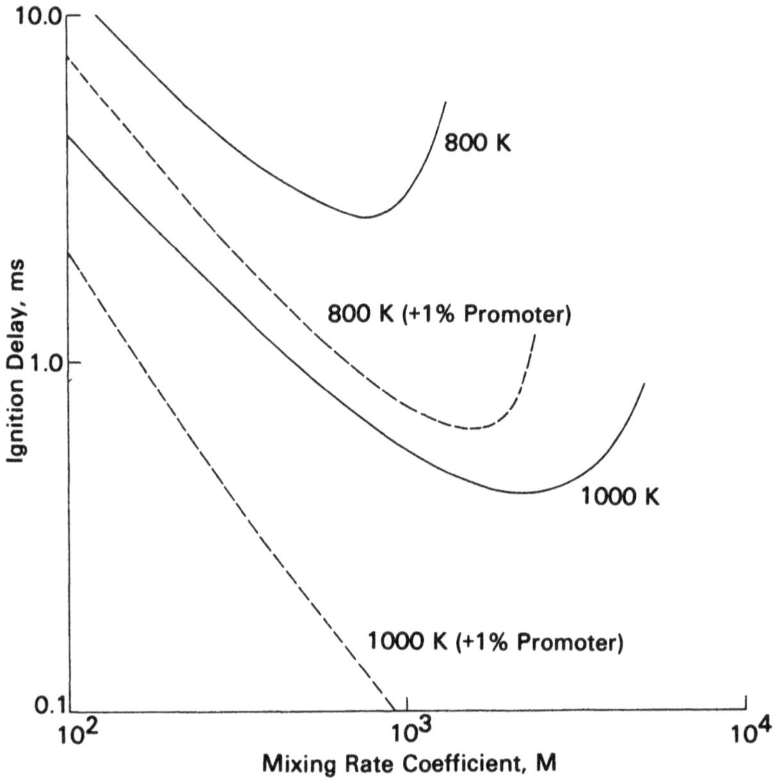

Fig. 15. Simulated ignition delays for the homogeneous mixing model.

fuel spray as a whole, one can envisage that a certain fuel element will experience a mixing profile approximating to the optimum, and that this will provide the source of autoignition. Encouragingly, the magnitude of the minimum ignition delays simulated in Fig. 15 corresponds approximately to those that might be expected for such a fuel under the appropriate gas conditions. Fig. 16 illustrates the progress of the reaction under the conditions for minimum ignition delay. Under these rapid mixing conditions, the existence of cool flame or two-stage ignitions is masked.

It is interesting to note that the optimum mixing rate is not the same at the two temperatures. The effective mixing rate across the spray is, of course, a function of the injection characteristics, and these results suggest that the optimization of an injection system for reliable ignition may be a sensitive function of operating temperature.

The broken lines in Fig. 15 illustrate one final application of the chemical model in this context, namely to simulate the effect of ignition promoters. For these simulations, the chemical model has been modified by introducing the reaction:

$$P \rightarrow 2R$$

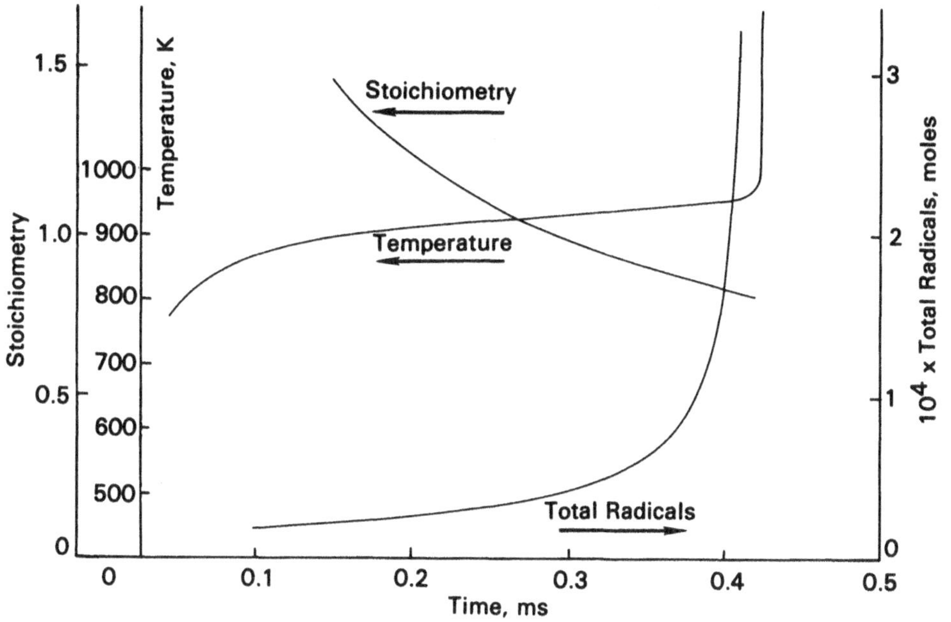

Fig. 16. Simulation of ignition by the homogeneous mixing model. Mixing rate coefficient = 2×10^3.

which describes the unimolecular decomposition of a molecular species releasing free radicals into the system. The Arrhenius parameters for this reaction were chosen to correspond to those for a known ignition promoter, di-t-butyl peroxide. The introduction of this additive shortens the ignition delays and raises the optimum mixing rate at a given temperature.

These are admittedly very primitive calculations, but they are illustrative of the kind of understanding of a complex physico-chemical process that can be achieved by the application of modeling techniques. Further development of the diesel spray-ignition modeling concept requires coordination of experimental and modeling programs in a similar manner to that used in the development of our knock model. Combustion rigs, such as the rapid compression machine, provide a more controlled environment for this kind of study than is possible in engines. At Thornton, a program has therefore been initiated to carry out a detailed high-speed photographic characterization of diesel ignition in the rapid compression machine which, in conjunction with appropriate modeling studies, will help to develop a fuller understanding of the complex processes involved.

CONCLUSIONS

It has been our main intention in this paper to demonstrate how our present understanding of hydrocarbon oxidation systems can be condensed into a form suitable for practical modeling applications. Modeling of autoignition in the end-

References pp. 222-223.

gas of a gasoline engine has confirmed the validity of this theory of knock, and detailed evaluation against engine measurements has demonstrated the potential of such a model as a working tool. The autoignition model provides a useful vehicle for studying modifications to the oxidation mechanism introduced by additives such as antiknock agents or ignition promoters. Preliminary studies have illustrated the potential value of the chemical model in understanding the processes of diesel ignition, and this promises to be a fruitful avenue for future development.

ACKNOWLEDGMENT

The authors wish to acknowledge the contribution of many colleagues towards the research described in this paper: Dr. C. P. Quinn and Dr. M. P. Halstead, who conceived the knock modeling idea and whose fundamental autoignition studies were of primary importance; Mr. A. Prothero and Mr. M. J. Hinde for development of and assistance with the mathematical techniques employed; Mr. D. B. Pye, Mr. R. Summers and Mr. J. C. Nield for their technical assistance.

REFERENCES

1. *H. R. Ricardo, "The Progress of the Internal Combustion Engine and the Fuel," Melchett Lecture, Institute of Fuel, 1935.*
2. *S. Curry, "A Three-Dimensional Study of Flame Propagation in a Spark Ignition Engine," SAE Trans., Vol. 71, Paper No. 452B, pp. 628–650, 1963.*
3. *E. M. Barber, L. E. Endsley and T. H. Randall, "Some Characteristics of Combustion in the ASTM-CFR Knock-Test Engine," API Proc III, pp. 103–115, 1947.*
4. *L. J. Kirsch and D. B. Pye, "Modelling the Action of Antiknock Agents," Second European Symposium on Combustion, Orleans,.France, Vol. 2, Paper 136, pp. 806–811, 1975.*
5. *D. C. Bull, "Autoignition at Hot Surfaces," Proc. Second International Symposium on Loss Prevention and Safety Promotion in the Process Industries, Heidelberg, pp. 165–172, September 1977.*
6. *R. W. Walker, "Rate Constants for Gas Phase Hydrocarbon Oxidation," Reaction Kinetics, Vol. 1, Specialist Periodical Reports, Ed. P. G. Ashmore, Chem. Soc. London, pp. 161–211, 1975.*
7. *R. W. Walker, "Rate Constants for Reactions in Gas Phase Hydrocarbon Oxidation," Gas Kinetics and Energy Transfer, Vol. 2, Specialist Periodical Reports, Ed. P. G. Ashmore and R. J. Donovan, Chem. Soc. London, pp. 296–330, 1977.*
8. *G. McKay, "Gas Phase Oxidation of Hydrocarbons," Progress in Energy and Combustion Sciences, Vol. 3, pp. 105–126, 1977.*
9. *M. P. Halstead, L. J. Kirsch and C. P. Quinn, "The Autoignition of Hydrocarbon Fuels at High Temperatures and Pressures—Fitting of a Mathematical Model," Comb. and Flame, Vol. 30, pp. 45–60, 1977.*
10. *A. Prothero and A. Robinson, "On the Stability and Accuracy of One-Step Methods for Solving Stiff Systems of Ordinary Differential Equations," Math. Comp., Vol. 28, pp. 145–162, 1974.*
11. *W. S. Affleck and A. Thomas, "An Opposed Piston Rapid Compression Machine for Preflame Reaction Studies," Proc. I. Mech. E., Vol. 183, pp. 365–385, 1969.*
12. *M. P. Halstead, L. J. Kirsch, A. Prothero and C. P. Quinn, "A Mathematical Model for Hydrocarbon Autoignition at High Pressures," Proc. Roy. Soc., A346, pp. 515–538, 1975.*
13. *L. J. Kirsch and C. P. Quinn, "A Fundamentally Based Model of Knock in the Gasoline Engine," Sixteenth Symposium (International) on Combustion, The Combustion Institute, Pittsburgh, Pennsylvania, pp. 233–244, 1976.*

14. A. G. Bell, "The Relationship between Octane Quality and Octane Requirement," SAE Paper No. 750935, 1975.
15. W. J. D. Annand, "Heat Transfer in the Cylinders of Reciprocating Internal Combustion Engines," Proc. I. Mech. E., Vol. 177, pp. 973-990, 1963.
16. ASTM Manual, Part 47, "Test Methods for Rating Motor, Diesel and Aviation Fuels," 1978.
17. H. C. Gerrish and J. L. Meem, "The Measurement of Fuel/Air Ratios by Analysis of the Oxidized Exhaust Gas," NACA Report 757, 1943.
18. J. C. Dent, "A Basis for Comparison of Various Experimental Methods for Studying Spray Penetration," SAE Trans, Vol. 80, Paper No. 710571, pp. 1881-1884, 1971.

DISCUSSION

P. Eyzat *(Institut Francais du Pétrole, France)*

You mentioned that you used the results from the rapid compression machine experiments to determine the temperature and pressure coefficients. As far as I am aware, the rapid compression machine cannot be used with industrial gasoline or diesel fuel. Did you encounter any such problems with the operation of the rapid compression machine?

L. J. Kirsch

First of all, the rapid compression machine can be used with gasoline and we have done so. It can't be used with diesel fuel. Of course, the compression

L. J. Kirsch

machine we use can be heated up to 150°C so that we can use fuels which are volatile below that temperature.

P. Eyzat

But this temperature level is too low. That is, the temperature should be at least 200°C. The problem with the rapid compression machine is that if you start using a temperature which is too high, the time at which the kinetic rates begin to become appreciable occurs nearer to bottom dead center. So it is difficult to analyze the results. This is a problem with fuels which are evaporation limited. Did you experience this problem?

L. J. Kirsch

First of all, the present work was carried out with reference fuels, which are completely volatile under the conditions employed. Other people ask the same question that you have asked, namely, can it be done with gasolines? We have done similar experiments with gasolines, which of course, contain a mixture of fuels having a range of boiling points. Considering the small proportions of the higher boiling elements in a stoichiometric mixture, at 100°C they can be essentially completely volatized with carefully selected gasolines. In all our experiments fuels are pre-injected into the evacuated and heated cylinder to assist evaporation and preparation of a homogeneous mixture.

P. Eyzat

While we, too, heat our rapid compression machine, our experience has been that it can't be used with industrial gasolines. If the unit temperature is not high enough, one has the volatility problem, and if the temperature is too high, say 200°C, the early reaction rate problem occurs.

L. J. Kirsch

We didn't need to go that high in temperature. The majority of this work was based on reference fuels. Moreover I should point out that the use of inert gas diluents to vary mixture specific heat, together with the nature of the compression stroke in our rapid compression machine give very rapid rates of temperature rise at the end of compression, particularly at the highest experimental temperatures. This overcomes the kinetic problem you refer to without the use of very high precompression temperatures.

F. F. Pischinger *(Institute of Applied Thermodynamics, Germany)*

Before I ask some questions, I would like to mention briefly what we at the Aachen Institute have done to calculate knock. We used methane because the mechanism of methane is simpler, and also because we had a methane engine. We found that the whole knocking mechanism is very sensitive to the termination reactions as well as to some starting reactions. Because today's kinetics data are not sufficiently reliable, one can easily adjust each knocking point by changing various kinetics data.

The second problem we observed was that the statistical character of the engine process made it nearly impossible to compare computations with experimental data. There was very little correlation between the pressure trace and knock. As is well known, if knocking occurs, it doesn't mean that every cycle is knocking. Rather, only several cycles are knocking, and those cycles are not always the ones with the highest pressures. We finally stopped this work because we found it to be so intricate.

We also tried to evaluate compounds affecting engine knock. Lead compounds like tetraethyl lead have, of course, been found to prevent knock with gasoline. However, if this compound is used with methanol, it has an adverse effect of promoting knock. There were so many phenomena which we could not predict.

Finally, my questions are, "Did you use a mean pressure trace for the engine in your modeling, or did you take into account the cyclic variations of the pressure, which means a burning function? Second, were you able to make any prediction regarding the action of the lead compounds in the knocking cycle?"

L. J. Kirsch

First let me address one of your early comments—the one regarding work you did with methane. Now, I made a sweeping statement at the beginning when I said that the behavior of all hydrocarbons is essentially the same. That holds true all the way down to propane, but ethane and methane are very different from all the others. They don't exhibit this rather complex pattern of behavior that is the essence of this model, though, of course, the detailed kinetics are much better known. You also mentioned the sensitivity to the termination process. This is certainly the case, and I would point out that parameters of our model, as we derived them, are not uniquely determined. Although two termination parameters are built into our model, they are very critical but not uniquely determined. What we simply have done is to devise a tool based on a sound but generalized kinetics framework which predicts the form of the observations and then to apply it in a definite environment.

With regard to your question of whether we used single pressure measurements, the measurements were averages of several pressure traces. These measurements were repeated on a day-to-day basis to ensure reproducibility.

You asked one other question, about the action of lead compounds. We have looked at the action of lead on ignition delays, not in a full knock model, but rather in the compression machine, where one can observe the same essential trends that are observed in an engine. We've used the model to try to rationalize the way in which lead acts. The basic conclusion is that it has to act catalytically, and it has to act in particulate form. That conclusion, of course, is very general.

H. K. Newhall *(Chevron Research Company)*

I would like to make a couple of remarks which also bear on Dr. Pischinger's comments. The first relates to the experimental CFR engine, where I understand that you used mean pressure traces to arrive at the crank angle value at which knock occurred. In comparing those data with your calculations, one can't really draw any conclusions without being given some statistics regarding the range of values surrounding those mean values. I would urge that in future work the 90%

or 95% confidence intervals around each of those experimental points be identified so we can tell where the computations fall and assess the differences between calculated and measured data.

The second question I have concerns fuel sensitivity. We know that there are other compounds besides aromatics that increase fuel sensitivity. The important ones, of course, are olefins. In exercising your model, did you put any olefins into the gasoline to determine their effect?

L. J. Kirsch

On the first point please remember that for each engine condition the normal combustion/heat transfer parameters of the model were fitted to a single, albeit averaged pressure curve. This aspect of the model is not predictive, and statistical deviations in the sense that you imply them are not strictly applicable. We are predicting the development of end-gas reactions when combustion follows that particular, but representative, pattern.

While we did not examine olefins in gasoline, we did look at compression machine data for olefins and concluded that they are not quite the same as aromatics. In fact, there are two factors which lead to sensitivity. These factors are, first, temperature and, second, charge density. If one increases temperature, then one tends to degrade sensitive fuels. This is particularly important with regard to olefins. Furthermore if one decreases charge density, one again tends to degrade sensitive fuels. This is less significant with the olefins and more significant with the aromatics.

H. K. Newhall

One final question. Did you at any time look at full-boiling-range fuels in the situation where a lot of the octane quality is in the heavier hydrocarbons?

L. J. Kirsch

We did not do so in the full knock model. We have looked at full-boiling-range fuels in the compression machine, and, in terms of ignition delay, they behaved rather as one might expect. For example, a full-boiling-range gasoline with a given sensitivity shows ignition delays and trends in ignition delay intermediate between those for the toluene reference fuel and the primary reference fuel of corresponding research octane number, which have very high sensitivity and zero sensitivity, respectively.

U. Montalenti *(National Research Council, Italy)*

Have you found a correlation between sensitivity to ignition delay and the surface tension of the fuel? If so, how does it depend on temperature?

L. J. Kirsch

No, we didn't get into the physical aspects of this at all. The model is purely for a homogeneous gas phase. Aspects such as surface tension are not taken into account.

P. N. Blumberg *(Ford Motor Company)*

You showed pretty good agreement between the predicted and measured crank angle at which knock occurs. The model assumes that the unburned charge is uniform in temperature and yet the kinetics scheme shows a negative temperature coefficient. I wonder what the effect of including a boundary layer treatment in which the temperature is allowed to vary between the wall temperature and the bulk unburned gas temperature would have on the predicted results. Do you think that knock would be predicted to occur preferentially at some point in the boundary layer?

L. J. Kirsch

That's a very pertinent question. People who use compression machines are interested in knowing gas temperatures. Because, of course, there will be a boundary layer, the temperature in the middle of the chamber will be higher than the temperature at the walls. To cut a long story short, this decision wasn't taken lightly. We did a lot of work looking at the temperature inside the compression machine to determine the best criterion for working this out. In the end we decided to assume that it was entirely homogeneous and used simple perfect gas laws to estimate a bulk temperature, which was then assumed to be uniform across the chamber. That was the assumption that went into the fitting of the temperatures I plotted on that graph (Fig. 2 of paper). Having made that assumption, of course, there is no point in making more sophisticated assumptions in the engine cycle simulation. To answer your question constructively and quantitatively, one would have to go back and re-analyze the compression machine data, since the model is calibrated to that assumption. As a qualitative comment, under low pressure laboratory conditions ignitions can certainly be observed propagating from cool to hot regions of an unstirred reactor. Whether this would occur under the high pressure, turbulent, short time scales of an engine is another question.

G. A. Karim *(University of Calgary, Canada)*

The data from compression ignition machines are free from residual effects. Our computations show that if you use one fuel, like methane, the autoignition of methane in the engine is extremely sensitive to the kinetic and thermal effects of the residuals of exhaust gases from the previous cycle. Therefore, for the same data obtained from compression ignition machines, if you alter the contribution of the residuals you should get substantially different behavior. Therefore, it is important for any model to take account of these residuals.

L. J. Kirsch

Maybe this is true with methane. However, I would be surprised if the exhaust gas residuals carried out of one cycle to another have any more than a purely physical effect in terms of the specific heat of the mixture. This effect, of course, is taken into account in the model. If you have other evidence I would be interested in seeing it.

I. M. Khan *(Renault, France)*

Just one small comment regarding the effect of mixing rate on ignition delay in diesels. One may find experimentally that the mixing rate may be too slow, leading to very long delays. This has been observed to be the reason for cold starting problems. One therefore increases the delay to increase the mixing rates. When you are mixing at a fast rate, you do not see an increase in delay. Have you any thoughts about that?

L. J. Kirsch

I think that there is an implied assumption in what you say. What you inferred was that the mixing rate is the same for every element of fuel that is injected into the mixture. However, when the fuel is injected it breaks up, and I suppose the answer is that different elements of fuel have different mixing rates, and one of them will have just the right mixing rate. Thus the implication of our simple simulation is that the optimum 'site' of ignition within the spray may vary with such factors as injection rate.

S. C. Sorenson *(University of Illinois)*

What is the maximum temperature limit that you would expect this sort of mechanism to be valid for; does it have any relation to the kinetics that actually occur in a flame?

L. J. Kirsch

The maximum temperature that we get to in the compression machine is about 900 K, which, of course, is way below flame temperatures. The fact is that the mechanism gets simpler in conceptual terms as you go up in temperature. The unfortunate thing, as far as modeling knock is concerned, is that you're in this region where you have this very complex behavior in that the temperature coefficient changes as a function of temperature. But as the temperature increases, the mechanism simplifies and you eventually get to a straight-line Arrhenius plot which is much simpler to handle. This indicates that perhaps you don't need such a complex model to model the generalized kinetics of flames.

G. L. Borman *(University of Wisconsin)*

In using the model to analyze ignition in a diesel engine, it appeared that you presupposed the coupling between temperature and air-fuel ratio. Did you uncouple these parameters to determine their separate effects? What I am thinking is that these parameters are not necessarily coupled so simply in a diesel spray.

L. J. Kirsch

No. We haven't looked into that area. Until the point where self reactions generate a significant heat release, the temperature of the fuel-air mixture is determined by a simple heat balance as cold fuel mixes with hot air.

TOWARD A COMPREHENSIVE MODEL FOR COMBUSTION IN A DIRECT-INJECTION STRATIFIED-CHARGE ENGINE

T. D. BUTLER, L. D. CLOUTMAN, J. K. DUKOWICZ and J. D. RAMSHAW

Los Alamos Scientific Laboratory, Los Alamos, New Mexico

R. B. KRIEGER

General Motors Research Laboratories, Warren, Michigan

ABSTRACT

A description is given of a numerical model that calculates the fluid dynamics in an axisymmetric engine cylinder. The model is used to simulate the compression and expansion strokes, including the effects of fuel injection, mixing, ignition, combustion, and nitric oxide formation. The fluid dynamics computation is based on the ICED-ALE finite difference technique for flows of arbitrary Mach number using a combined Lagrangian-Eulerian description. Some advantages of this approach include (i) the ability to use locally fine zoning, (ii) the ability to treat arbitrarily shaped and moving boundaries (such as a cupped piston), and (iii) the reduction of numerical diffusion that arises in purely Eulerian difference methods. A new numerical scheme is described for the spray injection of fuel droplets into the engine cylinder. The method treats a spectrum of droplet sizes in the fuel spray and includes vaporization. The combustion of the turbulent stratified fuel-air mixture is calculated by a procedure similar to that used in the RICE and APACHE computer codes, which solve the governing equations for the species transport, mixing, chemical reactions and accompanying heat release in a multicomponent system. The combustion chemistry is simplified by the use of global kinetic equations. Calculational results are presented for a cupped piston geometry under operating conditions typical of those contemplated for direct-injection stratified-charge engines. Particular emphasis is placed on the spatial and temporal distribution of fuel following injection, and on nitric oxide production during combustion.

NOTATION

β	reaction rate preexponential
C_D	particle (droplet) drag coefficient
C_{pk}	specific heat of particle (droplet) at constant pressure k
C_v	specific heat at constant volume
$C_{v\alpha}$	energy temperature ratio for species α
D	species diffusivity
D_k	particle (droplet) drag function
$\underline{\underline{\mathbf{e}}}$	rate of strain tensor
η	heat-transfer parameter defined by Eq. 24
λ	second coefficient of viscosity
h	specific enthalpy
h^*	heat-transfer coefficient defined by Eq. 23
h_α	specific enthalpy of species α
h_{pk}	specific enthalpy of particle (droplet) k
h_v	specific enthalpy of vapor leaving droplet surface
I	specific internal energy
$\underline{\underline{\mathbf{I}}}$	unit tensor
K	thermal conductivity
M_α	molecular weight of species α
m_k	mass of particle k
μ	first coefficient of viscosity
p	pressure
ϕ	fuel-air equivalence ratio
Q	effective heat of reaction
q_k	heat transfer to droplet k

$\dot{Q}_c$	rate of heat release by chemical reactions
R	universal gas constant
r	radial coordinate
r_k	particle (droplet) radius
ρ	density
ρ_α	density of species α
$(\dot{\rho}_\alpha)_c$	rate of change of density of species α from chemical reactions
$(\dot{\rho}_\alpha)_p$	rate of change of density of species α from evaporation
T	temperature
T_s	droplet wet-bulb temperature
T_k	temperature of particle (droplet) k
T_0	droplet temperature at which no evaporation occurs
t	time
θ	void fraction in spray equations
$\mathbf{u}$	velocity
$\mathbf{u}_g$	velocity of control volume surface in Eqs. 8–10
$\mathbf{u}_{pk}$	velocity of particle (droplet) k
V, S, n	volume, surface, and normal to surface for integration volume in Eqs. 8–10
z	axial coordinate

INTRODUCTION

Increased concern with long-term energy reserves and environmental quality has given impetus to efforts to enhance the efficiency and reduce the undesirable emissions of the automotive internal combustion engine. One engine concept that holds promise for improvement in these areas is the direct-injection stratified-charge (DISC) engine. Ideally, this concept will combine simplicity of design and the favorable ignition characteristics of a rich mixture with the higher efficiency and lower emissions that characterize lean-mixture operation.

In an attempt to gain understanding of and insight into the complex coupled

References pp. 259-260.

processes that occur in such an engine, we are developing computational models that simulate the single-cylinder dynamics of an engine cycle. Among the important physical effects that need to be represented are fluid dynamics at low Mach number, two-phase flow (spray injection of fuel droplets), turbulence, the ignition process, chemistry (turbulent flame propagation, pollutant formation), wall heat transfer and the associated phenomenon of wall quenching, and moving-boundary surfaces (valves and piston). Although emphasis is presently placed on the DISC engine, most of these effects are also of importance in other spark-ignition or diesel engines. The associated models are therefore quite generally applicable.

The overall approach of our modeling effort is to solve the governing partial differential equations describing the above processes using multidimensional finite-difference techniques. These techniques provide detailed spatial and temporal descriptions of the dynamical processes. An essential part of our approach is the ICED-ALE method [1, 2] for solving fluid flow problems at all flow speeds using an arbitrary Lagrangian-Eulerian computing mesh. This implicit method is naturally suited to engine simulations because it avoids the Courant sound-speed time-step limit to which explicit methods are subject, and it also provides the capability to represent curved and/or moving boundary surfaces. We have extended the basic ICED-ALE method to include species transport, mixing, and chemical reactions. This extension was implemented in a manner similar to that of the RICE [3] and APACHE [4] codes.

Crucial to the operation of a DISC engine is the distribution of fuel in the cylinder at the time of ignition. To predict this distribution, we have developed a new model for fuel-spray behavior [5]. The model represents the spray as discrete particles that exchange mass, momentum and energy with the gas. This approach has the advantages of allowing the convenient representation of a spectrum of particle sizes and eliminating Eulerian diffusion of the particles. The coupling terms between the particles and gas are treated in an implicit manner similar to that of the IMF method [6].

The important effects of turbulence have so far been modeled in a rudimentary fashion, using a scalar eddy viscosity. We have not yet developed models for wall heat transfer or wall quenching.

Practical engine configurations will invariably involve fully three-dimensional (3D) dynamics, and the models are being developed with this in mind. Our ultimate goal is the development of a comprehensive 3D computer program that simultaneously treats all the important physical effects. Those models that have been developed to date are currently embodied in axisymmetric two-dimensional (2D) pilot codes. The ICED-ALE reactive fluid dynamics method is contained in a computer code that is a simplified version of CHOLLA, while the particle-fluid spray model is contained in the SOLA-SPRAY code. These will soon be combined into a single unified 2D code, which will serve as a prototype for the final 3D program.

In the remainder of this paper we describe the current status of our modeling effort in greater detail. The next section describes the ICED-ALE method, extended to include species transport and chemistry. Following that, the particle-fluid fuel spray model and the chemical reactions currently in use are described.

The primary fuel consumption is represented by a single-step "global" reaction, while NO formation is estimated using extended Zeldovich kinetics. Species concentrations for the NO calculations are obtained from a modified form of the equilibrium model of Olikara and Borman [7]. The subsequent section is devoted to a discussion of some of the limitations of the models as they stand today, and prospects for their improvement. Illustrative calculations are then described in which we present a comparison between homogeneous and stratified operation in an axisymmetric cylinder with a cupped piston. Finally, concluding remarks are given.

REACTIVE FLUID DYNAMICS

The partial differential equations that govern multicomponent gas dynamics, species mixing, and chemical reactions are described in this section. A full description of the ICED-ALE computing method used to solve these equation is not given in this paper; more complete descriptions are found in Refs. 1, 2 and 8. However, a brief outline of the procedure is presented in order to familiarize the reader with the basic concepts of the method.

The equations given below apply either to laminar flow or to the mean flow in a turbulent medium. In the former case the transport coefficients have their molecular values, while in the latter case they must be regarded as turbulent eddy diffusivities. In the turbulent case, these equations are justified by separating the flow variables into mean and fluctuating components. The mean values are defined by the familiar mass-weighted averaging procedure, and the gradient-flux approximation is used to model the fluctuating terms. This leads to equations of the same form as in the laminar case, except that the turbulent energy flux is proportional to the internal energy gradient rather than the temperature gradient. However, if gradients in specific heat are neglected these two gradients are proportional, so that the turbulent equations become identical in form to the laminar ones; only the transport coefficients are modified.

Governing Equations—The continuity equation for species α [9] is

$$\frac{\partial \rho_\alpha}{\partial t} + \underline{\nabla}\cdot(\rho_\alpha \underline{u}) = \underline{\nabla}\cdot[\rho D \underline{\nabla}(\rho_\alpha/\rho)] + (\dot{\rho}_\alpha)_c \tag{1}$$

in which ρ_α is the mass density of species α, $\underline{u}$ is the mixture velocity, D is the species diffusivity, and $(\dot{\rho}_\alpha)_c$ represents the rate of change of density of species α as a result of chemical reactions. The expressions that we use for $(\dot{\rho}_\alpha)_c$ are given in a following section. The global density ρ is given by

$$\rho = \sum_\alpha \rho_\alpha, \tag{2}$$

in which the indicated summation is over all species α.

Momentum conservation is expressed by

$$\frac{\partial \rho \underline{u}}{\partial t} + \underline{\nabla}\cdot(\rho \underline{u}\underline{u}) = -\underline{\nabla} p + \underline{\nabla}\cdot(\mu \underline{\underline{e}} + \lambda \underline{\underline{I}}\,\underline{\nabla}\cdot\underline{u}) \tag{3}$$

References pp. 259-260.

in which p is the pressure, $\underline{\mathbf{I}}$ is the unit tensor, μ and λ are the first and second coefficients of viscosity, respectively, and $\underline{\mathbf{e}}$ is the rate-of-strain tensor. In 2D cylindrical coordinates the non-zero components of $\underline{\mathbf{e}}$ for non-swirling flows are given by

$$
\begin{aligned}
e_{rr} &= 2\frac{\partial u}{\partial r} \\
e_{\theta\theta} &= \frac{2u}{r} \\
e_{zz} &= 2\frac{\partial v}{\partial z} \\
e_{zr} &= \frac{\partial u}{\partial z} + \frac{\partial v}{\partial r}
\end{aligned}
\tag{4}
$$

where u and v are the radial and axial components of velocity, respectively.

The internal energy equation is

$$\frac{\partial \rho I}{\partial t} + \underline{\nabla}\cdot(\rho I \underline{\mathbf{u}}) + p\underline{\nabla}\cdot\underline{\mathbf{u}} = \underline{\nabla}\cdot K\underline{\nabla}T + \frac{\mu}{2}\underline{\mathbf{e}}:\underline{\mathbf{e}} + \lambda(\underline{\nabla}\cdot\underline{\mathbf{u}})^2 + \dot{Q}_c \tag{5}$$

where I is the specific internal energy, K is the thermal conductivity, T is the temperature and $\dot{Q}_c$ is the rate of heat release by chemical reactions. T and I are related by

$$I = C_v T \tag{6}$$

and

$$C_v \equiv \frac{1}{\rho}\sum_\alpha C_{v\alpha}(T)\rho_\alpha \tag{7}$$

The quantity $C_{v\alpha}(T)$ is not actually a specific heat; it is rather the energy-temperature ratio for species α. This ratio is useful because it is a weaker function of temperature than the partial energy of species α. The temperature dependence of $C_{v\alpha}(T)$ accounts for both dissociation and temperature-dependent specific heat. Notice that I is the total specific internal energy less the energy of chemical bonding. The enthalpy diffusion contribution to the heat flux has been neglected in Eq. 5 for simplicity, since it is ordinarily very small. However, this contribution will be included for completeness in the unified 2D fluid-particle code.

The total pressure is computed as the sum of partial pressures:

$$p = \sum_\alpha \frac{R}{M_\alpha}\rho_\alpha T$$

where R is the universal gas constant and M_α is the molecular weight of species α.

Method of Solution—The ICED-ALE scheme forms the basis for the solution of Eqs. 1 through 7. Following Ref. 1, Eqs. 1, 3 and 5 are more conveniently expressed in the ICED-ALE method when integrated over a volume V, whose surface may be moving with an arbitrarily prescribed velocity, $\mathbf{u}_g$. Denoting the surface of V by S and the outward normal on S by $\mathbf{n}$, these equations become

$$\frac{d}{dt}\int_V \rho_\alpha dV - \int_S \rho_\alpha(\underline{\mathbf{u}}_g - \underline{\mathbf{u}})\cdot\underline{\mathbf{n}}dS = \int_S (\rho D\underline{\nabla}\rho_\alpha/\rho)\cdot\underline{\mathbf{n}}dS + \int_V (\dot{\rho}_\alpha)_c dV \quad (8)$$

$$\frac{d}{dt}\int_V \rho\underline{\mathbf{u}}dV - \int_S \rho\underline{\mathbf{u}}(\underline{\mathbf{u}}_g - \underline{\mathbf{u}})\cdot\underline{\mathbf{n}}dS = -\int_S p\underline{\mathbf{n}}dS + \int_S (\mu\underline{\underline{\mathbf{e}}} + \lambda\underline{\underline{\mathbf{I}}}\,\underline{\nabla}\cdot\underline{\mathbf{u}})\cdot\underline{\mathbf{n}}dS \quad (9)$$

and

$$\frac{d}{dt}\int_V \rho I dV - \int_S \rho I(\underline{\mathbf{u}}_g - \underline{\mathbf{u}})\cdot\underline{\mathbf{n}}dS + \int_V p\underline{\nabla}\cdot\underline{\mathbf{u}}dV$$

$$= \int_S (K\underline{\nabla T})\cdot\underline{\mathbf{n}}dS + \int_V \left(\frac{\mu}{2}\underline{\underline{\mathbf{e}}}:\underline{\underline{\mathbf{e}}} + \lambda(\underline{\nabla}\cdot\underline{\mathbf{u}})^2 + \dot{Q}_c\right) dV \quad (10)$$

We note that if $\underline{\mathbf{u}}_g = \underline{\mathbf{u}}$, the equations are expressed in their Lagrangian form. If $\underline{\mathbf{u}}_g = 0$, the equations take the Eulerian form.

The finite difference approximations to Eqs. 8 through 10 are obtained by identifying the integration volumes with the cells of a finite-difference mesh. The computational mesh is composed of cells whose cross sections are arbitrarily shaped quadrilaterals, as shown in Fig. 1. The velocity components, u and v, and spatial coordinates, r and z, are defined at cell vertices. The quantities ρ, ρ_α, I, T and p are cell-centered quantities. The quantity V in Eqs. 8 and 10 is the cell volume. Because velocities are located on cell vertices, the volume V in Eq. 9 is defined by connecting the diagonals of the four cells adjacent to the vertex.

$\underline{\mathbf{u}}_g$ is the velocity of the grid and provides the flexibility to permit arbitrary movement of the grid with respect to the fluid. The significance of this in an engine calculation is that the computing mesh can move with the piston in a

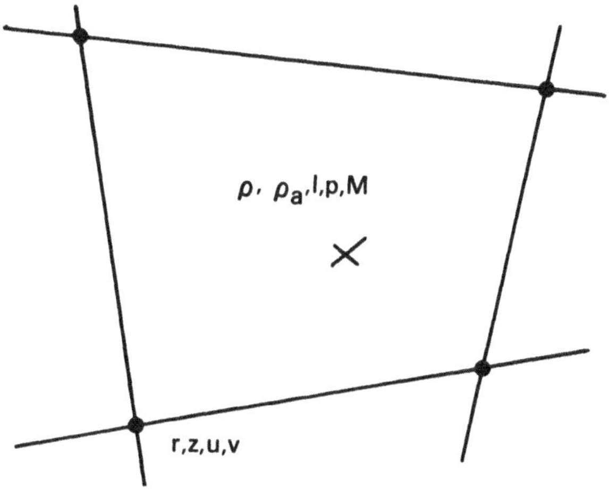

Fig. 1. The assignment of variables in a typical ICED-ALE computing zone.

References pp. 259-260.

mostly Lagrangian manner, thereby minimizing the deleterious effects of numerical diffusion. Some deviation of the grid velocity from the local fluid velocity is required in order to keep the mesh from distorting severely as the fluid distorts. This can happen in regions of shear along the rigid boundaries or in recirculating flow regions.

Each computational cycle consists of three phases. Phase I is a complete explicit Lagrangian time step, except that the vertices are not moved. Phase II consists of an iterative procedure that determines the pressure gradients that produce a self-consistent set of velocities, densities and pressures. This phase eliminates the sound speed from the Courant stability condition, thereby permitting the use of a much larger time step than is allowed in purely explicit methods. Phase III is the rezone phase that moves the grid to its new configuration.

Specifically, the calculation procedure is as follows:

(i) The viscous contributions to the momentum equations are computed, together with the diffusion fluxes of energy and chemical species.
(ii) The local chemical reaction rates are calculated and the ρ_α and I are updated accordingly.
(iii) Changes to the velocity components arising from gradients in the pressures from the previous time steps are computed.
(iv) Explicit changes to the internal energy are made to account for pressure work terms and viscous dissipation.
(v) We iterate the pressure, density, and velocity fields with a point-relaxation scheme to obtain values consistent with Eqs. 8 and 9 and the equation of state.
(vi) Cell vertices are moved to their new positions.
(vii) The fluxes of mass, momentum, and internal energy resulting from the motion of the grid relative to the fluid are computed using a mixture of centered and donor-cell differencing. Using these fluxes, final values of species mass, momentum, internal energy, and pressure are determined for the cycle.

SPRAY DYNAMICS

The DISC concept involves the injection of a fuel spray to produce the desired stratification of fuel and air at the time of ignition. Because it is so important to the success of the concept, it is necessary to predict the fuel-air mixture distribution accurately.

Modeling of fuel sprays has been largely empirical. Recent developments in modeling multidimensional two-phase flow [6, 10] offer a new approach for modeling sprays. We attempted to use these techniques to calculate the dynamics of nonevaporating sprays but we were discouraged for the following reasons. The behavior of sprays is strongly influenced by the distribution of droplet sizes composing the spray. The two-fluid model, however, is inherently limited to a single "effective" droplet size. This limitation might be removed by going to a multifluid model, but this would be very costly in terms of computer time.

Furthermore, practical injectors have orifice sizes that are exceedingly small in comparison to typical dimensions of an engine cylinder. To resolve the spray shape in the vicinity of the orifice with a finite difference grid would require prohibitively small mesh zoning. Finally, the continuum representation of the spray in these models usually requires that computation of the spray equations be performed throughout the mesh, even in cells that contain no droplets. To overcome these various difficulties would make excessively large demands on computer time and storage.

These considerations led to the development of an alternative procedure which represents the spray by discrete particles, rather than by continuous distributions [5]. This amounts to a statistical (Monte Carlo) formulation of the problem, since the finite number of particles used represents a sample of the total population of particles. Each computational particle is considered to represent a group of particles possessing the same characteristics, such as size, composition, etc. The use of discrete particles eliminates the problems of a single effective droplet size, Eulerian numerical diffusion and lack of resolution in the vicinity of the injector. We have found that the number of particles required to achieve satisfactory accuracy is not excessive, and since frequently the particles are limited to a part of the mesh, the computing requirements are less severe. An important characteristic of this development is its adaptability for use in different multidimensional fluid dynamics codes, especially in 3D and ICED-ALE codes.

The application of this technique to nonevaporating sprays has been successful [5], as illustrated in Fig. 2. In this figure are summarized the results of a number of calculations for a spray from a single-orifice injector. Agreement between computations and the experimental data of Hiroyasu and Kadota [11] has been obtained for a wide range of gas pressures. The figure also illustrates the resulting spray shape. The shape of the droplet-size distribution function was determined from the experiments, while the mean droplet size was not the experimental value but was inferred from a Weber number criterion for droplet stability.

The particles interact with the continuum phase by exchanging mass, momentum and energy, as well as by volume displacement of the gas. The implicit numerical formalism used in our technique permits computation of the strong coupling between the droplets and gas which frequently occurs in fine sprays.

There are certain limitations in this technique. We assume that droplets are spherical, and we neglect small effects such as the Basset force, virtual mass contributions, and non-uniform temperature within each droplet. Collective effects between droplets, which are expected to be important in the close vicinity of the injector orifice, are also ignored. For a given injector, measured droplet size and velocity distributions near the injector orifice are required as boundary conditions by this method. This information makes it unnecessary to compute the details of the atomization process and the inter-particle interactions in the thickest part of the spray. However, the required information (injector droplet sizes and velocities) is generally not available, although efforts are currently underway at the General Motors Research Laboratories to alleviate this problem. In addition, we currently rely upon available empirical correlations for the phase-change rate and the particle-drag function.

References pp. 259-260.

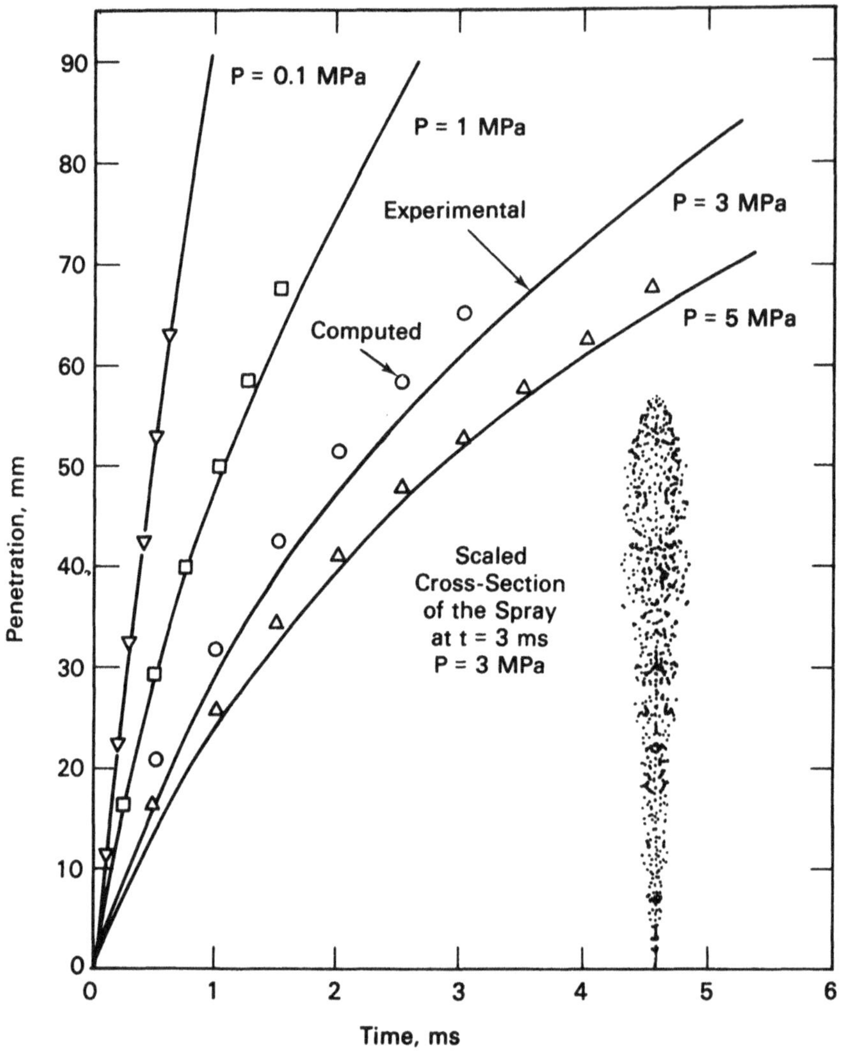

Fig. 2. Comparison between theory and experiment for the spray tip penetration versus time for various gas pressures. The inset shows a scaled cross section of the particle plot, indicating the spray shape.

The effects of internal droplet circulation are neglected. In the Stokes' regime, theory [12] predicts that the change in drag due to circulation is of the order of the ratio of gas to liquid viscosities, which is very small. The effect on evaporation, if any, should be incorporated in the empirical heat-transfer correlation. In any case, experimental evidence [13] suggests that droplets of the size considered here do not have internal circulation, possibly due to surface-tension effects.

Governing Equations—Omitting chemical reactions, the continuity equation for species α is

$$\frac{\partial \theta \rho_\alpha}{\partial t} + \underline{\nabla} \cdot (\theta \rho_\alpha \underline{u}) = \underline{\nabla} \cdot \theta \rho D \underline{\nabla} \left(\frac{\rho_\alpha}{\rho} \right) + (\dot{\rho}_\alpha)_p \tag{11}$$

where θ is the void fraction, or the fraction of the volume occupied by the gas. The presence of the void fraction in this and the following equations accounts for the displacement effect of the particles. The last term $(\dot{\rho}_\alpha)_p$ is the source term giving the rate of change of density of species α due to change of phase (evaporation) of the particles. We assume that the droplets are composed of a single component so that

$$\begin{aligned} (\dot{\rho}_\alpha)_p &= 0; \ \alpha \neq v \\ (\dot{\rho}_v)_p &= -\frac{1}{V} \sum_k \frac{dm_k}{dt} \end{aligned} \tag{12}$$

where the subscript v represents the fuel vapor. The indicated summation is over all the particles in a sub-volume V, which is taken to be the volume of a computational cell in the finite difference solution; m_k is the mass of particle k.

The fluid momentum equation is given by

$$\frac{\partial \theta \rho \underline{u}}{\partial t} + \underline{\nabla} \cdot (\theta \rho \underline{u}\underline{u}) = -\theta \underline{\nabla} p + \underline{\nabla} \cdot \theta (\mu \underline{\underline{e}} + \lambda \underline{\underline{I}} \underline{\nabla} \cdot \underline{u})$$

$$- \frac{1}{V} \sum_k D_k (\underline{u} - \underline{u}_{pk}) \tag{13}$$

in which $\underline{u}_{pk}$ is the velocity of particle k, and D_k is the particle drag function given by

$$D_k = 6\pi \mu r_k + \tfrac{1}{2}\pi r_k^2 \rho C_D \left| \underline{u} - \underline{u}_{pk} \right| \tag{14}$$

where r_k is the particle radius and C_D is the drag coefficient. This form of the drag function assumes that the drag force is the sum of the Stokes' drag and the form drag.

The fluid internal energy equation is written as

$$\theta \rho \left[\frac{\partial I}{\partial t} + \underline{u} \cdot \underline{\nabla} I \right] = \frac{\theta p}{\rho} \left[\frac{\partial \rho}{\partial t} + \underline{u} \cdot \underline{\nabla} \rho \right]$$

$$+ \underline{\nabla} \cdot \theta \left[K \underline{\nabla} T + \rho D \sum_\alpha h_\alpha \underline{\nabla} \left(\frac{\rho_\alpha}{\rho} \right) \right] + \theta \left[\frac{\mu}{2} \underline{\underline{e}} : \underline{\underline{e}} + \lambda (\underline{\nabla} \cdot \underline{u})^2 \right]$$

$$+ \frac{1}{V} \sum_k \left[\left(D_k - \frac{1}{2} \frac{dm_k}{dt} \right) \left| \underline{u} - \underline{u}_{pk} \right| - q_k + (h - h_v) \frac{dm_k}{dt} \right] \tag{15}$$

in which q_k is the heat-transfer rate from the gas to particle k, and h_α is the specific enthalpy of species α. The total gas enthalpy h is defined by

$$h = \frac{1}{\rho} \sum_{\alpha} \rho_\alpha h_\alpha \tag{16}$$

The first term on the right-hand side of Eq. 15 represents the work due to the compression of the gas (the PdV work). The following terms take into account thermal conduction, enthalpy diffusion and viscous dissipation within the gas. The remaining terms account for the rate of energy change due to the presence of particles. The first part of the term involving the relative velocity between the particles and the gas accounts for particle friction. The second part of this term accounts for the fact that fuel vapor comes off at particle velocity and must be accelerated to gas velocity. The last term in the particle contribution is the enthalpy change due to the mixing of the vapor whose enthalpy must be brought from its initial enthalpy at evaporation to the local gas enthalpy.

The particle equations are

$$\frac{d}{dt} \underline{\mathbf{x}}_{\mathrm{k}} = \underline{\mathbf{u}}_{\mathrm{pk}} \tag{17}$$

$$\frac{d}{dt} m_k \underline{\mathbf{u}}_{\mathrm{pk}} = -\frac{m_k}{\rho_p} \underline{\nabla} p + D_k(\underline{\mathbf{u}} - \underline{\mathbf{u}}_{\mathrm{pk}}) \tag{18}$$

and

$$m_k \frac{dh_{pk}}{dt} = q_k + (h_v - h_{pk}) \frac{dm_k}{dt} \tag{19}$$

where ρ_k is the particle density and h_{pk} is the specific enthalpy of particle k. The pressure-gradient term in the momentum equation is usually small but it is retained for consistency with the corresponding term in the two-fluid equations [6]. The last term in the energy equation represents the energy required for phase change (the latent heat of evaporation).

To complete this set of equations we need a model to specify the evaporation rate. We start from the assumption that in thermal equilibrium the droplet is at its wet-bulb temperature, T_s. The equilibrium is a balance between heat transfer to the droplet and the latent heat of evaporation carried away by the vapor

$$(h_v - h_{pk}) \frac{dm_k}{dt} = -q_k \tag{20}$$

It can be expected that a large portion of the droplet lifetime is spent while in this equilibrium. However, the droplet must be heated from its initial temperature to wet-bulb temperature. For this portion of its lifetime we assume that

$$m_k \frac{d}{dt} (C_{pk} T_k) = \eta (T_s - T_k) \tag{21}$$

$$(h_v - h_{pk}) \frac{dm_k}{dt} = -q_k + \eta (T_s - T_k) \tag{22}$$

where T_k is the droplet temperature, assumed to be uniform within the droplet, C_{pk} is its specific heat, and η defines a characteristic time constant for droplet heating. The heat transfer, q_k, is obtained from an empirical correlation [14]; it can be expressed in the form

$$q_k = h^*(T - T_k) \tag{23}$$

where T is the local gas temperature. The parameter η is then deduced from Eq. 22:

$$\eta = h^* \left(\frac{T_0 - T}{T_0 - T_s} \right) \tag{24}$$

where T_0 is that droplet temperature at which no evaporation takes place. (More precisely, it is that temperature for which $\rho_v/\rho \mid_{\text{surf}} = \rho_v/\rho \mid_\infty$.)

Method of Solution—Briefly, the solution procedure through one time cycle is accomplished in the following way:

(i) Intermediate values of specific internal energy are obtained from Eq. 15 omitting the conduction, enthalpy diffusion, and particle contributions.
(ii) New particles are injected into the computing region.
(iii) Characteristic evaporation temperatures T_0 and T_s, together with the rate of change of T_s with gas temperature $(\partial T_s/\partial T)$, are computed.
(iv) The evaporation of the particles is calculated using a predictor-corrector scheme to allow accurate calculation for cases with strong evaporation. In the predictor phase, evaporation of particles is calculated using previous values of T and T_s, and this is used to predict new values of T and T_s.
(v) In the corrector phase, final values for the particle radii, temperatures, and phase change contributions to mass and energy are calculated.
(vi) The specific internal energy is then updated for conduction, enthalpy diffusion, and all contributions due to the presence of particles such as evaporation, heat transfer between phases, enthalpy of mixing between the vapor and the gas, and particle friction.
(vii) Eq. 17 is used to update particle positions using velocities from the previous time step. The new void fraction is then computed.
(viii) Again using a predictor-corrector method, the particle drag function is evaluated, and intermediate particle and gas velocities are obtained using a linearly implicit technique.
(ix) The final advanced-time velocities, pressures, and total densities are obtained by iteration using a technique similar to that used in RICE [3].
(x) The species densities are then obtained using Eq. 11.
(xi) Finally, the particle velocities are updated to account for changes in gas velocities and pressures obtained in the iteration.

CHEMISTRY

For the calculations reported in this paper, we used a single-step global chemical kinetics equation to model the combustion of the fuel-air mixture. The overall

References pp. 259-260.

reaction is represented by

$$2\ C_8H_{18} + 25\ O_2 \rightarrow 16\ CO_2 + 18\ H_2O \tag{25}$$

The reaction proceeds according to the equation

$$(\dot{\rho}_f)_c = -\beta \rho_f \rho_{O_2} \exp\left(\frac{-T_A}{T}\right) \tag{26}$$

where the subscript f denotes the fuel (n-octane), T_A = 15 780 K, and β = 9.38 $\times$ 10^{11} mL/(g s). The effective heat of reaction Q = 4.46 $\times$ 10^{11} ergs per gram of fuel.

Nitric oxide (NO) formation is modeled by the extended Zeldovich mechanism. Since this mechanism was applied to engines by Lavoie, Heywood and Keck [15], its use has become almost universal and is supported by extensive comparisons with experiment [16–20]. The mechanism consists of three reactions:

$$N_2 + O \rightleftharpoons NO + N \tag{27}$$

$$O_2 + N \rightleftharpoons NO + O \tag{28}$$

$$N + OH \rightleftharpoons NO + H \tag{29}$$

with N atoms assumed to be in steady state. Since reactions 27–29 are slow relative to the hydrodynamic and energy-release processes, the rate expression corresponding to reactions 27–29 is evaluated only once per time step.

To minimize the increase in computation time necessary to include NO kinetics, several simplifications were made. First, reaction 29 was omitted for computational cells which are lean. This omission introduces negligible error. For lean cells, the only species concentration required for the NO rate expression which is not already available in the code is the oxygen-atom concentration. To compute this, an approximate method is used. First, an estimate is obtained from the $O_2 - O$ equilibrium constant. Second, a correction is applied to the estimate. The correction is a regression expression obtained by fitting the ratio of the O atom concentration obtained from an equilibrium routine [7] to the values computed from the $O_2 - O$ equilibrium constant. The regression expression is accurate in the range $0.82 \leqslant \phi < 1.0$, 2200 K $\leqslant T \leqslant$ 3200 K, 1.0 MPa $\leqslant p \leqslant$ 3.0 MPa. For $\phi < 0.82$, the equilibrium constant-based O atom concentration is accurate. This method provides a fast alternative to calling the equilibrium routine.

For stoichiometric and rich cells, reaction 29 contributes significantly to NO formation and is therefore included. To evaluate the NO rate expression when reaction 29 is included, values for O_2, O, OH and H concentrations are required. Here, the equilibrium routine [7] is called. Although this approach is more time consuming than that used for lean cells, there is no obvious alternative. The rate constants used for reactions 27 and 28 are from Ref. 21 and that for reaction 29 is from Ref. 22.

CURRENT MODEL LIMITATIONS

In this section we discuss the principal limitations of the computational model at present, together with some possible approaches for alleviating them.

One of the principal limitations of our model in its present form is the use of a scalar eddy diffusivity to represent the effects of turbulence. This diffusivity, of course, affects the flame propagation as well as the turbulent transport, and its value must be adjusted in conjunction with the chemical reaction rate to give the correct flame speed. This procedure yields surprisingly good results on a case-by-case basis, but the diffusivity and reaction rate must be readjusted for each case, depriving the model of a truly predictive capability. To remedy this deficiency, it will be necessary to represent the physics of turbulent flame propagation more faithfully. An essential aspect of this physics is the fact that the different turbulent length scales have entirely different effects on the flame, the smaller length scales acting as an effective increase in the molecular diffusivities, while the larger length scales act to wrinkle the flame [9]. To represent these effects in a fundamental way it would be necessary to use a turbulence model that distinguishes between the different turbulent length scales. This level of description is not needed in most other contexts, and most available turbulence models do not provide it. For example, the two-parameter turbulence models that have become so widely used in the past decade [23–26] contain, either explicitly or implicitly, only a single dominant turbulent length scale. These models are therefore unlikely to be directly applicable to the prediction of turbulent flame propagation.

It may, however, be possible to utilize such models in conjunction with existing empirical correlations for turbulent flame speeds [27]. One might proceed as follows. The use of the turbulent diffusivity predicted by the two-parameter model, with no other changes, would give the wrong flame speed, because this diffusivity includes the effects of all the turbulent length scales. To compensate for this, the chemical reaction rate would have to be altered locally so that the flame propagates at the speed determined by the correlation. If successful, this approach would result in a predictive capability whose scope coincides with that of the correlation used.

A more fundamental approach to the problem would be the use of the sub-grid-scale (or "large eddy simulation") method of turbulence modeling [23–26, 28–30]. In this method, only the turbulent length scales that are too small to resolve in the finite-difference mesh are modeled; the larger length scales are simply calculated along with the mean flow. The sub-grid-scale method, therefore, automatically distinguishes between the different turbulent length scales in the required manner, and there is hope that the irregularities present in the large-scale motions will produce the flame-wrinkling effect. Unfortunately, this approach will probably require somewhat finer zoning than is economically feasible for full engine cylinder calculations on present-day computers. (Moreover, it is strictly applicable only in three dimensions.) In the near future, the usefulness of the sub-grid-scale approach is likely to be limited to simpler geometries, where

References pp. 259-260.

fine zoning is feasible. Such calculations could yield useful insight into turbulent flame propagation that will influence the more phenomenological approaches.

Alternative approaches to the modeling of turbulent flames are currently being pursued elsewhere [31-33]. Some of this work may ultimately prove useful in modeling in-cylinder engine dynamics.

The use of simplified chemical kinetics to represent the overall combustion of the fuel is another rather severe limitation. Unfortunately, computer time and storage constraints preclude the inclusion, in multidimensional models, of the large numbers of chemical species and reactions that would be required to represent the reaction mechanism in detail. Even if this were not the case, the complete reaction mechanism and the associated rate expressions are largely unknown for practical fuels. For these reasons, the use of simplified chemical kinetics in multidimensional calculations will be necessary for the foreseeable future. Efforts to develop such simplified schemes are being pursued [34-36]. These schemes can give remarkably good results in limited regions of parameter space, but tend to lack universality over a wide range of pressures and compositions. This is another respect in which the present model lacks a truly predictive capability.

We have not yet developed models for wall heat transfer and wall quenching. The necessarily coarse resolution does not allow the calculation of wall heat transfer directly from the temperature gradient at the wall. It may be necessary to specify the wall heat flux from appropriate empirical correlations. This procedure, however, is not expected to be sufficient for an accurate prediction of wall quenching because of the inadequate spatial resolution and the detailed chemical kinetic considerations involved.

TABLE 1

Engine Configuration and Operating Conditions

Bore	98.4 mm
Stroke	95.5 mm
Cup diameter	49.2 mm
Cup depth	33.4 mm
Compression ratio	10:1
Speed	1600 r/min
Period of Revolution	37.5 ms
Swirl Ratio	0.
Fuel	n-octane
Timing:	
Injection	15.4 ms (32° BTDC)
Ignition—Stratified	17.1 ms (15.8° BTDC)
Ignition—Homogeneous	17.4 ms (13° BTDC)
Initial conditions (BDC):	
Temperature	314 K
Pressure	0.1 MPa

NUMERICAL EXAMPLES

Two numerical examples are presented in this section to contrast homogeneous and stratified-charge combustion in a cupped piston geometry. The overall equivalence ratio of the homogeneous charge is approximately stoichiometric (ϕ = 0.95), typical of premixed operation, while the stratified-charge case has an overall equivalence ratio of 0.5, typical of DISC engine operation. Both cases use the same engine geometry and the operating parameters listed in Table 1.

Chemistry parameters used in these calculations are those presented earlier, except that β was increased by a factor of two in the stratified case. This was done to account partially for increased turbulence due to the spray. An overall eddy diffusivity of 10^4 mm^2/s is assumed, together with unity values of turbulent Prandtl and Schmidt numbers. We have not modeled valve motion and we have neglected wall heat transfer. Scavenging efficiency is assumed to be 90%.

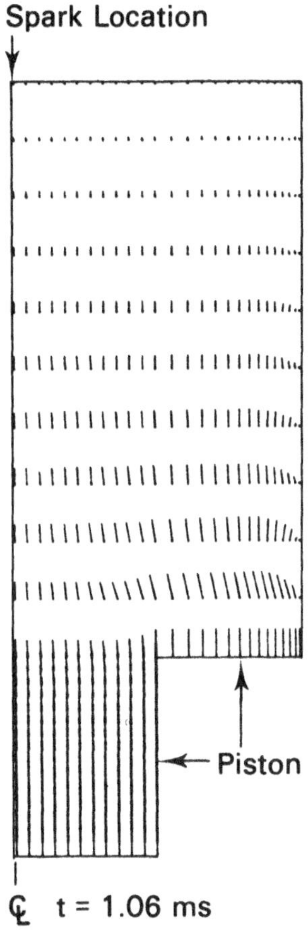

Fig. 3. Velocity vectors early in the compression stroke. Time = 1.06 ms. Maximum velocity = 1.42 m/s. Piston velocity = 1.42 m/s.

No-slip velocity boundary conditions are assumed on all walls including the piston, which moves with its prescribed sinusoidal velocity. The initial computing mesh is composed of an axisymmetric region of rectangular cross section with 25 zones in each direction. An internal obstacle of 14 × 15 zones is positioned in the lower right-hand corner of the mesh to form the cup geometry. As the piston approaches TDC, the top 10 rows of cells are compressed. For computational efficiency we rezone the computing mesh to eliminate the unnecessary zones in the squish region. At TDC the number of axial zones is reduced to three, while retaining the original number of zones in the cup. We ignite the fuel-air mixture in one zone in the upper left-hand corner of the mesh by rapidly raising the gas temperature to 1600 K.

Uniform Charge—The calculated flow patterns generated within the cylinder are similar during the compression stroke for both the uniform and stratified

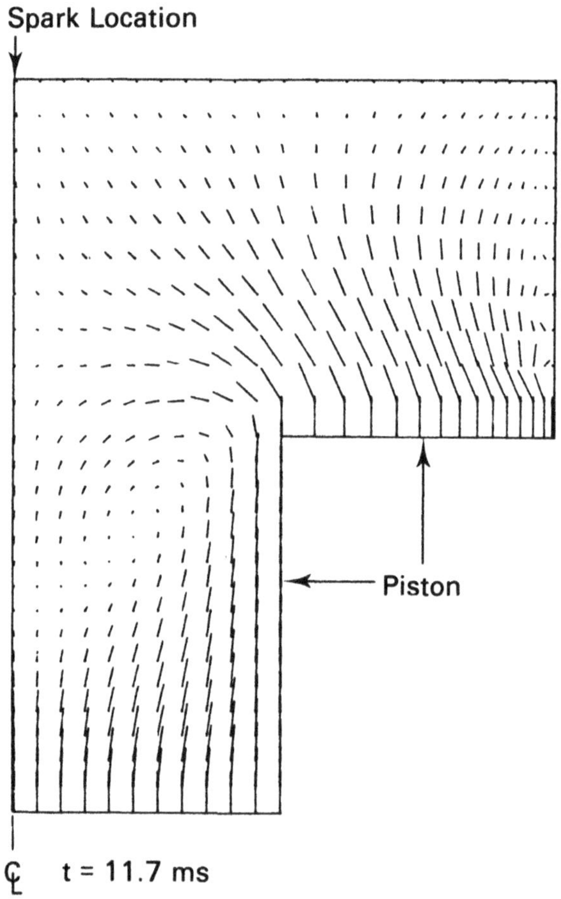

Fig. 4. Velocity vectors at an intermediate time during compression. Time = 11.7 ms. Maximum velocity = 7.40 m/s. Piston velocity = 7.40 m/s.

cases. For this reason, we restrict our discussion of the flow field to the uniform-charge case. Fig. 3 illustrates the velocity vectors early in the compression stroke. The flow field can be conveniently divided into several regions with distinct properties. In the core region the velocity field is essentially uniform in the radial direction and varies linearly from the value of velocity at the piston face to zero velocity at the cylinder head. Boundary layers are formed along the vertical sides of the cup and the cylinder wall. The boundary layer is most pronounced along the cylinder wall near the piston. A vortex appears to be formed in this corner, which would be most apparent in a frame fixed to the piston. At a later time (68° BTDC), Fig. 4 shows the initial influence of squish. A recirculating flow is established in the upper portion of the cup with a net inflow of mass into the cup. Up to this time the process has been an adiabatic compression with essentially uniform density, temperature and pressure. The next figure (Fig. 5) shows the velocity field at the time of ignition (13° BTDC). Here the squish region has been rezoned to be four cells thick. The flow field is dominated by the squish, which produces a strong jet as a result of the rapid decrease of the squish volume.

The velocity vectors in these figures are scaled such that the maximum velocity vector length does not exceed two times the cell dimensions. The maximum velocities, as well as the piston velocities, are noted in the figure captions. We

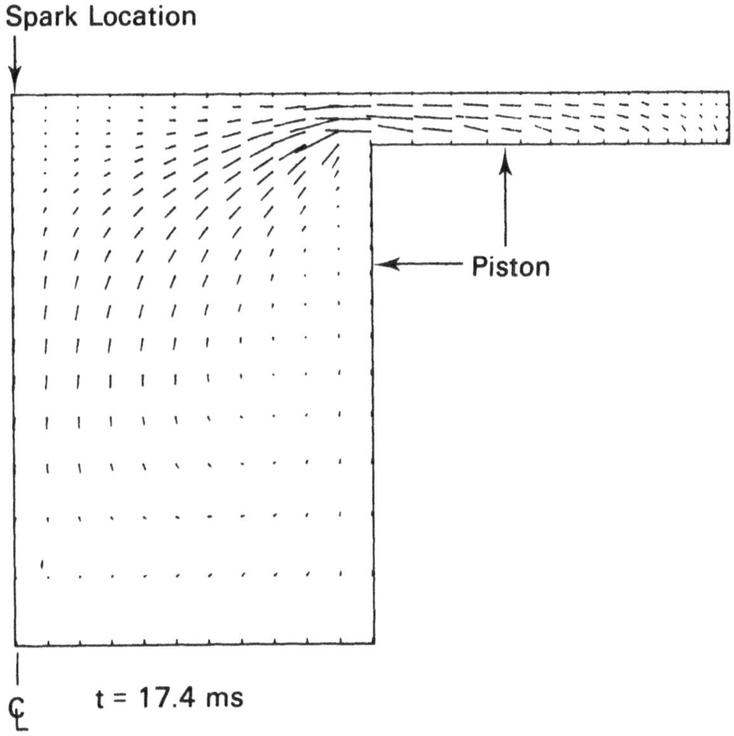

Fig. 5. Velocity vectors at the time of ignition. Time = 17.4 ms. Maximum velocity = 21.25 m/s. Piston velocity = 1.80 m/s.

References pp. 259-260.

note that the squish velocities in Fig. 5 are in excess of 2.5 times the maximum piston speed of 8 m/s.

The progress of the flame propagating in the cylinder can be illustrated by contour plots of gas temperature and *NO* density. Fig. 6 shows these plots at 19.7 ms (9.5° ATDC). The flame thickness is approximately two cell widths (~4 mm). The maximum temperature behind the flame is about 2500 K. The unburned gas temperature is approximately 680 K. The flame front remains continuous until it reaches the rim of the cup. It then splits into two independent segments, one in the squish region and the other in the cup, as shown in Fig. 7. The burn is completed at 21.1 ms (22.6° ATDC). The maximum gas temperature of 2920 K is reached just after the burn is finished.

The *NO* contours in Figs. 6 and 7 show a small amount of *NO* produced in the flame, with significantly larger amounts generated behind the flame. *NO* production continues even after the combustion is completed, as will be discussed later.

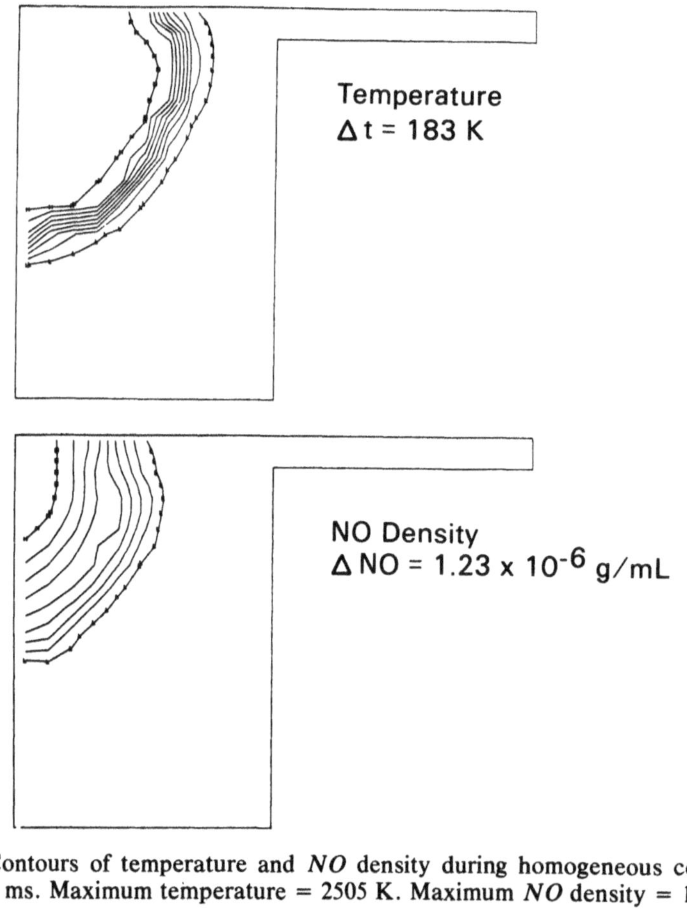

Fig. 6. Contours of temperature and *NO* density during homogeneous combustion. Time = 19.7 ms. Maximum temperature = 2505 K. Maximum *NO* density = 1.24×10^{-5} g/mL.

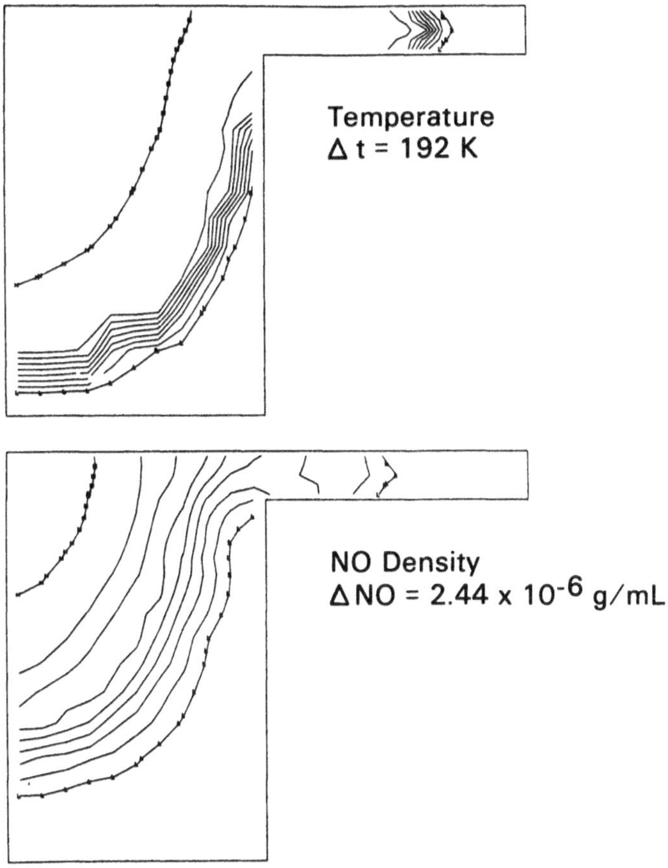

Fig. 7. Contours of temperature and *NO* density during homogeneous combustion, after the flame enters the squish region. Time = 20.6 ms. Maximum temperature = 2692 K. Maximum NO density = 2.46×10^{-5} g/mL.

Stratified Charge—The stratified charge calculation was made in three parts. The first segment was calculated using the ICED-ALE code, without fuel, for the period from BDC to injection time (15.4 ms, 32° BTDC). The second segment was performed with the SOLA-SPRAY code for the injection period from 32° BTDC to shortly before ignition (16.2° BTDC). The SOLA-SPRAY solution provided the initial conditions for the ICED-ALE code to perform the third segment of the calculation. This procedure was necessary in the absence of a unified code. We expect to incorporate the particle-spray model into the ICED-ALE code as the next step in our development of a comprehensive program.

Prior to injection, the solution is similar to the uniform-charge case. At the start of injection, this solution is used to supply initial conditions for velocity, density and temperature to the SOLA-SPRAY code. The SOLA-SPRAY computing mesh is similar to the ICED-ALE mesh except that the squish region is omitted. The effects of squish are supplied as time-varying mass flux boundary

References pp. 259-260.

TABLE 2

Injection Characteristics

Fuel	n-Octane
Spray Included Angle	16°
Droplet Sauter Mean Diameter	6 μ
Injection Pressure	14.7 MPa
Max. Spray Velocity	135 m/s
Mass of Injected Fuel	0.0225 g
Duration of Injection	1.5 ms
Initial Fuel Temperature	367 K

conditions obtained from continuing the ICED-ALE solution through the period of injection.

Injector characteristics are listed in Table 2. Details of the spray model used are given in an earlier section. The injection lasts for 1.5 ms and terminates at 16.9 ms. The droplets are then allowed to evaporate until 17.1 ms. Fig. 8 shows the resulting SOLA-SPRAY velocity vectors, particle distribution, gas tempera-

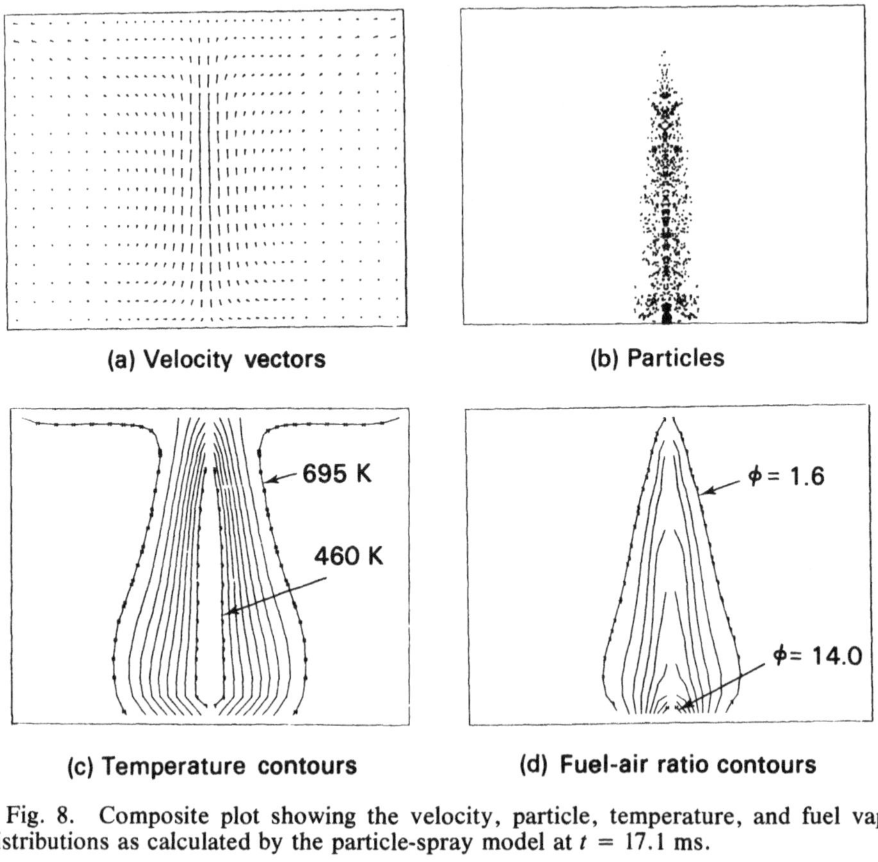

Fig. 8. Composite plot showing the velocity, particle, temperature, and fuel vapor distributions as calculated by the particle-spray model at $t = 17.1$ ms.

ture, and evaporated fuel-air mass-ratio contours at this time. The velocity induced by the spray is very pronounced in the core of the spray. Vapor density is highest along the axis near the head of the spray. The evaporation of the droplets has produced a region of low temperature along the center of the spray. The high vapor density and low temperature in this region retard evaporation of the remaining droplets. Approximately 20% of the total mass of fuel is still in the form of droplets at this time.

At 17.1 ms (16.2° BTDC), the remaining droplets are instantaneously converted to vapor, the resulting fuel distribution and velocity field are mapped onto the ICED-ALE mesh, and the solution is continued. The fuel is ignited shortly after this time (15.8° BTDC). Ignition procedure is the same as in the uniform-charge case. Fig. 9 shows the fuel mass-fraction contours at the time the solution is transferred from SOLA-SPRAY, and at 18.8 ms, which is 1.7 ms after ignition. The flame propagates down the column of fuel, consuming the leaner portions of the mixture. The flame is slowed when it reaches the richer portions of the mixture near the spray tip at the bottom of the cup. The flame front remains

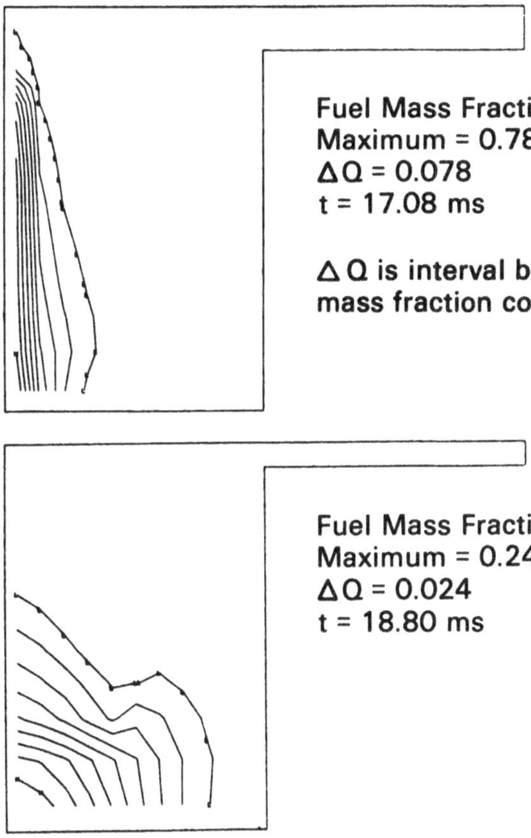

Fig. 9. Contours of the fuel mass fraction during stratified combustion at the time of ignition and during the premixed portion of the burn.

References pp. 259-260.

nearly stationary while fuel and oxygen diffuse into the flame from opposite sides. The flame thus changes in character from a pre-mixed to a diffusion flame. In Fig. 10 are shown contours of gas temperature and *NO* density at 18.8 ms. At this time the flame front has penetrated about three-quarters of the way into the cup. The flame propagates no further into the fuel-rich core region, but continues to propagate at the periphery of the fuel-rich region. As the piston withdraws, the hot gases are drawn partially back into the expanding squish region, as shown in Fig. 11. The maximum temperature behind the flame at 18.8 ms is about 2600 K. This temperature decreases as the expansion continues, becoming about 2400 K by 22.4 ms. During the premixed portion of the burn, the *NO* production occurs immediately behind the flame in the hottest portion of the gas. During the later diffusion-controlled portion of the burn, the region of appreciable *NO* production is less localized and tends to be on the side of the flame where oxygen is available (see Fig. 11).

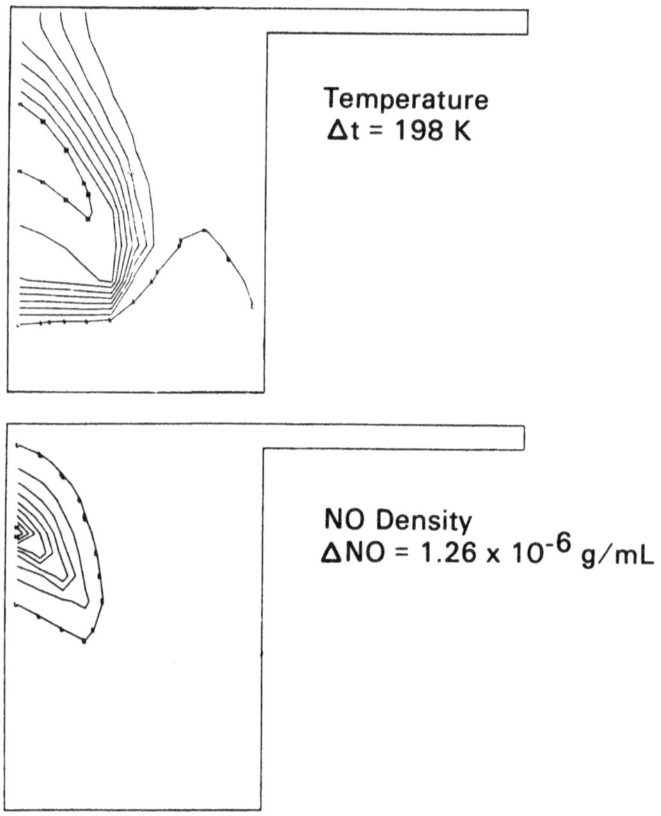

Fig. 10. Contours of temperature and *NO* density during the combustion of the stratified mixture in the premixed portion of the burn. Time = 18.8 ms. Maximum temperature = 2596 K. Maximum *NO* density = 1.27×10^{-5} g/mL.

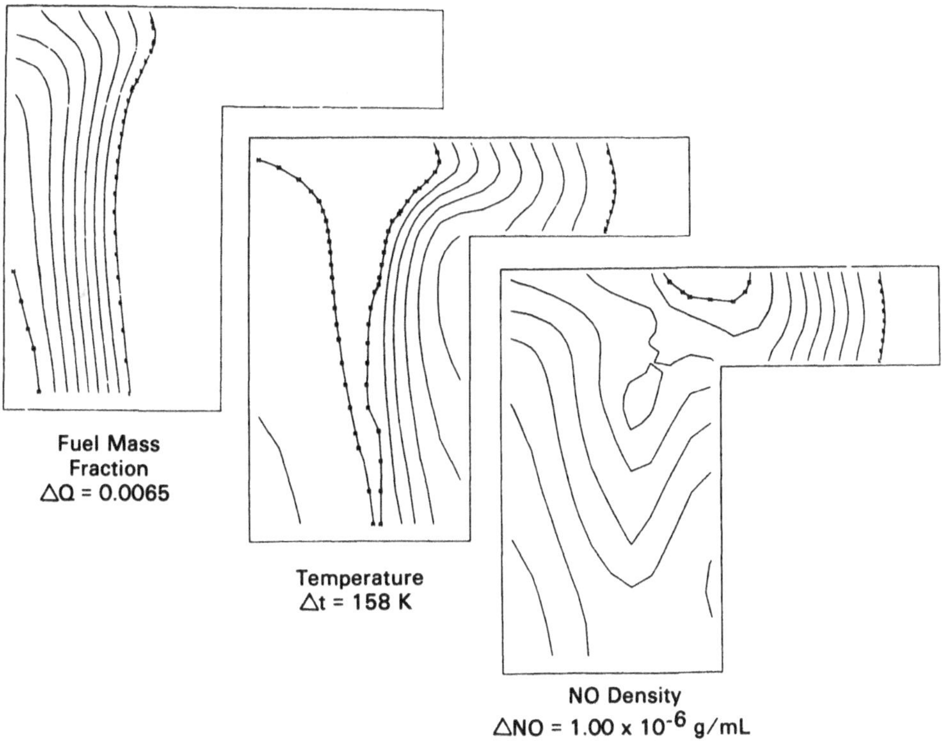

Fig. 11. Contours of fuel mass fraction, temperature, and *NO* density in the diffusion-limited portion of the stratified combustion. Time = 22.4 ms. Maximum fuel mass fraction = 0.0647. Maximum temperature = 2417 K. Maximum *NO* density = 1.04×10^{-5} g/mL.

Discussion of Results—The difference between homogeneous and stratified-charge combustion is clearly illustrated in Fig. 12, which shows the mass of unburned fuel as a function of time. Immediately after ignition, the flame kernel in the homogeneous case propagates slowly. As the combustion proceeds, the unburned gases are compressed, and the combustion rate increases exponentially. The burn terminates suddenly as the remaining unburned gases are consumed. The initial period of the stratified-charge combustion is very similar to the homogeneous case while the portion of the charge within the flammability limits is burned. Following this initial period, the combustion abruptly changes character as the flame becomes controlled by diffusion of oxygen from one side and fuel from the other. This produces a relatively slow rate of fuel consumption. When the calculation was terminated at 26 ms, approximately 5% of the original fuel remained unburned, but combustion was continuing.

The pressure histories of the two cases are shown in Fig. 13. These pressures are taken from a location on the cylinder wall in the squish region. They are representative of overall cylinder pressures because pressure variations are typically 1%. The initial portion of the curves shows the adiabatic compression due

References pp. 259-260.

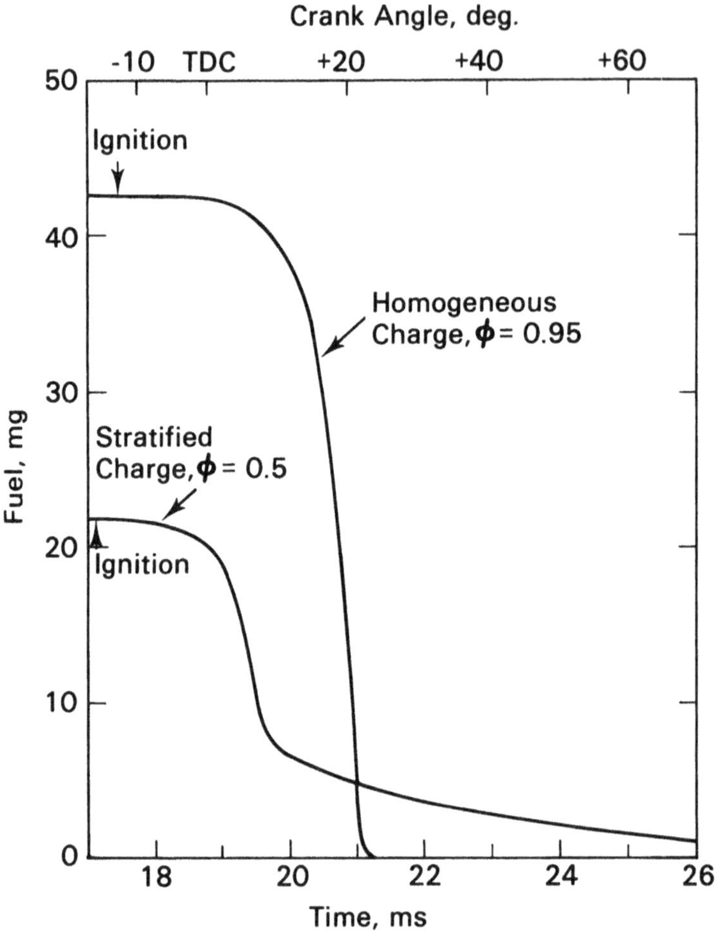

Fig. 12. Comparison of the unburned fuel time histories for the homogeneous and stratified cases.

to piston motion. The steep rise in pressure due to the burn correlates with the rapid rate of fuel consumption shown in Fig. 12. Following the burn, the pressure drops due to the withdrawal of the piston. These pressure histories show small oscillations which are due to weak pressure waves generated within the cylinder. The fluctuations near 17 ms are caused by the injection and/or ignition process. We believe that the oscillations near the peak of the pressure curve in the homogeneous case are resonant oscillations of the gas in the squish region, excited by the pressure pulse at the end of combustion. The oscillations in the stratified-charge pressure curve occur near TDC and are correlated with the end of the rapid premixed-burn phase.

The total mass of NO in the cylinder as a function of time is plotted in Fig. 14. The rate of NO production is most rapid during the rapid-burn phase of the homogeneous case. The production of NO continues even after the burn is

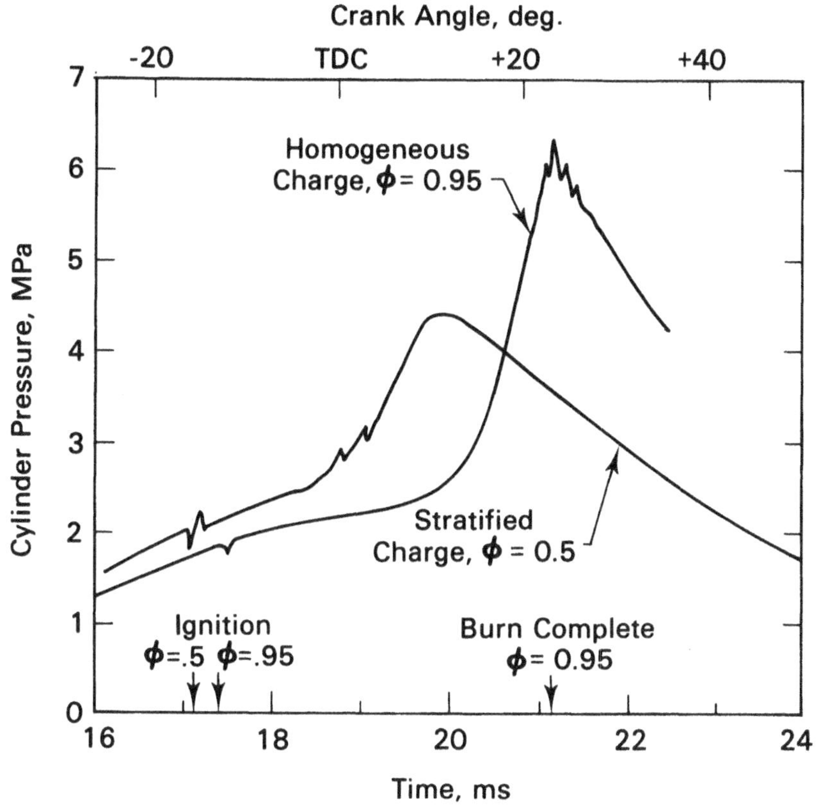

Fig. 13 Comparison of the cylinder pressure time histories for the homogeneous and stratified cases.

completed, reaching a maximum mass of *NO* of approximately 3.5 mg. Toward the end of the calculation, the reverse *NO* reactions dominate, reducing the total mass of *NO* in the cylinder. We expect that these reactions would freeze out later in the expansion, but we did not carry the calculation out far enough in time to verify this phenomenon. The stratified-charge case shows that the total mass of *NO* reaches a maximum of approximately 0.7 mg and appears to be frozen out at this value. The relative amounts of *NO* produced are in qualitative agreement with experiment.

Although the numerical examples reported in this paper do not incorporate the effects of swirl, we have made other calculations in which swirl was included. The principal difference observed is the presence of Ekman layers on the piston face and on the cylinder head. The Ekman layers produce an inward radial flow along these surfaces and therefore promote mixing.

CONCLUDING REMARKS

The satisfactory operation of DISC engines hinges on the ability to achieve the desired stratified fuel-air distribution, and this distribution must be accurately

References pp. 259-260.

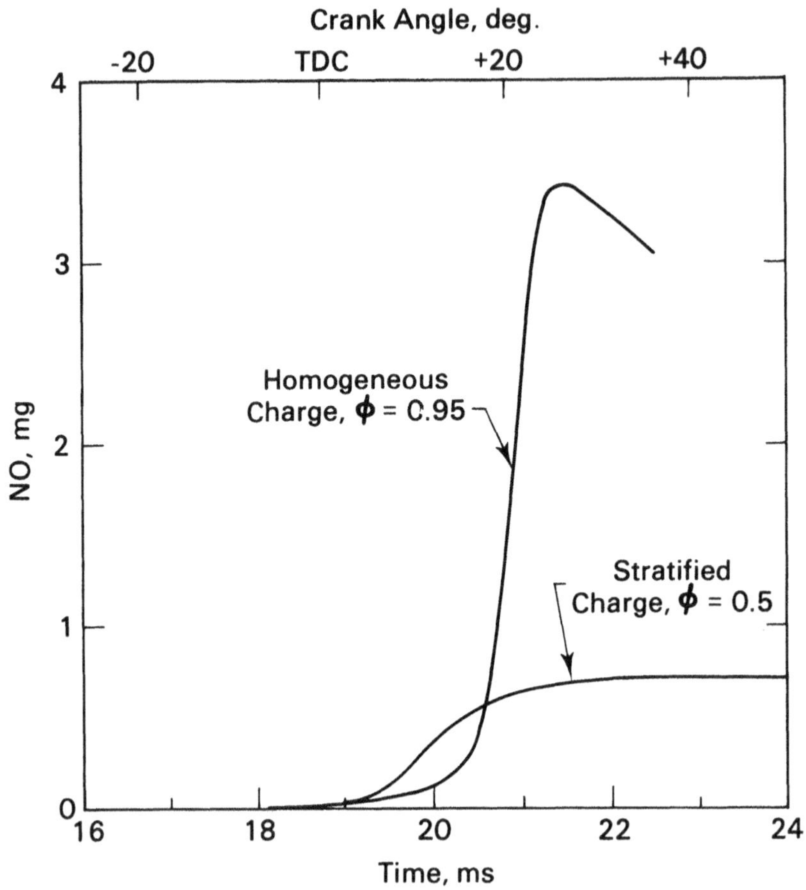

Fig. 14. Comparison of the time histories of the total mass of *NO* for the homogeneous and stratified cases.

calculated if a numerical model is to be useful as a predictive tool. The primary advantage of a comprehensive multidimensional model is that it allows one to calculate the detailed spatial and temporal distributions within the cylinder. These details are frequently difficult to obtain experimentally and are not available from simple theories. For example, Fig. 8 shows that a single-hole axisymmetric injector produces a fuel distribution that is very rich in the core, which tends to inhibit further droplet evaporation in this region. Since such a distribution is undesirable and will produce very slow burning, a detailed 3D model can be used to study the effects of design modifications or changes in operating conditions in an effort to improve the fuel distribution and to promote the evaporation of the remaining liquid droplets.

ACKNOWLEDGMENT

We would like to express our thanks to our colleagues, A. A. Amsden and H. M. Ruppel, for their efforts to make the ICED-ALE code more efficient. This work was performed under the auspices of the U. S. Department of Energy Contract W-7405-ENG-36.

REFERENCES

1. C. W. Hirt, A. A. Amsden and J. L. Cook, "An Arbitrary Lagrangian-Eulerian Computing Method for All Flow Speeds," J. Comp. Phys., Vol. 14, pp. 227-253, 1974.
2. A. A. Amsden and C. W. Hirt, "YAQUI: An Arbitrary Lagrangian-Eulerian Computer Program for Fluid Flow at All Speeds," Los Alamos Scientific Laboratory Report LA-5100, 1973.
3. W. C. Rivard, O. A. Farmer, T. D. Butler and P. J. O'Rourke, "RICE: A Computer Program for Multicomponent Chemically Reactive Flows at All Speeds," Los Alamos Scientific Laboratory Report LA-5812, 1975.
4. J. D. Ramshaw and J. K. Dukowicz, "APACHE—A Generalized-Mesh Eulerian Computer Code for Multicomponent Chemically Reactive Fluid Flow," Los Alamos Scientific Laboratory Report LA-7427, 1979.
5. J. K. Dukowicz, "A Particle-Fluid Numerical Model for Liquid Sprays," J. Comp. Phys. (accepted for publication).
6. F. H. Harlow and A. A. Amsden, "Numerical Calculation of Multiphase Fluid Flow," J. Comp. Phys., Vol. 17, pp. 19-52, 1975.
7. C. Olikara and G. L. Borman, "A Computer Program for Calculating Properties of Equilibrium Combustion Products with Some Applications to I. C. Engines," SAE Paper No. 750468, 1975.
8. J. L. Norton and H. M. Ruppel, "YAQUI User's Manual for Fireball Calculations," Los Alamos Scientific Laboratory Report LA-6261-M, 1976.
9. F. A. Williams, "Combustion Theory," Addison-Wesley, Reading, Mass., 1965.
10. J. R. Travis, F. H. Harlow and A. A. Amsden, "Numerical Calculation of Two-Phase Flow," Nuc. Sci. Eng., Vol. 61, pp. 1-10, 1976.
11. H. Hiroyasu and T. Kadota, "Fuel Droplet Size Distribution in Diesel Combustion Chamber," SAE Trans., Vol. 83, Paper No. 740715, pp. 2615-2624, 1974.
12. H. Lamb, "Hydrodynamics," 6th ed., Dover, New York, 1945.
13. W. N. Bond and D. A. Newton, "Bubbles, Drops, and Stokes' Law (Paper 2)," Phil. Mag., Vol. 5, pp. 794-800, 1928.
14. M. C. Yuen and L. W. Chen, "Heat-Transfer Measurements of Evaporating Liquid Droplets," Intern. J. Heat Mass Transfer, Vol. 21, pp. 537-542, 1978.
15. G. A. Lavoie, J. B. Heywood and J. C. Keck, "Experimental and Theoretical Study of Nitric Oxide Formaion in Internal Combustion Engines," Comb. Sci. and Tech., Vol. 1, pp. 313-326, 1970.
16. P. N. Blumberg and J. T. Kummer, "Prediction of NO Formation in Spark-Ignited Engines—An Analysis of Methods of Control," Comb. Sci. and Tech., Vol. 4, pp. 73-95, 1971.
17. K. K. Chen and R. B. Krieger, "A Statistical Analysis of the Influence of Cyclic Variation on the Formation of Nitric Oxide in Spark Ignition Engines," Comb. Sci. and Tech., Vol. 12, pp. 125-134, 1976.
18. J. B. Heywood, S. M. Mathews and B. Owen, "Predictions of Nitric Oxide Concentrations in a Spark-Ignition Engine Compared with Exhaust Measurements," SAE Paper No. 710011, 1971.
19. M. Watfa, D. E. Fuller and H. Daneshyar, "The Effects of Charge Stratification on Nitric Oxide Emission from Spark Ignition Engines," SAE Paper No. 741175, 1974.
20. G. A. Lavoie and P. N. Blumberg, "Measurements of NO Emissions from a Stratified Charge Engine: Comparison of Theory and Experiment," Comb. Sci. and Tech., Vol. 8, pp. 25-37, 1973.
21. K. L. Wray and J. D. Teare, "Shock-Tube Study of the Kinetics of Nitric Oxide at High Temperatures," J. Chem. Phys., Vol. 36, pp. 2582-2596, 1962.
22. T. M. Campbell and B. A. Thrush, "Reactivity of Hydrogen to Atomic Nitrogen and Atomic Oxygen," Trans. Faraday Soc., Vol. 64, pp. 1265-1274, 1968.
23. B. E. Launder and D. B. Spalding, "Mathematical Models of Turbulence," Academic, London, 1972.
24. F. H. Harlow, ed., "Turbulence Transport Modeling," Vol. XIV, AIAA Selected Reprint Series, AIAA, New York, 1973.

25. W. C. Reynolds and T. Cebeci, "Calculation of Turbulent Flows," in Turbulence, edited by P. Bradshaw, Springer-Verlag, Berlin, 1976.
26. W. C. Reynolds, "Computation of Turbulent Flows," Ann. Rev. Fluid Mech., Vol. 8, pp. 183–208, 1976.
27. R. G. Abdel-Gayed and D. Bradley, "Dependence of Turbulent Burning Velocity on Turbulent Reynolds Number and Ratio of Laminar Burning Velocity to R.M.S. Turbulent Velocity," Sixteenth Symposium (International) on Combustion, The Combustion Institute, Pittsburgh, Pennsylvania, pp. 1725–1736, 1976.
28. J. W. Deardorff, "A Numerical Study of Three-Dimensional Turbulent Channel Flow at Large Reynolds Numbers," J. Fluid Mech., Vol. 41, pp. 453–480, 1970.
29. J. W. Deardorff, "On the Magnitude of the Subgrid Scale Eddy Coefficient," J. Comp. Phys., Vol. 7, pp. 120–133, 1971.
30. J. W. Deardorff, "Numerical Investigation on Neutral and Unsteady Planetary Boundary Layers," J. Atmos. Sci., Vol. 29, pp. 91–115, 1972.
31. F. V. Bracco, ed., "Turbulent Reactive Flows," Comb. Sci. and Tech., Vol. 13, pp. 1–275, 1976.
32. F. C. Lockwood, "The Modeling of Turbulent Premixed and Diffusion Combustion in the Computation of Engineering Flows," Comb. and Flame, Vol. 29, pp. 111–122, 1977.
33. J. Chomiak, "Dissipation Fluctuations and the Structure and Propagation of Turbulent Flames in Premixed Gases at High Reynolds Numbers," Sixteenth Symposium (International) on Combustion, The Combustion Institute, Pittsburgh, Pennsylvania, pp. 1665–1674, 1976.
34. F. L. Dryer and I. Glassman, "The High Temperature Oxidation of Carbon Monoxide and Methane," Fourteenth Symposium (International) on Combustion, The Combustion Institute, Pittsburgh, Pennsylvania, pp. 987–1003, 1973.
35. J. R. Creighton, "A Two Reaction Model of Methane Combustion for Rapid Numerical Calculations," Preprint UCRL-79669, Rev. 1, Lawrence Livermore Laboratory, 1977.
36. C. K. Westbrook, "Fuel Motion and Pollutant Formation in Stratified Charge Combustion," Preprint UCRL-81115, Lawrence Livermore Laboratory, 1978.

DISCUSSION

W. A. Sirignano *(Princeton University)*

With regard to your spray model, I think that there is one area where improvement is desired. It is not only important to track the particle velocity and the particle size, but also the temperature in order to determine the transient vaporization rate correctly. Tracking the temperature is not a trivial thing, but I think it should be done. Another point I might make is that this particle and cell method which you have is quite interesting, but I think it's unfair for you to have made the statement that the continuum approach, the standard two-phase flow approach, automatically leads to smearing. The partial differential equations which govern the droplet characteristics are hyperbolic, and it's possible to use something like Glimm's method, as put forth by Chorin,* to remove the smearing effect. There are also corrective means with other approaches. I might make a third point. With regard to the eddy diffusivity, it seems to me that when you have strong swirl, as you seem to have, the turbulence is no longer isotropic, due to the centrifugal effects. I believe that Bradshaw pointed this out. Streamline

* A. J. Chorin, "Random Choice Methods with Applications to Reacting Gas Flows," J. Comp. Phys., Vol. 25, No. 3, pp. 253-272, 1977.

curvature could lead to that situation. That is something that should be accounted for in your turbulence model when combined with swirl.

T. D. Butler

I would like to try to address some of those points. With respect to your first point, each of those particles has a temperature associated with it, and we solved the transport equation associated with each particle for the temperature of the drop. So the temperature does change as a function of time and strongly influences the rate of evaporation. In fact in the numerical example that you saw in our film, the droplets make it all the way to the bottom of the cupped piston and tend to sit there. Because of vaporization and heat transfer between the droplets and the gas, there was a temperature gradient in the center core of the spray, resulting in a 200°C difference between the spray and ambient gas temperatures. So the mechanism is there.

Your second point, concerning smearing, is quite right. The continuum equations have convective terms giving rise to the offensive diffusion errors in an ordinary finite-difference scheme. To accurately calculate the dynamics you have to do something very special about them. A particle model handles it. You recommended one. There's a flux-corrected transport theory that might handle it. There are a number of ways of trying to get away from those effects, but this model was developed on the basis of how we were doing the overall fluid dynamics and coupling it to that in the three-dimensional code. We did not want to use the other methods for correcting the convective terms.

W. A. Sirignano

I'm not being critical of your method.

T. D. Butler

Yes, I know what you mean. We should have said that the particle model is more accurate than usual finite difference methods that do not take special care to control numerical diffusion.

Regarding the third point about the streamline curvature and eddy diffusivity, there is no way to defend our turbulence model. Suffice to say, we have to be more sophisticated.

W. A. Woods *(University of Liverpool, England)*

I'd like to make two points. One is about some experiments we did at Liverpool some years ago. This picks up on the point Tony Oppenheim made earlier about creating a novel physical model of a real engine. We made a machine which didn't have any valves. It was just a compression and expansion machine with a transparent cylinder. We arranged to dump some talcum powder into the engine to try to visualize the flow without disturbing it. Anyway, the point I'm getting at is that on one of the films we did detect this corner vortex—not the rollup vortex that we've heard quite a lot about, but one at the top of the cylinder on the downward stroke. Now, I think this vortex appeared on one film.

T. D. Butler

Did you have swirl?

W. A. Woods

No, there was no swirl. What I'm saying is that this visualization hasn't appeared in the published literature, but it's on one of the films. I must confess that you have to look very hard and be in the right frame of mind. After attending this conference, it may be much easier for me to see it now.

The second point, which is a bit more important, is that we also had some hot wires in the cylinder to measure the squish velocity. We used these at various immersions. We also did some simple calculations and found the measurements were very much lower than the calculations. Leakage past the piston rings was included in very simple computations. We found that this leakage had a predominant influence on the motion in this gap and changed the flow field radically.

D. T. Pratt *(University of Michigan)*

We developed a similar two-phase model, which we called "particle source in the cell," in the early 70's. We applied it first to droplet penetration in the gas turbine combustor. Next, we found it was remarkably good at predicting flow in cyclone separators, and we got quite excited about it. Then we decided to apply it to combustion in pulverized coal problems—and got burned when we found out that the turbulent diffusion problems were enormous. The small particles don't follow the mean motion at all, of course. They're deflected out with the gas motion in turbulent flows. You have that to look forward to. In addition to that, you sort of brushed off the thin-spray/thick-spray problem, but I hope you're not trying to tell us that your model deals with droplet/droplet interactions.

T. D. Butler

No, it does not at all, and we should have made that clear. I think Larry Cloutman mentioned that the droplets do not interact with one another. That brings up an important point. A lot of very detailed measurements are required in order to characterize an injector such that the information can be used in a model like this one. We are not dealing with jet breakup or atomization from first principles in any way. Rather, the approach is very empirical. But again, this model is being developed with the intention of using it with a three-dimensional code. We just can't go to the level of detail that's required, even if we knew how to do the whole jet breakup problem. We simply don't know the physics that well.

J. W. Daily *(University of California)*

I want to know what the philosophical implications are of plotting the calculations as points and the experiment as a solid line.

L. D. Cloutman

Probably there are more experimental numbers than calculated ones, and we drew a line through them for clarity of the graph.

L. D. Cloutman

T. D. Butler

As a matter of fact, though, the droplet penetration tip, which is the measured quantity in the calculations, evolves very much the way it does in the experiment. We do not have a nice analytical fit for the penetration. It's a matter of finding that lead particle at a given time and saying, "You're the tip." So it does come to us in discrete ways—but no philosophical problem.

T. Morel *(General Motors Research Laboratories)*

I have a question which is a practical one to people trying to model and calculate ignition. You showed the ignition point as one single point. Did you really succeed with igniting one point, or did you have to take more points?

L. D. Cloutman

We ignited all of those calculations in one computational cell. It was a finite area, but it was a single cell located at the wall.

W. G. Agnew *(General Motors Research Laboratories)*

Regarding that calculation with the DISC engine, if I gave you a lean flammability limit of 0.5, let's say, for the equivalence ratio, can you tell me how much fuel there is in the cylinder that's beyond that limit and therefore presumably wouldn't burn when the flame is ignited? Then if I can delay the spark timing relative to injection, can you tell me how much that unburned fuel quantity will change?

T. D. Butler

We do keep track of the vapor density in every computational cell, and so we can tell you the mass of fuel that's outside the equivalence ratio equal to 0.5 range. Yes, you can obtain that kind of information. Other examples of information coming from the model include nitric oxide as a function of time, and unburned fuel left in the cylinder after combustion.

SUPPLEMENTARY COMMENTS: POTENTIAL APPLICATIONS OF HOLOGRAPHIC INTERFEROMETRY TO ENGINE COMBUSTION RESEARCH

J. C. DENT

Loughborough University of Technology, Leicestershire, England

ABSTRACT

The principles of holographic interferometry are discussed in some detail and applied to the measurement of local instantaneous air-fuel ratio in an axisymmetric evaporating fuel spray injected into the quiescent combustion space of a motored engine cylinder.

Application of similarity principles indicates how the technique might be applied to the scaling of actual engine processes, thereby facilitating their study on smaller experimental engines and isothermal "bombs".

Extension of the techniques developed to include effects of wall impingement and air swirl, which results in an asymetric fuel spray development, are discussed, and further application to air utilization measurements during combustion of the injected fuel spray is considered.

NOTATION

B	transfer number
C	velocity of light in vacuum
C_p	specific thermal capacity
D_i	mass diffusion coefficient
d	orifice diameter
d'	orifice equivalent diameter
E, E_0	displacement and amplitude of electric wave
f	frequency

References pp. 290–291.

h_{fg}	enthalpy of vaporization
I	intensity of light
k	thermal conductivity
K	Gladstone-Dale constant
L	optical path length or length scale
m	mass fraction
M	molecular weight
n	refractive index
N	number of fringes
P	partial or total pressure
R	jet radius
S	fringe count
t	time
T	temperature
$\bar{U}$	spatial or temporal mean velocity
V	velocity of light in a medium
x, y, z	cartesian coordinates
α	thermal diffusivity
γ	droplet diameter
Δ	difference
λ	wavelength
μ	absolute viscosity
σ	surface tension
ρ	fluid density
ω	circular frequency

INTRODUCTION

Developments in laser technology over the past decade have resulted in a growth of new diagnostic techniques applicable to fluid dynamics and combustion research, and a revival and improvement of previously well known methods. The object of this paper is to indicate some potential uses for laser interferometry in engine combustion research.

REVIEW OF PRINCIPLES

A comprehensive exposition of optical interference is beyond the scope of this review. However, the following explanation will form a basis on which the principles of interferometry may be understood.

Nature of Light—Light is an electromagnetic wave phenomenon by which energy is transported from one point to another. The physics of electromagnetic fields is summarized in Maxwell's equations, the simplest solution of which is for the case of plane transverse waves.

The transverse electric and magnetic waves are in phase and mutually perpendicular. It is usual to discuss the electric wave in detail, because it is this wave which is of interest in optics.

The electric wave is a vector quantity normal to the direction of travel of the wave. The description of the vector direction defines the polarization state of the light, with the simplest polarization state, i.e., linear, being the state in which the direction of the electric vector remains unchanged. For the present purpose, discussion is confined to linearly polarized light. The concepts are summarized schematically in Fig. 1.

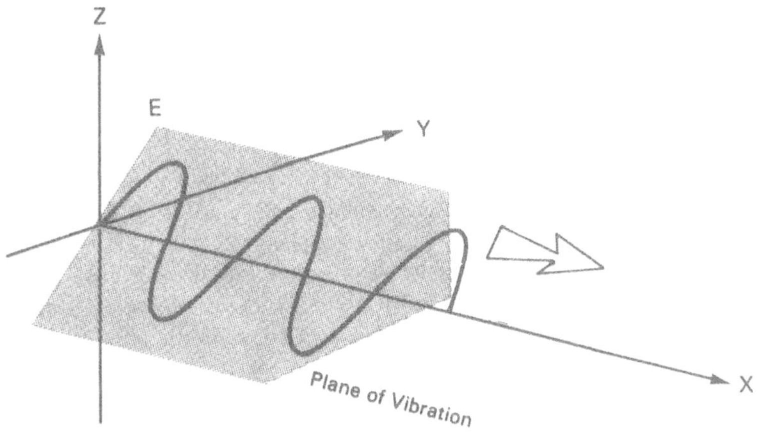

Fig. 1. Linearly polarized transverse electric wave.

References pp. 290–291.

Interference—When beams of light intersect, they obey the principle of superposition. Because of the wave-like nature of light, this summation produces patterns of varying intensity of illumination known as interference.

A beam of light can be represented approximately by wave trains of the form

$$E = E_o \sin \frac{2\pi}{\lambda} (x - Ct) \tag{1}$$

where E is the displacement, E_o the amplitude, λ the wavelength and C the velocity of the sinusoidal waveform moving in a direction x in a time t. The intensity of illumination, I, is a measure of the energy flux (per unit area and time) and is proportional to the square of the amplitude. It is more usual to express Eq. 1 in the form

$$E = E_o \sin \left(2\pi \frac{x}{\lambda} - \omega t \right) \tag{2}$$

where $\omega = 2\pi f$ and $c/\lambda = f$, the frequency of the wave. The term $2\pi(x/\lambda)$ is a "phase" associated with the wavetrain.

The superposition of two wavetrains, assuming for simplicity that each has the same frequency, is given by the simple addition

$$\begin{aligned} E &= E_1 + E_2 \\ &= E_{o_1} \sin \left[2\pi \left(\frac{x_1}{\lambda_1} \right) - \omega t \right] + E_{o_2} \sin \left[2\pi \left(\frac{x_2}{\lambda_2} \right) - \omega t \right] \end{aligned} \tag{3}$$

Through the use of standard trigonometric identities,Eq. 3 can be expressed as

$$E = \left\{ E_{o_1}^2 + E_{o_2}^2 + 2E_{o_1}E_{o_2} \cos 2\pi \left[\left(\frac{x_1}{\lambda_1} \right) - \left(\frac{x_2}{\lambda_2} \right) \right] \right\}^{1/2} \sin(\alpha - \omega t) \tag{4a}$$

where

$$\tan \alpha = \frac{E_{o_1} \sin 2\pi \left(\frac{x_1}{\lambda_1} \right) + E_{o_2} \sin 2\pi \left(\frac{x_2}{\lambda_2} \right)}{E_{o_1} \cos 2\pi \left(\frac{x_1}{\lambda_1} \right) + E_{o_2} \cos 2\pi \left(\frac{x_2}{\lambda_2} \right)} \tag{4b}$$

From Eq. 4a it can be seen that superposition has resulted in a wavetrain similar to those of the component waves but with the amplitude now a function of the phase relationship and amplitudes of the component wavetrains. The phase relationship $\cos 2\pi[(x_1/\lambda_1) - (x_2/\lambda_2)]$ depends on the distances x_1 and x_2 traveled by the two wavetrains and their wavelengths in the media traversed. The phase relationship is $+1$ when the wavetrains are "in phase", and the resultant amplitude from Eq. 4a is $(E_{o_1} + E_{o_2})$. When the wavetrains are "out of phase", the phase relationship is -1 and the resultant amplitude is $(E_{o_1} - E_{o_2})$. Between these limits, the resultant amplitude varies in a sinusoidal manner. If the wavetrains are superimposed on a screen in such a way that the difference in optical path length $(x_1 - x_2)$ traveled by the wavetrains varies over the screen, bright and dark fringes of light will be seen, these being determined by the phase relationship. These fringes are known as interference effects. It is the phase relationship between the two wavetrains that is made use of in interferometry, which will be discussed later.

An important condition for interference which can also be used to discuss the coherence length of a light source may be derived from Eq. 4a. Consider the resultant amplitude given by the superposition of N wavetrains:

$$E^2 = \sum^{N} E_{0_n}^2 + \sum_{A}^{N} \sum_{B}^{N} E_{0_A} E_{0_B} \cos 2\pi \left(\frac{x_A}{\lambda_A} - \frac{x_B}{\lambda_B} \right) \tag{5}$$

It can be seen that with N large and a random distribution of phase relationships, these will on the average tend to produce as many positive as negative values of the cosine term so that

$$\sum_{A}^{N} \sum_{B}^{N} E_{0_A} E_{0_B} \cos 2\pi \left(\frac{x_A}{\lambda_A} - \frac{x_B}{\lambda_B} \right) = 0 \tag{6}$$

and the resultant intensity will be given by the sum of the individual intensities. Thus the phase information in the light beam is lost, and hence there is no interference. This demonstrates the important point concerning the coherence of a light source. In an ordinary light source such as a tungsten filament lamp the atoms radiate independently, producing light waves which are incoherent both spatially (wavetrains have random phasing) and temporally (wavetrains emitted with differing wavelengths). Temporal coherence can be achieved using a monochromatic light source such as a mercury lamp with a green line filter (0.53 μm wavelength), and spatial coherence can be obtained by using a pinhole (to produce a plane wavetrain). The result is a spatial coherence length of the order of 0.001 m. In contrast, all the light produced by a laser is both spatially and temporally coherent. For example the spatial coherence length of a stabilized helium-neon laser is approximately 10 m, while that of a pulsed ruby laser is of the order of 1 m. With interferometry the design and quality of the optical system must be such that it operates within the coherence length of the light source to be used. The advantage of the laser source is readily apparent.

The Mach-Zehnder Interferometer—Most applications of interferometry in fluid dynamics and combustion have utilized the Mach-Zehnder interferometer, shown schematically in Fig. 2. This is an amplitude division system in which a collimated source beam is divided into an object and reference beam by use of the beam splitter M_1. The object beam, after reflection at M_3, passes through the test section and is reflected by the beam splitter M_4 into receiving optics and onto a photographic plate. The reference beam is reflected at M_2, passes through compensating windows (to compensate for the test-chamber windows) and on through the beam splitter M_4, and is combined with the object beam at the photographic plate of the camera.

With mirrors M_1 and M_4 adjusted and truly parallel, the path length of the object (x_{obj}) and reference (x_{ref}) beams will be the same, and the distance over

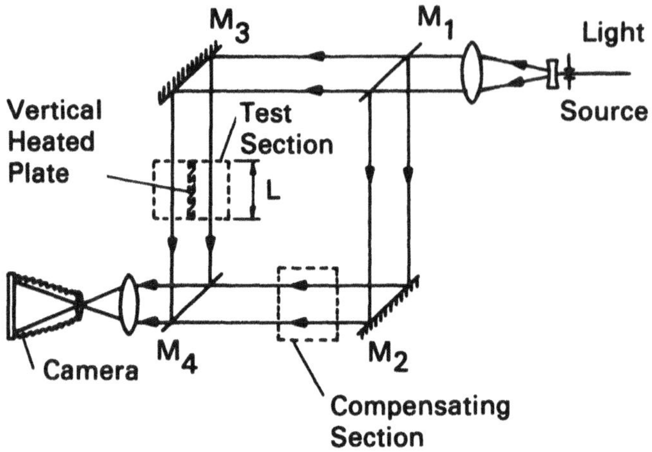

Fig. 2. Schematic of Mach-Zehnder interferometer.

which any phase change between the object and reference beams will be observed is the length of the test section, L. Therefore the bracketed term of the phase relationship will have the form

$$\left(\frac{L}{\lambda_{obj}} - \frac{L}{\lambda_{ref}}\right) \tag{7}$$

But L/λ_{obj} and L/λ_{ref} are the numbers of object and reference beam fringes N_{obj} and N_{ref} in the length L. From the definition of refractive index, n,

$$n = \frac{C}{V} = \frac{\lambda_o}{\lambda} \tag{8}$$

where V and λ are the velocity and wavelength of light in the medium under study,respectively, and C and λ_o are the velocity and wavelength of light in a vacuum. Now $L/\lambda_{obj} = n_{obj}(L/\lambda_o)$ and $L/\lambda_{ref} = n_{ref}(L/\lambda_o)$. Therefore

$$(N_{obj} - N_{ref})\lambda_o = L(n_{obj} - n_{ref}) \tag{9}$$

With the test section undisturbed, $n_{obj} = n_{ref}$, and there is uniform illumination at the camera plane. This setting is known as the "infinite fringe field". If a heated vertical flat plate is now introduced into the test section, temperature, and hence density and refractive index variations, will be set up normal to the plate. From the phase relationship and Eq. 9 it will be seen that destructive interference, i.e., dark fringes, will occur at odd multiples of $\lambda_o/2$. Hence a series of fringes will be set up adjacent to the heated plate (Fig. 3). From Eq. 9 the general fringe-shift equation for the Sth dark fringe may be written as

$$S\frac{\lambda_o}{2} = L(n_s - n_{ref}) \tag{10}$$

where $(N_s - N_{ref}) = S = 1, 3, 5, 7$, etc. n_s is the only unknown in Eq. 10 and can be established from a fringe count. The ability to measure refractive index directly with the Mach-Zehnder interferometer is apparent. In most engineering applications it is the density, ρ, or temperature, T, which is of interest.

The Gladstone-Dale equation relating fluid density, ρ, and refractive index is

$$\frac{n-1}{\rho} = K \tag{11}$$

where K is the Gladstone-Dale constant for the fluid under study. Combining Eqs. 10 and 11 yields

$$KL(\rho_s - \rho_{\text{ref}}) = \frac{S\lambda_o}{2} \tag{12}$$

From Eq. 12 we note that the number of fringes is directly proportional to the optical path length of the test section, L, and to the difference in density between the reference state and the Sth fringe.

The basic ideas of interferometry, including the interpretation of interference fringe data, have been illustrated with the well known Mach-Zehnder system. Other types of interferometer and their applications and limitations are discussed in the literature [1–4].

HOLOGRAPHIC INTERFEROMETRY

Holographic Recording—Holography is based entirely upon the excellent coherence properties of laser light sources. In the off-axis system shown sche-

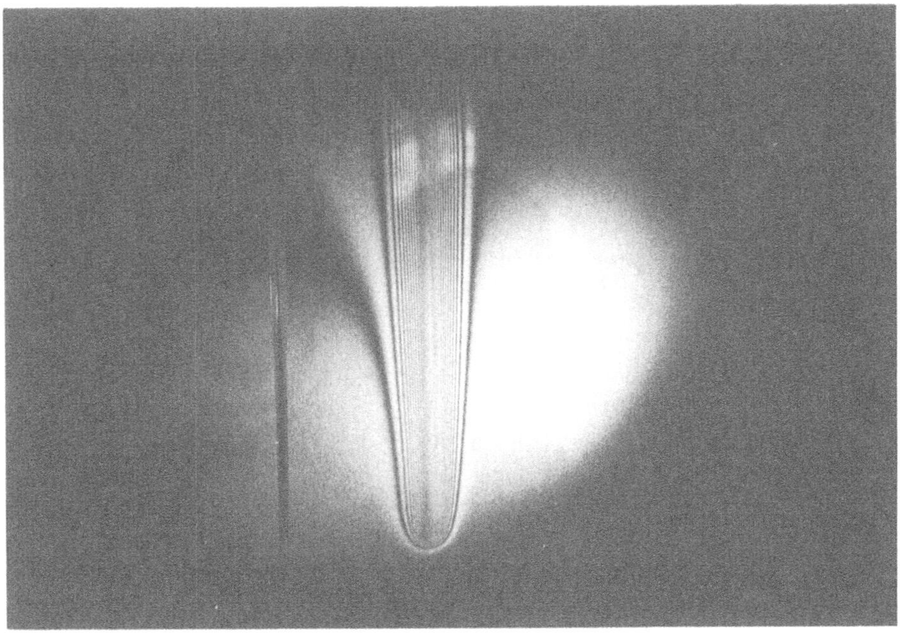

Fig. 3. Interference fringe pattern about a heated vertical flat plate.

References pp. 290–291.

matically in Fig. 4, the photographic plate is simultaneously exposed to the object and reference beams. The object beam is directed at the spray plume through a diffuser, i.e., a frosted glass plate. Each point of the diffuser acts as a discrete source of light. The light passing through every point in the spray is impeded in its progress through it by differing amounts and is therefore "out of phase" with the reference beam by differing amounts. The object and reference beams interfere at the photographic plate, and these interference patterns are recorded as variations in light intensity on the photographic emulsion. These patterns relate to droplets at differing distances from the plate.

The photographic plate records only intensity I, which is a constant multiplied by the sum of the squares of the amplitudes of the object and reference beams. From Eq. 4a

$$I = E^2_{0_{\text{obj}}} + E^2_{0_{\text{ref}}} + 2E_{0_{\text{obj}}} \cdot E_{0_{\text{ref}}} \cos 2\pi \left[\left(\frac{x}{\lambda}\right)_{\text{obj}} - \left(\frac{x}{\lambda}\right)_{\text{ref}} \right] \tag{13}$$

The amplitude transmittance of the developed photographic negative can be made proportional to I. When this negative is illuminated with the beam from a laser E_{rec} (here assumed to be of the same frequency and in the same direction as the reference beam), the photographic negative, i.e., the hologram, behaves as a complex diffraction grating. Thus the emerging wavefronts are dispersed to reconstruct the virtual and real images of the object volume. The emergent wavefront E_F is proportional to the product $I \times E_{\text{rec}}$, where E_{rec} has the form

$$E_{\text{rec}} = E_{0_{\text{rec}}} \cdot \sin \left[2\pi \left(\frac{x}{\lambda}\right)_{\text{rec}} - \omega t \right] \tag{14}$$

Combining Eqs. 13 and 14, rearranging, and noting that $(x/\lambda)_{\text{rec}} = (x/\lambda)_{\text{ref}}$ results in

$$\begin{aligned} E_F &= I \times E_{\text{rec}} \\ &= \underbrace{[E^2_{0_{\text{obj}}} + E^2_{0_{\text{ref}}}] E_{0_{\text{rec}}} \sin \left[2\pi \left(\frac{x}{\lambda}\right)_{\text{ref}} - \omega t \right]}_{I} \\ &\quad + E_{0_{\text{obj}}} E_{0_{\text{ref}}} E_{0_{\text{rec}}} \\ &\quad \times \Bigg\{ \underbrace{\sin \left[2\pi \left(\frac{x}{\lambda}\right)_{\text{obj}} - \omega t \right]}_{II} + \underbrace{\sin \left[4\pi \left(\frac{x}{\lambda}\right)_{\text{ref}} - \omega t - 2\pi \left(\frac{x}{\lambda}\right)_{\text{obj}} \right]}_{III} \Bigg\} \end{aligned} \tag{15}$$

Term I contains no information about the phase of the object and is in fact the zero-order undeflected beam.

Term II contains information about the phase of the object and except for the multiplicative constant is identical in form with the object beam. It is responsible

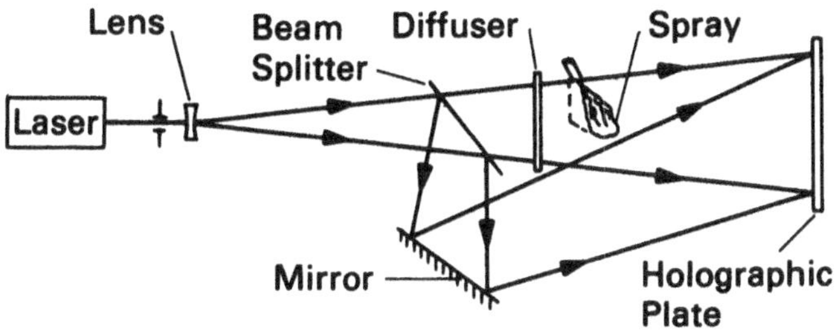

Fig. 4. Schematic of "off-axis" holographic recording.

for the reconstruction of the virtual image of the object, which is located behind the hologram in the position of the actual object, and is the positive first-order diffracted beam.

Term *III* contains information about the object in a pseudoscopic form and results in the real image at a location in front of the hologram. It is the negative first-order diffracted beam.

These ideas are summarized in Figs. 5 and 6, the latter showing a sequence of three photographs of a fuel spray obtained with differing depths of focus in the virtual-image volume created by a single hologram.

It should be noted that the reconstruction of the virtual image behind the hologram is similar to the viewing of the object field through a window in that complete three dimensionality and parallax effects are present. Also, the field of view is limited by the aperture of the window, a large holographic plate providing a wider angle of view of the object than a smaller one.

Holographic Interferometry—In the double-exposure technique, a hologram is made of the test volume in the absence of the event to be studied, giving a

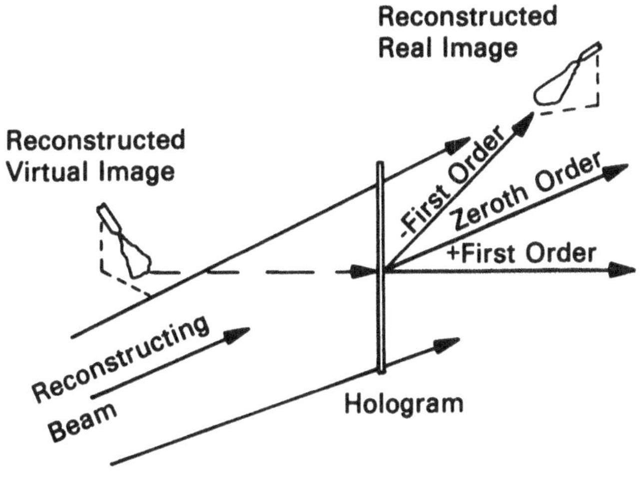

Fig. 5. Schematic of holographic reconstruction.

References pp. 290–291.

Fig. 6. (a)

Fig. 6. (b)

Fig. 6. Illustration of the depth of field available in holographic recording. Single holographic record of a diesel fuel spray in air. (Angle between spray axis and holographic plate = 30°, spray length = 150 mm). a) Focus at injector tip, b) Focus at mid-length of spray, c) Focus near spray tip.

Fig. 6. (c)

reference state exposure. Before processing the plate, a second exposure is made with the plate undisturbed but during which the test volume has been modified by the event under study, thus giving the object state exposure. The result is two overlapping wavefront systems which form an interference pattern indicative of the modifications to the test volume by the event under study in a manner similar to that discussed in connection with the Mach-Zehnder interferometer (Fig. 2). The double-exposure technique is usually undertaken with a pulsed laser light source in engine studies.

In the real-time or live-fringe technique, a reference state hologram is made of the test volume as in the double-exposure method described above. The hologram is then illuminated with a continuous-wave laser source to reconstruct the virtual image of the test volume, which is then made to overlap the actual test volume precisely. Any modification to the actual test volume by the event under study will appear as a continuously changing fringe pattern, which may be recorded with a motion picture camera.

The main advantage of holographic interferometry for engine applications is the use of a common optical path for the reference and object beams. This eliminates the need for high-quality optical components and cell windows. The rapid switching and short pulse duration (30 ns) of pulsed lasers enables precise control of laser firing in synchronism with engine events and overcomes problems associated with movement during the exposure of the plate.

The live-fringe technique is more suited to bomb studies, where precise align-

References pp. 290–291.

ment of the test bomb and its virtual image presents less of a problem than an engine system would using the same method.

An Engine-Based Interferometry System—The test facility is based on a 4-cylinder in-line spark-ignition engine of 86-mm bore and 64-mm stroke. One cylinder was modified in a manner similar to that outlined by Bowditch [5] to enable optical access to it via a 50-mm diameter window in the piston crown. A second co-axial window of the same size is accommodated in the cylinder head, allowing transmission of the light beam through the cylinder cavity. Because of the large diameter of the cylinder-head window, conventional valves could not be accommodated in the head. Instead, three reed valves were located adjacent to the cylinder-head window. The valves are held open against stiff "leaf" stops by a simple spring tensioner which controls the instant of opening and closing of the reed valves during the expansion and compression strokes of the engine cycle. The "breathing" action of the test cylinder follows a two-stroke cycle. The test cylinder was motored by two of the remaining cylinders of the engine for balance purposes and had a maximum rotational speed of 3500 r/min.

A schematic of the engine optics is shown in Fig. 7, while Fig. 8 shows a photograph of the cylinder head and piston detail. The clearance space of the test cylinder has a disc configuration with a 16-mm depth (spacer ring Fig. 8).

Liquid fuel injection to the test cylinder was from a standard diesel injector and jerk pump, the latter driven from the engine crankshaft by a toothed belt.

The peak air pressure in the test cylinder is variable between 0.4 and 1.0 MPa by use of a combination of intake throttling and pressure boosting from the laboratory air supply. Facility for the electrical heating of the inlet air supply was also available.

A 300-mJ pulsed ruby laser by J. K. Lasers, Limited, was used. It had variable pulse separation (2 − 1000 μs) and a coherence length of 1 m.

Optical components were mounted on a vibration-isolated steel table surrounding the engine by means of magnetic bases. Optics were used to obtain an enlarged image of the test volume at the holographic plate. This was necessary because of the long optical path from the test chamber volume (approximately *TDC* compression) to the holographic plate, and the limited aperture of the piston window. In the absence of the imaging optics, the effect on holographic reconstruction was of peering down a tube to observe the image of the injection process in actual size. This was inadequate for fringe analysis.

Firing of the laser at precise times of the fuel injection cycle with a high degree of repeatability was achieved by an electronic logic control system actuated from the engine crankshaft marker. Allowance was made for the buildup of power ($\simeq$ 1.5 ms) following the firing of the laser flash tube before the electro-optic shutter was energized to allow transmission of the light pulse from the laser.

Interferograms were obtained using the double-exposure method. The electro-optic shutter was energized at the point of interest of the injection cycle but in the absence of fuel injection. Hence a hologram of the test volume was obtained, giving the reference condition. A second exposure of the holographic plate at the same point in the injection cycle but with fuel injection present was then made

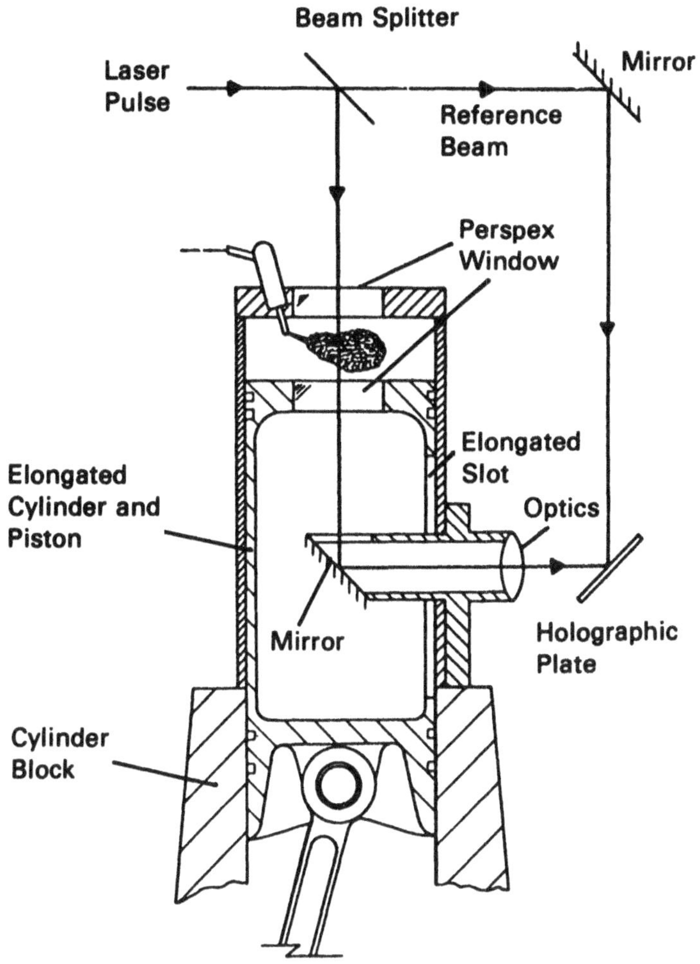

Fig. 7. Schematic of engine optics.

on a later engine cycle. The developed holographic plate was re-illuminated on an optical table with a continuous-wave helium-neon laser source. Closer study of interference fringes, jet structure, etc, was made by imaging the reconstructed virtual image with a television camera and appropriate vidicon tube onto a video monitor. At this stage, data to be stored for further reference were recorded on video tape using a conventional video recorder.

A typical interferogram for the injection of a liquid pentane jet into the test cylinder is shown in Fig. 9. The fringe structure is clearly evident. It should be noted that as the axis of the jet is approached, the fringe density increases rapidly due to the presence of liquid fuel. This is further illustrated by a hologram of a liquid pentane injection, Fig. 10, at identical conditions to that used for the interferogram. The recording was made with a single exposure of the plate with injection present, and is effectively a photograph of the jet. Here the dense liquid

References pp. 290–291.

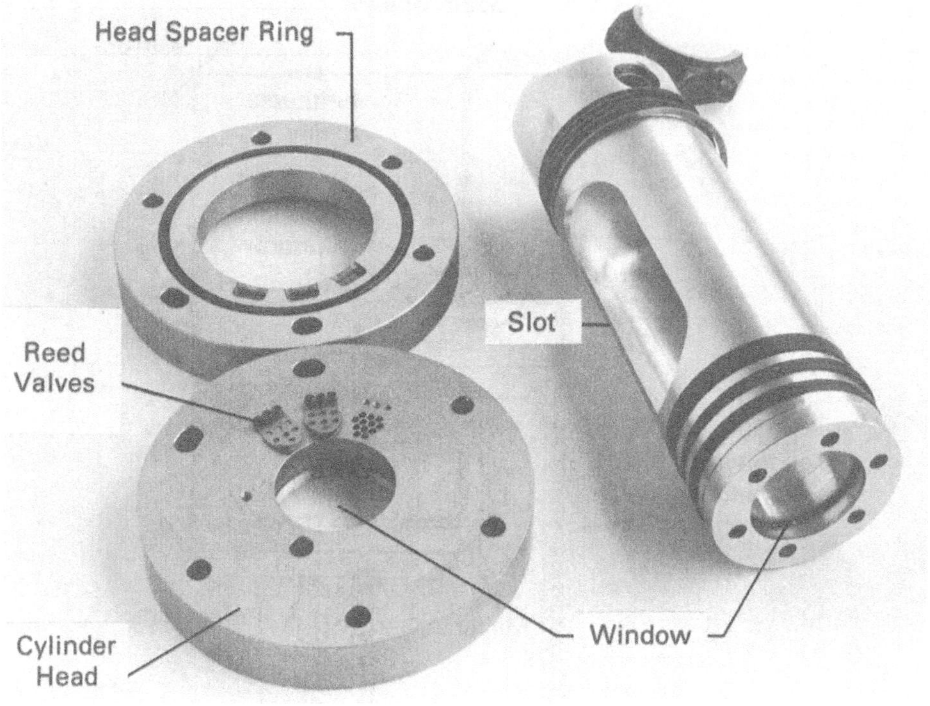

Fig. 8. Piston and cylinder-head details.

core can be seen, the fuel vapor is transparent to the laser beam, and small droplets (< 5 μm) are not resolved. The similarity in the core structure of the two jets is very apparent.

Inversion of Fringe Count – Axisymmetric Case—The fringe-shift equation (Eq. 10) derived for the Mach-Zehnder interferometer is for the one-dimensional variation of refractive index normal to the optical path length, n_y, and integrated along it. Using suffix "o" for the reference condition instead of "ref", Eq. 10 can be expressed as

$$S(y) = \frac{2}{\lambda_o} \int_0^L (n_y - n_o)\, dL \tag{16}$$

For an axisymmetric jet, assuming refraction effects are negligible, the observed fringe system is the integrated result of interference caused by the passage of the object and reference beams through the test volume. Fig. 11 shows a slice, dz, across a vaporizing fuel jet for which the refractive index variation is axisymmetric.

Applying a simple transformation of coordinates to Eq. 16 for the axisymmetric geometry of Fig. 11 results in

$$S(y) = \frac{4}{\lambda_o} \int_y^R \frac{(n_r - n_o) r \, dr}{(r^2 - y^2)^{1/2}} \tag{17}$$

Eq. 17 gives the fringe number, $S(y)$, at y for the integrated effect along the optical path shown in Fig. 11. n_r is the refractive index of the fluid within the jet at radius r, and n_o is the refractive index at a reference condition, i.e., state of the air outside the jet. The variation of the fringe number, $S(y)$, with y has the form shown in Fig. 12.

Eq. 17 is a form of the Abel integral equation and, provided $(n_r - n_o) = 0$ for $r > R$, can be inverted to yield

$$(n_r - n_o) = -\frac{\lambda}{\pi} \int_r^R \frac{[dS(y)/dy] \, dy}{[y^2 - r^2]^{1/2}} \tag{18}$$

Fringe counts were undertaken at a number of axial stations along the fuel jet. This was carried out manually off the video monitor screen, using the camera focus and monitor contrast to sharpen local areas of the fringe system. At each axial station, Eq. 18 was evaluated numerically [6] to yield $(n_r - n_o)$.

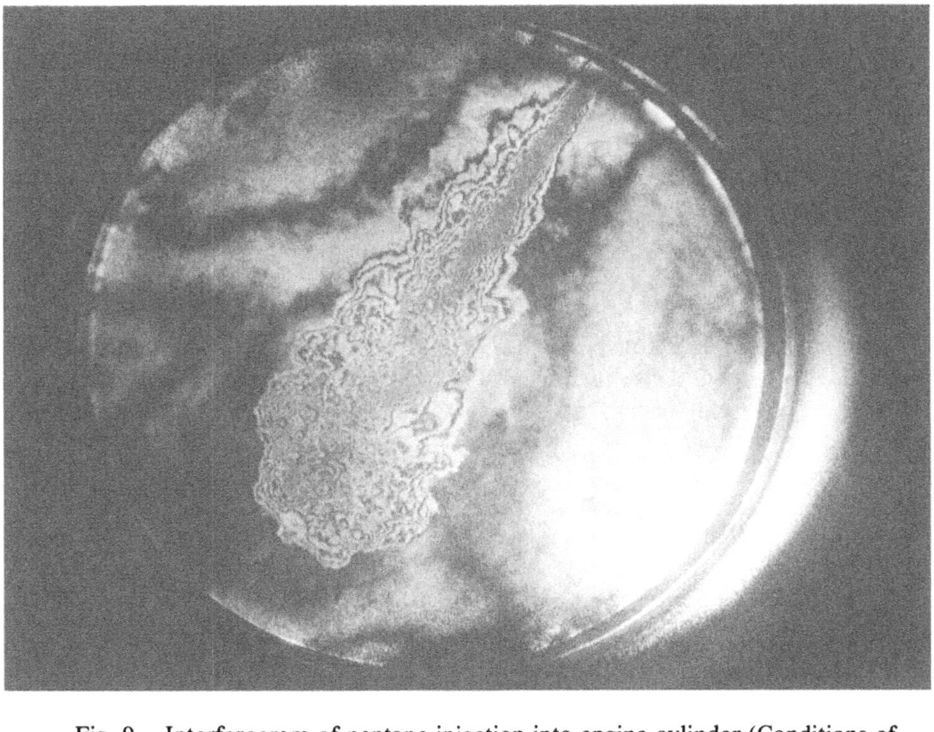

Fig. 9. Interferogram of pentane injection into engine cylinder (Conditions of Table 1).

References pp. 290–291.

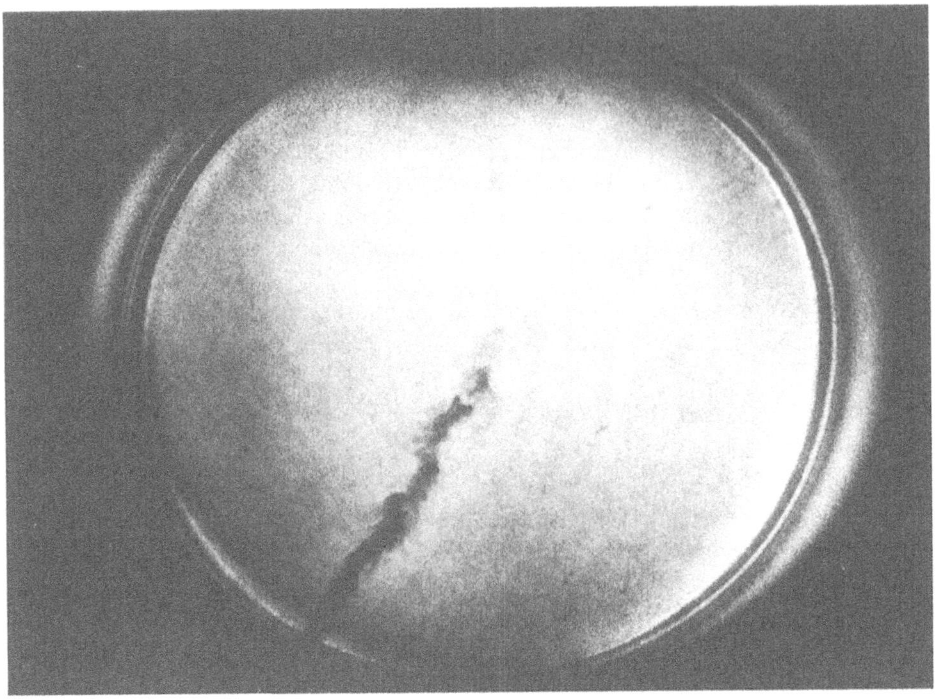

Fig. 10. Hologram of pentane injection into engine cylinder (Conditions of Table 1).

EVALUATION OF LOCAL AIR-FUEL RATIO

The Gladstone-Dale equation (Eq. 11) relates n_r to mixture density, ρ_r, at location r so that

$$n_r = \rho_r K_r + 1 \tag{19}$$

The Gladstone-Dale constant, K_r, for the binary mixture of air and fuel vapor (assumed to behave as an ideal gas) will be dependent on the mass fractions of fuel vapor, m_j, and air, m_o, at location r. Hence Eq. 19 can be expressed as

$$n_r = \rho_r[m_{j_r} K_j + m_{o_r} K_o] + 1 \tag{20}$$

where $m_{j_r} + m_{o_r} = 1$. Therefore

$$\frac{n_r - n_o}{\rho_o K_o} = \frac{\rho_r}{\rho_o}\left[m_{j_r}\left(\frac{K_j}{K_o} - 1\right) + 1\right] - 1 \tag{21}$$

Applying the equation of state to the mixture at r and the reference state (air) condition "o", and noting that the pressure is constant, results in

$$\frac{\rho_r}{\rho_o} = \left(\frac{T_o}{T_r}\right)\left(\frac{M_r}{M_o}\right) \tag{22}$$

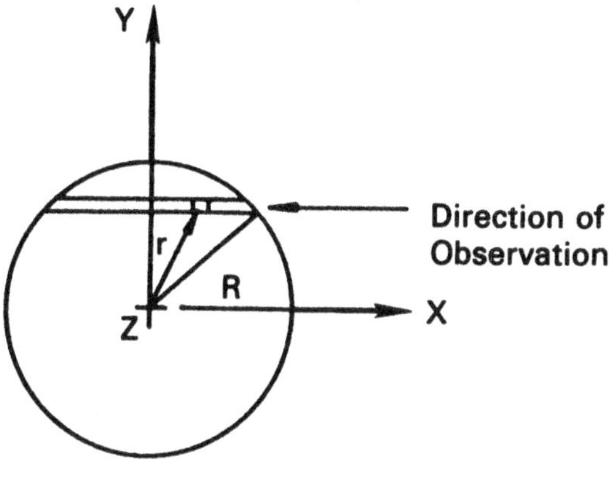

Fig. 11. Elemental transverse slice dz through the fuel spray.

Also, the mixture volume at r can be expressed as the sum of the partial volumes of air and fuel vapor at the mean temperature, T_r, and constant pressure P_o, yielding

$$\frac{1}{M_r} = \frac{m_{j_r}}{M_j} + \frac{m_{o_r}}{M_o} \tag{23}$$

From Eqs. 22 and 23

$$\frac{\rho_r}{\rho_o} = \frac{T_o}{T_r}\left[m_{j_r}\left(\frac{M_o}{M_j} - 1\right) + 1\right]^{-1} \tag{24}$$

Combining Eqs. 21 and 24 yields the mass fraction of fuel vapor at r, m_{j_r}, which is

$$m_{j_r} = \frac{\left(\frac{T_o}{T_r} - 1\right) - \left(\frac{n_r - n_o}{\rho_o K_o}\right)}{\left[\left(\frac{M_o}{M_j} - 1\right)\left(\frac{n_r - n_o}{\rho_o K_o} + 1\right) - \frac{T_o}{T_r}\left(\frac{K_j}{K_o} - 1\right)\right]} \tag{25}$$

In Eq. 25, $(n_r - n_o)$ is obtained from the inversion of the fringe-shift equation (Eq. 18); M_o, M_j, ρ_o and T_o are known. K_j and K_o are calculated from tabulated [7] values of n_j and n_o at a known state and the Gladstone-Dale relationship (Eq. 11). T_r is evaluated as the mass mean of the temperature of a vaporizing droplet, T_s, in the surrounding reference state, T_o, in the following manner.

In considering the vaporization from the surface of a liquid "j" which is at a saturation state "s" into air which is at state "o", energy and mass balances between the liquid and gaseous phases yield the transfer number B [8]. With the assumption of a Lewis number (D_i/α) of unity, B can be expressed as:

$$B = \frac{m_{o_j} - m_{s_j}}{m_{s_j} - 1} = C_{p_{j-o}} \frac{(T_o - T_s)}{h_{fg_s}} \tag{26}$$

References pp. 290–291.

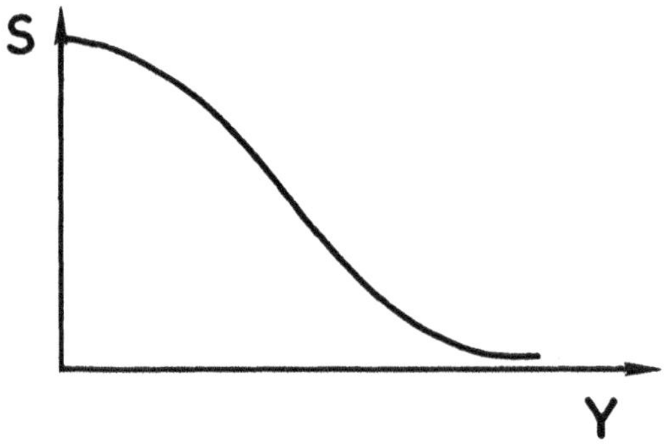

Fig. 12. Variation of fringe count with Y.

m_{o_j} and m_{s_j} are the mass fractions of fuel vapor in the "o" and "s" states per unit mass of air-fuel mixture. h_{fg_s} is the enthalpy of vaporization at the saturation temperature, T_s. At the reference air state "o", $m_{o_j} = 0$. Assuming that the vapor-air mixture behaves as an ideal gas, it can be shown that

$$M_{s_j} = \frac{\left(\frac{M_j}{M_o}\right)\left(\frac{P_{s_j}}{P_T - P_{s_j}}\right)}{1 + \left(\frac{M_j}{M_o}\right)\left(\frac{P_{s_j}}{P_T - P_{s_j}}\right)} \tag{27}$$

where P_{s_j} and P_T are the saturation vapor pressures of j and the total cylinder pressure, respectively.

Using a mass-averaged value for $C_{p_{j-o}}$, the specific thermal capacity of the air-vapor mixture, and thermodynamic property data [7] for the saturated vapor, j, Eq. 26 may be solved for T_s by iteration. Hence

$$T_r = m_{s_j} T_s + (1 - m_{s_j}) T_o \tag{28}$$

Results for local air-fuel ratio in the pentane jet are presented at the end of the next section.

SCALING OF ENGINE PROCESSES

From the previous section it is apparent that adequate scaling relationships must exist for the processes of air-fuel mixing and vaporization in the actual engine and in the "model" used to simulate it for the interferometry experiment, if meaningful quantitative results for local air-fuel ratio are to be obtained. This section, based on previously published work by the author and his co-workers [9, 10], briefly indicates the scaling relationships used and presents experimental results to test the ideas put forward.

Similarity between the actual engine and its "model" needs to be established for the following processes:

(i) Injection of fuel into air and their subsequent mixing, both when the air is in motion (swirl) and when it is not (quiescent chamber).
(ii) Injection period.
(iii) Evaporation rate from the fuel spray.

In previous work [9, 10] it was shown that the dynamics of the air-fuel mixing process with air swirl present is controlled by the ratio of fuel-jet to air-swirl momenta, so that

$$\frac{\bar{U}_f^2 \rho_f d_f^2}{\bar{U}_{a_E}^2 \rho_{a_E} L_E^2} = \frac{\bar{U}_g^2 \rho_g d_g^2}{\bar{U}_{a_M}^2 \rho_{a_M} L_M^2} \tag{29}$$

In the absence of air swirl, $\bar{U}_{a_E}$ and $\bar{U}_{a_M}$ are proportional [11] to $\bar{U}_f$ and $\bar{U}_g$ respectively. If we now consider that $\bar{U}_g$ is made equal to $\bar{U}_f$, Eq. 29 reduces to

$$\left(\frac{d_E'}{L_E}\right) = \left(\frac{d_M'}{L_M}\right) \tag{30}$$

where

$$d_E' = d_f\left(\frac{\rho_f}{\rho_{a_E}}\right)^{1/2}$$

and

$$d_M' = d_g\left(\frac{\rho_g}{\rho_{a_M}}\right)^{1/2}.$$

The injection period was shown [10] to be proportional to $(L_E/\bar{U}_f)$. Therefore considering the equality of $\bar{U}_f$ and $\bar{U}_g$

$$\left(\frac{\Delta t_E}{L_E}\right) = \left(\frac{\Delta t_M}{L_M}\right) \tag{31}$$

Spray vaporization was considered [10] to be based on the mass transfer from an isolated droplet in air at a high temperature and pressure using the relation [12]

$$\dot{m}'' = 1.46\left(\frac{k}{C_p}\right)_{\text{air-fuel}} Re_\gamma^{1/2} \frac{B}{(1+B)\gamma} \tag{32}$$

However, to incorporate jet effects into Eq. 32, the fuel jet velocity, $\bar{U}_f$, was used in the Reynolds number Re_γ and the droplet diameter, γ, was replaced with the Sauter Mean Diameter of the spray (S.M.D.).

The S.M.D. can itself be related to the fuel injection process through the

References pp. 290–291.

empirical equation of Knight [13]:

$$\text{S.M.D.} \propto \left(\frac{d_f{}^{0.41}}{\Delta P^{0.46}}\right)\left(\frac{\mu_f{}^{0.22}}{\rho_f{}^{0.318}}\right)(\sigma_f{}^{0.115})\left(\frac{\mu_{a_E}^{0.257}}{\rho_{a_E}^{0.372}}\right) \tag{33}$$

Now $\bar{U}_f \propto \Delta P_E$, and the collective effects of σ_f, μ_f and ρ_f were shown to be small in the scaling [10]. The similarity between engine and "model" vaporization processes is therefore controlled by

$$\left(\frac{k}{C_p}\right)_{a_{E-f}}\left(\frac{\rho_{a_E}^{0.69}}{\mu_{a_E}^{0.63}}\right)\left(\frac{1}{d_E{}^{0.21}}\right)\left(\frac{B_E}{1+B_E}\right) = \left(\frac{k}{C_p}\right)_{a_{M-g}}\left(\frac{\rho_{a_M}^{0.69}}{\mu_{a_M}^{0.63}}\right)\left(\frac{1}{d_M{}^{0.21}}\right)\left(\frac{B_M}{1+B_M}\right) \tag{34}$$

where B is the transfer number defined in Eq. 26 and is evaluated after iteration to determine T_s.

The engine geometry and operating conditions fix the left-hand side of Eqs. 30, 31 and 34. In the "model", choice of the scale, L_M, and the pure hydrocarbon fuel liquid density, ρ_g, are made through equipment availability and suitable boiling point characteristics of the fuel with pressure. These choices reduce the unknowns to be determined for the "model" to four, namely, the orifice diameter, d_M, the air pressure, P_{a_M} and temperature, T_{a_M} during injection, and T_{s_M}, the saturation temperature in B_M. These unknowns may be determined by iteration in Eqs. 26, 30 and 34, using the equation of state for an ideal gas and the thermodynamic properties for the liquid chosen to relate pressure, temperature and density.

The evaluation of (k/C_p) for air-fuel mixtures is based on the relations [14]

$$k_{\text{mix}} = \frac{\sum_{\text{All } n} x_n k_n M_n{}^{1/3}}{\sum_{\text{All } n} x_n M_n{}^{1/3}} \tag{35}$$

and

$$C_{p\,\text{mix}} = \frac{\sum_{\text{All } n} x_n C_{pn} M_n}{\sum_{\text{All } n} x_n M_n} \tag{36}$$

where

$$x_n = \frac{\text{Saturation vapor pressure at } T_s}{\text{Cylinder pressure } P_T}$$

EXPERIMENTS WITH PENTANE AND HEXANE INJECTION

To test the basis on which the scaling relationships were developed, they were applied to provide similarity between two fuel injection processes, one using pentane and the other hexane, injected into different gas states (pressure and temperature) in the test cylinder of the engine described. Local air-fuel ratio in the vaporizing pentane and hexane jets was determined using the interferometric technique discussed earlier.

Table 1 summarizes the orifice sizes and the engine operating conditions necessary to achieve similarity. Fig. 13 provides the results for radial variation of air-fuel ratio across the vaporizing sprays at three axial locations. Good agreement for approximately 40% of the half-width of the jet is noted. A small lateral displacement is noted due to inaccuracy in the determination of the jet edges. If the results are plotted on the basis of radius ratio, Fig. 14, the good agreement of local air-fuel ratio is very apparent. The inability to resolve fringes clearly as the liquid core of the fuel spray is approached is a limitation of the manual method of fringe counting. Better resolution can be achieved using a densitometer [15].

The method described here relies upon an empirical input for temperature. This restriction can be lifted if the experiment can be simultaneously executed with two beams having differing wavelengths (producing two interferograms). The result would be two sets of fringe-shift equations similar to Eq. 17 and their inversions (Eq. 18). Hence, two equations in two unknowns, ρ_r and m_{j_r}, can be solved, and with the equation of state and measured pressure yield m_{j_r} and T_r, the local mass fraction of fuel and temperature, directly.

The concept can be realized by using a setup similar to that described here. The output beam from the pulsed ruby laser is divided into two beams with a beam splitter. One beam passes through a frequency shifter, e.g., frequency doubling cell, resulting in an output beam of 0.3472 μm (blue) wavelength. The other beam will have the original 0.6943 μm (red) wavelength. Because of the losses involved, a high-powered pulsed ruby laser would be necessary. A similar approach would be used with a "bomb" configuration and a live-fringe technique. Here an argon-ion continuous-wave laser would be necessary.

TABLE 1

Conditions for Similarity
Engine Speed 1100 r/min

Injected Fluid	Nozzle Orifice Dia. (mm)	Injection Timing (°CA)	Laser Pulse Timing (°CA)	Peak Cylinder Pressure (bar)	Peak Cylinder Temp. (°C)
Pentane (liquid)	0.305	24 BTDC	20 BTDC	4.0	150
Hexane (liquid)	0.400	4 BTDC	TDC	7.0	230

References pp. 290–291.

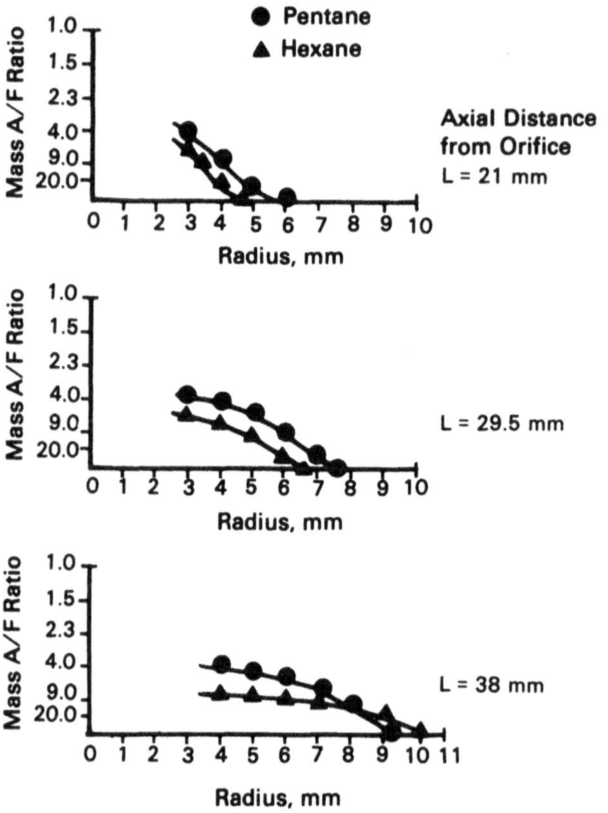

Fig. 13. Comparison of air-fuel ratio from dynamically similar evaporating jets of pentane and hexane (conditions of similarity in Table 1).

CONSIDERATION OF THREE-DIMENSIONAL FLOWS

The engine applications of interferometry discussed in the foregoing sections and elsewhere [10] have been limited to the study of axisymmetric fuel-spray development. In actual engines, the fuel-spray development departs considerably from the idealized axisymmetric jet configuration due to:

(i) Wall impingement of the fuel jet and the proximity of cylinder head and piston surfaces, causing recirculation effects.
(ii) Effects of bulk air motion, i.e., swirl, in high-speed direct-injection engines. Here the effect is the deflection of the fuel-spray trajectory and the creation of vortex regions downstream of the spray [9]. The combined effects of impingement and swirl also produce a strong flow of a fuel-rich mixture along the combustion chamber wall [9].

Extension of double exposure holographic interferometry to these three-dimensional flow situations needs to be considered.

Application of holographic interferometry and inversion of data to yield density in an asymmetric three-dimensional density field of an inclined free jet has been undertaken by Matulka [16] and Matulka and Collins [17], while Sweeney [18] and Sweeney and Vest [19] have made extensive studies of various schemes for data inversion.

Here, the essential principle of the holographic recording technique will be discussed in the context of engine measurement, the aim being to illustrate the potential application. The discussion of data inversion is confined to a grid method as this relates directly to the technique discussed earlier. Sweeney and Vest [20] should be consulted for a comparative study of methods of inversion.

In the earlier discussion of holographic recording and reconstruction, it was seen that actual holograms have a field of view limited by the size of the photographic plate used for recording. This, coupled with the need to avoid distortion of the fringe field during analysis by viewing at angles that do not deviate significantly from the object-beam direction, requires that sufficient angular coverage of the object field be obtained during holographic recording (180° when no planes of symmetry are present). In the engine context this would require simultaneous recording of the double-exposure interferograms grouped around the test volume. This is illustrated schematically in Fig. 15.

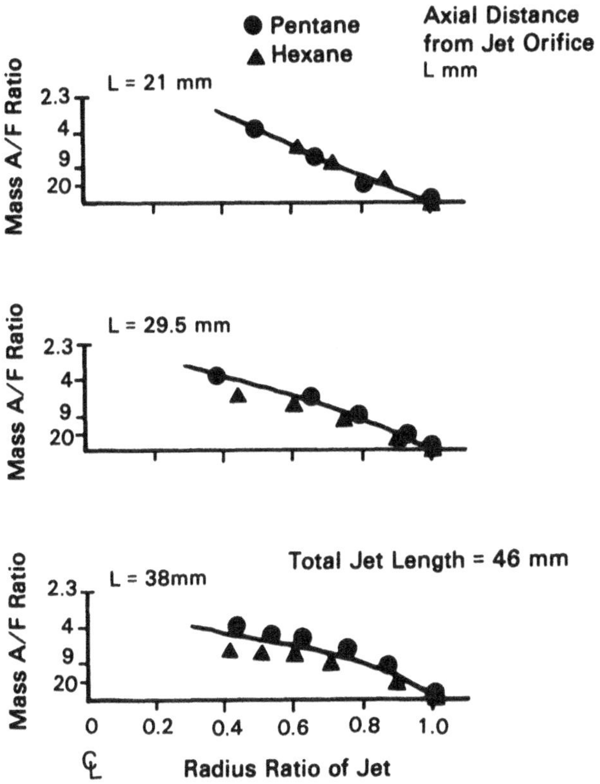

Fig. 14. Data of Fig. 13 plotted on basis of radius ratio.

References pp. 290–291.

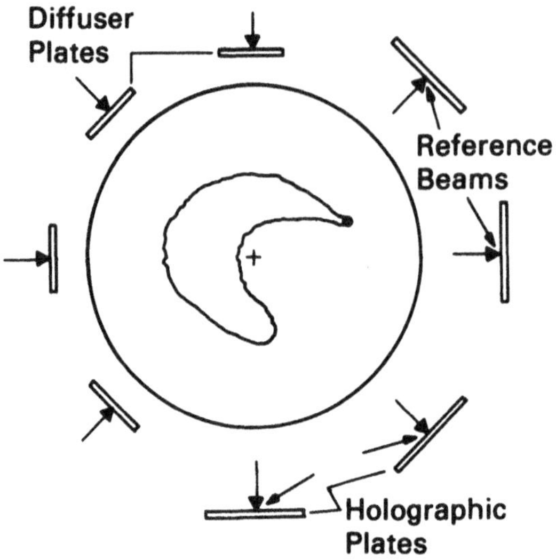

Fig. 15. Schematic of arrangement for recording three-dimensional flow field in engine or bomb.

For the case of a three-dimensional object field, the fringe-shift equation can be written as

$$S_i = \frac{2}{\lambda_o} \int_{L_i} [n_{(x,y,z)} - n_{o(x,y,z)}] \, dL_i \tag{37}$$

The line integral is evaluated along the ray path L_i, and S_i is the number of fractional wavelengths measured at the interferogram. In the grid-method of data inversion, data are analysed at a number of planes (x, y) of the object field—z locations. Each (x, y) plane is divided into a rectangular array $(o \times p)$ of finite elements, and within each element of the array the refractive index is assumed to have a uniform value, e.g., n_j, relative to a reference state "o" outside the array. If a segment of a ray path of length ΔL_i lies within an element of refractive index n_j, the contribution to the total optical path length is $n_j \Delta L_i$ and the total optical path length is the sum of the contributions from all elements through which the ray L_i passes. It will be equal to S_i. This reduces Eq. 37 to

$$S_i = \frac{2}{\lambda_o} \sum_{L_i} n_j \, \Delta L_i \tag{38}$$

By considering $(o \times p)$ independent ray paths, a set of algebraic equations such as Eq. 38 is generated which can be solved for $(o \times p)$ values of n_j in terms of measured ΔL_i's and S_i's, the latter from fringe counts for each ray path.

Engine application of the above technique could be accommodated in principle with little difficulty through the use of a transparent spacer ring (Fig. 8), thus enabling holographic recording around approximately 180° of the cylindrical test volume. Because of the need to illuminate each holographic plate simultaneously with its own reference beam and the fact that the reference beam intensity should be about four times that of the object beam for the production of clear holograms, increased laser power is necessary in comparison with the axisymmetric case studied here.

FURTHER APPLICATIONS

Extension of the ideas discussed here to the estimation of air utilization under a fired situation, either engine or bomb, can be achieved in principle as follows, using either the pulsed double-exposure technique or a live-fringe technique. Using a reference state without fuel injection and the object state with combustion of the fuel spray, attention can be focused on the outer air-fuel vapor regions of the spray. Eq. 19 can now be applied to a small burning zone, r, and will have the form.

$$n_{r_B} = \rho_{r_B}[\sum_{\text{All } j} m_{j,r} K_j] + 1 \tag{39}$$

where $m_{j,r}$ is the mass fraction of the products j in the reaction at r, K_j is the Gladstone-Dale constant for j, and ρ_{r_B} is the gas density of the burning zone at r.

In an unburned gas (air) reference state at "o", well away from the fuel jet

$$n_o = \rho_o K_o + 1 \tag{40}$$

Hence

$$\frac{n_{r_B} - n_o}{\rho_o K_o} = \frac{\rho_{r_B}}{\rho_o}\left[\sum_{\text{All } j} \frac{m_{j,r} K_j}{K_o}\right] - 1 \tag{41}$$

The quantity $(n_{r_B} - n_o)$ is obtained from a suitable inversion of fringe data. The assumption of axisymmetry need not be made in the inversion process. The gas density, ρ_{r_B}, will depend on the system pressure (measured), the burning zone temperature, T_{r_B}, and the mass fractions of the products. An equilibrium composition for the reaction process is assumed in which the major products are CO_2, H_2O, N_2, O_2 and CO, along with dissociation products arising from the equilibria [21].

$$\begin{array}{ll} CO_2 \rightleftarrows CO + \tfrac{1}{2}O_2 & \text{(a)} \\ H_2O \rightleftarrows H_2 + \tfrac{1}{2}O_2 & \text{(b)} \\ H_2O \rightleftarrows \tfrac{1}{2}H_2 + OH & \text{(c)} \\ \tfrac{1}{2}H_2 \rightleftarrows H & \text{(d)} \\ \tfrac{1}{2}O_2 \rightleftarrows O & \text{(e)} \\ \tfrac{1}{2}N_2 + \tfrac{1}{2}O_2 \rightleftarrows NO & \text{(f)} \end{array} \tag{42}$$

The reaction equations can be solved for T_{r_B} for a given pressure once the initial stoichiometry is specified. Therefore, iteration between Eq. 41 and the equilibrium reaction equations, using the equation of state, will yield the initial stoichi-

References pp. 290–291.

ometry and T_{r_B}. However, difficulty arises in the determination of the species refractive index, n_j, and hence K_j for OH. This can be estimated from methods outlined by Alpher and White [22].

Varde [23], in an interferometric determination of flame-front temperature in a propane-air mixture in a two dimensional spark-ignited 'bomb', neglected dissociation of molecules to free atoms and radicals (Eqs. 42c to f). His results indicate an adequate representation of temperature profile for a known initial stoichiometry. It would appear, therefore, that the scheme proposed here is feasible.

CONCLUSIONS

Application of laser interferometry to air-fuel ratio determination in unfired and fired diesel engine and "bomb" studies has been surveyed in some detail. The major advantage of the technique is the ability to obtain data over the whole combustion chamber volume in a single run, with the ability to analyze locally during the reconstruction of the holograms. In this respect the live-fringe technique with the combustion bomb provides an excellent way to study the complete fuel injection, ignition delay and combustion processes in a series of single-run experiments. A disadvantage of the technique is an inability to operate at pressures much in excess of about eight bar under motoring, or "cold bomb" conditions, due to increased fringe density (count) with compression-air density. This problem has been circumvented to some extent by the use of the scaling method discussed earlier. Under fired conditions there will be a decrease in the optical system sensitivity, due to the much higher combustion temperatures and therefore a decrease in fringe density. Vibration resulting in distortion of the fringe structure is a problem, but with suitable precautions it can be minimized.

Holographic interferometry as applied to engine combustion research merits further consideration. It is hoped that this survey will provide stimulus to that end.

ACKNOWLEDGMENT

I wish to express my thanks to my associate Chris De Boer in obtaining Figs. 3 and 6.

REFERENCES

1. J. D. Trolinger, *"Laser Instrumentation for Flowfield Diagnostics," Agardograph No. 186 (Agard-AG-186), March 1974.*
2. F. J. Weinberg, *"Optics of Flames," Butterworths, London, 1963.*
3. R. W. Ladenburg (Ed), *"Physical Measurements in Gas Dynamics," Part I of Vol IX, "High Speed Aerodynamics and Jet Propulsion," Oxford University Press, 1955.*
4. C. Veret, *"Review of Optical Techniques with Respect to Aero Engine Applications," Agard Lecture Series No. 90, (Agard-LS-90), pp. 2-1 to 2-6, July 1977.*
5. F. W. Bowditch, *"A New Tool for Combustion Research, A Quartz Piston Engine," SAE Trans., Vol. 69, Paper No. 150B, pp. 17–23, 1960.*

6. C. Park and D. Moore, "A Polynomial Method for Determining Local Emission Intensity by Abel Inversion," NASA, TN D-5677, Feb. 1970.
7. "International Critical Tables," McGraw-Hill, New York, Vol. 7, p. 10, 1930.
8. D. B. Spalding, "Convective Mass Transfer," Edward Arnold, London, pp. 56-98, 1963.
9. C. J. Morris and J. C. Dent, "The Simulation of Air-Fuel Mixing in High Swirl Open Chamber Diesel Engines," Proc. I. Mech. E., Vol. 190, C 47/76, pp. 503-513, 1976.
10. J. C. Dent, J. H. Keightley and C. D. De Boer, "The Application of Interferometry to Air Fuel Ratio Measurement in Quiescent Chamber Diesel Engines," SAE Trans., Vol. 86, Paper No. 770825, pp. 2858-2869, 1977.
11. F. P. Ricou and D. B. Spalding, "Measurements of Entrainment by Axisymmetrical Turbulent Jets," Jour. of Fluid Mech., Vol. 9, pp. 21-32, 1961.
12. T. Natarajan and T. S. Brzustowski, "Some New Observations on the Combustion of Hydrocarbon Droplets at Elevated Pressures," Comb. Sci. and Tech., Vol. 2, pp. 259-269, 1970.
13. B. E. Knight, written discussion to: A. Radcliffe "The Performance of a Type of Swirl Atomizer," Proc. I. Mech. E., Vol. 169, pp. 104-105, 1955.
14. R. H. Perry and C. H. Chilton (Eds), "Chemical Engineers Handbook," 5th Edition, New York, McGraw-Hill, pp. 3-244-3-238, 1973.
15. A. B. Witte and R. F. Wuerker, "Laser Holographic Interferometry Study of High Speed Flow Fields," AIAA Paper No. 69-347, presented at the Fourth Aerodynamics Testing Conference, Cincinnati, Ohio, April 1969.
16. R. D. Matulka, "The Application of Holographic Interferometry to the Determination of Asymmetric Three Dimensional Density Fields in Free Jet Flows," PhD Thesis, U.S. Naval Postgraduate School, 1970.
17. R. D. Matulka and D. J. Collins, "Determination of Three Dimensional Density Fields from Holographic Interferograms," Jour. App. Phys., Vol. 42, No. 3, pp. 1109-1119, 1971.
18. D. W. Sweeney, "Interferometric Measurement of Three Dimensional Temperature Fields," PhD Thesis, Univ. of Michigan, 1972.
19. D. W. Sweeney and C. M. Vest, "Measurement of Three Dimensional Temperature Fields above Heated Surfaces by Holographic Interferometry," Int. Jour. Heat and Mass Transfer, Vol. 17, pp. 1443-1454, 1974.
20. D. W. Sweeney and C. M. Vest, "Reconstruction of Three Dimensional Refractive Index Fields from Multi-directional Interferometric Data," Applied Optics, Vol. 12, No. 11, pp. 2649-2663, 1973.
21. A. G. Gaydon and H. G. Wolfhard, "Flames—Their Structure, Radiation and Temperature," Chapman-Hall Ltd., London, p. 264, 1953.
22. R. A. Alpher and D. R. White, "Optical Refractivity of High Temperature Gases," I—"Effects Resulting from Dissociation of Diatomic Gases," II—"Effects Resulting from Ionization of Monatomic Gases," Physics of Fluids, Vol. 2, No. 2, pp. 153-161 and 162-169, 1959.
23. K. S. Varde, "Optical Fringe Displacement as a Measure of Temperature Profile," ASME Paper 77-WA/TH-5, presented at the Winter Annual Meeting, Atlanta, Georgia, November 1977.

DISCUSSION

F. V. Bracco *(Princeton University)*

That's a very interesting experiment. This type of information is very useful and practical both from the point of view of designing the engine and of checking the models. What kind of confidence would you attach to that technique for determining equivalence ratio?

J. C. Dent

We compared* our interferometric measurements with the results from a sampling-valve study that we did in parallel, and the agreement was fairly good. In

J. C. Dent

this case an ethane jet was injected into an engine using conventional valves and a slightly different optical arrangement. The agreement was on the order of 20%. We have quite a fair amount of confidence in the sampling-valve study.

A. K. Oppenheim *(University of California)*

I had the good fortune of having Dr. Dent's paper before I came here. I would like to take this opportunity to tell you that the paper is really remarkable. It's very erudite, extremely clear, and really comprehensive. I recommend it to anyone who would like to use this type of interferometry.

Towards the end of his paper Dr. Dent pointed out some of the advantages one can derive from shear interferometry. This technique should be of particular service to the diagnostics of high-speed fluid mechanic processes occurring in internal combustion engines, and for this reason I am offering the following comments.

Shear interferometry, developed by Professor Felix Weinberg of Imperial College**, is basically two-shot photography: first an exposure of the test section

---

* *J. C. Dent, J. H. Keightley and C. D. DeBoer, "The Application of Interferometry to Air Fuel Ratio Measurement in Quiescent Chamber Diesel Engines," SAE Trans., Vol. 86, Paper No. 770825, pp. 2858–2869, 1977.*

** *M. J. R. Schwar and F. J. Weinberg, "Coherent Light Sources and Refractive Index Fields," Phys. Bull., Vol. 21, pp. 490–492, 1970, and A. R. Jones, M. J* R. Schwar, and F. J. Weinberg, "Generalizing Variable Shear Interferometry for the Study of Stationary and Moving Refractive Index Fields with the Use of Laser Light," Proc. Roy. Soc., Vol. 322, pp. 119–135, 1971.

without the event and, second, with the event. The resulting record "subtracts" the effects of one from the other, thus eliminating all the optical imperfections of the system. Hence windows do not have to be made of high-quality glass and, what is more relevant, their distortion with use does not affect the quality of the record.

The essential elements of the optical system are presented in Fig. 16*. They consist of a diffraction grating used for the separation of the beam, and a collimating lens employed for the production of the image on the film plane with an overlap or "shear", yielding by interference a diffraction record of the refractive index field existing in the test section. The actual interferogram is obtained subsequently by reconstruction of two superposed diffraction records, one of the undisturbed and the other of the disturbed field under study. The optical set-up used for this purpose is shown in Fig. 17. It is sufficient here to use a white point light source for illuminating the diffraction records from the back and to introduce a stop that singles out only one diffracted beam for admission into the film plane of the camera focused upon these records. The interference fringes in the latter are, in effect, Moiré patterns formed between the two superposed diffraction gratings, i.e., the patterns of points where the diffraction fringes cross each other, since their diffractive effects are obliterated so that the light passing through them is blocked from the camera by the stop.

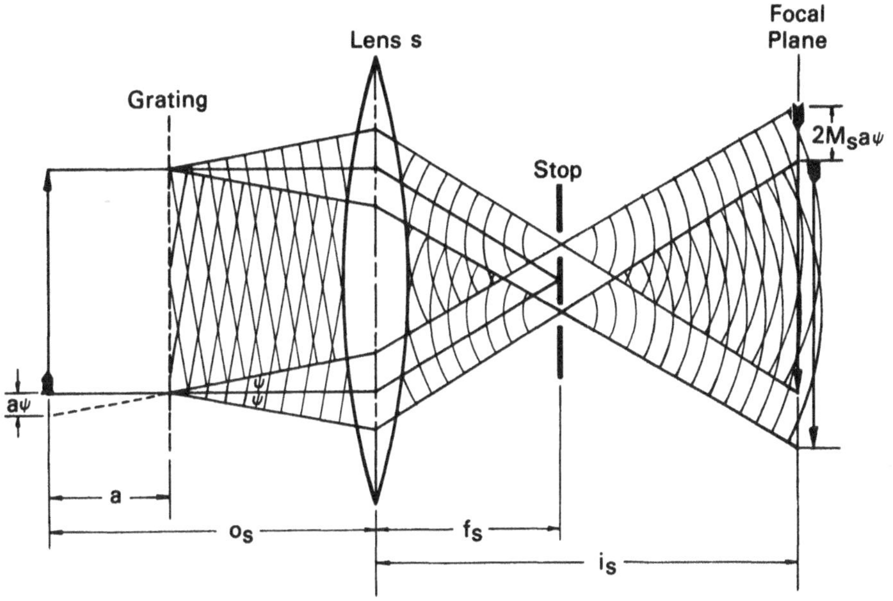

Fig. 16. Shear interferometer.

* *For more details see A. K. Oppenheim and M. M. Kamel, "Laser Cinematography of Explosions," Springer-Verlag, Wien-New York, 1971.*

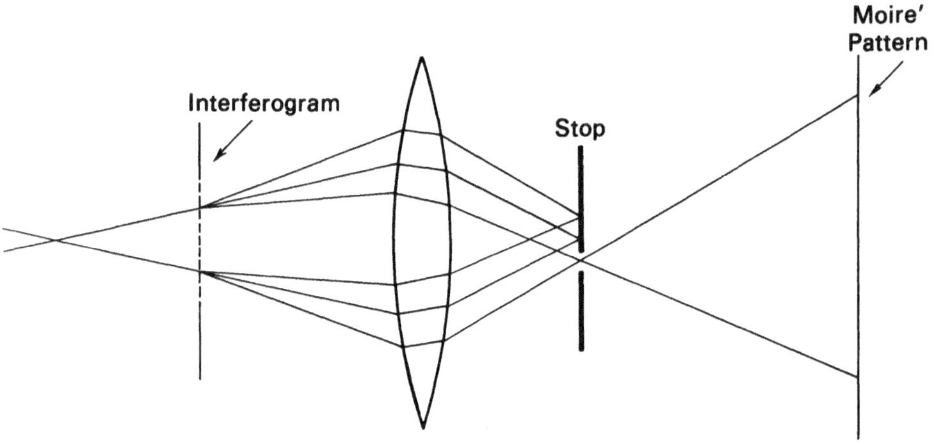

Fig. 17. Diffraction record reconstruction optics.

On this basis a high-speed cinematographic technique has been developed*. Its practical success was due primarily to the fact that the conventional infrared film has been found to have a sufficiently high grain density (equivalent to about 80 lines per mm) to produce satisfactory diffraction records (requiring not more than 20 lines per mm). For this purpose two reference records are taken, each with the diffraction grating tilted in its own plane by a small angle, first in one direction and then in the other. The record of the event is subsequently obtained while the diffraction grating is left in the last position. A direct print of the latter yields a schlieren photograph. The reconstructed image produced by superposing it with the first reference record is a finite-fringe interferogram, while that obtained with the use of the second reference record is an infinite fringe interferogram.

Such three records obtained at a frequency of 500,000 frames per second are shown in Figs. 18 and 19, the latter providing the enlargement of four frames in the third column of the former to demonstrate the remarkable space resolution one can also achieve by this technique. The test section is square, and the interferogram displays its full width of 32 mm next to the back wall. The particular phenomenon presented by the records is the combustion wave initiated behind a reflected shock in an argon-diluted methane-oxygen mixture—an autoignition process akin to knock in a spark-ignited internal combustion engine. The cinematographic strip on the left of Fig. 18 is the schlieren record, in the middle is the infinite fringe interferogram and on the right is the finite fringe interferogram.

The schlieren record provides information on the trajectory of the combustion-driven shock front, the infinite fringe interferogram represents loci of constant

* *A. K. Oppenheim, L. M. Cohen,, J. M. Short, R. K. Cheng, and K. Hom, "Shock Tube Studies of Exothermic Processes in Combustion," Modern Developments in Shock Tube Research, Edited by Goro Kamimoto, Shock Tube Research Society, Japan, pp. 557–574, 1975.*

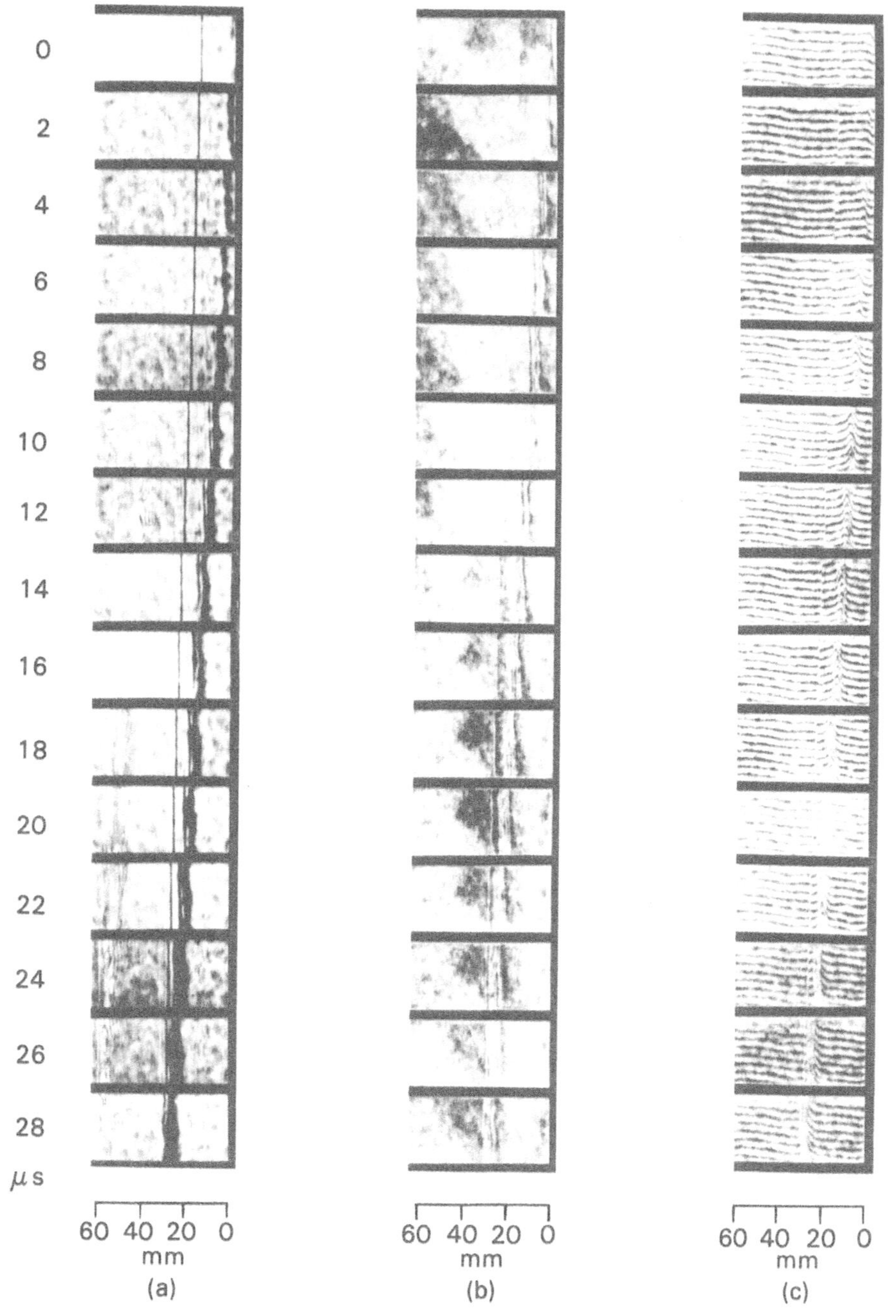

Fig. 18. Typical results obtained by shear interferometry. (a) Schlieren record, (b) Infinite-fringe interferogram, (c) Finite-fringe interferogram.

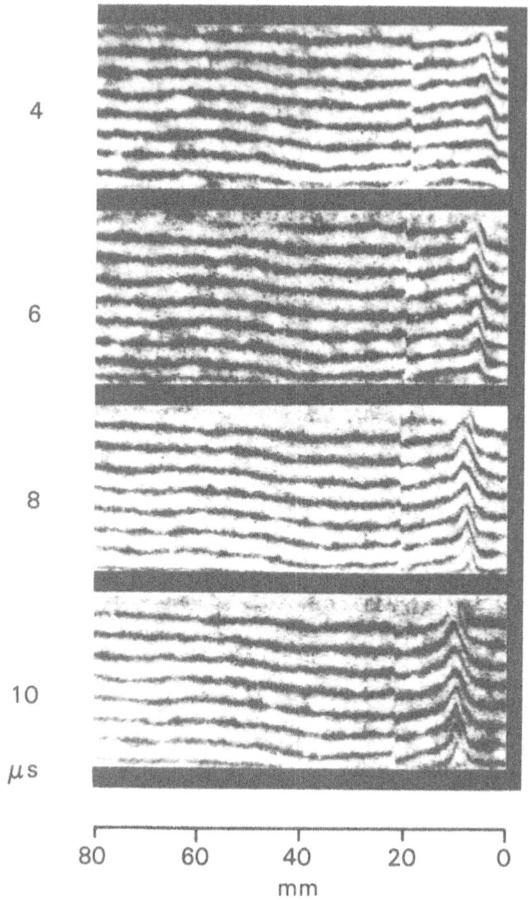

Fig. 19. Enlarged finite-fringe interferogram from Fig. 18.

density, and the finite fringe interferogram depicts the variation of the refractive index, from which one can deduce the density profiles. With pressure profiles evaluated independently at the same time, one can thus obtain data on the amount and the rate of deposition of energy in the combustible medium—information equivalent to the evolution of heat release by combustion and its power profile traced as a function of time at a sufficiently high resolution to reveal the essential dynamic features of the process.

J. C. Dent

A point I didn't make when discussing the shearing interferometer is that because it's a one-shot operation, you have no comparison beam that's coming around the outside (the reference beam). Therefore it is less susceptible to vibration, which can be a strong advantage for engine applications.

F. C. Gouldin *(Cornell University)*

I gather that in interpreting your data, you assume this axisymmetric jet. I wonder how good that assumption is in turbulent jet flow. Also, in removing the need to infer a temperature, you suggest measuring at two different wavelengths. Does that depend on the variation of the Gladstone-Dale constant between the two wavelengths, and if so, how sensitive is it to small variations?

J. C. Dent

The assumption of an axisymmetric jet can be a restriction in certain engine applications. However, there is a considerable body of published work on the interpretation of three-dimensional refractive index fields, which I feel can account adequately for the turbulent impinging jet flow with recirculation which is encountered in the high-swirl direct-injection diesel engine. Work in this direction is progressing in my laboratory.

In answer to your second question, some preliminary work on the variation of the Gladstone-Dale constant was done by Olsen*. He generated a procedure for looking at the composition in a flame by using a multi-wavelength experiment, and very conclusively proved the point using a binary mixture (an oxygen-air system.)

I. M. Khan *(Renault, France)*

That is very interesting work. What I would like to ask you is, "Could one look at a fluid with a full-boiling-point range?"

J. C. Dent

Yes, I think one could do it with a very simple mixture. For example, I think one could look at something like a cetane-hexadecane mixture, because there one can represent the vaporization characteristics fairly accurately. One could then interact the optical information with a theoretical two-component vaporization study in order to obtain another input for the binary liquid vaporization.

T. M. Dyer *(Sandia Livermore Laboratories)*

If I were trying to guess from intuition what the holographic interferogram of your fuel spray would look like, I would have expected to see fringes more nearly radial from the source, with perhaps some gradation along the axis as a function of distance from the orifice. What I'd like to ask you to comment on is the fact that maybe the signal you're seeing is due to index-of-refraction changes, that is, the light passing through the fuel spray being deflected, so that you no longer

---

* *H. L. Olsen, "An Interferometric Method of Gas Analysis, Third Symposium (International) on Combustion, The Combustion Institute, Pittsburgh, Pennsylvania, pp. 663–674, 1949.*

have a reasonable representation of the actual spray dimensions. Such an effect could also cause the noise that seems to be apparent inside the fuel spray.

J. C. Dent

I'm quite aware of the point that you made. In fact, Professor Vest* at Michigan has published quite a lengthy paper (not directed specifically at sprays) on inversion techniques, and has shown that the Abel inversion does not give the high distortions that some of the other inversion techniques do. He shows that this inversion gives a good representation in a fairly dense optical field. In counting fringes we located a disc (having equally spaced concentric rings ruled on its surface) in the test volume corresponding to the point in the engine cycle at which the injection process was to be studied, and photographed it from the position of the holographic plate. The subsequent fringe count and jet radius evaluation during reconstruction was carried out in relation to a projection of the ruled disc. I think that at the tip of the spray we are seeing the effect of quite a lot of turbulent mixing. If one took multiple directional views, one could isolate the local refractive index more precisely.

* *C. M. Vest, "Interferometry of Strongly Refractive Axisymmetric Phase Objects," Applied Optics, Vol. 14, No. 7, pp. 1601–1606, 1975.*

SCALING OF SOME FLOWFIELDS WITH FLAMES

F. V. BRACCO

Princeton University, Princeton, New Jersey

ABSTRACT

Consider a charge in a variable-volume chamber. If (i) the thicknesses of the flame and wall boundary layers are much smaller than the dimensions of the chamber, (ii) the speed of the gas is much less than that of sound, (iii) the charge is initially nearly uniform, (iv) the geometry of the chamber is compact and unobstructive, (v) the ratio of the rate of wall heat transfer to the rate of energy release is independent of flame speed, and (vi) the ratio of the wall velocity to the flame speed is also independent of flame speed, then the local and instantaneous values of all thermodynamic variables are very insensitive to flame speed and thickness, except within the flame front itself and near the walls, and they depend almost exclusively on flame position, chamber geometry, and elements of the initial and boundary conditions. Also, the local and instantaneous gas velocity is very nearly proportional to the instantaneous flame speed. Equivalently, the flowfield scales in time as the inverse of the flame speed, except within the flame and wall layers. The properties hold for any number of species, any forms of the diffusivities, and any number of space dimensions. Uniform-charge spark-ignited open-chamber engine combustion tends to respect the conditions and to exhibit the properties, but non-scaling flowfields are often found in divided-chamber engines. The scaling should also apply to some degrees of swirl and some forms of charge stratification, but species not in equilibrium, e.g., emissions, are excluded. The scaling extends to geometrically similar engines of different displacement if the flame speed increases proportionally to the length scale of the combustion chamber. The properties can be used to generalize results of multidimensional computations and to guide experiments. Since the flowfield is controlled by diffusion and reaction processes, but only from thin isolated flame and wall layers, measurements outside of those layers are not nearly as informative as those within them. On the other hand, when the similarity breaks down and diffusion and reaction processes are controlling everywhere, time- and space-resolved measurements at all locations are informative. But then the flowfield is also fully three-dimensional and almost impossible to compute at present. Thus engine flowfields that are likely to exhibit scaling properties and to be amenable to being computed are difficult to resolve experimentally, and those that do not scale and cannot be computed lend themselves more easily to informative measurements.

References p. 323.

NOTATION

C_v	specific heat at constant volume, erg/(g K)
D	$u_f\delta_f$ definition of effective diffusivity. For a premixed laminar flame, it is about equal to the mass diffusion coefficient. For a premixed turbulent flame, it depends on its presumed structure. cm^2/s
e	thermal internal energy, erg/g
h_i^0	enthalpy of formation of species i, erg/g
K	thermal conductivity, erg/(cm s K)
L	characteristic length of the chamber, $\sim V^{1/3}$, cm
M	Mach number. Also mass, g
$\dot{M}_b$	mass burning rate, g/s
N	engine speed, r/min
$\underline{n}$	outward unit vector
p	pressure, dyne/cm^2
q	$\bar{W}\,\Sigma_l \hat{q}_l$ thermal enthalpy change due to all reactions, erg/g
$\hat{q}_l$	heat of reaction $l \equiv -\Sigma_i h_i^0 W_i(\nu''_{i,l} - \nu'_{i,l})$, erg/mole
$\dot{q}_r$	$\Sigma_l R_l \hat{q}_l$, rate of thermal energy change due to all reactions, erg/(cm^3 s)
$\dot{Q}_r$	total rate of thermal energy change due to all reactions, erg/s
Q_w	total rate of thermal energy change due to wall heat transfer, erg/s
r	radial position, cm
R	gas constant, erg/(g K)
R_C	chamber radius, cm
R_l	rate of reaction l, mole/(cm^3 s)
S	chamber surface area, cm^2
S_f	flame surface area, cm^2
t	time, s
T	temperature, K

$\underline{u}$	gas velocity, cm/s
u_f	flame speed = $(Dw/\rho)^{1/2}$ from definitions of D and w, cm/s
V	chamber volume, cm^3
w	$M_b/\delta_f S_f = \rho u_f/\delta_f$ = definition of effective reactant-to-product conversion (or reaction) rate. For a premixed laminar flame, $w = \bar{w}'\delta_f'/\delta_f$. For a premixed turbulent flame it depends on its presumed structure. g/(cm³ s)
$\bar{w}'$	$\bar{\dot{q}}_r'/q = (\overline{\Sigma_l R_l \hat{q}_l})'/\bar{W}\,\Sigma_l \hat{q}_l$ = average reactant-to-product conversion rate in the reaction zone of a premixed laminar flame, g/(cm³ s)
W_i	molecular weight of species i, g/mole
$\bar{W}$	average molecular weight, g/mole
x	**space dimension, cm**
Y	mass fraction
α	parameter of the α-transformation
β	parameter of the β-transformation
γ	ratio of specific heats
δ_f	flame thickness = $(D\rho/w)^{1/2}$ from the definitions of D and w, cm
δ_f'	thickness of the reaction zone within a premixed laminar flame, cm
δ	characteristic length of any flowfield nonuniformity: δ_{ρ_i} of nonuniformity of mass distribution; $\delta_{\rho u}$ of nonuniformity of momentum distribution; δ_e of nonuniformity of energy distribution; δ_w of wall boundary layers, cm
θ	crank angle, deg
λ	first viscosity coefficient, g/(cm s)
μ	second viscosity coefficient (for nil bulk viscosity, $\lambda = -2\mu/3$), g/(cm s)
$\nu'_{i,l}$	stoichiometric index of species i on the LHS of Reaction l
$\nu''_{i,l}$	stoichiometric index of species i on the RHS of Reaction l
ρ	density, g/cm³
ρ_i	density of species i, g/cm³

Subscripts

b	burned

References p. 323.

f	flame or final condition
fl	fuel
o	initial condition
ox	oxidant
u	unburned

SOME QUESTIONS

What do the four items discussed in the following paragraphs have in common?

In Fig. 1 three flames are represented that have propagated through a uniform stoichiometric air-fuel mixture for identical periods of time [1]. The pressures are almost the same, and so are the gas velocity, temperature, density and mass fraction distributions everywhere except within the flame fronts, which are seen

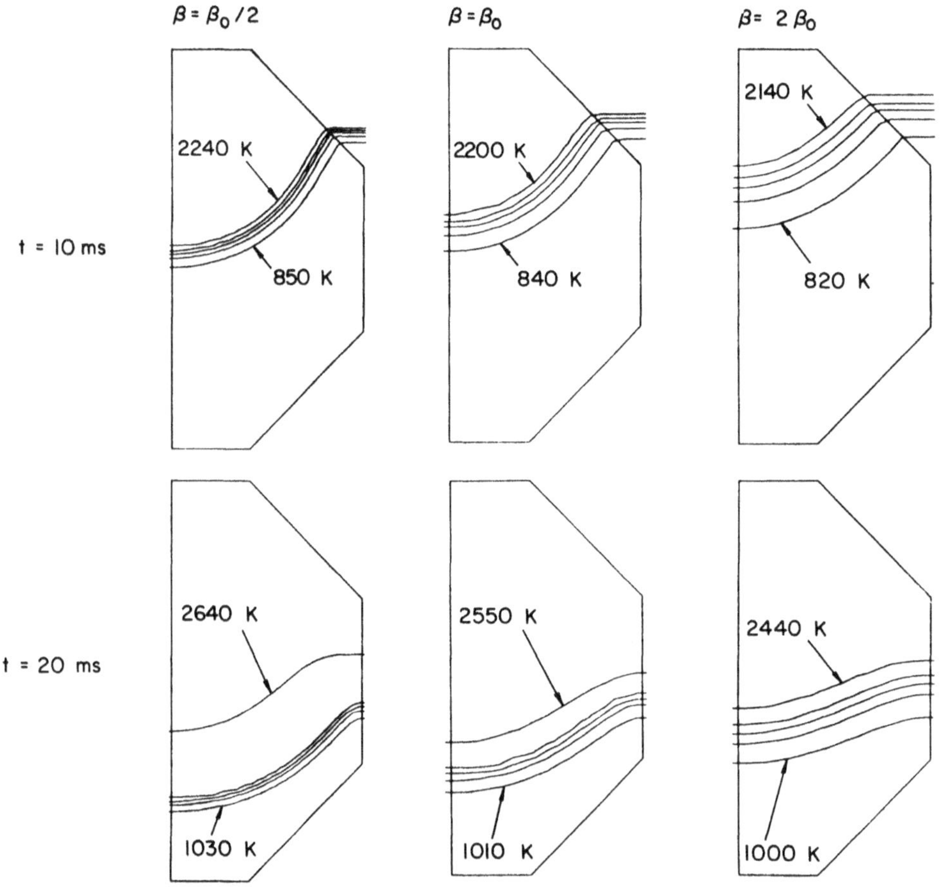

Fig. 1. Comparisons of temperature contour plots at two different times for three calculated flames with approximately equal velocities but different thicknesses.

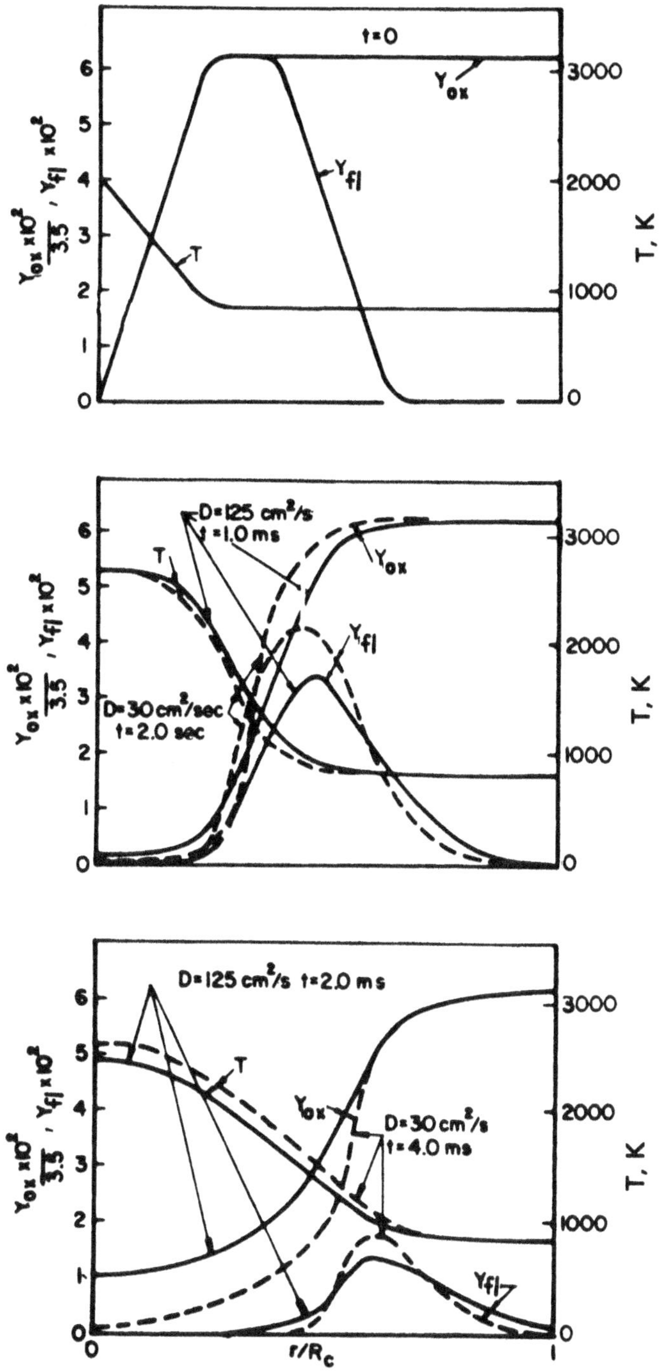

Fig. 2. Effect of diffusivity on maintenance of stratification and flame propagation. Fuel and oxygen mass fractions and temperature versus radial position and time for $D = 125\ cm^2/sec$ and $D = 30\ cm^2/sec$.

References p. 323.

to be of different thicknesses. The three flames are the results of three computations that differ only in diffusivities and reaction rates. These quantities were changed in such a way that, according to the simplest of flame speed and thickness equations ($u_f^2 \sim Dw/\rho$; $\delta_f^2 \sim D\rho/w$), the flame speeds should be the same, and the thicknesses should differ by factors of two. Under what conditions are the solutions of the complete two-dimensional unsteady conservation equations almost the same everywhere and at all times except within the flame front?

The second item is presented in Fig. 2. Computed results are shown for two flames moving through a stratified charge in a constant-volume cylindrical chamber [2]. Initially, a stoichiometric fuel-air mixture is located near the axis, and air is located away from it. The top diagram shows the temperature and the fuel and oxygen mass fractions shortly after an idealized ignition all along the axis. Only the values of the diffusivities differ in the two computations, corresponding to turbulence fields of different intensities. The continuous and dashed lines show results for D = 125 cm²/sec and D = 30 cm²/sec, respectively, and at different times from ignition. The data of the low diffusivity case are at times that are twice as long as those of the high diffusivity computation, as if the flame speed, which goes as the square root of the diffusivity, were the only important parameter of the problem. It is seen that the two flow fields do not overlap exactly at corresponding times, but they do come surprisingly close considering that the two fields are different solutions of the complete one-dimensional conservation equations. Under what conditions do flow fields scale in time almost exactly and due to the flame speed alone?

The two flames of Fig. 3, taken from a transparent-head divided-chamber engine, constitute the third item [3]. Within normal cyclic variations, they look identical at the same crankangles, and so do their measured pressure histories. But they are not two flames from different cycles of the same engine configuration. In fact, the two configurations differ considerably in engine speeds (1000 r/min versus 1500 r/min), throat areas (1.1 cm² versus 0.75 cm²), and overall equivalence ratios (1.26 versus 0.62). Are the similarities coincidental? Or, phrased differently, under what conditions is engine combustion expected to scale? And, does such scaling extend to emissions? Or equivalently, consider a set of detailed results obtained through an expensive three-dimensional computation and for a specific set of engine conditions. Can they be scaled to obtain solutions corresponding to other conditions?

The fourth item concerns the experimental evaluation of the accuracy of the results of multidimensional computations of combustion such as those of Fig. 2. Is the measurement of any of the dependent variables at any location and any time equally useful for the purpose? If not, which criteria should be used for the selection?

In the following paragraphs, a perspective is presented that shows the relationship among the above items and allows answers to be given to some of the questions. However, it is admitted from the onset that the puzzle is a complex one, and that some of its pieces are still missing. But the subjects are important, and by opening discussion of them we hope to stimulate others to consider them.

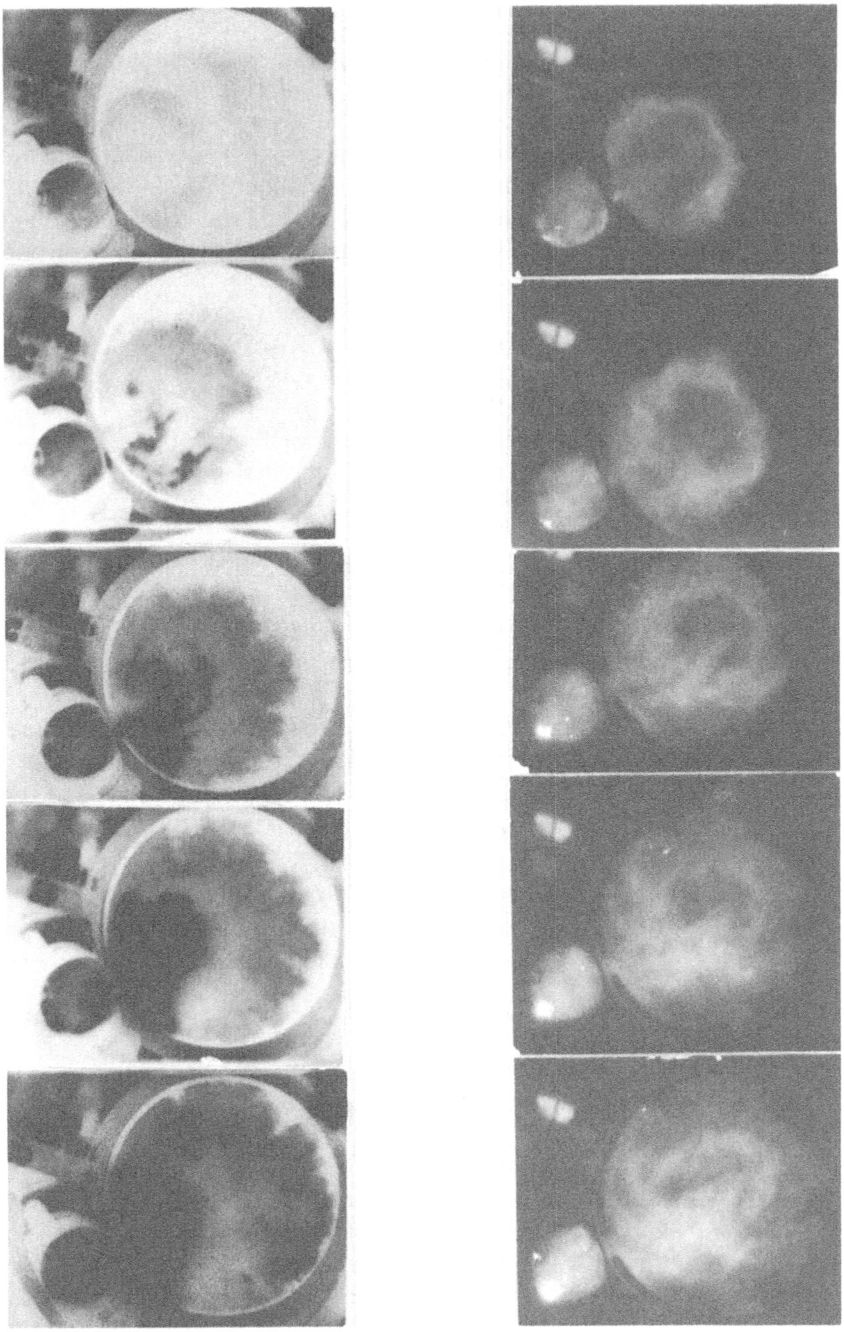

Fig. 3. Filmed flames at about the same crankangle in a divided-chamber engine operated at two different conditions.

References p. 323.

USING ZERO-DIMENSIONAL EQUATIONS

Consider a premixed uniform charge in laminar or turbulent motion in a closed vessel whose volume may change with time. If, during the propagation of the flame, the speed of the gas is everywhere and always much smaller than the speed of sound, then the pressure in the chamber can be considered uniform in space and varying only with time. If, in addition, the flame is thin, the burned and unburned fractions of the charge are distinct, and the following well-known equations expressing conservation of mass and isentropic compression of the reactants apply:

$$M_u + M_b = M \tag{1}$$

$$M_u = pV_u/RT_u \qquad M = p_0V_0/RT_0 \tag{2}$$

$$T_u/T_0 = (p/p_0)^{(\bar{\gamma}_u-1)/\bar{\gamma}_u} \tag{3}$$

$$\frac{M_b}{M} = 1 - \left(\frac{V}{V_0} - \frac{V_b}{V_0}\right)\left(\frac{p}{p_0}\right)^{1/\bar{\gamma}_u} \tag{4}$$

where the fourth equation follows from the previous three. The assumption that the unburned charge is compressed isentropically is often acceptable for most of the flame propagation, even in the presence of nonadiabatic walls, since most of the heat is lost by the high-temperature combustion products to the necessarily cooler walls.

The principle of conservation of energy can also be used to obtain the following equations that are valid under progressively more restrictive assumptions:

$$\frac{d}{dt}\int_{V(t)} \rho e dV = -p\frac{dV}{dt} + \int_{V(t)} \dot{q}_r dV + \int_{S(t)} K\nabla T\cdot \underline{n} dS \tag{5a}$$

$$\frac{d}{dt}\left[\frac{p}{p_0}\left(\frac{V_b/V_0}{\bar{\gamma}_b - 1} + \frac{V/V_0 - V_b/V_0}{\bar{\gamma}_u - 1}\right)\right] = -\frac{p}{p_0}\frac{dV/V_0}{dt} + \frac{\dot{Q}_r}{p_0V_0} - \frac{\dot{Q}_w}{p_0V_0} \tag{5b}$$

where

$$\dot{Q}_r \equiv \int_{V(t)} \dot{q}_r dV$$

and

$$\dot{Q}_w \equiv -\int_{S(t)} K\nabla T\cdot \underline{n} dS$$

$$\frac{p}{p_0}\left[\frac{V_b/V_0}{\bar{\gamma}_b - 1} + \frac{V/V_0 - V_b/V_0}{\bar{\gamma}_u - 1}\right] + \frac{q\rho_0}{p_0}\left(\frac{V}{V_0} - \frac{V_b}{V_0}\right)\left(\frac{p}{p_0}\right)^{1/\bar{\gamma}_u}$$
$$-\left[\frac{1}{\bar{\gamma}_u - 1} + \frac{q\rho_0}{p_0}\right] = -\int_1^{V/V_0} \frac{p}{p_0} d\frac{V}{V_0} - \frac{Q_w}{p_0V_0} \tag{5c}$$

where

$$\dot{Q}_r = q\dot{M}_b = -qM\frac{d}{dt}\left[\left(\frac{V - V_b}{V_0}\right)\left(\frac{p}{p_0}\right)^{1/\bar{\gamma}_u}\right]$$

and

$$Q_w \equiv \int_0^t \dot{Q}_w dt$$

$$\left[\frac{p_f V_f}{\bar{\gamma}_b - 1} - \frac{p_0 V_0}{\bar{\gamma}_u - 1}\right] + Q_{w_f} + \int_{V_0}^{V_f} p dV = \rho_0 V_0 q \tag{5d}$$

$$\frac{p - p_0}{p_f - p_0} = \frac{M_b}{M} \tag{5e}$$

Eq. 5a, which applies also for turbulent flows, follows directly from the general form of the energy conservation equation neglecting only kinetic energy and work of viscous forces, in accordance with the assumption of low Mach-number flows. Eq. 5b is obtained by setting $\rho e = p/(\bar{\gamma} - 1)$, using the assumption of a thin flame, and selecting two average values of γ corresponding to burned and unburned fractions of the charge. It is important that appropriate averages for γ be used, and they are obtained by assuring that the correct thermal energies are evaluated: $\bar{\gamma} = R/\bar{C}_v + 1$, $\bar{C}_v = (\int_0^T C_v dT)/T$. Notice that, even though there may be large temperature and density nonuniformities behind the flame and in the vicinity of cold walls, the energy per unit volume, ρe, is rather uniform everywhere as long as the pressure is uniform. That is, acoustic waves quickly redistribute over the entire volume the local pressure nonuniformities generated by the increase or decrease of thermal energy. On account of the spatial uniformity of ρe, the volume integral on the right-hand side of Eq. 5a can be carried out, thus obtaining Eq. 5b. Specification that the energy released by the combustion be proportional to the mass burned and integration over time of Eq. 5b give Eq. 5c which, when applied at the initial and final stages of the flame propagation, yields Eq. 5d. Eq. 5d clearly shows the redistribution of the energy released by the combustion as work, increased thermal energy, and wall heat losses. Finally, Eq. 5e, which is often applied to engine combustion [4], is obtained from Eqs. 4 and 5c by setting $\bar{\gamma}_b = \bar{\gamma}_u = \bar{\gamma}$, $V =$ constant, $Q_w = 0$.

If the volume is constant and the walls are adiabatic, the mass and energy equations (Eqs. 4 and 5c) contain only three variables (M_b/M, V_b/V, p/p_0) and three constants ($\bar{\gamma}_b$, $\bar{\gamma}_u$, $q\rho_0/p_0$). It can then be stated that for a given gas ($\bar{\gamma}_b$, $\bar{\gamma}_u$) and set of initial conditions ($q\rho_0/p_0$), the relationship among burned mass fraction, its volume, and pressure increase is universal. In particular, such a relationship is independent of the size and shape of the volume and of the speed and thickness of the flame, provided the latter two are much smaller than the speed of sound and the size of the vessel, respectively. Actually, a unique relationship among M_b/M, V_b/V, and p/p_0 exists also for a family of configurations for which volumes are not constant and walls are not adiabatic as long as the terms that account for these processes on the right-hand side of Eq. 5c are the same for all members of the family. Hence, the degree of departure from a universal relationship is due to *differences* in the magnitudes of the contributions from changing volumes and wall heat transfer during combustion and not to the magnitudes themselves. Indeed, in later sections it will be argued that scaling

References p. 323.

flowfields can exist even in the presence of moving boundaries and wall heat transfer when certain conditions are satisfied. In engines, the contributions from the two effects are not negligible but are small with respect to the heat of combustion, expecially for fast combustion around top dead center (TDC). Hence, departures from a unique relationship among M_b/M, V_b/V, and p/p_0, for a given value of qp_0/p_0 and for different engines or operating conditions, are due to differences in quantities that are already small and therefore not likely to be accurately measurable or very important. This trend and the tendency of both heat release and wall heat-transfer rates to be proportional to engine speed contribute to the explanation of the relative success of Eq. 5e in engine applications. Indeed, one can obtain Eq. 5e from Eq. 5b also by defining an effective heat of reaction, q^*, that includes the relatively constant corrections for changing volume and wall heat transfer (and by setting $\bar{\gamma}_b = \bar{\gamma}_u = \bar{\gamma}$).

Another interesting use of Eqs. 4 and 5c is the determination of the instantaneous wall heat-transfer rate and M_b/M if p/p_0, V_b/V_0 and $V(t)/V_0$ are measured in an engine or a bomb. Such an exercise was carried out using flame-position and pressure data that were obtained by Volkswagen with a quiescent stoichiometric methane-air mixture, initially at $p_0 = 0.35$ MPa and $T_0 = 350$ K. The charge was spark ignited at the periphery of a cylindrical bomb that has flat transparent top and bottom plates [5] (the same experiment for which the computations of Fig. 1 were made). The result is shown in Fig. 4, in which the ratio of the energy transferred to the walls to that released by the combustion is given versus time during the propagation of the laminar flame. At the beginning, the

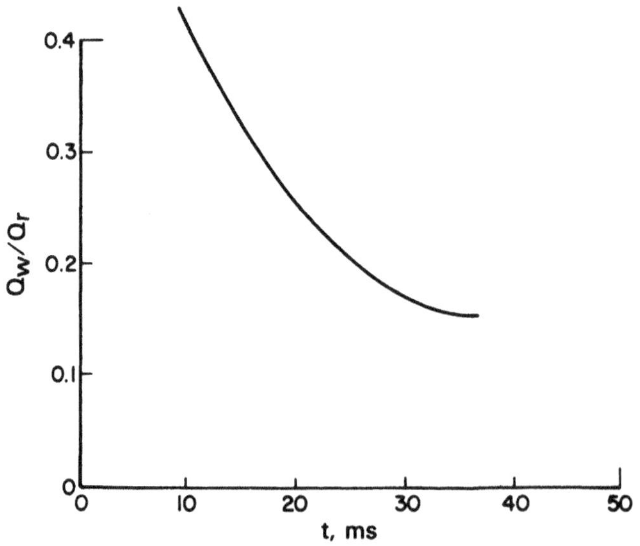

Fig. 4. Ratio of the heat transferred to the walls to that released by combustion during the propagation of a laminar flame in a cylindrical vessel.

fractional heat loss is very large, due to the high surface-to-volume ratio of the space occupied by the products. During the propagation it drops to a reasonable 15%. Initial and final values are not reported because they are subject to large errors, due to difficulties in measuring accurately V_b and V_u, respectively. Nor is it clear whether the ratio increases or decreases toward the end of the burn, since both the rate of energy release and that of energy loss tend to increase. But the final value of the ratio can be estimated independently using Eq. 5d, the measured final pressure, and the constant-volume heat of reaction. It was found to be about 15%. It should be noted that the intermediate values of the ratio are not affected by difficulties at the extremes, since they depend only on the instantaneous magnitudes of p/p_0 and V_b/V.

Another common engine application of Eqs. 4 and 5c, or equivalent ones, is the computation of p/p_0 and V_b/V, given M_b/M. The latter can be specified as a rate of combustion through a selected combustion angle or through a correlation for the flame speed. In the flame-speed approach, the flame front is assumed to be spherical, and an Eularian or Lagrangian coordinate is introduced to compute the nonuniform distribution of temperature behind the flame front if more than pressure-time histories are of interest. Similar approaches are reconsidered in later sections, but first additional properties are discussed that are exhibited by unsteady flowfields with slow and thin flames, even when such fields are completely three-dimensional.

USING MULTIDIMENSIONAL EQUATIONS

The same conclusions and several more can also be reached with a different approach. Given a confined charge with its initial and boundary conditions, it is found from the complete set of conservation equations that when certain conditions are satisfied, an artificial increase in flame speed almost exclusively brings about a proportionate acceleration of all events such that the flowfield remains the same at corresponding scaled times, except for the gas velocity, which increases proportionally to the flame speed. When other conditions are satisfied, it is found that an artificial increase in flame thickness brings about little change in the flowfield, except within the flame front itself. These two properties are useful when numerical solutions are sought, since computation time decreases drastically as flame speed and thickness are artificially increased. The physically meaningful solution is then recovered simply by scaling back the numerical one [1]. But there may be other useful applications of these scaling properties.

Scaling of the artificial solutions occurs because terms in the equations, and their initial and boundary conditions, are made to vary in specific ways. Hence, if there are physical situations in which the terms happen to vary in the same way, the physical flowfield will exhibit the same scaling properties. In general, practical configurations do not readily meet this requirement. But some of them do so approximately, and that could be the reason combustion in them scales to the extent that it does. But before getting to possible applications, it is first necessary to outline the two scaling transformations [1].

Consider the complete set of three-dimensional unsteady conservation equa-

References p. 323.

tions for laminar and turbulent flows with an unrestricted number of species and elementary reactions, and make in them and in their boundary and initial conditions the following substitutions:

α-Transformation for $M^2 << 1$	β-Transformation for $\delta_f << \delta$
$\mu^* = \alpha\mu$	$\mu^* = \beta\mu$
$\lambda^* = \alpha\lambda$	$\lambda^* = \beta\lambda$
$K^* = \alpha K$	$K^* = \beta K$
$D^* = \alpha D$	$D^* = \beta D$
$R_l^* = \alpha R_l$	$R_l^* = R_l/\beta$
$t^* = t/\alpha$	$\underline{x}^* = \beta \underline{x}$
$\underline{u}^* = \alpha \underline{u}$	

Under the α-transformation, transport coefficients (μ, λ, K, D), reaction rates (R_l) and velocities ($\underline{u}$) are multiplied by α, and time is divided by it. Under the β-transformation, the transport coefficients and the space dimensions are multiplied by β, and the reaction rates are divided by it. Notice that the thermodynamic properites (p, ρ_i, T, e) are not changed by either transformation.

Upon substitution into the general conservation equations and their initial and boundary conditions, it is found for the α-transformation that if terms of order $\alpha^2 M^2$ are neglected with respect to terms of order one, the parameter α drops out from the transformed equations and their initial and boundary conditions. Hence, for low Mach-number flows, the solution of the transformed equations does not depend explicitly on the value of α, and all of the thermodynamic properties and $\underline{u}^*$ are functions of $\underline{x}$, t^*, scaled diffusivities and reaction rates, vessel geometry, and initial and boundary conditions. Since the flame speed is proportional to α, $u_f^{*2} \sim \alpha^2 Dw/\rho$, it can also be stated that under appropriate conditions the space dependence of p, ρ_i, T and e does not change with flame speed, and the gas velocity increases proportionally to it. But combustion flowfields of practical interest do not readily respect the necessary conditions.

The most comprehensive field may exhibit nonuniformities in the initial distribution of chemical species (e.g., charge stratification), momentum (e.g., swirl), and energy (e.g., unmixedness of fresh and residual charges); may be enclosed by chemically active, dragging, conductive and moving walls; and may be controlled by diffusive and reactive processes everywhere and at all times. For two such fields to α-scale into each other, the values of all diffusivities, all reaction rates, and initial and boundary velocities of one must be α times those of the other, and everything else must be the same. These are requirements that are difficult to meet, but there are some limited cases of practical interest.

In the simplest of flowfields,* diffusion and reaction processes are negligible

---

* *The following definitions are used here and elsewhere: an initially uniform flowfield is a field in which the initial distribution of mass, momentum, and energy is uniform within the volume; a perfect boundary, or wall, is impermeable, chemically inert, frictionless, and adiabatic; a vessel of simple geometry is compact and unobstructive to the flow; a nearly perfect flowfield, or the simplest flowfield, is a field in which diffusion of mass, momentum, and energy, and reactions*

everywhere except within the flame fronts. Then any effect that arbitrary changes of diffusivities and reaction rates may have on such a field must be through changes that may occur within the flame front because the flame region is the only one in which such processes are controlling. For such a field, the α-transformation allows one to conclude that, for $M^2 << 1$, the space dependence of all thermodynamic variables is insensitive to flame speed and is determined by the geometry of the vessel, the initial conditions, and the flame position, and that the instantaneous gas velocity is proportional to the instantaneous flame speed.

In the β-transformation, diffusivities and length scales are multiplied by β, and reaction rates are divided by the same factor. Since the flame speed is not changed and length scales are increased by β, the duration of the combustion event is lengthened by the factor β. When the transformed variables are substituted into the general conservation equations and their initial and boundary conditions, one finds that the parameter β drops out of them if the time derivatives are neglected with respect to the other terms of the equations. Thus, in unsteady flowfields the β-transformation is applicable only when nonuniform regions are separated by uniform ones and the properties of the uniform regions change significantly only on a time scale that is much longer than the characteristic diffusion and/or reaction times of the nonuniform regions. Then the flowfield is seen as a steady field by the nonuniform regions, and within them the time derivatives are small in comparison to some other terms of the conservation equations. Thus, for example, if the nonuniform region is a flame in a uniform charge in a closed vessel, the characteristic time for significant changes of the flowfield is L/u_f, and that for diffusion and reaction within the flame is δ_f/u_f, so that the condition for the flame to be quasi-steady is $\delta_f << L$. But the flame could be imbedded in a nonuniform region that, in turn, is imbedded in the flowfield. Then the conditions for quasi-steadiness of the nonuniformities within the field of the flame within the nonuniformities are $\delta_f << \delta_{\rho_i, \rho u, e} << L$, where $\delta_{\rho_i, \rho u, e}$ are the characteristic lengths of regions of spatial nonuniformity of mass, momentum, and energy, respectively. Consequently, wherever a flame front and a wall boundary layer merge, the quasi-steady assumption, in general, breaks down ($\delta_f \approx \delta_w$).

When the stated conditions are satisified, β drops out of the transformed equations, and their solution no longer depends on it explicitly. All variables become unique functions of $\underline{x}^*$, t, scaled diffusivities and reaction rates, vessel geometry, and initial and boundary conditions. Then the flame thickness, the size of nonuniform regions and that of the vessel, and all other dimensions increase proportionally to β, and the entire solution scales with the size of the container. That is, solutions obtained for geometrically similar vessels of different sizes

that exchange energies comparable to the local thermal energy, are significant only within the flame region. It requires that the field be initially nearly uniform, the boundaries nearly perfect, the geometry adequately simple, and the wall motion properly restricted. Momentum nonuniformities are induced by the flame even in initially uniform charges. But the field can remain nearly perfect throughout the flame propagation if the geometry of the vessel is adequately simple.

References p. 323.

overlap if plotted versus scaled lengths when the flames are at corresponding scaled locations.

The required proportionate decrease of all reaction rates and increase of all diffusivities with increasing vessel size again is difficult to match in practice, but for restricted configurations, the necessary conditions are reduced and met more easily. At the extreme, there is once more the simplest flowfield, one in which any effect that arbitrary changes of diffusivities and reaction rates may have on it can occur only through changes that may occur within the flame front. Then the β-transformation allows one to conclude that, for $\delta_f << L$, the space dependence of all properties is insensitive to flame thickness, except within the flame front itself, and is determined by the geometry of the vessel, the initial conditions and the instantaneous position of the flame.

A less extreme case in which the physics of the problem seems to match naturally the requirements for the β-similarity is the behavior of laminar flames with respect to the pressure of the charge. Classical experiments have shown that, at constant temperature and for several practically interesting mixtures, the flame speed is insensitive to pressure, whereas the flame thickness is inversely proportional to it. In turn, this is often explained by letting $D \sim p_0^{-1}$, $\rho \sim p_0$, and $w \sim p_0^2$ so that $u_f^2 \sim Dw/\rho$ = constant, whereas $\delta_f^2 \sim D\rho/w \sim p_0^{-2}$. This appears to be a case of β-similarity in which $\beta \propto p_0^{-1}$ so that $D \sim \beta$ and $w/\rho \sim \beta^{-1}$. Then having computed the solution at one pressure, i.e., having the spatial distribution of p/p_0, ρ_i/ρ_{i_0}, T/T_0, e/e_0, and $\underline{u}/\underline{u}_0$, solutions at other pressures should be obtainable simply by applying the β-scaling. Computations for this interesting application have not yet been made.

Finally, since the two transformations are independent of each other, they can be applied simultaneously, and the conclusion is reached for the simplest flowfield that for low Mach numbers and thin flames, the space distributions of all thermodynamic variables, p, ρ_i, T, e, etc., depend only on initial conditions, geometry of the vessel, and instantaneous position of the flame, except within the flame front, whereas the gas velocity, $\underline{u}$, is proportional to the flame speed.

The above conclusion confirms and complements the one obtained earlier with the zero-dimensional equations for the simplest of flowfields. Then it was found that the space integrals of mass and energy are independent of flame speed and thickness, but nothing was said about the gas speed because the momentum conservation equation had not been used. Now it is concluded that the space distributions of all thermodynamic variables, not just their volume integrals, are independent of flame speed and thickness (except within the flame front itself) and that the gas speed is proportional to the flame speed. Since the zero-dimensional Eqs. 1 through 5 are volume integrals of the three-dimensional ones, it is not surprising that properties of spatial distributions were obtained with the latter and corresponding integral properties were obtained with the former.

THE SIMPLEST FLOWFIELD

The computations of Fig. 1 correspond to the most idealized case of propagation of flames in closed vessels, as discussed in the previous section. The walls are stationary and perfect, the charge is initially uniform, the geometry is compact and unobstructive, the Mach number is small compared to unity, and the flame is thin compared to the dimensions of the vessel.

The results were obtained solving the complete two-dimensional conservation equations without dropping any of their terms, even where the magnitude of some of them was expected to be small. Thus ahead of and behind the flame front, diffusive and reactive processes were allowed to occur, but their contributions were found to be small. The complete solutions exhibit the expected trends and confirm the validity of the similarity arguments. All thermodynamic properties ahead of and behind the three flames of different thicknesses are nearly the same and so are the dynamic ones, since in this case the three flames have the same speed. Moreover, within the flame fronts, the spatial distributions of properties scale with the flame thickness. Away from the walls the curvature of the three flames is constant along their fronts and is the same. Near the walls, the flame fronts become perpendicular to the boundary by bending locally with a radius of curvature about equal to the flame thickness. When the flow induced by the flame is locally parallel to the walls, i.e., when walls and stream surfaces locally coincide, as nearly occurs at $t = 10$ ms, the only force interaction between the wall and the flow would be through friction, and the free-slip boundary condition eliminates this. But when the direction of the flow tends to have a component normal to the wall, as occurs at $t = 20$ ms, the wall also exerts a normal pressure that forces the flow to run parallel to it.

The other observable differences are due mostly to the three computations having been initiated from the same flame kernel instead of from flames of different thicknesses, as should have been done. Consequently, at the same times, the flames are not exactly at the same positions. It can be seen that the lower temperatures are exhibited by the flames that have traveled less, i.e., whose burned products occupy smaller volumes. Had the comparisons been made at corresponding burned volumes, instead of at the same time, the flames would have been at the same position, and the fields ahead and behind the flame fronts would have been even more similar.

Thus it has been confirmed that, under the stated conditions, the flowfield: is insensitive to the thickness of the flame, except within the flame front; scales in time as the inverse of the flame-speed, a characteristic suggested only indirectly by the results of Figs. 1 and 2 but verified by other computations; and, for a given flame position, depends almost exclusively on the geometry of the vessel and the initial conditions. Then for a given geometry and set of initial conditions, one multi-dimensional unsteady solution gives the field for all flame thicknesses and speeds. Within the constraints of a nearly perfect flowfield, the latter quantities can be changed, for example, by changing the initial intensity of turbulence while keeping the initial mean velocity negligible, or by the addition of trace species to increase or decrease the flame speed, while keeping energetic reactions confined to the flame region. In the particular case in which the geometry of the vessel is so simple that a spherical flame front can be predicted during some part of the burn, the relationship between flame surface area and volume of burned gas can be determined from geometrical considerations, and the multidimen-

References p. 323.

sional computations are not needed. The zero-dimensional equations and a correlation for the flame speed are sufficient. The zero-dimensional equations give pressure, burned volume and mass fractions, and the conditions ahead of the flame as functions of the initial conditions, while a flame-speed correlation relates the burned mass fraction to the flame surface area and the conditions ahead of the flame: $\dot{M}_b = S_b \rho_u u_f$. (If only pressure-time histories are of interest, not even the resolution of the nonuniform temperature field behind the flame is necessary, since the pressure is uniform and can be calculated without it.) The multidimensional computations, or when applicable, the zero-dimensional ones, can be used to study the pressure-time history versus the chamber geometry, for example, and can be extended to include the calculation of trace species ahead of and behind the flame front. But such applications are of greater practical value when walls are allowed to transfer energy and to move as considered next.

CHARGE STRATIFICATION

The configuration discussed in the previous section is so idealized that its practical usefulness is considerably limited. Therefore, it is appropriate to ask if scaling properties may also be exhibited by at least some configurations in which the charge is nonuniform or in which the walls are not adiabatic, frictionless, and stationary. Some possibilities are discussed in this and following sections. They were surmised from equations and available data but have not been tested with specific computations. However, the important point at this time is not how exact the predictions are but that it is indeed likely that configurations of practical interest possess scaling properties, at least approximately.

The results of Fig. 2, which were obtained prior to consideration of scaling properties and requirements, pertain to two cases in which the condition of the initial uniformity of the charge is violated. As previously stated, in the two computations only the diffusivities are different, which is roughly equivalent to changing the speed of an engine without changing anything else. Since the reaction rates are the same and the diffusivities differ by a factor of four, it is instructive to compare the two sets of results with what one would have obtained from two similarity solutions in which $\alpha = \beta = 1$ and $\alpha = \beta = 2$, thus combining the same reaction rates with diffusivities that differ by a factor of four. Then one would predict that in half the time, the faster flame achieves a configuration everywhere identical to that of the slower one, except within the flame front, which would be twice as thick for the faster flame. However, for such a similarity to apply, the charge should have been uniform, initiation of combustion should have been from kernels of different sizes (or from a point source), and the flames should always have been much thinner than the size of the vessel. In spite of the violations of what appear to be necessary conditions, the two flames of Fig. 2 still scale reasonably well in time. Thus, it is surmised that the similarity holds even in the presence of some initial nonuniformities provided that the flame thickness is much smaller than the characteristic length of the nonuniform regions, that all slopes change proportionally and that the characteristic length of the expanded nonuniform regions still remains much smaller than that of the container: $\beta\delta_f << \beta\delta << L$. In an earlier section, analogous conditions were stated to be sufficient to maintain similarity if the size of the vessel is also increased proportionally to the flame thickness. The validity of the proposition can be checked readily with computations similar to those of Fig. 2, but they have not been made yet.

WALL HEAT TRANSFER AND MOVING BOUNDARIES

If one removes the restriction that the walls are adiabatic and frictionless, in general one finds non-scaling solutions of the conservation equations for different flame speeds and thicknesses. This is because the changes in diffusivities that alter the rate of heat release by changing the flame speed also alter the rate of wall heat loss by changing the wall boundary layers, but not necessarily in the same proportions. But suppose that wall heat transfer is dominated by such a type of forced convection that the local rate of heat loss is proportional to the local gas speed. Then if the gas speed is proportional to the flame speed and the mass burning rate is proportional to the flame speed, the rate of wall heat loss is also proportional to the rate of energy release. Hence, the thermodynamic variables of the flowfield again become independent of flame speed and dependent only on flame position. Indeed, such a configuration appears to be compatible with the conditions for the existence of the α-scaling. Similarly, if the walls are free to move but confine burns differing in flame speed, with the ratio of the wall to the flame speed being the same (but not necessarily constant in time), then it would appear that scaling should be maintained, and all thermodynamic variables should depend only on flame position. This configuration is again compatible with the conditions for the existence of the α-scaling, which still requires $M^2 << 1$ everywhere and at all times. Numerical solutions of the complete two-dimensional equations could again verify that the entire field scales under the above special cases of wall heat transfer and wall motion, but they have not yet been performed.

ENGINE COMBUSTION

The information of the previous sections can now be brought together and applied to engine combustion. First, consider a spark-ignition engine of arbitrary combustion-chamber geometry, but with nearly uniform charge and for which $M^2 << 1$ and $\delta_f << L$. Further assume that both the rate of heat release (or, equivalently, the flame speed) and the rate of wall heat transfer are proportional to the engine speed. Since the piston speed is also proportional to the engine speed, the indicated work becomes independent of it. This is shown by Eq. 5b of the zero-dimensional model (substitute $dt = d\theta/6$ r/min, $\dot{Q}_r \propto N$, $\dot{Q}_w \propto N$ and notice that N drops out of the equation), which also indicates that the volume integrals of the thermal energy and of all thermodynamic variables ahead of and behind the flame are independent of engine speed.

We can now add that the local and instantaneous values of all thermodynamic variables are independent of flame speed, i.e., engine speed, and thickness,

References p. 323.

except within the flame front itself and near the walls, and that they depend only on initial and boundary conditions, chamber geometry, and flame position. Also, the local and instantaneous gas velocity is proportional to the instantaneous flame speed, which implies in particular that the wall heat-transfer rate is proportional to the gas speed. But where flame front and wall boundary layers merge, the indicated behavior of the flowfield, in general, breaks down because $\delta_f \approx \delta_w$. The conclusions should be valid for any number of species and elementary reactions and any forms of the diffusivities for mass, momentum, and energy, but a few special cases are interesting. If the effective reactant-to-product conversion rate is postulated to be independent of engine speed, i.e., of turbulence level, then within the flame front the effective diffusivities tend to increase as the square of the engine speed and flame thickness as the engine speed ($u_f \propto N$ and $w/\rho \approx$ constant, then $D \sim u_f^2\, \rho/w \sim N^2$ and $\delta_f \sim D/u_f \sim N$).* Alternatively, if the flame thickness is postulated to be independent of the engine speed, then within the flame front both effective diffusivities and effective reaction rates tend to increase proportionally to the engine speed ($u_f \propto N$ and $\delta_f =$ constant, then $D \sim u_f\delta_f \sim N$ and $w/\rho \sim u_f/\delta_f \sim N$). Other combinations are allowed as well.

In the section on the simplest configuration, some uses were discussed of the result of one computation on account of the scaling properties at that flowfield. Since the introduction of restricted forms of boundary motion and wall heat transfer has not changed the nature of the flowfield, those considerations are still applicable. Moreover, it would appear that the scaling should also hold for some degree of swirl if the swirl velocity is proportional to the engine speed, and for some forms of charge stratification, as discussed in a previous section. Parenthetically, if geometrically similar chambers of different displacement are considered, and if it is assumed that the turbulence level can be increased so as to increase the flame speed proportionally to the characteristic length of the chamber, then the flowfield also would be the same for geometrically similar engines and simply scale in space and time. Such a flowfield still would be a function of the initial conditions of the charge, i.e., load, degree of swirl, degree of stratification, boundary conditions (magnitude of the wall heat and momentum transfers) and the geometry of the combustion chamber. Then one solution must be obtained for each combination of such quantities, the solution scaling with engine speed

* *The interpretation of D and w in the expressions $u_f^2 \approx Dw/\rho$, $\delta_f^2 \approx D\rho/w$ is clear for laminar flames, in which case D is about equal to the mass diffusion coefficient evaluated ahead of the flame (where ρ is also evaluated), and $w = \bar{w}'\delta_f'/\delta_f$ is an effective reactant-to-product conversion rate which can be estimated by computing the average reaction rate, $\bar{w}'$, in the high temperature region, δ_f', and translating it into an effective (lower) rate spread over the entire flame thickness, δ_f. For a turbulent premixed flame, D is an effective diffusivity, and w is an effective conversion rate defined by $\delta_f u_f = D$ and $w = \dot{M}_b/\delta_f S_f = \rho u_f/\delta_f$, where $\dot{M}_b$, ρ, u_f, and δ_f are measurable quantities. But the mechanism through which a specific turbulent flame exhibits specific values of D and w depends on one's interpretation of the events that occur within the turbulent flame front. For example, in some cases ρ/w may be interpreted as a turbulent mixing time instead of a chemical reation time. However, the conclusions of this section are not limited by any specific interpretation of D and w, and are valid as long as $\delta_f << L$, $M^2 << 1$ and break down in those regions of the flowfield in which $\delta_f \approx \delta$.*

and possibly with displacement. But it would appear that eventually one should also be able to scale the result of one computation to obtain flowfields that correspond, at least approximately, to somewhat different loads, degrees of swirl, degrees of stratification and magnitudes of wall heat and momentum transfer.

The end product of such computations is the evaluation of the indicated efficiency and its changes with respect to all of the mentioned parameters. But it could also be used to compute concentrations of trace species ahead of and behind the flame front as functions of the same parameters in a two-step approach. First the dimensionless flowfield is computed and stored. Next the species conservation equations are solved for the trace species, using the stored flowfield information applied to the specific set of the above parameters which is of interest. This approach is needed because the concentrations of trace species ahead of and behind the flame front are not expected to scale as the rest of the flowfield ($\delta_{\rho_i} \approx L$), and it is possible because such species do not influence the flowfield significantly. Indeed, the procedure is valid only if the local reactions that pertain to the trace species involve energies which are negligible with respect to the local thermal energy. Radicals ahead of the flame, and NOx and particulates behind it, should satisfy the requirement. Carbon monoxide and unburned hydrocarbon will satisfy the requirement, but only at sufficiently low concentrations.

Finally, it may be possible to devise techniques to scale solutions in time and space, even for small departures from the assumed proportionalities of flame speed and wall heat-transfer rate to engine speed and displacement. But the type of scaling discussed in this paper does break down if the conditions $M^2 << 1$ and $\delta_f << L$ are not satisfied.

Indeed, as the Mach number becomes comparable to unity, the pressure is no longer spatially uniform, and waves are established of characteristic time close to the acoustic time of the chamber. Scaling of the flowfield in time as the engine speed varies is no longer possible because the combustion time varies as $1/N$ whereas the acoustic time of the chamber is about constant. Hence, synchronization is excluded, and so is addition of contributions since neither deflagration waves nor high-amplitude pressure waves are linear.

Similarly, wherever the flame thickness approaches the characteristic length of the local flowfield, as at walls, any outcome of the interaction, such as the reaction rate, depends on the local and instantaneous details of the flowfield, and there is no assurance that the outcome will scale in space and time as the rest of the field. Thus when δ_f tends to L, the entire field is coupled, and the discussed type of scaling does not hold. But it is not clear at this point exactly *how large* the Mach number, and/or *how thick* the flame can be before the solutions deviate from each other unacceptably. Even for open chamber engines, the question is relevant for high-speed operation (> 5000 r/min) and/or in the presence of squish.

Examples of clear breakdown of either one or both conditions are often found in divided-chamber engines in which the throat Mach number can approach unity and the flame thickness can be of the order of the equivalent radius of the throat. For instance, it has been established that when the throat Mach number approaches unity, even for very short times, pressure oscillations are set up whose

References p. 323.

magnitude not only does not stay constant with engine speed, but does not even change monotonically with it [6]. Moreover, as previously stated, the ratio of combustion to acoustic time is not constant with engine speed, so that the flowfield does not scale in time as $1/N$ at all.

Yet the flames of Fig. 3, that were photographed at different engine speeds, appear to scale in time, and other flames under different conditions exhibited strong visual similarities. Several positions can be taken about this: (i) that the apparent similarity is fortuitous, particularly since it is not expected, even independently of the arguments about M and δ_f, because the two flames are for different initial conditions and chamber geometries and proper resolution of the flowfield in throat and prechamber regions would have evidenced differences; (ii) that mild, local and/or short departures from the conditions for similarity do not alter the *entire* field permanently and drastically and that the present similarity can be extended to include families of initial conditions and of chamber geometries; (iii) that other similarities exist. These new questions cannot be answered with the available information. Therefore the current conclusion must remain that when the conditions $M^2 << 1$ and $\delta_f << \delta$ are violated, scaling in time with engine speed is not possible.

When the conditions are satisfied, the flowfield scales ultimately because of its simplicity. Diffusion and reaction processes control it but only from thin isolated layers of rapidly changing properties along the walls and flame front. In the broad spaces separating the layers, their contributions are small, much milder gradients prevail, and acoustic waves travel rapidly to redistribute energy changes originated within the layers. Moreover the rates of the energy changes that set the time scale of the process conveniently increase or decrease together in the same proportions as the engine speed varies. But the geometry of the field may be fully three-dimensional, and the thin layers may be curved almost at will.

A particular case of some interest is when the geometry of the chamber and the location of the ignition source are such that the flame front can be assumed to be spherical. Then a quasi-one-dimensional model for the bulk flow based on the assumption of spherical symmetry, matched along the walls to submodels for the two-dimensional unsteady wall boundary layers, may be considered as an alternative to two- and three-dimensional models. For that situation to hold, the flame and the wall boundary layers must still be thin ($\delta_f << L$, $\delta_w << L$), otherwise the field is fully three-dimensional. The condition $M^2 << 1$ could be removed, and expectation of scaling with speed abandoned, but an engine field that exhibits a spherical flame is very likely to satisfy it. So this field becomes a particular case of the previous one. The advantages of the one-dimensional matched model with respect to a fully two-dimensional one is that in certain chamber configurations it may yield more accurate results, both by reproducing the bulk flow better and by making possible greater resolution of flame and wall layers. With respect to a three-dimensional model, the one-dimensional model may yield results of comparable accuracy at a fraction of the cost. Its disadvantages include inability to account for most forms of swirl and charge stratification, and some difficulties in accounting for the piston motion properly. Moreover, the development of an appropriate wall boundary-layer submodel and its match to the bulk flow and the piston motion may not be simple matters.

The structure of the field also suggests the location and nature of the quantities that should be measured to elucidate its details and check its models. When diffusion and reaction processes occur in thin layers, the flowfield is not sensitive to their details. It is sensitive only to the resulting net rate of energy change, which in turn can be achieved by infinite combinations of diffusion and reaction rates, each yielding different flame speeds and thicknesses and different wall boundary layers. Hence, measurements ahead of and behind the flame front and outside of the wall layers are not as informative as those within them, i.e., the measurements should be made where the controlling processes are most active.

On the other hand, when diffusion and reaction processes are active and significant everywhere, the field is made up of thick layers that spread over the entire volume and merge continuously into each other. The flame front is not clearly distinguishable, and separation of the charge into burned and unburned fractions is not meaningful. But now time- and space-resolved measurements of most quantities at most locations within the combustion chamber would constitute valuable information. Parenthetically, this flowfield, that is inherently three-dimensional ($\delta_f \approx \delta_w \approx L$), could still scale with engine speed, as engine combustion often does, if $M^2 << 1$ and if the increasing engine turbulence resulted in spatially uniform proportionate increases of both diffusivities and reaction rates, i.e., if all conditions for the validity of the α-scaling were respected. But such a proposition is not nearly as likely as expecting $\delta_f << L$.

Thus we have come to the conclusion that flowfields that are likely to exhibit scaling properties and to be amenable to computation are difficult to resolve experimentally, and those that do not scale and cannot be computed lend themselves more easily to informative measurements.

It would appear that flames and wall boundary layers in engines are adequately thin ($\delta_f \approx 0.1L$?), and that combustion in them does scale to a good degree, at least with engine speed. Hence the most informative measurements will be difficult to make. The use of completely nonuniform combustion flowfields (as can be obtained with gaseous fuel injection and very high swirl, for example) to facilitate measurements to test the models for simpler charges would appear to be a possibility. But the three-dimensional computations cannot be made routinely, and the mechanism of control of more complex fields is not necessarily the same as that of simpler ones. So the dilemma persists and compromises are probably in order.

SOME ANSWERS

Consider a set of flames of different thicknesses but having the same speed. Under what conditions do the corresponding flowfields, i.e., the corresponding solutions of the complete three-dimensional unsteady conservation equations, scale up in space as the flame thickness, and under what conditions are the flowfields almost the same everywhere at all times except within the flame front?

References p. 323.

The flame thickness must be much smaller than the dimension of the vessel ($\delta_f << L$). Then, in general, the solutions scale up in space as the flame thickness, if the flowfields are identical except for proportionate increases of vessel size and all diffusivities and proportionate decreases of all reaction rates. In particular, as the flame thickness varies, the solutions are almost the same everywhere at all times except within the flame front if diffusion and reaction processes are negligible outside of it. Then all variables and their space distributions are determined by the geometry of the vessel, the initial conditions, and the position of the flame. The property is expected to hold also in the presence of some initial nonuniformities in the distribution of the chemical species, i.e., even if diffusion processes are also controlling outside of the flame and wall layers, provided that the flame thickness is much smaller than the size of the nonuniform regions, that all gradients change proportionally, and that the size of the nonuniform regions remains much smaller than that of the vessel ($\beta\delta_f << \beta\delta_{\rho_i} << L$).

Consider a set of flames of different speeds but having the same thickness. Under what conditions do the flowfields scale in time almost exactly and due to the flame speed alone?

The Mach number must be much less than unity everywhere and at all times ($M^2 << 1$). Then, in general, solutions scale in time as the inverse of the flame speed if the flowfields are identical except for proportional increases of all diffusivities and reaction rates. In particular, solutions also scale in time as the inverse of the flame speed, except possibly within the flame front, if diffusion and reaction processes are negligible outside of it. Then all thermodynamic variables and their space distributions are determined by the geometry of the vessel and the initial conditions and the position of the flame, and the instantaneous gas velocity is proportional to the instantaneous flame speed.

The properties exhibited by the flowfield with respect to changes in flame thickness at constant flame speed and in flame speed at constant flame thickness can be combined to conclude that if $\delta_f << L$, $M^2 << 1$, and diffusion and reaction processes are negligible everywhere except within the flame front, then all thermodynamic variables and their space distributions are determined by the geometry of the vessel, the initial conditions and the position of the flame (except within the flame front), and the instantaneous gas velocity if proportional to the instantaneous flame speed.

Under what conditions is engine combustion expected to scale, and does such scaling extend to emissions?

Consider a spark-ignition engine of arbitrary combustion-chamber geometry but with nearly uniform charge and for which $M^2 << 1$ and $\delta_f << L$. Further, assume that both the rate of heat release (or equivalently, the flame speed) and the rate of wall heat transfer are proportional to the engine speed. Then the local and instantaneous values of all thermodynamic variables are independent of flame speed (i.e., engine speed) and thickness, except within the flame front itself and near the walls, and depend only on initial and boundary conditions, chamber geometry and flame position. Also, the local and instantaneous gas velocity is proportional to the instantaneous flame speed. But in general, where flame front and wall boundary layers meet, the indicated behavior of the flowfield breaks down. Moreover, the scaling does not extend to emissions. The properties are expected to hold for any number of species and elementary reactions and any forms of the diffusivities for mass, momentum, and energy.

Although calculations have not yet been made to verify any of the statements of this paragraph, it would appear that the scaling should be valid also for some degree of swirl if the swirl velocity is proportional to the engine speed, and for some forms of charge stratification. If geometrically similar chambers of different displacement are considered, and if it is assumed that the turbulence level can be increased so as to increase the flame speed proportionally to the characteristic length of the chamber, then the flowfield would also be the same for geometrically similar engines and would scale in space and time. Such a flowfield would still be a function of the initial conditions of the charge, e.g., load, degree of swirl, degree of stratification, boundary conditions (extent of volume change and the magnitude of the wall heat and momentum transfers), and of the geometry of the combustion chamber, so that one solution must be obtained for each combination of such quantities, which then scales with engine speed, and possibly displacement. But it is conceivable that the result of one computation can also be scaled to obtain flowfields that correspond to somewhat different loads, degrees of swirl, degrees of stratification, and magnitudes of wall heat and momentum transfer. It may even be possible to devise techniques to scale solutions in time and space, even for small departures from the assumed proportionalities of flame speed and wall heat-transfer rate to engine speed and displacement. But the scaling does break down if the conditions $M^2 << 1$ and $\delta_f << L$ are not satisfied. As the Mach number becomes comparable to unity, the pressure is no longer spatially uniform, and waves are set up of characteristic time close to the acoustic time of the chamber. Scaling of the flowfield in time as the engine speed varies is no longer possible because the combustion time varies as $1/N$ whereas the acoustic time of the chamber is about constant. Hence, synchronization is excluded and so is addition of contributions, since neither deflagration waves nor high-amplitude pressure waves are linear. Similarly, wherever the flame thickness approaches the characteristic length of the local flowfield, as at walls, the outcome of the interaction depends on the local and instantaneous details of the flowfield, and there is no assurance that the outcome will scale in space and time as the rest of the field. Thus, when δ_f tends to L, the entire field is coupled, and the discussed type of scaling does not hold. But it is not clear at this point exactly *how large* the Mach number, and/or *how thick* the flame, can be before solutions deviate from each other unacceptably.

Examples of violations of either one or of both conditions are often found in divided-chamber engines in which the throat Mach number can approach unity and the flame thickness can be of the order of the equivalent radius of the throat. Hence, when similarities are observed among flames in divided-chamber engines, several positions can be taken: that the apparent similarity is fortuitous and that proper resolution of the flowfield in throat and prechamber regions would have evidenced differences; that mild, local and/or short departures from the conditions for similarity do not alter the *entire* field permanently and drastically; that other

References p. 323.

similarities exist. These new questions cannot be answered with the available information, so that the current conclusion must remain that when the conditions $M^2 << 1$ and $\delta_f << \delta$ are violated, scaling in time with the inverse of the engine speed is not possible.

In the case of engine combustion, is the measurement of any of the dependent variables, at any location and any time, equally informative and useful in checking models? If not, which criteria should be used for the selection?

The structure of the flowfield suggests the location and nature of the quantities that should be measured to elucidate its details and check its models. When $M^2 << 1$ and $\delta_f << L$, the field is ultimately a simple one, even if it is three-dimensional and unsteady. Diffusion and reaction processes control it, but only from thin isolated layers of rapidly changing properties along the walls and flame front. In the broad spaces separating the layers, their contributions are small, properties change much more gradually, and acoustic waves travel rapidly to redistribute energy changes originated within the layers. Thus most of the flowfield is not sensitive to the details of the controlling diffusion and reaction processes, but only to the resulting net rate of energy change that, in turn, can be achieved by infinite combinations of diffusion and reaction rates, each yielding different flame thicknesses and wall boundary layers. Hence measurements ahead of and behind the flame front and outside of the wall layers are not nearly as informative as those within them, i.e., the measurements should be taken where the controlling processes are most active.

On the other hand, when diffusion and reaction processes are active and significant everywhere ($\delta_f \approx \delta_f \approx L$), no clear flame front is distinguishable, separation of the charge into burned and unburned fractions is not meaningful, and time- and space-resolved measurements of most quantities at most locations within the chamber constitute valuable information. But this flowfield is also completely three-dimensional and almost impossible to compute at present. Moreover, for it to scale with engine speed, as engine combustion often does, conditions must be verified that are much more unlikely than $\delta_f << L$.

Thus we have come to the conclusion that engine flowfields that are likely to exhibit scaling properties and to be amenable to computation are difficult to resolve experimentally, and those that do not scale and cannot be computed lend themselves more easily to informative measurements.

ACKNOWLEDGMENT

It is a pleasure to acknowledge that the computations of Figs. 1 and 2 were performed by Mr. P. J. O'Rourke, and those of Fig. 4 were performed by Dr. S. A. Syed. The experiments of Fig. 3 were conducted by Dr. R. L. Steinberger. Support for the engine combustion research is being provided by the U.S. Department of Energy, Volkswagenwerk AG of West Germany, the U.S. National Science Foundation, the General Motors Research Laboratories, and FIAT of Italy.

REFERENCES

1. *P. J. O'Rourke and F. V. Bracco, "Two Scaling Transformations for the Numerical Computation of Multi-Dimensional Unsteady Flames," to appear in the Journal of Computational Physics.*
2. *F. V. Bracco, "Modeling of Two-Phase, Two-Dimensional, Unsteady Combustion for Internal Combustion Engines," Stratified Charge Engine Conference, Proc. I. Mech. E. C171/77, pp. 167–187, 1976.*
3. *H. C. Gupta, R. L. Steinberger and F. V. Bracco, "Combustion in a Divided Chamber, Stratified Charge, Reciprocating Engine: Initial Comparisons of Calculated and Measured Flame Propagation," presented at Seventeenth Symposium (International) on Combustion, The Combustion Institute, Pittsburgh, Pennsylvania, 1978, to appear in Comb. Sci. and Tech.*
4. *E. F. Obert, "Internal Combustion Engines," International Textbook Company, Cambridge, Massachusetts, 1977.*
5. *W. R. Brandstetter and R, Decker, "Fundamental Studies of the Volkswagen Stratified Charge Combustion Process," Combustion and Flame, Vol. 25, pp. 15–23, 1975.*
6. *H. C. Gupta and F. V. Bracco, "On the Origin of Pressure Oscillations in Divided Chamber Engines," presented at the 1978 Society of Automotive Engineers Congress, Detroit, Michigan, February 1978, and submitted for publication to Combustion and Flame.*

DISCUSSION

J. C. Keck *(Massachusetts Institute of Technology)*

Using your model, you have indeed produced a very thick flame front. Furthermore, you showed a smooth temperature variation in which the temperature rises over a fairly large thickness. The character of that thick front is qualitatively different from the character in the front that you showed in your combustion pictures. In addition to your own pictures, there are many other combustion pictures that show what the turbulent flame front looks like. They show that it's a widely fluctuating region in which the leading edge is quite well defined. The temperature jumps and then fluctuates. If you look at an ionization probe as the flame front passes over it, you will see wildly fluctuating ionization. Furthermore, if you looked at the radiation emitted, you would find wildly fluctuating radiation associated, for example, with *OH* radicals and the like. This behavior suggests that the flame is not really a smooth transition between the unburned and the burned gases, but rather a region in which the flame front is highly convoluted. There are wild temperature fluctuations and wild concentration fluctuations. I don't see how you can reconcile your type of model of turbulent flame propagation with what is actually seen in engines.

F. V. Bracco

Current engine computations give only average values of all variables, not their turbulent fluctuations. As far as the structure of the flame is concerned, when the charge is uniform and the flame is thin, say less than 10% of the bore, the specific mechanism by which reactants are converting to products within it influence the flowfield only through the flame speed, which in turn is often

F. V. Bracco

correlated well with very simple expressions. However, if reaction and diffusion processes are controlling everywhere, the concept of a flame region is misleading, and the details of all reaction and diffusion processes everywhere influence all the variables anywhere. But it would appear that open-chamber uniform-charge engines exhibit adequately thin flames.

J. C. Keck

I think the one parameter that you didn't bring into your equations is the ratio of a laminar flame thickness to the turbulent scale. It's quite clear that if the laminar flame thickness is thin compared to the scale of turbulence, you will in fact have a complicated wrinkled structure which may be multiple connected at the start of the flame front. When you start calculating the coupling of this sort of a flow field to the chemistry, you're going to get very different results because of the highly nonlinear character of that coupling. For example, if you assume one of these very smoothly varying temperature profiles with a highly convoluted flame front, what you have done in effect is to multiply the effective flame front area by a large factor, namely, the ratio of the turbulent flame speed to the laminar flame speed. In engines, that number can be ten. So, in effect, you have something like ten times the effective flame front area if you do have a complicated wrinkled situation with multiple connected islands of burned gas. I think that these effects are important, for example, in determining *NO* formation rates. You might get very different results in the two cases.

F. V. Bracco

I refer to the previous answer as far as the influence of the details of the flame structure on the field is concerned. But non-equilibrium species, such as *NO*, generally do not scale as the rest of the flowfield and may be influenced everywhere by the details of the turbulent structure.

D. P. Hoult *(Massachusetts Institute of Technology)*

The measured spectral properties of turbulence near top dead center support the idea that there is a range of length scales that are both larger and smaller than the thickness of the turbulent flame front, so I don't know how to apply your criteria.

F. V. Bracco

To evaluate whether a flowfield with turbulence may scale with respect to flame speed, or flame thickness, or other parameters, one should define equations that represent the essence of what one thinks is the contribution of turbulence to the specific flowfield and see if the complete set of equations with their initial and boundary conditions allow scaling solutions. Such an approach is rather unequivocal, but without it, it is difficult to argue one way or the other. However, for the sake of the discussion, turbulence of scale equal to, or smaller than, the flame thickness should influence the field only through the flame speed. If the convective velocities of the larger eddies are smaller than those induced by the flame, the large eddies should not influence the field much; if the velocities are of the same order, they should produce distortion of the flame front of size comparable to the scale of the eddy; if much larger, they should produce random, separated flame fronts and, if the eddy scale is as large as the bore, they should produce random swirl. While there appears to be no reason to believe that any one of these configurations can be achieved, it would appear that most engines exhibit rather reproducible reasonably thin flame fronts, which should indicate that large-scale turbulence is not controlling.

SESSION II WRAP-UP

G. L. BORMAN

University of Wisconsin, Madison, Wisconsin

It seems to me that we're making great progress in modeling, although perhaps not for the most practical engine cases, but rather for simplified cases. We are also making great progress in experimental techniques to test these models. Those investigators who are experimentalists look at the great detail demanded by the models and are awed by the great detail of the measurements that are required to test them. So we need both sets of progress. What has occurred to me is that the combustion community is famous for all of us wanting to go our own way and do our own things. In one sense, this is fine in that it leads to a variety of approaches, experiments and results. But I wonder if there aren't a few key experiments that should be considered by all of us. Such experiments should be simplified. I don't think they can be engine experiments at the moment, or at least exclusively engine experiments. To evaluate such models both experimentally and theoretically we must first agree upon our boundary conditions and our initial conditions in some detail. There's some of this cooperation already existing, but I think there could be more. If I see any recurrent theme in my mind after hearing these papers, it is that such cooperation might result in something better than that resulting from independent efforts.

SESSION III

MODELING OF ENGINE EXHAUST EMISSIONS

Session Chairman
H. K. NEWHALL

Chevron Research Company
Richmond, California

MODELING OF ENGINE EXHAUST EMISSIONS—AN INTRODUCTORY OVERVIEW

H. K. NEWHALL

Chevron Research Company, Richmond, California

ABSTRACT

Exhaust emissions modeling, perhaps more than any other application of engine cycle simulation, has relied upon what is sometimes called a "phenomenological" approach. Broadly, this approach is characterized by disaggregation of the engine cycle and/or engine system into discrete components with detailed attention focused on those areas most closely related to the phenomenon or phenomena of interest.

Historically, the phenomenological approach has been associated with nondimensional or so-called zero-dimensional combustion models. The implied empiricism has resulted in criticism of this approach on the grounds that predictive capability is severely limited.

At the other extreme, several recent modeling efforts have been aimed at characterization of engine cycle events through solution of the governing conservation equations—including conservation of momentum—in one or more physical dimensions. The objective is more rigorous characterization of the combustion process, leading hopefully to enhanced predictive capabilities. This is particularly important for engine cycle processes dominated by fluid motion or mixing.

While these detailed models will undoubtedly add much to the understanding of engine processes, they suffer definite limitations specific to the emissions problem. Most important is the difficulty in handling detailed chemical kinetic computations. Because these multidimensional procedures become intractable when attempting to incorporate all but the simplest kinetic mechanisms, many of the real phenomena fundamental to the exhaust emissions problem are currently beyond their reach. Further difficulties stem from the localized and at the same time complex flow processes that are critical to pollutant formation. An example is the apparent contribution of the "roll-up vortex" in spark-ignition engine hydrocarbon emissions.

Faced with the computational constraints currently associated with detailed multidimensional models on the one hand, and on the other hand the need to minimize empiricism in existing phenomenological models, the modeler must make important decisions regarding future directions of his work.

This overview paper reviews the areas where phenomenological models have been successful. The respective levels of success are related to existent understanding of the governing physical and chemical processes. One of the major conclusions is that a strong need for fundamental experimental information relevant to the formation and emission of pollutants continues to exist. Important areas needing further experimental study are identified.

Finally, the increasing need for an interactive relationship between modeler and experimenter is reiterated. While this idea is not unique to the emissions problem—nor to this author by any means—it is of key importance in unravelling and managing the persisting complexity of exhaust emissions processes.

INTRODUCTION

The next four Symposium papers will deal specifically with the question of modeling engine exhaust emissions. These papers represent the state of the art in this field, and we are fortunate in that the authors all are at the forefronts of their respective modeling areas.

This introductory overview paper is not intended as a detailed technical treatment or review of the subject. Rather, it is planned to offer a perspective from which to view the more detailed work of the session's key authors. Also, rather than providing a balanced overview of the modeling field, it is intended to emphasize those aspects that are critical to the emissions modeling problem. In some instances, these considerations and the suggested directions for further work are divergent from the directions and goals of work covered in other sessions of this Symposium.

One of the major points in this regard is the treatment of chemical kinetics. In the topic areas covered in previous sessions, an objective has been to eliminate or decouple detailed chemical kinetic mechanisms from the major modeling formulations. This, of course, permits more detailed computation of fluid motions and transport processes that may be key to flame propagation, heat release, and mixing processes.

In the case of exhaust emissions, however, the objective usually is to predict the concentrations of pertinent chemical species appearing in engine exhaust gases. In most cases, these are controlled by kinetic or rate-limited processes. As a consequence, in the area of emissions modeling, in contrast to other modeling areas, it is incumbent on the researcher to avail himself of the best and most up-to-date kinetics information. This point will be amplified in the ensuing discussion.

GOALS OF EMISSIONS MODELING

Important specific goals of emissions modeling are listed in Table 1. Recently, much emphasis has been placed on the predictive capability of emissions models. Clearly, development of models capable of predicting relationships between engine design, operating conditions, performance, and pollutant emissions is a worthy goal. However, due to the complexity of the emissions problem—particularly when combinations of engines with advanced control systems are consid-

TABLE 1

Emissions Modeling Goals

Predict Engine Design, Performance, Emissions Relationships
Explain Engine Design, Performance, Emissions Relationships
Provide Framework for Improving Detailed Understanding of Specific Chemical and Physical Processes Affecting Emissions
Provide (System) for Storing, Managing and Rendering in Understandable Form the Many Complex Interactions Among Engine Operating Variables, Performance Parameters and Emissions Levels

ered—there is a legitimate place for models of limited predictive capability designed primarily to explain observed relationships between pollutant emissions and other parameters in fundamental terms.

A further objective of modeling in the emissions area is improved understanding of detailed chemical and physical processes controlling formation and destruction of pollutant species. What is envisioned here is the use of small submodels developed to explore specific chemical and physical phenomena related to processes and species of particular importance.

Finally, when considering the entire vehicle system, interactions among engine design and operating variables, performance, fuel economy and emissions have become increasingly complex. From this perspective, it seems that modeling has a role to play in storing and, more importantly, in rendering in useful and understandable form these complex interactions.

CHRONOLOGY OF ENGINE CYCLE, COMBUSTION AND EMISSIONS MODELING

It is interesting to view current emissions modeling efforts from a historical perspective. In this regard, Table 2 presents a brief chronology of engine cycle analysis starting with Carnot's treatise of 1824. It is evident that from that time until about 1960, major advances in engine cycle or combustion analysis occurred at roughly 20 to 50 year intervals. However, during the early 1960's, widespread availability of high-speed digital computers led to increased interest in engine cycle simulation. Simultaneously, the emerging exhaust emissions problem provided incentive for more detailed simulation or modeling of engine cycle events with the objective of predicting pollutant species concentrations or emissions rates.

During the years 1965 to 1975, predictive modeling of *NOx* emissions from spark-ignition engines became a reality. The efforts that led to this will be reviewed in detail in a following section of this paper. In the early 1970's, models for *NOx* emissions from diesel engines began to appear, and this continues to be an active area of investigation.

Models for unburned hydrocarbon emissions from spark-ignition engines have

References pp. 341–343.

TABLE 2

Chronology of Engine Cycle, Combustion and Emissions Modeling

1824	Carnot Cycle—Carnot [1]
1882	Air Standard Cycle Analysis—Clerk [2]
1927	Fuel-Air Cycle Analysis—Goodenough and Baker [3]
1936	Equilibrium Thermodynamic Charts—Hershey, Eberhardt, and Hottel [4]
1960–1965	First Digital Computer Simulations of Engine Cycle Events with Predictions of Performance and Fuel Economy [5]
1965–1975	Development of Models for Predicting *CO* and *NOx* Emissions from Spark Ignition (Homogeneous) Engines
1970	Models for Diesel (Heterogeneous) Engine *NOx* Emissions
1978	Models for Spark-Ignition Engine Hydrocarbon Emissions
19??	Models for Heterogeneous Engine Organic and Particulate Emissions

just recently been formulated. In fact, two of the papers to be presented during this Symposium session represent the most complete publications on this subject to date of which the author is aware.

Activity in the foregoing modeling areas is expected to continue for the foreseeable future. In addition, it is anticipated that attempts to model organic and particulate emissions from diesel and other heterogeneous combustion engines will be evidenced shortly. Each of these areas will be considered in more detail.

APPROACHES TO EMISSIONS MODELING

It seems appropriate to discuss briefly two distinct modeling approaches that have evolved during the past several years—"phenomenological modeling" and "detailed multidimensional modeling" (Table 3).

TABLE 3

Approaches to Emissions Modeling

Phenomenological Modeling
Disaggregation of the Engine Cycle and/or Engine System into Discrete Components
Detailed Computation of Chemical or Physical Phenomena Key to the Pollutant Species of Interest
Detailed Multidimensional Modeling
Solution of the Fundamental Partial Differential Equations Governing Physical and Chemical Processes

Major recent efforts in combustion modeling have been aimed at detailed multidimensional characterization of engine processes. This involves solution of the fundamental partial differential equations governing pertinent physical and chemical processes. Detailed modeling has been particularly important in attempting to characterize bulk fluid motions and mixing processes that control the behavior of heterogeneous-combustion engines.

In the case of emissions modeling, the detailed approach, in principle, appears to be of value. However, in a practical sense, inclusion of the complex kinetic mechanisms that control species emissions rates can increase computer storage and time demands to intractable levels. For this reason, the phenomenological approach appears to be more realistic. Here we benefit from a disaggregation of the engine cycle and/or system into discrete components, with attention focused on those areas most closely related to the phenomenon of interest. At present, it appears that emissions work must rely heavily, if not exclusively, on phenomenological modeling. All papers for this Symposium session are, in fact, based on this approach.

In the future, as experience with detailed modeling develops, it is possible that hybrid methods involving combinations of detailed and phenomenological computation will evolve as the best long-term approach to modeling the emissions problem.

EVOLUTION OF SUCCESSFUL EMISSIONS MODELS—*NOx* EMISSIONS FROM SI ENGINES

As mentioned previously, accurate predictive modeling of *NOx* emissions from spark-ignition engines is now a reality. At the present time, it is possible to predict with reasonable accuracy the influence of engine design and operating variables on *NOx* emissions. In view of the level of success achieved in this case, it is of interest to review the historic evolution of SI-engine *NOx* modeling. Hopefully, the resulting chronology can be generalized to reveal the ingredients necessary to the development of successful models in new or emerging emissions areas.

Events leading ultimately to formulation of successful *NOx* emissions models actually date back to the mid-1940's with extensive fundamental study of the chemical kinetics of *NO* formation (and destruction). Table 4 lists in chronological sequence key fundamental kinetics studies, all of which have ultimately impacted on *NOx* modeling. These efforts continued for about a 20-year period. A wide variety of experimental methods was used, and a broad range of physical conditions is represented.

These fundamental studies were followed by experiments concerned with application of kinetics to engine processes. A number of the significant applied studies are listed in Table 5. These investigations, each of which represented a major experimental effort, were concerned for the most part with temporal and spatial resolution of species concentrations during the engine cycle and their interpretation with regard to previously proposed kinetic mechanisms. Again, a variety of experimental methods provided a solid background of applicable data.

References pp. 341–343.

TABLE 4

Evolution of S.I.-Engine *NOx* Modeling
I. Fundamental Chemical Kinetic Studies

Zeldovich et al., 1946, 47 [6, 7]	Combustion Vessel
Wise and Frech, 1952 [8]	Flow Reactor
Kaufman and Kelso, 1955 [9]	Static Reactor
Kistiakowsky et al., 1957 [10]	Stirred Reactor
Ford and Endow, 1957 [11]	Photolysis
Glick, Klein and Squire, 1957 [12]	Shock Tube
Fennimore and Jones, 1958 [13]	Flat Flame Burner
Kaufman and Decker, 1958 [14]	Static Reactor
Harteck et al., 1958 [15]	Flow Reactor
Kaufman and Kelso, 1958 [16]	Microwave Discharge
Duff and Davidson, 1959 [17]	Shock Tube
Ogryzlo and Schiff, 1959 [18]	Shock Tube
Kaufman and Decker, 1960 [19]	Static Reactor
Clyne and Thrush, 1961 [20]	R. F. Discharge
Freedman and Daiber, 1961 [21]	Shock Tube
Fennimore and Jones, 1962 [22]	Burner
Wray and Teare, 1962 [23]	Shock Tube
Kretchmer and Peterson, 1963 [24]	Microwave Discharge
Modica, 1965 [25]	Shock Tube

It is apparent at this point that spark-ignition engine *NOx* modeling was preceded by extensive fundamental and applied experimental work related to chemical kinetics. The summation of evidence from these studies led ultimately to the simplified kinetic schemes that have been used successfully in modeling engine-cycle *NO* formation.

When compared with the level of work required in developing an experimental background, as represented by Tables 4 and 5, the effort involved in actual model construction, as shown by Table 6, appears relatively small. In Table 6, several SI-engine *NOx* modeling schemes appearing in the published literature are cited.

TABLE 5

Evolution of S.I.-Engine *NOx* Modeling
II. Experimental Study of Applicability of Kinetics to Engine Combustion

Zeldovich, 1946–1947 [6, 7]	Combustion Vessel
Alperstein and Bradow, 1966 [26]	S.I. Engine—Timed Gas Sampling
Starkman et al., 1967 [27]	S.I. Engine—I.R. Emission Spectroscopy
Starkman, Stewart and Zvonow, 1969 [28]	S.I. Engine—Timed Gas Sampling
Shahed et al., 1970 [29]	Combustion Vessel—U.V. Absorption Spectroscopy
Lavoie, Heywood and Keck, 1970 [30]	S.I. Engine—U.V. and Visible Emission Spectroscopy
Shahed et al., 1971 [31]	Combustion Vessel—U.V. Absorption Spectroscopy
Muzio, Starkman and Caretto, 1971 [32]	S.I. Engine—Timed Gas Sampling

TABLE 6

Evolution of S.I.-Engine *NOx* Modeling
III. Model Construction and IV. Validation

Blumberg and Kummer, 1971 [33]
Heywood, Mathews and Owen, 1971 [34]
Komiyama and Heywood, 1973 [35]
Hiroyasu and Kadota, 1974 [36]

Also included in Table 6 are experimental model validation studies. Historically, such studies have been conducted primarily by the respective modelers. Two points here probably deserve attention. First, it will become apparent in ensuing discussion that the modeling field would benefit from increased emphasis on model validation experiments. Second, a broader group of experimentalists, some independent of the model construction process, should be involved in validation work.

The intent of the review provided by Tables 4, 5 and 6 has been to place in perspective the general sequence of events that has led to a highly successful field of emissions modeling. This perspective is consolidated very simply in Table 7, in which the four steps involved in successful emissions model development are summarized. The major role played by experimental work at both fundamental and applied levels is one of the major conclusions drawn from this summary and the foregoing discussion. In the case of SI-engine *NOx* modeling, much of the key experimental work predated initial modeling efforts and thus provided necessary guidance during the formative stages of model development. Again, the importance of detailed chemical kinetics knowledge is evident.

DIESEL (HETEROGENEOUS COMBUSTION) ENGINE *NOx* EMISSIONS

The technical papers to be presented in this session deal with two developing areas in emissions modeling—diesel engine *NOx* emissions and spark-ignition engine hydrocarbon emissions.

Papers authored by Shahed et al. and by Hiroyasu et al. represent the state of the art in diesel engine emissions modeling. Both are characterized by the phen-

TABLE 7

Evolution of Successful Emissions Models

I.	Delineate Fundamental Chemical Kinetics Key to the Pollutant of Concern	Experimental
II.	Verify Applicability of Known Kinetics to Engine Cycle Processes	Experimental
III.	Construct Model	Conceptual
IV.	Validate Model	Experimental

References pp. 341–343.

TABLE 8

NOx Emissions from Diesel (Heterogeneous) Engines

I. Chemical Kinetics—Excellent Background Since 1965	
II. Experimental Study of Kinetics Applicability to Engine Processes	
Bennethum, Mattavi and Toepel, 1975 [37]	Timed Gas Sampling
Carver et al., 1975 [38]	Timed Gas Sampling
Voiculescu and Borman, 1978 [39]	Cylinder Dumping
Rhee, Uyehara, and Myers, 1978 [40]	Timed Gas Sampling
III. Model Construction and IV. Validation	
Bastress et al., 1971 [41]	
Shahed et al., 1973 and 1975 [42, 43]	
Kau, Heap, Tyson and Wilson, 1976 [44]	
Hiroyasu and Kadota, 1976 [45]	

omenological approach, as are all other diesel emissions models published to date.

It is of interest to assess the development of current diesel *NOx* models (Table 8) within the sequential framework provided by Table 7. First, concerning the chemical kinetics of *NO* formation and destruction, it appears that most of the fundamental work previously cited is applicable to the diesel engine, and it is doubtful that further effort in this area is warranted. However, experimental work aimed at application of *NO* kinetics to diesel engine processes is much less extensive than has been the case for SI engines. In fact, in contrast to SI-engine modeling, detailed diesel engine combustion studies have followed rather than preceded initial modeling efforts. Thus, modeling efforts have proceeded on a less informed basis than was the case for the SI engine.

Closely allied to the implied need for detailed experimental study of diesel engine cycle processes is a parallel need for model validation experiments. Because the experimental basis for model construction is not extensive at present, the need for carefully designed and exhaustive model validation experiments is particularly acute.

SPARK IGNITION (HOMOGENEOUS COMBUSTION) ENGINE HYDROCARBON EMISSIONS

Engine cycle modeling efforts aimed at predicting hydrocarbon emissions have only recently been undertaken. Two of the papers to be presented at today's session represent the most recent advances in this field and have been authored by researchers who are among the most knowledgeable on the subject.

If the status of hydrocarbon modeling is reviewed in the framework of Table 7, it becomes apparent first that a wealth of chemical kinetics data fundamental to hydrocarbon reactions exists (Table 9). However, efforts to consolidate and systematize this information, thereby casting it in a form relevant to the engine emissions problem. would be beneficial to further modeling attempts.

TABLE 9

Spark Ignition (Homogeneous) Engine Hydrocarbon Emissions

I.	Chemical Kinetics	
	A Wealth of Fundamental Information Exists	
	Information Should be Pulled Together in a Form Relevant to Emissions Problem	
II.	Experimental Study of Kinetics Applicability to Engine Processes	
	Daniel, 1956 [46]	Optical Study of Engine Quench Layer
	Daniel and Wentworth, 1962 [47]	Timed Sampling of Exhaust Gases
	Daniel, 1967 [48]	Timed Sampling of Quench Layer Gases
	Tabaczynski, Heywood and Keck, 1972	Timed-Resolved Exhaust Hydrocarbon Mass Flow
III.	Model Construction and IV. Validation	
	Patterson et al., 1972 (Exhaust Reactions) [50]	
	Hiroyasu and Kadota, 1974 [36]	
	Sakai et al., 1977 (Exhaust Reactions) [51]	
	Lavoie and Blumberg, 1977 [52]	
	Herrin et al., 1978 (Exhaust Reactions) [53]	

At the same time, there is a strong need for experimental work related to application of hydrocarbon kinetics to engine cycle processes. While initial exploratory work of this type was reported as early as 1957, there does not appear to have been a consistent effort leading to increasingly detailed knowledge of the hydrocarbon emissions process. Thus, as in the case of diesel emissions, SI-engine hydrocarbon modeling is beginning from a position of relative ignorance regarding details of fundamental engine cycle processes. Again, in this situation, validation experiments become critical to successful modeling efforts.

It is pertinent again to emphasize the importance of incorporating fundamental kinetics knowledge in emissions models. Where hydrocarbon emissions are concerned, there has been a tendency to use global reaction mechanisms providing single rate expressions for complete oxidative destruction of fuel species. Analysis of engine exhaust gases, however, typically reveals a hydrocarbon composition significantly different from that of the parent fuel. A great majority of the hydrocarbon species appearing in exhaust gases are synthesized during the engine cycle. As a consequence, hydrocarbon composition varies throughout the cycle, and a single oxidation mechanism cannot be expected to be representative. Considerable experimental work aimed at identifying key reaction processes within the engine can be justified, in the author's view.

DIESEL (HETEROGENEOUS COMBUSTION) ENGINE ORGANIC AND PARTICULATE EMISSIONS

Particulate emissions from diesel and other heterogeneous combustion engines are receiving increasing attention. To date, the author is aware of no significant attempts at modeling the particulate emissions process. However, with the current level of interest in this problem and the apparent technical difficulties as-

TABLE 10

Diesel-Engine Organic and Particulate Emissions

I.	Chemical Kinetics
	Large Existing Literature on Particulate Formation
	Need to Consolidate from Standpoint of Relevance to Engine Processes
II.	Experimental Study of Kinetics Applicability to Engine Processes
	No Significant Work in Published Literature
III.	Model Construction and IV. Validation
	No Significant Published Work beyond Empirical Correlations

sociated with its control, it is reasonable to expect that such attempts will be forthcoming.

With the prospect of emerging modeling efforts in the particulate emissions area, it is of interest to consider the status of knowledge (Table 10) in terms of the framework previously developed (Table 7). First, there exists a large, if somewhat diffuse, literature on the fundamental kinetics of particle formation in diffusion flames and related combustion processes. Consolidation of this information from the standpoint of relevance to engine processes and fuels would be a worthwhile initial step toward better understanding of the emissions problem.

Detailed experimental knowledge of particle formation (or destruction) in engine cycle processes is virtually nonexistent. Thus, at present, there exists no basis for rational application of kinetic models to the engine cycle. With the significant advances in experimental techniques—particularly in the field of optical diagnostics—that have occurred recently, it should be possible to gain significant new insight into particle formation and destruction processes. Guided by the present understanding of fundamental kinetic processes, such studies should provide the basis ultimately for successful model development. In the author's view, diagnostic experimental work with these objectives should receive a high priority for the next several years.

SUMMARY

The intent of this overview has been to focus on currently developing emissions modeling areas from the historical perspective provided by previously successful efforts in SI engine *NOx* modeling. Generalizing from this evaluation, the following sequence of steps appears to be the key to development of successful emissions models.

(i) Delineation of fundamental chemical kinetics relevant to species of interest.
(ii) Application of kinetics to engine cycle processes
(iii) Model construction
(iv) Model validation

Two of the major points that derive from this train of thought are, first, the importance of detailed chemical kinetics knowledge relevant to the pollutant species of interest and, second, the major commitment to experimental work required to support successful model construction.

When viewed from this perspective, it appears that currently developing emissions modeling areas would benefit substantially from increased experimental study aimed at increased understanding of engine cycle processes, as well as from more detailed and exhaustive model validation. The case for experimental study will be stronger still if modeling activity should emerge in the area of particulate emissions from heterogeneous combustion engines.

REFERENCES

1. S. Carnot, "Reflexions sur la Puissance Motrice du Few," edited by E. Mendoza, 1960.
2. D. Clerk, "Theory of the Gas-Engine," Proc. Inst. of Civil Engrs., Vol. 69, pp. 220-250, 1882.
3. G. A. Goodenough and J. B. Baker, "A Thermodynamic Analysis of Internal Combustion Engine Cycles," Univ. of Illinois, EES Bull. 160, 1927.
4. R. L. Hersey, J. E. Eberhardt and H. C. Hottel, "Thermodynamic Properties of the Working Fluid in Internal-Combustion Engines," SAE Trans., Vol. 31, pp. 409-424, 1936.
5. E. S. Starkman (Editor), "Digital Calculations of Engine Cycles," SAE Progress in Technology Series, Vol. 7, 1964.
6. Y. B. Zeldovich, "The Oxidation of Nitrogen in Combustion Explosions," Acta Physicochimica USSR, Vol. 21, pp. 577-628, 1946.
7. Y. B. Zeldovitch, P. Y. Sadovnikov and D. A. Frank-Kamenetskii, "Oxidation of Nitrogen in Combustion," Academy of Sciences of USSR, Institute of Chemical Physics, Moscow-Leningrad (Translated by M. Shelef), 1947.
8. H. Wise and M. F. Frech, "Kinetics of Decomposition of Nitric Oxide at Elevated Temperatures. I. Rate Measurements in a Quartz Vessel, II. The Effect of Reaction Products and the Mechanism of Decomposition," J. Chem. Phys., Vol. 20, pp. 22-24, 1724-1727, 1952.
9. F. Haufman and J. R. Kelso, "Thermal Decomposition of Nitric Oxide," J. Chem. Phys., Vol. 23, pp. 1702-1707, 1955.
10. G. B. Kistiakowsky and G. G. Volpi, "Reactions of Nitrogen Atoms, I. Oxygen and Oxides of Nitrogen," J. Chem. Phys., Vol. 27, pp. 1141-1149, 1957.
11. H. W. Ford and N. Endow, "Rate Constants at Low Concentrations. IV. Reactions of Atomic Oxygen with Various Hydrocarbons," J. Chem. Phys., Vol. 27, pp. 1277-1279, 1957.
12. H. S. Glick, J. J. Klein and W. Squire, "Single Pulse Shock Tube Studies of the Kinetics of the Reaction $N_2 + O_2 \rightleftharpoons 2NO$ Between 2000-3000 K, Cornell Aeronautical Laboratory Report AD-959-A-1, 1957.
13. C. P. Fenimore and G. W. Jones, "Determination of Oxygen Atoms in Lean, Flat, Premixed Flames by Reaction with Nitrous Oxide," J. Phys. Chem., Vol. 62, pp. 178-183, 1958.
14. F. Kaufman and L. J. Decker, "Effect of Oxygen on Thermal Decomposition of Nitric Oxide at High Temperatures," Seventh Symposium (International) on Combustion, The Combustion Institute, Pittsburgh, Pennsylvania, pp. 57-60, 1958.
15. P. Harteck, R. R. Reeves and G. Mannella, "Reaction of Oxygen Atoms with Nitric Oxide," J. Chem. Phys., Vol. 29, pp. 1333-1335, 1958.
16. F. Kaufman and J. R. Kelso, "Reactions of Atomic Oxygen and Atomic Nitrogen with Oxides of Nitrogen," Seventh Symposium (International) on Combustion, The Combustion Institute, Pittsburgh, Pennsylvania, pp. 53-57, 1958.
17. R. E. Duff and N. Davidson, "Calculation of Reaction Profiles Behind Steady State Shock Waves. II. The Dissociation of Air," J. Chem. Phys., Vol. 31, pp. 1018-1027, 1959.
18. E. A. Ogryzlo and H. I. Schiff, "The Reaction of Oxygen Atoms with NO," Can. J. Chem., Vol. 37, pp. 1690-1695, 1959.
19. F. Kaufman and L. J. Decker, "High Temperature Gas Kinetics with the Use of the Logarithmic

References pp. 341-343.

Photometer," Eighth Symposium (International) of Combustion, The Combustion Institute, Pittsburgh, Pennsylvania, pp. 133–139, 1960.

20. *M. A. Clyne and B. A. Thrush, "Rates of the Reactions of Nitrogen Atoms with Oxygen and with Nitric Oxide," Nature, Vol. 189, pp. 56–57, 1961.*
21. *E. Freedman and J. W. Daiber, "Decomposition Rate of Nitric Oxide Between 3000 and 4300 K. J. Chem. Phys., Vol. 34, pp. 1271–1278, 1961.*
22. *C. P. Fenimore and G. W. Jones, "Rate of the Reaction, $O + N_2O \rightarrow 2NO$," Eighth Symposium (International) on Combustion, The Combustion Institute, Pittsburgh, Pennsylvania, pp. 127–133, 1962.*
23. *K. L. Wray and J. D. Teare, "Shock-Tube Study of the Kinetics of Nitric Oxide at High Temperatures," J. of Chem. Phys., Vol. 36, pp. 2582–2596, 1962.*
24. *C. B. Kretschmer and H. L. Peterson, "Kinetics of Three-Body Atom Recombination," J. Chem. Phys., Vol. 39, pp. 1772–1778, 1963.*
25. *A. P. Modica, "Kinetics of the Nitrous Oxide Decomposition by Mass Spectrometry. A Study to Evaluate Gas-Sampling Methods Behind Reflected Shock Waves," J. Phys. Chem., Vol. 69, pp. 2111–2116, 1965.*
26. *M. Alperstein and R. L. Bradow, "Exhaust Emissions Related to Engine Combustion Reactions," SAE Trans., Vol. 75, Paper No. 660781, pp. 876–884, 1966.*
27. *H. K. Newhall and E. S. Starkman, "Direct Spectroscopic Determination of Nitric Oxide in Reciprocating Engine Cylinders," SAE Trans., Vol. 76, Paper No. 670122, pp. 743–762, 1967.*
28. *E. S. Starkman, H. E. Stewart and V. A. Zvonow, "An Investigation into the Formation and Modification of Emission Precursors," SAE Paper No. 690020, 1969.*
29. *H. K. Newhall and S. M. Shahed, "Kinetics of Nitric Oxide Formation in High Pressure Flames," Thirteenth Symposium (International) on Combustion, The Combustion Institute, Pittsburgh, Pennsylvania, pp. 381–389, 1970.*
30. *G. A. Lavoie, J. B. Heywood and J. C. Keck, "Experimental and Theoretical Study of Nitric Oxide Formation in Internal Combustion Engines," Comb. Sci. and Tech., Vol. 1, pp. 313–326, 1970.*
31. *S. M. Shahed and H. K. Newhall, "Kinetics of Nitric Oxide Formation in Propane-Air and Hydrogen-Air-Diluent Flames," Combustion and Flame, Vol. 17, No. 2, pp. 131–137, 1971.*
32. *L. J. Muzio, E. S. Starkman and L. S. Caretto, "The Effect of Temperature Variations in the Engine Combustion Chamber on Formation and Emission of Nitric Oxides," SAE Trans., Vol. 80, Paper No. 710158, pp. 652–662, 1971.*
33. *P. N. Blumberg and J. T. Kummer, "Prediction of NO Formation in Spark-Ignited Engines—An Analysis of Methods of Control," Comb. Sci. and Tech., Vol. 4, pp. 73–95, 1971.*
34. *J. B. Heywood, S. M. Mathews and B. Owen, "Predictions of Nitric Oxide Concentrations in a Spark-Ignition Engine Compared with Exhaust Measurements," SAE Paper No. 710011, 1971.*
35. *L. Komiyama and J. B. Heywood, "Predicting NOx Emissions and Effects of Exhaust Gas Recirculation in Spark-Ignition Engines," SAE Trans., Vol. 82, Paper No. 730475, pp. 1458–1476, 1973.*
36. *H. Hiroyasu and T. Kadota, "Combustion Simulation for Combustion and Exhaust Emissions in Spark Ignition Engine," Fifteenth Symposium (International) on Combustion, The Combustion Institute, Pittsburgh, Pennsylvania, pp. 1213–1223, 1974.*
37. *J. E. Bennethum, J. N. Mattavi and R. R. Toepel, "Diesel Combustion Chamber Sampling-Hardware, Procedures, and Data Interpretation," SAE Trans., Vol. 84, Paper No. 750849, pp. 2213–2240, 1975.*
38. *C. P. Carver, M. P. Heap, C. McComis and T. J. Tyson, "Investigation of Diesel Combustion by Direct In-cylinder Sampling," SAE Paper No. 750850, 1975.*
39. *I. A. Voiculescu and G. L. Borman, "An Experimental Study of Diesel Engine Cylinder-Averaged NOx Histories," SAE Paper No. 780228, 1978.*
40. *K. T. Rhee, P. S. Myers and O. A. Uyehara, "Time and Space-Resolved Species Determination in Diesel Combustion Using Continuous-Flow Gas-Sampling," SAE Paper No. 780226, 1978.*
41. *E. K. Bastress, K. M. Chng and D. M. Dix, "Models of Combustion and Nitrogen Oxide Formation in Direct and Indirect Injection Compression-Ignition Engines," IECEC Meeting Paper No. 719053, 1971.*

42. S. M. Shahed, W. S. Chiu and V. S. Yumlu, "A Preliminary Model for the Formation of Nitric Oxide in DI Diesel Engines and Its Application in Parametric Studies," SAE Trans., Vol. 82, pp. 338-351, 1973.
43. S. M. Shahed, W. S. Chiu and W. T. Lyn, "A Mathematical Model of Diesel Combustion,—I. Mech. E. Conf. on Combustion in Engines," C94/75, Cranfield, England, pp. 119-128, 1975.
44. C. J. Kau, M. P. Heap, T. J. Tyson and R. P. Wilson, "The Prediction of Nitric Oxide Formation in a Direct Injection Diesel Engine," Sixteenth Symposium (International) on Combustion, The Combustion Institute, Pittsburgh, Pennsylvania, pp. 337-350, 1976.
45. H. Hiroyasu and T. Kadota, "Models for Combustion and Formation of Nitric Oxide and Soot in Direct Injection Diesel Engines," SAE Trans., Vol. 85, Paper No. 760129, pp. 513-526, 1976.
46. W. A. Daniel, "Flame Quenching at the Walls of an Internal Combustion Engine," Sixth Symposium (International) on Combustion, The Combustion Institute, Pittsburgh, Pennsylvania, pp. 886-893, 1956.
47. W. A. Daniel and J. T. Wentworth, "Exhaust Gas Hydrocarbons—Genesis and Exodus," SAE Paper 486B, 1962.
48. W. A. Daniel, "Engine Variable Effects on Exhaust Hydrocarbon Composition (A Single-Cylinder Engine Study with Propane as Fuel)," SAE Trans., Vol. 76, Paper No. 670124, pp. 774-795, 1967.
49. R. J. Tabaczynski, J. B. Heywood and J. C. Keck, "Time Resolved Measurements of Hydrocarbon Mass Flow Rate in the Exhaust of a Spark Ignition Engine," SAE Trans., Vol. 81, Paper No. 720112, pp. 379-391, 1972.
50. D. J. Patterson, R. H. Kadlec, B. Carnahan, H. A. Lord, J. J. Martin, W. Mirsky and E. A. Sondreal, "Kinetics of Oxidation and Quenching of Combustibles in Exhaust Systems of Gasoline Engines," Annual Progress Report No. 3 to CRC, 1971-1972.
51. Y. Sakai, Y. Nakagawa, S. Tange and R. Maruyama, "Fundamental Study of Oxidation in a Lean Thermal Reactor," SAE Paper No. 770297, 1977.
52. G. A. Lavoie and P. N. Blumberg, "A Fundamental Model for Predicting Emissions and Fuel Consumption for the Conventional Spark-Ignition Engine," Eastern States Section, The Combustion Institute, Hartford, Connecticut, November 1977.
53. R. J. Herrin, D. J. Patterson and R. H. Kadlec, "Modeling Hydrocarbon Disappearance in Reciprocating-Engine Exhaust," General Motors Research Laboratories Symposium on Combustion Modeling in Reciprocating Engines, Warren, Michigan, November 1978.

H. K. Newhall

A MODEL FOR THE FORMATION OF EMISSIONS IN A DIRECT-INJECTION DIESEL ENGINE

S. M. SHAHED, P. F. FLYNN and W. T. LYN

Cummins Engine Company, Columbus, Indiana

ABSTRACT

A mathematical model of diesel spray combustion in direct-injection diesel engines is described.

The model is based on concepts of quasi-homogeneous multiple-zone combustion. Combustion zones evolve progressively as a result of the fuel-air mixing process. The thermodynamic and chemical kinetics model is flexibly designed without geometrical constraints to accept any fuel-air mixing model. Calculation of mixing rates is based on transient spray-plume-growth equations together with concentration profiles within the spray plume.

Unlike concepts of apparent heat-release rates, the model combines, in a consistent set of equations, the fuel-air mixing process, local chemical energy release, pressure evolution and pollutant-formation kinetics. The paper illustrates the use of the model in a fundamental study of the effect of high injection pressure on the combustion and pollutant-formation process. A fundamental tradeoff between cycle work and nitric oxide results when fuel-air mixing rate is used as a controlling parameter. Experimental results are presented to show this tradeoff, and the model is used in an analysis of the fundamental causes.

Engine-development experience shows a similar strong tradeoff between particulate and nitric-oxide emission. The potential use of the model in studying the fundamental causes of this tradeoff is illustrated.

NOTATION

A	zone of fresh air
B_i	combustion zone i
b	spray half-width
C	rich core zone
c, c_m	mass fraction of fuel in spray and on the spray axis, respectively

References pp. 362–363.

H absolute enthalpy (subscripts a, b_i, c refer to respective zone)

M, M_a, M_e, M_f total mass, mass of air, mass entrained and mass of fuel respectively (subscripts a, b_i, c)

P instantaneous cylinder pressure

Q heat transfer (subscripts a, b_i, c with subscript ch refers to total combustion chamber)

R_a, R_f gas constants for air and fuel vapor respectively

t time

T temperature (subscripts a, b_i, c, ch)

U absolute internal energy (subscripts a, b_i, c, ch)

V volume (subscripts a, b_i, c, ch)

X_t spray tip penetration

X_L spray tail penetration

X coordinate direction along spray axis

Y coordinate direction normal to spray axis

α transient fuel distribution parameter

θ crank angle

ρ density of jet

$\Phi_L, \Phi_R, \Phi_i, \Phi_{i+1}$ equivalence ratios at lean limit of combustion, rich limit of combustion, and boundaries of zone B_i, respectively

INTRODUCTION

Diesel combustion modeling has lagged behind modeling of spark-ignition engine and gas turbine engine combustion. One of the reasons is the complex heterogenous as well as transient nature of the processes involved. In addition, three phases, namely particulate carbon, liquid fuel, and fuel vapor and other gaseous species, exist in the combustion chamber at various times and affect in a major way the fundamental rate processes taking place during combustion. Discriminating in-cylinder information on fuel-air distribution, charge motion and heat-transfer rate has also been sparse, leading to substantially differing concepts of combustion.

The complexity of the problem and the corresponding effort required calls for an explicit understanding of the expectations from a mathematical model and the purpose of modeling. A model is considered a mathematical catalogue of the most current physical information, providing a framework of disciplined thinking about fundamental variables and mechanisms affecting engine performance. Where such fundamental information is not available, the model necessarily, and hopefully temporarily, bridges the gap with empirical and semi-empirical correlations. By its very nature, this process identifies areas where fundamental information is needed. The model should therefore be designed to be receptive to new information without a change in its basic framework. A model can further serve to provide a detailed analysis of experimental results where a quantitative description of the intermediate steps is desired. Finally, in advanced stages of development the model may serve as a predictive design tool.

The present paper illustrates such features of the diesel combustion model described.

COMBUSTION MODEL

Physical Description—Diesel combustion is strongly controlled by mixing of fuel and air. During the ignition-delay period a part of the injected fuel mixes with air and forms a combustible mixture. Ignition begins at multiple nucleii and rapidly engulfs this already mixed part of the fuel. The remainder of the fuel in the spray core is too rich to burn. This and any further injected fuel entrains air and burns gradually as it mixes with enough oxygen to form a combustible mixture near the rich limit. Further, the process of combustion of this rich mixture is not an instantaneous release of all its heat of combustion. Rather, it may be considered to go through phases of partial oxidation to shifting equilibrium products dictated by local oxygen availability. Thus, the mixing of raw fuel and air, as well as the mixing of products of rich combustion and air, dictate the chemical-energy-release rate.

The rate of mixing and the local conditions of temperature and composition under which energy is released are important considerations in modeling the combustion process. Both the total energy release and, even more so, pollutant formation are strong functions of local properties. Hot combustion zones are a source of nitric oxide as they pass through the appropriate composition range. Hot rich zones are a source of soot resulting from fuel pyrolysis, with subsequent burning upon mixing with oxygen.

Combustion Zones—The model described here is based on the concept of quasihomogeneous combustion zones. In considering the thermodynamic and combustion process, a zone does not have to be defined geometrically, nor does it necessarily have to be contiguous. A zone is defined by the fuel-air mixture prepared for combustion during a given crank-angle interval. The rate process feeding the fuel-air mixture to the zone can be either jet momentum exchange or

References pp. 362–363.

droplet evaporation. The source of fuel-air mixture can be concentrated, such as the core of a jet, or distributed, such as droplets.

The temperature and composition of each zone is followed from its inception to exhaust-valve opening as it goes through its mixing and oxidation history. Internal energy release and pollutant kinetics calculations are carried out simultaneously.

The model described in this paper is based on semiempirical jet-mixing calculations in which the spray is assumed to be essentially vaporous. However, it is flexible enough to accept refined or even radically different mixing rate processes.

Jet Mixing—Jet-mixing rate processes are calculated based on a dimensional analysis of jet-penetration and jet-width data acquired over a wide range of injection system and chamber parameters. The data were obtained in a constant-volume chamber equipped with quartz windows and an injection system designed to obtain a single shot of injection. High-resolution shadowgraphs and high-speed schlieren movies were used as means of data acquisition.

The details of jet penetration, jet deflection under swirling conditions, jet-width growth, effect of swirl on jet width and similar considerations have been the subject of an earlier publication [1] and will not be described here. A brief outline of the calculation of air entrainment into the jet is given below for the sake of completeness.

From similarity analyses of steady-state sprays [2] the concentration distribution of fuel along the spray axis is assumed to be hyperbolic, while the concentration across it is assumed to follow a normal distribution curve. Thus

$$C_m = \frac{1}{\alpha(t)x + 1} \quad x_L < x < x_t \tag{1}$$

$$= 0 \quad x < x_L \text{ and } x > x_t$$

$$\frac{C(x, y, t)}{C_m(x, o, t)} = 1 - \left(\frac{y}{b}\right)^{1.5} \tag{2}$$

These concentration profiles are shown in Fig. 1. The use of steady-state concentration profiles for an unsteady jet has lately received partial justification from the work of Dent [3]. The parameter $\alpha(t)$ in Eq. 1 is obtained from an overall mass balance of the fuel within the spray plume. Thus

$$\int_0^t \frac{dM_f}{dt}\, dt = 2\pi \int_{x_L}^{x_t} \int_0^b c\rho y\, dy\, dx \tag{3}$$

where dM_f/dt is the fuel-injection rate and the local density ρ is

$$\rho = \frac{P_a}{[(1 - c)R_a + cR_f]T_a} \tag{4}$$

Numerical iteration is employed to evaluate $\alpha(t)$. $\alpha(t)$ is a fundamental parameter representing the transient nature of the spray. It is constant for a steady-state

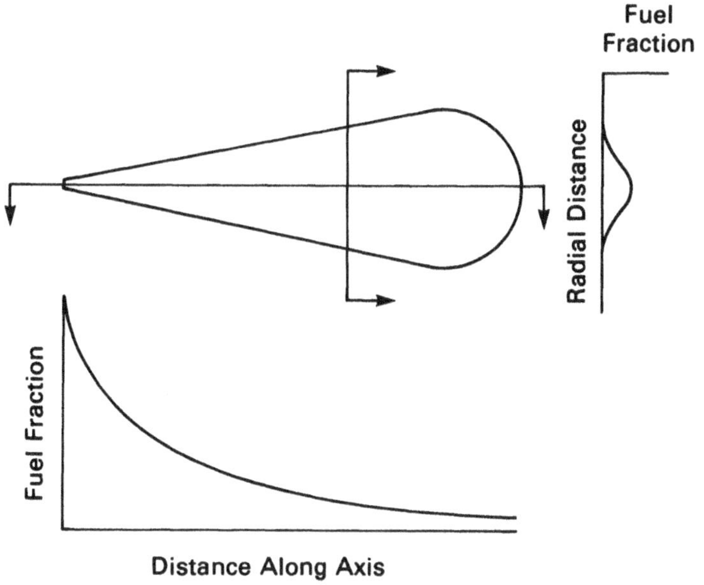

Fig. 1. Concentration distribution in a spray.

jet. Attempts have been made in the literature to calculate air-entrainment rate into the jet using steady-state jet velocity profiles [4, 5, 6, 7]. The wide variation of $\alpha(t)$ over the injection duration and a significant portion of the mixing duration clearly illustrates the inadequacy of the steady-state approach.

As discussed earlier, a set of discrete combustion zones progressively evolving along with the developing jet is superimposed on this continuous fuel-air distribution, giving the quasi-homogenous zones referred to earlier. Each zone is determined by following and accounting for a fixed amount of fuel.

At incipient ignition, the chamber is considered to be divided into $(n + 2)$ zones, as shown in Fig. 2a: a zone 'A' of air, a zone 'C' forming the rich core of the spray, and 'n' zones 'B_i' of mixture, mixed to combustible proportions during the ignition-delay period. The combustion zones 'B_i' are bounded by the rich and lean limits of combustion and by arbitrarily chosen intermediate equivalence ratios, Φ_i, such that

$$\Phi_R \geq \Phi_i \geq \Phi_L$$

The mass of fuel in zone B_i is

$$M_{fbi} = 2\pi \int_{x_L}^{x_t} \int_{y(\Phi_{i+1})}^{y(\Phi_i)} c\rho y dy dx \tag{5}$$

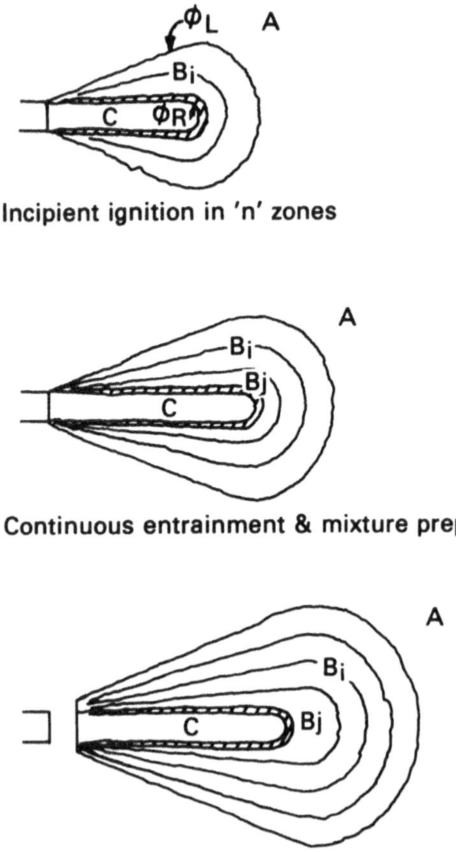

(a) Incipient ignition in 'n' zones

(b) Continuous entrainment & mixture preparation

(c) End of injection. Continuous mixing

Fig. 2. Evolution of combustion zones.

and of air is

$$M_{abi} = 2\pi \int_{x_L}^{x_t} \int_{y(\phi_{i+1})}^{y(\phi_i)} (1 - c)\rho y dy dx \tag{6}$$

where the iso-equivalence ratio boundary, $y(\phi_i)$, is calculated from Eq. 2. The mass of zone C is computed from equations similar to Eqs. 5 and 6 by considering integration limits from $y(\Phi = \infty)$ to $y(\Phi = \Phi_R)$. The mixture leaner than the lean limit, Φ_L, is too lean to burn and is considered part of zone A. The integration limits for zone A are $y(\Phi = 0)$ to $y(\Phi = \Phi_L)$, thus giving M_{fa} from an equation similar to Eq. 5.

Fig. 2a also depicts the "skin" of the core that is the source of further prepared mixture which develops into more zones, B_i. This development is shown in Figs. 2b and 2c. Fresh mixture is considered to develop into a zone "B_J" until the mass of fuel in that zone equals the zone average. At this point, while air entrainment into the zone continues, "J" is updated to the next higher number, creating a new zone into which freshly prepared mixture is fed. Thus, fuel mass in each zone except the newest zone, B_J, is fixed and nearly equal to the zone average.

At each computation step an updated spray geometry and distribution coefficient, $\alpha(t)$, are obtained. From a knowledge of the mass of fuel, M_{fa} and M_{fbi}, an iterative calculation is performed progressively on Eqs. 5 and 6 to calculate the updated values of $y(\Phi i)$, starting with zone A at $y(\Phi = 0)$ and ending at $J + 1$, the inner boundary of B_J. The calculation near the skin of the core gives the mixture prepared for combustion during the updated interval. Thus

$$\Delta M_{mp} = 2\pi \int_{x_L}^{x_t} \int_{y(\phi_{J+1})}^{y(\phi_R)} c\rho y dy dx + 2\pi \int_{x_L}^{x_t} \int_{y(\phi_{J+1})}^{y(\phi_R)} (1 - c)\rho y dy dx \tag{7}$$

This prepared mixture is fed to the existing or an updated zone, B_J, depending on the mass of fuel already present in it. Further details of the jet mixing calculation are given in Ref. 1.

Thermodynamic Calculations—The jet-mixing equations described above give the updated mass, fuel-air ratio and entrainment rates into each zone. Conservation of energy is next applied to the zones in order to calculate the thermodynamic variables, namely pressure, volume, temperature and equilibrium-combustion-product composition. The details of this calculation are given in Ref. 8. Only a brief outline is given here.

$$\frac{d(M_c U_c)}{d\theta} = -P\frac{dV_c}{d\theta} + \frac{dQ_c}{d\theta} + \frac{dM_f}{d\theta}H_f + \frac{dM_{ec}}{d\theta}H_a - \frac{dM_{mp}}{d\theta}H_m \tag{8}$$

$$\frac{d(M_a U_a)}{d\theta} = -P\frac{dV_a}{d\theta} + \frac{dQ_a}{d\theta} - \frac{dM_a}{d\theta}H_a \tag{9}$$

$$\frac{d(M_{bi} U_{bi})}{d\theta} = -P\frac{dV_{bi}}{d\theta} + \frac{dQ_{bi}}{d\theta} + \frac{dM_{ebi}}{d\theta}H_a \quad i = 1, \ldots (J - 1) \tag{10}$$

$$\frac{d(M_{bj} U_{bj})}{d\theta} = -P\frac{dV_{bj}}{d\theta} + \frac{dQ_j}{d\theta} + \frac{dM_{mp}}{d\theta}H_m - \frac{dM_{ebj}}{d\theta}H_a \tag{11}$$

Eq. 11 applies only to the zone B_j immediately surrounding the core, since it is the only zone receiving prepared fuel-air mixture for combustion. Heat transfer is calculated from a correlation similar to Annand's work [9]. It is based on a bulk average temperature defined by

$$T_{ch} = \frac{\sum M_{bi}T_{bi} + M_a T_a + M_c T_c}{\sum M_{bi} + M_a + M_c} \tag{12}$$

The total heat transfer is apportioned between the various zones based on their mass and temperature.

References pp. 362–363.

$$\frac{dQ_{bi}}{d\theta} = \frac{M_{bi}T_{bi}}{\sum M_{bi}T_{bi} + M_aT_a + M_cT_c}\frac{dQ_{ch}}{d\theta} \tag{13}$$

The equation of state for each zone

$$PV = MRT \tag{14}$$

and the overall volume constraint

$$\sum V_{bi} + V_a + V_c = V_{ch}(\theta) \tag{15}$$

complete the set of equations necessary to calculate pressure and local temperature. The energies, U, and enthalpies, H, are absolute values, including those of formation. Each combustion zone, B_i, is considered to be at chemical equilibrium composition determined by the local properties. It is seen that an iterative solution to these equations is required. The detailed flow chart for the solution is given in Ref. 8.

One significant point needs to be made before leaving the subject of thermodynamic analysis. From conservation-of-mass equations implied in the jet mixing calculations

$$\sum_i \frac{dM_{ebi}}{d\theta} + \frac{dM_{ec}}{d\theta} = -\frac{dM_a}{d\theta} \tag{16}$$

In other words, air entrained into the jet comes from the air zone. This fact, combined with the volume constraint of Eq. 15, enables the summation of Eqs. 8 through 11 to give the overall energy balance for the chamber:

$$\frac{d(M_{ch}U_{ch})}{d\theta} = -P\frac{dV_{ch}}{d\theta} + \frac{dQ_{ch}}{d\theta} + \frac{dM_f}{d\theta}H_f \tag{17}$$

Diesel combustion models based on the concepts of apparent heat-release rate involve the solution of Eq. 17. However, such solutions assume homogeneous combustion, with instantaneous equilibrium at increasing (overall lean) fuel-air ratios, as fuel is added at an imaginary rate corresponding to the apparent heat-release rate [5, 6, 7, 10, 11, 12]. Unfortunately in subsequent combustion literature not only has the qualification "apparent" been dropped, but also when Eq. 17 is solved for $dM_f/d\theta$ from an experimental measurement of $P(\theta)$, the result has been called "experimental" heat-release rate. Significant differences between local and average thermodynamic properties exist because of their strong dependence on fuel-air ratio and temperature. While it may be possible to adjust empirical constants to overcome this difficulty and obtain a fair correlation for cylinder pressure, no amount of adjustment at the mean temperatures encountered can yield reasonable nitric-oxide correlations. Modelers have therefore been forced to adopt a different model of combustion for nitric-oxide formation than for pressure development [12], thus leading to an inconsistent set of equations. Many combustion models in the literature [7, 12, 13] suffer from this inconsistency. The simultaneous and consistent calculation of thermodynamic and chemical kinetic variables, together with transient jet-mixing calculations, are the distinguishing features of the model presented here. Such consistency, however, has not been achieved in heat-transfer modeling. A more detailed discussion is given later.

Pollutant Kinetics—The solution of Eqs. 8 through 11 yields values of local temperature and equilibrium specie concentration. The equilibrium concentrations of atomic and molecular oxygen and nitrogen are used with the Zeldovich mechanism and appropriate rate constants [14] to calculate the rate of formation of nitric oxide in each combustion zone. The nitric-oxide contribution of each zone is summed up to obtain the instantaneous chamber total. Integration of the rate equation over time thus yields the exhaust concentration.

The model was exercised for a variety of direct-injection engines over a wide range of operating conditions. Validation of the model was based on cylinder pressure history and exhaust concentration of nitric oxide. Some of these comparisons have been published earlier [1, 8]. However, validation of a complex formation history on the basis of a single exhaust value is inadequate. Figs. 3 and 4 are reproduced from the work of Voiculescu and Borman [15]. They measured cylinder-averaged concentration of nitric oxide as a function of crank angle, by quenching the total cylinder contents at various stages of the combustion expansion process. The present model was used to calculate nitric-oxide history using engine design and operating conditions as input. It should be emphasized that these conditions are far removed from the conditions of the initial validation of the model [1, 8], and that no further "refinement" of the model was done for this calculation. The agreement between experiment and theory on the shape of the nitric-oxide-concentration history is very encouraging.

Plans are underway to treat the kinetics of soot formation and oxidation on the basis of local temperature and equivalence ratio in much the same way. MacFarlane [16] and Radcliffe and Appleton [17] have measured threshold equivalence ratios under which carbon particle precipitation occurs. While the rates of

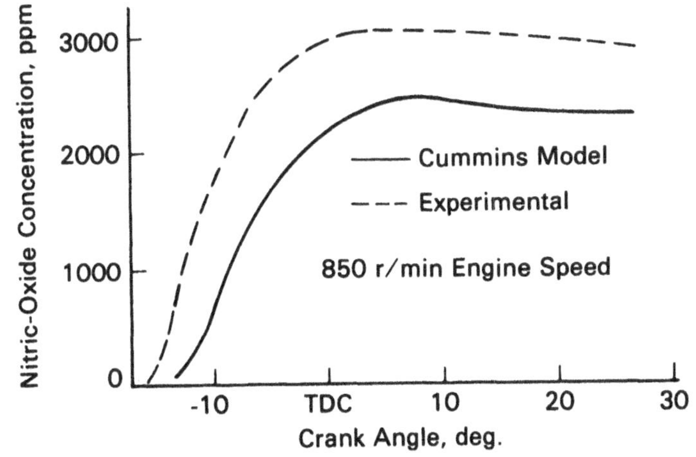

Fig. 3. Nitric oxide formation history at 850 r/min [15].

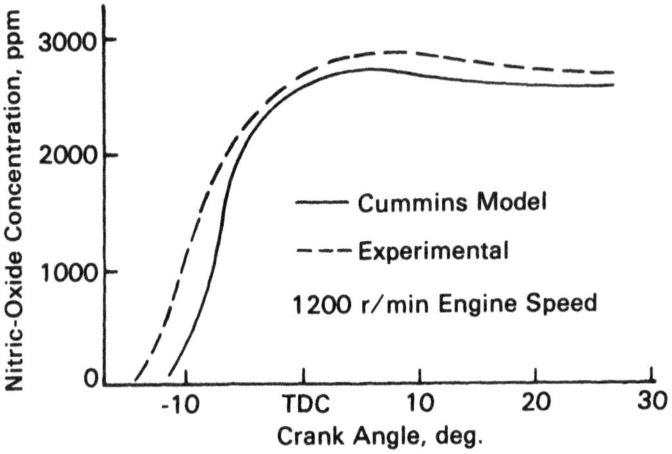

Fig. 4. Nitric oxide formation history at 1200 r/min [15].

carbon formation are not fully understood, equilibrium composition [18] may be assumed as a starting point. Subsequent oxidation of carbon particles may be calculated based on rate equations such as those in Ref. 19. This provides a potent framework for the extension of the capabilities of the model.

Heat Transfer—The heat-transfer calculation in the present model is not based on local temperatures, but rather on a bulk average temperature per Eq. 12. Both convection and radiation heat transfer are based on this value. Fig. 5 shows the thermodynamic mean temperature and mean combustion-zone-temperature histories for a typical turbocharged full-load condition. The mean-temperature history is in agreement with calculations based on an overall thermodynamic cycle simulation. Such a history indicates that calculated peak heat transfer occurs later than 30° ATDC. This is in sharp contrast to experimental measurements of radiation and total heat-transfer histories [20, 21]. A comparison of the possible radiating temperature given by the combustion zone average (Fig. 5) with the mean temperature indicates that radiant heat transfer based on the former may be several factors higher than that based on the latter in the early stages of combustion. Figs. 3 and 4 show that this is also the period of significant activity as far as nitric-oxide formation is concerned. This points out a serious inadequacy of the model. During the period when nitric-oxide formation occurs, radiant heat transfer is a significant part of the total [20]. In the model, total heat transfer is apportioned as a linear function of temperature while the radiant part depends on the fourth power. Hence, a lesser proportion of the total heat transfer is attributed to the hotter zone, resulting in an overestimation of its temperature, hence of nitric oxide. In order to compensate for this, a higher reverse rate coefficient was necessary in the kinetics rate equations.

Such problems arise because of the inconsistency in basing the primary heat-transfer calculation on the bulk average temperature, as opposed to the local temperature. This represents one of the areas requiring major improvement in diesel combustion modeling.

MODEL APPLICATIONS

Effect of High Injection Pressure—The preceding sections of the paper have described the diesel combustion model, illustrating its distinguishing features of a simple consistent framework capable of accepting improved formulations and/or updated information. The following sections give an example of the use of the model in a fundamental analysis of experimental results in order to understand the intermediate steps involved.

Retarded injection timing has been one of the standard techniques used to reduce the emission of nitric oxide from diesel engines. Unfortunately, this has invariably been accompanied by increased fuel consumption and smoke. Several workers have discovered the beneficial effects of high injection pressure in controlling smoke at retarded timing [22, 23]. Work at the authors' laboratory on an

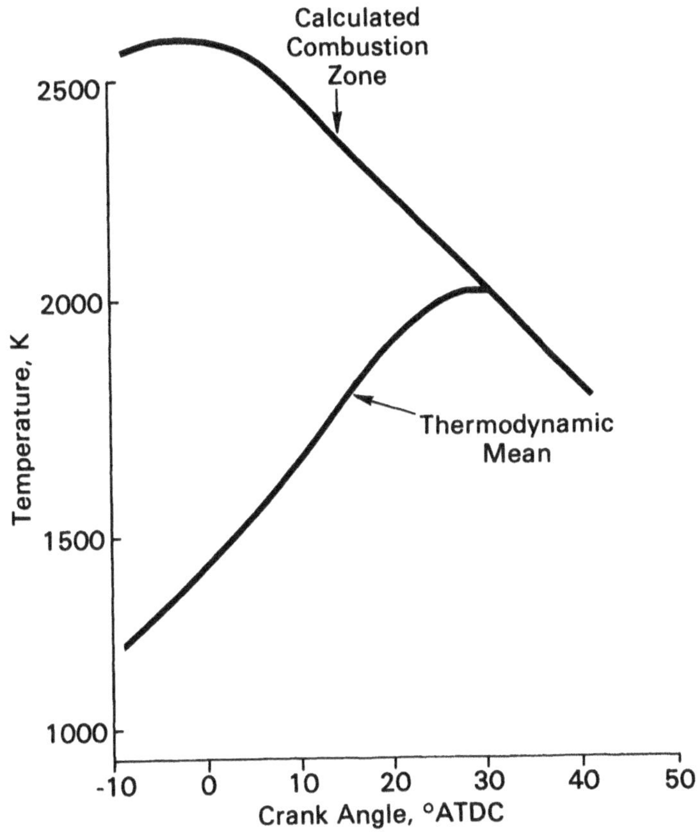

Fig. 5. Combustion zone and bulk average temperature under typical full-load turbocharged conditions.

References pp. 362-363.

experimental single-cylinder engine with an externally operated experimental fuel-injection systems also gave similar results, shown in Fig. 6, indicating the possibility of using very retarded injection timing with good smoke control. Fig. 7 shows the effect of injection pressure on fuel consumption, indicating once again that very retarded operation is possible, with loss of fuel consumption appropriately recovered by using high injection pressure. Fig. 8 shows the corresponding effect on nitric oxide. It is seen that at a given timing an increase in injection pressure results in increased nitric-oxide emissions, thus defeating the original purpose of timing retard. The fuel-consumption versus nitric-oxide tradeoff, shown in Fig. 9 for three different injection pressures, indicates that there is little advantage in using high injection pressure to break the tradeoff, especially considering that injection work is not included in the data. This effect was also observed by Parker [22]. Various shapes of injection rate control have been suggested [23] as having possible desirable effects. The combustion model described above was exercised over the range of conditions represented by the data of Figs. 6 through 9 in order to understand the intermediate fundamental processes that control fuel-consumption and nitric-oxide tradeoff. Further, an attempt was made to predict the effect of injection-rate shape on such tradeoff.

Analysis of Fuel-Air and Product-Air Mixing Rates—The effect of mixing rates is best followed by considering the history of a typical combustion zone. This is done for a retarded timing (injection at TDC) moderate injection pressure (108

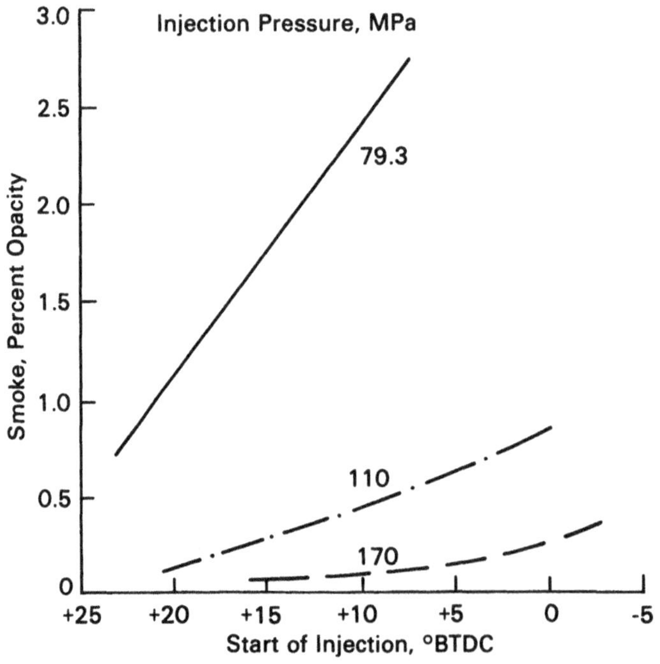

Fig. 6. Effect of injection pressure and timing on smoke.

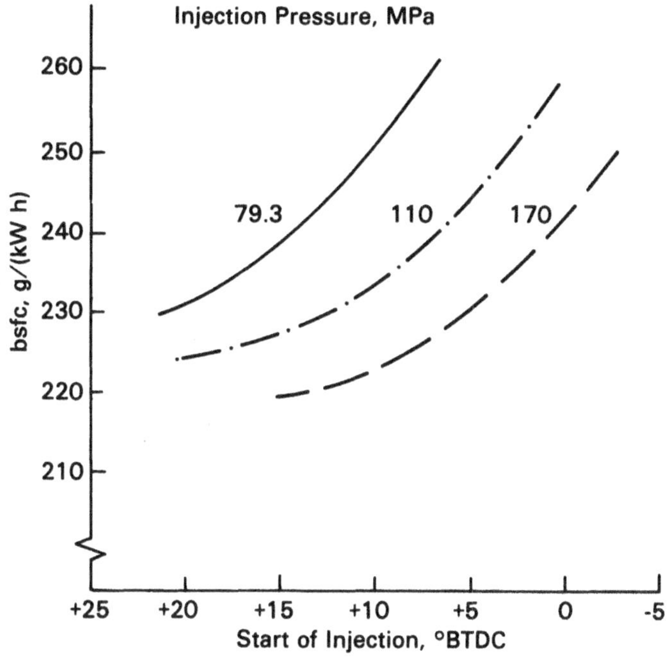

Fig. 7. Effect of injection pressure and timing on *bsfc*.

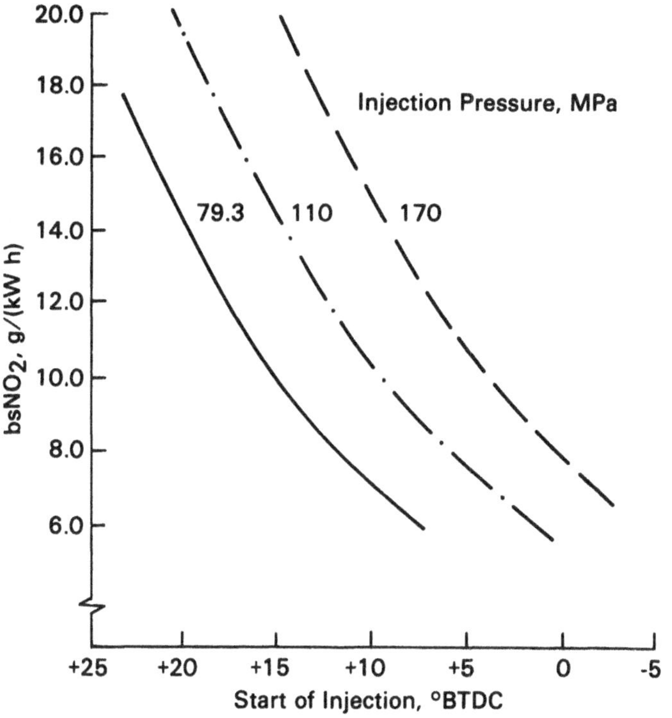

Fig. 8. Effect of injection pressure and timing on $bsNO_2$.

References pp. 362–363.

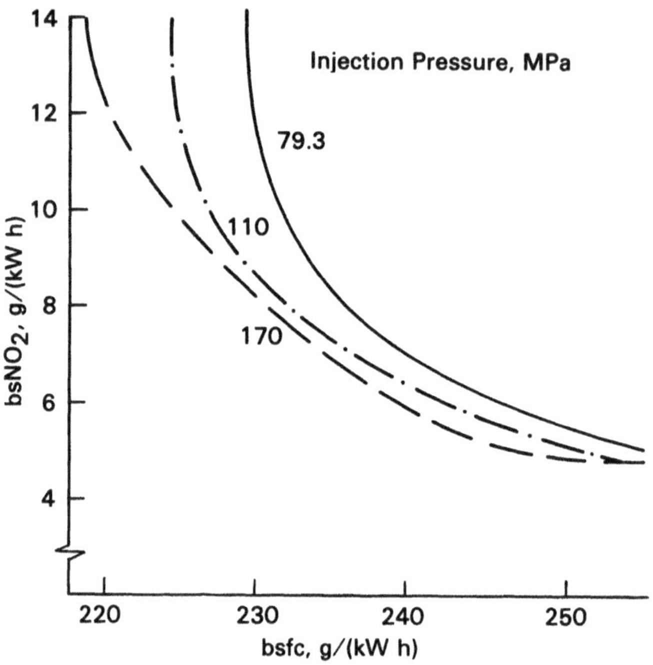

Fig. 9. Effect of injection pressure on *bsfc-bsNO*$_2$ tradeoff.

MPa) case, shown in Fig. 10. The temperature and composition history are illustrated. It is seen that the first assignment of prepared mixture to this zone occurs at around 8°BTDC, with the zone progressively leaning out to the average fuel-air ratio. Peak temperature as high as 2600 K is observed as the zone passes through the stoichiometric range of its history. The sharp peak in nitric oxide formation rate is shown. Fig. 11 shows the calculated and measured cylinder-pressure history associated with this condition and serves as an indication of the cycle work done.

The local temperature and composition history is shown in Fig. 12 for the same zone under the same operating conditions but a higher injection pressure. The same qualitative picture is seen. However, it is evident that critical stoichiometric mixture ratio is reached much faster. It is accompanied by higher local maximum temperature, resulting in nearly double the peak nitric-oxide formation rate. Fig. 13 shows the corresponding measured and calculated cylinder-pressure histories, which indicate that the cycle work (hence efficiency) is higher than that for the lower-injection-pressure case shown in Fig. 11. Detailed analysis of mixing rate under various injection pressure conditions shows that nitric-oxide formation is not a simple function of injection timing alone, but also depends on the fuel-air and product-air mixing rates. Any enhancement of this mixing rate to improve efficiency seems to result inevitably in an increase in nitric-oxide formation rate. It must be pointed out that such a tradeoff can be studied only when the same basic mixing-rate processes determine both pressure evolution and nitric-oxide formation. When one is calculated based on an overall apparent heat release and

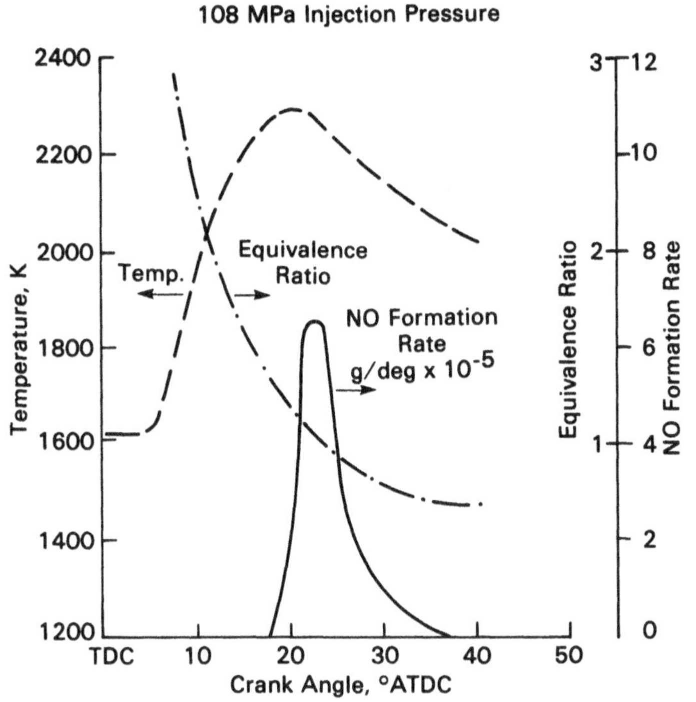

Fig. 10. Typical combustion zone history under moderate injection pressure conditions.

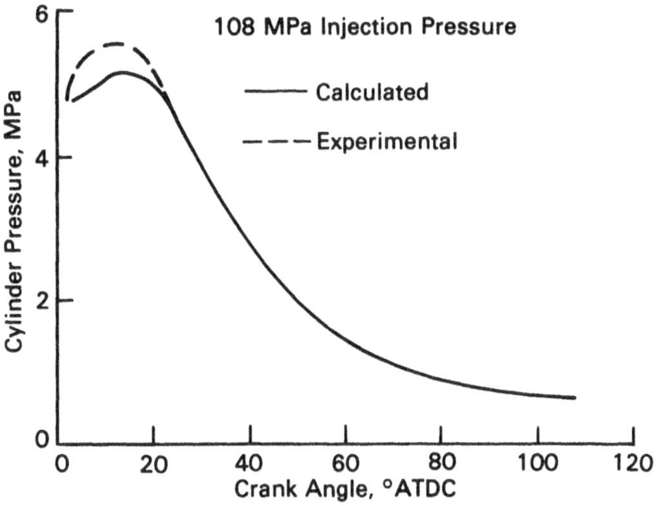

Fig. 11. Typical cylinder pressure history under moderate injection pressure conditions.

References pp. 362–363.

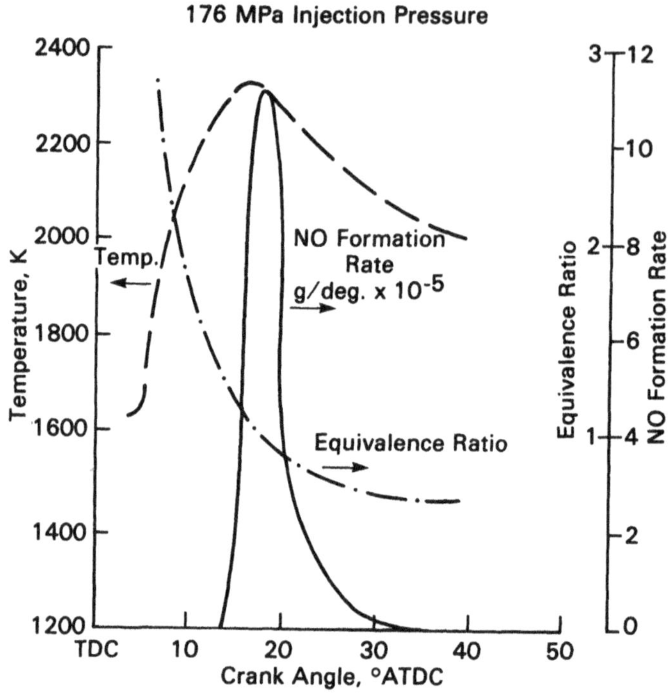

Fig. 12. Typical combustion zone history under high injection pressure conditions.

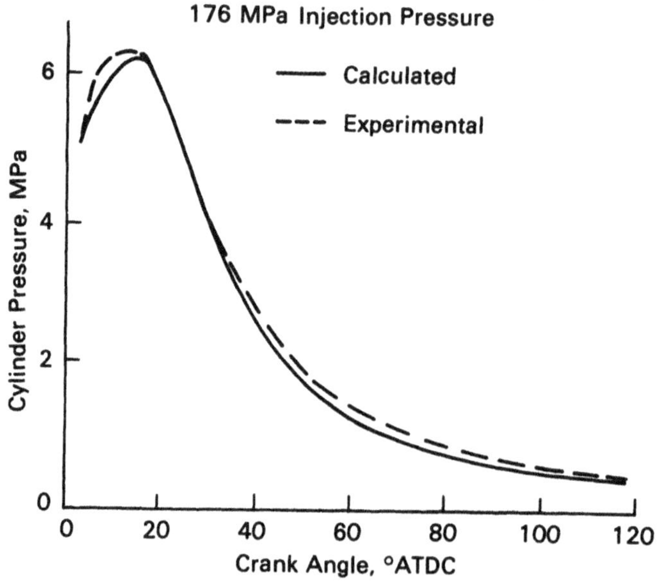

Fig. 13. Typical cylinder pressure history under high injection pressure conditions.

the other superimposed on it by empiricism, then an analysis of the fundamental intermediate processes is not possible.

Effect of Injection-Rate Shape—Before the significance of the effect of mixing rate could be fully understood, it was thought that the advantages of high injection pressure could be better realized by optimizing the rate of injection. Rectangular, rising, and falling triangular injection rates were approximated on the experimental system, and an analytical investigation was conducted. Both investigations showed that essentially the same nitric-oxide versus efficiency tradeoff is obtained by these methods. The results of the calculation are shown in Fig. 14. The injection-rate shapes are shown by the shape of the symbols used. The tradeoff line represents an injection-timing excursion at otherwise constant conditions. The results clearly indicate that the simultaneous effect of mixing rate on pressure evolution and nitric-oxide formation result in no net advantage in the tradeoff.

Particulate Emissions—Engine development experience shows a strong tradeoff between particulate and nitric-oxide emissions. The concept of particulate emissions presented in this model involves a rapid carbon-particle precipitation rate, followed by a mixing-rate-limited oxidation. Emitted particles correspond to those still left unmixed or whose oxidation was quenched during expansion. Such a concept represents a competition between rapid mixing and expansion. The strong effect of injection pressure on smoke, even at very retarded timings,

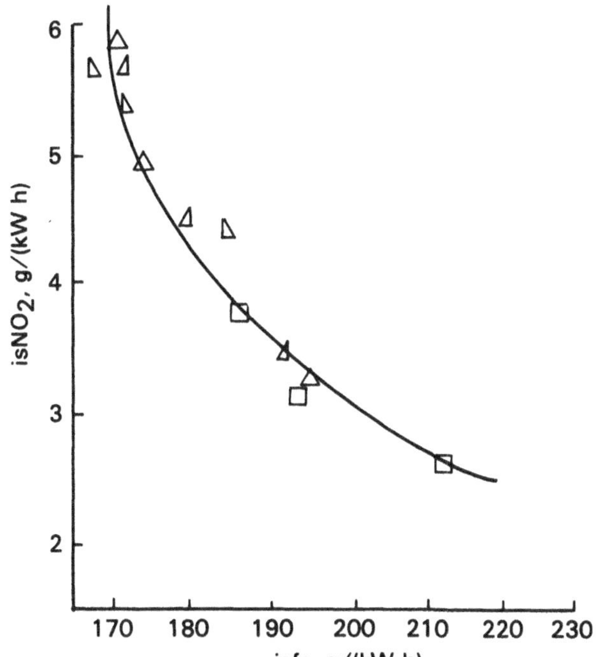

Fig. 14. Effect of injection rate shape on *bsfc*-$bsNO_2$ tradeoff.

References pp. 362–363.

appears to lend strength to it. It is suspected that the tradeoff between particulate and nitric-oxide emission is another manifestation of the strong dependence of both these processes on the fuel-air and product-air mixing rates. It is expected that when the carbon-particle-formation model outlined earlier is completed, it will yield a more quantitative estimation of the effects of mixing rates.

SUMMARY

The combustion and emissions model described in the paper represents a departure from the purely empirical attempt at modeling, while being far away from fundamental three-dimensional transient Navier-Stokes analyses. The paper illustrates the use of the model as a mathematical catalogue and its use in detailed analysis of experimental results. The need for more information, particularly in the area of heat transfer, has been identified, and the model is shown capable of accepting new information as it becomes available. It is hoped that as this is incorporated, the model will come nearer the stages of development when it can reliably be used as a predictive tool.

REFERENCES

1. W. S. Chiu, S. M. Shahed and W. T. Lyn, "A Transient Spray Mixing Model of Diesel Combustion," SAE Trans., Vol. 85, Paper No. 760128, pp. 502–512, 1976.
2. G. N. Abramovich, "The Theory of Turbulent Jets," MIT Press 1963.
3. J. C. Dent, J. H. Keightley, C. D. Deboer, "The Application of Interferometry to Air-Fuel Ratio Measurements in Quiescent Chamber Diesel Engines," SAE Trans., Vol. 86, Paper No. 770825, pp. 2858–2869, 1977.
4. D. Adler and W. T. Lyn, "The Evaporation and Mixing of Diesel Fuel Spray in a Diesel Air Swirl," Proc. I. Mech. E., Vol. 184-3J, No. 16, pp. 171–180, 1970.
5. J. M. Rife and J. B. Heywood, "Photographic and Performance Studies of Diesel Combustion Using a Rapid Compression Machine," SAE Trans., Vol. 83, Paper No. 740948, pp. 2942–2961, 1974.
6. H. C. Grigg and M. H. Syed, "The Problem of Predicting Rate of Heat Release in Diesel Engines," Proc. I. Mech. E., Vol. 184-3J, No. 18, pp. 192–202, 1970.
7. C. J. Kau, M. P. Heap, T. J. Tyson and R. P. Wilson, "The Prediction of Nitric Oxide Formation in a Direct Injection Diesel Engine," Sixteenth Symposium (International) on Combustion, The Combustion Institute, Pittsburgh, Pennsylvania, pp. 337–350, 1976.
8. S. M. Shahed, W. S. Chiu and W. T. Lyn, "A Mathematical Model of Diesel Combustion," Proc. I. Mech. E., C94/75, pp. 119–128, 1975.
9. W. J. D. Annand, "Heat Transfer in the Cylinders of Reciprocating Internal Combustion Engines," Proc. I. Mech. E., Vol. 177, pp. 973–996, 1963.
10. W. T. Lyn, "Study of Burning Rate and Nature of Combustion in Diesel Engines," Ninth Symposium (International) on Combustion, The Combustion Institute, Pittsburgh, Pennsylvania, pp. 1069–1082, 1962.
11. R. B. Krieger and G. L. Borman, "The Computation of Apparent Heat Release for Internal Combustion Engines," ASME. Paper No. 66-WA/OGP-4, 1966.
12. S. M. Shahed, W. S. Chiu and V. S. Yumlu, "A Preliminary Model for the Formation of Nitric Oxide in DI Diesel Engines and its Application in Parametric Studies," SAE Trans., Vol. 82, Paper No. 730, pp. 338–351, 1973.
13. I. M. Khan, G. Greeves and D. M. Probert, "Prediction of Soot and Nitric Oxide Concentration in Diesel Engine Exhaust," Proc. I. Mech. E., pp. 205–217, 1971.
14. H. K. Newhall and S. M. Shahed, "Kinetics of Nitric Oxide Formation in High Pressure

Flames," *Thirteenth Symposium (International) on Combustion, The Combustion Institute, Pittsburgh, Pennsylvania, pp. 381-389, 1970.*
15. I. A. Voiculescu and G. L. Borman, "An Experimental Study of Diesel Engine Cylinder-Averaged NOx Histories," SAE Paper No. 780228, 1978.
16. J. J. MacFarlane, "Carbon Formation in Premixed Methane Oxygen Flames Under Constant Volume Conditions," Combustion & Flame, Vol. 14, pp. 67-72, 1970.
17. S. W. Radcliffe and J. P. Appleton, "Shock Tube Measurement of Carbon to Oxygen Atom Ratios for Incipient Soot Formation with C_2H_2, C_2H_4 and C_2H_6 Fuels," MIT Report, 1971.
18. S. Gordon, "Complex Chemical Equilibrium Calculations," NASA SP-239 (1970).
19. S. W. Radcliffe and J. P. Appleton, "Soot Oxidation Rates in Gas Turbine Engines," Comb. Sci. & Tech., Vol. 4, pp. 171-175, 1971.
20. P. F. Flynn, M. Mizusawa, O. A. Uyehara and P. S. Myers, "An Experimental Determination of Instantaneous Potential Radiant Heat Transfer within an Operating Diesel Engine," SAE Trans., Vol. 81, Paper No. 720022, pp. 95-126, 1972.
21. J. C. Dent and S. I. Sulaiman, "Convective and Radiative Heat Transfer in DI Diesel Engines," SAE Trans., Vol. 86, Paper No. 770407, pp. 1758-1783, 1977.
22. R. F. Parker, "Future Fuel Injection System Requirements of Diesel Engines for Mobile Power," SAE Trans., Vol. 85, Paper No. 760125, pp. 482-490, 1976.
23. J. A. Kimberly and R. A. DiDomenico, "UFIS—A New Diesel Injection System," SAE Paper No. 770084, 1977.

DISCUSSION

M. L. Monaghan *(Ricardo Consulting Engineers, England)*

I have in fact discussed the evolution of this model with Dr. Shahed from time to time over the years and would like to make one or two points about it. First, I must congratulate him, as I hope I have done in the past, on the way in which the model has been structured as to allow new concepts and new input data to be added to enhance its validation. This is one of the features that certain other models lack, and one area where I believe Cummins' combination of practical and academic work will prove to be of great advantage. Having said all that, let me make a few comments. Cummins' main thrust has in fact been in the field of relatively quiescent diesel engines, where the air motion is not quite so well constrained, if I may use that word, as 90% of the other diesel engine makes (perhaps not numbers) in the world. As far as the indicated *NOx* versus indicated fuel consumption tradeoff goes, have you tried to feed into your model varying mixing rates for swirling engines and also varying initial conditions of air motion in so far as it concerns both the gross air motion, and indeed turbulence? One of the thorny questions at the moment is precisely how the initial conditions during compression and at the end of compression affect the fuel conditioning. Have you done any, at least preliminary, correlation work here?

S. M. Shahed

Turbulence, no. Gross air motion, yes. I did not dwell at all on the fuel jet-air mixing model that is associated with us, but that model allows for a deflection of the spray as a result of interaction with swirl. In fact, one of the validation slides

S. M. Shahed

that you saw had air swirl as an independent variable. All of this was indeed done on Cummins' experimental single-cylinder engines where the swirl was obtained by using masked valves. The swirl levels that were used were probably lower than the kinds of swirl levels that you are referring to. They were typically 6000 to 7000 r/min swirl on a 2000 r/min engine. Swirl was modeled as a gross solid-body rotation. We treat the very gross effect of spray deflection due to the swirl.

F. C. Gouldin *(Cornell University)*

I would like to suggest that there's possibly another shortcoming to your model, namely the neglect of the effect of temperature fluctuations and species fluctuations in your *NO* calculations. I don't believe that it is possible, even in simple turbulent diffusion flames, to calculate *NO* formation rates *a priori*. There are also suggestions regarding the success of *NO* prediction models in premixed engines, where you might think that the environment is turbulent. I'd like to suggest it's not, at least in the post-flame gases where most of the *NO* is formed. So I would say that you should look into adding some sort of turbulent mixing effect and temperature fluctuation effect in the model if you want to get good quantitative comparisons. Whether it impacts the relative studies, for example, the effect of injection pressure changes, is a different question, but I'd certainly be careful if the changes affected the turbulence structure, particularly turbulent time scales relative to the *NO* time scales.

S. M. Shahed

I'm quite cognizant of the fact that we do not have any interaction between turbulence and mixing rate. We have started with the assumption that the jet mixing rate, the momentum exchange associated between a jet injected at high injection pressure in air, is far greater than local turbulence. That may or may not be true. The correlations are very gross and cannot take into account the kinds of refinements that you're talking about. I do not think that I can claim that the model is in any way close to being a predictive tool. Again, it provides the kind of things that we were talking about—a mathematical catalog of decipherent thinking and a tool to analyze some of the results. Hopefully someday it will be a predictive tool, but it's not one at this time.

J. B. Heywood *(Massachusetts Institute of Technology)*

I have some comments about model validation in the light of your statement that your agreement between theory and measurements is plus or minus 25%. There is a tendency to think that experiments are right and models are wrong, and I just want to make some general comments, if it's that simple. Obviously there are inadequacies in models that lead to differences between predictions and experimental data. But there is also scatter, sometimes substantial in the emissions area, in experimental data. And there's a third aspect that comes in that isn't often mentioned, that is, getting correspondence between a model calculation and an experiment. It's often not that easy, and there is a source of error often introduced by not getting that appropriate correspondence. That is particularly important in getting the timing of the combustion process the same in both the calculations and the experiment. *NO* emissions, for example, are very sensitive to timing the combustion process around the optimum timing. Concerning Fred Gouldin's point about the *NO* kinetics, I understand why you adjust your kinetic rate constants to use the model as a development tool. But sometime I would like to see a comparison of predictions with experimental data made with nonadjusted kinetics. At any rate, I have a couple of suggestions of ways that you might improve your model in a manner that might improve the matching, without having to adjust the kinetic constants. One involves the introduction of concentration fluctuations. We have looked at that in gas turbine combustors. It significantly changes the *NOx* formation-rate profile with equivalence ratio. It drops and broadens the peak which you show on your slides. Secondly, you might consider a more sophisticated heat-transfer model. We have found it is necessary in spark-ignition engines to go to some kind of boundary-layer heat-transfer model because the bulk gases are not cooled, and that's where most of the *NO* comes from. In addition, radiation is presumably occurring along the soot burnout process region, which is thought to be an interface between rich elements and lean elements in the flow. That would of course affect the subsequent temperature of leaner elements as they grow leaner. I think that perhaps those are two areas of important sources of discrepancy between predictions of *NOx* and experiment.

S. M. Shahed

These are all very worthwhile comments, and indeed we feel that heat-transfer considerations should be one of our major efforts to try to improve our model.

H. K. Newhall *(Chevron Research Company)*

I'd like to make one response to John Heywood's first statement regarding the validity of experimental versus computer data. The problem was scatter in experimental data. It's not a real problem; it's nothing that cannot be handled with simple statistics. I don't really think that's a crucial point. Experiments done properly and analyzed properly have to provide the framework for validation of computer models. I would be interested in any of the other people's viewpoints on this.

W. T. Lyn

I want to make a comment about turbulence. The effect of turbulence is to produce a time-varying history of concentration. The other aspect of turbulence is to increase mixing. In our model we have decoupled the hydrocarbon equilibrium from the *NO* kinetics. I recently learned from Dr. Pischinger about some interesting experiments conducted at Aachen. They found in working with a homogeneous mixture that the turbulence had an effect on *NOx* production.

F. F. Pischinger *(Institute of Applied Thermodynamics, Germany)*

What we discussed was that with quick injection and very rapid mixing, one gets the same tradeoff between *NOx* emissions and fuel consumption. The same disappointment occurs with a spark-ignited engine. That is, if you would invoke high turbulence, you can lean out the mixture, but you get nearly the same tradeoff between fuel consumption and *NOx* emissions. That's what we discussed. But it's not really the same effect we are now discussing. Obviously, mixing history is important, and I think that by changing this history, it is possible to influence the *NOx* versus fuel consumption tradeoff. For instance, the precombustion chamber engine has a different mixing history, and it's well known that you can alter the dependence of this relationship. Perhaps I could end with the question, "Did you investigate synthetic or prescribed mixing rates in your model? Did you try to determine how the spray should mix for the best tradeoff?"

S. M. Shahed

No, we did not try any artifically prescribed mixing rates. We did try higher and higher injection pressures in our calculations, but not in our experiments. But the injection pressures do specify the mixing history. Obviously, the kind of mixing rate that we would like to have is initially a kind of mixing that is rapid enough to avoid carbon particle formation but not fast enough that it will get into nitric oxide formation regimes, followed by mixing that is so rapid that it outraces nitric oxide kinetics. If you can design an injection system that will do that, it's fine.

G. L. Borman *(University of Wisconsin)*

Regarding heat transfer, I was wondering if you had compared your modeling results with your experimental work on adiabatic engines. It seems that in the adiabatic engine, the radiation would be about the same but the conduction would be reduced.

S. M. Shahed

We have not. We should and we will.

N. A. Henein *(Wayne State University)*

We are aware of the assumptions in the model. We find that it is very sensitive to many parameters. One of these is the spray angle. We found that by changing the spray angle by only 20 percent, you can increase the *NO* formed significantly. How did you determine the spray angle and the effect of the increased injection pressure on the spray angle? A second series of questions is, "What were the physical properties of the fuel you used? Did you use a distillate or a pure compound? If you did use a distillate, how did you determine its physical properties?" In our work we are running tests with pure compounds in order to be able to determine physical properties at some of the points.

S. M. Shahed

The spray mixing equations were developed independently of engine experiments. The phenomenological approach, or the quasi-dimensional modeling if you like, takes bits and pieces of the model and validates them on an experimental reference for which the spray mixing equations were developed. Once having developed the spray geometry and spray angle it would not be quite right to go back to the engine and adjust that angle back and forth until one gets a nitric oxide correlation. We are not using engine experimental results to study fundamental nitric oxide kinetics or fundamental spray fluid mechanics. So the spray mixing equations that we have published in the past give the effect of injection pressure, air swirl, and the chamber pressure on spray geometry and the mixing calculations. And those remain constant. The fuel that we have used is No. 2 diesel fuel. There is nothing very magical about it. We have tried to use an average carbon to hydrogen ratio in our equilibrium calculations to calculate the equilibrium composition. I again don't think that it is very sensitive to those calculations.

D. J. Patterson *(University of Michigan)*

Any model should be able to survive the test where you introduce some *NO* into the inlet of the engine and observe the response in both the engine and the model. Such a test was done by Aiman* here at General Motors.

---

* *W. R. Aiman, "A Critical Test for Models of the Nitric Oxide Formation Process in Spark Ignition Engines," Fourteenth Symposium (International) on Combustion, The Combustion Institute, Pittsburgh, Pennsylvania, pp. 861–868, 1972.*

S. M. Shahed

That's a very good suggestions. We did follow Aiman's work and tried to repeat it. In our experiments we examined the effect of injecting into the engine CO_2, nitrogen, argon, and a number of other gases, and we were able to explain those results with the model. Unfortunately, by the time that we got around to injecting nitric oxide, the experimental system broke down. Therefore we do not have those results. But you're quite right, we don't have that validation in hand.

SUPPLEMENTARY COMMENTS: FUEL SPRAY CHARACTERIZATION IN DIESEL ENGINES

H. HIROYASU, T. KADOTA and M. ARAI

Hiroshima University, Hiroshima, Japan

ABSTRACT

A fundamental study has been made of the spray characterization in diesel engines. Spray penetration, spray angle, droplet size, vaporization rate, and ignition delay are measured in the constant-volume chamber with several injection conditions. The wide range of such measured data has been expressed in explicit equations, with numerical values for the computation of diesel-engine combustion. These equations for spray characteristics have been applied to compute the cylinder pressure and concentration of nitric oxide and soot in the direct-injection diesel engine, using the mathematical model which is developed by the authors.

NOTATION

A	constant
B	amount of fuel delivered, m^3/stroke
C	constant
c	discharge coefficient
c_{pa}	specific heat of air, J/(kg K)
c_{pf}	specific heat of vaporized fuel, J/(kg K)
c_{pl}	specific heat of liquid fuel, J/(kg K)
C_s	correlation factor of penetration associated with air flow
C_w	coefficient associated with the air entrainment of an impinging spray

References pp. 401–402.

C_θ	correlation factor of spray angle associated with air flow
D	apparent activation energy
$\mathcal{D}$	binary diffusivity, m^2/s
d_o	nozzle orifice diameter, m
H	heat of evaporation, J/kg
h^*	heat-transfer coefficient at high mass transfer, $W/(m^2\ K)$
k^*	mass-transfer coefficient at high mass transfer, $kg/(m^2\ K)$
L	breakup length, m
m_l	mass of fuel droplet, kg
N	swirler speed, r/min
Nu	Nusselt number
n	engine speed, r/min
P	cylinder pressure, Pa
P_a	ambient pressure, Pa
P_o	opening pressure of injection nozzle, Pa
P_{O_2}	partial pressure of oxygen, Pa
ΔP	difference between injection pressure and ambient pressure, Pa
Q	net rate of heat release, J/s
Q_d^*	heat-transfer rate to droplet, W/s
R	rack position, m
r_s	swirl ratio $=N/n$
Sh	Sherwood number
S_N	needle lift of injection nozzle, m
s	spray tip penetration, m
s_s	spray tip penetration with air flow, m
T_a	ambient temperature, K

T_l	droplet temperature, K
t	time, s
t_{break}	transition time, s
v	droplet volume, m^3
x	droplet size, m
$\bar{x}_{32}$	Sauter mean diameter, m
$\bar{x}_m$	median mean diameter, m
Y_{fo}	mass fraction of vaporized fuel at the droplet surface
u	spray tip velocity, m/s
u_o	initial fuel jet velocity, m/s
u_θ	swirl velocity, m/s
α	constant
β	constant
δ	constant
ζ	flow rate ratio of air to fuel at droplet surface
μ_a	viscosity of air, Pa s
θ	spray angle or crank angle, deg
θ_s	spray angle with air flow, deg
ϕ	equivalence ratio of the vaporized fuel
λ	thermal conductivity, W/(m K)
ρ_a	density of air, kg/m^3
ρ_l	density of liquid fuel, kg/m^3
τ	ignition delay, ms

INTRODUCTION

Diesel combustion is strongly controlled by the character of the spray formed by the fuel injected into the combustion chamber. Concentrations of smoke and

References pp. 401–402.

nitric oxide in the engine exhaust depend, to a great extent, upon the spray form, droplet size distribution, and the air temperature, pressure, and motion in the combustion chamber.

Several combustion models for predicting the effect of different parameters on trends in performance and pollutant formation, and for helping in the design and development of the engine, have been reported during recent years [1-7]. In spite of the fact that these combustion models need the characterization of the fuel spray as input data, such information is lacking. This lack of knowledge concerning the spray characteristics has been a stumbling block in the path of progress in diesel combustion analysis.

The objective of this work is to characterize the fuel spray in diesel engines and to obtain explicit equations with numerical values for the computation of diesel engine combustion.

SPRAY PENETRATION AND SPRAY ANGLE

Many expressions have been developed for the calculation of spray penetration in diesel engines and are reviewed by Hay and Johnes [8]. Contributing research includes that of Schweitzer [9], Lyshevskiy [10, 11], Wakuri [12], Rusinov [13], Sitkei [14], Parks [15], Ogasawara [16], Oz [17], Burt [18], Taylor [19] and Dent [20]. Scott [21] found that the Schweitzer [9] and Wakuri [12] correlations agreed well with his observations. Hay and Johnes [8] evaluated these twelve correlations and recommended the use of the Dent correlation under all conditions except for large chamber densities, where the Dent correlation tends to overpredict penetrations. Correlations of Schweitzer, Wakuri and Dent show that spray penetration is proportional to the square root of time from the start of injection. All experimental works are poor in the initial part of injection, i.e., the period up to about 1 ms after the start of injection. The penetration during the period approximately equivalent to the ignition delay period in an engine is, however, very important. Recently, Rife and Heywood [22] compared experimental data on the geometry of a jet in a quiescent combustion chamber with results from a two-phase jet model and concluded that the centerline velocity of the fuel spray at the initiation of injection is constant.

Reitz [23], based on his extensive photographic study of jet atomization, reported that the velocity of the fuel jet upon initial emergence is constant. An extensive study of fuel-jet penetration has also been reported by Hiroyasu et al. [24].

Experimental Apparatus—Two pieces of experimental apparatus were used in this work. First, a cylindrical constant-volume bomb (180-mm diameter × 540-mm height) with glass windows on two opposite sides was used to measure the spray tip penetration and spray angle in quiescent air. The gas remained at constant temperature and pressure during each test. The combustion bomb was pressurized using nitrogen, with maximum attainable pressure being 10 MPa. A cylindrical electric furnace was contained in the pressure bomb. Temperature in the furnace was varied by controlling the voltage supply to the electric heating elements. Fuel was injected using a specially designed injection system which consisted of a fuel accumulator and an injection nozzle operated with an electromagnetic valve. With the aid of a photographic technique, spray penetration and spray angle were measured at various ambient temperatures and pressures, injection pressures and injection durations.

The second experimental apparatus was used to determine the swirling effect of combustion air on the spray formation. This apparatus, which again involved a constant-volume combustion chamber (Fig. 1), was designed to simulate an open-chamber diesel engine with central injection. The high-pressure combustion air was heated up to 500°C with electric heating elements. Fuel was injected from a nozzle with a single-shot injection unit constructed by remodeling an injection system of an actual engine. Swirling motion of air in the combustion bomb was induced by a rotating swirler installed in the bomb. The swirler was driven by a motor. The velocity profile in the bomb was measured beforehand with a constant-temperature hot-wire anemometer. The results showed the existence of a solid vortex in the bomb and no slip between the swirler and the air.

Results and Discussions—First, spray tip penetration and spray angle were measured with a constant rate of fuel injection into quiescent air. This injection rate was measured using the Bosch injection-rate measuring technique.

Fig. 2 shows the spray tip penetration at various ambient gas pressures, where the logarithmic penetration on the ordinate is correlated with the logarithmic time from the injection start on the abscissa. The illustration shows a linear relationship with two different slopes. At the spray initiation the slope is 1, but after a short period of time, the slope changes to 0.5. This shows that the spray velocity at the initiation of injection is constant, and then afterwards, the spray develops into a steady jet. The spray tip penetration decreased with an increase in ambient gas pressure.

Fig. 3 shows the spray angle at various ambient gas pressures. The spray angle increases with an increase in ambient gas pressure.

Fig. 4 shows the effect of the injection pressure on the spray tip penetration with constant ambient pressure and constant injection duration. It is indicated that the relationship between the penetration and time is linear with two different slopes in a similar manner to that shown in Fig. 2. The time at the point of intersection of the two lines decreases with an increase in the injection pressure. The spray angle increases with an increase in injection pressure, as shown in Fig. 5.

Figs. 6 and 7 show the effect of ambient temperature on the spray tip penetration and spray angle. Even if the ambient gas temperature changes from room temperature to 320°C, the spray tip penetration is not significantly changed, but spray angle is decreased with an increase in ambient gas temperature. This indicates that the peripheral portion of the spray is quickly vaporized in a high-temperature gas, but the center portion of the spray is not so quickly vaporized.

As shown in the figures mentioned above, the spray penetration is directly proportional to time during the early stage of injection, followed by a period when the penetration is proportional to the square root of time. Fig. 8 shows the

References pp. 401–402.

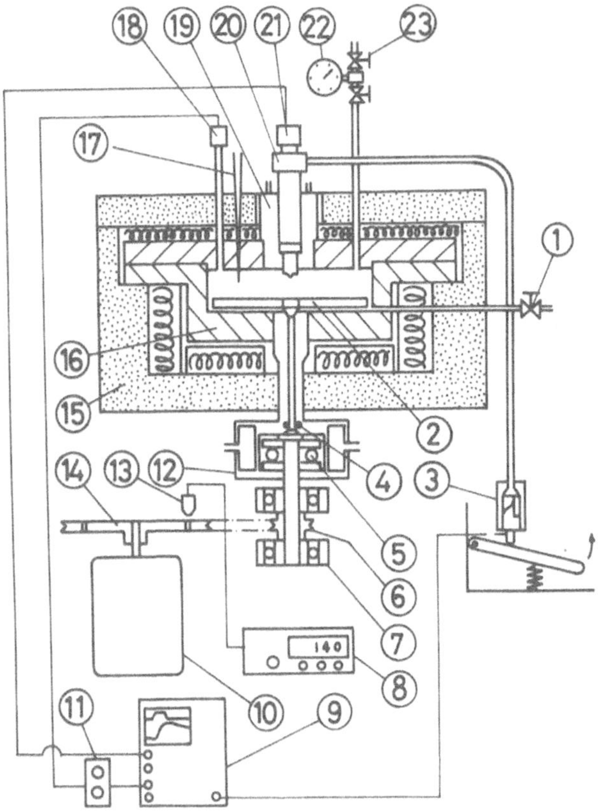

Fig. 1. Constant volume combustion chamber.

1. Exhaust valve
2. Swirler
3. Injection pump
4. O-ring
5. Bearing
6. Pulley
7. Bearing
8. Tachometer
9. Oscilloscope
10. Motor
11. Amplifier
12. Cooling water jacket
13. Detector of tachometer
14. Pulley
15. Electric furnace
16. Combustion chamber
17. Thermocouple
18. Pressure transducer
19. Cooling water jacket
20. Fuel injection nozzle
21. Light beem choppering type needle lift detector
22. Pressure gauge
23. Inlet valve

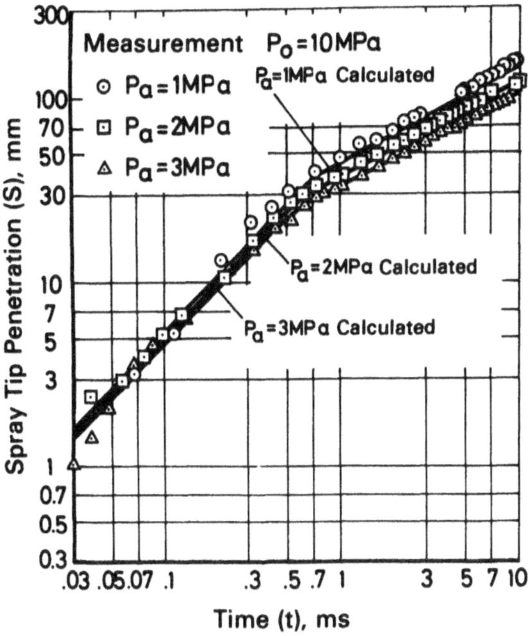

Fig. 2. Spray tip penetration at various ambient pressures.

relationship between injection pressure and the transition time at which the slope changes from a value of 1 to a value of 0.5.

Mathematical Expressions for Spray Tip Penetration—The observations in the previous section indicate that the jet is divided into two regions, namely, a

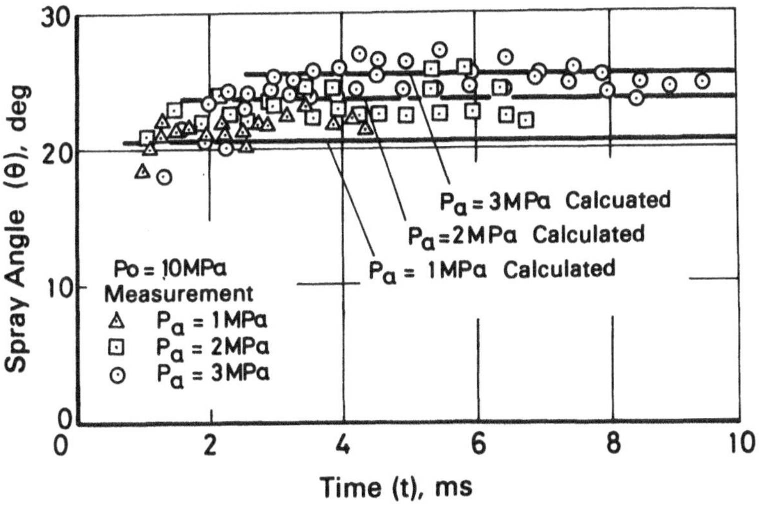

Fig. 3. Spray angle at various ambient pressures.

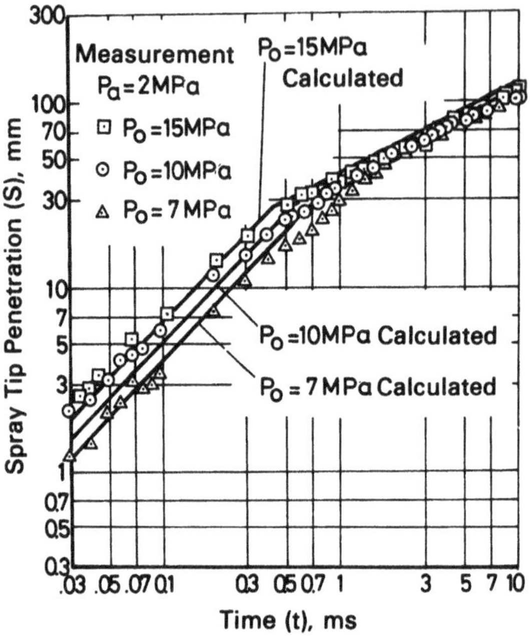

Fig. 4. Spray tip penetration at various injection pressures.

developing region from the nozzle orifice to transition, and a fully developed region beyond the transition point. The developing region is in the 0.3 to 1.0 ms time interval, depending on the injection condition. These time durations are approximately equivalent to the ignition delay period. The following expressions were derived from the data obtained during this investigation and from applying

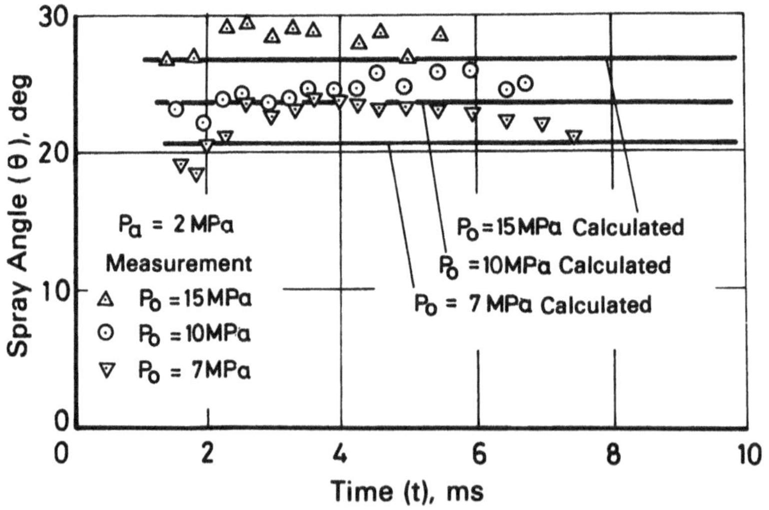

Fig. 5. Spray angle at various injection pressures.

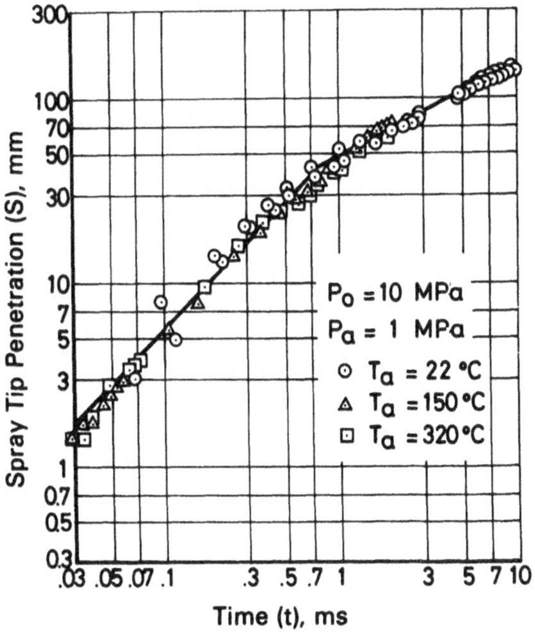

Fig. 6. Effect of ambient temperature on spray tip penetration.

the jet disintegration theory of Levich [25]. The derivation of the jet model is given in Appendix A.

$$0 < t < t_{break}$$

$$s = 0.39 \sqrt{\frac{2\,\Delta P}{\rho_l}}\, t \tag{1}$$

$$t \geq t_{break}$$

$$s = 2.95 \left(\frac{\Delta P}{\rho_a}\right)^{1/4} \sqrt{d_o\, t} \tag{2}$$

Furthermore, the following empirical expression for spray angle, relating effective injection pressure, air density, etc., was obtained (Appendix B) from the data measured in experimental runs without swirl:

$$\theta = 0.05 \left(\frac{d_o^2\, \rho_a\, \Delta P}{\mu_a^2}\right)^{1/4} \tag{3}$$

Computed spray penetration and spray angle data are compared with measured values in Figs. 2 through 8.

Spray Form with Crossflow—The high-speed direct-injection diesel engine may have substantial air motion in its combustion chamber. The spray form injected into the combustion chamber is strongly affected by air motion. Therefore, an

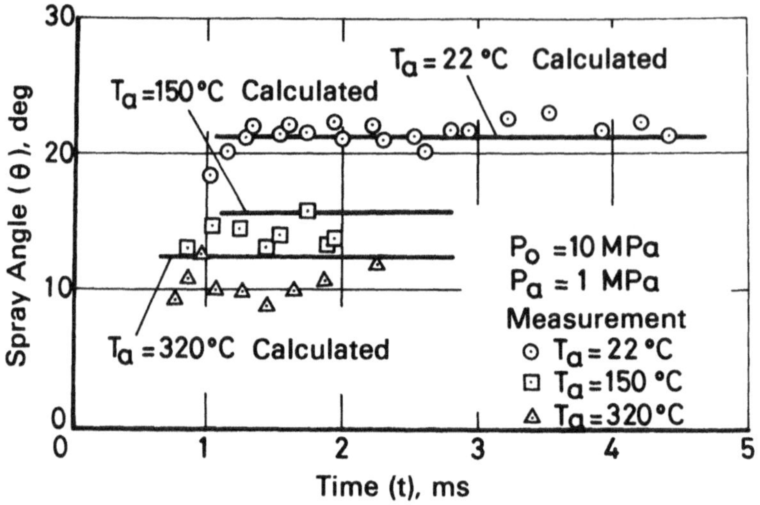

Fig. 7. Effect of ambient temperature on spray angle.

experimental study was made of the effect of swirling motion of the combustion air on the spray form. The constant-volume combustion bomb shown in Fig. 1 was used. Swirling motion of air in the combustion bomb was induced by a rotating swirler installed in the bomb. The swirler was driven by a motor. The velocity profile in the bomb is that associated with a solid vortex in the bomb, with no slip between the swirler and the air. The spray development was photographed using a high-speed camera. Film records of air motion obtained at several speeds are shown in Fig. 9. The nozzle had six holes, but the diagrams show only two jets from each test so that the results may be compared.

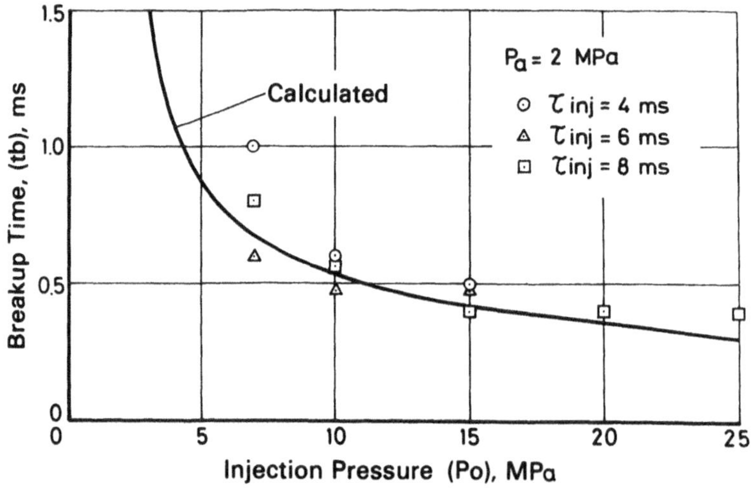

Fig. 8. Relationship between the transition time and injection pressure.

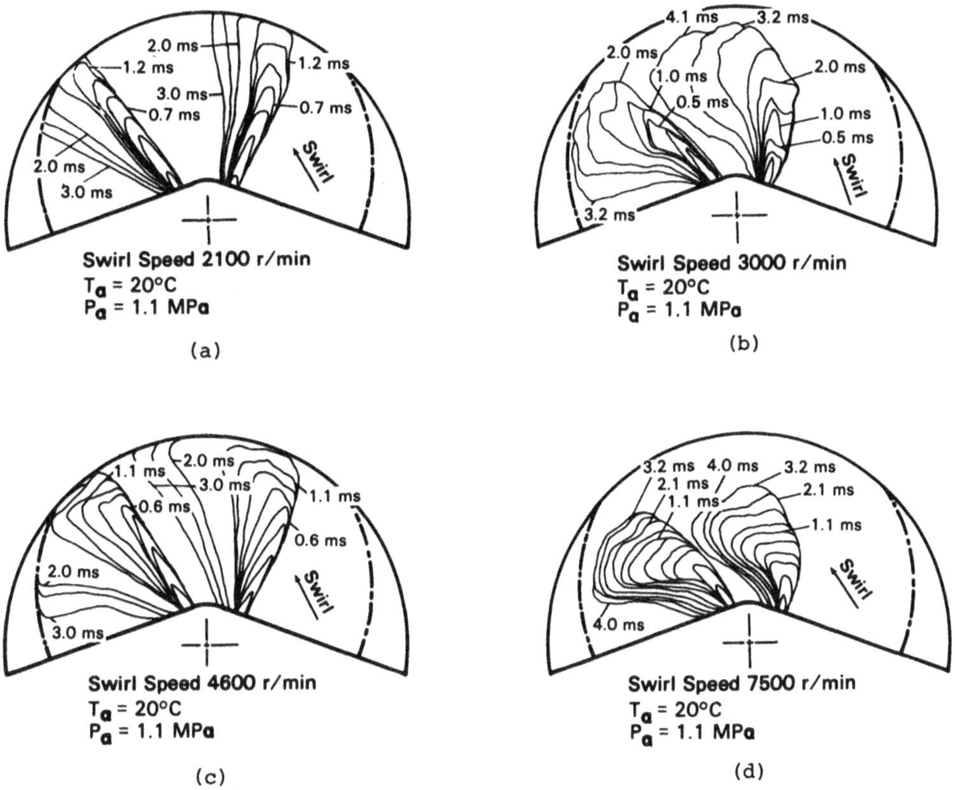

Fig. 9. Spray form with cross airflow.

Empirical correlations for analyzing a fuel jet with air motion normal to the axis of the jet are quite complicated. We used the following correlation factors in conjunction with the equation for the quiescent air situation to obtain the equation of the spray form with air motion. The definition of spray angle with air swirl is quite difficult because the spray is bent by air swirl, and the angle changes with the distance from the orifice. So, we defined the spray angle as the mean value of the maximum angle and the minimum angle:

$$s_s = C_s s \tag{4}$$

$$\theta_s = C_\theta \theta \tag{5}$$

where

$$C_s = \left(1 + \frac{\pi r_s n s}{30 u_0}\right)^{-1} \tag{6}$$

$$C_\theta = \left(1 + \frac{\pi r_s n s}{30 u_0}\right)^{2} \tag{7}$$

Derivations of these correlation factors are given in Appendix C. Experimental

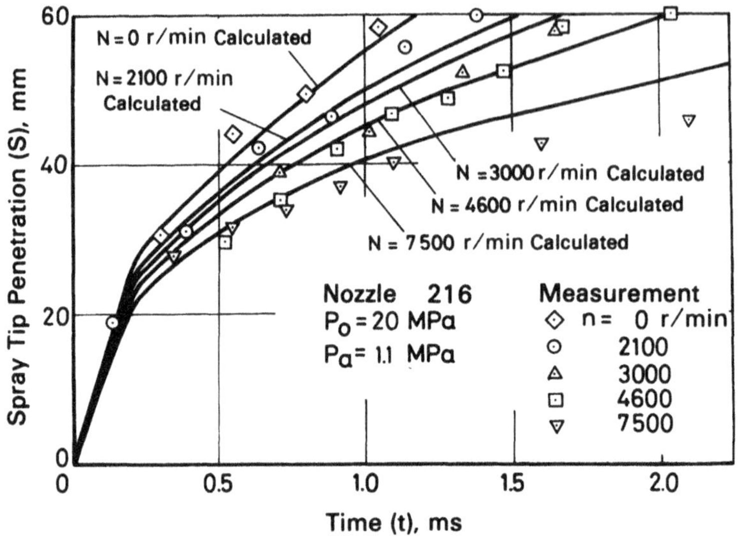

Fig. 10. Comparison of measured and computed tip penetration.

data for the spray tip penetration and spray angle are compared with computed results in Fig. 10.

MEAN DROPLET SIZE AND DROPLET SIZE DISTRIBUTION

Good spray atomization is very important for the proper operation of high-speed open-chamber diesel engines because the time available for combustion is limited. The aim of the atomization is to divide a liquid fuel into a multitude of droplets, hence greatly increasing the total surface area of the fuel in preparation for its subsequent evaporation and combustion.

The experimental measurement of the droplet sizes and distribution in diesel engines is extremely difficult. Numerous measuring techniques have been developed for determining the droplet size distribution. Each method has its advantages and disadvantages, but none is entirely satisfactory. In this study, droplet size was measured using the liquid immersion sampling technique. The method that is most commonly employed consists of collecting a sample of the spray in a cell filled with immersion liquid and making a microscopic size count of thousands of droplets. The transient characteristics of the droplet size and the time-averaged results during an injection period, especially under the effects of ambient pressure, rack position, and pump speed, are measured.

Transient Characteristics of Droplet Size [26]—In an injection period, the injection conditions, such as injection pressure, nozzle orifice area, and injection rate, fluctuate. Consequently, droplet size will also change with time during the injection period. Therefore, the objective of the present work was to obtain fundamental knowledge of the transient characteristics of the droplet size distribution in the diesel engine.

The experimental apparatus is shown schematically in Fig. 11. An element of the spray injected during a small increment of time was sampled with an instantaneously sliding shutter (12) located just under the nozzle tip. The shutter, which was operated with a DC solenoid (7), had a stroke of 16 mm. The start of sampling and the sampling period were adjusted individually by varying the width of the shutter opening and its velocity. Residual fuel which did not pass through the slit was evacuated with a vacuum pump (16) through a teflon tube (19) into a vessel (17). In order to obtain the synchronous signal, a pulse generator, which consisted of a disc (23) with a thin slit, a phototransistor (22), and a lamp (24) was installed on the pump axis. The immersion liquid, a 0.05% water solution of FC128 [27] (a kind of fluoric surface active agent), filled a small transparent cell (18) about 2 mm in thickness. Injection pressure and nozzle needle lift were measured to determine the state of the injection system. Pressure at the inlet of the nozzle was sensed with a semiconductor strain gauge transducer (9) and was displayed on an oscilloscope (13). Nozzle needle lift was measured with a light-beam chopper-type of needle lift detector (10) at the nozzle holder.

As pump speed and rack position were adjusted to a desired value, fuel was injected consecutively. When the switching circuit of the DC solenoid was activated, the syncronous signal allowed the shutter to move right under the nozzle orifice to sample a small element of a spray among consecutive fuel sprays issuing from the nozzle. Droplets of the spray which passed through the slit were sampled in the immersion liquid. Photographed images of droplets on films were measured and counted with a scale.

Results of the Transient Characteristics of Droplet Size—The variation of the frequency-distribution diagram of the droplet volume with sampling time is shown in Fig. 12 for the hole nozzle with a single orifice. In the figure, parameter t (ms) represents the lapse of time from the start of injection to the time when the droplets were sampled. The duration for sampling droplets was about 1 ms in the present experiment. Other parameters, P_o, R and n, represent the valve opening pressure, the rack position which determines the quantity of fuel injected per stroke of the injection pump, and pump speed, respectively. It is evident from the figure that the droplet size varies considerably with time during an injection period. Fig. 13a shows simultaneously the pressure at the inlet of the nozzle, P_N, the rate of fuel injection, dB/dt, and the nozzle needle lift, S_N. In addition, the Sauter mean diameter, X_{32}, is plotted versus the lapse of time from the start of injection, t. It is seen from the figure that in general, higher injection pressure produces smaller droplets. Figs. 13b and c also show the results for a throttling nozzle and a pintle nozzle. These results agree with those derived from the empirical correlation which will be described later, and show that droplet size decreases with increasing velocity of discharge.

Time-Averaged Characteristics of Droplet Size [28]—In the present study, fuel was injected into high-pressure room-temperature gaseous environments with a diesel-engine injection system, and droplet size was measured using the liquid-

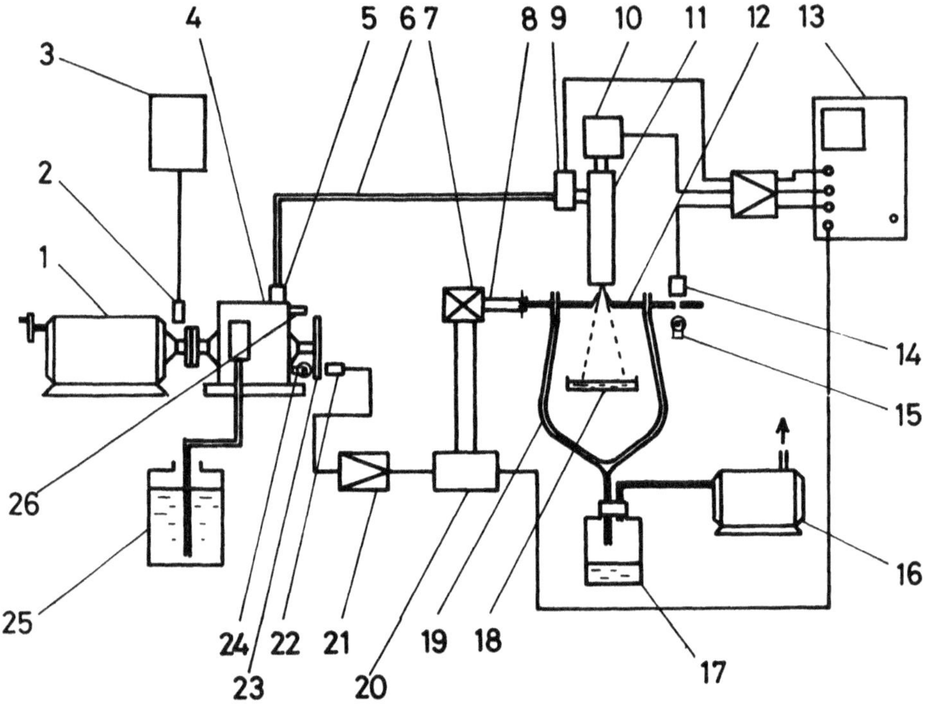

Fig. 11. Experimental apparatus for measuring transient characteristics of droplet size.

1. Motor
2. Detector of tachometer
3. Tachometer
4. Injection pump (Bosch Type)
5. Delivery chamber
6. Fuel pipe
7. D.C. solenoid coil
8. D.C. solenoid plunger
9. Semi-conductor strain gauge transducer
10. Light beam choppering type needle lift detector
11. Fuel injection nozzle
12. Sliding shutter
13. Oscilloscope
14. Photo-transistor
15. Lamp
16. Vacuum pump
17. Vessel
18. Sampling cell
19. Teflon tube
20. S.C.R. switching element
21. Step-signal generator
22. Photo-transistor for pulse generator
23. Disc for pulse generator
24. Lamp for pulse generator
25. Fuel tank
26. Fuel control rack

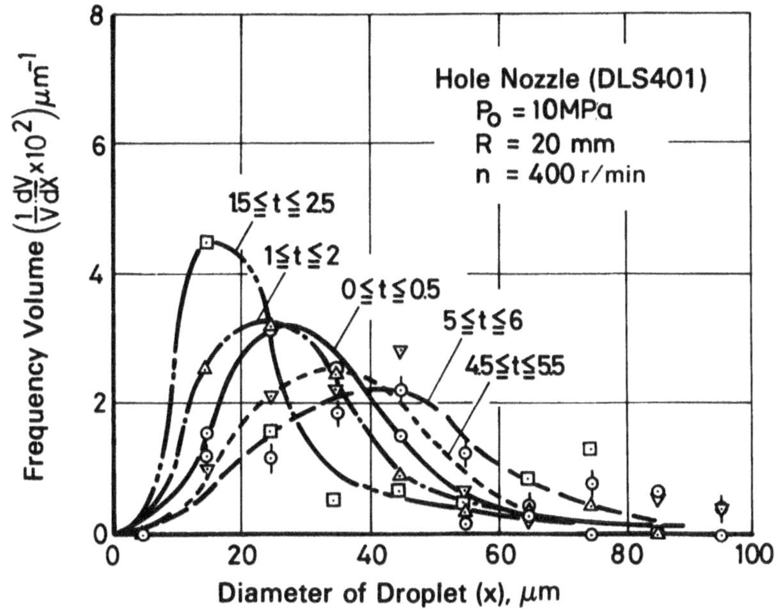

Fig. 12. Droplet size distribution at different sampling times.

immersion sampling technique to obtain the effect of ambient pressure, pump speed, and rack position of the fuel pump.

Effect of Sampling Position—Droplets sampled at various radial distances from the nozzle axis were measured. An example of the droplet size distributions at

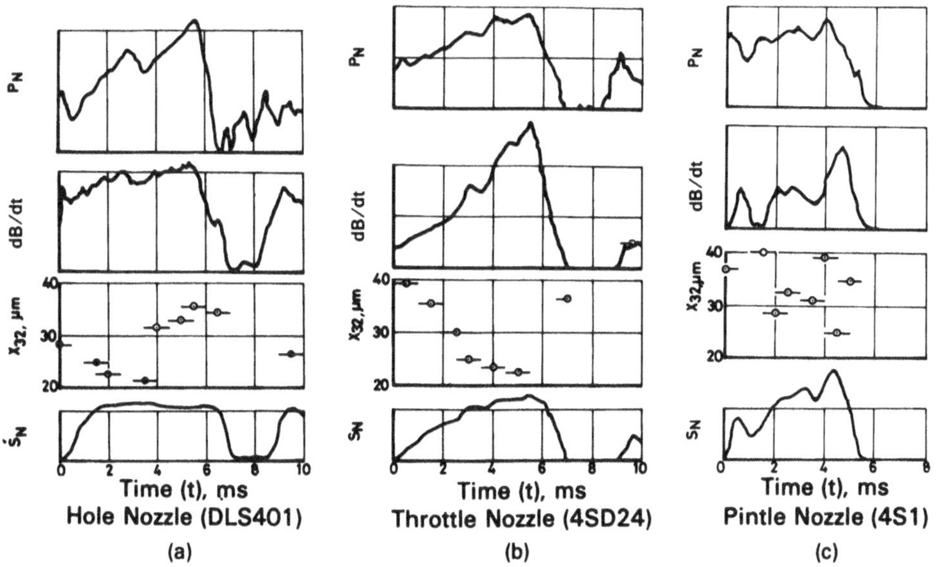

Fig. 13. Relation between the Sauter mean diameter and injection conditions.

References pp. 401–402.

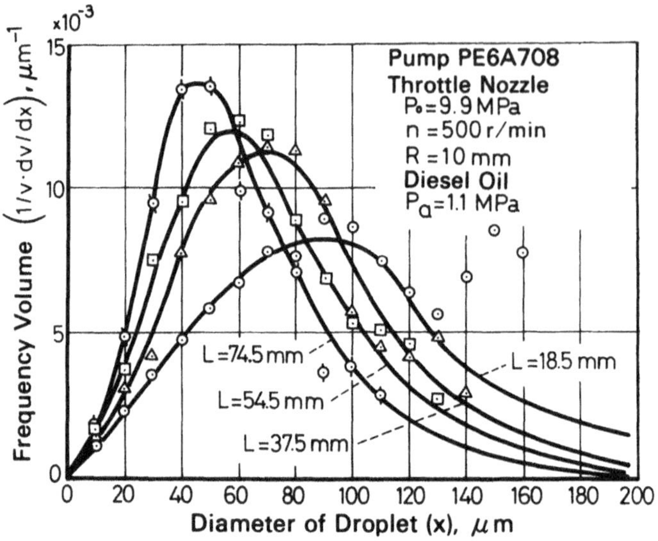

Fig. 14. Droplet size distribution at various sampling positions.

various sampling positions is shown in Fig. 14 for a back pressure of 1.1 MPa. The peak of the distribution diagrams shifts to the side of large droplet diameter with a decrease in radial distance. This means that more small droplets are in the periphery of the spray.

The relation between the Sauter mean diameter and the radial distance at various back pressures is shown in Fig. 15. The Sauter mean diameter decreases with an increase in the radial distance of the sampling position from the nozzle

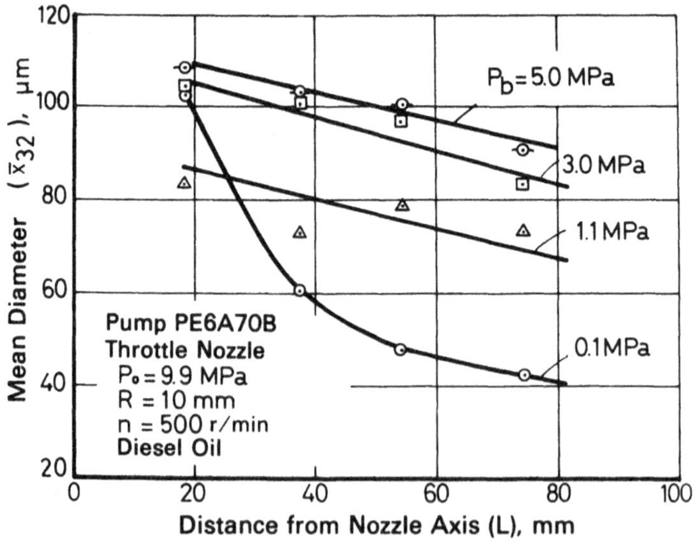

Fig. 15. Sauter mean diameter at various sampling positions.

axis. The relation differs with the back pressure. The slope of the curve is rather gentle for the data at high back pressure. This may be due to the variation in spray angle with back pressure.

Effect of Ambient Gas Pressure—In order to study the effect of ambient gas pressure on the Sauter mean diameter, the air pressure into which fuel was injected was varied from 0.1 to 5.0 MPa. Injection conditions such as valve opening pressure were held constant. A frequency diagram of the volume of droplets is shown in Fig. 16. The great influence of ambient gas pressure on droplet size distribution is clear. The curve in Fig. 16 does not, however, express sufficiently the effect of ambient gas pressure, since the droplets were sampled at a single position in the spray. Since the Sauter mean diameter varies with the sampling position, as shown previously, the Sauter mean diameter of the whole spray must be obtained. For this purpose, the fuel dispersion in a spray was measured.

The effect of the ambient gas pressure on the Sauter mean diameter, considering total droplet size distribution in a whole spray, is shown in Fig. 17. The Sauter mean diameter in a whole spray increases with an increase in ambient gas pressure.

Increasing back pressure results in a decrease in spray velocity and therefore in an increase in droplet size. On the other hand, there is a decrease in droplet size due to the increase in specific weight and viscosity of the ambient gas. In this way, the back pressure has opposing effects on the mean diameter. Another effect, which is probably of primary significance, is that the coalescence of droplets due to a small penetration of the spray at high back pressures results in large droplet size.

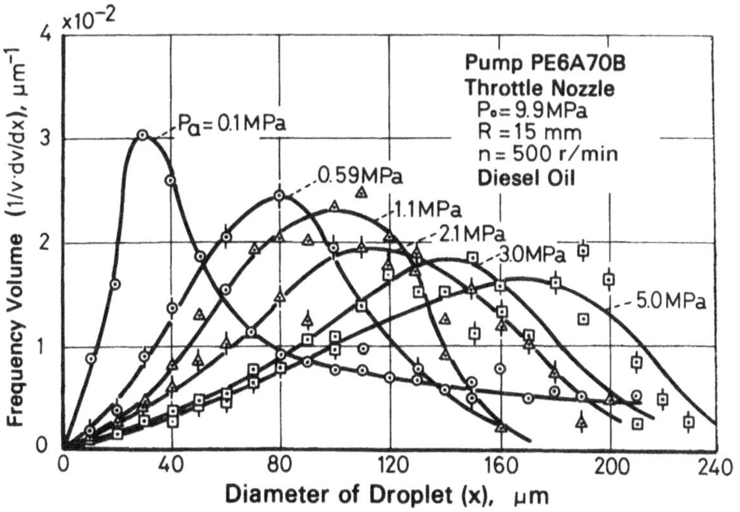

Fig. 16. Droplet size distribution at various back pressures.

References pp. 401–402.

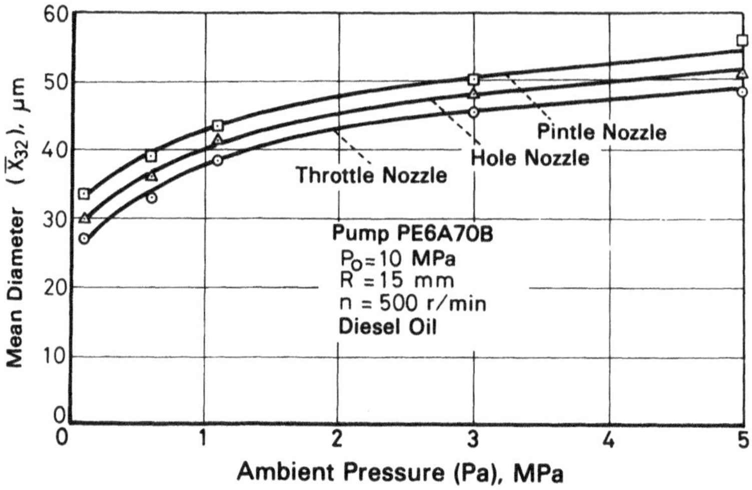

Fig. 17. Effect of ambient pressure on the Sauter mean diameter.

Effect of Pump Speed—Fig. 18 shows the variation of the Sauter mean diameter with pump speed. Mean diameter decreases slightly with an increase in pump speed for three types of nozzles. The increase in the injection pressure under constant back pressure results in an increase in spray velocity at the outlet of the nozzle and thus produces, presumably, the decrease in the Sauter mean diameter.

Effect of Rack Position—The amount of fuel in a spray is regulated with the rack. The relation between the Sauter mean diameter and the rack position is shown in Fig. 19. A large amount of fuel in a spray produces a high probability of coalescence among neighboring droplets. This is considered to be the main cause of the increase in mean diameter.

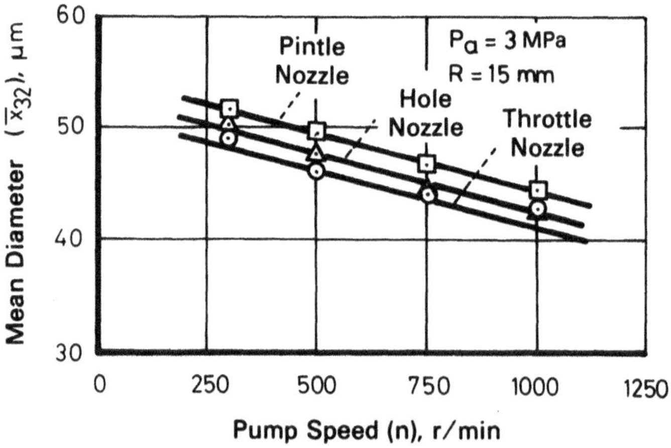

Fig. 18. Effect of pump speed on the Sauter mean diameter.

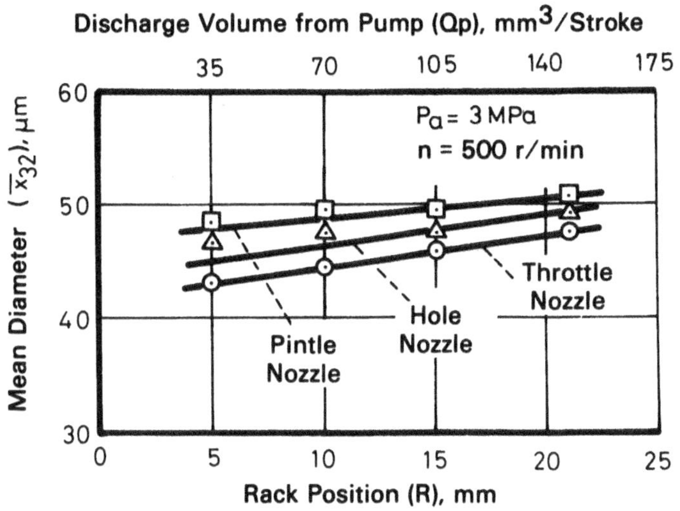

Fig. 19. Effect of rack position on the Sauter mean diameter.

Comparison with Different Types of Nozzles—In this investigation, three types of nozzles, namely a hole nozzle, a pintle nozzle, and a throttling pintle nozzle, are used under the same operating conditions, which include opening pressure, fuel delivery, and pump speed. As shown in Figs. 18 and 19, the measured results of the mean diameter are not very different for the three types of nozzles.

The effect of the nozzle orifice diameter was studied using 0.20, 0.25, 0.30 and 0.50-mm orifice diameters. When measurements were made using different orifice nozzle diameters, no remarkable difference was noted in the Sauter mean diameter.

Mathematical Expressions for Droplet Size Distribution and the Sauter Mean Diameter—In order to analyze the combustion phenomena of a diesel engine mathematical expressions for droplet size distribution and mean diameter, which include the physical conditions, are desirable. Many mathematical expressions have been proposed for the droplet size distribution in liquid sprays. We found, however, that one nondimensional expression can give the droplet size distribution for sprays injected through three types of nozzles, namely, the hole nozzle, the pintle nozzle and the throttling pintle nozzle, under conditions simulating diesel engines:

$$\frac{dv}{v} = 13.5\left(\frac{x}{\bar{x}_{32}}\right)^3 exp[-3.0(x/\bar{x}_{32})] \cdot d(x/\bar{x}_{32}) \quad (8)$$

and

$$\frac{dv}{v} = 30.3\left(\frac{x}{\bar{x}_m}\right)^3 exp[-3.67(x/\bar{x}_m)] \cdot d(x/\bar{x}_m) \quad (9)$$

References pp. 401–402.

These expressions are independent of the operating conditions and the kind of nozzle used. If we can get the mean diameter, i.e., the Sauter mean diameter or the median diameter, which is dependent of operating conditions, total droplet size distribution can be obtained immediately. One more desired item of information is an expression for the Sauter mean diameter as a function of air back pressure, speed of the fuel pump, and rack position. An attempt was made to develop such an expression from the data obtained during this investigation.

The empirical relationship relating effective injection pressure, air density, quantity of fuel delivered, and the Sauter mean diameter of spray droplets is

$$\bar{x}_{32} = A(\Delta P)^{-0.135}(\rho_a)^{0.121}(B)^{0.131} \tag{10}$$

where A is a constant, which equals 2.33×10^{-3} for the hole nozzle, 2.18×10^{-3} for the throttling pintle nozzle, and 2.45×10^{-3} for the pintle nozzle.

THE VAPORIZATION OF FUEL DROPLETS

The vaporization of the liquid fuel droplets is the stage following atomization and penetration of the injected fuel. Vaporization of the fuel is essential, since combustion occurs only in the vapor phase of a fuel-air mixture. Extensive studies concerning the evaporation of a single droplet have been reported. Ranz and Marshall [29], Spalding [30], El-Wakil et al. [31, 32], Nishiwaki [33] and others tried to elucidate these phenomena experimentally. Almost all of them were limited to a droplet at atmospheric or near atmospheric pressure. The investigation of the mechanism of droplet vaporization at elevated pressures and temperatures is a necessary requirement for the simulation of diesel engine combustion. Savery et al. [34] and Matlosz et al. [35] conducted experiments on droplet evaporation at fairly high pressure, but low gas temperature. The effects of temperature and pressure in the gaseous environments were not made sufficiently clear by these works. The objective of our work, therefore, was to obtain information on liquid droplet evaporation at high pressures and temperatures corresponding to the supercritical state of a droplet.

Experimental Methods—As shown in Fig. 20, the main apparatus in the experimental study consisted of a pressure vessel and a small cubic electric furnace which moved. In order to make the measurements in high-pressure gaseous environments, a cylindrical pressure vessel (1) was manufactured. The pressure in the vessel is increased by feeding gaseous nitrogen from a bomb (10). The vessel contains an electric furnace (16) which is supported by a rack and pinion mechanism. A droplet hanging on a 0.4-mm diameter quartz thread is exposed to the hot gas by the displacement of the furnace, thereby initiating evaporation. In order to suspend a droplet at the tip of the quartz thread, a droplet maker was installed at the pressure vessel. It is operated out of the vessel and designed to prevent the leakage of liquid fuel, even at high pressure. A copper-constantan thermocouple of 50 μm diameter is fixed on the tip of the fine quartz thread for the measurement of the droplet temperature.

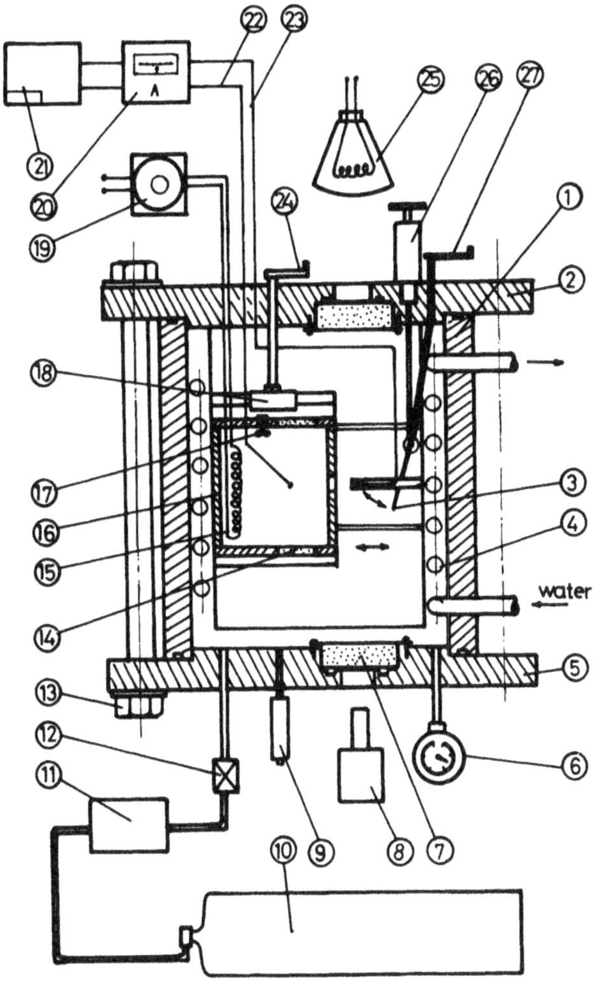

Fig. 20. Experimental apparatus for droplet vaporization.

1. Cylindrical pressure vessel
2. Upper flange
3. Needle of droplet maker
4. Cooling water tube
5. Bottom flange
6. Pressure gauge
7. Pyrex glass windows
8. 16mm movie camera
9. Exhaust valve
10. Nitrogen cylinder
11. Pressure regulator
12. Stop valve
13. Bolts
14. Quartz glass windows
15. Heating elements
16. Electric furnace
17. Fan
18. Rack and pinion mechanism
19. Volt slider
20. Milli-voltmeter
21. Recorder
22. Thermocouple
23. Thermocouple
24. Handle of rack and pinion mechanism
25. Lamp
26. Droplet maker
27. Handle for droplet maker

References pp. 401–402.

After the temperature and the pressure is increased to the desired value, a droplet is suspended on a fine quartz thread with the droplet maker. A droplet hanging on the quartz thread is quickly subjected to hot gas by the movement of the furnace and begins to evaporate. A silhouetted droplet image is recorded through the glass window, using a 16mm movie camera. The time history of the temperature of an evaporating droplet is recorded with an oscilloscope. The liquids studied included *n*-heptane, iso-octane, *n*-hexadecane, ethanol, benzene, kerosene and light oil. The experimental range of conditions consisted of ambient gas pressures from 0.1 MPa to 5 MPa and ambient gas temperatures from 100°C to 500°C, which corresponds to the subcritical, critical and supercritical state of an evaporating droplet.

Experimental Results—The wet-bulb temperature rises with an increase in ambient gas temperature and pressure. The critical point is a state in the process where gas and liquid phases coexist and simultaneously transport heat and mass at high temperature and pressure. Droplet temperatures are compared with the critical temperature of the liquid in Fig. 21. As shown, the droplet and wet-bulb temperatures have been nondimensionalized by dividing each by the critical temperature of the liquid. Although the wet-bulb temperature approaches the critical temperature of a liquid with an increase in gaseous temperature and pressure, it does not reach the critical temperature, even at supercritical ambient conditions. A droplet was not heated up to the critical temperature in environments whose reduced temperature and pressure were 1.9 and 1.4, respectively. Time histories of droplet size are shown in Fig. 22 for various liquids. Large expansion and small evaporated mass sometimes make the size of a droplet exceed its initial size.

The rate at which the square of the diameter of an evaporating droplet decreases with time at quasi-steady state, i.e., the gradient of the line in an x^2 versus t diagram such as Fig. 22, is called the evaporation rate ($K = -dx^2/dt$) and is an important parameter. The effects of the ambient gas temperature and pressure

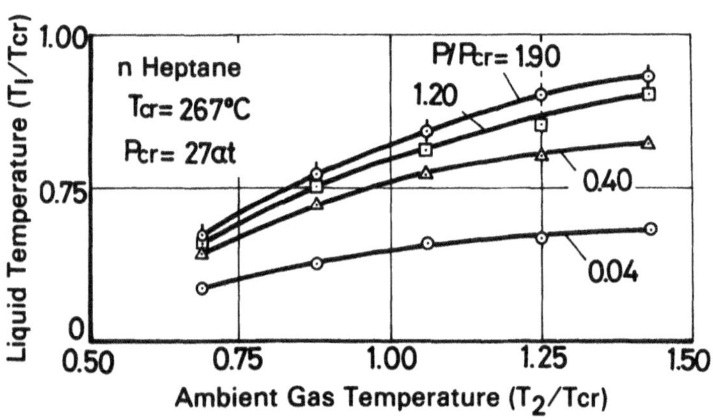

Fig. 21. Reduced wet-bulb temperature.

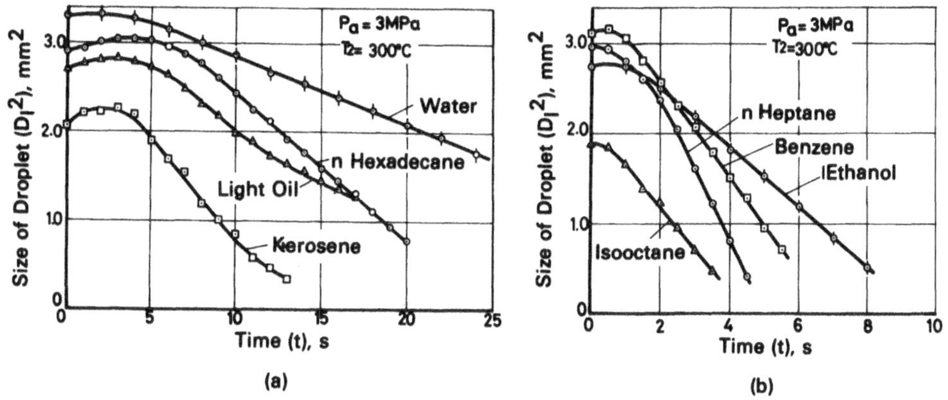

Fig. 22. Time history of droplet size.

on the evaporation rate are similar to their effects on the life time. That is to say, with the rise in gas temperature, the evaporation rates increase at all gas pressures.

The Mathematical Model—Using the approach of our previous work [36], the following equations can be set up to enable the extension of single-droplet evaporation in high-pressure and high-temperature gaseous environments. The calculations cover the unsteady and steady state of droplet evaporation, considering the effect of non-ideal mixtures, the fuel-vapor concentration at the droplet surface, and the non-ideality of the enthalpy of vaporization. The differential equation for the droplet temperature is determined from the overall heat balance between the droplet and the gaseous environment:

$$\frac{dT_l}{dt} = \frac{1}{m_l c_{pl}} \left(Q_d{}^* + H \frac{dm_l}{dt} \right) \tag{11}$$

The heat-transfer rate to the droplet is

$$Q_d{}^* = x^2 h^* (T_a - T_l) \tag{12}$$

where h^* is the coefficient of heat transfer affected by the mass transfer rate and is expressed as

$$h^* = \frac{-\dfrac{dm}{dt}(c_{pf} + \zeta c_{pa})}{\pi x^2 \left[\exp \left\{ \dfrac{-dm/dt(c_{pf} + \zeta c_{pa})}{\pi x \lambda} \dfrac{1}{Nu} \right\} - 1 \right]} \tag{13}$$

The following differential equation gives the rate of change of droplet diameter:

$$\frac{dx}{dt} = \frac{2}{\pi x^2 \rho_l} \left(\frac{dm_l}{dt} - \frac{\pi x^3}{6} \frac{d\rho_l}{dT_l} \frac{dT_l}{dt} \right) \tag{14}$$

References pp. 401–402.

The mass-transfer rate of fuel at the surface is

$$-\frac{dm_l}{dt} = \pi x^2 k^* \frac{Y_{fo}}{1-(1+\zeta)Y_{fo}} \tag{15}$$

The mass-transfer coefficient, k^*, is affected by the mass transfer rate, $-dm/dt$, and expressed as

$$k^* = \frac{\rho_a \mathscr{D}}{x} \frac{1-(1+\zeta)Y_{fo}}{(1+\zeta)Y_{fo}} Sh\, ln \frac{1}{1-(1+\zeta)Y_{fo}} \tag{16}$$

where

$$\zeta = -\frac{\rho_a(1-Y_{fo})}{\rho_l + \frac{x}{6}\frac{d\rho_l}{dt}\Big/\frac{dx}{dt}} \tag{17}$$

The above equations determine the history of the diameter and the temperature of the droplet. The correlations of Ranz and Marshall [29] for the Nusselt and Sherwood number are used. Theoretical results are compared with experimental results in Figs. 23 and 24, where the broken lines or dots indicate the experimental results and the solid lines show the theoretical results.

SPONTANEOUS IGNITION DELAY OF FUEL SPRAY

In diesel engines, ignition delay is one of the major factors affecting the heat-release rate. The increasing ignition delay in diesel engines leads to an increase

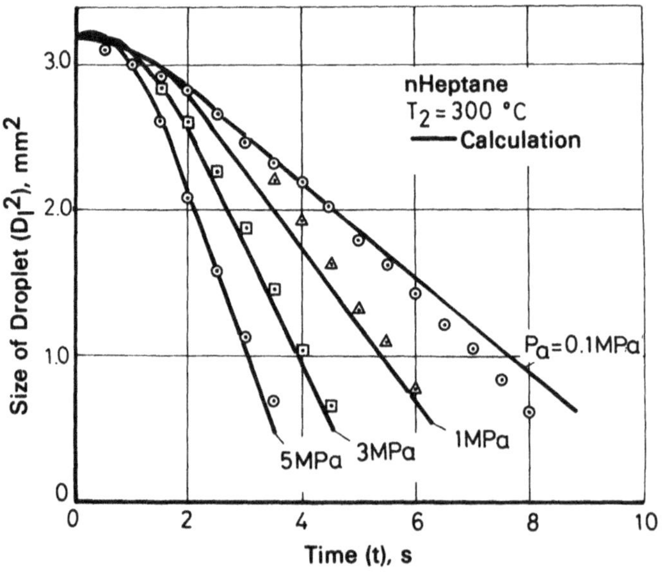

Fig. 23. Comparison of measured and computed time history of droplet size.

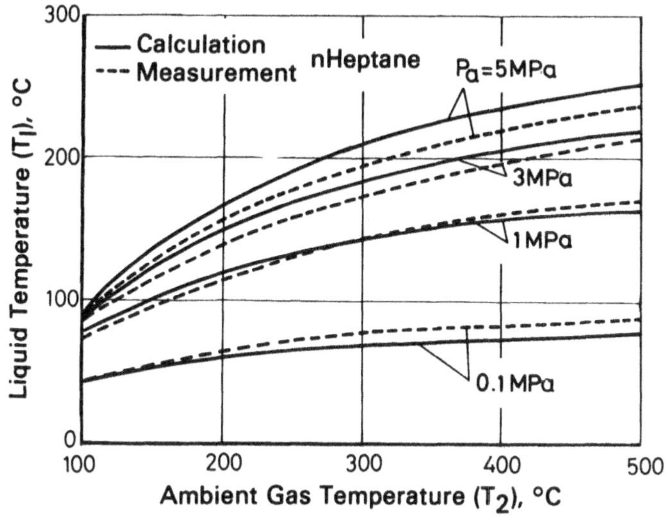

Fig. 24. Comparison of measured and computed final droplet temperature.

in the amount of fuel vaporized preceding autoignition of fuel, and therefore also leads to an increase in the portion of premixed combustion, which sometimes results in a diesel knock. Several investigations on the ignition delay of liquid fuel sprays have been made with actual engines, constant-volume bombs, and combustion tubes in which fuel is injected into an air stream. A motored research engine technique is useful to obtain understanding of the influence of operating conditions, such as injection timing, engine speed, and swirling air velocity. However, in an engine it is not easy to correlate the ignition delay with primary and fundamental factors such as temperature and pressure because these parameters vary simultaneously with time and it is sometimes difficult to estimate their quantity. In contrast, the constant-volume bomb is well suited for fundamental understanding of ignition delay. Constant environmental conditions can be maintained in this apparatus.

Apparatus and Procedure—The experimental apparatus is shown in Fig. 25. A cylindrical electric furnace is contained in the pressure vessel. Temperature in the furnace is varied by controlling the voltage supply to the electric heating elements. The pressure vessel is also equipped with a single-shot injection unit for supplying a liquid fuel spray into the vessel.

There are various ways of determining the ignition delay. They are based on either the temperature rise, pressure increase, or luminescence at combustion. The last method was used in the present study. Ignition delay was defined as the period of time from the start of injection to the first appearance of a small luminous nucleus in the spray. The instant of the start of spray injection was determined from the movement of the nozzle needle. A small luminous nucleus in the spray was detected with a phototransistor (12) which was located at a glass window of the pressure vessel. Both signals were displayed simultaneously on

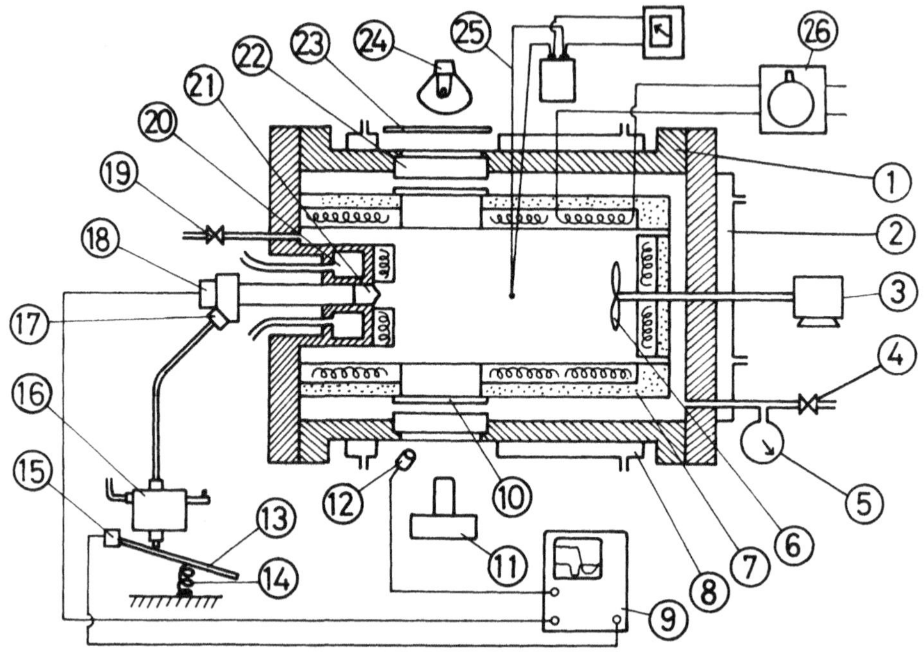

Fig. 25. Constant-volume combustion bomb.

1. Cylindrical pressure vessel
2. Cooling water jacket
3. Motor
4. Feeding valve
5. Pressure gauge
6. Fan
7. Electric furnance
8. Cooling water jacket
9. Oscilloscope
10. Quartz glass window
11. Camera
12. Photo-transistor
13. Pushing lever
14. Spring
15. Triggering switch
16. Bosch type injection system
17. Semi-conductor strain gauge type of pressure transducer
18. Light beam choppering type needle lift detector
19. Exhaust valve
20. Cooling water jacket
21. Fuel injection nozzle
22. Quartz glass window
23. Screen
24. Lamp
25. Thermocouple
26. Volt slider

an oscilloscope to determine the ignition delay. A spray evaporating and burning in the furnace was photographed with a 16-mm high-speed movie camera (4000 ~ 6000 frames per second) (11). The ignition delay determined from this method was compared with the results from the phototransistor method, as mentioned previously. The comparison showed good agreement between them. Consequently, the ignition delay was determined exclusively by the phototransistor method.

Results and Discussions—High-speed photographs of spray combustion showed that the evaporation of fuel ended before the luminous nucleus appeared. Soon after the first appearance of the luminous nucleus, consecutive appearances of small luminous nuclei occurred at different positions in the spray. Simultaneously, the size of the flame nuclei increased, resulting in an abrupt spread of flame over the spray. At the first stage of spray combustion, flame development acted like a premixed flame.

Ignition delay was measured at ambient gas pressures ranging from 0.1 MPa to 3.0 MPa, temperatures ranging from 400°C to 700°C, and oxygen concentration varying from 0.5 to 1.0. Oxygen concentration, ϕ, in the present study was defined as $\phi = P_{O_2}/(0.21P)$. P_{O_2} is the partial pressure of the oxygen molecules in ambient gas, and P is the total pressure in the furnace.

Summarizing the experimental results for the effects of ambient gas on the spray ignition delay, the following representation was obtained:

$$\tau = AP^{B}\phi^{C}\exp(D/T) \tag{18}$$

The values of constants A, B, C and D in the above equation are summarized in Table 1. D in the above equation is the apparent activation energy, which includes such physical factors as atomization and evaporation of liquid fuels, as well as chemical factors. Ambient gas pressure produces both chemical and physical effects on the spray ignition delay. Physical processes affected by gas pressure are the atomization of fuel and the evaporation of droplets. The spray penetration decreases and spray angle increases with an increase in gas pressure. The droplet size distribution also varies with pressure, and the evaporation rate of a single droplet increases with an increase in gas pressure. This indicates a high rate of increase in fuel concentration of the ambient gaseous mixture. The chemical effect of ambient gas pressure, excluding the rate constant of the kinetics, is the increase in oxygen molecules of the mixture, which results in the short ignition

TABLE 1

Values for Ignition Delay

($\tau = AP^{B}\phi^{C}e^{D/T}$) ($\tau$: ms, p: atm, T: K)

Fuel	A	B	C	D
Light Oil	2.76×10^{-1}	−1.23	−1.60	7280
n-heptane	7.48×10^{-1}	−1.44	−1.39	5270
n-dodecane	8.45×10^{-1}	−1.31	−2.02	4350
n-hexadecane	8.72×10^{-1}	−1.24	−2.10	4050

References pp. 401–402.

delay since the rate of chemical reaction is proportional to the product of the number of fuel and oxygen molecules in the mixture.

The relationship between injection conditions and ignition delay is contained only in the value of A in Eq. 18. Also examined was how the injection conditions affect the ignition delay. Injection conditions produce a minor effect on the ignition delay compared with the effect of ambient gas conditions. Ignition delay increases with an increase in the amount of fuel sprayed, which is particularly significant at low gas temperatures. Ignition delay is almost independent of nozzle opening pressure, nozzle orifice diameter, and plunger speed.

Ignition Delay when Spray Impinges on the Wall [37]—In the previous section, ignition delays are described when spray is injected into free compressed air. In the small high-speed diesel engine, however, there are many cases in which the fuel spray hits the chamber wall and then ignites. This section describes measurements of the ignition delay when the spray impinges on the chamber wall. The same measuring apparatus and constant-volume combustion chamber was used. The wall, whose temperature can be controlled, was oriented perpendicular to the spray axis of the combustion chamber. Fig. 26 shows the results of the ignition delay with wall impingement of the fuel. The distance between the nozzle tip and the wall was 100 mm, and wall temperature was equal to the gas temperature. The dotted line and its continuation are ignition delay data without the

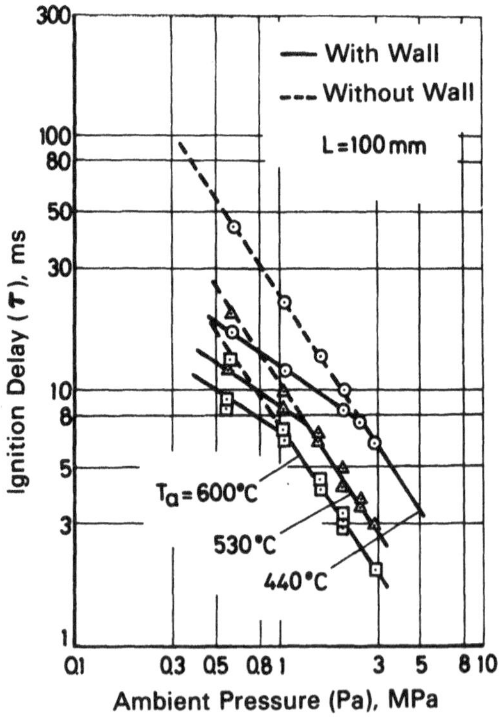

Fig. 26. Ignition delay when spray impinges on the wall at various ambient pressures.

wall. The dots and solid line represent the ignition delay data when the spray impinges on the wall. These results show a constant slope for the variation of ignition delay with ambient pressure. When the ambient gas pressure is high, the data are the same with and without the wall. But when the ambient pressure is below a specific pressure, the ignition delay with wall impingement is shorter than the ignition delay without the wall. When the ambient temperature is increased, ignition delay is decreased. These effects of temperature are the same, with and without the wall. Fig. 27 shows the data when the wall temperature is varied from 200°C to 520°C. The gas temperature is kept at 440°C. It is noticed that ignition delay decreases with the wall present, as compared to the case without the wall, when the wall temperature is lower than the gas temperature.

SPRAY FORMATION AND HEAT-RELEASE MODELS

The model described herein is based on the fundamental experimental results and mathematical formulas which are introduced in previous sections for a free or wall jet, in quiescent or swirling air. The fundamental concept and detailed description of this model are reported elsewhere [5].

The fuel is assumed to form an axisymmetric spray pattern without air swirl. If air swirl is present, the spray is bent in the direction of swirl. The spray is divided into 250 isolated packages without mixing. Axial division of the spray

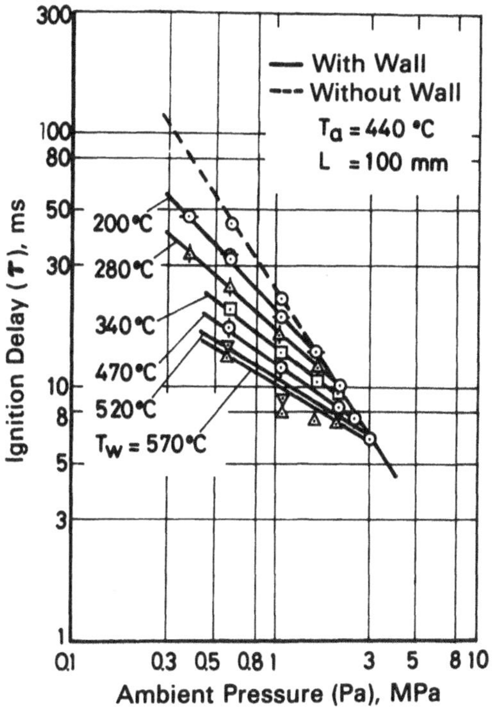

Fig. 27. Ignition delay when spray impinges on the wall at various wall temperatures.

TABLE 2

Direct-Injection Engine Used in Model Calculation

Bore	0.135 m
Stroke	0.13 m
Length of Connecting Rod	0.23 m
Clearance volume	0.1283×10^{-3} m^3
Swept Volume	1.8×10^{-3} m^3
Compression Ratio	15.5
Opening Timing of Inlet Valve	113° BTDC
Opening Timing of Exhaust Valve	107° ATDC
Injection Pump	Bosch AD
Plunger Bore	10 mm
Diameter of Nozzle Hole	0.35 mm
Number of Nozzle Holes	4

coincides with the location of the fuel elements injected during successive small increments of time. The radial division produces cones of equal solid angles with their vertices at the nozzle orifice. Each package consists of air, a group of liquid droplets, and fuel vapor. Penetration of all packages is postulated to be the same. The amount of fuel in each package is determined from the injection rate, assuming a normal distribution among the packages. The droplet size distribution in each package is assumed to follow an empirical correlation in terms of the Sauter mean diameter. The mass of fuel evaporated in each package is calculated from single-droplet evaporation equations. The local equivalence ratio is calculated from the fuel vapor and air in each package. The combustion of each package

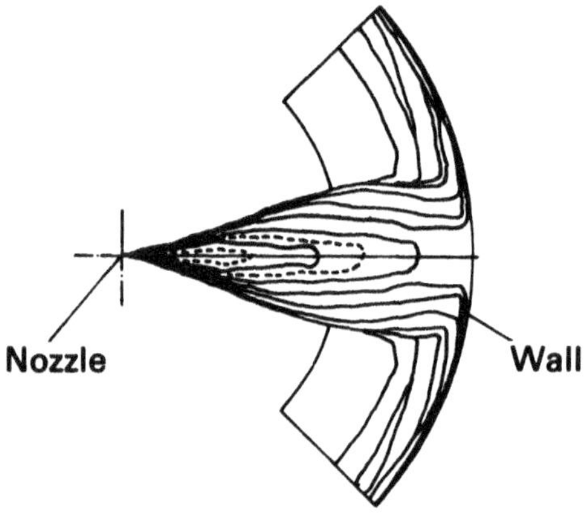

$n = 1250$ r/min

$\theta_{inj} = -17°$

$B = 90$ mg

$r_s = 0.0$

Fig. 28. Computed spray form with quiescent air.

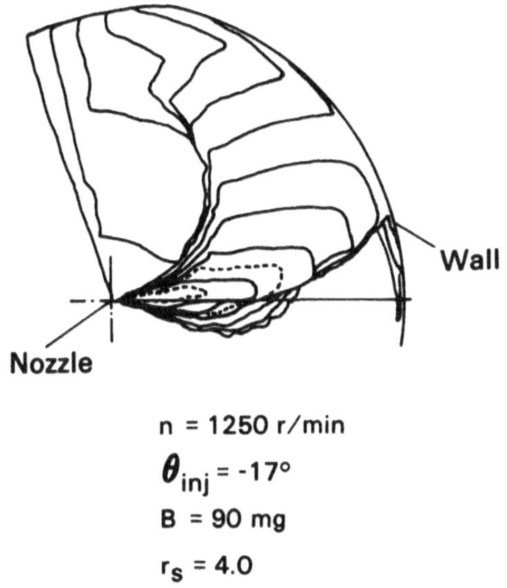

Fig. 29. Computed spray form with swirl ratio r_s = 4.0.

starts after an ignition delay which is calculated from an empirical equation for homogeneous mixtures. The net heat release is calculated from the summation of the heat of combustion of the fuel in all packages after subtracting the heat losses to the combustion chamber walls. Each package has a different temperature, which varies with time and space, and is different from the mass-average

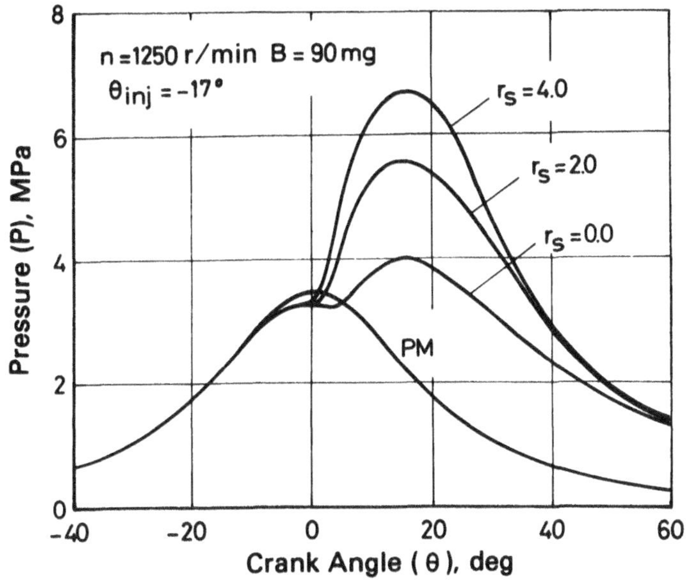

Fig. 30. Effect of swirl ratio on computed cylinder pressure.

References pp. 401–402.

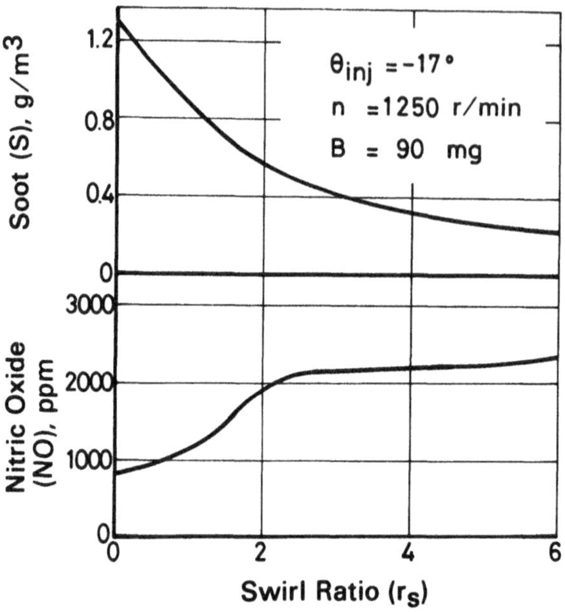

Fig. 31. Effect of swirl ratio on computed soot and *NO* concentration.

temperature calculated from the cylinder pressure. The *NO* calculations were according to the extended Zeldovich mechanism, assuming a homogeneous mixture in each package. Soot formation and combustion were also calculated from empirical equations.

Fig. 28 represents the spray behavior obtained from computed results. The operating condition is given in Table 2. The fuel was injected into quiescent air, and although the nozzle had four holes, only one spray was used in the calculations. After some period of time from the beginning of the fuel injection, the ignition occurs at the peripheral portion of the spray, followed by rapid expansion of the combustion mixture. Since the temperature and the pressure in the cylinder increase with the lapse of time, the ignitability of the spray package increases. This means that the ignition delay of each succeeding package decreases. Consequently, the flame spreads rapidly and the spray impinges the cylinder wall.

Fig. 29 shows the spray form at a swirl ratio, r_s, of 4.0. The spray is bent with air swirl and spreads. Figs. 30 and 31 show the effect of swirl ratio on the pressure history in the cylinder and the concentration of *NO* and soot in the exhaust gas.

CONCLUSIONS

The spray characteristics in diesel engines have been examined experimentally, and equations describing the spray characteristics have been developed for use in the computation of diesel engine combustion.

The spray penetration and spray angle have been measured with and without air swirl for several injection conditions. Experiments were conducted in a constant-volume chamber equipped with observation windows. At various swirl levels, a single shot of fuel was injected into a high-pressure environment. The growth of the spray was followed with high-speed movies. Experimental equations of spray penetration and spray angle were obtained with and without the air swirl.

The mean droplet size and droplet size distribution were measured for several injection conditions using a liquid immersion sampling technique. Droplet size distribution function and Sauter mean diameter data were obtained.

An analytical model of a single droplet evaporating in a high-pressure and high-temperature gaseous environment was investigated and showed good agreement with the experimental results for single-droplet evaporation.

Ignition delay was also measured with and without wall impingement in a constant-volume chamber.

Equations obtained for spray characteristics were applied to compute the cylinder pressure and the concentrations of nitric oxide and soot in a direct-injection diesel engine using the mathematical model developed by the authors.

REFERENCES

1. *E. K. Bastress, K. M. Chng and D. M. Dix, "Models of Combustion and Nitrogen Oxide Formation in Direct and Indirect Injection Compression Ignition Engines," Proc. Intersoc. Energy Conversion Eng. Conf., Paper No. 719053, pp. 364-375, 1971.*
2. *I. M. Khan, G. Greeves and D. M. Probert, "Prediction of Soot and Nitric Oxide Concentrations in Diesel Engine Exhaust," Proc. I. Mech. E., C142, pp. 295-217, 1971.*
3. *S. M. Shahed, W. S. Chiu and V. S. Yumlu, "A Preliminary Model for the Formation of Nitric Oxide in Direct Injection Diesel Engines and Its Application in Parametric Studies," SAE Trans., Vol. 82, Paper No. 730083, pp. 338-351, 1973.*
4. *W. S. Chiu, S. M. Shahed and W. T. Lyn, "A Transient Spray Mixing Model for Diesel Combustion," SAE Trans., Vol. 85, Paper No. 760128, pp. 502-512, 1976.*
5. *H. Hiroyasu and T. Kadota, "Models for Combustion and Formation of Nitric Oxide and Soot in Direct Injection Diesel Engines," SAE Trans., Vol. 85, Paper No. 760129, pp. 513-526, 1976.*
6. *C. J. Kau, M. P. Heap, T. J. Tyson and R. P. Wilson, "The Prediction of Nitric Oxide Formation in a Direct Injection Diesel Engine," Sixteenth Symposium (International) on Combustion, The Combustion Institute, Pittsburgh, Pennsylvania, pp. 337-350, 1976.*
7. *M. Meguerdichian and N. Watson, "Prediction of Mixture Formation and Heat Release in Diesel Engines," SAE Paper No. 780225, 1978.*
8. *N. Hay and P. L. Johnes, "Comparison of the Various Correlations for Spray Penetration," SAE Paper No. 720776, 1972.*
9. *P. H. Schweitzer, "Penetration of Oil Sprays," Pennsylvania State University Bulletin No. 46, 1937.*
10. *A. S. Lyshevskiy, "Determination of the Length of an Atomized Fuel Spray," Trudy Novocherkasskogo, Politekhnicheskogo Instituta, Vol. 26, pp. 391-401, 1955.*
11. *A. S. Lyshevskiy, "The Coefficient of Free Turbulence in a Jet of Atomized Liquid Fuel," NASA TT-F351, 1956.*
12. *Y. Wakuri, M. Fujii, T. Amitani and R. Tsuneya, "Studies of the Penetration of a Fuel Spray in a Diesel Engine," Bull. JSME, Vol. 3, No. 9, pp. 123-130, 1960.*

13. R. V. Rusinov, "Length of Atomized Fuel Jet in a Diesel Engine," Russian Engrg. Jrn., Vol. 43, 1963.
14. G. Sitkei, "Kraftstoffaufbereitung und Verbrennung bei Dieselmotoren," Springer-Verlag, 1964.
15. M. Parks, C. Polonski, and R. Toye, "Penetration of Diesel Fuel Sprays in Gases," Paper No. 660747, 1966.
16. M. Ogasawara and H. Sami, "Study on the Behavior of a Fuel Droplet Injected into the Combustion Chamber of a Diesel Engine," SAE Trans., Vol. 78, Paper No. 670468, pp. 1690-1707, 1967.
17. I. Hakki Öz, "Calculation of Spray Penetration in Diesel Engines," SAE Trans., Vol. 78, Paper No. 690254, pp. 1107-1116, 1969.
18. R. Burt and K. Troth, "Penetration and Vaporization of Diesel Fuel Sprays," Proc. I. Mech. E., Vol. 184, Part 3J, pp. 147-170, 1970.
19. D. H. Taylor and B. E. Walsham, "Combustion Processes in a Medium Speed Diesel Engine," Proc. I. Mech. E., Vol. 184, Part 3J, pp. 67-76, 1970.
20. J. C. Dent, "A Basis for the Comparison of Various Experimental Methods for Studying Spray Penetration," SAE Trans., Vol. 80, Paper No. 710571, pp. 1881-1884, 1971.
21. W. M. Scott, "Looking in on Diesel Combustion," SAE Paper No. 690002, SP-345, 1969.
22. Joe Rife and John B. Heywood, "Photographic and Performance Studies of Diesel Combustion with a Rapid Compression Machine," SAE Trans., Vol. 83, Paper No. 740948, pp. 2942-2961, 1974.
23. Rolf D. Reitz, "Atomization and Other Breakup Regimes of a Liquid Jet," Ph.D. Thesis, Princeton University, 1978.
24. H. Hiroyasu, T. Kadota and S. Tasaka, "Penetration of Diesel Fuel Sprays," JSME Trans., Vol. 44, No. 385, pp. 3208-3219, 1978.
25. Veniamin G. Levich, "Physicochemical Hydrodynamics," Prentice-Hall Inc., Englewood Cliffs, New Jersey, pp. 639-650, 1962.
26. H. Hiroyasu, Y. Toyota and T. Kadota, "Transient Characteristics of Droplet Size Distribution in Diesel Sprays," Paper No. 5-3, presented at the First International Conference on Liquid Atomization and Spray Systems, Tokyo, 1978.
27. T. Kamimoto, S. Matsuoka and S. Shiga, "Penetration of Fuel Spray at the Initial Stage of Injection in Diesel Engines," JSME Trans., Vol. 41, No. 342, pp. 672-684, 1975.
28. H. Hiroyasu and T. Kadota, "Droplet Size Distributions in Diesel Engines," SAE Trans., Vol. 83, Paper No. 740715, pp. 2615-2624, 1974.
29. W. E. Ranz and W. R. Marshall, Jr., "Evaporation from Drops," Chem. Eng. Progs., Vol. 48, No. 3, p. 141, 1952, and Vol. 48, No. 4, p. 173, 1952.
30. D. B. Spalding, "The Combustion of Liquid Fuels," Fourth Symposium (International) on Combustion, The Combustion Institute, Pittsburgh, Pennsylvania, pp. 847-864, 1952.
31. M. M. El-Wakil, O. A. Uyehara and P. S. Myers, "A Theoretical Investigation of the Heating-up Period of Injected Fuel Droplets Vaporizing in Air," NACA TN 3179, 1954.
32. M. M. El-Wakil, R. J. Priem, H. J. Brikouski, P. S. Myers and O. A. Uyehara, "Experimental and Calculated Temperature and Mass Histories of Vaporizing Fuel Drops," NACA TN 3490, 1956.
33. N. Nishiwaki, "Kinetics of Liquid Combustion Processes: Evaporation and Ignition Lag of Fuel Droplets," Fifth Symposium (International) on Combustion, The Combustion Institute, Pittsburgh, Pennsylvania, pp. 148-158, 1955.
34. C. W. Savery, D. L. Juedes and G. L. Borman, "n-Heptane, Carbon Dioxide and Chlorotrifluoromethane Droplet Vaporization Measurements at Supercritical Pressures," Ind. Eng. Chem. Fundam., Vol. 10, No. 4, pp. 543-553, 1971.
35. R. L. Matlosz and S. Leipziger, "Investigation of Liquid Drop Evaporation in a High Temperature and High Pressure Environment," Int. J. Heat and Mass Transfer, Vol. 15, pp. 831-852, 1972.
36. T. Kadota and H. Hiroyasu, "Evaporation of a Single Droplet at Elevated Pressures and Temperatures," JSME Trans., Vol. 42, No. 356, p. 1216-1222, 1976.
37. H. Hiroyasu, T. Kadota and S. Tanaka, "Spontaneous Ignition Delay of Diesel Sprays Impinging on a Wall," JSME Paper No. 760-16, pp. 222-224, 1976.

APPENDIX A—SPRAY TIP PENETRATION

Levich [25] obtained the breakup length of a liquid jet under the following assumptions:

(i) The liquid jet of density ρ_l is moving in a gas medium with a density $\rho_a \ll \rho_l$.
(ii) The relative velocity between the liquid jet and the gas medium is large.
(iii) The amplitude of the jet surface disturbance is accelerated by the pressure disturbance of the gas.
(iv) As the amplitude of the jet increases, the jet tends to be unstable, and finally, the jet may break up into droplets.
(v) The breakup Length, L, is calculated from the breakup time, t_{break}.

$$L \approx u_o t_{break} \approx \alpha \sqrt{\frac{\rho_l}{\rho_a}} d_o \tag{A1}$$

Furthermore we assumed that

(vi) Jet velocity within the intact length is equal to the initial jet velocity.
(vii) The spray tip velocity is proportional to $\sqrt{t}$ based on continuous jet theory.

From these assumptions, the jet velocity before jet breakup to droplets is

$$u_o = c \sqrt{\frac{2\Delta P}{\rho_l}} \tag{A2}$$

The transient time from the start of injection to jet breakup is obtained from u_o and L:

$$t_{break} = \frac{\alpha \rho_l d_o}{\sqrt{2c^2 \rho_a \Delta P}} \tag{A3}$$

Since we assume the spray tip penetration is proportional to $\sqrt{t}$ after breakup to droplets, spray tip penetration can be expressed as follows:

$$s = \beta\sqrt{t} \tag{A4}$$

We can obtain the value of β from the conditions of $s = L$ and $t = t_{break}$:

$$\beta = 2^{1/4}(\alpha c d_o)^{1/2} \left(\frac{\Delta P}{\rho_a}\right)^{1/4} \tag{A5}$$

The values of the constants can be determined from the experiment described in previous sections:

$$\alpha = 15.8$$
$$c = 0.39 \tag{A6}$$

Therefore, the spray tip penetration is

$$0 < t < t_{break}$$

$$s = 0.39 \frac{2}{\rho_l} \Delta P t \tag{A7}$$

$$t \geq t_{break}$$

$$s = 2.95 \left(\frac{\Delta P}{\rho_a}\right)^{1/4} \sqrt{d_o t} \tag{A8}$$

where

$$t_{break} = 28.65 \frac{\rho_l d_o}{\sqrt{\rho_a \Delta}} \tag{A9}$$

APPENDIX B—SPRAY ANGLE

Wakuri et al. [12] obtained the following function for the spray angle from their dimensional analysis:

$$\theta = f\left(\frac{\rho_l}{\rho_a}, \frac{d_o u_o \rho_l}{\mu_a}\right) \tag{B1}$$

If we assume that the spray angle is constant after the jet breakup to droplets, the following equation can be written from Eq. B1:

$$\theta = \delta \left(\frac{\rho_l}{\rho_a}\right)^m \left(\sqrt{\frac{2 \cdot \Delta P}{\rho_l}} \frac{d_o \rho_l}{\mu_a}\right)^n \tag{B2}$$

δ, m and n are constants obtained from experimental results. From the experimental data

$$\delta = 0.00413, \quad m = -\tfrac{1}{4} \quad \text{and} \quad n = \tfrac{1}{2} \tag{B3}$$

Then

$$\theta = 0.05 \left(\frac{d_o^2 \rho_a \Delta P}{\mu_a^2}\right)^{1/4} \tag{B4}$$

APPENDIX C—PENETRATION AND SPRAY ANGLE WITH CROSS FLOW

A jet of initial velocity, u_o, is injected into a crossflow of velocity, $u_\theta(s)$. The penetration with air swirl s_s is determined by applying a correlation factor, c_s, to the penetration without swirl, s.

$$s_s = c_s s \tag{C1}$$

The correlation factor, C_s, is given by

$$C_s = \left(1 + \frac{\text{swirl velocity at } s}{\text{jet velocity}}\right)^{\alpha} \qquad \text{(C2)}$$

where, for simplicity, we use the initial jet velocity, u_o, for the jet velocity, u. The swirl velocity at s is given by

$$u_{\theta}(s) = \frac{\pi r_s n s}{30} \qquad \text{(C3)}$$

Therefore, the correlation factor, c_s, is

$$C_s = \left(1 + \frac{\pi r_s n s}{30 u_o}\right)^{\alpha} \qquad \text{(C4)}$$

where

r_s = the swirl ratio

n = engine speed

α = constant

From the experimental results, α was found to be equal to -1.

The spray tip penetration with air swirl is

$$s_s = \left(1 + \frac{\pi r_s n s}{30 u_o}\right)^{-1} s \qquad \text{(C5)}$$

The spray angle with air swirl can be expressed in the same manner as the penetration, s_s:

$$\theta_s = C_{\theta}\theta \qquad \text{(C6)}$$

where

θ_s = spray angle with air swirl

θ = spray angle without air swirl

C_{θ} = correlation factor

The correlation factor, C_{θ}, is expressed as follows:

$$C_{\theta} = \left(1 + \frac{\pi r_s n s}{30 u_o}\right)^{\beta} \qquad \text{(C7)}$$

β, based on experiments, equals 2. Therefore

$$\theta_s = \left(1 + \frac{\pi r_s n s}{30 u_o}\right)^{2} \theta \qquad \text{(C8)}$$

DISCUSSION

V. W. Wong *(Cummins Engine Company)*

I have a couple of questions about the correlation between your experimental data on penetration and drop size. You showed a graph [Figs. 2 and 4 of paper] of penetration versus time that had a kink in it, and you referred to that kink as "breakup." You also showed a graph [Fig. 12] of droplet size distribution, but I didn't see any apparent discontinuity in the distribution of sizes around the time of the kink. That's the first question. The second question concerns the droplet size distribution. In your illustrated droplet size distribution data [Fig. 12], the droplet size appeared, first, to go to lower values with increasing time. This was followed with a time period in which the droplets became larger with increasing time. Is this behavior caused by condensation or coagulation of droplets at the later stage of injection?

H. Hiroyasu

Spray penetration of diesel injection refers to a fixed position in space, but drop size is related to time. The velocity of every droplet, or every parcel of liquid fuel at the initiation of the injection, is constant. Afterwards the spray develops into a steady jet, as shown in Figs. 2 and 4. But the droplets in the spray were measured at a position downstream of that location at which the spray develops into a steady jet. Therefore the droplet size is not dependent on the

H. Hiroyasu

spray penetration. Regarding the second question, droplet size is mainly affected by injection pressure. The high injection pressure produces small droplets. Furthermore, I think there is coagulation of droplets during the later stage of injection. But the main cause of the change of droplet size is the change of the injection pressure.

F. V. Bracco *(Princeton University)*

I am an admirer of Professor Hiroyasu's experimental tricks in gathering very difficult experimental information. I also agree with several of his conclusions, but I disagree with others. In particular, I agree with the conclusion that the initial penetration tip velocity can be approximated as a linear function. I don't think there is any need for a discontinuity. That could be reconciled simply by an exponential function that changes gradually. If a discontinuity is observed, it may be related to the unsteadiness of the pressure upstream. Now, continuing with the conclusions with which I agree, I also would like to suggest that the difference in the distribution function of droplet sizes downstream can be correlated to two factors. One is the instantaneous pressure at the time at which a parcel of fluid was going through the nozzle itself. That's the quasi-steady assumption that we have proved reasonably well in our own experiments. The situation is further complicated by the phenomenon that was described by the Los Alamos calculations, that is, the parcel of droplets which is at the tip is not the one that came out of the nozzle at the corresponding time. Rather it's continuously changing. Therefore, if one measures downstream, one cannot relate directly to the upstream pressure in a one-to-one time-velocity correlation.

Now the part on which I disagree. You have the correlation for the breakup distance or for the intact length, whatever you prefer to call it. That correlation includes only one geometrical parameter, the diameter of the nozzle. I think we have proved, beyond reasonable doubt, that more than just the diameter of the nozzle is required. The length-to-diameter ratio and the radius of curvature of the inlet part of the nozzle make a drastic difference between having an intact length and not having any intact length at all. For the same diameter, having a long length can produce a very intact jet; having a very short length results in no intact length. Therefore, one geometrical parameter to describe the breakup length is not adequate. Similarly, the same parameter is used to describe the spray angle. I submit to you that that parameter also is not adequate. We have shown that for the same diameter, you can get spray angles varying by a factor of two or three, simply by changing the nozzle geometry. In fact, you can get an intact jet as well as a completely shattered one. So this is item one of two disagreements, namely, that we must be concerned with more than just the diameter. The other item of some contention is that your correlations are in terms of the density ratios, gas versus liquid. This is exactly the parameter that we have found to be important, by changing the molecular weight of the inert gas so as to keep the same pressure and temperature but with different densities. However, many of the results are presented in terms of gas pressure. I submit that the gas pressure is not the important parameter, but rather the gas density. So in this case I agree with your correlation, but I disagree with your presentation of the numbers.

H. Hiroyasu

Thank you very much for your comments, Professor Bracco. I agree with your comments, but the mechanism of disintegration of liquid is quite a complicated phenomenon. At this time, I believe, nobody fully understands this intermittent liquid spray phenomenon. Of course, we know that the spray disintegration mechanism, including the droplet size, spray penetration and spray angle, is affected by nozzle diameter, length-to-diameter ratio of the nozzle, inside roughness of the nozzle, turbulence of the issuing liquid, injection pressure, gas density, etc. We need more detailed experimental data. But if we apply the penetration and droplet size equation to the diesel combustion simulation, I believe a simpler equation is better than a complicated equation.

J. C. Dent *(Loughborough University of Technology, England)*

I would just like to ask if, in the course of your experiments, you determined any radial size distribution of the droplets, and if so, could you give us some idea about the variation of droplet size with radius?

H. Hiroyasu

Yes, we measured radial size distribution of the spray droplets. In order to make this measurement, the sampling position was changed with the radial distance of the spray. The detailed technique and results are given in our paper.

G. L. Borman *(University of Wisconsin)*

If you look at your data on droplet sizes, it appears to a lot of us that the Sauter mean diameter correlation you have presented gives Sauter mean diameters that are too large. The explanation could be that because of the long length there was agglomeration. Have you had any further thoughts about how one might analyze that or get at that problem?

H. Hiroyasu

The immersion liquid method is a simple method for measuring drop size, but this method also has a disadvantage in that it requires a sufficiently large distance from the nozzle tip to the sampling position. In such a case, there are many chances for the droplets to coalesce. To avoid this effect, optical or laser optical methods can be used. However, even if we used those methods, we cannot measure the droplet size in the dense-spray region, that is, near the nozzle tip. Obtaining correct droplet size data is quite difficult.

HYDROCARBON EMISSIONS MODELING FOR SPARK IGNITION ENGINES

G. A. LAVOIE, J. A. LORUSSO and A. A. ADAMCZYK

Ford Motor Company, Dearborn, Michigan

ABSTRACT

The Ford hydrocarbon model for conventional, homogeneous-charge engines is reviewed. The model is phenomenological in that it contains no formal spatial dependence in the description of burning gas elements within the cylinder. However near the wall, an integral boundary-layer approach is employed to calculate the growth of a thermal boundary layer and the evolution of a chemically reacting quench layer. The initial thickness of the quench layer is assumed to be proportional to the laminar two-plate quench thickness. Following oxidation in the cylinder, additional burnup of hydrocarbons in the exhaust port is calculated by means of a plug-flow analysis for each exhaust-gas element, according to its time of exit from the cylinder. Included in the exhaust gas is a contribution from the ring-crevice hydrocarbons, part of which leave the cylinder with the bulk gas during blowdown and exhaust, while the remainder are entrained in a roll-up vortex which leaves the cylinder, in part, later in the exhaust stroke.

In all, the model has seven adjustable parameters. Two of these define the heat transfer in the cylinder and in the exhaust port. The remaining five parameters relate to the hydrocarbon emission process and define the effective temperature at which quenching takes place, the ratio of one-wall to two-plate quench thickness, the rate of hydrocarbon burnup, the distribution of ring-crevice hydrocarbons in the exhaust, and the exiting fraction of the roll-up vortex.

Several modifications have been made to the model since it was originally presented. These changes relate to the surface area assigned to wall quenching, and to the distribution of ring-crevice hydrocarbons in the exhaust gas. The latter change resulted in a significant improvement of the predicted trend of hydrocarbons versus engine load. Additional validation experiments have been performed to test the model's ability to predict the effects of wall-temperature and compression-ratio variations. The model showed the proper trend with wall temperature, but the predicted change in hydrocarbon emissions was less than observed experimentally. Good agreement was obtained in the case of compression-ratio changes. As reported earlier, good predictions were obtained in the case of equivalence-ratio variation for stoichiometric and

References pp. 440–441.

lean mixtures, for changes in spark timing and in *EGR* rate, while poor agreement was obtained in the case of variations in engine speed.

Strengths and weaknesses of the current model are discussed with a view to identifying the critical areas where new knowledge is essential for establishing more accurate predictive engine models. One such area is that of in-cylinder fluid-mechanics processes governing the exhaust distribution of quench-layer and ring-crevice hydrocarbons. Another area is the quenching process itself. The detailed nature of the quenching process, including the post-quench burnup and structure of the layer, remains unclear at this time. This has important ramifications regarding the effects of fuel type and turbulence on quenching, and the distribution of individual hydrocarbon species within the layer itself.

NOTATION

A	pre-exponential factor in global rate expression
$A(\alpha^*)$	total surface area per unit mass of the element burning at mass fraction burned α^*.
C_1	quenching constant of proportionality, dq_1/dq_2
c_p	specific heat at constant pressure
C_R	non-dimensional oxidation parameter
c_v	specific heat at constant volume
dq_1	single-wall quench distance
dq_2	two-plate quench distance
E	activation energy in global rate expression
f	in-cylinder heat-transfer friction factor
f_p	exhaust-port heat-transfer friction factor
f_{res}	residual mass fraction
f_v	fraction of the roll-up vortex exiting during exhaust
g_{rc}	parameter defining fraction of ring crevice hydrocarbons that exit with the bulk gas
h	convective heat-transfer coefficient
$[HC]$	unburned hydrocarbon concentration
k	burnup ratio

m	mass fraction coordinate with origin at the wall and encompassing the total mass of the element in question when $m = 1$
$m_{BL}(\alpha,\theta)$	boundary-layer mass fraction in element α at crank angle θ.
m^i_{BL}	initial boundary-layer mass fraction of element α at the moment of burning
$m_s(\alpha,\theta)$	thermal boundary layer mass fraction of mass element α at crank angle θ
m_z	total unburned material in the boundary layer
$[O_2]$	oxygen concentration
P	cylinder pressure
$(q/A)_{cham}$	cycle-average heat flux in the combustion chamber
$(q/A)_{port}$	cycle-average heat flux in the exhaust port
R	universal gas constant
Re_i	engine "throughput" Reynolds number based on mass flow, cylinder average viscosity and cylinder bore.
$S/V_{HC}(\alpha^*)$	effective surface-to-volume ratio for quenching at mass fraction burned α^*
$S/V(\alpha^*)$	total surface-to-volume ratio at mass fraction burned α^*
t	time
T	temperature
T_B	bulk-gas temperature
T_{bq}	adiabatic burned-gas temperature in the quench layer
T_c	chamber coolant temperature
T_{cp}	exhaust-port coolant temperature
T_{lm}	log-mean temperature through the quench layer
T_u	effective adiabatic unburned compression temperature
T_{uq}	effective unburned-gas temperature in the quench layer
T_w	average chamber wall temperature
T_{wp}	average exhaust-port wall temperature

References pp. 440–441.

U_{cham}	heat-transfer coefficient through the wall and coolant film in the combustion chamber
$\overline{U}_p$	mean piston speed
U_{port}	heat-transfer coefficient through the wall and coolant film in the exhaust port
V	total cylinder volume
$v_b(\alpha^*)$	specific volume of element α just after it burns, i.e., when $\alpha^* = \alpha$
$V_b(\alpha^*)$	total volume of burned gas behind the flame front when mass fraction burned equals α^*
X_T	constant defining the effective quenching temperature, T_{uq}
y	distance from wall
y_{02}	local oxygen mass fraction
$y_{02,q}$	local oxygen mass fraction in the pure unburned quench mixture
$y_{02,b}$	local oxygen mass fraction in the bulk gas
z	local mass fraction of unburned material in the quench layer
α	mass fraction coordinate, ordering elements from 0 to 1 in sequence of burning.
$\alpha^*(\theta)$	mass-fraction burned
η	non-dimensional boundary-layer coordinate
γ	specific heat ratio $= c_p/c_v$
$\bar{\gamma}_b$	average burned-gas specific heat ratio
ω	non-dimensional temperature
ϕ	fuel-air equivalence ratio
ρ	average element gas density
θ	crank angle degrees after top dead center
θ_i	crank angle at which combustion begins
θ_f	crank angle at which combustion ends

INTRODUCTION

Mathematical models of homogeneous-charge engine performance and emissions offer great potential as aids in understanding the complex tradeoffs between fuel economy and emission control that are encountered in conventional engine design and calibration. Furthermore they provide a useful testing ground for the physical submodels of emissions formation that will be needed in future models both of homogeneous- and heterogeneous-charge engines.

Currently, performance characteristics, *NOx* and *CO* emissions can be modeled with reasonable accuracy for the conventional homogeneous-charge engine, provided that a minimum amount of experimental data is available to enable determination of empirical model parameters such as flame speed, heat-transfer coefficient, etc. These models are generally phenomenological in nature in that they do not involve the solution of the full Navier-Stokes equations of fluid motion, but instead rely on empirical knowledge of the flow patterns in the engine. This has meant a great reduction in complexity and computational requirements, which has enabled the successful application of these models to practical programs in engine design and control. Blumberg et al. [1] have documented the development of these models both for conventional- and stratified-charge engines.

In contrast to the relatively successful modeling of performance, *NOx* and *CO* emissions, modeling of hydrocarbon emissions has been slow to develop because of the complexity and number of physical and chemical mechanisms involved, many of which remain poorly understood at this time. Although a great deal is known experimentally about the mechanism of hydrocarbon emissions in engines and combustion bombs [2–11], and about the variations of emissions with engine operating conditions and design parameters [12–14], in the past only three attempts at modeling hydrocarbon emissions in engines have been reported in the literature [13, 15, 16], and these have been to a large extent incomplete or limited in applicability [1].

Recently, Lavoie and Blumberg [17, 18] have presented a major extension of earlier Ford work on *NOx* modeling [19–21] to include many of the known features of the hydrocarbon emission process, within the framework of a comprehensive engine-simulation model. The purposes of the present paper are to outline briefly the structure of the Ford hydrocarbon model, which is reported in detail in Ref. 18, to report on the results of some recent modifications to the model and on the outcome of new validation experiments that have been performed, and finally to indicate areas where future development and refinement of hydrocarbon models are most needed.

THERMODYNAMIC MODEL

The organization of the thermodynamic model is shown in Fig. 1. The term "thermodynamic" refers in this instance to all the computational elements necessary to calculate the basic gas states throughout the engine cycle and the overall balance of energy for the engine system. In terms of engine events, the

References pp. 440–441.

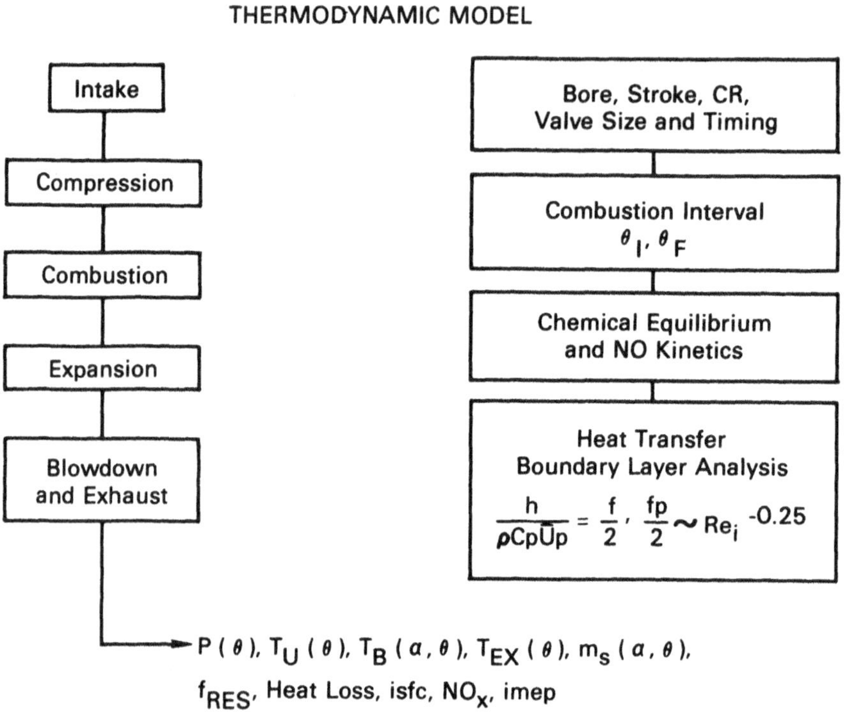

Fig. 1. Organization of thermodynamic model, showing major computational elements on the left and important submodels and input data on the right.

major elements are shown on the left side of Fig. 1. The output of this sequence of computations includes cylinder pressure, unburned temperature, burned-gas temperature, etc., as functions of crank angle and element mass fraction, α, (where applicable) as well as integrated quantities such as heat loss, *imep*, *isfc*, etc.

On the right side of Fig. 1 are the main submodels and empirical information necessary to carry out these thermodynamic calculations. Included are geometric quantities such as engine bore, stroke and compression ratio. Valve size and empirical lift curves are employed to calculate intake and exhaust flow by means of a quasi-steady-flow analysis, as described by Sherman and Blumberg [22]. The energy-release schedule must be specified from experimental information or from a flame-propagation model. Although such models exist [23–25], the current model uses a simple cosine burning law:

$$\alpha^*(\theta) = \tfrac{1}{2}\left[1 - cos\left(\frac{180^\circ\,(\theta - \theta_i)}{(\theta_f - \theta_i)}\right)\right] \tag{1}$$

where α^* is the mass fraction burned at crank angle θ. The initial and final crank angles of combustion, θ_i and θ_f, are input parameters and may be determined from experiment or specified *a priori* for parametric studies. When supported by accurate pressure or PV^γ data, the cosine law has been shown to give a satisfactory representation of the energy-release schedule [18].

The NO kinetics calculations are carried out as described by Lavoie and Blumberg [18] and employ the extended Zeldovich mechanism with an empirical correction for "prompt NO". The rate constants employed for the Zeldovich mechanism are as recommended by Baulch et al. [26], except for that of the formation reaction $O + N_2 \rightarrow NO + N$, which has been adjusted downward by 35% in order to obtain a better fit to the experimental data [18]. This adjustment is justified on the basis of the wide scatter in the experimental kinetics data used to generate the recommended rate constant.

The heat-transfer calculation is based on a Reynolds-analogy formulation, as indicated in Fig. 1:

$$\frac{h}{\rho c_p \overline{U}_p} = \frac{f}{2}, \frac{f_p}{2} \tag{2a}$$

$$f = f_o(Re_i/Re_{io})^{-0.25} \tag{2b}$$

$$f_p = f_{op}(Re_i/Re_{io})^{-0.25} \tag{2c}$$

where h, c_p and $\overline{U}_p$ are the heat-transfer coefficient, the specific heat and the mean piston speed, respectively, and ρ is the instantaneous average density of the mass element in question. The terms f and f_p are the friction factors in the cylinder and in the exhaust port respectively, and are determined at an arbitrary engine condition (denoted by subscript 'o') by empirically matching predicted to experimental specific fuel consumption and exhaust temperature. The friction factors scale with engine throughput Reynolds number, Re_i, in a similar manner to that found in turbulent pipe flow, as indicated in Eqs. 2b and 2c. For the purposes of matching to experimentally obtained exhaust temperatures, a predicted "Nusselt"-averaged temperature, T_{Nu}, is calculated, which takes into account the time variation of both the exhaust temperature and the heat-transfer coefficient to the experimental probe (typically a thermocouple).

An integral boundary-layer analysis is employed to calculate the growth of a thermal boundary layer, which is assumed to have a linear temperature profile from the wall temperature to the temperature of the central adiabatic core of the bulk cylinder gases. This separation of the cylinder contents into a boundary layer and an adiabatic core gives improved predictions of NO formation [18], compared with a uniform-temperature model. Equally important, the calculated boundary-layer mass, $m_s(\alpha,\theta)$, for each gas element, α, as a function of crank angle, θ, is employed as input to the hydrocarbon model to define the rate at which the quench layer is diluted by diffusing into the bulk gas.

Wall-Temperature Calculation—Since the original documentation of the model [18], several modifications have been made, one of which is employed to calculate the average surface temperature of the chamber, T_w, and the exhaust port, T_{wp}, as a function of the calculated average heat loading:

References pp. 440–441.

$$T_w = T_c + \frac{(q/A)_{\text{cham}}}{U_{\text{cham}}} \tag{3}$$

$$T_{wp} = T_{cp} + \frac{(q/A)_{\text{port}}}{U_{\text{port}}} \tag{4}$$

where (q/A) represents the calculated cycle-average heat flux and U the total heat-transfer coefficient through the wall and coolant film. Subscripts "cham" and "port" refer to the chamber and exhaust port, respectively. T_c and T_{cp} denote coolant temperatures for the chamber and port, respectively. The model is based on a simple steady-state analysis of the heat transfer through the wall and coolant film, and has as input the overall heat-transfer coefficients and coolant temperatures for the chamber and exhaust-port thermal systems. The calculations to be described later were carried out with an estimated heat-transfer coefficient of 1.89 kW/(m^2 s K) and gave an overall temperature drop of approximately 47°C through the chamber wall/coolant film and 23°C temperature drop through the exhaust-port wall/coolant film at medium load and speed conditions. Temperatures were not available to verify the model, but the predicted temperature drop across the chamber wall is similar to the values measured by Wentworth [7] under similar engine conditions. Although not yet validated in detail, this portion of the model provides a simple means of incorporating the known dependence of wall temperature on engine speed and load.

HYDROCARBON MODEL

The basic elements of the hydrocarbon model are shown in Fig. 2. Information from the thermodynamic model feeds into the major computational segments of the model, which include the assignment of the initial quench-layer content, in-cylinder diffusion and burnup, and oxidation of ring-crevice and wall-quench hydrocarbons in the exhaust port, exhaust manifold and tailpipe. Exhaust-reactor and tailpipe calculations have been omitted since the single-cylinder engine configuration used in the experiments reported here is not representative of these processes. A more detailed description of these submodels can be found in Ref. 18. On the right side of Fig. 2 are shown the key relationships defining the major sources of phenomenological input to the model, and the associated input parameters. They include a number of relationships for defining the initial quench-layer content as well as an empirical global rate expression for hydrocarbon oxidation.

Initial Quench Layer—The initial hydrocarbon quench-content of each element of gas is dependent on a number of relationships. The first of these is the two-plate laminar quench-distance correlation of Lavoie [27], which prescribes the two-plate quench distance, dq_2, as a function of the local thermodynamic variables, i.e.:

$$dq_2 = f(P, \phi, f_{res}, T_{uq}, T_{bq}) \tag{5}$$

where P, ϕ, and f_{res} denote the pressure, equivalence ratio and residual fraction,

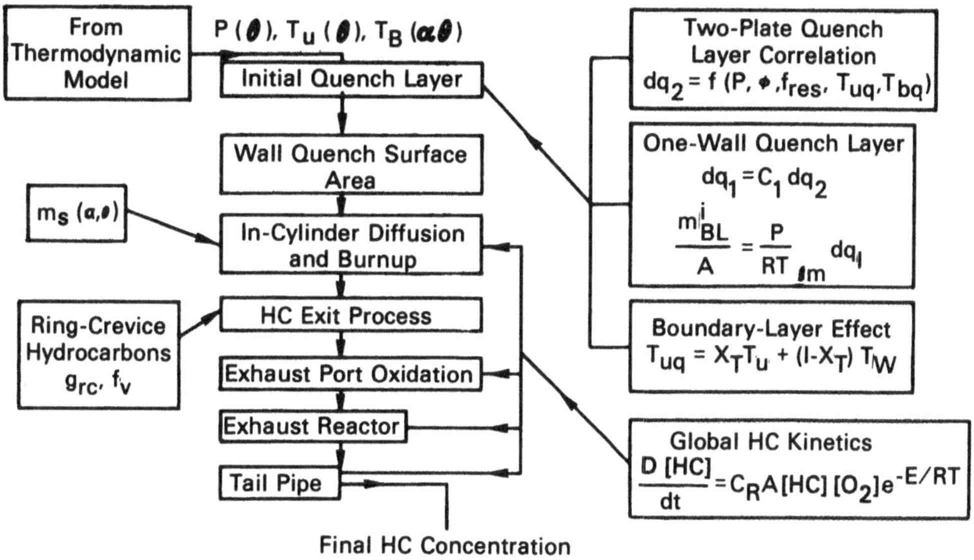

Fig. 2. Organization of the hydrocarbon model, with computational sequence outlined on the left and input information on the right.

respectively, while T_{uq}, and T_{bq} indicate the unburned- and burned-gas temperatures in the quench layer.

The second relationship defines the single-wall quench distance, dq_1, as a constant fraction of the two-plate quench distance, i.e.,

$$dq_1 = C_1 \, dq_2 \tag{6}$$

where C_1 is the constant of proportionality between dq_1 and dq_2, and is a primary input parameter of the hydrocarbon model.

Based on the results of a one-dimensional transient flame model presented by Adamczyk and Lavoie [28], it was determined that the mass in the quench layer per unit area at the moment of quenching can be approximated by the following expression:

$$m^i_{BL} \simeq \frac{P}{RT_{lm}} \, dq_1 \tag{7}$$

where the superscript "i" refers to the initial time of quench, and where dq_1 refers specifically to the distance of closest approach to the wall of the position of the maximum reaction rate in the flame. T_{lm} is the log-mean temperature through the layer and is defined by

$$T_{lm} \equiv (T_{bq} - T_{uq})/\log\,(T_{bq}/T_{uq}) \tag{8}$$

This expression takes into account the variation of density through the layer in an assumed linear temperature profile from the unburned quench temperature, T_{uq}, to the adiabatic burned temperature of the quench gas, T_{bq}.

References pp. 440–441.

In order to define the initial quench-layer content, it is also necessary to consider the effect that the thermal boundary layer, which develops during compression, could have on the effective unburned quench temperature, T_{uq}, in the region near the wall. Depending on the relative sizes of the thermal boundary layer and quench layer, T_{uq} could lie anywhere between the wall temperature, T_w, and the adiabatic unburned compression temperature, T_u. To acknowledge the existence of this effect, an empirical parameter, X_T, is introduced which determines T_{uq} as follows:

$$T_{uq} = X_T T_u + (1 - X_T) T_w \tag{9}$$

where X_T is a value between 0.0 and 1.0. As presented in Ref. 18, the constant X_T has been set equal to 0.0 so that $T_{uq} = T_w$, thus providing an implicit dependence of quench-hydrocarbon content on wall temperature, in agreement with known experimental facts.

Wall Quench Surface Area—In order to assign a specific quantity of unburned hydrocarbons to a particular mass element, it is necessary to define the amount of surface area that is associated with the element. In the previously reported version of the model [18], the surface-to-volume ratio of each mass element was assumed to be equal to the instantaneous value for the entire chamber. This is approximately correct for the case of a cylindrical flame moving in a pancake-type combustion chamber and is adequate for heat-transfer calculations. As will be seen, however, the assumption leads to a significant overestimation of the surface area for quenching, due to the compression and expansion of the burned gas behind the flame front and the fact that the combustion analysis is carried out by means of a Lagrangian mass-based calculation. Accordingly, the method of assigning quench surface to the elements has been revised to take these effects into account.

Fig. 3 shows an idealized thin flame front propagating between two parallel walls of a combustion chamber at the moment when the mass fraction burned is α^* and when a new mass element, $d\alpha$, is in the process of burning. The doubly cross-hatched region, $ABCD$, indicates the mass element just before it burns, when the flame is at position 1, while the singly cross-hatched region, $A'B'C'D'$, shows the element just after burning, when the flame is in position 2. Due to the pressure rise which accompanies the burning of element $d\alpha$ and the resulting compression of the burned gas volume, $V_b(\alpha^*)$, the new quenching surface ($AD' + BC'$) is less than the total surface area of the element ($A'D' + B'C'$). By assigning the total geometric surface-to-volume ratio to each element for quench-layer calculations, as was done in Ref. 18, the surface area utilized for quenching can be overestimated by as much as 50%. For heat-transfer calculations, however, the total surface area may be used without incurring significant error because the areas are continuously updated throughout the cylinder such that the apparent gain in area of element $d\alpha$ is balanced by a collective loss in the area of the previously burned gas.

Based on these arguments and assuming that the burned gas behind the flame behaves isentropically, it can be shown that the effective surface-to-volume ratio

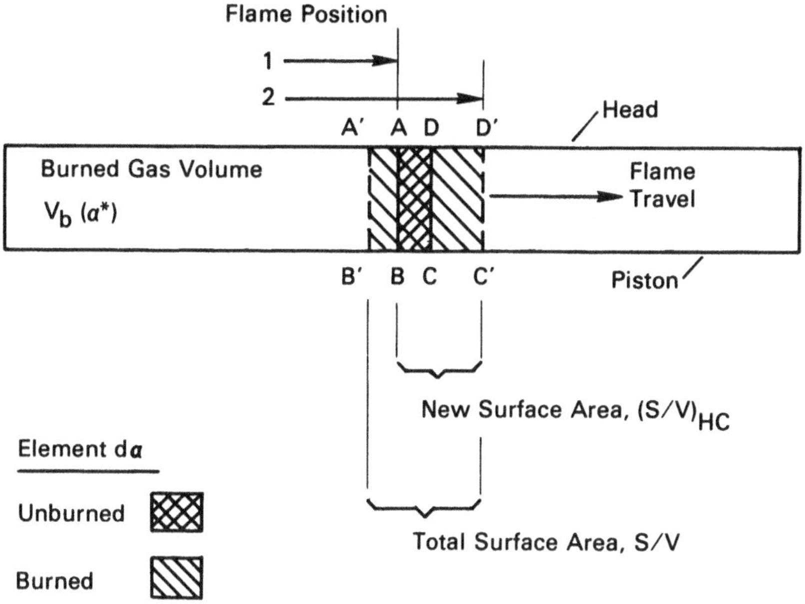

Fig. 3. Sketch of flame front before and after combustion of element $d\alpha$, showing total surface area and new surface area for heat transfer and quench calculations, respectively.

for quenching, $(S/V)_{HC}(\alpha^*)$, of the element which burns at mass fraction α^* is given by:

$$\frac{(S/V)_{HC}(\alpha^*)}{S/V\ (\alpha^*)} = 1 - \frac{V_b(\alpha^*)}{m_t v_b(\alpha^*)}\left[\frac{d}{d\alpha^*}\log\left(\frac{P(\alpha^*)^{1/\bar{\gamma}_b}}{S/V\ (\alpha^*)}\right)\right] \tag{10}$$

where $v_b(\alpha^*)$ is the specific volume of the element at α^* in the burned state, m_t is the total mass of gas in the chamber, and $\bar{\gamma}_b$ is the average burned-gas value of the ratio of specific heats, c_p/c_v. Inspection of Eq. 10 reveals that under conditions of rising pressure and approximately constant $S/V(\alpha^*)$, the quench area is less than the total area used for the heat-transfer calculation, due to the compression of the earlier burned gas behind the flame. After peak cylinder pressure is reached, the burned gas will expand with the result that $(S/V)_{HC}$ will be greater than S/V, i.e., there is a net gain in surface area exposed to burned gas.

Taking the wall quench-area effect into account, the initial quantity of mass in the boundary layer at the moment of quench may be written as:

$$m^i_{BL}(\alpha^*) = A(\alpha^*)\frac{(S/V)_{HC}(\alpha^*)}{S/V\ (\alpha^*)}\frac{P}{RT_{lm}}\,dq_1 \tag{11}$$

where $A(\alpha^*)$ is the total surface area per unit mass of the element and m^i_{BL} is the mass fraction of material in the boundary layer of element α^* at the moment of quenching.

References pp. 440–441.

The major effect of implementing the revised quenching-area calculation was generally to lower the effective surface area by about 30% and hence the hydrocarbon emissions for a given set of quench-thickness parameters. The correction was greatest for combustion near TDC, where pressure rise was a maximum. For long or late combustion conditions the correction was less, with the result that some changes in trends were observed, but they were generally negligible (less than 10%).

In-Cylinder Diffusion and Burnup—The process of in-cylinder diffusion and burnup of the quenched hydrocarbons is calculated by means of an integral boundary-layer analysis described in detail in Ref. 18. Suitable profile shapes are assumed for the distribution of unburned material and temperature in the boundary layer for each mass element $d\alpha$. The evolution of these profiles is calculated based on the overall diffusional growth of the layer, which is related to the growth of the thermal boundary-layer mass, $m_s(\alpha, \theta)$, and the average rate of hydrocarbon burnup, which is obtained by integrating the global kinetics rate expression across the layer.

The profile shapes chosen for the analysis are found in the solution to the one-dimensional diffusion problem defined by a sudden step change in wall temperature and a simultaneous injection of unburned material at the wall. For simplicity, a Lewis number of unity is assumed so that the species and thermal boundary layers diffuse away from the wall at the same rate.

When the appropriate matching of initial unburned-material content and temperature deficit in the layer is carried out, the resulting profiles for the non-dimensional temperature, ω, and unburned-mass fraction, z, are as follows:

$$\omega \equiv \frac{T - T_w}{T_b - T_w} = \operatorname{erf}(\eta) \tag{12}$$

$$z = \frac{2}{\pi}\left(\frac{m^i_{BL}}{m_{BL}}\right) \exp(-k^2\eta^2) \tag{13}$$

where T_b and T_w are the temperatures in the bulk gas and at the wall, respectively, and η is a non-dimensional boundary-layer coordinate defined by

$$\eta \equiv \frac{2}{\sqrt{\pi}} \frac{m}{m_{BL}} \tag{14}$$

In Eq. 14, m is a mass-fraction coordinate with origin at the quenching surface, and m_{BL} is the boundary-layer-mass thickness, equal to the mass fraction contained in a linear-profile boundary layer, with equivalent thermal deficit as the profile given by Eq. 12.

The total boundary-layer mass at any time is given by the sum of the initial mass, m^i_{BL}, and the thermal-layer mass, m_s, which is calculated as part of the thermodynamic model:

$$m_{BL} = m^i_{BL} + m_s \tag{15}$$

The burnup ratio, k, is the inverse fraction of the initial unburned material that remains in the layer at any time.

Fig. 4 shows the profiles of non-dimensional temperature and unburned-mass fraction as a function of the boundary-layer coordinate, η, for a number of possible conditions in the quench-layer evolution. The curves labeled $k = 1$ correspond to the situation with diffusion only. Two such curves are shown: one at the initial quench condition ($m_{BL}/m^i_{BL} = 1$), and the other at a condition when the boundary layer has doubled in size due to diffusion ($m_{BL}/m^i_{BL} = 2$).

The total quantity of unburned material, m_z, is given by the integral of z through the layer and initially is equal to one-half the boundary-layer mass, i.e.:

$$m_z^i = \left(\int_0^1 z\,dm\right)_{init} = 0.5\, m^i_{BL} \tag{16}$$

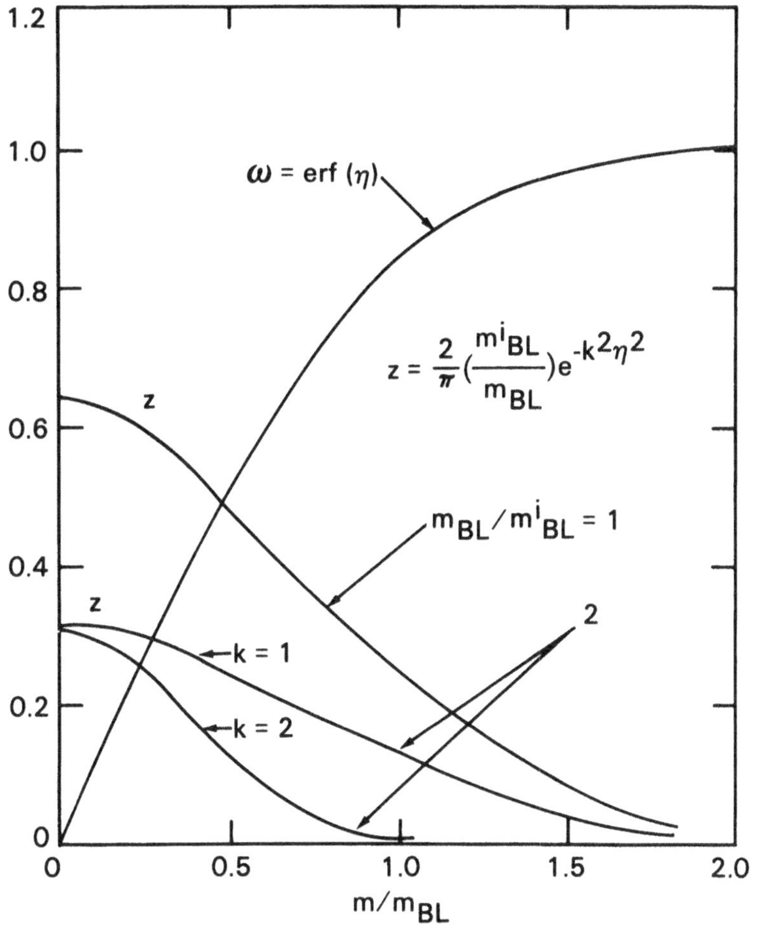

Fig. 4. Assumed profiles for non-dimensional temperature, $\omega = (T - T_w)/(T_B - T_w)$, and unburned mass fraction, z, for various conditions of dilution and burnup.

References pp. 440–441.

Mathematically, the factor of 0.5 in Eq. 16 is due to the particular form of the species and temperature profiles that have been used. Physically, the factor represents the effect of diffusion in the quench zone just prior to quenching and signifies that one-half of the gas in the quench layer is burned gas that has diffused toward the wall from the flame zone.

The calculation of burnup within the layer is carried out with an empirical global rate expression for the oxidation of hydrocarbons developed by Lavoie [27]:

$$\frac{d[HC]}{dt} = -C_R A[HC][O_2]e^{-E/RT}\ (moles/cm^3 \cdot s) \tag{17}$$

where the brackets denote concentrations in moles/cm^3. The activation energy, E, and pre-exponential factor, A, are 37,230 cal/mole and 6.7×10^{15} cm^3/(mole·s), respectively. C_R is a dimensionless parameter introduced to aid in fitting the model to experimental data.

With the above global reaction rate, the mass-average decay rate of unburned material over the entire boundary layer is calculated as follows:

$$\frac{1}{m_z}\frac{d\,m_z}{dt} = -C_R A \frac{\int_0^\infty z y_{02}(P/RT)e^{-E/RT}\,d\eta}{\int_0^\infty z\,d\eta} \tag{18}$$

where the local mass fraction of oxygen, y_{02}, is determined by linearly weighting the mass fraction in the unburned quench gas, $y_{02,q}$, and the bulk-gas mass fraction of oxygen, $y_{02,b}$, by the local mass fraction of unburned material, z, i.e.,

$$Y_{02} = y_{02,q}z + y_{02,b}(1 - z) \tag{19}$$

A typical unburned-material profile is shown in Fig. 4 for the case of 50% burnup and dilution of 2:1, i.e., $k = 2$ and $m_{BL}/m^i_{BL} = 2$.

Actual results for a burnup calculation are shown in Fig. 5, where the computed temperature and unburned-mass-fraction profiles are displayed as a function of mass fraction within the element that burns midway in the combustion process ($\alpha = 0.5$), under engine conditions of 1250 r/min, 380 kPa, $\phi = 0.9$, and MBT spark timing. Burnup is very rapid initially but eventually ceases for all practical purposes when about 85% of the initial material has been consumed, due to the fact that the remaining unburned material is in a region of the boundary layer that is relatively cold. The parameter values employed in this calculation were $C_1 = 0.8$, $X_T = 0.0$, and $C_R = 0.1$, and were chosen by a matching technique to be described in a later section.

Hydrocarbon-Exit Processes—In tests of the hydrocarbon model versus CFR engine data it has become apparent that the contribution of the ring-crevice hydrocarbons is greatly underestimated by the version of the model reported by Lavoie and Blumberg [18]. In that model, the entire ring-crevice content is assumed to be contained in the roll-up vortex and leaves the cylinder only to the

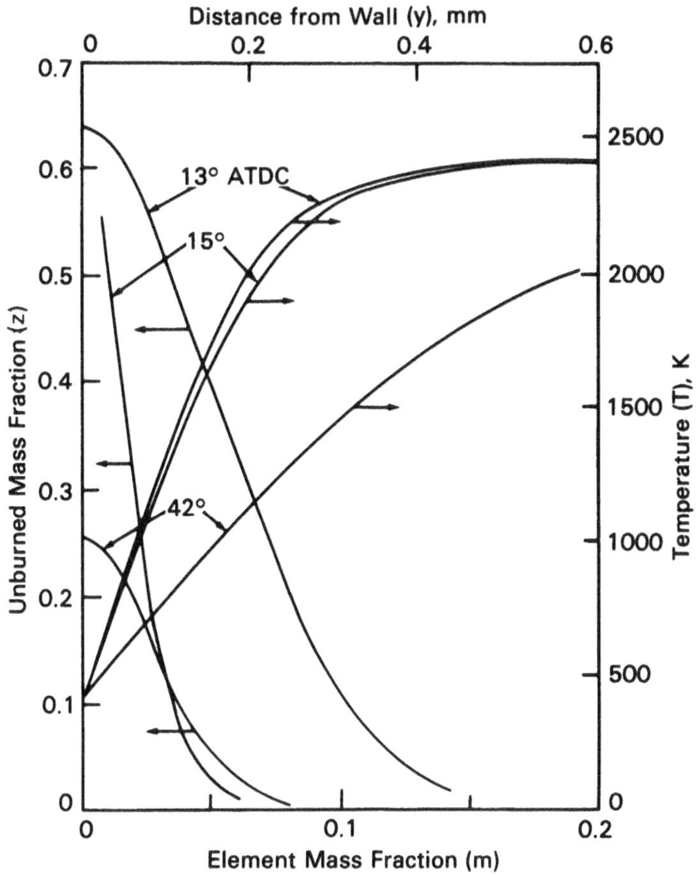

Fig. 5. Calculated boundary-layer profiles of temperature and unburned mass fraction vs. elemental mass fraction, m, for middle element ($\alpha = 0.5$), burning at 13° ATDC. Approximate scale at top of figure for distance, y, normal to the wall.

extent that it geometrically overlaps the lower face of the exhaust valve near TDC. In the case of the CFR engine, the pancake combustion chamber produces relatively large clearance heights, so that under most operating conditions only a small fraction (less than 10%) of the roll-up vortex is calculated to leave the cylinder. This is contrary to the available experimental data: first, in that the high measured concentrations of hydrocarbons near the end of the exhaust stroke are not predicted by the above mentioned model for the CFR engine geometry, and second, the well known effect of ring crevice volume on hydrocarbon emissions is not accounted for.

Based on fluid-mechanics arguments, a number of flow effects have been postulated to increase the proportion of ring-crevice hydrocarbons that leave the cylinder, as shown in Fig. 6. In Fig. 6a the engine cylinder is shown after expansion, just as the exhaust valve opens. At this time the unburned mixture from the ring crevice has expanded from its original state and is distributed along

References pp. 440–441.

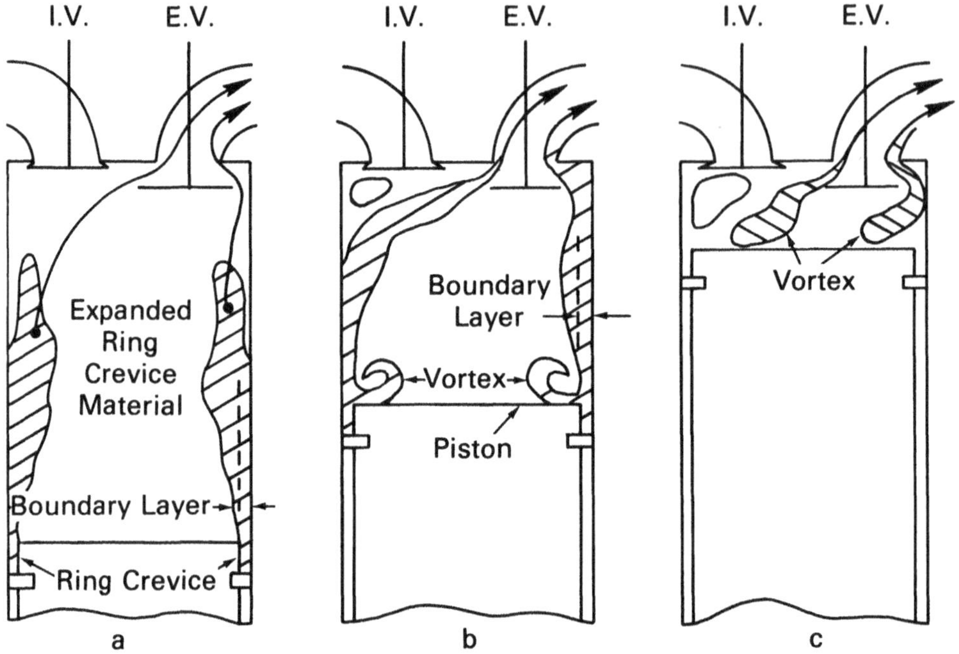

Fig. 6. Sketch of processes for ring-crevice *HC* exhaust; (a) blowdown, (b) exhaust, and (c) end of exhaust.

the cylinder wall, as indicated by the cross-hatched region. When the exhaust valve opens, the gas begins to flow out of the chamber. In the region near the walls, the flow is restrained by the buildup of a viscous boundary layer which begins to grow out from the walls at the moment that the flow is initiated. To the extent that the boundary layer is thinner than the layer of ring-crevice material, and this is certainly true initially, some of the ring-crevice material will move with the bulk gas and may be transported out during the blowdown process.

During the exhaust stroke a roll-up vortex begins to form, as shown in Fig. 6b. Provided the remaining ring-crevice material is outside the viscous boundary layer, it will be transported towards the exhaust valve ahead of the vortex. Thus a significant portion of the ring-crevice hydrocarbons can actually leave the cylinder during blowdown and exhaust.

The ring-crevice material that is situated within the viscous boundary layer will be swept into the vortex and will be pushed up toward the top of the cylinder by the piston. As shown in Fig. 6c, a recirculation flow is likely to build up in the upper corner of the cylinder away from the exhaust valve, so that the vortex will be detached from the wall and will tend to be swept out with the bulk gas. Some evidence of this is seen in the photographs of the exhaust process obtained in a rapid compression machine by Ishikawa and Daily [29]. In the corner nearest the exhaust valve, the flow is deflected around the valve, also tending to pull the vortex out of the chamber. In this way it is possible for a large part of the vortex to leave the cylinder despite the large clearance heights that are encountered in a CFR engine.

Based on this reasoning, the portion of the hydrocarbon model dealing with the ring-crevice exhaust process has been modified, and two additional input parameters have been introduced into the model. The first of these, g_{rc}, defines the fraction of ring-crevice hydrocarbons that exits with the bulk gas during blowdown and exhaust. The second parameter, f_v, defines the fraction of the roll-up vortex, containing the remaining ring-crevice hydrocarbons, that exits the cylinder. For these calculations, g_{rc} and f_v have been assigned the values 0.25 and 0.33, respectively. Thus it is assumed that one-fourth of the ring-crevice material is scavenged from the cylinder during blowdown and exhaust, and is uniformly distributed throughout the exhaust gas (representing the processes in Figs. 6a and 6b). Then, the remaining three-fourths of the material is assumed to be rolled up in the vortex, of which one-third actually exits, thus containing an additional one-fourth of the original ring-crevice material (representing process 6c). Once in the port, the vortex is assumed, as in Ref. 18, to be diluted 5:1 with the bulk gas. Because of the small cylinder-bore clearances involved, it is further assumed that the original material in the ring crevice is maintained at or near the wall temperature rather than at the adiabatic compression temperature.

Also, because of the in-cylinder flow pattern and boundary conditions, it is unlikely that wall-quench material on the piston face will be exhausted from the cylinder. For this reason, the fraction of exiting wall-quench hydrocarbons is assumed to be equal to the proportion of chamber surface area taken up by the cylinder head and walls at the end of combustion. This is on the order of 70% under normal operating conditions. Note that, as in the earlier version of the model, the hydrocarbon-quench and ring-crevice material that is mixed with the bulk gas (i.e., not in the vortex) exits the cylinder in proportion to the amount of total charge that leaves, i.e., $1-f_{res}$.

The above picture of the ring-crevice exhaust process is necessarily qualitative at this time. The details have not been verified directly by experiment or by multidimensional fluid-flow models. However, the effect of implementing the indicated changes is to produce results which are in better agreement with experiment than were obtained previously.

Fig. 7 shows a typical profile of exhaust hydrocarbon concentration versus exiting mass fraction for a CFR engine at 1250 r/min and 380 kPa *imep*. Notice the peak of hydrocarbons due to the vortex, which is in good qualitative agreement with the experimental evidence of previous investigators (Daniel and Wentworth [4] and Tabaczynski et al. [8]). In overall terms the ring-crevice contribution constitutes roughly 30% of the total emissions, in agreement with the relatively large effects that are known to occur with changes in ring-crevice volume (Wentworth [6], and Haskell and Legate [29]).

As will be seen in a later section, the revised model gives greatly improved results for trends of hydrocarbon emissions versus load, which had been a source of difficulty in the previous version of the model. The results for trends of hydrocarbons with variations in other engine variables are unaffected by the changes.

References pp. 440–441.

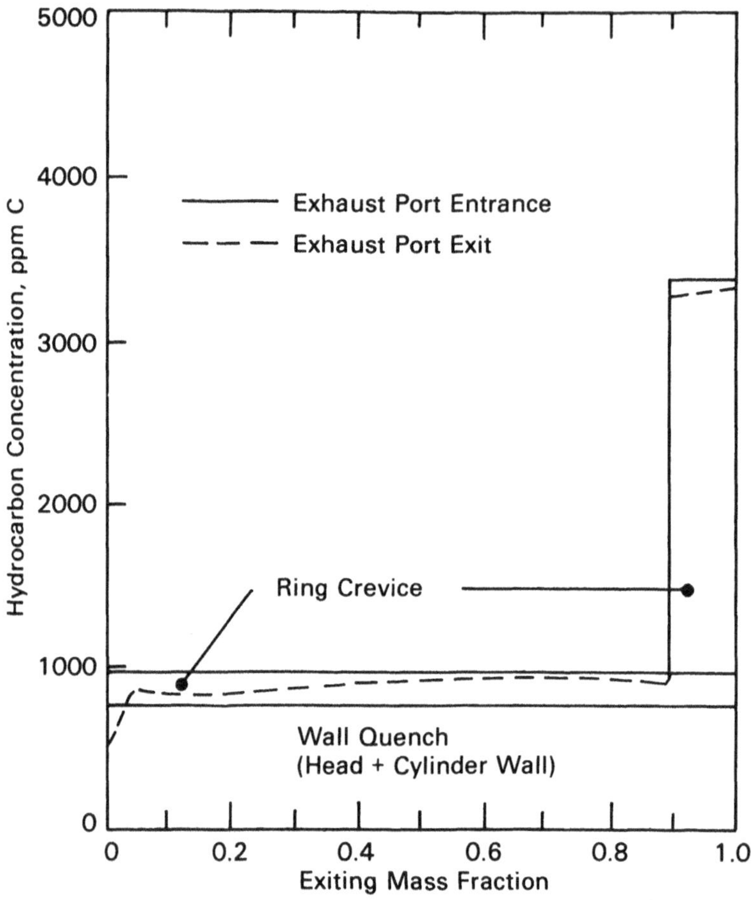

Fig. 7. Predicted hydrocarbon distribution in exhaust gas at entrance of exhaust port (solid lines) and at exit of port (dashed line). 1250 *r/min*, 380 kPa *imep*, MBT spark, $\phi = 0.9$, $C_1 = 0.80$, $C_R = 0.10$.

Exhaust-System Burnup—Oxidation of unburned hydrocarbons in the exhaust port is calculated in the same manner as has been previously reported [18]. Each exiting gas element is assumed to enter the exhaust port with an average hydrocarbon concentration determined by the model of the exit processes. Complete mixing of the unburned hydrocarbons with the bulk gas is assumed to be accomplished by the turbulence generated in passing through the exhaust valve, so that oxidation begins at the exhaust valve and is computed by means of a plug-flow analysis for each gas element. Initial temperature and residence time varies among the elements depending on the crank angle of exit and on the mass flow rate in the port at that time. The chemical kinetics calculations are carried out by means of the global reaction rate given in Eq. 23. Typical results of exhaust-port hydrocarbon concentrations as a function of exiting mass fraction are shown in

Fig. 7 for the engine conditions of that figure, and are indicated by the dashed line.

Further oxidation in the exhaust manifold and tailpipe can be calculated, if desired, by means of a well-stirred reactor model (manifold), and a plug-flow analysis (tailpipe), both of which employ the global rate expression of Eq. 17.

Hydrocarbon Model Parameters—As in the case of the thermodynamic model parameters, f_o and f_{op}, the parameters of the hydrocarbon model are chosen by a technique which matches the predicted hydrocarbon emission data to experimental data at a limited number of engine operating conditions. For reasons given earlier, the boundary-layer parameter, X_T, is fixed at the value 0.0, leaving only C_1 and C_R undetermined. As shown in Ref. 18, at a given stoichiometry, load and speed, the predicted final in-cylinder level of hydrocarbons is sensitive primarily to the value of C_1, while the amount of burnup in the exhaust port is dependent on the effect of exhaust temperature on C_R. With the exhaust temperature fixed by the choice of f_p, C_1 and C_R may be determined by adjusting both parameters so as to match exhaust-port hydrocarbon emissions at two points of greatly differing exhaust temperature, e.g., at MBT spark timing and a retarded condition. Once the parameters are fixed in this way, the model may be used at other engine conditions without any further adjustments.

TEST APPARATUS

A schematic of the test apparatus used to generate the experimental data to be presented is shown in Fig. 8. The engine used in this experiment was a CFR (Cooperative Fuels Research) engine, which is a single-cylinder valve-in-head variable-compression-ratio engine with six access ports to the combustion chamber. In order to minimize hydrocarbon-data scatter due to varying ring-crevice storage effects, the four compression rings were pinned with end gaps staggered 180° apart. Tables 1 and 2 give the engine specifications and the base point parameters used in the model calculations, respectively.

TABLE 1
CFR Engine Specifications*

Bore	82.5 mm
Stroke	114.3 mm
Compression Ratio	6.92
Crank *L/R*	4.44
S/V @ TDC	0.184 mm^{-1}
Topland Ring Crevice Volume	280 mm^3
Exhaust-Port Diameter	32 mm
Exhaust-Port Length	211 mm
Intake Valve	Shrouded
Fuel Type	Isooctane

* *Unless otherwise specified*

References pp. 440–441.

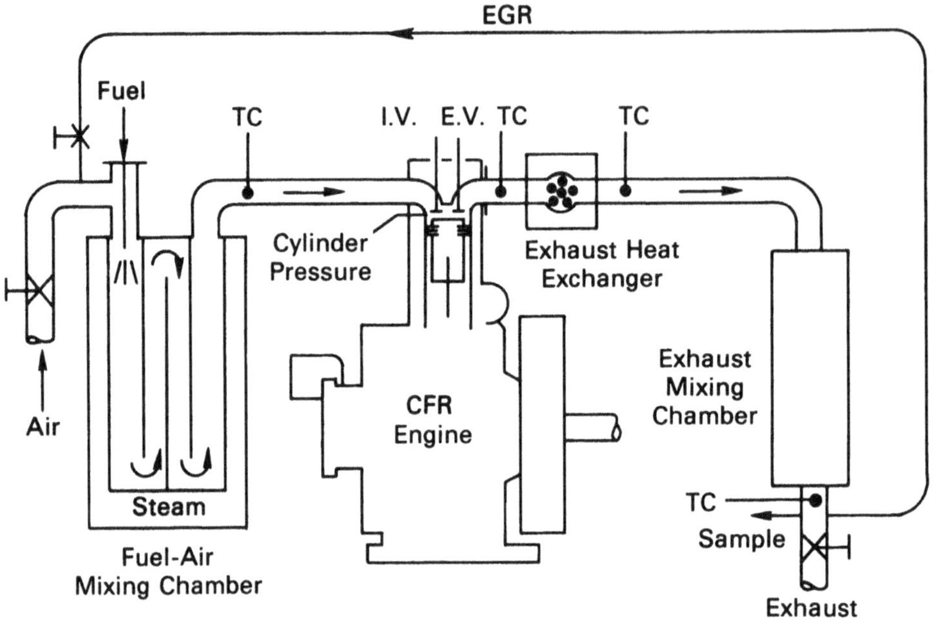

Fig. 8. Schematic of experimental test setup.

Referring to Fig. 8, air and atomized fuel are mixed in a constant temperature (358 K) mixing chamber to ensure that the inlet charge is vaporized and homogeneous. In order to reduce complexities in exhaust-port oxidation modeling, a water-cooled cross-flow heat exchanger was mounted close to the exhaust port to cool the exhaust gases and quench the oxidation of hydrocarbons downstream of that point. The 1.59-mm (tube diameter) hex-pack tube bundle in the heat exchanger was designed to reduce transient exhaust temperatures to 700 K or

TABLE 2

Model Parameters at the Base Point*

Speed	1250 r/min
imep	379 kPa
Equivalence Ratio	0.9
Spark Timing	− 22.5 CA° (MBT)
θ_i	− 2.5 CA°
θ_f	28 CA°
f_o	0.0472
f_{op}	0.0032
C_1	0.48
C_R	0.10
g_{rc}	0.25
f_v	0.33
X_T	0.00

* *Unless otherwise specified*

lower under the maximum heat loads measured in a similar CFR engine by Tabaczynski et al. [8]. After the exhaust heat exchanger, an exhaust mixing chamber was used to reduce the effects of stratification of hydrocarbons and other emitted species in the exhaust before sampling. Care was taken during the experiments to maintain exit mixing-chamber and sampling-system temperatures at a minimum of 422 K to prevent condensation of higher molecular-weight hydrocarbons.

The important measurements carried out during the experiments include: cylinder pressure, PV^γ (viz., combustion rate), major exhaust specie concentrations (CO, CO_2, O_2, NOx, and HC) and exhaust temperature. Cylinder pressure was measured with an AVL 8QP500C water-cooled transducer coated with RTV 108, as described in Ref. 31, to minimize transient thermal errors. Exhaust-specie-concentration measurements of O_2, NOx, CO, and CO_2 were made "dry", with standard instrumentation, and unburned hydrocarbons were measured "wet" with a heated flame-ionization detector. In some cases, a gas chromatograph was used for a more-detailed analysis of exhaust hydrocarbon species. The chemical equivalence ratio calculated from the exhaust products was always maintained to within ±3.5% of the equivalence ratio calculated from independent measurements of air and fuel flow rates. Exhaust temperature was measured with a half-shielded, chromel-alumel thermocouple, corrected for radiation losses.

RESULTS

Presented in Figs. 9 through 15 are comparisons of the hydrocarbon-model calculations and experimental engine data. Plotted are exhaust hydrocarbon concentrations as a function of various critical engine operating or design parameters.

As seen in Fig. 9, the model very accurately predicts the variation of hydrocarbon material as a function of indicated specific fuel consumption. Experiments were performed by retarding spark from MBT (−22°) to +4° ATDC, thus incurring a 30% efficiency penalty. The dashed lines indicate the HC levels at the entrance to the exhaust port, while the solid lines show the levels at the port exit, after partial burnup. It is interesting to note that the initial ring-crevice contribution (dashed lines) falls off with increasing fuel consumption because the density of the gases in the ring crevice at the end of combustion is diminished as the spark is retarded. The initial wall-quench contribution is relatively constant across the *isfc* range.

As explained earlier, spark-retard data are used in tuning the HC kinetic reaction rates in the model to experimental data. The values of C_1 and C_R, determined by matching to these data, are equal to 0.48 and 0.10, respectively. Depending on the magnitude of the ring-crevice contribution, determined by the parameters f_v and g_{rc}, the amount of exhaust-port burnup required to match the experimental data will vary to some degree and affect the matched values of C_1 and C_R. The fact that the ring-crevice model has not been verified directly introduces a source of uncertainty in determining the model parameters and the detailed behavior of the model. In particular, if the contribution of the ring crevice is reduced, the values of C_1 and C_R required to match the data will be

References pp. 440–441.

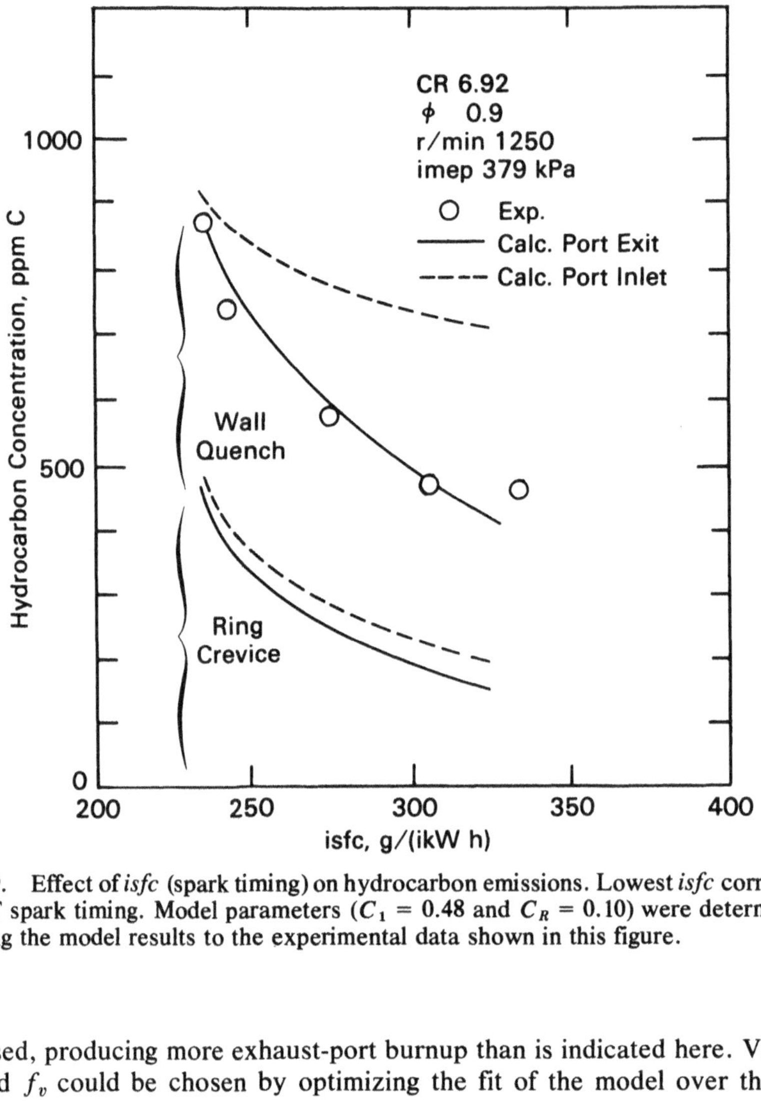

Fig. 9. Effect of *isfc* (spark timing) on hydrocarbon emissions. Lowest *isfc* corresponds to MBT spark timing. Model parameters (C_1 = 0.48 and C_R = 0.10) were determined by matching the model results to the experimental data shown in this figure.

increased, producing more exhaust-port burnup than is indicated here. Values of g_{rc} and f_v could be chosen by optimizing the fit of the model over the entire data base. To date this has not been done. Thus, the actual parameter values used should not be considered as absolute, and detailed interpretation of the model results using these parameters should be considered with some degree of flexibility.

In Fig. 10, hydrocarbon concentration is plotted as a function of engine load, with the engine operating at 1250 r/min, at MBT spark timing and fuel-air equivalence ratio ϕ = 0.9. Also plotted are the corresponding model predictions for total exhaust hydrocarbon emissions broken up into portions due to ring-crevice and wall-quench hydrocarbons, using the hydrocarbon mass-fraction distribution presented in Fig. 7. As can be seen, the model predictions are substantially improved from the earlier versions of the model, as presented by Lavoie and Blumberg [18]. This is due primarily to the increased contribution of ring-crevice hydrocarbons with load because the ring-crevice mass fraction of hydrocarbons scales linearly with the total chamber volume fraction of the ring-crevice, which remains approximately constant as a function of load. The exposed wall-quench distance decreases with increasing pressure, so that the quench zone occupies a smaller volume fraction of the chamber at higher loads, causing the fraction of hydrocarbons due to wall quench to decrease with increasing load, as described in the earlier model. The larger contribution of ring-crevice hydrocarbons reduces the dependence of total exhaust hydrocarbons on load, in agreement with experiment.

Fig. 11 presents the hydrocarbon predictions of the model and the corresponding experimental data obtained for variations in compression ratio at a constant *imep* of 379 kPa and with MBT spark timing. The functional dependence of hydrocarbons, as calculated by the model, is primarily due to the changes in ring-

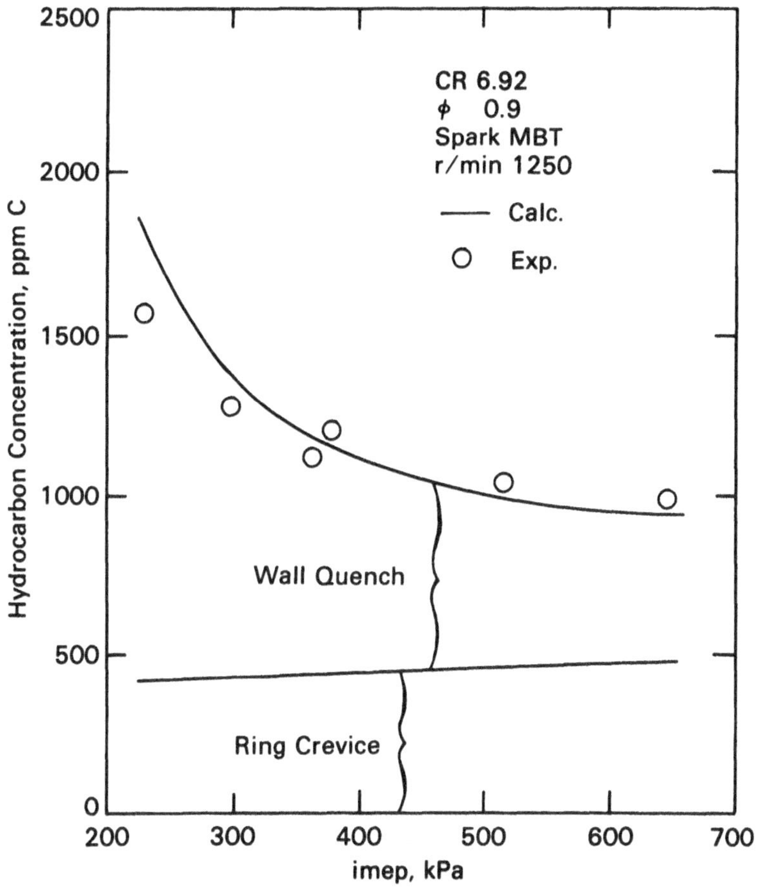

Fig. 10. Predicted and experimental hydrocarbon emissions as a function of engine load, showing breakdown of ring-crevice and wall-quench contributions to total predicted emissions. $C_1 = 0.8$, $C_R = 0.10$.

References pp. 440–441.

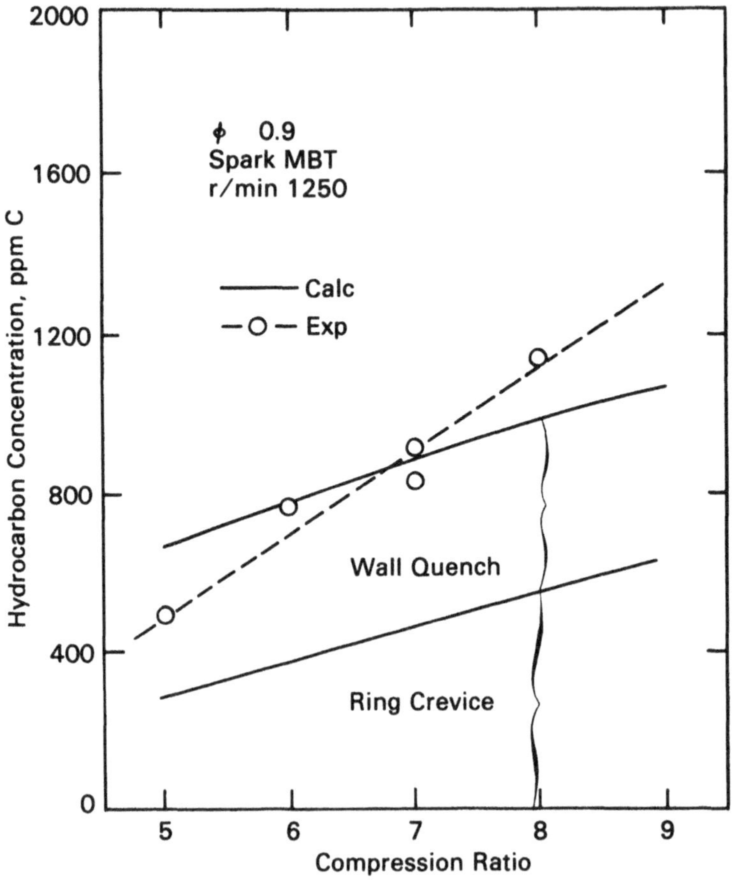

Fig. 11. Effect of compression ratio on predicted and experimental hydrocarbon emissions.

crevice hydrocarbons which arise from geometry considerations. Production of ring-crevice hydrocarbons increases as compression ratio increases, since the volume fraction of the ring-crevice hydrocarbons increases with increasing compression ratio, thereby producing a net overall increase in exhaust hydrocarbons. The model predicts only a small increase (~15%) in wall-quench hydrocarbons across the compression-ratio range, due to a number of competing effects. The increased pressure level in the cylinder which occurs with increasing compression ratio has a relatively small effect on the actual quantity of quench hydrocarbons, per unit area, due to decreasing quench-layer thickness and increasing density. Also, although the surface-to-volume ratio increases dramatically with increasing compression ratio, the actual surface area decreases slightly. Furthermore, under these conditions and for the particular values of parameters employed, the model predicts little exhaust-port burnup (~5 to 10%), so that the exhaust-temperature change due to compression-ratio variation had only a small effect on final *HC* levels. Interpretation of these results is subject to the uncertainty mentioned earlier regarding the ring-crevice contribution and the choice of parameters. For example, an increased amount of exhaust-port burnup at low compression ratios could be obtained by assuming a reduced amount of ring-crevice hydrocarbons and adjusting the parameter C_R upwards. The trade off of these two effects has not been investigated in detail and could be responsible for the difference in slope between the predicted- and experimental-curve trends on the hydrocarbon vs. compression-ratio plot.

Model predictions of hydrocarbon-level variations as a function of coolant temperature are shown in Fig. 12, along with the experimental data. Wall temperatures were calculated by means of Eqs. 3 and 4 and were evaluated to be approximately 47°C higher than the coolant temperatures shown. The amount of exhaust-port burnup predicted by the model under these conditions is small, so that the results plotted are indicative of the in-cylinder processes associated with wall-quench and ring-crevice hydrocarbons. The experimental results show a

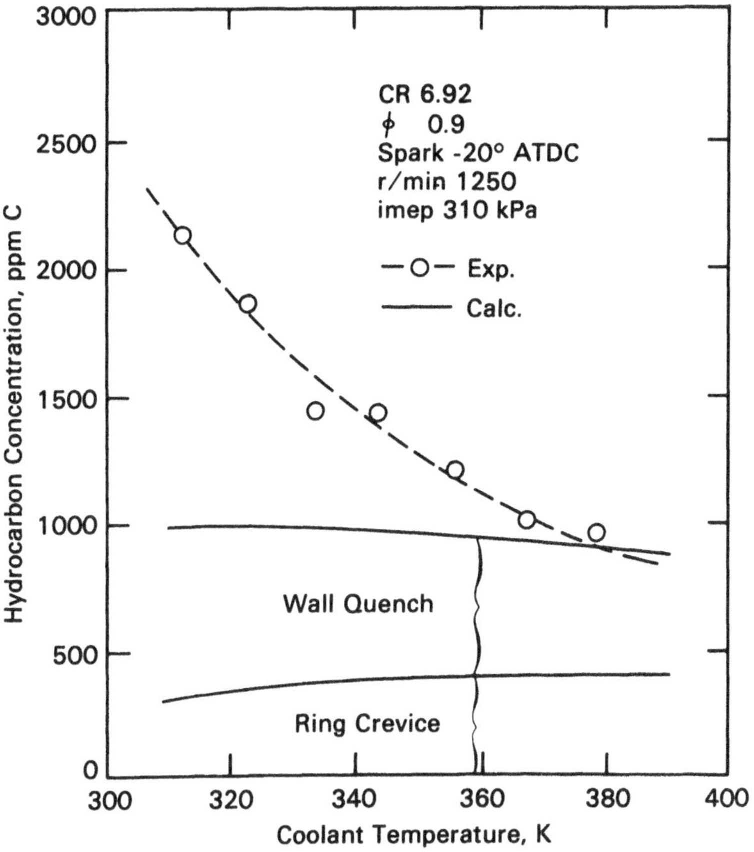

Fig. 12. Predicted and experimental hydrocarbon emissions as a function of coolant temperature for constant spark timing.

References pp. 440–441.

relatively sharp drop in hydrocarbons with increasing coolant temperature, in agreement with the measured results of Wentworth [7]. Although the in-cylinder quenching model is dependent on both the wall temperature and the bulk-gas temperature, the predicted change of hydrocarbons is less than observed experimentally. Similar results have been obtained under rich conditions (ϕ = 1.24) but have been omitted in the interest of brevity. These results suggest that the assumed global oxidation mechanism may be incomplete or incorrect, or possibly that the *HC*-entrainment process during the exhaust stroke is sensitive to the fluid viscosity and the temperature profile near the wall.

As noted in Ref. 18, the hydrocarbon model gives relatively poor predictions for engine-speed variations. A typical comparison of predictions and experimental results is shown in Fig. 13. The model predicts a rise in wall temperature of 35° C across the speed range, due to increased heat loading; however, the predicted effect of this rise on wall-quench hydrocarbons is small, as has been shown in

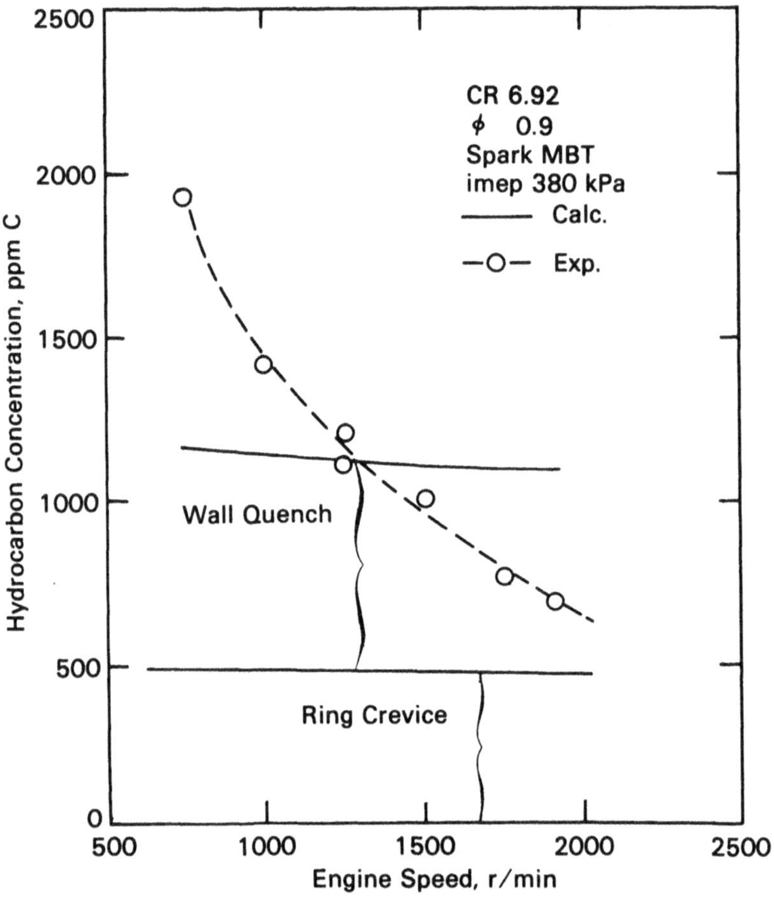

Fig. 13. Comparison of predicted and experimental hydrocarbon emissions for varying engine speed. C_1 = 0.80, C_R = 0.10.

the discussion on wall-temperature effects, with the result that little change is observed in the wall-quench hydrocarbons with changes in speed. Based on the experimental results of Fig. 12, it can be seen that the wall-temperature effect, if correctly accounted for, could explain roughly 50% of the observed speed discrepancy. In agreement with Wentworth [7] and Myers and Alkidas [32], apparently there are additional effects of speed which are not explained by wall-temperature changes alone. Since this model determines hydrocarbon concentration primarily by a laminar thermal-quenching process (wall quench), it has no internal mechanism to determine the effects of turbulence on the initial quench-layer thickness, or the effects of viscous boundary-layer fluid motions on the hydrocarbon scavenging process determining the quantity of hydrocarbons that leave the cylinder. Thus, the submodels incorporated into the calculation procedure cannot be expected to predict the hydrocarbon-speed variations accurately.

Fig. 14 depicts hydrocarbon variation with stoichiometry. At lean conditions, the model quite accurately predicts trends and absolute levels of produced hydrocarbons. However, under rich conditions the model underestimates both wall-quench and ring-crevice hydrocarbons by a factor of three at an equivalence ratio of 1.2. In addition to the measured total hydrocarbon levels, a gas chromatograph was incorporated into the exhaust line to determine amounts of various hydrocarbon constituents in the exiting gaseous products. From this analysis, levels of methane, ethylene and ethane were determined as a function of stoichiometry. Since these species are consumed at a slower rate than the original fuel molecule, it is reasonable to assume that the model predictions apply just to unburned fuel, and one may add the measured fraction of methane, ethylene and ethane to the predictions. By this procedure, a corrected curve (dashed) is obtained which is in better agreement with experimental results on the rich side of stoichiometric and only slightly changed for lean mixtures. It should be noted that other intermediate hydrocarbon species exist in the exhaust products, and in reality the corrected curve could possibly be raised even further, thereby approaching the experimentally determined curve for rich mixtures. These studies suggest that either the oxidation chemistry is sufficiently different on each side of the stoichiometric ratio to warrant substantial differences in the global reaction mechanism or that the simple global reaction mechanism represents only the disappearance of original fuel-type hydrocarbons.

Fig. 15 indicates hydrocarbon levels for various amounts of EGR dilution using Indolene clear as fuel in a CFR engine of different design with an unshrouded valve. These experimental data were presented in Ref. 18 but are included for completeness. As seen, the model predicts with great accuracy the hydrocarbon variation as EGR is varied, due entirely to the change in wall-quench hydrocarbons. It should be noted that when the EGR value was equal to approximately 20% of the gas volume, misfire began to occur in the engine, and from pressure traces it was estimated that the number of misfires equalled approximately 2% of the firing cycles. This misfiring constituted approximately a 2200 *ppm C* increase in the hydrocarbon level, and the calculated result almost exactly predicts the experimental data point. These experimental results differ considerably

References pp. 440–441.

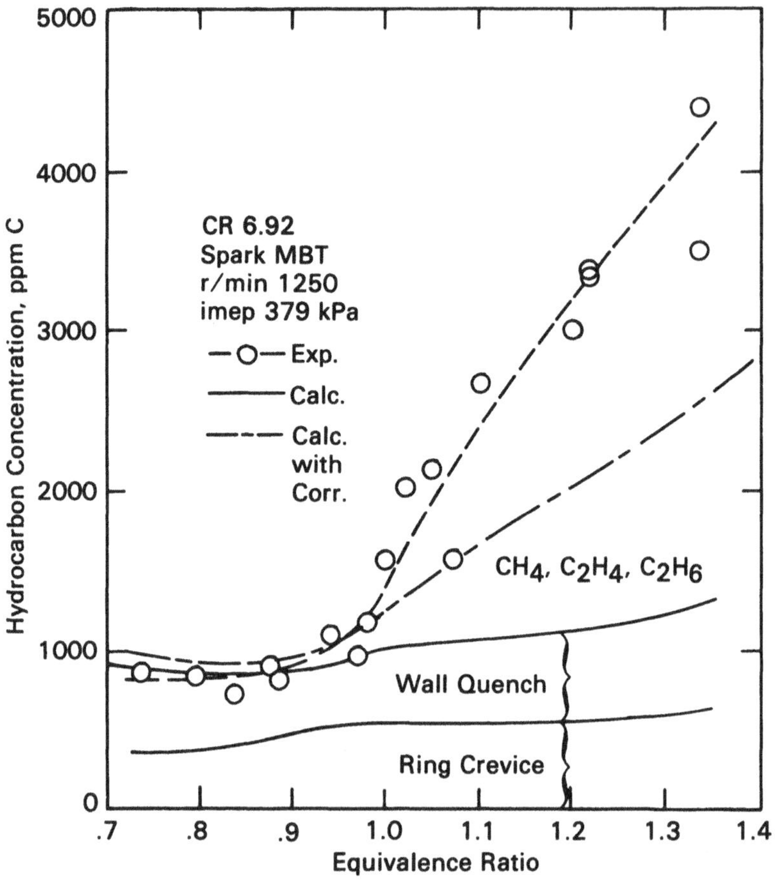

Fig. 14. Effect of equivalence ratio on predicted and experimental hydrocarbon emissions. The "corrected" curve shows predictions obtained by adding contribution of CH_4, C_2H_4 and C_2H_6 to the predicted curves, assuming the model results pertain only to unburned fuel.

from those shown in Figs. 9–14 due to the fact that the engine was operated on Indolene clear fuel and had an unshrouded intake valve. As a result, the model parameters for these conditions are also different, i.e., $C_1 = 2.4$ and $C_R = 0.5$. Because of the previously mentioned uncertainties in the model calibration, these results must be considered as qualitative. Nevertheless, the fact that C_1 for the unshrouded case is higher than the corresponding shrouded-valve case ($C_1 = 0.48$) is in agreement with the combustion-bomb experiments of Gottenberg et al. [10], in which turbulence was found to decrease the effective quench-layer thickness.

DISCUSSION

Based on the results presented here, the current hydrocarbon model is seen to give accurate predictions of exhaust hydrocarbons as a function of a number of

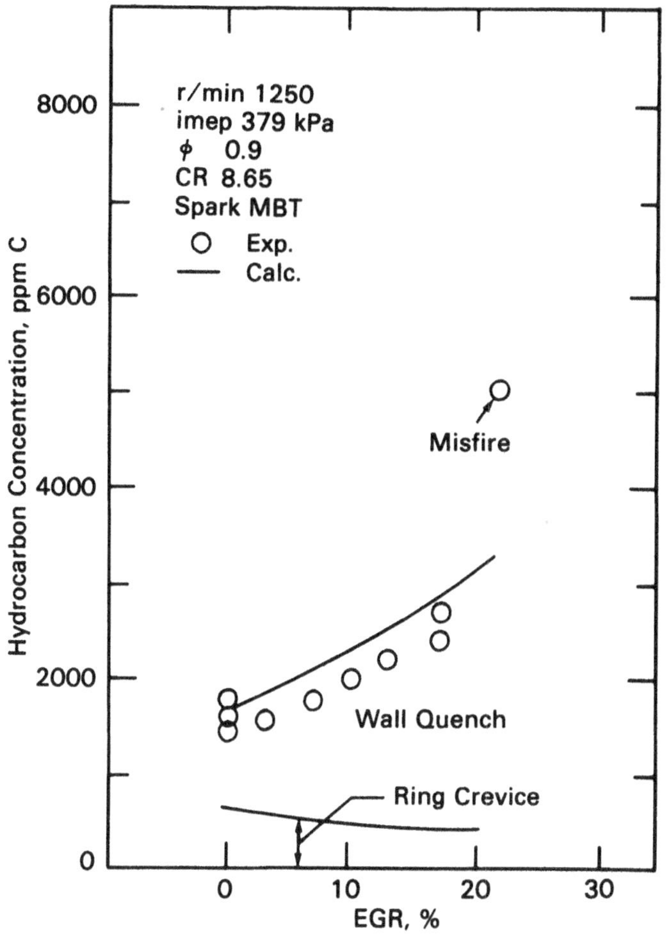

Fig. 15. Plot of hydrocarbon emissions as a function of EGR. Experimental data is from Ref. 18, and was obtained on an unshrouded CFR engine with Indolene clear fuel. Model parameter for these calculations are parameters $C_1 = 2.4$, $C_R = 0.5$.

important engine operating and design variables. These include spark timing, load, equivalence ratio (under stoichiometric and lean conditions), EGR and compression ratio. With regard to other variables, such as wall temperature, engine speed, and equivalence ratio under rich conditions, the model is less predictive. Since the model is essentially phenomenological in nature, the behavior of the model depends to a large degree on the choice and adequacy of the submodels that have been included in the overall formulation. For example, the model cannot be expected to predict the effects of blowby since this submodel has not been included in the formulation. Other critical phenomena, such as oxidation in the exhaust port, are included explicitly, thus ensuring the success of the model in predicting hydrocarbons as a function of spark retard. Beyond this, a great deal can be learned about the assumed mechanisms included in the

References pp. 440–441.

model by analyzing the results for other variables not explicitly tied to the parameterization of the individual submodels and by examining the values of the parameters that are necessary to fit the data.

With regard to the wall-quenching phenomenon, it is significant that the model gives excellent agreement with experiment for variations of engine load and EGR. These trends depend primarily on the variation of the wall-quench contribution through the influence of the laminar quench distance, which is obtained from two-plate quench data. Except in the rich region, where methane and other slow-burning hydrocarbons are significant, the laminar-quench approach gives good results for equivalence-ratio variations.

The absolute level of the single-wall quench thickness determined by fitting the model to experimental data ranges from 0.48 to 0.80 times the two-plate quench thickness for the shrouded valve, isooctane data, and is 2.4 times the two-plate distance for the unshrouded, Indolene clear experiments. These values of the parameter C_1 depend to some extent on the assumptions made regarding the influence of ring-crevice hydrocarbons and also on the accuracy of the exhaust-temperature measurement and matching procedure. It should also be noted that the quench-layer correlation is based on propane data, while the experiments have been performed with other fuels. Although somewhat different in detailed behavior, these fuels have quench distances similar to propane in the region of interest, i.e., near the stoichiometric ratio. Considering these uncertainties, the values of C_1 must be considered only as a rough estimate of the effective quench thickness actually occurring in the engine. Despite these uncertainties and the fact that the values of C_1 are somewhat larger than the expected values of approximately 0.2 (see Ref. 27), the evidence does not contradict the concept of a quench layer that is strongly controlled by laminar flow and diffusion processes.

In view of the laminar quenching approach used in the model and the fact that the wall temperature was used to calculate the quench distance, it is surprising that the predicted influence of wall temperature was so much less than that observed experimentally. This is a clear indication that the assumed global mechanism for post-quench oxidation is inaccurate or incomplete. It is possible, for example, that the reactions controlling the post-quench burnup near the wall have higher activation energies than the value currently used in the overall global reaction rate, which was determined from exhaust-reactor studies (see Ref. 27). Further work is needed in this area to elucidate the nature of the chemistry governing the quenching process, especially under fuel-rich conditions. Not only is this needed to understand the gross behavior of the total quantity of quenched hydrocarbons, but also it will be useful in understanding the detailed breakdown of the hydrocarbons into other constituent species.

The model shows little or no dependence on engine speed, in contrast to the experiments, and more insight is needed in this area. In part, the difficulty is due to the demonstrated insensitivity of the model to wall-temperature variations which are known to occur with increasing engine speed. However, as observed by other investigators [7, 32], there are other phenomena that play an important role in determining the trend of hydrocarbon emissions with speed. One of these is in-cylinder turbulence, which could influence the quench layer through changes in the structure of the mixing and diffusion layers near the wall. This is also suggested by the different values of C_1 obtained in the shrouded- and unshrouded-valve experiments. Little is known about the effects of turbulence on the quenching process, and it is an area where new information would be of great value.

Two additional areas can be suggested for future work in hydrocarbon modeling. Both involve the effects of in-cylinder fluid dynamics on the fate of unburned hydrocarbons. The first area concerns the ring-crevice hydrocarbons. Of obvious importance, the hydrocarbons from this source are responsible in large part for the improved results demonstrated in the case of load variations, as well as the good trends observed for compression-ratio changes. The present model employs a simple scheme which recognizes only the top-land crevice volume and does not take blowby or storage of hydrocarbons between the rings into account. The storage effect has been suggested by Daniel [15] as being the major cause of hydrocarbon variations with engine speed, although Wentworth [6] has demonstrated with sealed-ring experiments that it is not the only cause of the speed behavior. This is clearly an area where more experimental information is needed and where improved submodels could have a major impact on the usefulness of hydrocarbon models.

The second area where extensive work is needed concerns the hydrocarbon scavenging processes that control how much of the in-cylinder hydrocarbons actually leave the cylinder and their distribution in the exhaust gas. This involves the boundary-layer processes governing the exiting of the wall-quench hydrocarbons, as considered by Hicks et al. [33], as well as the vortex exit mechanisms that have been discussed in this paper and elsewhere [8]. Both experimental and modeling work is needed in this area.

CONCLUSIONS

Based on the preceding comparison of experimental data with the predictions of the Ford hydrocarbon model, the following conclusions may be stated regarding the current state of the model and areas where new insight is needed:

(i) The model gives excellent predictions of hydrocarbon emissions for variations in spark retard, EGR, load, compression ratio, and for stoichiometric and lean equivalence ratios.

(ii) Inadequate predictions are obtained for variations in wall temperature and engine speed, and for rich equivalence ratios.

(iii) The results suggest that both turbulent and laminar processes influence the quench layer and the amount of post-quench burnup.

(iv) The contribution of ring-crevice hydrocarbons has been found to be an important factor in determining the behavior of hydrocarbons as a function of load and compression ratio.

(v) Future work is needed, both in modeling and in basic experiments, to elucidate the chemical and fluid-mechanical processes influencing the wall-quenching process, ring-crevice storage and blowby effects, and

References pp. 440–441.

the scavenging processes governing the quantity and temporal distribution of hydrocarbons in the exhaust gas.

ACKNOWLEDGMENT

The authors wish to thank P. N. Blumberg, J. M. Novak and R. J. Tabaczynski for many helpful discussions regarding the development of the model and the experimental tests. In addition, the assistance of Steve Pardee and Dennis Quenneville in conducting the experiments is gratefully acknowledged.

REFERENCES

1. *P. N. Blumberg, G. A. Lavoie and R. J. Tabaczynski, "Phenomenological Models for Reciprocating Internal Combustion Engines," Department of Energy/Division of Power Systems Workshop on Modeling of Combustion in Practical Systems, Los Angeles, California, January 4–6, 1978. Also to be published in Progress in Energy and Combustion Science.*
2. *W. A. Daniel, "Flame Quenching at the Walls of an Internal Combustion Engine," Sixth Symposium (International) on Combustion, The Combustion Institute, Pittsburgh, Pennsylvania, pp. 886–893, 1956.*
3. *W. A. Daniel, "Engine Variable Effects on Exhaust Hydrocarbon Composition (A single-cylinder engine study with propane as the fuel)," SAE Trans., Vol. 76, Paper No. 670124, pp. 774–795, 1967.*
4. *W. A. Daniel and J. T. Wentworth, "Exhaust Gas Hydrocarbons-Genesis and Exodus," SAE Paper 486B, 1962.*
5. *J. T. Wentworth, "Piston and Ring Variables Affect Exhaust Hydrocarbon Emissions," SAE Trans., Vol. 77, Paper No. 680109, pp. 402–416, 1968.*
6. *J. T. Wentworth, "The Piston Crevice Volume Effect on Exhaust Hydrocarbon Emissions," Comb. Sci. and Tech., Vol. 4, pp. 97–100, 1971.*
7. *J. T. Wentworth, "Effect of Combustion Chamber Surface Temperature on Exhaust Hydrocarbon Concentration," SAE Trans., Vol. 80, Paper No. 710587, pp. 2003–2019, 1971.*
8. *R. J. Tabaczynski, J. B. Heywood and J. C. Keck, "Time Resolved Measurements of Hydrocarbon Mass Flow Rate in the Exhaust of a Spark Ignition Engine," SAE Trans., Vol. 81, Paper No. 720112, pp. 379–391, 1972.*
9. *J. N. Shinn and D. R. Olson, "Some Factors Affecting Unburned Hydrocarbons in Engine Combustion Products," SAE Paper No. 146, 1957.*
10. *W. G. Gottenberg, D. R. Olson and H. W. Best, "Flame Quenching During High Pressure, High Turbulence Combustion," Combustion and Flame, Vol. 7, pp. 9–13, 1963.*
11. *J. T. Agnew, "Unburned Hydrocarbons in Closed Vessel Explosions, Theory vs. Experiment; Applications to Spark Ignition Engine Exhaust," SAE Trans., Vol. 76, Paper No. 670125, pp. 796–810, 1967.*
12. *D. F. Hagen and G. W. Holiday, "The Effects of Engine Operating and Design Variables on Exhaust Emissions," SAE Paper 486C, 1962.*
13. *T. A. Huls, P. S. Myers and O. A. Uyehara, "Spark Ignition Engine Operation and Design for Minimum Exhaust Emission," SAE Trans., Vol. 75, Paper No. 660405, pp. 669–719, 1967.*
14. *D. J. Patterson and N. A. Henein, "Emissions from Combustion Engines and Their Control," Ann Arbor Science Publishers, Inc., Ann Arbor, Michigan, 1972.*
15. *W. A. Daniel, "Why Engine Variables Affect Exhaust Hydrocarbon Emission," SAE Trans., Vol. 79, Paper No. 700108, pp. 400–426, 1970.*
16. *H. Hiroyasu and T. Kadota, "Computer Simulation for Combustion and Exhaust Emissions in Spark Ignition Engine," Fifteenth Symposium (International) on Combustion, The Combustion Institute, Pittsburgh, Pennsylvania, pp. 1213–1223, 1974.*
17. *G. A. Lavoie and P. N. Blumberg, "A Fundamental Model for Predicting Emissions and Fuel*

Consumption for the Conventional Spark-Ignition Engine," Paper presented at the Fall Technical Meeting, Eastern States Section, The Combustion Institute, Hartford, Connecticut, November 1977.

18. *G. A. Lavoie and P. N. Blumberg, "A Fundamental Model for Predicting Fuel Consumption, NOx and HC Emissions of the Conventional Spark-Ignited Engine," To be published in Comb. Sci. and Tech.*
19. *P. N. Blumberg and J. Kummer, "Prediction of Nitric Oxide Formation in Spark Ignited Engines—An Analysis of Methods of Control," Comb. Sci. and Tech., Vol. 4, pp. 73–95, 1971.*
20. *P. N. Blumberg, "Nitric Oxide Emissions from Stratified Charge Engines: Prediction and Control," Comb. Sci. and Tech., Vol. 8, pp. 5–24, 1973.*
21. *G. A. Lavoie, and P. N. Blumberg, "Measurements of NO Emissions from a Stratified Charge Engine: Comparison of Theory and Experiment," Comb. Sci. and Tech., Vol. 8, pp. 25–37, 1973.*
22. *R. H. Sherman and P. N. Blumberg, "The Influence of Induction and Exhaust Processes on Emissions and Fuel Consumption in the Spark Ignition Engine," SAE Trans., Vol. 86, Paper No. 770880, pp. 3025–3040, 1977.*
23. *R. J. Tabaczynski, C. R. Ferguson and K. Radhakrishnan, "A Turbulent Entrainment Model for Spark-Ignition Engine Combustion," SAE Trans., Vol. 86, Paper No. 770647, pp. 2414–2433, 1977.*
24. *S. D. Hires, R. J. Tabaczynski and J. M. Novak, "The Prediction of Ignition Delay and Combustion Intervals for a Homogeneous Charge Spark-Ignition Engine," SAE Paper No. 780232, 1978.*
25. *B. S. Samaga and B. S. Murthy, "On the Problem of Predicting Burning Rates in a Spark Ignition Engine," SAE Trans., Vol. 84, Paper No. 750688, pp. 1660–1674, 1975.*
26. *D. L. Baulch, D. D. Drysdale and D. G. Horne, "Evaluated Kinetic Data for High Temperature Reactions," Vol. 2, Butterworth and Co., Ltd., London, 1973.*
27. *G. A. Lavoie, "Correlations of Combustion Data for S. I. Engine Calculations: Laminar Flame Speed, Quench Distance, and Global Reaction Rates," SAE Paper No. 780229, 1978.*
28. *A. A. Adamczyk and G. A. Lavoie, "Laminar Head-On Flame Quenching: A Theoretical Study," SAE Paper No. 780969, 1978.*
29. *N. Ishikawa and J. W. Daily, "Observation of Flow Characteristics in a Model I. C. Engine Cylinder," SAE Paper No. 780230, 1978.*
30. *W. W. Haskell and C. E. Legate, "Exhaust Hydrocarbon Emissions From Gasoline Engines—Surface Phenomena," SAE Paper No. 720255, 1972.*
31. *D. R. Lancaster, R. B. Krieger and J. H. Lienesch, "Measurement and Analysis of Engine Pressure Data," SAE Trans., Vol. 84, Paper No. 750026, pp. 155–172, 1975.*
32. *J. P. Myers and A. C. Alkidas, "Effects of Combustion Chamber Surface Temperature on the Exhaust Emissions of a Single-Cylinder Spark Ignition Engine," SAE Paper No. 780642, 1978.*
33. *R. E. Hicks, R. F. Probstein and J. C. Keck, "A Model of Quench Layer Entrainment During Blowdown and Exhaust of the Cylinder of an Internal Combustion Engine," SAE Trans., Vol. 84, Paper No. 750009, pp. 71–82, 1975.*

DISCUSSION

C. R. Ferguson *(Purdue University)*

I would like to offer two hypotheses. These would relate to your conclusion regarding the dependence of hydrocarbons on engine speed and wall temperature. For example, we know that blowby probably allows no more than one percent of the mass to leak out of the engine. But that mass is all going past the rings where the crevice hydrocarbons are, so it could be a very important effect. In fact, it may be possible to control hydrocarbons by allowing them to leak out of the chamber. Blowby leakage, of course, is reduced at higher engine speeds, which is exactly what I observed on your graph. I believe that the thermal expansion of the crevice volume needs to be taken into account as one changes coolant temperature. In trying to explain the results observed by Wentworth*, I've made some calculations for an engine in which the piston and liner were made of slightly different materials. I observed that for a rather small change in temperature, such as that in your study, the hydrocarbons changed by a factor of two.

G. A. Lavoie

With respect to your last question, I'm aware that that kind of idea has been proposed before. We make no claims, really, as to the validity of our hypothesis. The problem with this kind of modeling, as you can see, is that it is very difficult to get experimental data to test any of these ideas. Until these data can be provided, it really doesn't seem worthwhile to go ahead with that kind of a model. Regarding your question with respect to blowby, doesn't your idea mean that as you increase engine speed, there would be less blowby? Wouldn't that mean that there would be an increased hydrocarbon contribution from the ring crevice?

G. A. Lavoie

C. R. Ferguson

Yes.

G. A. Lavoie

So the way I see it, the dependence would go the wrong way in that hydrocarbons would tend to increase at higher engine speeds. I think the blowby effects

---

* *J. T. Wentworth, "Piston and Ring Variables Affect Exhaust Hydrocarbon Emissions," SAE Trans., Vol. 77, Paper No. 680109, pp. 402–416, 1968.*

are more complicated than you suggest, and Wayne Daniel* alluded to that in one of his early papers having to do with storage between the first two rings. We haven't made any attempts to put that into our model. But that's what I mean. There are a lot of things that need to be done.

J. W. Daily *(University of California)*

How do you measure the exhaust hydrocarbons? Do you do it at the tailpipe? Let me just say that from our own experience, putting physics into a problem for which you are measuring a single one-lump parameter is a bit dangerous.

G. A. Lavoie

I didn't really have time to show the experimental setup, but we make an attempt to get rid of most of the post-engine oxidation. We have a stainless steel tube-bundle heat exchanger as close to the port as we can get it, and we attempt to quench all of the oxidation beyond that point. Within the cylinder, I agree with you, it's really difficult to isolate individual effects. However, I think that within the framework we have done these experiments in, our conclusions are valid.

J. T. Wentworth *(General Motors Research Laboratories)*

I think you will find that blowby is typically between about two and three percent of the engine air flow rate. It doesn't change a whole lot with flow. However, hydrocarbons, other things being equal, will decrease markedly with an increase in blowby rate if you just artificially open up the gap. On another matter, I admire George's courage in trying to separate hydrocarbons into crevice and quench contributions. I would like to suggest an approach that I have used and published**, that is, to redesign the piston to completely eliminate the crevice and then to measure the hydrocarbons to see how much was due to the crevice. The result is that you will have a better idea of the separation of the two sources, and it simplifies the problem quite a bit. To do this successfully I have found that you have to pay attention to three things. First of all, you really have to eliminate the crevice volume, and I've done this by moving the top ring up closer to the top of the piston and then severely chamfering the corner of the piston. You also want to seal the top compression ring into the piston in order to eliminate any breathing past the top rings. I've done this with O-rings and rubber seals. Of course it eliminates blowby at the same time. Finally, you have to maintain excellent oil control because we know that oil ingestion will increase hydrocarbons. When I've done these three things, I've found that hydrocarbons are less

---

* *W. A. Daniel, "Why Engine Variables Affect Exhaust Hydrocarbon Emissions," SAE Progress in Technology, Vol. 14, Vehicle Emissions, Part 3, Paper No. 700108, pp. 341–367, 1971.*

** *J. T. Wentworth, "The Piston Crevice Volume Effect on Exhaust Hydrocarbon Emissions," Comb. Sci. and Tech., Vol. 4, pp. 97–100, 1971.*

than half that associated with a conventional piston. Of course the hydrocarbons that are left are those from the quench zone, that is, from the walls facing into the open chamber. I think that will make your modeling problems much simpler, and after you've managed that, you can go on to more complicated systems which include the ring crevices, oil ingestion, ring-belt breathing, deposits, and such things.

S. C. Sorenson *(University of Illinois)*

Have you made any comparison of your predictions with timed sampling results? There have been a few port sampling experiments published*, and I thought that might be a good technique to use.

With regard to the global kinetics model, we've done some emissions test work, and we've reviewed** work done on hydrocarbon oxidation. How well does your correlation compare with the data we've obtained under somewhat idealized conditions? I have another question. There have been data showing an interaction between nitric oxide formation and hydrocarbon oxidation. At least, we've shown that with ethylene. Your correlation doesn't show any effect of nitric oxide in the global kinetics. Would you care to comment on what you might expect?

G. A. Lavoie

We're aware that some people have observed an effect of NO on the kinetics. I think it's a question of our not wanting to put in anything more complex than we really had to. I think the uncertainties with regard to mixing of hydrocarbons and the temporal distribution far outweigh any advantages that you would gain by putting in a more accurate oxidation mechanism. I think this is another area where our kind of modeling really needs more experimental data in order to proceed. Regarding our global reaction rate correlation*** and how it compares with your experimental data: The rate we used to make the calculations (including the matched factor $C_R = 0.10$) is roughly a factor of 30 below your experiments for ethane. This is possibly because your rate refers only to the disappearance of ethane and not the burnup of total hydrocarbons. It also may reflect the role of finite rate mixing in the exhaust port which would reduce the effective burnup rate. I think your suggestion about port sampling is a good one. In fact we are embarking on a project to do in-cylinder sampling to measure the quench layer itself. We feel this is the best place to start. We hope to be able to get some results on that in the near future.

---

* *W. A. Daniel and J. T. Wentworth, "Exhaust Gas Hydrocarbons—Genesis and Exodus," SAE Paper 486B, 1962.*

W. A. Daniel, "Engine Variable Effects on Exhaust Hydrocarbon Composition (A Single-cylinder engine study with propane as the fuel)," SAE Trans., Vol. 76, Paper No. 670124, pp. 774–795, 1967.

** *R. W. Deller and S. C. Sorenson, "Use of Isothermal Plug Flow Reactors for Exhaust Hydrocarbon Reaction Studies," SAE Trans. Vol. 86, Paper No. 770638, pp. 2350–2362, 1977.*

*** *G. A. Lavoie, "Correlations of Combustion Data for S.I. Engine Calculations: Laminar Flame Speed Quench Distance and Global Reaction Rates," SAE Paper No. 780229, 1978.*

J. I. Gumaer *(Chrysler Corporation)*

You said that you calculated wall temperature by taking into account the heat transfer loss and the coolant temperature. Have you considered the temperature swing through the various stages of the cycle rather than considering the wall temperature as a global temperature? I would think that this would become particularly important in the upper part of the cylinder, where deposits would accumulate and where you have a local thermal inertia.

G. A. Lavoie

No, we haven't taken that into account.

J. P. Myers *(General Motors Research Laboratories)*

I want to offer a comment about the quenching process. First, I think you should be commended for attacking a very complex problem and doing a credible job. We have also looked at the quenching process. Our model does not include the array of submodels or the comprehensive approach that you just presented, but it has offered some insight and I think some agreement in the conclusions that you have reached. I would like to add my support.

I mentioned our model deals specifically with the quench layer. We use an approach whereby the quench thickness is based on an inverse proportionality to the turbulent wall heat flux. Then we determine a quench mass using the support of a heat-release model that includes spherical flame propagation. This approach is somewhat different from yours.

Our results show some similarities and some differences in comparison to your results. I would like to address specifically the engine speed data and coolant temperature data where you showed some disagreement with experimental work. First, we get fair agreement with coolant temperature, and we get fair agreement with engine speed. While not as good as it might be, it has provided us with the following insight. I believe that the slightly better agreement we get with surface temperature is due to the fact that our correlation includes surface temperature directly in the turbulent heat-flux calculation. We attempt to account for turbulence through a turbulent heat-flux calculation at the wall as well as through the heat-release model, in which flame propagation includes the effects of turbulence.

To summarize, I think it's important that one not overlook some of these basic effects which we have the tools available to study. I'd especially like to point out the importance of turbulence, which you mentioned very strongly in your conclusions. I heartily agree with them. I'd like to also suggest that possibly the laminar approach which has been so characteristically taken to study the quenching process may not be the entire explanation. I think there's some turbulence there that we can't overlook.

G. A. Lavoie

I agree on the last point very much. I just want to say that I haven't seen your model in detail, and I hope it is forthcoming soon. It sounds like you've made some progress in evaluating engine speed effects.

MODELING HYDROCARBON DISAPPEARANCE IN RECIPROCATING-ENGINE EXHAUST

R. J. HERRIN

General Motors Research Laboratories, Warren, Michigan

D. J. PATTERSON and R. H. KADLEC

University of Michigan, Ann Arbor, Michigan

ABSTRACT

Dilute combustion processes occurring in the exhaust from homogeneous-charge reciprocating engines influence the levels of some pollutants emitted to the atmosphere. This paper deals exclusively with hydrocarbon emissions and examines the modeling of hydrocarbon disappearance in engine exhaust. The chemical aspects of this modeling are addressed in a study to determine exhaust hydrocarbon kinetics during fuel-lean engine operation. The results of the study are presented and compared with results from other kinetics studies reported in the open literature.

The various kinetics expressions are then coupled with both classical and contemporary flow/mixing models to simulate hydrocarbon-disappearance data from two experimental studies reported in the literature. Comparisons of experiment and prediction allow selection of the most representative kinetics model and the most applicable flow/mixing model.

NOTATION

A	frequency factor
a	HC order
b	O_2 order
c	NO order
CO	carbon monoxide

References pp. 472–473.

CO_2	carbon dioxide
$CSTR$	continuous stirred-tank reactor
d	CO order
E	activation energy
F_{HC_i}	inlet HC molar flow rate
f_p	flow fraction through plug-flow element
HC	hydrocarbon
H_2O	water
NO	oxides of nitrogen
O_2	oxygen
P_x	Partial pressure of species X
P_{HC_i}	inlet HC partial pressure
P_{HC_o}	outlet HC partial pressure
P_{HC_p}	HC partial pressure at outlet of plug-flow element
P_{HC_s}	HC partial pressure at outlet of $CSTR$ element
POF	"Patterns-of-Flow" reactor simulation
R	ideal gas law constant
r_{HC}	HC reaction rate
T	gas temperature
τ	residence time
$\bar{\tau}$	mean residence time
V	volume
$\dot{V}$	volumetric flow rate
V_c	central plenum volume
V_t	total volume
X_f	final extent of conversion

X_p	plug-flow extent of conversion
X_s	*CSTR* extent of conversion
X_T	total extent of conversion
$[X]$	concentration of species X (C_6 equivalent for *HC*)

INTRODUCTION

Combustion in the cylinder of a homogeneous-charge reciprocating engine fails to consume all inducted fuel. The unburned-fuel residual is generally negligible in terms of engine efficiency, but is of major concern as a pollutant. Flame extinction in the cylinder marks the end of significant chemical-energy release, but non-flame, dilute oxidation processes continue as the spent charge travels through the engine exhaust system. Although these gas-phase oxidation reactions are relatively slow, they are nonetheless important because they alter the level of some pollutants during the journey from the engine cylinder to the atmosphere.

In this paper, we describe some of our efforts to model the chemical kinetics and flow dynamics necessary for simulating the disappearance of hydrocarbons (*HC*) in an engine exhaust system. The first portion of the paper deals with an experimental study designed to quantify exhaust *HC*-oxidation kinetics. In the latter portion of the paper, our kinetics models and others from the literature are coupled with available flow/mixing models. Simulated *HC* disappearances are compared with results from several experimental studies in order to evaluate predictive capabilities.

MODELING THE KINETICS OF EXHAUST *HC* OXIDATION

Several investigators have examined the kinetics of *HC* oxidation in automotive engine exhaust. However, these studies considered single *HC* species [1, 2, 3] or gas-stream constituents respresentative of fuel-rich engine operation [4, 5, 6]. None provide kinetics expressions for the complex blend of exhaust *HC* at fuel-lean engine conditions. Accordingly, a new experimental kinetics study was deemed appropriate.

Approach—In the conceptual stage of the kinetics experiment, three of the most crucial decisions involved choosing an analytical model, an experimental reactor, and a source of exhaust gas.

Analytical Model—Mathematical representation of a chemical process can take one of three basic forms:

(i) mechanistic, where all species and all reactions are accounted for,
(ii) quasi-global (partially lumped), where only significant species and controlling reactions are included, and

(iii) global (totally lumped), where a complex process is represented by a single empirical expression.

The exhaust of a gasoline-fueled engine contains scores of stable *HC* species, not to mention numerous short-lived intermediate products [7]. Furthermore, the complex mix of species varies with fuel, engine and operating condition. The successful application of a true mechanistic approach to such a system is all but impossible. The quasi-global approach would be more feasible, but accurate identification and representation of all controlling reactions promises to be a substantial undertaking. The global approach, on the other hand, is more easily implemented but less likely to be accurate if extrapolated beyond the range of available kinetics data.

The choice of kinetics model was tempered by knowledge of the flow model which must be included in order to simulate oxidation in the exhaust system. For multicylinder, reciprocating engines, that flow model is quite complex. The union of even a simplified quasi-global kinetics model and a complex flow model would result in an expensive and unwieldy simulation.

Our choice, then, was a global correlation where

$$r_{HC} = f[T, P_{HC}, P_{O_2}, P_{NO}, P_{CO}] \tag{1}$$

The need for extrapolation was minimized by taking kinetics data over a broad range of all pertinent variables.

Experimental Reactor—Most kinetics studies of this type are conducted with either a continuous stirred-tank reactor (*CSTR*) or a plug-flow reactor. The residence time distribution of a *CSTR* (Fig. 1) is difficult to approach in a real reactor, but experimentation and data reduction for a *CSTR* are quite simple. A plug-flow reactor, on the other hand, appears easily fabricated. In addition, a plug-flow reactor allows the examination of intermediate species, which are obscured by the back-mixing in a *CSTR*. However, external heating is required to obtain isothermal conditions, and true plug flow is difficult to achieve in practice.

Our choice of reactor was the *CSTR*. Availability of a *CSTR* kinetics reactor used in an earlier study [5] was an important factor in this selection. Also, the lack of knowledge of intermediate species in a *CSTR* does not preclude the use of a global kinetics model.

Gas Source—The source of exhaust gas for the kinetics reactor could be a burner [8] or a spark-ignition engine [1-6]. The burner approach has the advantages of a simplified apparatus and fairly good control over flow, temperature and constituents. However, the "synthetic" exhaust from a burner differs from "real" engine exhaust gas in myriad ways, some of which could be important. On the whole, it seemed much more desirable to feed the kinetics reactor with exhaust gas directly from an engine so that all effects and interactions, however subtle, are represented in the kinetics data. Unfortunately, this cannot be done. Feasible analysis of data from the kinetics reactor hinges on the fact that the gas

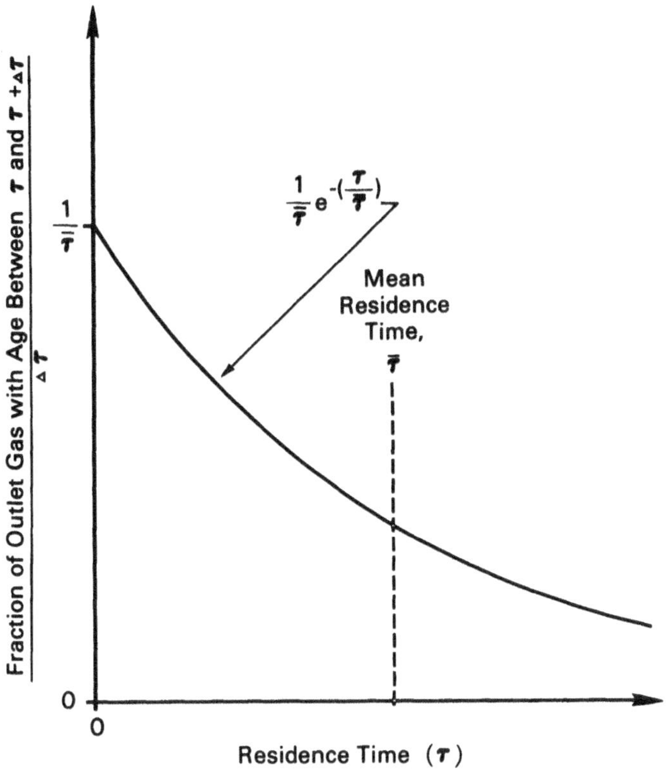

Fig. 1. CSTR residence time distribution.

properties at the reactor inlet are uniform and steady with time. The exhaust from a reciprocating engine is, however, anything but steady (Fig. 2). To make matters worse, the engine cycle time is the same order of magnitude as the mean residence time of the gas in an adequately sized reactor. Thus, engine exhaust can be used only if the various profiles are smoothed by back-mixing in a surge tank before entering the reactor. This presents no particular problem at fuel-rich engine conditions because the combustion products are close to chemical equilibrium. At fuel-lean engine conditions, however, the oxidation reactions proceed while the exhaust is being backmixed, and the reactants may be nearly consumed in the surge tank.

To overcome this problem, an experiment using a blend of real exhaust and synthetic gases was conceived. An engine running slightly richer than stoichiometric supplied exhaust gas to a large, insulated surge tank in which back-mixing eliminated temporal gradients in composition and temperature. Oxidation in the surge tank resulted in outlet gas essentially free of hydrocarbons and oxygen, but containing a selected level of carbon monoxide determined by the engine air-fuel ratio. At the inlet to the reactor, preheated air and a blend of hydrocarbons were injected into the exhaust gas to obtain an overall gas composition closely resembling the average composition of fuel-lean engine exhaust.

References pp. 472–473.

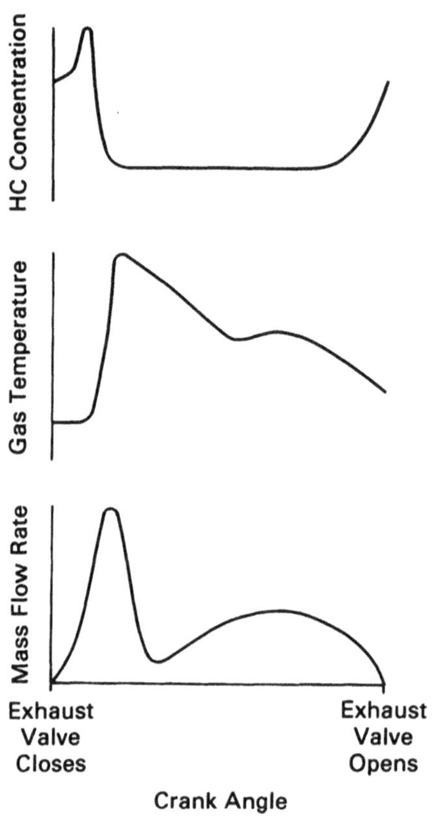

Fig. 2. Typical gas characteristics at outlet of reciprocating engine exhaust port.

The injected hydrocarbon gas was a blend of the nine most abundant species in the quenched exhaust from a lean-mixture multicylinder engine, as measured by gas chromatographic (GC) analysis. Table 1 lists the principal HC species in the engine exhaust gas (average of two operating conditions), as well as a subsequent analysis of the blended hydrocarbons injected for the kinetics experiment.

The above approach provided a uniform reactor inlet gas containing concentrations of HC, CO, O_2, NO, CO_2, H_2O and N_2 appropriate for fuel-lean engine operation. The inlet gas did not, however, contain the numerous trace HC species present in "real" exhaust, although many of those species may have appeared within the reactor from pyrolysis and reaction of the heavier injected species. Furthermore, the inlet gas probably did not contain any radical species which might influence oxidation reactions in a "real" exhaust system close-coupled to the engine. We believe that these shortcomings are secondary and feel that this approach provides an acceptable balance between the chemical and flow-dynamical requirements of a successful kinetics study.

Experimental Details—The following section contains a brief description of the kinetics experiment. Additional information can be found in Appendix A.

TABLE 1

Composition of Quenched Exhaust and Synthetic Injection Gas (Molar Basis)

	Quenched Exhaust		Injection Gas
Species	% of Total *HC*	% of Listed *HC*	% of Total *HC*
Ethylene	27.7	33.3	33.6
Acetylene	12.5	15.0	15.0
Propylene	12.2	14.7	14.6
Isobutene	7.7	9.3	9.2
Toluene	7.3	8.8	8.8
Methane	6.6	7.9	7.9
2,2,4 TMP	3.0	3.6	6.4
3,3 DMH	2.3	2.8	
Benzene	2.2	2.6	2.6
Isopentane	1.6	1.9	1.9
	83.1	100.0	100.0

Apparatus—A single-cylinder CFR engine was used as the source of exhaust gas. A 22.1-L surge/mixing tank was mounted at the engine exhaust port, followed by a 0.98-L reactor (Fig. 3). A bypass, branching off between the surge tank and reactor, allowed mean residence time in the reactor to be lengthened by diverting

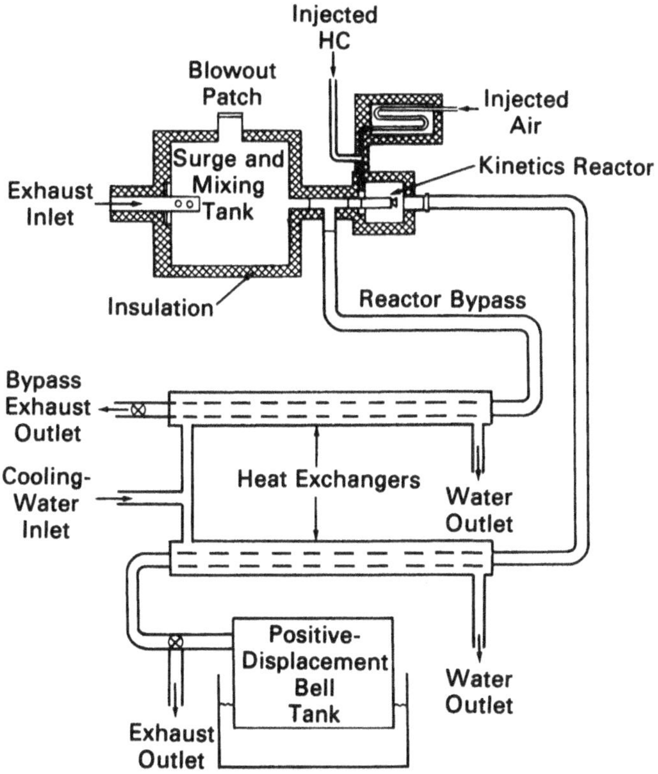

Fig. 3. Experimental apparatus.

part of the flow. The bypass and reactor outlet flows entered two concentric-tube counter-flow heat exchangers, where tap water cooled the exhaust gas. The reactor exhaust then entered a positive-displacement bell tank used to measure flow rate.

A more detailed cross section of the reactor is shown in Fig. 4. A cylindrical plug in the reactor inlet/sparger tube resulted in a small annular flow area and high gas velocities at the point where exhaust gas and injected gases merged. Injected air was preheated by electrical elements to minimize the temperature difference between the two streams and to trim the final reactor temperature. Additional details on the reactor and its development can be found in Refs. 5 and 9.

The injected-*HC* blend was made by placing 20 mL of proportioned liquid constituents (C_5 − C_8) in a nitrogen-purged 34.9-L tank and then pressurizing the tank to about 480 kPa with a hydrocarbon and nitrogen blend containing the proper proportion of C_4 and lighter hydrocarbons. Prior to injection, the blend was completely vaporized by heating the tank to about 95°C on a hot-plate.

Experimental Control—The apparatus assembled for this experiment allowed independent control of almost every major variable. Table 2 itemizes the individual control mechanisms. The *NO* concentration and gas temperature were not independent because they tended to vary inversely as the engine operating condition was changed.

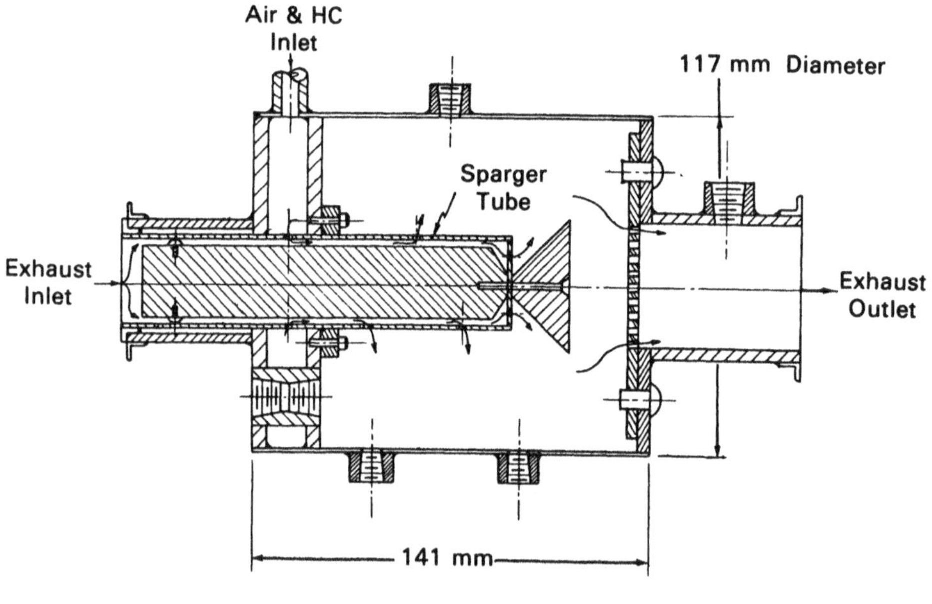

Fig. 4. Kinetics reactor.

TABLE 2

Means of Controlling Variables

Variable	Control Mechanism
HC	Blended Hydrocarbon Injection Rate
CO	Engine Air-Fuel Ratio
O_2	Air Injection Rate
NO	Engine Load, Speed, and Spark Timing
Gas Temperature	Engine Load, Speed, and Spark Timing
Mean Residence Time	Engine Load and Speed, and Exhaust Bypass

TABLE 3

Ranges of Variables

Variable	Minimum	Maximum	Mean
Inlet *HC* (ppm C_6)	158	829	406
Inlet *CO* (ppm)	208	1351	727
Inlet O_2 (%)	0.63	4.41	2.31
Inlet *NO* (ppm)	258	1089	522
Gas Temperature (°C)	582	823	701
Mean Residence Time (ms)	89	216	134
HC Conversion Efficiency (%)	29.0	99.8	84.0

Experimental Domain—A total of 109 individual runs were made, covering a broad spectrum of the possible combinations of the independent variables. Table 3 lists the minimum, maximum and mean values for each variable.

Analysis and Results—Analysis techniques employing both linear and nonlinear regressions were used in seeking the most representative global kinetics model.

Linear Models—A power-law form of Eq. 1 was chosen initially to correlate the kinetics data. This form hypothesizes that *HC*-disappearance rate depends on temperature and the partial pressures of *HC*, O_2, *NO* and *CO*:

$$r_{HC} = A\ e^{-E/RT} P_{HC}^{a}\ P_{O_2}^{b}\ P_{NO}^{c}\ P_{CO}^{d} \tag{2}$$

The linear form of Eq. 2 is obtained by taking logarithms:

$$ln\ r_{HC} = ln\ A - E/RT + a\ ln\ P_{HC} + b\ ln\ P_{O_2} + c\ ln\ P_{NO} + d\ ln\ P_{CO} \tag{3}$$

A multiple linear regression program employing the leaps and bounds technique for isolating "best" subset regressions [10] was used to fit Eq. 3 to the kinetics data. The results from this regression are shown on line 1 of Table 4. These results were disconcerting for a number of reasons. First of all, a recent plug-flow reactor study [11] found that *HC* conversion was essentially independent of *CO* level. Our regressed expression, however, shows the *HC* rate to depend strongly on *CO* partial pressure. Secondly, analysis of the data in Ref. 11 yields an apparent *HC* order of about 0.8. This value is in general agreement with an

References pp. 472–473.

TABLE 4

Results from Multiple Linear Regressions

	Rate Expression	Number of Observations	Multiple Correlation Coefficient Squared	Data Range
1.	$r_{HC} = 1.017\, e^{-25390/RT} P_{HC}^{0.28} P_{O_2}^{0.22} P_{NO}^{0.39} P_{CO}^{0.69}$	109	0.81	All
2a.	$r_{HC} = 4.196 \times 10^{-4} e^{-16030/RT} P_{HC}^{0.15} P_{CO}^{0.35}$	42	0.80	$100 < [HC]_{in} < 300$
2b.	$r_{HC} = 7.169 \times 10^{-4} e^{-16290/RT} P_{HC}^{0.16} P_{O_2}^{0.26} P_{CO}^{0.30}$	53	0.81	$350 < [HC]_{in} < 600$

Variable	Units
r_{HC}	gmol/(mL s)
R	1.987 cal/(gmol K)
T	K
P	kPa

earlier kinetics study [6], which found the *HC* order to be 1.0. Neither of these orders compares favorably with the very low value of 0.28 from our regression. Finally, in spite of extensive efforts to maintain tight experimental control, the multiple correlation coefficient squared (r^2) for the regression was a rather mediocre 0.81.

We interpreted these results as an indication of one or more of the following problems: (i) an experimental error, (ii) an inappropriate regression technique, or (iii) the wrong analytical model. The possibility of an experimental error was explored by re-evaluating the experimental technique and by double-checking the data acquisition and analysis. No significant errors were found.

A potential problem in the area of regression technique stems from significant numerical correlations (non-orthogonality) between "independent" variables in the experimental data set. In situations such as this, ridge regression [12] has sometimes improved on the coefficient estimates provided by least-squares regression. Accordingly, the kinetics data were analyzed using a comprehensive ridge-regression package [13]. In this case, ridge regression offered no significant improvement. The analysis did indicate that a strong correlation between gas temperature and outlet *HC* concentration in the experimental data caused the least-square estimates of activation energy and *HC* order to be inflated. However, these effects were subtle enough that ridge-regression coefficient estimates significantly different from those of the least-squares regression were difficult to justify.

One of the most prominent failings of a global model occurs when there is a major shift in the controlling mechanisms as one or more independent variables are changed. To check that this was not contributing to the apparent failure of the model, numerous linear regressions were performed on subsets of the data formed by restricting the ranges of certain variables. Since controlling mechanisms often change with temperature, regressions were done with several data subsets based on specific temperature ranges. We found no major differences among the various temperature ranges. However, restricting the range of inlet *HC* concentration had a remarkable effect. Lines 2a and 2b in Table 4 are the results of regressions for low and high inlet *HC*, respectively. Note that these two regressions are, with the exception of oxygen dependence, very similar. However, they differ substantially from the results on line 1 in Table 4, where all levels of inlet *HC* were lumped together. Of particular interest is the large decrease in the *CO* order. The fact that limiting the range of inlet *HC* had such a marked effect on *CO* order raised serious questions about the compatibility of the kinetics reactor and the analytical model.

It is well recognized that conditions in fuel-lean engine exhaust are conducive for partial oxidation of *HC* to *CO* [9, 11, 14, 15, 16, 17]. It follows, then, that outlet *CO* is indirectly related to inlet *HC*. The differences between the expressions on lines 1 and 2 of Table 4 suggest an influence of inlet *HC*, which the reactor model could quantify only through the *CO* order. The dependence of the rate expression on inlet concentrations tends to suggest non-*CSTR* behavior and caused us to question the mixing and residence-time-distribution characteristics of our experimental reactor. One means of checking these concerns is to add to

References pp. 472–473.

the regression a parameter which can quantify residence-time distribution. This required a considerable extension of the analytical model, forcing us to turn to nonlinear regression techniques.

Nonlinear Models—A model encompassing the effects of both kinetics and residence-time distribution was formed by considering a parallel-flow network of plug-flow and *CSTR* reactors (Fig. 5). By altering the fraction of total flow to each reactor element, the characteristics of the total network could be varied between ideal *CSTR* and ideal plug flow. Thus, flow fraction was added as a regression variable. The development of this analytical model is included as Appendix B.

The parallel-flow model was fit to the kinetics data using a finite-difference, Levenberg-Marquardt [18] routine designed to solve nonlinear least-squares problems [19]. The results from this regression are shown on line 1 of Table 5, with the results from the linear *CSTR* model (Table 4, line 1) reproduced on line 2 for comparison. The parallel-flow-model results on line 1 look quite favorable, having both a higher r^2 and a *HC* order more in line with results in Ref. 6 and our

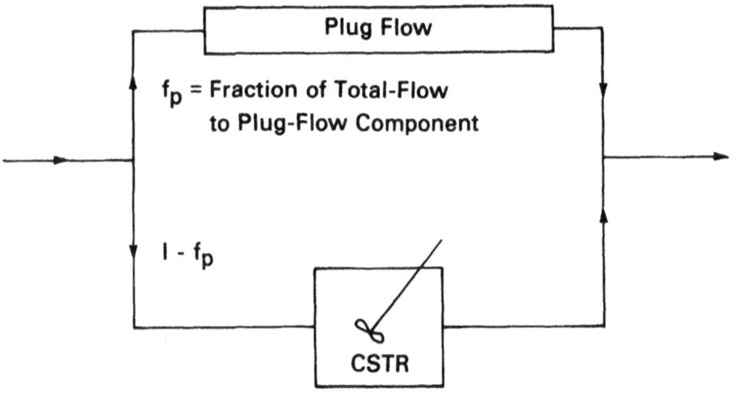

Fig. 5. Parallel-flow plug/CSTR network.

analysis of the data in Ref. 11. The *NO* order is shifted negative by the nonlinear regression, indicating that *NO* suppresses rather than enhances *HC* oxidation.

Perhaps the most unexpected result from the parallel-flow model involved the calculated flow split between the two branches. Although a departure from *CSTR* behavior was anticipated, we were surprised to find that the least-squares routine assigned 95% of the flow to the plug-flow branch. These results suggest that the residence-time distribution of the kinetics reactor is substantially different than originally determined from a coldflow test [5]. Additional experiments to explore this possibility are in order.

Comparisons of Kinetics Models—Three global *HC* kinetics expressions which have been reported in the literature are listed in Table 5, lines 3 through 5. For

TABLE 5

Comparison of Kinetic Expressions

Source	Expression	Number of Observations	Multiple Correlation Coefficient Squared
1. Current Study — Nonlinear Regression	$r_{HC} = 2.689 \times 10^{-2} e^{-20860/RT} P_{HC}^{0.94} P_{O_2}^{0.23} P_{NO}^{-0.35}$	109	0.90
2. Current Study— Linear Regression	$r_{HC} = 1.017\ e^{-25390/RT} P_{HC}^{0.28} P_{O_2}^{0.22} P_{NO}^{0.39} P_{CO}^{0.69}$	109	0.81
3. Ref. 6	$r_{HC} = 1.713 \times 10^{5} T^{-1.25} e^{-31000/RT} P_{HC}^{1.0} P_{O_2}^{0.25}$	181	—
4. Ref. 5	$r_{HC} = 1.213\ e^{-29800/RT} P_{HC}^{0.24} P_{O_2}^{0.54} P_{NO}^{0.42} P_{CO}^{0.51}$	105	0.80
5. Ref. 9	$r_{HC} = 4.630\ e^{-30500/RT} P_{HC}^{0.43} P_{O_2}^{0.33} P_{NO}^{0.34}$	76	0.58

comparative purposes, all expressions in Table 5 have been converted to the same units. Expressions 3 through 5 were obtained from data taken predominantly at fuel-rich engine conditions, so the ranges of some of the variables (most notably *CO* concentration) lie outside the area of interest for this study. Nonetheless, these expressions are potentially applicable to fuel-lean engine operation, and a comparison of all of the expressions in Table 5 reveals some interesting trends. First of all, each of the expressions, with the exception of line 4, exhibits an oxygen order fairly close to one-quarter. Secondly, activation energies tend to demark fuel-rich and fuel-lean engine operation, with the higher energies associated with rich operation. Finally, expressions 1 and 3 exhibit first-order *HC* dependence, which is in general agreement with the data in Ref. 11. The remaining expressions have *HC* orders clustering between one-quarter and one-half.

The traditional Arrhenius plot cannot capture all of the complex tradeoffs between the various expressions in Table 5. To provide some order-of-magnitude comparisons, however, Fig. 6 contains Arrhenius plots of these expressions at two "representative" conditions. Condition I is typical of exhaust gas just leaving an engine, whereas the low *HC* level in Condition II is more representative of gas with enough residence time in the exhaust system for considerable *HC* oxidation to have occurred. The significance of a high *HC* order is evident from a comparison of the plots in Fig. 6. The two expressions with high *HC* orders (lines 1 and 3, Table 5) exhibit a much greater drop in rate as *HC* level is decreased than do the expressions with low *HC* order. At either condition, the five expressions span nearly a ten-to-one range of predicted rates. The merits of these various expressions can be assessed by comparing predictions made with them to published experimental results.

PREDICTING EXHAUST *HC* OXIDATION

Modeling the disappearance of *HC* in an actual engine exhaust system requires a quantification of exhaust flow/mixing characteristics as well as chemical kinetics. In this section, we discuss results obtained by marrying the various kinetics models discussed in the previous section with classical and contemporary flow/mixing models.

Available Flow/Mixing Models—The chemical industry has evolved analytical models to describe the residence-time-distribution effects of several types of reactor vessels [20]. Investigators dealing with chemical reactions in automotive exhaust systems have adapted and expanded these basic models to accommodate unsteady flow and complex geometries [9, 21, 22]. We have used two basic classical models and one comprehensive contempory model in our efforts to simulate the disappearance of *HC* in exhaust systems.

Classical—Two ideal-flow-reactor concepts, i.e., plug-flow and *CSTR*, are useful in evaluating performance limits because they represent extremes in mixing. An ideal plug-flow reactor has no axial mixing, which results in composition

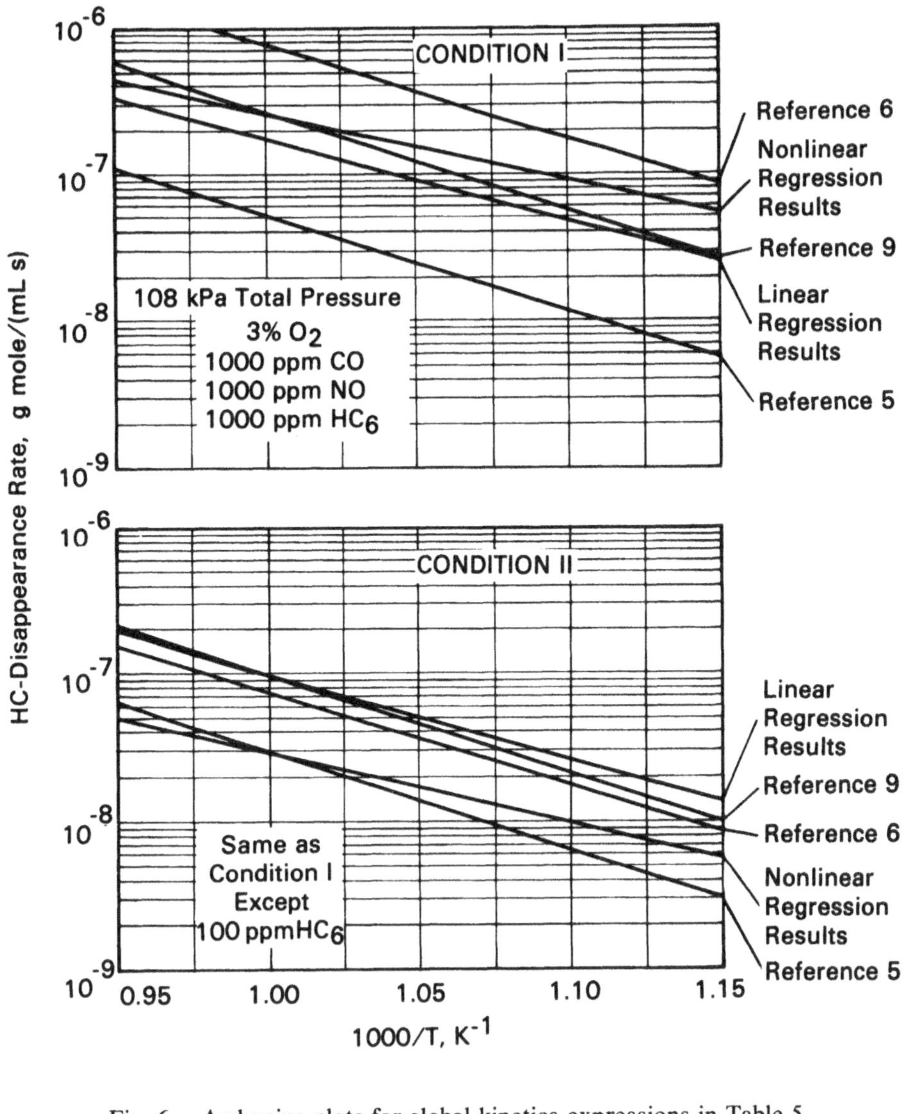

Fig. 6. Arrhenius plots for global kinetics expressions in Table 5.

gradients between the inlet and outlet. An ideal *CSTR*, on the other hand, provides complete back-mixing so that gas in the reactor is uniform and identical in composition to the outlet gas. While neither of these ideals is totally achieved in any engine exhaust system, they provide building blocks for more realistic representations. We have also found them useful for bracketing the performance predictions expected from a given kinetics model.

References pp. 472–473.

Contemporary—Significant extensions of the ideal flow models emerged from a thermal-reactor study reported in Refs. 9 and 22. Non-ideal *CSTR* and plug-flow-reactor models were formulated by segmenting the exhaust gas into discrete cells. With these models, microscopic mixing is controlled by selecting a frequency at which random pairs of cells are coalesced and redispersed [23]. Coalescence and redispersion are defined to occur instantly, and at that instant the properties of the two cells are averaged. Between coalescences, individual cells act as batch reactors. The basic distinctions between plug-flow and *CSTR* residence-time distributions are retained by controlling the choice of cells to be exited from the reactor. For a *CSTR*, exiting cells are chosen at random from all cells within the reactor. In contrast, the cells in a plug-flow-reactor model are segregated into "slugs" which progress sequentially through the reactor, and exiting cells are chosen at random only from the slug at the reactor outlet.

The cellular approach of these non-ideal reactors allows all properties of the inlet stream to be varied with time, so reciprocating engine exhaust characteristics (Fig. 2) can be closely approximated. The random-coalescence scheme allows the mixing implications of a non-uniform reactor feed to be represented realistically. The two non-ideal reactor models were included in a master flow simulation which allows the user to construct a series/parallel network of any combination of *CSTR* and plug-flow reactors. This simulation, called Patterns of Flow (*POF*), provides extensive flexibility in establishing both micro- and macro-scopic mixing, as well as bulk flow characteristics. Detailed discussions concerning the theory and development of this simulation can be found in Refs. 9 and 22. An improved, streamlined version of this simulation has been prepared at the University of Michigan. We have used this latest version in our attempts to predict *HC* disappearance in complex exhaust systems.

Experimental Data Bases—Two recent studies have obtained *HC*-disappearance data at fuel-lean engine conditions in a parametric format convenient for simulation/experiment comparisons. In one of these studies [11], exhaust gas from a four-cylinder engine was ducted into a long, well insulated tube which approximated ideal plug-flow-reactor conditions. Numerous gas-sample taps located axially down the pipe allowed *HC* disappearance to be correlated with residence time.

The second study [17] utilized a V-8 engine and a cylinder-head-mounted compact-reactor system which provided independent control of volume, configuration and heat loss. *HC* disappearance was correlated with mean residence time and temperature for both low and high levels of heat loss. Also, the influence of residence-time distribution on *HC* disappearance was qualitatively assessed by varying reactor internal configuration.

Assessing the Predictive Capabilities—A three-step approach was used in evaluating our ability to predict *HC* disappearance in an engine exhaust system. In the first step, the various kinetics expressions listed in Table 5 were screened by predicting performance characteristics of the least-complex device—the tubular reactor [11]. In the second step, additional kinetics-model screening was accomplished by bracketing the compact-reactor performance [17] between ideal *CSTR* and plug-flow-reactor performance predictions. In the final step, the comprehensive reactor simulation was used in an attempt to predict the subtleties of heat loss and residence-time-distribution effects for the compact reactor.

Tubular Reactors—The ideal plug-flow-reactor model was used to predict the performance of the tubular reactor used in Ref. 11. Since this reactor was fed from the exhaust manifold of a 4-cylinder engine, inlet flow, temperature and species concentrations undoubtedly exhibited high-frequency oscillations which might cause departure from ideal plug-flow performance. The investigators in Ref. 11 checked for this, however, by testing at different engine speeds and exhaust flow rates while maintaining constant levels of time-averaged inlet temperature and species concentrations. They found negligible changes in *HC* disappearance versus residence time for these tests, which indicates that mixing limitations were insignificant. Thus, application of the ideal plug-flow model was deemed appropriate.

Predictions of *HC* concentration versus residence time using the five kinetics expressions in Table 5 are shown in Figs. 7 and 8. The experimental data for three temperature conditions are replotted from Ref. 11. The analytical/experimental comparisons in Fig. 7 show that the kinetics expression from our nonlinear

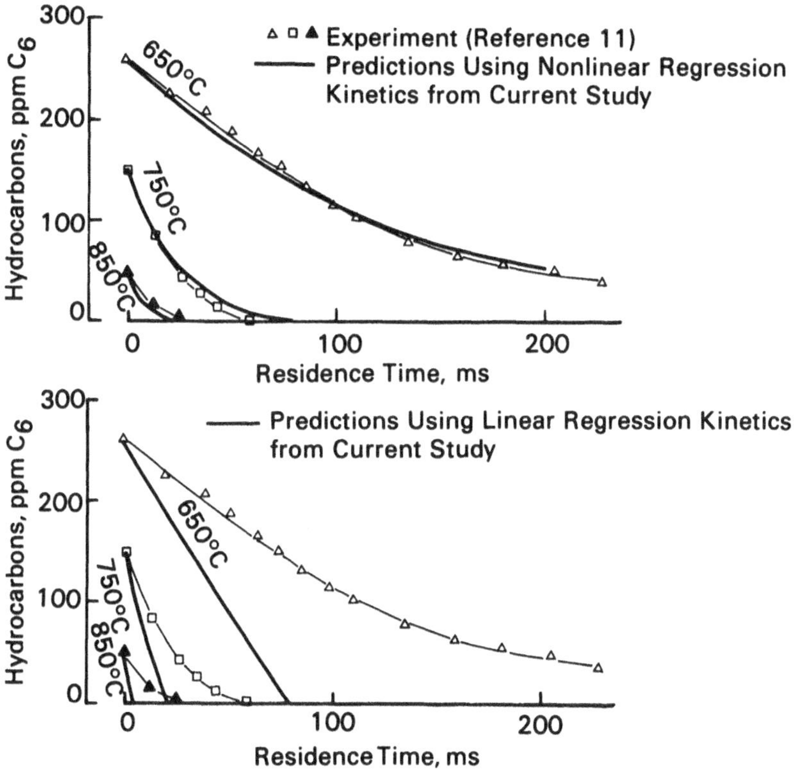

Fig. 7. Experimental and predicted tubular reactor performance.

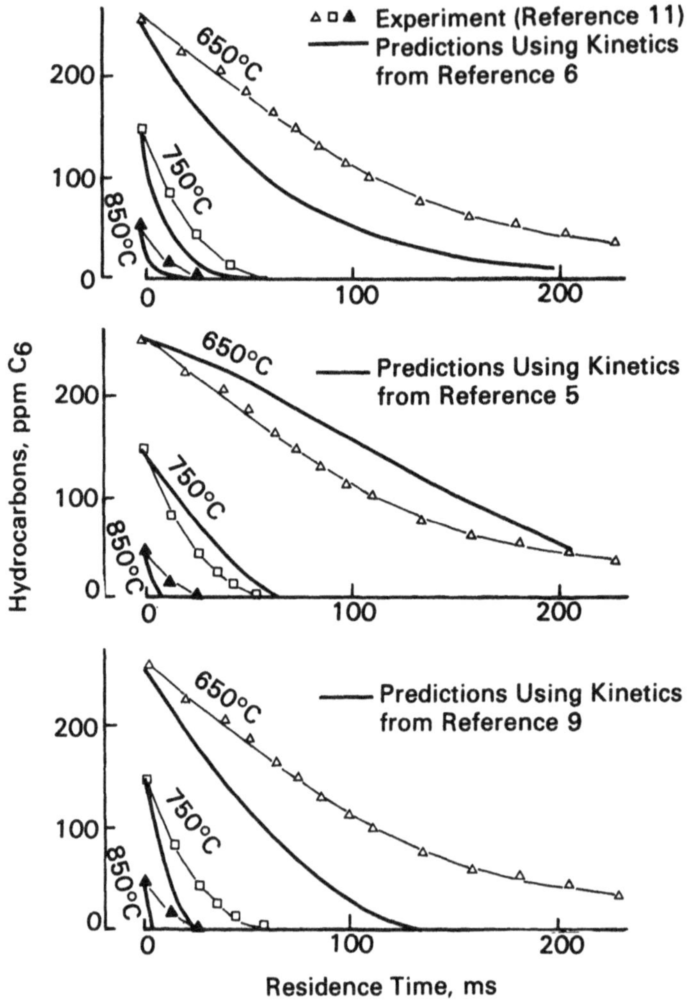

Fig. 8. Experimental and predicted tubular reactor performance.

regression (line 1, Table 5) provides good agreement, whereas the expression from our linear regression (line 2, Table 5) results in very poor agreement. The three kinetics expressions from the literature (Fig. 8) exhibit agreement somewhere between these extremes, with *HC* disappearance being consistently overpredicted by two of the three expressions. Overall, the kinetics expression from our nonlinear regression does the best job of matching these experimental data.

Compact Reactors/Ideal Models—Unlike the tubular reactor in Ref. 11, the cylinder-head-mounted compact reactors used in Ref. 17 cannot realistically be represented by an ideal reactor model. However, if micro-mixing limitations are not severe, the compact-reactor performance should fall between the performance predictions of ideal *CSTR* and plug-flow-reactor models. Thus, we can cross-check the data in Refs. 11 and 17 indirectly by determining if the kinetics expression which best fits the tubular-reactor data will also bracket the compact-reactor data. Results from *CSTR* and plug-flow predictions using our nonlinear regression kinetics (line 1, Table 5) are overlaid on the compact-reactor data for three operating conditions in Fig. 9, and we find that the predictions do bracket the data in almost all cases. A similar comparison is made in Fig. 10 for predictions using the Ref. 9 kinetics (line 5, Table 5), and here we find that the conversion predictions from both ideal models lie well above the experimental data. These comparisons indicate a consistent trend between the data in Refs. 11 and 17 and in the predictive ability of the various kinetics expressions.

Compact Reactors/Contemporary Model—The final step was to use the most promising kinetics expression in the *POF* reactor simulation in order to model some subtleties of compact-reactor performance which were reported in Ref. 17. In Fig. 11 are shown the schematic of a two-pass compact reactor and the flow network chosen for its simulation. The use of three *CSTR* modules to represent the central plenum of the reactor allowed spatial separation of the reactor inlet ports.

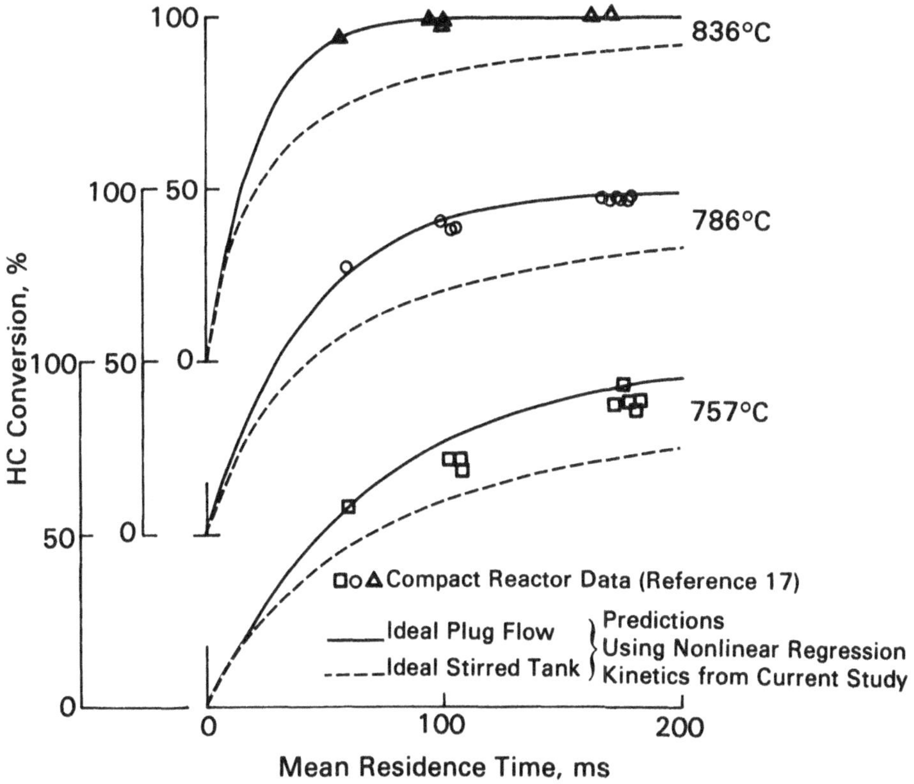

Fig. 9. Ideal reactor predictions of multiple-pass compact-reactor performance using nonlinear regression kinetics from current study.

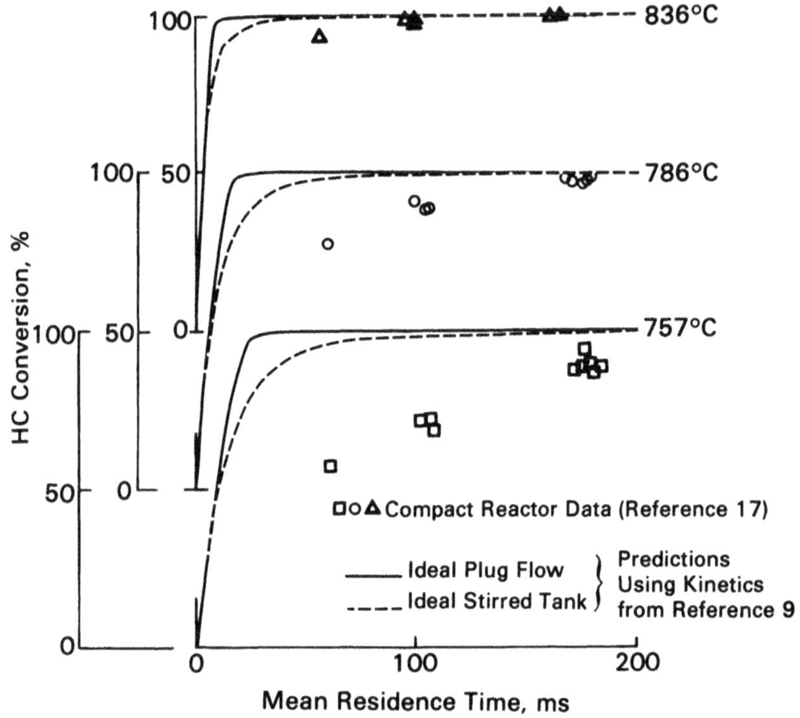

Fig. 10. Ideal reactor predictions of multiple-pass compact-reactor performance using kinetics from Ref. 9.

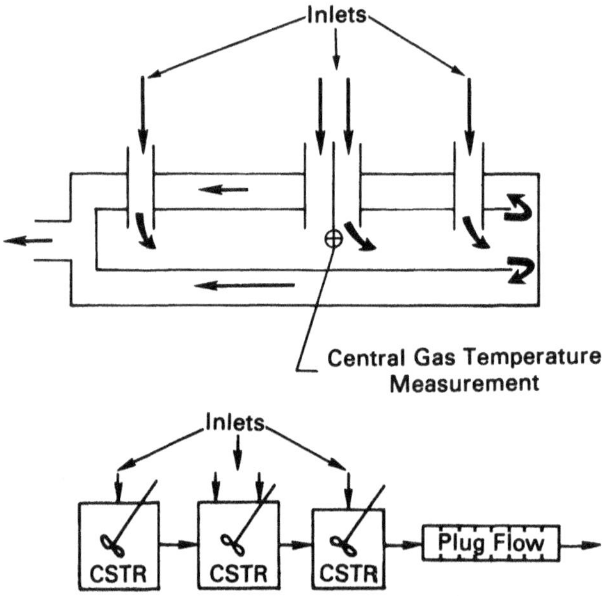

Fig. 11. Two-pass reactor schematic and POF network.

A series of simulations was made at three inlet conditions and two levels of heat loss for the reactor configuration shown in Fig. 11. Experimentally, the two levels of heat loss were obtained by operating with and without external insulation. In the simulation, heat loss was established by adjusting heat-transfer coefficients until gas-temperature predictions at the reactor central plenum and outlet agreed with experimental values. Fig. 12 shows a comparison between experimental data and the results of simulations using our kinetics expression from line 1, Table 5. The agreement between simulation and experiment is less than perfect but acceptable considering the complexity involved.

A more demanding test of the simulation was made by attempting to quantify configurational effects reported in Ref. 17. In that experimental study, it was shown that the *HC* conversion in a two-pass reactor varies inversely with the fraction of reactor volume contained in the central "collector" plenum. Bulk-flow considerations were used to demonstrate that a plug-flow residence-time

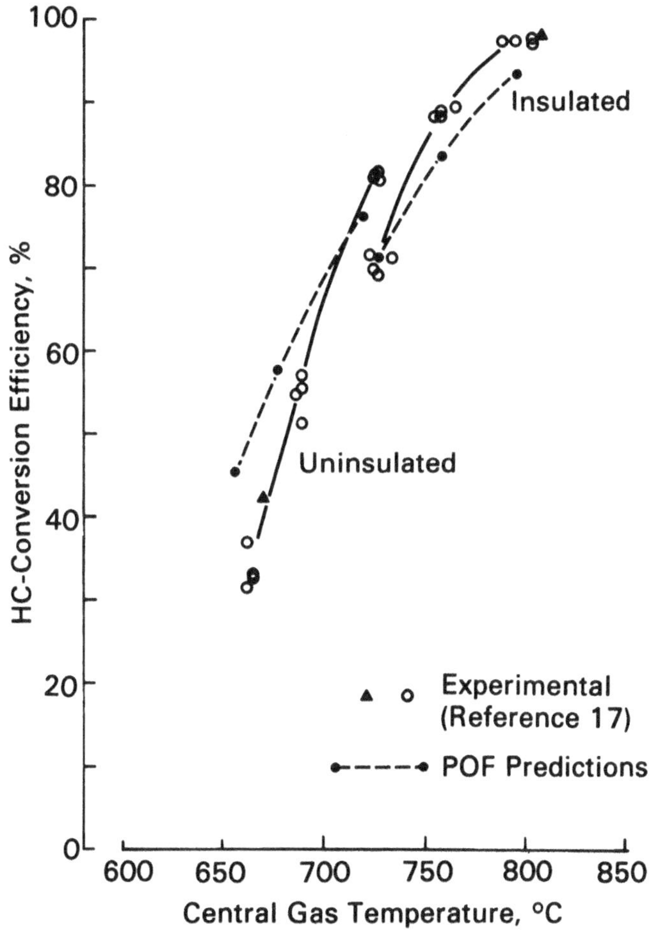

Fig. 12. Compact reactor experimental performance and POF predictions using non-linear regression kinetics from current study.

References pp. 472–473.

distribution is approached as the central plenum volume becomes small. The opposite situation occurs when the central plenum volume approaches the total volume; in the limit, the reactor becomes a one-pass reactor with a broad residence-time distribution and impaired performance. A schematic of a one-pass reactor and the *POF* flow network used to simulate it are included in Fig. 13. The flow network shown previously (Fig. 11) was used to simulate two-pass configurations, and the central plenum/total reactor volume ratio was adjusted by changing the size of the various *CSTR* and plug-flow modules.

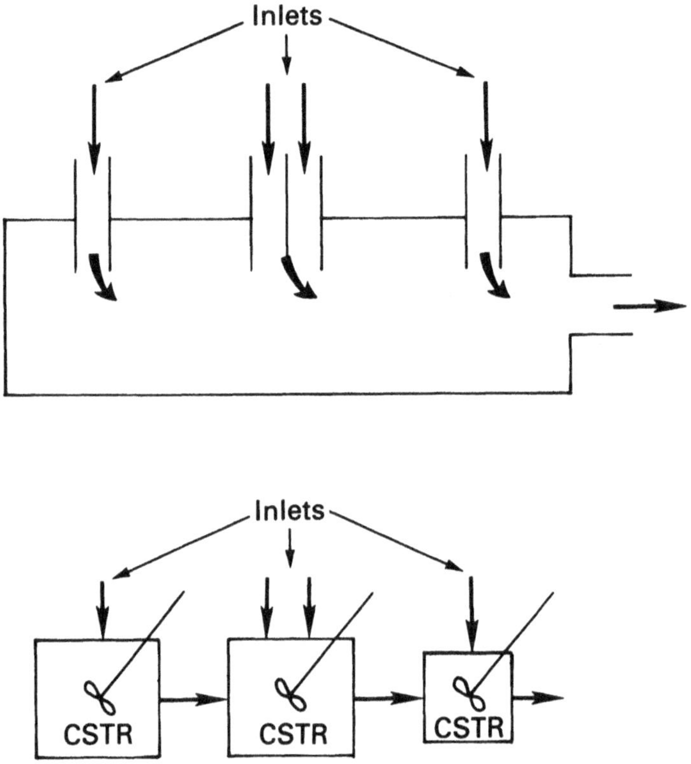

Fig. 13. One-pass reactor schematic and POF network.

The experimental and predicted effects of residence-time distribution for a typical inlet condition and a low level of heat loss are shown in Fig. 14. The simulation predictions of residence-time distribution for the four volume ratios of Fig. 14 are included in Fig. 15. These distributions appear to be in line with intuition, and both the predicted level and the trend of *HC* conversion in Fig. 14 are quite acceptable.

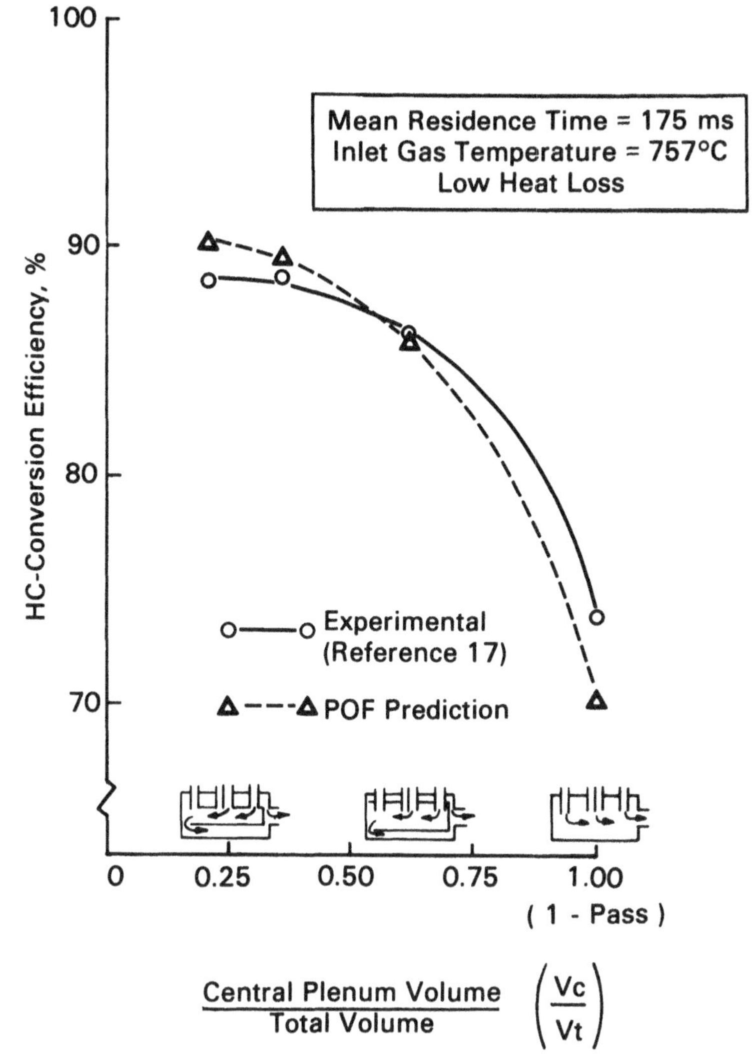

Fig. 14. Experimental and predicted residence-time-distribution effects.

Problem Areas—The previous sections have demonstrated that our kinetics expression from the nonlinear regression provides good predictions of *HC* disappearance when coupled with an appropriate flow/mixing model. However, this success is tempered by the knowledge that our kinetics have definite limitations.

A primary area of concern involves the *HC* rate dependence on *NO*. It was shown in Ref. 11 that *NO* concentration has a complex effect on *HC* disappearance. Some of the data on this effect are reproduced from Ref. 11 in Fig. 16, together with the theoretical effects predicted by our "successful" kinetics expression and the kinetics expression from Ref. 9. Obviously, our kinetics are

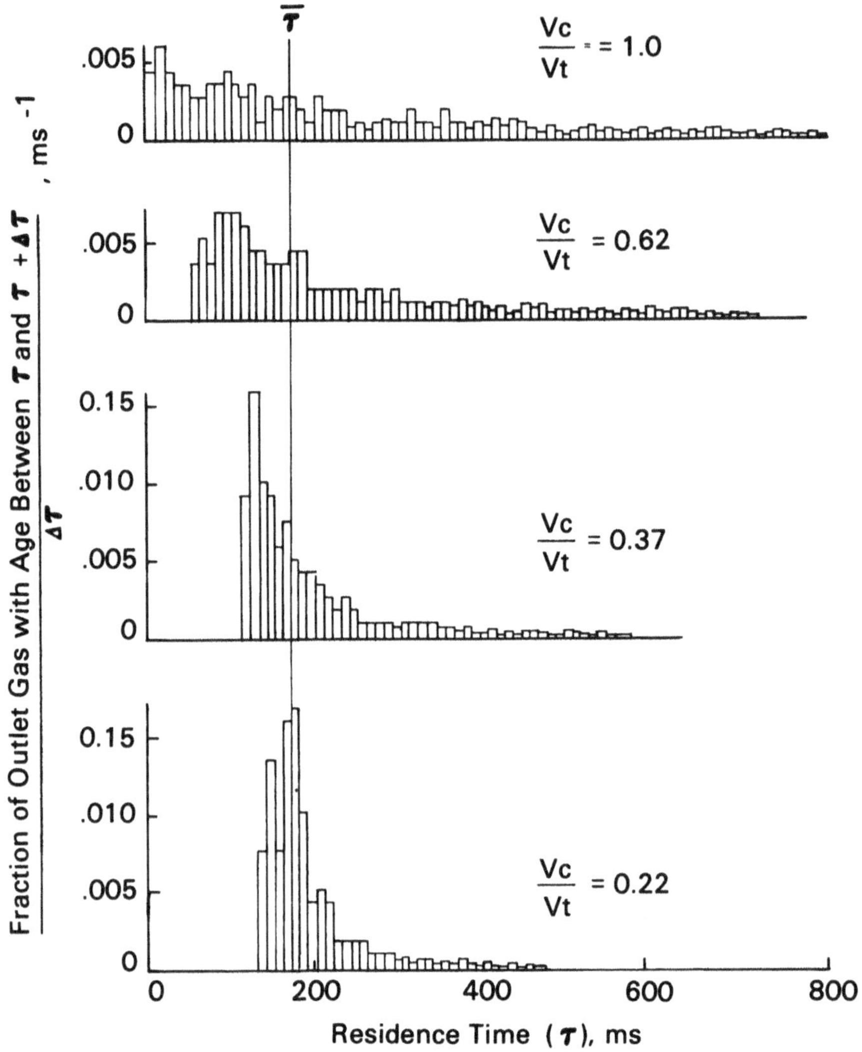

Fig. 15. POF predictions of residence time distributions for the four ratios of central plenum volume to total volume shown in Fig. 14.

not valid at low concentrations of *NO*, whereas the Ref. 9 kinetics are not valid at high *NO* concentrations. It is equally obvious that a simple, power-law kinetics model cannot duplicate the complex behavior of the experimental data shown in Fig. 16. More-complicated global models can be envisioned which could emulate these experimental findings and, indeed, a number of such models were tried in the nonlinear regression. Unfortunately, none of these models, when applied over the entire range of *NO* values, provided agreement which was even as good as that shown in Fig. 16. We were unaware of this *NO* effect when the kinetics experiment was executed and consequently made no special effort to ensure

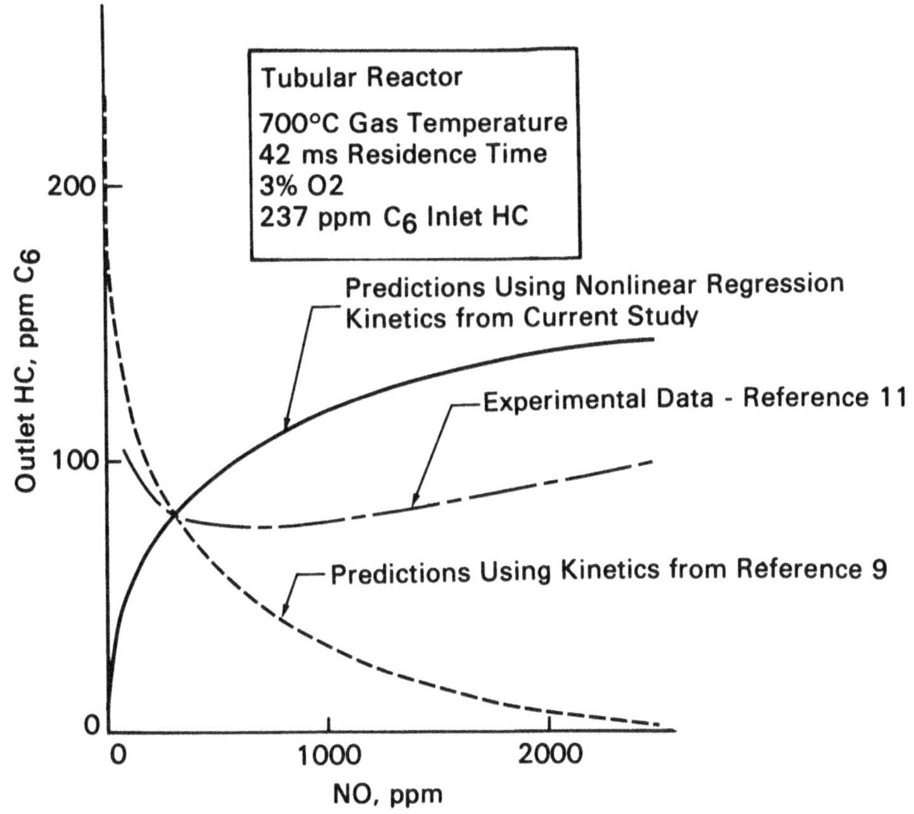

Fig. 16. Experimental and predicted effects of *NO* on *HC* disappearance.

orthogonality between *NO* and the other variables. We feel that the resulting numerical correlation between *NO* and gas temperature in our data set precludes any further refinement of our kinetics model. It would be wise in any future exhaust *HC* kinetics studies to vary *NO* independently.

A second area of concern involves the uncertainty about the residence-time distribution of the kinetics reactor. It is certainly difficult to achieve a *CSTR* residence-time distribution in the real world. However, it is also difficult to believe (as our nonlinear regression suggests) that a reactor designed and cold tested as a *CSTR* behaves like a plug-flow device. Experimental determination of the residence-time distribution is needed to clarify this situation.

SUMMARY

A global expression for the rate of exhaust *HC* disappearance was obtained for conditions representative of fuel-lean spark-ignition engine operation. Although this kinetics expression has known limitations, it has been shown to predict *HC* disappearance better than several other global expressions from the literature.

References pp. 472–473.

With regard to flow and mixing characterization in the exhaust system, it was found that both ideal and comprehensive models can be useful, depending on the complexity of the exhaust-system component being considered and on the resolution desired. For isothermal pipe sections, the ideal plug-flow model was satisfactory. For complex components such as an exhaust manifold, *HC* disappearance was bracketed by ideal *CSTR* and ideal plug-flow-reactor predictions. More complicated (and more realistic) cases involving both significant heat loss and non-ideal residence-time distributions were found to be handled satisfactorily by a comprehensive flow/mixing model reported in the literature.

REFERENCES

1. *H. W. Sigworth, Jr., P. S. Myers and O. A Uyehara, "The Disappearance of Ethylene, Proplyene, n-Butane and 1-Butane in Spark-Ignition Engine Exhaust," SAE Paper No. 700472, 1970.*
2. *S. C. Sorenson, P. S. Myers and O. A. Uyehara, "Ethane Kinetics in Spark-Ignition Engine Exhaust Gases," Thirteenth Symposium (International) on Combustion, The Combustion Institute, Pittsburgh, Pennsylvania, pp. 451-459, 1971.*
3. *R. W. Deller and S. C. Sorenson, "The Use of Isothermal Plug Flow Reactors for Exhaust Hydrocarbon Reaction Studies," SAE Trans., Vol. 86, Paper No. 770638, pp. 2350-2362, 1977.*
4. *R. C. Schwing, "An Analytical Framework for the Study of Exhaust Manifold Reactor Oxidation," Vehicle Emissions—Part III, SAE Progress in Technology Series, Vol. 14, Paper No. 700109, pp. 368-382, 1971.*
5. *H. A. Lord, E. A. Sondreal, R. H. Kadlec and D. J. Patterson, "Reactor Studies for Exhaust Oxidation Rates," SAE Paper No. 730203, 1973.*
6. *J. L. Bascunana, J. Skibinski and E. E. Weaver, "Rates of Exhaust Gas-Air Reactions," SAR Trans., Vol. 86, Paper No. 770639, pp. 2362-2372, 1977.*
7. *M. W. Jackson, "Effects of Some Engine Variables and Control Systems on Composition and Reactivity of Exhaust Hydrocarbons," Vehicle Emissions—Part II, SAE Progress in Technology Series, Vol. 12, pp. 241-267, 1967.*
8. *B. W. Gerhold and A. M. Mellor, "Operation of a Homogeneous Exhaust Manifold Reactor," Fifteenth Symposium (International) on Combustion, The Combustion Institute, Pittsburgh, Pennsylvania, pp. 1225-1232, 1974.*
9. *D. J. Patterson, R. H. Kadlec, B. Carnahan, H. A. Lord, J. J. Martin, W. Mirsky and E. A. Sondreal, "Kinetics of Oxidation and Quenching of Combustibles in Exhaust Systems of Gasoline Engines," Annual Progress Report No. 3 to CRC, 1971-1972.*
10. *G. M. Furnival, "All Possible Regressions With Less Computation," Technometrics, Vol. 13, pp. 403-408, 1971.*
11. *Y. Sakai, Y. Nakagawa, S. Tange and R. Maruyama, "Fundamental Study of Oxidation in a Lean Thermal Reactor," SAE Paper No. 770297, 1977.*
12. *A. E. Hoerl and R. W. Kennard, "Ridge Regression: Biased Estimation for Nonorthogonal Problems," Technometrics, Vol. 12, pp. 56-67, 1970.*
13. *D. I. Gibbons, "A Simulation Study of Some Ridge Estimators," Report GMR-2659, General Motors Research Laboratories, Warren, Michigan, March 1978.*
14. *R. J. Herrin, "Lean Thermal Reactor Performance Characteristics—A Screening Study," SAE Trans., Vol. 85, Paper No. 760319, pp. 1262-1273, 1976.*
15. *J. E. Lahiff and W. C. Albertson, "Volume and Temperature Influences on the Effectiveness of Lean Thermal Reactors," SAE Trans., Vol. 83, Paper No. 741168, pp. 3513-3521, 1974.*
16. *D. J. Pozniak, "A Spark Ignition, Lean-Homogeneous Combustion, Engine Emission Control System for a Small Vehicle," SAE Paper No. 760225, 1976.*
17. *R. J. Herrin, "Emissions Performance of Lean Thermal Reactors—Effects of Volume, Configuration, and Heat Loss," SAE Paper No. 780008, 1978.*
18. *D. W. Marquardt, "An Algorithm for Least-Squares Estimation of Nonlinear Parameters," J. SIAM, Vol. II, No. 2, June 1963.*

19. "The IMSL Library," International Mathematical & Statistical Libraries Inc., Houston, Texas, pp. ZXSSQ-1-ZXSSQ-6, 1975.
20. O. Levenspiel, "Chemical Reaction Engineering," John Wiley & Sons, Inc., New York, p. 97, 1972.
21. M. H. Blenk and R. G. E. Franks, "Math Modeling of an Exhaust Reactor," SAE Trans. Vol. 80, Paper No. 710607, pp. 2120-2147, 1971.
22. R. H. Kadlec, E. A. Sondreal, D. J. Patterson and M. W. Graves, Jr., "Limiting Factors on Steady-State Thermal Reactor Performance," SAE Paper No. 730202, 1973.
23. R. L. Curl, "Dispersed Phase Mixing: I. Theory and Effects in Simple Reactors," AIChE Journal, 9, No. 2, pp. 175-181, March 1963.

APPENDIX A—KINETICS EXPERIMENT DETAILS

Reactor

Configuration	Cylindrical
Volume	0.98 L
Material	Hastelloy – X
Inlet sparger	11 holes, 2.4 mm diameter

Engine

Type	ASTM – CFR
Displacement	0.61 L
No. of cylinders	1
Compression ratio	8:1
Bore	82.6 mm
Stroke	114.3 mm
Fuel	Propane
Speed	945 – 1220 r/min
Ignition timing	25° BTDC – 17° ATDC

Emissions Instrumentation

Hydrocarbons	Beckman FID Model 109A
Oxygen	Beckman Amperometric Model 715
Nitric Oxide	Thermo Electron Corp. CLA
Carbon Monoxide	Beckman NDIR Model 315A
Carbon Dioxide	Beckman NDIR Model 315

Basic Test Matrix (Values are approximate)

$\bar{\tau}$ (ms)	O_2 (%)	T (°C) / CO (ppm) / HC (ppm)	620: 500	620: 1000	620: 1500	705: 500	705: 1000	705: 1500	790: 500	790: 1000	790: 1500
100	1	200	X	X		X	X	X	X	X	
		400	X	X	X			X	X		X
		600		X	X	X	X		X	X	X
	3	200	X	X	X	X		X	X		X
		400		X	X	X	X	X	X	X	X
		600	X	X		X			X	X	
	5	200	X	X	X		X	X			X
		400	X	X	X	X			X	X	
		600	X		X	X	X	X	X		X
170	1	200		X	X	X	X	X	X	X	XX
		400	X	X		X	X	X	X	X	X
		600	X		X						
	3	200	X	X		X	X	X	X	X	X
		400	X	X	X	X	X		X	X	X
		600		X	X			X			
	5	200	X		X	X		X			
		400		X			X				
		600	X	X	X	X	X	X			

APPENDIX B—DEVELOPMENT OF PARALLEL-FLOW PLUG/CSTR MODEL

An analytical reactor model featuring a variable residence-time distribution was developed from the flow network shown in Fig. B1. Assuming constant total pressure, the extent of conversion for the entire network is given by

$$X_T = \frac{P_{HC_i} - P_{HC_o}}{P_{HC_i}} \tag{B1}$$

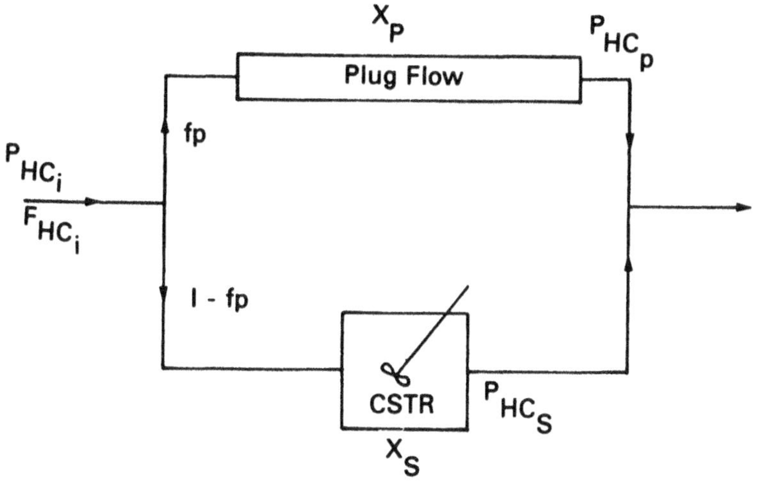

Fig. B1. Flow network for analytical model.

The outlet hydrocarbon partial pressure for the entire network and the two individual modules can be expressed as follows:

$$P_{HC_o} = P_{HC_p} f_p + P_{HC_s}(1 - f_p) \tag{B2}$$

$$P_{HC_p} = P_{HC_i}(1 - X_p) \tag{B3}$$

$$P_{HC_s} = P_{HC_i}(1 - X_s) \tag{B4}$$

Combining Eqs. B1 through B4 yields

$$X_T = 1 - [f_p(1 - X_p) + (1 - f_p)(1 - X_s)] \tag{B5}$$

Assuming equal mean residence times for both flow branches, and a hydrocarbon rate expression of the form

$$r_{HC} = - Ae^{-E/RT} P^a_{HC} P^b_{O_2} P^c_{NO} = - kP^a_{HC} \tag{B6}$$

specific expressions for extent of conversion in both the plug-flow and stirred tank modules can be derived.

The performance of an ideal plug-flow reactor is given in the following expression [20]:

$$\frac{V}{F_{HC_i}} = \int_o^{X_f} \frac{dX}{-r_{HC}} \tag{B7}$$

The extent of conversion is given by

$$X = 1 - \frac{P_{HC_o}}{P_{HC_i}} \tag{B8}$$

which can be differentiated to yield

$$dX = -\frac{dP_{HC}}{P_{HC_i}} \tag{B9}$$

Substituting the ideal-gas law, Eqs. B6 and B9 into Eq. B7 results in

$$\frac{V}{F_{HC_i}} = \frac{\bar{\tau}\,RT}{P_{HC_i}} = \frac{-1}{kP_{HC_i}}\int_{P_{HC_i}}^{P_{HC_o}} \frac{dP_{HC}}{P_{HC}^{a}} \tag{B10}$$

For the special case of first order kinetics with respect to HC (a = 1),

$$\bar{\tau}RTk = -\ln P_{HC}\Big|_{P_{HC_i}}^{P_{HC_o}} \tag{B11}$$

or

$$\frac{P_{HC_o}}{P_{HC_i}} = e^{-\bar{\tau}RTk} \tag{B12}$$

Substituting Eq. B12 into Eq. B8 yields

$$X_p = 1 - e^{-\bar{\tau}RTk} \tag{B13}$$

Similarly, for non-first-order HC kinetics ($a \neq 1$),

$$\bar{\tau}RTk = -\frac{P_{HC}^{1-a}}{1-a}\Big|_{P_{HC}}^{P_{HC_o}} \tag{B14}$$

$$P_{HC_o} = [P_{HC_i}^{1-a} - \bar{\tau}RTk(1-a)]^{\frac{1}{1-a}} \tag{B15}$$

$$X_p = 1 - \frac{[P_{HC_i}^{1-a} - \bar{\tau}RTk(1-a)]^{\frac{1}{1-a}}}{P_{HC_i}} \tag{B16}$$

Ideal stirred-tank performance is given by[20]

$$X_s = \frac{-r_{HC}V}{F_{HC_i}} \tag{B17}$$

Substituting the ideal-gas law and Eq. B6 into Eq. B17 results in

$$X_s = \frac{-r_{HC}VRT}{P_{HC_i}\dot{V}} = -\frac{r_{HC}\bar{\tau}RT}{P_{HC_i}} = \frac{kP_{HC}{}^{a}\bar{\tau}RT}{P_{HC_i}} \tag{B18}$$

By definition,

$$P_{HC_{contents}} = P_{HC_o} = P_{HC_i}(1 - X_s) \tag{B19}$$

Substituting Eq. B19 into Eq. B18,

$$X_s = \frac{k[P_{HC_i}(1 - X_s)]^a \bar{\tau} RT}{P_{HC_i}} \tag{B20}$$

Rearranging,

$$\bar{\tau} RTkP_{HC_i}^{a-1}(1 - X_s)^a - X_s = 0 \tag{B21}$$

which can be solved for X_s via numerical root-finding methods. In turn, values for X_s from Eq. B21 and X_p from Eq. B13 or B16 are substituted into Eq. B5 in order to calculate the total extent of conversion.

DISCUSSION

C. T. Bowman *(Stanford University)*

I think modelers sometimes tend to forget the limitations of these quasi-global kinetics models. Your experiments and kinetics analysis point out one of the significant limitations. That is, in general, these models can be applied with confidence only to the specific configuration and to the specific conditions for which they were developed. If you try to take plug-flow data and then try to use that to interpret stirred-reactor data, or if you take stirred-reactor data to try to interpret plug-flow data, or shock-tube data to try to interpret any of those, you're going to be in real difficulty. I'm not altogether convinced of the validity or the significance of your proposed global model. It may be an indication of its shortcomings by the fact that you're having the difficulty with *NO*. So, I think that's one word of caution that modelers ought to keep in mind—that you cannot go in and arbitrarily take somebody's quasi-global model and expect it to work at all, much less to have any reality for any particular configuration that you're dealing with.

The second thing is a request to people who are going to do hydrocarbon modeling. It seems to me that the way we want to go here is not to look just at the global rates of hydrocarbon oxidation, because as we all know, some hydrocarbons are more reactive in the atmosphere than others. I really think that, sooner or later, the various regulatory agencies are going to recognize this too. And, they're going to start saying, "Okay, methane is good, but some of the higher hydrocarbons are bad." I think that we want to go in the general direction of identifying what these hydrocarbons are, determining when they come out in the exhaust, and developing kinetics models to allow us to get the various splits among the various hydrocarbon species.

R. J. Herrin

I agree with you, especially on that last item, up to a point, and I may disagree with Newhall a little in this area. I feel that global models have definite limitations, and I think a more rational way to go in the future would be toward partially lumped, or quasi-global, types of arrangements in which you can identify some

R. J. Herrin

of the more salient species. I think Dr. Newhall's implication was that we should try to model the engine exhaust mechanistically, and I really don't think that's feasible at this point in time. I think you're going to run into more reactions than you can possibly handle. But I do agree with you—we need to go into more depth.

S. C. Sorenson *(University of Illinois)*

We've done some work with plug-flow reactions, and as mentioned before, there is an *NO* effect. Our results* tend to agree with those of Sakai** that there is both an inhibition and a promotion in the *NO* effect. We used a similar technique, so maybe we should agree. So, as you pointed out, it looks as if the global kinetics just can't do anything in that situation.

---

* *R. W. Deller and S. C. Sorenson, "The Use of Isothermal Plug Flow Reactors for Exhaust Hydrocarbon Reaction Studies," SAE Trans., Vol. 86, Paper No. 770638, pp. 2350-2362, 1977.*

** *Y. Sakai, Y. Nakagawa, S. Tange and R. Maruyama, "Fundamental Study of Oxidation in a Lean Thermal Reactor, SAE Paper No. 770297, 1977.*

R. J. Herrin

Actually, you could model it globally. In fact, I tried to do that. You can come up with an expression other than a power-law expression (which is still global) which would represent that trend. However, we were unable to garner that out of this data, primarily because we had a strong correlation between gas temperature and *NO*, resulting from the way the data were taken. We did not realize the *NO* influence at that time, so we made no special efforts to make those two variables independent in our data set.

S. C. Sorenson

I also want to point out some of the problems with global models. For example, we showed a *NOx* dependency which changed. That is, above 5% oxygen, the *NOx* dependence disappeared, whereas before that it was comparatively linear. So if you have occasions in the final state of reaction to convert a paraffin to a large amount of olefin, for example, your global kinetics obviously fall apart. Harry Sigworth* pointed that out.

R. J. Herrin

That's right.

D. T. Pratt *(University of Michigan)*

That whole analysis parallels, strikingly, the application of chemical reactor theory to gas turbine combustion, going back over the past 20 years. If one would refer to the 1950's literature, or even to Chapter 8 of Levenspiel's** book on reactor theory, one can find elucidated what is called the Bragg criterion for maximizing combustion efficiency and analyzing periods of conversion. The criteria are that the stirred-tank section should be only big enough to cause good stabilization of the flame, and that one should maximize the plug-flow section to complete burnout. It looks like you've rediscovered that principle.

R. J. Herrin

Perhaps we have, although I think that there may be a difference. I believe you're talking about a process that's strongly mixing controlled, whereas in this case, I do not believe that we're *strongly* mixing controlled. Because we're talking about overall lean stoichiometries where there's always an excess of oxygen in any element of gas, I think our mixing is not that significant.

---

* *H. W. Sigworth, Jr., P. S. Myers and O. A. Uyehara, "The Disappearance of Ethylene, Propylene, n-Butane and 1-Butane in Spark-Ignition Engine Exhaust," SAE Paper No. 700472, 1970.*

** *O. Levenspiel, "Chemical Reaction Engineering," John Wiley and Sons, Inc., New York, 1972.*

D. T. Pratt

No, the analogy that I was referring to is a Continuous Stirred Tank Reactor (CSTR)/plug-flow reactor-theory analysis, and the model stands independent of the application.

R. J. Herrin

A CSTR can be more efficient than a plug flow if the inlet is not well mixed, but if you have a well mixed inlet, then the reverse is true. Correct?

D. T. Pratt

I think not, but I'll have to think about it.

P. N. Blumberg *(Ford Motor Company)*

I have a question about the Patterns of Flow analysis that you did, particularly with respect to the study of the variation of the volume of the central core to the annular area. It seems that in reality if you reduce the volume of the central core, you go to a situation where each pulse out of a particular cylinder into this exhaust reactor will be more of a plug flow since, in the limit of zero volume, each pulse has to exit instantaneously as a pulse from the central core. Yet, you kept your model constant in your analysis of the variation with this volume ratio, and you got good agreement.

R. J. Herrin

I didn't keep the volumes constant. I changed the volumes of the three individual modules to represent that.

P. N. Blumberg

Yes, but you had the same basic model, and you changed the volumes in it. I guess I'm uncomfortable with the physical significance of what you've shown, and I have the sneaking suspicion that there are enough adjustable parameters in this model that you could fit almost any curve.

R. J. Herrin

I respect your concerns there. I should enumerate that one of the things that you can tweak infinitely in this model is mixing intensity. In order to try to make a fair comparison, I selected a mixing intensity (in fact, I selected a total number of coalescences per unit time) of the cells in the cellular model and then did not change that throughout all of the studies. So, I did not tweak knobs to make the agreement better, but I admit there are a lot of knobs to tweak in this model. I disagree with you about the applicability of still using those three CSTR's though, independent of changing the volume, because you do want to separate your inlets spatially. That's really the only way you can do it. Even as the volumes get very small, you're still going to have a difference in the volume traversed by a port at one end compared to a port at the other end, and you have to represent that somehow. In the limit of going to a very small central core, each CSTR would have zero volume, so there would be no reaction. That is, the reactor degenerates to plug flow in the limit of zero volume in the central core.

A. M. Mellor *(Purdue University)*

Going back to your response to Dave Pratt, you said that in an overall lean system, flame stabilization would not be mixing controlled, and that's not right. What is controlling is the rate of mixing of heat as opposed to species.

R. J. Herrin

Yes, that is important in this instance.

A. M. Mellor

I think you've said that you wanted to maximize the plug-flow reactor for stability. Did you mean the well-stirred reactor?

R. J. Herrin

No, I said maximize the plug flow for combustion efficiency.

A. M. Mellor

Oh! Efficiency.

R. J. Herrin

You have raised a good point, however. If you look at the cyclic temperature and hydrocarbon profiles from a reciprocating engine, you have a high hydrocarbon peak which comes out in low-temperature gas. In order to optimize performance, ideally you'd like to mix that low-temperature high-hydrocarbon gas with the clean bulk gas in order to raise its temperature.

J. B. Heywood *(Massachusetts Institute of Technology)*

In the test of your model against the engine exhaust data, you talked about a test of the kinetics and volume division in your reactions. But in addition, you're really testing your heat-transfer and mixing-parameter models, and furthermore, you have a built-in assumption about the unsteady nature of instantaneous flow.

I wonder if you could say a little bit about the degree to which you examined the sensitivity to developing what you haven't mentioned in heat transfer/mixing, and what you assumed about the inputs? That is, it seems that you should say a little bit about the physics and about the heat-transfer models.

R. J. Herrin

The physics in the heat-transfer model is extremely simple. Basically, you specify an area and film-coefficient product for a module, and you specify a temperature to which it will lose or gain heat—a sink temperature, if you will. That's basically all there is to it. For the compact-reactor comparisons, I had extensive experimental temperature data, and I adjusted $\overline{hA}$ coefficients until the temperature throughout the reactor agreed with the experimentally measured temperature profiles.

To incorporate the unsteady character of the exhaust pulses into the system, I examined some data which has been published on instantaneous exhaust flow rate and hydrocarbon concentration, as well as some engine-simulation predictions of instantaneous temperatures. The reactor simulation has a provision to approximate those profiles with straight-line segments. It then creates, by quadrature, a cellular "train" entering the reactor which will approximate the cyclic characteristics. So, I tried to get data close to the conditions that I was running experimentally and then approximated those data using the line-segment approach.

J. W. Daily *(University of California)*

What would happen if you took the plug-flow scheme, ran it for a variety of residence times, and then you multiplied by the residence time distribution function? Would you be able to show the same drop in conversion efficiency?

R. J. Herrin

Yes, I would. That's exactly where the performance degradation is coming from. Because the conversion efficiency versus absolute residence time curve is concave downward, any time you increase a distribution on that curve, you will degrade the performance.

J. W. Daily

Then why do you need well stirred reactors in the model?

R. J. Herrin

I'm not sure I do need them. However, I felt that of the two possibilities (that is, plug flow or stirred tank) I had available, the stirred reactor was more representative of what was physically happening in the reactor interior.

SESSION IV

APPLICATIONS OF ENGINE COMBUSTION MODELS

Session Chairman
R. B. KRIEGER

General Motors Research Laboratories
Warren, Michigan

APPLICATIONS OF ENGINE COMBUSTION MODELS—AN INTRODUCTORY OVERVIEW

R. B. KRIEGER

General Motors Research Laboratories, Warren, Michigan

ABSTRACT

This paper presents an overview of the applications of engine combustion models (ECM's). In the context of this overview, applications are divided into two broad areas: *Applications for Design* and *Applications for Understanding*.

Applications for design purposes are divided into three areas: *engine structural design, engine flow-system design* and *engine combustion-system design*. ECM's can be used to aid engine structural design in four areas: mechanical stress analysis, thermal stress analysis, bearing analysis, and noise analysis. In mechanical stress analysis, ECM's are used to compute peak cylinder pressures which characterize the load applied to the engine structure. The peak and average heat fluxes necessary to carry out a thermal stress analysis of the engine can also be computed with an ECM. The form of the engine cylinder pressure-time function is required for an engine bearing analysis as well as being one of the key elements in the analysis of noise generation from the engine structure. ECM's can be used to generate the pressure-time information necessary for these analyses.

While engine flow-system design depends on engine combustion performance, flow-system analysis is somewhat less dependent on engine combustion than is engine structure design because engine intake and exhaust flow are not strongly dependent on details of the engine combustion process. Thus ECM's used for this purpose generally need not be extremely precise. One exception to this generalization is turbocharger-system design which depends on details of the engine combustion through the sensitivity of exhaust energy to combustion process rate and timing. Here, engine flow and combustion models have been used for configuring and sizing exhaust manifolds for optimum turbocharger use.

Currently ECM's can also be used in some cases to answer specific questions about design details of the combustion system. However, these models are not generally sufficiently developed to be used for predictive combustion system design. Thus, ECM's are currently more useful for answering combustion-system design questions through their ability to generate and enhance understanding.

The first category considered in the *applications for understanding* area is diagnosis. Here it is shown that ECM's are used extensively to analyze experimental data to transform it into a more fundamental form. For example, engine cylinder-pressure traces are routinely studied through a heat-release analysis to determine the mass burning rate, a more fundamental quantity than the cylinder-pressure trace. In addition, ECM's are used to compute quantities not easily measurable, such as the instantaneous nitric oxide concentration.

Prediction constitutes a second type of application for understanding. It can be used not only to identify more optimum values of operating variables and engine geometries, but also to elucidate unexpected features of the engine flow and combustion processes, such as backflow of residual cylinder gases into the engine intake system during part load operation.

The third area of applications for understanding which is discussed is synthesis. This application overlaps design applications somewhat, but the intent here is the generation of broad guidelines. A good example of this application is the question facing developers of direct-injection stratified-charge (DISC) engines: What are the optimum fuel and momentum distributions within the engine cylinder gases at the beginning of combustion? ECM's, when sufficiently developed, can be used to address questions like this to guide the synthesis of new configurations.

INTRODUCTION

While the literature on engine combustion modeling dates back to the 1920's, interest in combustion modeling has been growing rapidly since about 1974. Most of the literature in this area describes the development of engine combustion models. Papers describing applications of models, which are any uses of models which go beyond model development, constitute only 30 to 40% of those which have been published. This overview considers only a sampling from this 30 to 40%. Those papers within the sample were selected to illustrate a classification system for modeling applications.

This overview begins with a description of the classification system. Next, each of the two major categories of the classification system are discussed by describing published examples of each type. However, since this is an overview rather than a literature survey, where an obvious application appears to have been neglected in the literature, this application is suggested and discussed. It is hoped that this grouping and sampling will help clarify the applications area and, more importantly, highlight it, since applications are the obvious end toward which all modeling work is directed.

CLASSIFICATION SYSTEM FOR APPLICATIONS OF ENGINE COMBUSTION MODELS

Applications of engine combustion models have been divided into two broad areas: *Applications for Design* and *Applications for Understanding*. While these two subdivisions inevitably involve some overlap, it is helpful to think of engine combustion model *Applications for Design* as referring to specific questions about specific engine designs. Engine combustion model *Applications for Understanding* is meant to cover all other uses.

Applications for design purposes are discussed in the context of three design areas: *Engine Structural Design, Engine Flow-System Design*, and *Engine Combustion-System Design*. The areas considered as applications for understanding are *Diagnosis, Prediction, and Synthesis*. The classification system for modeling applications is shown in Fig. 1.

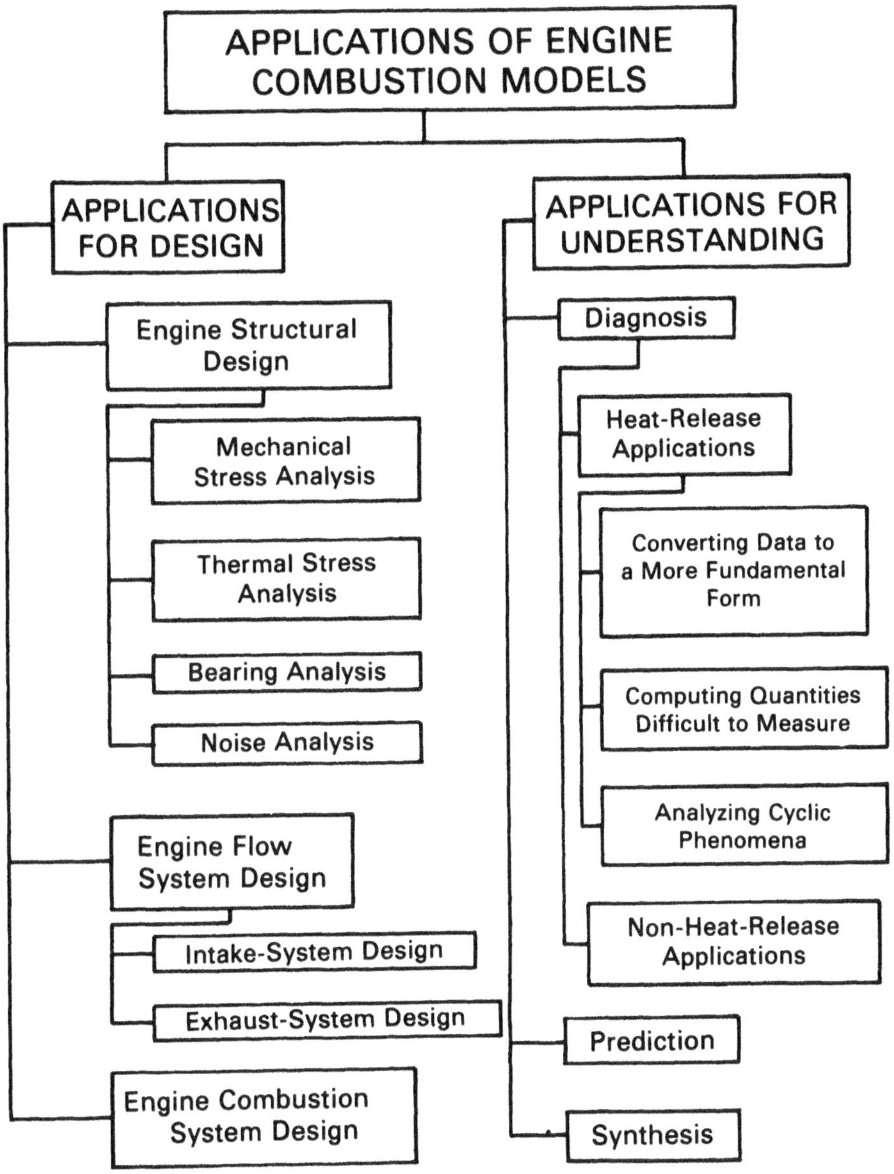

Fig. 1. Classification system for combustion modeling applications.

References pp. 501–503.

APPLICATIONS OF ENGINE COMBUSTION MODELS FOR DESIGN

The process of engine design can be broken down into three main areas: design of the engine structure, design of the engine gas-flow system, and design of the engine combustion system. The applications of engine combustion models in the design area are mostly indirect. That is, engine combustion models only provide input for design analyses. Thus, their discussion herein will be somewhat brief relative to the discussion of *Applications for Understanding*, which are mostly direct applications.

Engine Structural Design—Engine combustion models can be applied to four aspects of engine structural design: mechanical stress analysis, thermal stress analysis, bearing load analysis, and noise analysis.

In the analysis of mechanical stresses in engines, two loads are important: inertia loads and combustion-generated loads [1]. Of the combustion-generated loads in an engine, the most important category is structure loads, and these are directly proportional to peak cylinder pressure [2]. Engine combustion models have been used to predict peak cylinder pressures for mechanical stress analyses. Cameron et al. [3] describe such an application for the design of a 3000 kW locomotive engine. They illustrate the use of the cylinder pressure diagram for the analysis of connecting-rod forces and moments.

The engine combustion process is the driving force for the entire engine thermal-load distribution. Engine combustion models have been used to determine the driving heat fluxes for predictive thermal stress analyses of engines at the design stage. Janota et al. [4] used a computer code with an engine combustion model to study the effect of moving the ring band down the piston in a large diesel engine. This calculation enabled the designers to judge where the ring belt should be positioned, based on oil breakdown temperatures and maximum component temperatures. Cameron et al. [3] also generated a piston temperature distribution in their design study. This distribution allowed them to design the piston underside cooling provisions so that maximum temperature limits would not be exceeded.

Maximum engine bearing loads are normally generated by inertia forces. However, in highly loaded engines, particularly those which are turbocharged, combustion-pressure forces can dominate the inertia forces for part of the cycle. Here again, engine combustion models have been used to determine the cylinder pressure-versus-time function, which is required for an engine bearing analysis, and minimum-film-thickness calculation [3].

The noise generated by an engine structure is strongly dependent on the cylinder pressure-time function as discussed by Priede [5] and Chung [6]. In principle, engine combustion models could be used to generate the pressure-time function for an analysis of engine noise. However, this requires knowledge of the engine-noise transfer functions. Although these transfer functions are not known at the design stage, the engine combustion model could be used to determine the effect of different combustion rates on the noise from a particular engine structure with known noise transfer functions.

Engine Gas-Flow System Design—The engine gas-flow system is considered herein to consist of the intake flow system, including all regions between the entry of the air from the atmosphere and the inlet valve. The exhaust flow system is defined analogously. Thus these systems include the insides of pipes, plenums, turbochargers, manifolds and ports. These two systems are coupled together by the processes occurring in the engine cylinder during compression, combustion and expansion. Thus, engine combustion models have been used as part of comprehensive engine flow models. Such models have been described by Borman [7], Eberle [8], Benson [9, 10] and Brandstetter [11]. Since the coupling between the intake flow and exhaust flow is not always strongly influenced by the engine combustion process, the inclusion of an engine combustion model has sometimes been ignored, as in the models of Wright [12], and Novak [13].

The principal application of engine flow models which include combustion models appears to be in large diesel engines for turbocharger selection and matching, as illustrated by the work of Ryti [14], Eberle [8], and Streit and Borman [15]. For this application, the engine combustion model plays an important role through its influence on the exhaust energy. This is especially true when engine transients are considered, as in the work of Watson and Marzouk [16]. The other common application of engine flow models which include combustion models is in two-cycle, spark-ignition engine flow-system design. In such engines, flow maximization and power output are strongly dependent on exhaust pulse energy (Blair and Johnson [17], and Wallace and Langell [18]).

Engine Combustion-System Design—Of the three engine design areas, the design of the engine combustion system is the one for which it is the most difficult to distinguish between *Applications for Design* and *Applications for Understanding*. This stems from the fact that, within the three design areas, the least understood controlling process is the combustion process. Thus, somewhat ironically, the area of engine design where the engine combustion model is least employed is in the design of the engine combustion system. This is the case because until recently very little capability existed for computing the combustion rate on the basis of design variables. In fact the only publication found where the application of an engine combustion model was made to combustion-system design for a specific system is that of Mattavi et al. [19]. Nonetheless, engine combustion models have contributed extensively to design improvements through their ability to help increase understanding. Thus, their discussion fits more naturally into *Applications for Understanding*.

Summary of Design Applications of Engine Combustion Models—The extent to which engine combustion models are used for specific engine design tasks appears to depend on two factors: the unit cost of the engine being designed and the maximum loading which the engine will encounter. Thus predictive computer design using engine combustion models has generally been applied to larger engines in the diesel category [3, 4]. Computer design is probably used less for small, inexpensive engines because experimental costs are lower. Consequently, analysis is generally used less for such engines. One exception to this generality

References pp. 501–503.

is the area of high-performance spark-ignition two-stroke engines, where computer design is used frequently.

A good example of how a comprehensive engine design model depends on the engine combustion model is illustrated in Fig. 2 [3].

APPLICATIONS OF ENGINE COMBUSTION MODELS FOR UNDERSTANDING

The applications of engine combustion models for understanding found in the literature can be roughly divided into three areas: *Diagnosis, Prediction*, and *Synthesis*. Diagnosis is primarily concerned with enhancing the interpretation of experimental data. Many of the studies in the literature fall very naturally into this category. Prediction is concerned with the use of combustion models to extend understanding beyond what is known experimentally. In general, this involves parametric studies. Synthesis, as used herein, is the development of concepts or strategies which go beyond the direct output of predictive or diagnostic models. This is perhaps best illustrated by an example. If a detailed predictive model for the direct-injection stratified-charge (DISC) engine were available, a study calculating the effect of changing the spray angle or number of

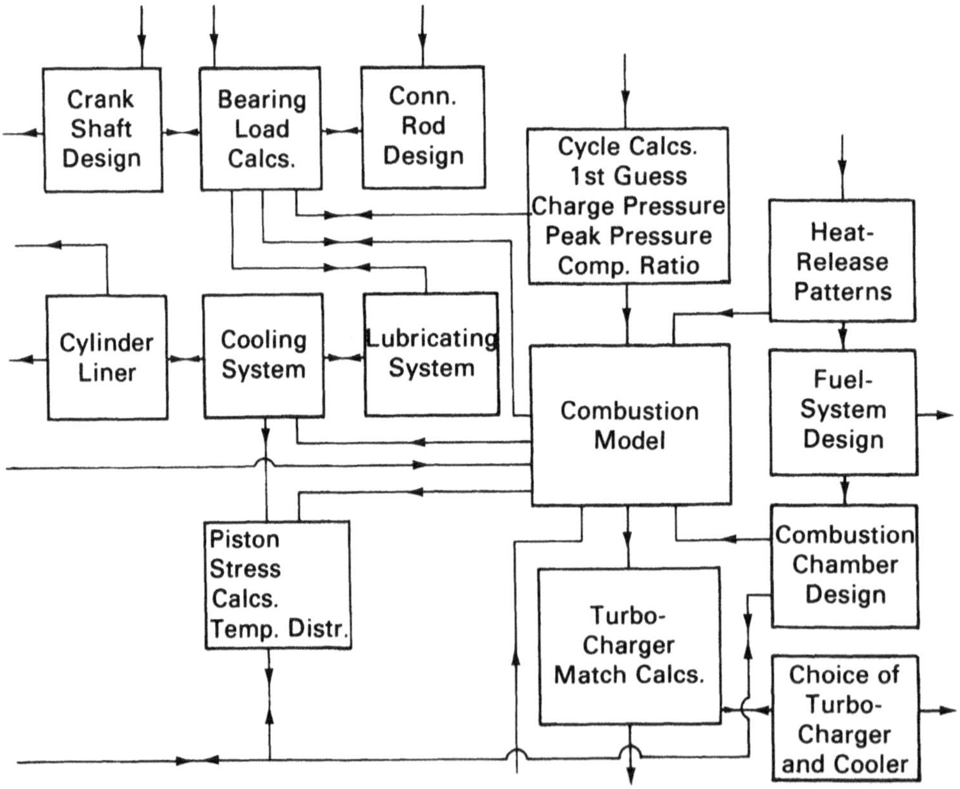

Fig. 2. Design flow diagram [3].

holes in the fuel-injection nozzle would be considered *prediction*. A broad study like the optimization of fuel and velocity distributions in such an engine and how to obtain them would be considered *synthesis*.

Diagnosis—The use of an engine combustion model as a diagnostic tool is defined herein as any application of the model whose primary purpose is to enhance interpretation of experiments or experimental data. Obviously then, the quality of the experimental data analyzed is one of the critical elements for successful diagnosis. Recent advances in the ability to measure cylinder pressure-time data accurately and quickly [20, 21] have considerably enhanced the utility of diagnosis.

Heat-Release Applications—The heat-release calculation is at the heart of most diagnostic applications. In this technique, measured engine cylinder pressure data are analyzed to determine how fast energy was released in order to create the measured pressure. Schweitzer [22] described a heat-release model for diesel engines more than 50 years ago. He lumped heat-transfer losses with the energy addition from combustion to derive a simple equation for the energy release rate. This equation could be evaluated completely with data from a pressure-volume (P-V) diagram. This model reproduced all of the important features of the diesel heat-release diagram, as shown by Obert [23]. Twelve years after Schweitzer's report, Rassweiler and Withrow [24] described a method which they developed for conventional, homogeneous spark-ignition (CHSI) engines. Based on experiments, they found that the fraction of mass burned was equal to the fractional pressure rise due to combustion. Thus, the fraction of mass burned was directly determined from the P-V diagram without any information about engine geometry, air-fuel ratio or trapped mass. This method was used for CHSI engines for many years.

The step to digital-computer codes to carry out heat-release calculations was taken by Austen and Lyn [25] with their model for direct-injected diesel engines. Not long after, Krieger and Borman [26] presented a computer-based thermodynamic model for CHSI heat-release calculations as well as an updated direct-injection diesel heat-release model. Their CHSI model assumed two zones, one corresponding to the unburned gases and the other to the burned gases. The flame front was assumed to be infinitesimally thin, and instantaneous heat-transfer computations were included. Their diesel heat-release model considered a single zone and assumed that injected fuel was immediately burned and mixed within the cylinder.

In the literature, heat-release models have been used: (i) to convert experimental data into a more fundamental form, (ii) to provide a means of computing or estimating quantities which are difficult to measure, and (iii) to evaluate the effects of cyclic variation.

Converting Experimental Data to a More Fundamental Form—In its simplest form, the heat-release model is used to convert cylinder pressure data into the more fundamental quantity, mass burning rate. Much insight into diesel and

References pp. 501–503.

spark-ignition combustion processes has been obtained from the use of heat-release models in this way.

Austen and Lyn [25] analyzed pressure data from a direct-injection diesel engine for a range of operating conditions. Their analysis led to a phenomenological description of the direct-injection diesel combustion process which has remained more or less intact since it first appeared. This description breaks the process into three phases: a very rapid combustion of the ignitable mixture prepared during the ignition-delay period (defined as the time between the start of injection and the first detectable pressure rise due to combustion), a period of falling combustion rates during which combustion is controlled by the rate of mixing, and a third period characterized by a long duration and low combustion rate. These features are identified in the heat-release diagram of Fig. 3.

In the case of the conventional, homogeneous spark-ignition engine, heat-release models have been used extensively to convert cylinder pressure data into mass burning rate. At the General Motors Research Laboratories, an engine diagnostic package based on a heat-release model is in routine use, as discussed by Young and Lienesch [27].

Lancaster et al. [28] extended the Krieger-Borman heat-release model by adding spherical flame geometry to compute instantaneous turbulent flame speed, a quantity one step more fundamental than mass burning rate. The additional information provided by the knowledge of the turbulent flame speed over knowledge of only the mass burning rate enabled Lancaster et al. to propose a phenomenological description of the combustion process in a CHSI engine. They identified four regimes of flame propagation during the combustion process: ignition and kernel growth, flame development, fully developed propagation, and termination. To obtain more information from the turbulent flame speed, Lancaster et al.

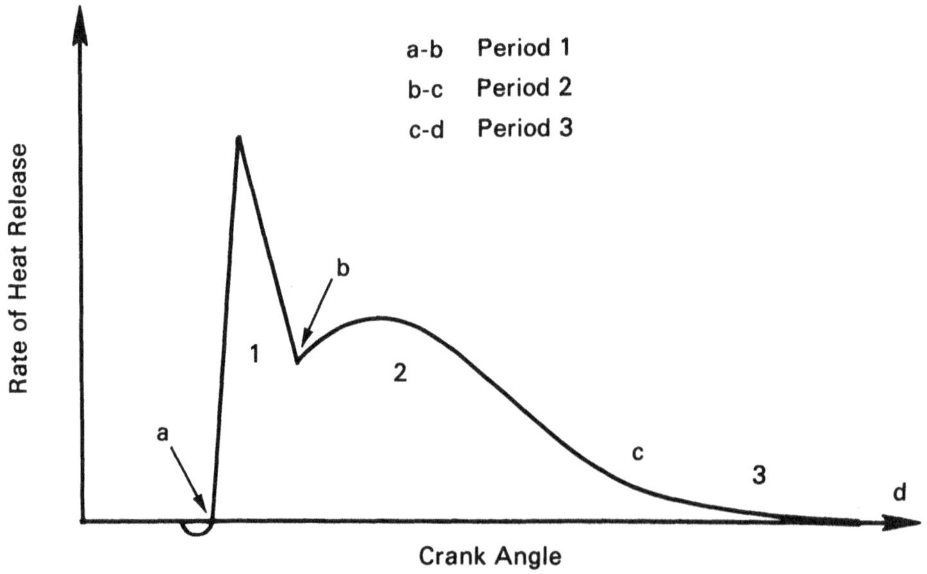

Fig. 3. Regimes of direct-injection diesel combustion process [25].

divided it by the instantaneous laminar flame speed and correlated this ratio with measured motored engine turbulence. For their engine and for the conditions considered, they found that the turbulent-to-laminar flame speed ratio during the fully developed flame-propagation period correlated with the relative volumetric flow rate through the intake valve.

Mattavi et al. [19] extended the approach of Lancaster et al. by incorporating more general combustion chamber geometry and by formulating the model in both predictive and diagnostic forms. This model was then used to analyze a specific engine geometry to propose a way of increasing combustion rate, as referred to earlier in the Combustion-System Design section.

While heat-release analyses have been used extensively for interpreting pressure data from both diesel and CHSI engines, this author is aware of no publications which illustrate the use of a heat-release analysis for stratified-charge engines.

Computing Quantities Difficult to Measure—Numerous investigators have used heat-release-type engine combustion models to compute quantities difficult to measure. Three examples of this for the CHSI engine are the computation of burned-gas temperature, nitric-oxide (*NO*) formation and "ignition delay" (defined for the spark-ignition engine as the time between the spark and the first detectable pressure rise due to combustion).

Eyzat and Guibet [29] computed *NO* formation histories from pressure histories of different engines at lean-mixture conditions. Lavoie et al. [30] used a more refined model to compute the burned-gas temperature gradient between the early-burned gases and the late-burned gases. This gradient has some significance in the interpretation of the spatial variation of *NO* formation. Komiyama and Heywood [31] used a refinement of the Lavoie et al. model to identify the *NO*-freezing differences between CHSI engines operating rich and those operating lean. They found that for rich mixtures, *NO* emission levels were controlled primarily by the *NO* decomposition rate while for lean mixtures, *NO* emissions were controlled primarily by the *NO* formation rate.

Raukis and McLean [32] used a heat-release model to study the effects of hydrogen addition on the combustion process in CHSI engines. Using a heat-release model, they were able to resolve ignition delay down to 0.1 ms. They found that added hydrogen reduced ignition delay significantly, especially in lean mixtures. Being able to separate ignition delay from flame-propagation time through the heat-release analysis, they found that the added hydrogen had little effect on flame propagation once a turbulent flame was well established.

Ignition delay is a fundamental factor in diesel combustion systems. There is no general way to determine ignition delay without the use of a heat-release model. Austen and Lyn [25], in their heat-release study, used ignition delay to examine rate-of-pressure rise in direct-injected diesel engines. They found that the rate-of-pressure rise, which must be controlled if engine-structure knock is to be avoided, correlated with the quantity of fuel which was injected during the ignition-delay period. This information gave them useful physical insight into the processes controlling the rate-of-pressure rise.

References pp. 501–503.

Analyzing Cyclic Phenomena—Cyclic variation in homogeneous spark-ignition engines is a phenomenon which is not well understood. However, some light has been shed on this phenomenon by heat-release calculations. For example, Peters and Borman [33] observed that cyclic variability of mass burning rates was higher early in the combustion process. This lent support to the often-stated hypothesis that cyclic variation is caused by variations in mixture motion near the spark plug at the time of ignition. Heywood et al. [34] and Chen and Krieger [35] examined the effects of cyclic variation on *NO* emissions by using heat-release models to analyze a range of pressure-time measurements at specific engine operating conditions. These studies, which were motivated by the temperature nonlinearity of the *NO* kinetics, had two objectives. First, they could establish if the reduction of cyclic variation could affect *NO* emissions. Second, they could show whether the use of an average pressure trace for a heat-release and *NO* calculation would yield an *NO* value corresponding to the average cycle or whether the nonlinearity would cause the *NO* to be underpredicted. Somewhat surprisingly, both studies found that the effects of cyclic variation on *NO* formation were not dramatic.

Non-Heat-Release Diagnostic Applications—Lavoie and Blumberg [36] used a predictive model in a diagnostic application. (However, it could be said that they used their predictive model as a heat-release model since their combustion-rate function was adjusted until observed pressure-time data were matched.) They observed that for premixed operation of a spark-ignition engine at low *NO* levels, their extended Zeldovitch kinetics mechanism could not account for corresponding measured *NO* values. They concluded that "flame-formed *NO*" must account for the difference and added a correction to their model for this effect. Thus, another diagnostic application is to aid in the synthesis and evaluation of model assumptions.

When Lavoie and Blumberg applied their model to stratified-charge-engine operation, a fuel-air stratification function was inferred for the engine through efforts at matching computed and measured *NO* emissions. They found that a highly stratified fuel-air-ratio distribution, with variations from the mean on the order of the mean itself, had to be used in the model to explain the data.

Summary of Diagnostic Applications—Engine combustion models are extensively used for diagnostic purposes in the engine field. For the most part, these applications are based on some form of heat-release model. Heat-release models have been extremely useful because models for predicting combustion rate from design variables are not available. Thus, experiments and the interpretation of experimental results have been the predominant methods used in the design and development of internal combustion engines. Accordingly, the success of diagnosis depends heavily on the quality of the experimental data analyzed.

Prediction—As discussed in this overview, prediction is, for the most part, the inverse of diagnosis. Prediction means to foretell in advance, before the experiment what will happen. Diagnosis is to go backwards in time to tell what happened during an experiment already completed. Successful application of combustion models requires a deep understanding of the strengths and weaknesses of a model as well as an understanding of the range of values of the model variables over which the model is valid. This is particularly true for prediction. For example, if one is to use a set of calculations to study the effect of changing a variable whose effects are not well modeled, the prediction will always be suspect. Likewise, if a set of calculations is used to consider values of a variable beyond the range for which the model is valid, other physics than those incorporated into the model may become important, causing an inaccurate prediction. Thus, unless the predictive model is very fundamental and extensively validated, and there has probably never been such a model for engine combustion, considerable judgment must be applied if the predictive calculations are to be successful.

The studies in the literature which fall into this category address problems for all three general types of reciprocating piston engines: conventional homogeneous spark-ignition (CHSI) engines, diesel engines, and stratified-charge (SC) engines.

Prediction for CHSI Engines—Thermodynamics-based cycle-efficiency calculations for CHSI engines to consider effects of parameters like compression ratio date back to the 1930's, when combustion charts became available. However, the first application of a physically realistic combustion model for predictive use appeared in the literature in the 1960's. For example, Walker [37] used a model which contained the major features of the CHSI combustion models in current use. These features were two zones (burned and unburned), heat transfer to the walls, and thermodynamic processes not confined to a series of isopleths of thermodynamic variables. However, the combustion rate was specified arbitrarily by functionally relating the fraction of mass burned to time. Walker used this model to study the effects of combustion duration, and of the shape of the mass-burned-fraction curve. He concluded that the principal difference between real-engine efficiencies and Otto-cycle efficiencies was attributable to heat-transfer effects rather than the lack of constant-volume (Otto-cycle) combustion. However, he found that shorter combustion durations, i.e., faster rates, gave higher efficiencies. This work illustrates a good use of a predictive model. Walker's model employed realistic physics for most of the important processes once combustion rate was specified. Thus the model responded reliably when combustion rate was changed, and useful guidance was generated even though nothing could be said about how to obtain the combustion rates he specified.

Blumberg and Kummer [38], using a predictive model with assumptions like the heat-release-type model of Lavoie et al. [30], carried out a thorough study of the effects of combustion parameters on NO formation. Blumberg and Kummer (B&K) considered the effects of changes in intake-air humidity, air-fuel ratio, exhaust gas recirculation, compression ratio, residual-gas fraction, inlet temperature, connecting-rod length-to-stroke ratio, and fuel type. For each variable studied, they varied the combustion timing and duration to examine tradeoffs. Although they did not include the effects of heat transfer, their calculations generated considerable insight into the CHSI combustion process with respect to NO emissions. For example, when operating at any load, speed and air-fuel combination, they found that there is a tradeoff between NO emissions and fuel

References pp. 501–503.

consumption. This is seen from Fig. 4, where each curve corresponds to a different combustion duration. As one moves from right to left on any given curve by retarding the spark, *NO* at first decreases rapidly for small increases in specific fuel consumption (*bsfc*). Eventually, however, *bsfc* begins to increase rapidly for small reductions in *NO*. In addition to this qualitative behavior, the results of B&K provide quantitative information about the magnitude of these changes.

Based partly on the B&K observations about the effects of combustion rate on the *NO* emissions/fuel consumption tradeoff, Mayo [39] undertook an experimental study to determine ways of increasing combustion rate by chamber ge-

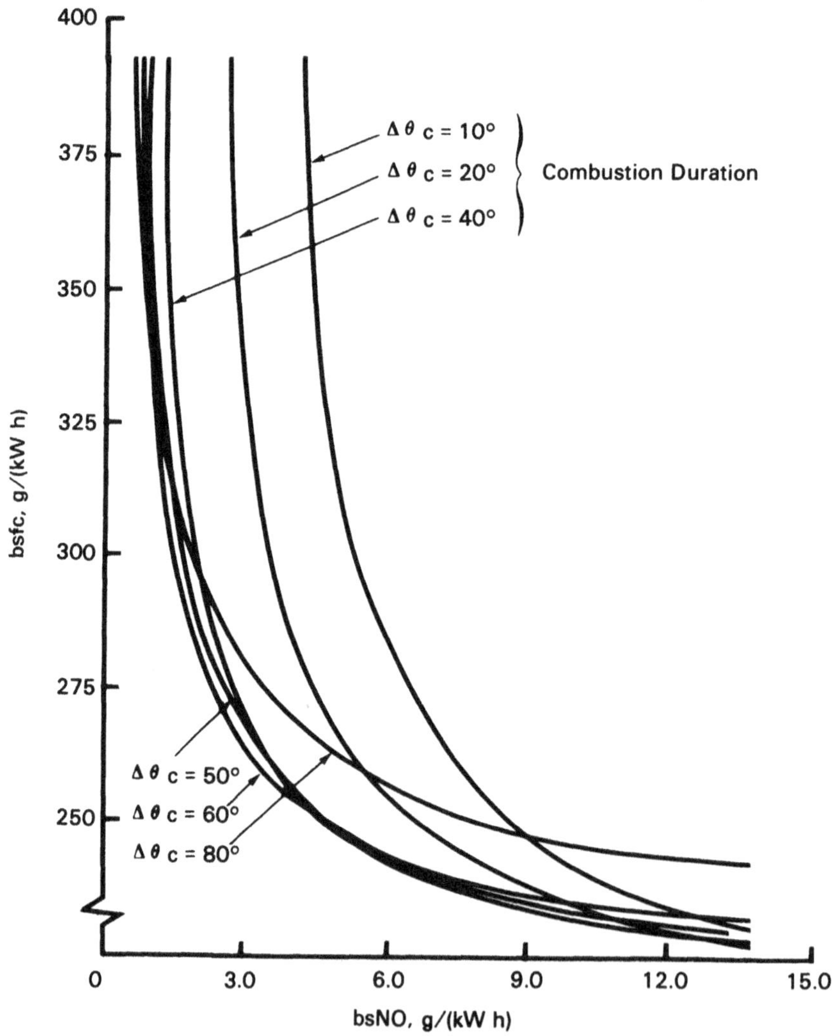

Fig. 4. Fuel consumption (*bsfc*/*bsNO*) tradeoff characteristics for different combustion durations [38].

ometry changes. This is an excellent example of how modeling studies can help guide experiments.

Prediction for Diesel Engines—On a somewhat parallel path, parametric studies using models have been carried out for diesel engines. Lyn [40] studied the effect of different triangularly shaped and variably timed heat-release-rate functions on cylinder pressures and cycle efficiencies for direct-injection diesel engines. Using these arbitrary heat-release-rate functions and considering different heat-release timings and compression ratios, Lyn concluded that a heat-release diagram of 40 degrees crank-angle duration, with the peak of the triangle in the middle, gave the best compromise between pressure-rise rate, a structure limitation, and cycle efficiency.

Shahed et al. [41] carried out a predictive parametric study of the effects of various operating parameters on *NO* formation in an open-chamber diesel engine, using a group of models which included a fuel-injection simulation, an ignition-delay program, a heat-release-rate program and a burning-law program, which was the model of Lyn [40]. In carrying out their study, Shahed et al. considered the effects of fuel cetane number, injection timing, air inlet temperatures and exhaust gas recirculation, all of which can be varied and studied experimentally. However, the fifth variable they studied, fuel injection rate, is not easily varied over a wide range experimentally. Such a study is thus an excellent application of a model. From this study, they found that there was a definite tradeoff between *NO* emissions and fuel consumption, just as for the CHSI engine. However, in their case *NO* was not strongly affected by changes in combustion rate schedule or model assumptions. Thus Shahed et al. concluded that from the standpoint of *NO* emissions, an experimental study of the effects of fuel injection rate on *NO* emissions was not warranted. This is another example of how modeling can guide experiments.

Prediction for Stratified-Charge Engines—Predictive parametric combustion-modeling studies have also been carried out for stratified-charge (SC) engines. Although there are many different forms and types of SC engines, the predominance of the literature on these engines falls into two areas: (single-chamber) direct-injection stratified-charge (DISC) engines, and three-valve divided-chamber carbureted stratified-charge (DCCSC) engines.

Even though it is probably a serious oversimplification, DCCSC engines are generally thought of, and certainly modeled, as being primarily geometrically stratified. That is, the important stratification is thought of as between chambers rather than within chambers. There are a group of similar models for this engine based on such an assumption, e.g., those of Davis et al. [42], Hires et al. [43] and Asanuma et al. [44].

Wall et al. [45] have carried out an extensive parametric study of the effects of changes to both geometric and operating variables on fuel consumption and exhaust temperature for the DCCSC engine using the model of Hires et al. [43]. They considered the effects of prechamber and main chamber combustion timing and duration, stratification between chambers, prechamber volume and orifice

References pp. 501–503.

size, engine load, and exhaust gas recirculation. Their study of the effects of prechamber connecting-orifice size illustrates one of the potential dangers of predictive modeling. Wall et al. found little effect of orifice size on NO emissions. However, they correctly rejected this result since they had not varied combustion duration when they varied the orifice size. It was known from experiments that orifice size has a strong effect on combustion duration. This example illustrates the importance of understanding the limits of validity of a model for a successful predictive study.

Summary of Prediction Applications—This aspect of modeling applications can be very valuable if used properly. It is often very helpful as a guide for selecting useful experiments, for carrying out studies of parameters not easy to vary experimentally, or conversely, as an indication of experiments which are not likely to be worthwhile. In addition, it can help quantify effects which are only known qualitatively. It is particularly useful for studying the effects of geometric changes which are generally costly and difficult to change experimentally, but simple to change in a modeling study. To be effective, the parametric variations chosen for study should produce effects which depend on aspects of the combustion process which are well modeled.

Synthesis—Synthesis is really an extension of prediction since synthesis would probably begin with a parametric study of some combustion process. However, prediction would generally involve a series of calculations for a set of parametric variations proposed before the calculations. Synthesis would start out like prediction but go beyond it toward an optimum based on some insight gained from the preplanned calculations. Thus the distinction between prediction and synthesis is a narrow one. Nonetheless, it is one which is worthy of consideration.

Synthesis would seem to be the most difficult of the three areas—diagnosis, prediction, and synthesis. It would seem to be more difficult than diagnosis because, in general, synthesis is more difficult than analysis. Synthesis is more difficult than prediction because it goes beyond it. It seems to this author that the likelihood that prediction will lead to synthesis during the study of a particular combustion process is directly proportional to the number of degrees of freedom in the control of the combustion process and inversely proportional to the number of people who have already spent their careers studying the concept experimentally.

Synthesis for CHSI Engines—During the period when relatively little was known about the effects of combustion rate on NO emissions in CHSI engines, thus making synthesis more probable, Eyzat and Guibet [29] made a study of the effect of combustion rate on NO formation. Although their model was rather simple, and slow bimolecular kinetics were used, their predictions were very close to measured exhaust concentrations because high values of specific heat ratios gave high temperatures, thus speeding up the kinetics. Using a mass-burned-fraction function similar to that of Walker [37], they concluded that a short combustion duration with high mass burning rates near the end of combustion with retarded spark timing gave high cycle efficiencies and relatively low *NO* formation. This optimization study was a good example of synthesis.

The combustion process of the stratified-charge engine seems to be a prime candidate for synthesis because it has several degrees of freedom which the conventional homogeneous engine does not have. Further, it has not yet been studied extensively experimentally.

Synthesis for DISC Engines—Blumberg [46], Tabaczynski and Klomp [47], and Watfa et al. [48] all carried out broad studies for DISC engines with simple models to find the optimum initial fuel distribution, which they called the stratification function. Blumberg considered three basic stratification functions: A linear distribution of fuel from the beginning of combustion to the end, a sinusoidal distribution, and a step distribution. Blumberg's model was quite simple since it did not consider fuel injection, vaporization, heat transfer and mixing. In addition, the study was carried out using arbitrary combustion rates. However, in spite of the model's simplicity, useful guidance resulted from the study. First, Blumberg concluded that stratification was more effective near the stoichiometric mixture ratio than on the rich or lean side of stoichiometric because no matter how poorly stratified at the overall stoichiometric condition, all elements of the charge for the unstratified case were replaced for the stratified case by elements producing less *NO*. Rich-to-lean stratification was found to be more effective than lean-to-rich. When rich-to-lean stratification was used, a tradeoff was observed between maintaining stratification long enough to minimize *NO* formation on the one hand and allowing sufficient time for *CO* burnup on the other. The *CO* remaining in the rich portion of the charge cannot start burning until destratification begins, thus providing the oxygen that *CO* burnup requires.

Synthesis for DISC Rotary Engines—Reitz and Bracco [49] have reported on a study which involved a direct-injection, stratified-charge rotary engine. For their calculations, Reitz and Bracco used a one-dimensional model which, by including the spray equation for an assumed monodisperse spray, also included droplet vaporization and droplet drag calculations. A time-dependent turbulent diffusivity was used as well as an empirical spray-penetration correlation. The motivation for the study was apparently the unexplained presence of misfire in various configurations of the DISC rotary engine. Reitz and Bracco began their calculation by computing fuel distributions at the time of spark for two-hole and four-hole injection nozzles. They observed that over a load range, the air-fuel ratio at the ignition site at the time of ignition was too lean to allow successful ignition, according to the model. This led them to ask: What type of fuel distribution would be conducive to maintaining combustible mixture at the spark plug at the time for ignition over a range of loads and speeds?

Reitz and Bracco proposed that a steep-fronted rich region should surround the ignition site. They also suggested that this region should grow in extent as more fuel was injected, rather than getting richer. These concepts are illustrated in Fig. 5. So far this hypothesis was not new. However, by examining the characteristics of multi-hole nozzles, they concluded that their hypothesized fuel

References pp. 501–503.

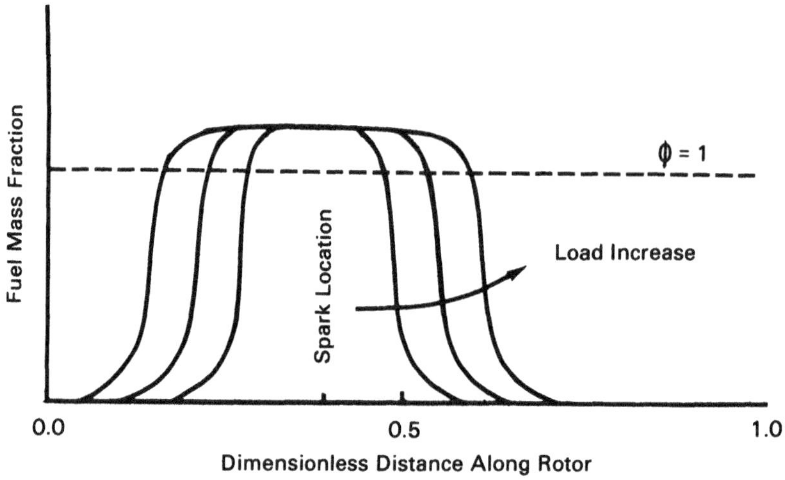

Fig. 5. Proposed ideal fuel distributions for disc rotary engine [49].

distribution was not attainable by variations in injection pressure with load. This observation was important because conventional pump and nozzle fuel injection systems change load predominantly through changes in injection pressure. Reitz and Bracco hypothesized that by varying nozzle area and direction of flow from the nozzle with load, the desired effect could be achieved. They demonstrated this effect by making calculations using a showerhead-type nozzle for which the number of holes and the angle between holes both increased with load. The calculated results suggested that with this approach, the ignitable region would tend to increase in size but not richness, thus supporting their hypothesis.

Summary of Use of Engine Combustion Models for Synthesis—Synthesis has not been extensively used to optimize engines. This probably stems from the fact that synthesis is the most difficult of the three (diagnosis, prediction, and synthesis) and thus requires the most judgment. This lack of extensive use also comes from the fact that models which inspire sufficient confidence to be used for synthesis have not been available.

Summary of Use of Engine Combustion Models for Understanding—The literature on *applications for understanding* overwhelms that on *applications for design*. Up until now, the diagnostic uses of models have had the greatest impact, and their extensive use will no doubt continue. The predictive use of models has trailed diagnostic use because, in general, predictive models require more accurate physics. This situation exists because the implication of prediction is extrapolation. Synthesis is a very useful application of models and has the potential for greater impact than either diagnosis or prediction, particularly with respect to stratified-charge engines.

REFERENCES

1. A. Scholes and D. J. Slater, "The Stress Analysis of Diesel Engine Frames by Computer," Proc. I. Mech. E., Vol. 182, Part 3-L, Paper 19, pp. 193–199, 1967–68.
2. B. Martin and G. Wright, "High Output Diesel Engine Design Philosophy," SAE Paper No. 770755, 1977.
3. H. Cameron, D. H. Freeston and D. J. Picken, "The Use of a Small Digital Computer in Diesel Engine Design," Proc. I. Mech. E., Vol. 182, Part 3-L, Paper 1, pp. 1–11, 1967-68.
4. M. S. Janota, A. J. Hallam, E. K. Brock and S. G. Dexter, "The Prediction of Diesel Engine Performance and Combustion Chamber Component Temperatures Using Digital Computers," Proc. I. Mech. E., Vol. 182, Part 3-L, Paper 7, pp. 61–73, 1967-68.
5. T. Priede, "Relation Between Form of Cylinder Pressure Diagram and Noise in Diesel Engines," Proc. I. Mech. E. (A.D.), No. 1, pp. 41–55, 1960-61.
6. J. Y. Chung, "The Use of Digital Fourier Transform Methods in Engine Noise Research," SAE Trans., Vol. 86, Paper No. 770010, pp. 1–10, 1977.
7. G. L. Borman, "Mathematical Simulation of Internal Combustion Engine Processes and Performance Including Comparisons with Experiment," Ph.D. Thesis, Mechanical Engineering Department, University of Wisconsin, 1964.
8. M. K. Eberle, "Computations of Scavenging and Supercharging of Internal Combustion Engines," Proc. I. Mech. E., Vol. 182, Part 3-L, Paper 12, pp. 123–132, 1967–68.
9. R. S. Benson, "A Comprehensive Digital Computer Program to Simulate a Compression Ignition Engine Including Intake and Exhaust Sytems," SAE Paper No. 710173, 1971.
10. R. S. Benson, W. J. D. Annand and P. C. Baruah, "A Simulation Model Including Intake and Exhaust Systems for Single Cylinder Four-Stroke Cycle Spark Ignition Engines," Int. J. Mech. Sci., Vol. 17, pp. 97–124, 1975.
11. W. R. Brandstetter, "Modeling of a Stratified Charge Engine with an Unscavenged Prechamber," General Motors Research Laboratories Symposium on Combustion Modeling in Reciprocating Engines, Warren, Michigan, November 1978.
12. E. J. Wright, "Computer Simulation of Engine Gas Dynamic Processes: A Design Package," SAE Paper No. 710174, 1971.
13. J. M. Novak, "Simulation of the Breathing Processes and Air-Fuel Distribution Characteristics of Three-Valve, Stratified Charge Engines," SAE Paper No. 770881, 1977.
14. M. Ryti, "Computing the Gas Exchange Process of Pressure Charged Internal Combustion Engines," Proc. I. Mech. E., Vol. 182, Part 3-L, Paper 11, pp. 113–122, 1967–68.
15. E. E. Streit and G. L. Borman, "Mathematical Simulation of a Large Turbocharged Two-Stroke Diesel Engine," SAE Trans., Vol. 80, Paper No. 710176, pp. 733–768, 1971.
16. N. Watson and M. Marzouk, "A Non-Linear Digital Simulation of Turbocharged Diesel Engines Under Transient Conditions," SAE Trans., Vol. 86, Paper No. 770123, pp. 491–508, 1977.
17. G. P. Blair and M. B. Johnson, "Unsteady Flow Effects in Exhaust Systems of Naturally Aspirated, Crankcase Compression Two-Cycle Internal Combustion Engines," SAE Trans., Vol. 77, Paper No. 680594, pp. 2423–2443, 1968.
18. F. J. Wallace and P. V. Langell, "Theoretical and Experimental Analysis of Air and Gas Flows in a Crankcase Scavenged Two-Stroke Engine Employing Boost Ports," SAE Paper No. 690134, 1969.
19. J. N. Mattavi, E. G. Groff, F. A. Matekunas, J. H. Lienesch and R. N. Noyes, "Engine Improvements Through Combustion Modeling," General Motors Research Laboratories Symposium on Combustion Modeling in Reciprocating Engines, Warren, Michigan, 1978.
20. R. V. Fisher and J. P. Macey, "Digital Data Acquisition with Emphasis on Measuring Pressure Synchronously with Crankangle," SAE Paper No. 750028, 1975.
21. D. R. Lancaster, R. B. Krieger and J. H. Lienesch, "Measurement and Analysis of Engine Pressure Data," SAE Trans., Vol. 84, Paper No. 750026, pp. 155–172, 1975.
22. P. Schweitzer, "The Tangent Method of Analysis of Indicator Cards of Internal Combustion Engines," Bulletin No. 35, Pennsylvania State Univ., September 1926.
23. E. F. Obert, "Internal Combustion Engines," International Textbook Co., Scranton, Pennsylvania, pp. 610–612, 1968.

24. G. M. Rassweiler and L. Withrow, "Motion Pictures of Engine Flames Correlated with Pressure Cards," SAE Trans., Vol. 42, pp. 185-204, May 1938.
25. A. E. W. Austen and W. T. Lyn, "Relation Between Fuel Injection and Heat Release in a Direct-Injection Engine and the Nature of the Combustion Processes," Proc. I. Mech. E. (A.D.), No. 1, pp. 47-62, 1960-61.
26. R. B. Krieger and G. L. Borman, "The Computation of Apparent Heat Release for Internal Combustion Engines," ASME Paper No. 66-WA/DGP-4, 1966.
27. M. B. Young and J. H. Lienesch, "An Engine Diagnostic Package (EDPAC)—Software for Analyzing Cylinder Pressure-Time Data," SAE Paper No. 780967, 1978.
28. D. R. Lancaster, R. B. Krieger, S. C. Sorenson and W. L. Hull, "Effects of Turbulence on Spark-Ignition Engine Combustion," SAE Trans., Vol. 85, Paper No. 760160, pp. 689-710, 1976.
29. P. Eyzat and J. C. Guibet, "A New Look at Nitrogen Oxides Formation in Internal Combustion Engines," SAE Trans., Vol. 77, Paper No. 680124, pp. 481-500, 1968.
30. G. A. Lavoie, J. B. Heywood and J. C. Keck, "Experimental and Theoretical Study of Nitric Oxide Formation in Internal Combustion Engines," Comb. Sci. and Tech., Vol. 1, pp. 313-326, 1970.
31. K. Komiyama and J. B. Heywood, "Predicting NO_x Emissions and Effects of Exhaust Gas Recirculation in Spark-Ignition Engines," SAE Trans., Vol. 82, Paper No. 730475, pp. 1458-1476, 1973.
32. M. J. Raukis and W. J. McLean, "The Effect of Hydrogen Addition on Ignition Delays and Flame Propagation in Spark Ignition Engines," Comb. Sci. and Tech., Vol. 19, No. 5-6, pp. 207-216, 1979.
33. B. D. Peters and G. L. Borman, "Cyclic Variations and Average Burning Rates in a S.I. Engine," SAE Paper No. 700064, 1970.
34. J. B. Heywood, S. M. Mathews, B. Owen, "Predictions of Nitric Oxide Concentrations in a Spark-Ignition Engine Compared with Exhaust Measurements," SAE Paper No. 710011, 1971.
35. K. K. Chen and R. B. Krieger, "A Statistical Analysis of the Influence of Cyclic Variation on the Formation of Nitric Oxide in Spark Ignition Engines," Comb. Sci. and Tech., Vol. 12, pp. 125-134, 1976.
36. G. A. Lavoie and P. N. Blumberg, "Measurements of NO Emissions From a Stratified Charge Engine: Comparison of Theory and Experiment," Comb. Sci. and Tech., Vol. 8, pp. 25-37, 1973.
37. G. Walker, "Effect of the Rate of Combustion on Gasoline Engine Performance," Journal of the Inst. of Fuel, Vol. 37, pp. 228-233, June 1964.
38. P. N. Blumberg and J. T. Kummer, "Prediction of NO Formation in Spark-Ignited Engines—An Analysis of Methods of Control," Comb. Sci. and Tech., Vol. 4, pp. 73-95, 1971.
39. J. Mayo, "The Effect of Engine Design Parameters on Combustion Rate in Spark-Ignited Engines," SAE Trans., Vol. 84, Paper No. 750355, pp. 869-888, 1975.
40. W. T. Lyn, "Calculations of the Effect of Rate of Heat Release on the Shape of Cylinder-Pressure Diagram and Cycle Efficiency," Proc. I. Mech. E. (A.D.), No. 1, pp. 12-24, 1960-61.
41. S. M. Shahed, W. S. Chiu and V. S. Yumlu, "A Preliminary Model for the Formation of Nitric Oxide in Direct Injection Diesel Engines and Its Application in Parametric Studies," SAE Trans., Vol. 82, Paper No. 730083, pp. 338-351, 1973.
42. G. C. Davis, R. B. Krieger and R. J. Tabaczynski, "Analysis of the Flow and Combustion Processes of a Three Valve Stratified Charge Engine with a Small Prechamber," SAE Trans., Vol. 83, Paper No. 741170, pp. 3534-3550, 1974.
43. S. D. Hires, A. Ekchian, J. B. Heywood, R. J. Tabaczynski and J. C. Wall, "Performance and NO_x Emissions Modeling of a Jet Ignition Prechamber Stratified Charge Engine," SAE Trans., Vol. 85, Paper No. 760161, pp. 711-738, 1976.
44. T. Asanuma, M. K. Ġ. Babu and S. Yagi, "Simulation of the Thermodynamic Cycle of a Three-Valve Stratified Charge Engine," SAE Paper No. 780319, 1978.
45. J. C. Wall, J. B. Heywood and W. A. Woods, "Parametric Studies of Performance and NO_x Emissions of the Three-Valve Stratified Charge Engine Using a Cycle Simulation," SAE Paper No. 780320, 1978.
46. P. N. Blumberg, "Nitric Oxide Emissions From Stratified Charge Engines: Prediction and Control," Comb. Sci. and Tech., Vol. 8, pp. 5-24, 1973.

47. *R. J. Tabaczynski and E. D. Klomp, "Calculated Nitric Oxide Emissions of an Unthrottled Spark Ignited, Internal Combustion Engine," SAE Paper No. 741171, 1974.*
48. *M. Watfa, D. E. Fuller and H. Daneshyar, "The Effects of Charge Stratification on Nitric Oxide Emission from Spark Ignition Engines," SAE Paper No. 741175, 1974.*
49. *R. D. Reitz and F. V. Bracco, "Studies Toward Optimal Charge Stratification in a Rotary Engine," Comb. Sci. and Tech., Vol. 12, pp. 63-74, 1976.*

DISCUSSION

J. B. Heywood *(Massachusetts Institute of Technology)*

That was very nice. I wanted to quarrel a little with your comment that you want as much physics as possible involved. I would rather rephrase that and suggest that you want the appropriate level of detail of physics and chemistry, as you put it. For different kinds of applications of these models, there are different degrees of detail required. My experience with different kinds of models is that as you gain in some areas, there are inevitably some losses associated with the model, so I feel that as much physics as possible is a little too crude a statement. If you would refine it, I'd be very happy.

R. B. Krieger

I was really trying to suggest a tradeoff. Probably the more physics you have in a model, the less insight is required to avoid falling into a trap. Insight doesn't always seem to be in great supply. I think that as we try to transfer models from use by the people who have created them to the users, who in general will understand less about the model physics and less about the traps that one might

R. B. Krieger

fall into, we should work toward incorporating more physics in them. The basic problem is that as a model goes from the creator to the user, sometimes the appreciation for the limitations is lost. I was trying to make that point, but I agree with your refinement.

G. Sovran *(General Motors Research Laboratories)*

This gives me an opportunity to make a comment, one I feel reasonably strong about. *Appropriate* physics is what is needed in all of these models, but *appropriate* physics does not necessarily mean more complexity. I have always reacted to the conclusion, for example, that something that is more complex is more sophisticated. I think that something that is more complex can be even less sophisticated. The prime examples for simplicity, elegance, and incorporation of all the physics that I like to think of are $F = ma$ and $E = mc^2$.

R. B. Krieger

I agree completely, and that's why I noted that more physics rather than more complication is required.

W. T. Lyn *(Cummins Engine Company)*

I would like to make two comments. First, in striving for more detailed physics, one should bear in mind that a model is only as good as its weakest link. For example, at the present time heat transfer is a weak link in the prediction of NO. Hence, there is no point in using a refined model of the mixing phenomenon in an engine cylinder unless an accurate heat-transfer model is also included. Second, the degree of refinement required depends upon the particular application of the model. A simple model is sufficient to generate a cylinder pressure-time diagram for mechanical development of an engine. However, a highly detailed model is probably required to predict hydrocarbon emissions or particulate emissions.

R. B. Krieger

There have been many comments about the lack of information on heat-transfer rates in engines, and this seems to me to be an area that has really been neglected. What is really needed is experimental information. It would be nice to see more research conducted to generate that information, because that's becoming more and more the next key element of models that has to be improved if we are to continue to make progress.

H. C. Watson *(University of Melbourne, Australia)*

I might just comment on diagnosis, particularly since you showed a slide of Lancaster's work* delineating the modes of flame propagation in the homogeneous engine. I think there's a great need to get evidence of replicability because of the large number of assumptions made, even in a relatively simple approach to obtain a single flame speed ratio. We attempted to repeat that sort of work. Rather than taking an ensemble average of cylinder pressures, we looked at it on a cycle-by-cycle basis. One finds that the characteristics shown in that slide are not apparent in a single-cycle path analysis. There seems to be quite a variety of complicated flame propagation processes, which is no doubt due to the time variations of the cycle-by-cycle variability of turbulence, and so on. So I think we need to see evidence of replicability from worker to worker before we can accept theories that are being devised on the basis of this sort of technology.

R. B. Krieger

The second paper in this afternoon's session extends some of the assumptions made in the work of Lancaster and co-authors, so we'll see then how good some of these assumptions are for other chambers.

G. L. Borman *(University of Wisconsin)*

I just wanted to go back to your first slide (Fig. 1 of paper) in which you showed a line going into the development phase and then not much into the application phase. I think something lacking here is the fact that often those who are developing these models and bringing them to the point where they could be applied don't exert the extra effort of introducing these models to people who could use them. That is, they don't do a teaching job, if you will, to bring them into the main stream of the designers who are really going to apply them. This is an industry problem, and it's also a problem for educators.

R. B. Krieger

One of the reasons why I structured my talk the way I did was to present applications to *users*. The modeling community knows very well what the applications are, and so what I have tried to do is illustrate how users and appliers can learn from models. I agree completely that this process has been neglected. I think the investigators I referred to are trying to say, "Here's a model that's ready to go. Somebody use it." But they have to do more to accomplish this goal.

---

* *D. R. Lancaster, R. B. Krieger, S. C. Sorenson and W. L. Hull, "Effects of Turbulence on Spark-Ignition Engine Combustion," SAE Trans., Vol. 85, Paper No. 760160, pp. 689–710, 1976.*

P. N. Blumberg *(Ford Motor Company)*

I think that was a very nice presentation, and I would like to take the opportunity to make a somewhat philosophical statement. You mentioned that many times insight is in short supply. But I think it is unrealistic to expect that a modeler will ever be able to hand over a model to a user and say, "Go ahead and do a design problem or do an optimization." It's been my experience that in the process of formulating a model, and John Heywood alluded to this yesterday, the person who is doing the model actually gains a great deal of insight about what's going on in the engine system or whatever system happens to be under consideration. He is in a far better position to exercise that model in a development-type application than perhaps anybody else. So I think the issue of technology transfer is one in which the person who develops the model really has to be the interface between the development of the model and the application of the model to practical problems. If I were a research manager in an engine manufacturing company or a research and development manager, I would take cognizance of that fact and really bring these two groups together in a consolidated way to achieve the desired result that we are all advocating.

R. B. Krieger

Yes, I agree completely. I struggled for many years trying to figure out a way to transfer the insight that one gains from formulating the model to somebody who's using the model. There probably isn't any way, and thus what you say is probably the only way to take advantage of that insight and yet apply the model.

G. L. Borman *(University of Wisconsin)*

I take issue with you on that because it seems to me you're saying that you can't educate people. I have some experience with this in cycle analysis. What we did was to bring together a class of people who were going to apply the model, but who didn't know much about it, and we exchanged information until they understood. Now, they're not going to have perfect understanding, and you're still going to have to nurse them along for a while. That is what Paul [Blumberg] seemed to think. So you can't just say, "Well, now you've had the lecture. Go apply it, but I don't want to see you again." What this means is that you have to put some effort into teaching and into showing and into making yourself available for awhile. But it doesn't mean that the people who are doing the designing can't use these things and that they can't be producing results.

P. N. Blumberg

This is not meant in a debating sense, but just one follow-up comment. I agree with Gary [Borman], and I think that the division between us is really perhaps semantic. I agree 100% with his comment, provided the model has certain features which are non-speculative—for example, features for diagnosis-type applications like heat-release models. But in areas where we are still dealing with speculative models, I believe that interaction has to exist on a very on-going basis because only the person who really formulated the model has the full grasp of what's speculative, what's real, and how to do all the juggling and interpretation that goes on. I think that those of us who have done this sort of modeling know how that goes.

R. B. Krieger

I'd like to add a cautionary note to that. Not only is the person who formulated and developed the model the one who knows the most about it, but he may also have the most fixed ideas and perhaps naivete in some areas. Overcoming that is something that is gained by transferring use of the model to someone else. Others don't have the same prejudices, and there may be some weaknesses in the model which the model developer is convinced aren't weaknesses, but which somebody else sees more clearly are.

MODELING OF A STRATIFIED-CHARGE ENGINE WITH AN UNSCAVENGED PRECHAMBER

W. R. BRANDSTETTER

Volkswagenwerk AG, Wolfsburg, West Germany

ABSTRACT

A mathematical model which simulates gas flow and combustion in a single-cylinder divided-chamber engine is described. The computation includes the flow across engine valves and in the intake and exhaust systems.

The main equations are given and explained. The system of ordinary differential equations is solved by a predictor-corrector method with variable step size. A set of dimensionless parameters which completely describe the problem is derived.

Application of the model is demonstrated for the case of a stratified-charge engine with an unscavenged prechamber, where a fraction of the total fuel is injected into the prechamber during the compression stroke. The resulting two-stage combustion process is simulated by two Wiebe functions.

The main purpose of the computational program is to study the effects of various significant parameters during the design stage and to aid interpretation of test results, mainly by obtaining information about quantities that are difficult to measure. Particular design and operating variables, such as ignition timing, combustion duration, prechamber volume, prechamber orifice size, degree of stratification, etc., are investigated and discussed.

NOTATION

c	speed of sound, specific heat
d	diameter
e	internal energy
h	heat-transfer coefficient
i	enthalpy

References pp. 535–536.

k	ratio of specific heats
l	valve lift
m	mass
p	pressure
q	heat
s	piston stroke
t	time
w	work, velocity, Wiebe exponent
x	mass fraction burned
z	exhaust gas fraction
B_1, B_2	fuel supplied by mixture formation device 1 or 2
D	cylinder bore
I	enthalpy of fuel
F	area
H_u	heat of combustion
R	heat of vaporization
T	temperature
V	volume
α	flow coefficient
ϵ	compression ratio
λ	relative air-fuel ratio, connecting-rod ratio
ρ	density
φ	crank angle

Subscripts

c	cycle
max	maximum

red	reduced
uv	unburned
v	constant volume
B	begin
C	compression
E	end, injection
EX	out
H	displacement
IN	in
K	piston
KA	connecting port
R	reaction
VK	prechamber
W	wall
Z	cylinder
1 2 3 4 5 6	reference points in flow model

INTRODUCTION

A number of mathematical models is available to simulate the engine cycle and engine charging process. Both problems have a large number of parameters. Depending on the special problem to be solved, the best suited simulation program is chosen. To use one universal model is not advisable.

The aims of the computer program discussed here are to shorten the optimization work during the development of a stratified-charge combustion process and to understand better the difficult phenomena involved. The simulation program has been used to assist interpretation of test results and to indicate quantitative trends, while carrying out experiments. It has also been applied to special case studies, with or without a prechamber, before the design phase.

The problem under investigation has a large number of variables. Certain parameter combinations are of interest from theoretical and practical points of view. However, experimental studies of these parameters require great expense and very sophisticated equipment. It is therefore desirable to apply analytical techniques to evaluate these parameters.

References pp. 535–536.

The program version described here deals with the processes during the entire engine cycle, including the intake and exhaust flows. It is a so-called zero-dimensional model, where a uniform state is assumed within a system, e.g., the main combustion chamber or the prechamber.

Program systems of another type are those being worked on at Princeton University [1, 2]. In that work, the geometrical shape of the combustion chamber is considered, and the solution depends not only on time, but also on space coordinates. Presently, processes in a divided combustion chamber can be solved in two dimensions. Results from this work are reported in Refs. 1 and 2.

In a number of mathematical models that are based on zero-dimensional treatment, reaction kinetics are the essential part of the calculation. Such computations, for example, were carried out at the Institute of Applied Thermodynamics at the Technical University of Aachen [3].

Recently, the operational characteristics of stratified-charge engines of the divided-chamber type have been studied intensively using simulation methods [4–8]. The majority of these publications deals with 3-valve stratified-charge engines. In these studies the combustion process was simulated in various ways. In Ref. 9, however, the burning law and flow in the connecting port are calculated from pressure indicator diagrams measured on a divided-chamber engine.

DESCRIPTION OF MODEL

A cross-section through the engine cylinder head and the engine block of a stratified-charge engine with prechamber injection (PCI combustion process) is shown in Fig. 1. The prechamber, with spark plug and injector arrangement, can be seen. Detailed specifications of the prototype engine can be found in Ref. 10. Flow and combustion phenomena are to be modeled in this engine. Emphasis is on the treatment of the main chamber and prechamber processes. Therefore, wave effects in the intake and exhaust systems are not described. The intake and exhaust pipe are represented by plenums of equivalent volume. A description of this mathematical model, as well as some results, is reported in Ref. 11.

A schematic of the single-cylinder four-stroke prechamber engine is given in Fig. 2. The system has three volumes of constant size (prechamber, V_{VK}, and intake and exhaust plenums, V_{IN} and V_{EX}) and one with variable size (main combustion chamber, V_Z).

Knowledge of the pressure and temperature of the gas and of the gas composition (mass fraction burned) in these volumes is of particular interest. Furthermore, gas velocities in significant flow areas (F_2, F_3, F_4, F_5, F_{KA}) are sought as a function of crank angle. In addition to these time-dependent variables, other quantities, such as the volumetric efficiency, indicated mean effective pressure, wall heat losses, relative air-fuel ratio of the prechamber, and the main chamber mixture at ignition, are computed.

Engine speed is expressed through cycle duration, t_c. Throttling of intake air at engine part-load operation is described by appropriate flow coefficients at cross-section F_2. The intake system is represented by volume V_{IN} and surface F_{IN}. To calculate the heat transfer, a mean wall temperature, $T_{W,IN}$, and a heat-

Fig. 1. Cross-section through stratified-charge engine with prechamber and prechamber fuel injection (PCI).

transfer coefficient are specified. The exhaust side is described in an analogous form. The main combustion chamber has one intake and one exhaust valve. Furthermore, valve timing, valve clearance, shape of the valve-lift curve, maximum valve lift and valve-head diameter, flow coefficients as a function of valve lift and flow direction, and the cross-section of the intake and exhaust ports must be specified. The geometry of the crank gear is determined by the piston stroke, s, and the ratio of the crank radius to the connecting-rod length, λ. Additional parameters are prechamber volume, V_{VK}, prechamber surface area, F_{VK}, connecting port cross-sectional area, F_{KA}, flow coefficients across the connecting port, α_{ZVK} and α_{VKZ}, compression ratio, ϵ, and average cylinder and prechamber wall temperatures, $T_{W,Z}$ and $T_{W,VK}$. The fuel quantity, B_1, is injected into the

References pp. 535–536.

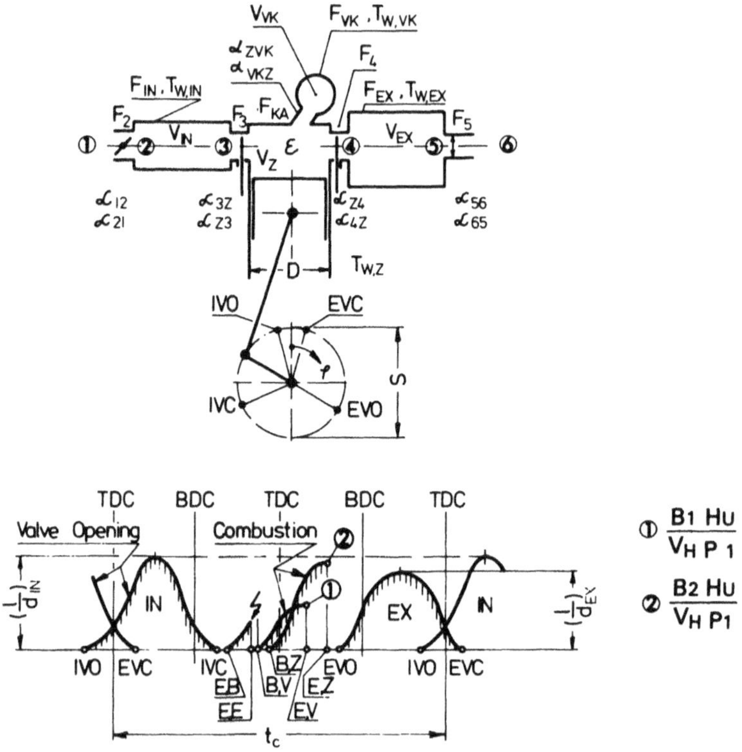

Fig. 2. Schematic of flow model.

prechamber during the injection time $\varphi_{E,B}$ to $\varphi_{E.E}$. In addition, fuel enters with the homogenous intake mixture of the main chamber. It is assumed that the energy release during the two-stage combustion is closely described by two Wiebe functions [12]. This requires specification of start of combustion, $\varphi_{B,VK}$ and $\varphi_{B,Z}$, end of combustion, $\varphi_{E,VK}$ and $\varphi_{E,Z}$, and the so-called Wiebe exponents, w_{VK} and w_Z.

The PCI combustion process differs from that of 3-valve stratified-charge engines. The three special mixture zones, namely, in the prechamber, in the main chamber, and in the vicinity of the connecting port, as described by Krieger and Davis [13] and particularly studied by Matsuoka et al. [14], do not exist. According to investigations by Furukawa et al. [8], main mixture and prechamber mixture burn without mixing with each other. It is assumed in this study that unburned mass leaving the prechamber with composition λ_{VK} reacts according to the Wiebe function of the prechamber, while charge of composition λ_Z leaving the cylinder reacts according to the cylinder Wiebe function, even when it is in the prechamber.

The mathematical model is designed for the simulation of processes in single-cylinder engines. The program is also suited to study special cases, such as an engine without a prechamber, motored engine operation, an adiabatic system, extremely large intake and exhaust plenums and many others.

MATHEMATICAL FORMULATIONS

Model Assumptions—The following assumptions have been made to formulate the mathematical equations:

- (i) Prechamber and main combustion chamber are separate thermodynamic systems.
- (ii) In each system the gas state is assumed to be homogeneous. Burned and unburned mass mixes completely and instantly (resulting in uniform temperatures and exhaust gas fraction).
- (iii) The flow across an area restriction is quasi-steady and adiabatic.
- (iv) The ideal gas law is applicable.
- (v) The heat-transfer coefficients for the engine cylinder and prechamber are computed with a Reynolds analogy, according to Borman [15].
- (vi) The burning law for the main chamber and the prechamber mixture is described by two independent Wiebe functions [12]. The mass of unburned air-fuel mixture in the prechamber and in the main chamber is calculated at ignition and monitored as mass is exchanged between the prechamber and main chamber.
- (vii) Intake and exhaust pipes are represented by equivalent plenums having the same volume.
- (viii) The specific heats for the gases in the prechamber and main chamber are calculated according to a correlation suggested by Krieger and Borman [16], which accounts for the dependence on temperature, pressure and air-fuel ratio.

Governing Equations—Gas pressure, p, mass, m, and exhaust gas mass fraction, z, were chosen as independent variables. For each of the four volumes, one can apply the first law of thermodynamics,

$$de = \sum i\,dm + dq - dw = d\,(mc_vT) \tag{1}$$

a mass balance,

$$dm = \sum \rho wF\,dt \tag{2}$$

and a mixing relation

$$m\,dz = \sum z\rho wF\,dt \tag{3}$$

Together with the gas equation in differentiated form, one obtains a set of first-order ordinary differential equations.

For reason of space, only the equations that apply for the main combustion chamber and which are the most universal ones are given below. Similarly, one can obtain the equations for the other three control volumes by changing the

References pp. 535–536.

indices and by dropping certain terms. All equations are written in dimensionless form.

After a number of substitutions one obtains

$$\frac{dp_Z/p_1}{d\varphi} = \left[k\left(\frac{F_3 w_3 p_3}{F_K c_1 p_1} - \frac{F_{KA} w_{KA} p_{KA}}{F_K c_1 p_1} - \frac{F_4 w_4 p_4}{F_K c_1 p_1}\right) \right.$$
$$- k\left(720 \frac{s}{c_1\, t_c} \frac{p_Z}{p_1} \frac{dV_Z/V_H}{d\varphi}\right)$$
$$+ (k-1)\left(720 \frac{s}{c_1\, t_c} \frac{dq_{R,Z}}{d\varphi}\right)$$
$$\left. - h_{red} \frac{F_Z}{F_K} \frac{h_Z}{h_{red}}\left(\frac{T_Z}{T_1} - \frac{T_{W,Z}}{T_1}\right)\right] \Big/ \left(720 \frac{s}{c_1\, t_c}\right) \frac{V_Z}{V_H} \tag{4}$$

From the conservation of mass equation

$$\frac{dm_Z/\rho_1 V_H}{d\varphi} = \left[\frac{F_3 w_3 \rho_3}{F_K c_1 \rho_1} - \frac{F_{KA}}{F_K} \frac{w_{KA}}{c_1} \frac{\rho_{KA}}{\rho_1} \right.$$
$$\left. - \frac{F_4 w_4 \rho_4}{F_K c_1 \rho_1}\right] \Big/ \left(720 \frac{s}{c_1\, t_c}\right) \tag{5}$$

The mixing equation for the period when no combustion takes place is as follows:

$$\frac{dz_Z}{d\varphi} = \left[(z_3 - z_Z) \frac{F_3 w_3 \rho_3}{F_K c_1 \rho_1} - (z_4 - z_Z) \frac{F_4 w_4 \rho_4}{F_K c_1 \rho_1} \right.$$
$$\left. - (z_{KA} - z_Z) \frac{F_{KA} w_{KA} \rho_{KA}}{F_K c_1 \rho_1}\right] \Big/ \left(720 \frac{s}{c_1\, t_c} \frac{m_Z}{\rho_1 V_H}\right) \tag{6}$$

Considering the terms within the brackets in Eq. 4, the three on the top line represent enthalpy exchange across the system boundaries, the term on the second line is the mechanical work on the piston, the term on the third line expresses the energy released during combustion, and the final term of the numerator represents the heat flow to the cylinder wall.

During the combustion periods and while both valves are closed, burned mass is formed according to a Wiebe law in which 99.9% combustion efficiency is assumed.

$$\frac{dx}{d\varphi} = 6.908\,(w+1)\left(\frac{\varphi - \varphi_B}{\varphi_E - \varphi_B}\right)^{w} e^{-6.908\left(\frac{\varphi - \varphi_B}{\varphi_E - \varphi_B}\right)^{w+1}} \Big/ (\varphi_E - \varphi_B) \tag{7}$$

Combustion is modeled as an instantaneous transformation of unburned charge into equilibrium burned products according to the appropriate Wiebe function.

Eq. 7 requires specification of the start of combustion, φ_B, end of combustion, φ_E, and the Wiebe exponent, w. It is further assumed, that fuel quantity B_1 is injected into the prechamber through an injector and that a fuel quantity B_2 per engine cycle is brought in with the intake air flow. The distribution of fuel enthalpy at ignition therefore depends on the distribution of the fresh charge at that time.

$$I_{VK} = \frac{B_1 H_u}{V_H p_1} + \frac{B_2 H_u}{V_H p_1} \frac{m_{VK,uv,B}}{m_{VK,uv,B} + m_{Z,uv,B}} \tag{8}$$

$$I_Z = \frac{B_2 H_u}{V_H p_1} \frac{m_{Z,uv,B}}{m_{VK,uv,B} + m_{Z,uv,B}} \tag{9}$$

During the combustion process, a fuel exchange between the main combustion chamber and the prechamber takes place. This results in a significant enthalpy exchange between the two volumes. The unburned mass that leaves the prechamber and flows into the main chamber, as well as the unburned mass from the main chamber that might be pushed into the prechamber, is calculated.

For the cylinder, the energy-release expression in Eq. 4 is

$$\frac{dq_{R,Z}}{d\varphi} = \frac{m_{Z,uv,B} - m_{uv,ZVK}}{m_{Z,uv,B}} I_Z \frac{dx_Z}{d\varphi} + \frac{m_{uv,VKZ}}{m_{VK,uv,B}} I_{VK} \frac{dx_{VK}}{d\varphi} \tag{10}$$

As explained above, the energy release in the cylinder results from a corrected mass of the main chamber mixture and, in part, from a fraction of the prechamber mixture.

Eq. 6 must be replaced during combustion by the following relation:

$$\frac{dz_Z}{d\varphi} = \left[720 \frac{s}{c_1 t_c} \frac{m_{Z,uv,B} - m_{uv,ZVK}}{\rho_1 V_H} \frac{dx_Z}{d\varphi} + \frac{m_{uv,VKZ}}{\rho_1 V_H} \frac{dx_{VK}}{d\varphi} - (z_{KA} - z_Z) \frac{F_{KA} w_{KA} \rho_{KA}}{F_K c_1 \rho_1} \right] \Big/ \left(720 \frac{s}{c_1 t_c} \frac{m_Z}{\rho_1 V_H} \right) \tag{11}$$

In the case of excess air it is assumed that products of combustion are formed only by the stoichiometric parts of air and fuel. The remainder is 'air'.

The application of two Wiebe functions for the two-stage combustion process has proven satisfactory. However, one should realize that both stages of combustion are coupled. The right choice of Wiebe parameters, depending on engine operating conditions, is not simple. The effects of the two exponents for the first and second stage of combustion, as well as the fraction of prechamber fuel enthalpy of the total energy release, have been investigated in a separate study. For some cases, the Wiebe parameters have been determined by comparing the theoretical energy release with that obtained from a thermodynamic pressure analysis.

The combustion duration used here represents the total time from ignition until the end of combustion, and not, as often done, the time for a 10 to 90% conversion.

Between the start of combustion in the prechamber and the start of combustion in the main chamber, a delay of 5 crank angle degrees has been assumed in most of the investigated cases.

Eqs. 4 through 6 and Eq. 11 contain the unknown velocities, w/c_1, pressures, p/p_1, and exhaust-gas fraction, z. Depending on the pressure ratio between the

References pp. 535–536.

two connected volumes, one can determine flow direction and check for critical flow. One must also differentiate between two cases, namely, where the velocity to be calculated is either upstream or downstream of a flow restriction. The derivation of the boundary equations is not presented here.

The system of linear first-order ordinary differential equations is solved with a modified Hamming predictor-corrector method with variable step size [17]. In addition, integrated quantities such as indicated mean effective pressure, volumetric efficiency, total wall heat loss, etc., are calculated.

At least two complete engine cycles were calculated in order to obtain a solution under periodic conditions. The calculations were carried out on an IBM 370/168 computer.

SIMILARITY PARAMETERS

A set of similarity parameters which uniquely describe the processes is obtained from the coefficients of the dimensionless equations. Therefore the computations have a certain degree of generality. Many design parameters and operating variables, however, can only be varied within narrow limits in practice.

These parameters (most of which are illustrated in Fig. 2) are summarized and defined in the following section. For the cases that are reported here, the specific values or ranges are given.

Engine speed parameter, $s/c_1 t_c$, = 0.0053 *to* 0.0117
(with piston stroke, s, = 80 mm)

Inflow area (with piston area, F_K, = 4960 mm^2 and cylinder bore, D, = 79.5 mm)	$F_2/F_K = 0.135$
Maximum intake port area	$F_3/F_K = 0.135$
Maximum exhaust port area	$F_4/F_K = 0.120$
Outflow area	$F_5/F_K = 0.151$
Connecting port area	$F_{KA}/F_K = 0.012$ *to* 0.042
Inlet system surface	$F_{IN}/F_K = 3.13$
Exhaust system surface	$F_{EX}/F_K = 3.84$
Prechamber surface	$F_{VK}/F_K = 0.29$ *to* 0.60
Intake system volume	$V_{IN}/V_H = 0.288$
Exhaust system volume	$V_{EX}/V_H = 1.77$
Prechamber volume	$V_{VK}/V_H = 0.013$ *to* 0.0395
Intake valve opening	φ_{IVO} = 73 deg BTDC
Intake valve closing	φ_{IVC} = 115 deg ABDC
Exhaust valve opening	φ_{EVO} = 122 deg BBDC
Exhaust valve closing	φ_{EVC} = 66 deg ATDC
Maximum inlet valve lift	$l/d_{IN} = 0.302$
Maximum exhaust valve lift	$l/d_{EX} = 0.332$
Start of combustion in cylinder	$\varphi_{B,Z}$ = 35 to −5 deg BTDC
End of combustion in cylinder	$\varphi_{E,Z}$ = 50 to 115 deg ATDC

Start of combustion in prechamber ..	$\varphi_{B,K}$ = 40 to −10 deg BTDC
End of combustion in prechamber ...	$\varphi_{E,VK}$ = 10 to −85 deg BTDC
Wiebe exponent for cylinder	$w_Z = 0.35$ *to* 2
Wiebe exponent for prechamber	$w_{VK} = 0.35$ *to* 4
Start of fuel injection	$\varphi_{E,B}$ = 70 deg BTDC
End of fuel injection	$\varphi_{E,E}$ = 60 deg BTDC
Enthalpy of prechamber injection fuel (1.7 corresponds to about 2 mm^3/ stroke)	$B_1H_u/V_Hp_1 = 0$ *to* 3.4
Enthalpy of main intake fuel	$B_2H_u/V_Hp_1 = 7.9$ *to* 30
Flow coefficient at inlet	$\alpha_{12} = 0.2$ *to* 0.9
	$\alpha_{21} = 0.2$ *to* 0.9
Flow coefficient at exit	$\alpha_{56} = 0.9$
	$\alpha_{65} = 0.9$
Flow coefficient at connecting port ..	$\alpha_{ZVK} = 0.7$
	$\alpha_{VKZ} = 0.85$
Compression ratio	$\epsilon = 8.5$
Connecting rod ratio	$\lambda = 0.294$
Specific heat ratio	$k = 1.35$ or $F(T,p,\lambda)$

For the specification of boundary conditions:

Relative heat of combustion	$H_u/c_vT_1 = 20.8$
Relative heat of vaporization	$R/H_u = 0.0076$
Inlet pressure	p_0/p_1 = 0.5 to 1.0
Exhaust back pressure	p_6/p_1 = 1.05 to 1.24
Inlet temperature	$T_0/T_1 = 1.0$
Exhaust temperature	T_6/T_1 = 3.0 to 4.0
Inlet-system wall temperature	$T_{W,IN}/T_1 = 1.1$
Exhaust-system wall temperature	$T_{W,EX}/T_1$ = 1.4 to 2.3
Cylinder wall temperature	$T_{W,Z}/T_1 = 1.6$
Prechamber wall temperature	$T_{W,VK}/T_1$ = 1.8 to 3.6
Exhaust-gas fraction in intake system	$z_1 = 0$
Exhaust-gas fraction in exhaust system	z_6 = 0.87 to 1.0

The reference pressure, p_1, is 0.1 MPa and the reference temperature, T_1, is 293 K (20°C).

ERROR SENSITIVITY ESTIMATE

Before practical calculations were made, the sensitivity of the computed results to uncertain parameters was checked. Heat transfer and combustion simulation parameters were regarded as most important.

Influence of Heat Transfer—The heat flow between the gas and cylinder wall or prechamber wall, respectively, depends on the heat-transfer coefficient and

References pp. 535–536.

the temperature difference. A large number of heat-transfer-coefficient correlations can be found in the literature. A specific correlation for the prechamber, accounting for inflow and outflow conditions, was not available. In Ref. 9 the same correlation is used for both combustion chambers. For an accurate description, time- and space-average wall temperatures are necessary, depending on the operating conditions. Initially, temperatures from conventional engines had to be used for a first approximation.

The sensitivity study showed that a 50% increase in the prechamber heat-transfer coefficient had a small effect in all investigated cases because of the relative small prechamber surface area in comparison with the total heat-transfer area. The peak prechamber gas temperature dropped by only 3%. Specific studies on the heat-flow pattern of a prechamber insert have shown that a fairly good agreement between experiment and calculation is obtained if the same heat-transfer correlation is used for the prechamber and main chamber. As expected, a modification of the heat-transfer coefficient for the main chamber shows a more pronounced effect.

Influence of Energy Release—The start of combustion is determined by ignition timing. Combustion duration and the Wiebe exponent determine the center of the energy release. Thermal efficiency decreased by only 1.5% when prechamber combustion duration was increased by 50%. The maximum gas velocity at the connecting port was significantly reduced and occurred later in the cycle. The maximum peak pressure was 18% less.

A reduction of both Wiebe exponents from 1.0 to 0.5 corresponds to a larger initial combustion rate and resulted in higher peak velocities at the connecting port (115 m/s versus 75 m/s), an increase in peak combustion pressure from 2.8 to 3.4 MPa, and an increase in indicated mean effective pressure by 6%.

These quantitative changes are valid only for the chosen geometric specifications.

PARAMETRIC STUDY

The following investigations were made for an engine operating condition which frequently occurs during the Federal emission test cycle, namely, an engine speed of 2700 r/min and an engine indicated mean effective pressure of approximately 0.3 MPa. Some calculations were carried out at full load and nominal speed. In the standard case, F_{KA}/F_K was 0.023 and V_{VK}/V_H was 0.023.

Influence of Ignition Timing—First, the effect of ignition timing for a given combustion duration was studied. In the standard case, it was assumed that the first stage of combustion lasts 75 crank angle degrees and the second stage starts 5 degrees later and lasts 110 degrees. These values may seem fairly large. However, they are based on a thermodynamic analysis and represent 0 to 100% combustion. In addition, calculations were also made for much shorter combustion periods. The outflow velocity from the prechamber usually reaches two maxima following ignition. This will be explained later with an example. The first velocity peak occurs shortly after ignition, even with the piston still ascending. During this event a prechamber torch will form and initiate the main chamber combustion. In Ref. 5 a delay of 6 crank angle degrees is suggested between the first and second stage of combustion. From the observations made with these calculations, it seems advisable to couple the start and the rate of the second stage of combustion with the timing and momentum of the torch jet. Such correlations have not been investigated thus far.

The dependency of the mass in the prechamber on ignition timing is shown in Fig. 3. The dimensionless mass in the cylinder and prechamber is approximately 0.40 in all cases. In the standard case, with long combustion periods, the mass leaving the prechamber after combustion is complete in the prechamber is fairly small. The change in prechamber mass during the exhaust valve flow is almost negligible. Because of the generally rich prechamber mixture and its importance to emissions of unburned hydrocarbons, this seems to be a beneficial result. In principle, this finding is supported by time-resolved measurements [18] of the exhaust.

Combustion simulation by Wiebe functions is also employed in the work reported in Refs. 4 and 6. However, the combustion periods for the main chamber and prechamber are much shorter. Results with a prechamber combustion duration of only 30 degrees and a main chamber combustion duration of 60 degrees are also plotted in Fig. 3. The mass which leaves the prechamber after prechamber

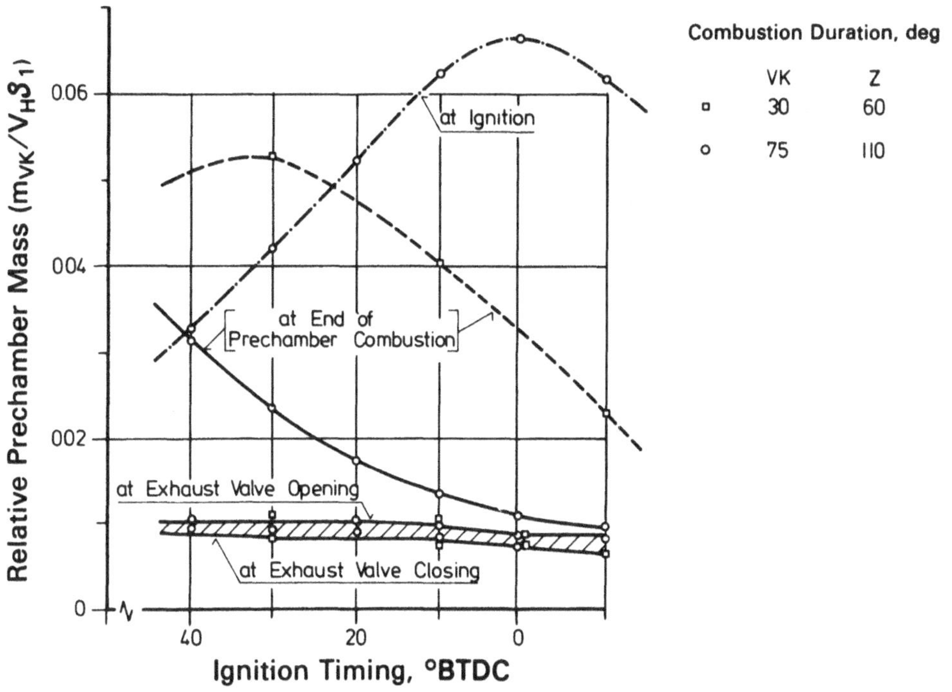

Fig. 3. Mass fraction in prechamber after ignition (part load).

References pp. 535–536.

combustion is terminated and which contains high concentrations of unburned products is much larger in this case. During the exhaust flow period, the mass change is equally as small as in the standard case.

Fig. 4 contains additional information from this cycle calculation. The peak combustion pressure is strongly reduced with late ignition. The average relative air-fuel ratio in the prechamber, λ_{VK}, at ignition also changes significantly because of the different degree of dilution with leaner main-chamber charge. The degree of stratification, which is defined as the ratio of fuel injected into the prechamber to total fuel, does not change with ignition timing. The maximum indicated mean effective pressure is obtained in the standard case with ignition at 20 degrees BTDC. Higher peak and mean effective pressures, as well as improved fuel consumption, are obtained with shorter combustion periods.

The cylinder pressure development for ignition occurring between 40 degrees BTDC and 10 degrees ATDC is shown in Fig. 5. The maximum peak pressure is 2.2 MPa for ignition at 40 degrees BTDC. With ignition 10 ATDC, it is lower than the peak compression pressure.

In these calculations, volumetric efficiency was 32% at a total relative air-fuel ratio of 1.23, which corresponds to a relative main chamber air-fuel ratio of 1.44. The residual gas fraction in the prechamber at ignition was 13%. The maximum velocity in the connecting port was in the range of 40 to 58 m/s. The peak value is obtained at late ignition and occurs generally 20 to 30 degrees ATDC.

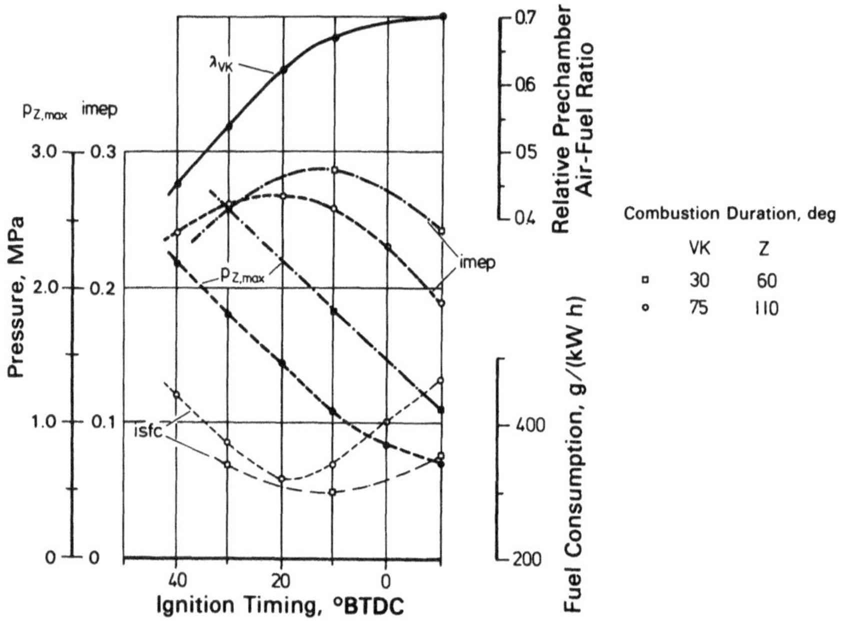

Fig. 4. Effect of ignition timing on indicated mean effective pressure, peak pressure, indicated specific fuel consumption and relative air-fuel ratio (part load).

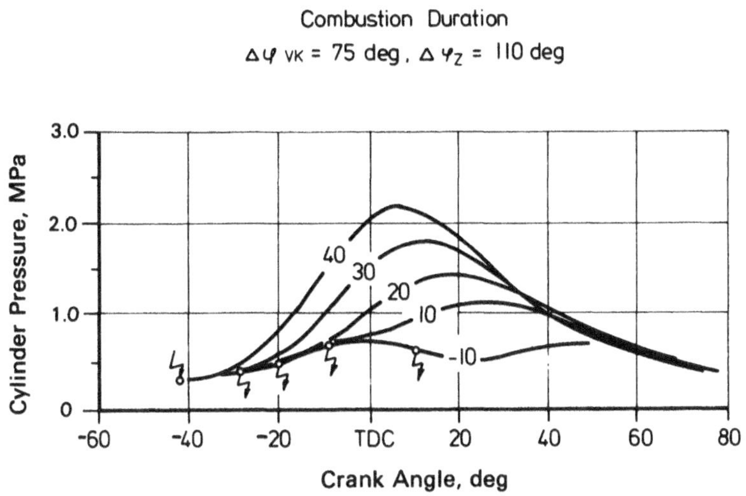

Fig. 5. Effect of ignition timing on cylinder pressure development (part load).

Influence of Main Chamber Combustion Duration—Experimental investigations were also carried out at the specified operating condition. The energy-release rate and combustion duration were computed from a thermodynamic analysis of indicated cylinder pressure development.

The total mass burned computed from measured cylinder pressure data is plotted in Fig. 6 for two levels of stratification and a relative air-fuel ratio of 1.15. In both cases ignition was set at 20 degrees BTDC, which was found to result in

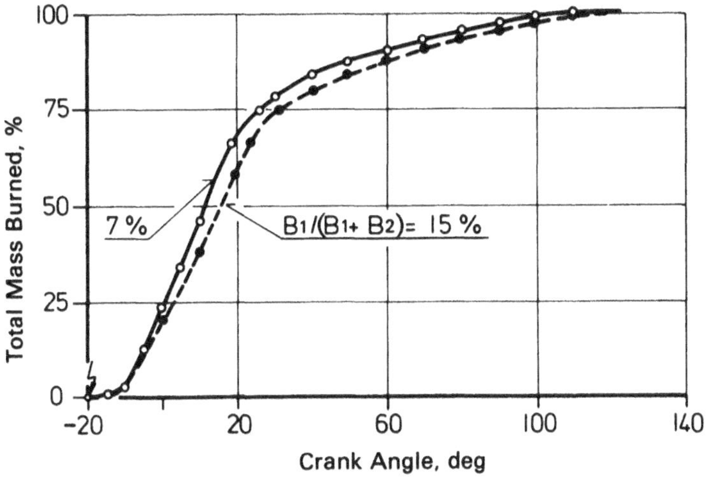

Fig. 6. Total mass burned for 7 and 15% stratification as computed from measured cylinder pressure development (part load).

References pp. 535–536.

minimum fuel consumption. Combustion is visibly accelerated 10 degrees later. This ignition delay is rather short when compared with typical homogeneous-charge engine values. After 40 crank angle degrees, approximately 60% of the total fuel is burned. If combustion were continued at this rate, excellent thermal efficiency could be expected. Unfortunately, combustion rates are significantly reduced during expansion. Approximately 80 crank angle degrees are necessary to burn the remaining 40% of the fuel. After completion of prechamber combustion (without torch assistance), the total combustion rate is significantly reduced. A reduction in the degree of stratification from 15 to 7% shows little improvement. Total combustion duration is 140 crank angle degrees, while 10 to 90% conversion requires 100 degrees. This slow burning phenomenon is probably the result of rich products from the prechamber mixing slowly and reacting with the excess oxygen in the main chamber.

Main combustion duration was varied at a fixed ignition timing (10 degrees BTDC). In one case, the prechamber combustion duration was 75 crank angle degrees and total relative air-fuel ratio was set at 1.24. In a second series of calculations, prechamber combustion duration was reduced to 30 crank angle degrees (as in the previous section) and total relative air-fuel ratio was simultaneously increased with main chamber combustion duration. Therefore throttling also changed slightly. A relationship between air-fuel ratio and combustion duration was taken from an investigation [19] for a conventional spark-ignition engine.

As can be seen in Fig. 7, in every case a delay in combustion results in a drop

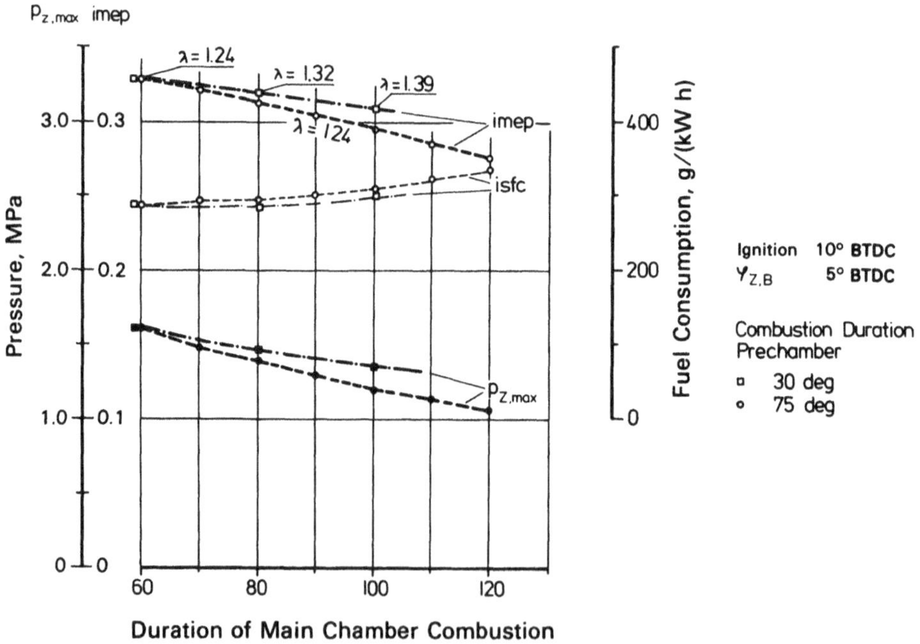

Fig. 7. Indicated mean effective pressure, peak pressure and indicated specific fuel consumption as a function of main chamber combustion duration (part load).

in peak pressure and indicated mean effective pressure and also an increase in fuel consumption. In the second case, peak pressures and mean effective pressures are slightly higher.

The temperature development is affected more significantly. Exhaust gas temperature at exhaust valve opening increases by 250°C if combustion is delayed by 50 crank angle degrees. The combined effects of changes for both Wiebe functions will be discussed later in more detail.

Influence of Prechamber Volume and Connecting Port Size—Two design parameters, prechamber volume and connecting-port cross-sectional area, characterize the divided-chamber engine. Both have a significant effect on the inflow and ignition process in the prechamber and on the intensity of the torch. Ignition of main-chamber charge and the rate of main-chamber heat release are determined, among other things, by the outflow from the prechamber.

First, the influence of the prechamber volume at a given connecting-port size (F_{KA}/F_K = 0.023, corresponding to a 12-mm diameter) is considered. With ignition at 10 degrees BTDC, a combustion duration of 75 crank angle degrees, and a compression ratio of 8.5, the prechamber volume was varied from 10 to 30% of the clearance volume. In practice one must expect the combustion process to change during such a design modification. However, the relationship is not known. Some results are summarized in Fig. 8. The relative air-fuel ratio of the prechamber increases with prechamber volume at a given degree of stratification (15% of the total fuel is injected into the prechamber) and at a constant total relative air-fuel ratio of 1.24. Also shown in Fig. 8 is the relative fuel enthalpy in the main chamber and prechamber at ignition. The maximum gas velocity at the connecting port, $w_{KA,max}$, changes approximately proportionally with prechamber volume.

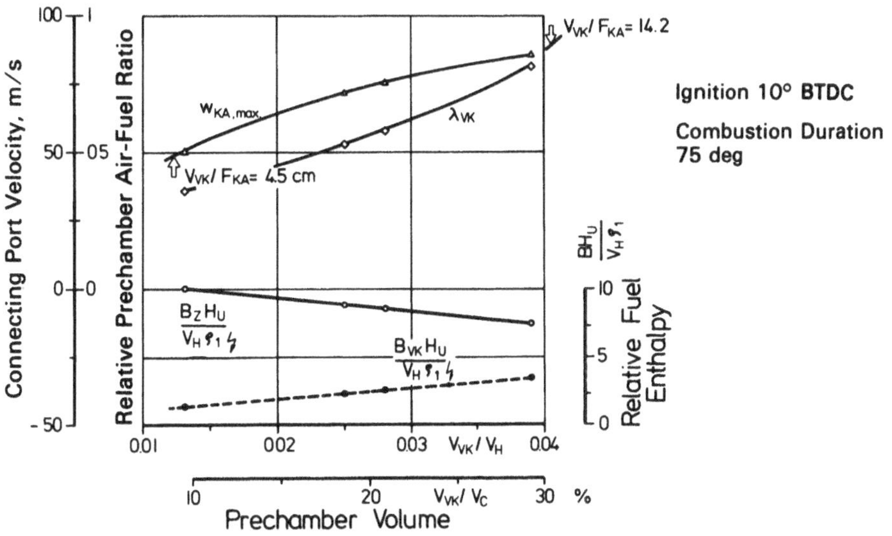

Fig. 8. Influence of prechamber volume for F_{KA}/F_K = 0.023 (part load).

References pp. 535–536.

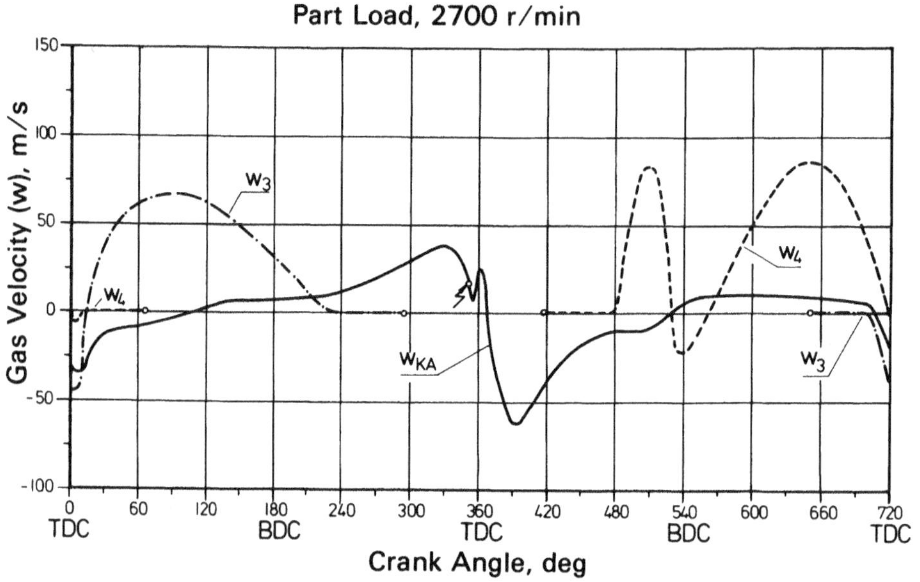

Fig. 9. Velocities at connecting port, intake and exhaust port (part load).

To demonstrate the processes in more detail, some characteristic quantities are plotted versus crank angle in Figs. 9, 10 and 11. They are valid for a prechamber size of 21% of the clearance volume. In Fig. 9 the history of gas velocities in the connecting port and in the intake and exhaust port can be seen. Because of part-load operation, the velocities are not very high. Following ignition, the inflow velocity is strongly reduced, but not completely reversed. Maximum velocity in

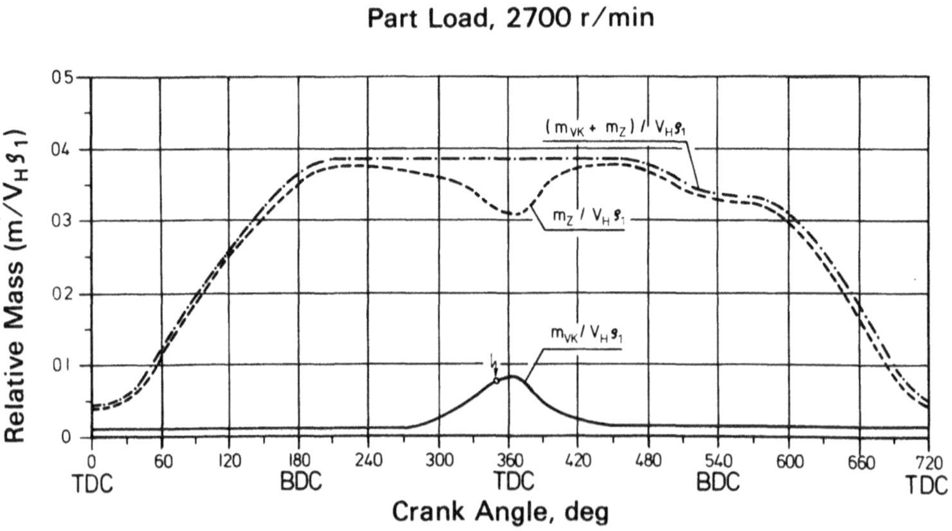

Fig. 10. Mass in prechamber and in main chamber (part load).

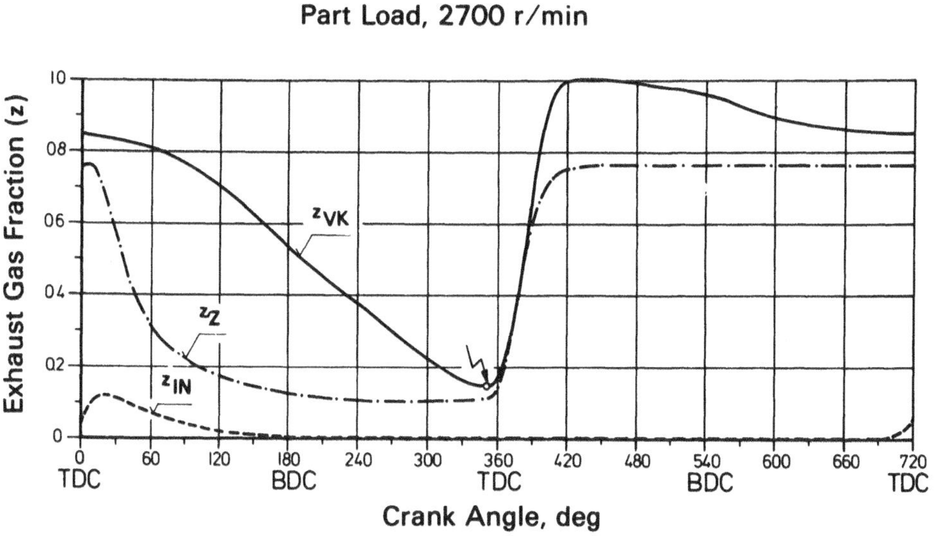

Fig. 11. Exhaust gas fraction in prechamber, main chamber and intake system (part load).

the intake port is 67 m/s and still positive after BDC. The exhaust process shows a pronounced blowdown effect with a short flow-reversal period. The maximum velocity is 85 m/s. The relative mass is plotted in Fig. 10. The prechamber mass is almost constant during 75% of cycle time. Because of an early pressure rise in the main chamber, the prechamber mass increases slightly, even after TDC. The exhaust gas fraction in the intake system, the cylinder, and the prechamber are plotted in Fig. 11. Because of flow reversal, an exhaust gas fraction of up to 12% is experienced in the intake system for a short period. The exhaust gas fraction in the cylinder, z_Z, drops to 10% during the intake process and reaches 75% at the end of combustion. The change in residual gas fraction in the prechamber is delayed and decreases only because of overflow of fresh charge from the main chamber. The residual gas fraction is 13% at ignition and 100% at the end of prechamber combustion. During the exhaust stroke, the prechamber residual gas fraction is reduced because of dilution by mass from the main chamber.

The influence of a constant connecting-port flow area is considered next. This parameter strongly affects the gas velocity and therefore the turbulence during the mass exchange between the prechamber and main chamber. Optimization for low exhaust emissions and low combustion noise has led to relatively large cross-sectional areas. At engine part-load operation, the influence of a parameter change in F_{KA}/F_K from 0.013 to 0.043 was studied. This corresponds to a connecting port diameter variation of 9 to 16.5 mm. Prechamber volume was 0.023, or 17% of the clearance volume. In Fig. 12 maximum gas velocities during the compression and power strokes are plotted. The velocity peak occurs at approximately the same crank angle position. Because of improved scavenging, the residual gas fraction in the prechamber at ignition is slightly reduced with increased orifice area.

References pp. 535–536.

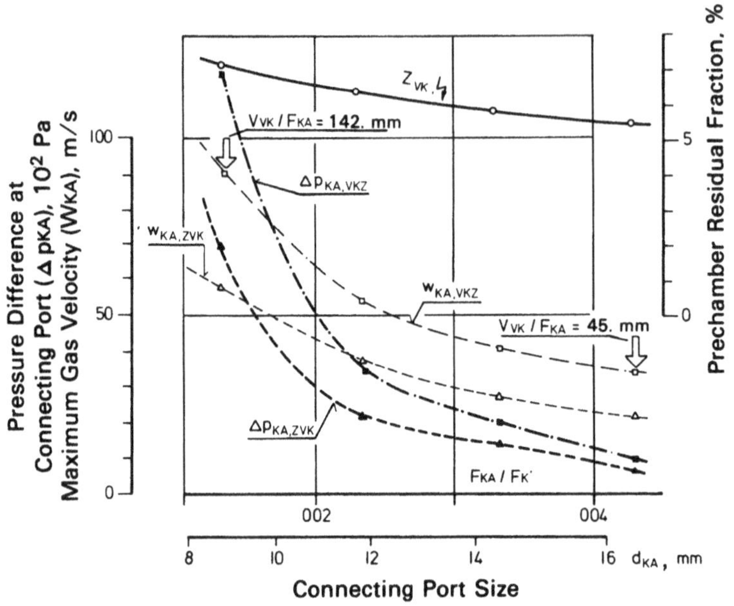

Fig. 12. Influence of connecting port size for $V_{VK}/V_H = 0.023$ (part load).

The pressure differential at the torch opening does not affect the indicated mean effective pressure. $\Delta P_{KA,VKZ}$ only rises with small orifice diameter. The combustion peak pressure is not noticeably affected.

Various authors, e.g., Ref. 20, use the ratio of prechamber volume to cross-sectional area as a characteristic parameter. In Figs. 8 and 12, two cases with V_{VK}/F_{KA} of 45 and 142 mm, one at constant volume and one at constant area, have been marked with arrows. A comparison shows that the peak outflow velocity at $V_{VK}/F_{KA} = 45$ mm is 35 or 48 m/s, respectively. At $V_{VK}/F_{KA} = 140$ mm, it is approximately 90 m/s in both cases.

Kuck and Brandstetter have investigated experimentally the dimensionless torch opening parameter $(F_{KA}/V_{VK})(V_H/F_K)$ on an engine having a scavenged prechamber.

As can be seen from their experimental data in Fig. 13, one can change the magnitude and position of minimum fuel consumption, depending on the relative air-fuel ratio. At a slightly lean mixture, low NOx and fuel consumption are obtained with wider openings, while with much leaner mixtures, small cross-sectional areas are required.

On a smaller engine with an unscavenged prechamber, maximum inflow and outflow velocities were calculated for a combination of connecting port and prechamber volume size such that V_{VK}/F_{KA} was 80 mm in all cases. Results, which are summarized in Table 1, were obtained for zero prechamber injection and for an engine speed of 6000 r/min.

Significant differences were found during inflow as well as during outflow for

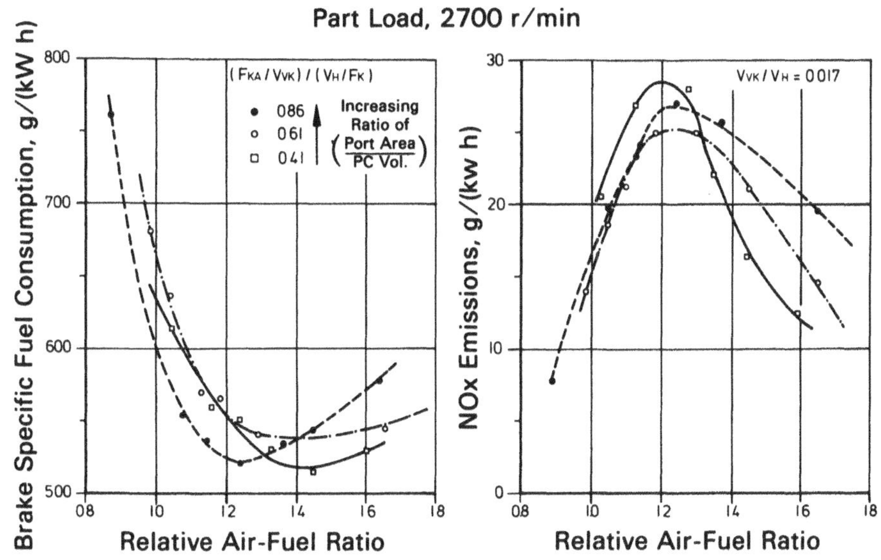

Fig. 13. Engine data showing the effect of torch opening parameter on fuel consumption and NOx emissions (part load).

the same V_{VK}/F_{KA}, which indicates that this parameter does not guarantee similarity.

Influence of Degree of Stratification—Another parameter of interest is the "degree of stratification," namely, the ratio of fuel injected into the prechamber to the total fuel. The simulation was carried out in two steps. First, the influence of the resulting increase in fuel enthalpy in the prechamber was studied. Everything else was kept constant. Four cases (with a stratification of 0, 7.5, 15 and 30%) were analyzed. With a total relative air-fuel ratio of 1.24, the average relative air-fuel ratio in the prechamber at ignition varies with increasing stratification from 1.24 (0%) to 0.5 (30%), while the relative air-fuel ratio of the main-chamber mixture increases simultaneously from 1.24 (0%) to 1.8 (30%). (See Fig. 14.) In cases with very rich prechamber mixtures, a strong increase in the

TABLE 1

Maximum Inflow and Outflow Velocity in the Connecting Port for Prechamber Volumes and Connecting Port Diameters with V_{VK}/F_{KA} = 80 mm (6000 r/min)

V_{VK}/V_C (%)	d_{KA} (mm)	$w_{max,ZVK}$ (m/s)	$w_{max,VKZ}$ (m/s)
5	5.7	100	57
10	8.1	95	66
20	11.5	75	60

References pp. 535–536.

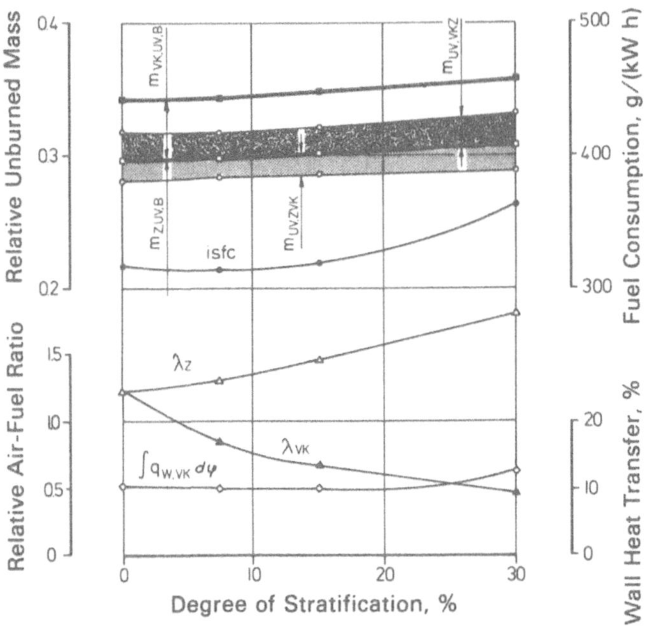

Fig. 14. Influence of degree of stratification (part load).

prechamber energy-release rate is expected, resulting in a rapid rise in temperature and pressure. Outflow from the prechamber, however, carries with it a significant fuel fraction, thereby reducing this effect.

In all cases studied, the residual gas fraction at ignition was approximately 10% in the main chamber and 11% in the prechamber. The volumetric efficiency was nearly constant, its slight increase with increasing stratification reflecting the replacement of fuel vapor with liquid fuel. When simulating part-load operation, backflow at the start of intake takes place. As a result, the residual gas fraction quickly reaches 10% in the intake plenum.

The experimentally determined influence of the degree of stratification on the burning rate has already been mentioned. (See Fig. 6.) From such a thermodynamic analysis of measured pressure data, Wiebe exponents were estimated such that a close fit for the simulation procedure could be expected. In a second series of calculations, the main-chamber burning duration was increased from 90 (0%) to 100 (7.5%), 110 (15%) and 130 crank angle degrees (30%). Furthermore, it was assumed that with no stratification, combustion duration in the prechamber and main chamber were equal. As the prechamber mixture became richer, shorter prechamber combustion durations were assumed. Under these assumptions, indicated mean effective pressure is affected little initially (0 to 10% stratification). However, a loss up to 15% was noticed at 30% stratification.

It has been discussed above that some unburned mass will leave the prechamber following ignition. A small pressure rise resulting from prechamber combustion will produce more displacement of mass than products of combustion formed. As mentioned earlier, these mass fractions are assumed to react in the main chamber according to the burning law of the prechamber, because of their original mixture ratio. It can be seen from Fig. 14 how much of the total unburned mass is in the prechamber ($m_{VK,uv,B}$) and how much is in the main chamber ($m_{Z,uv,B}$) at ignition. The variation with stratification is insignificant and within the computational accuracy. The mass fraction of the prechamber at ignition, which occurs at 20 degrees BTDC in all cases, is 14%, while the prechamber volume is 17% of the clearance volume. During the combustion process, 58% of the initially unburned prechamber mass ($m_{VK,uv,B}$) flows unburned into the main combustion chamber. On the other hand, because of piston motion and combustion events, approximately 8% of the unburned main chamber charge ($m_{Z,uv,B}$) is likely to react in the prechamber.

The ratio of wall heat flow in the prechamber to the total heat losses is also plotted in Fig. 14. For the prechamber size studied here, the fraction is within 10 to 13% and increases slightly with prechamber fuel quantity. Similar relative quantities were found at the full-load nominal-speed operating condition.

The calculated indicated specific fuel consumption (isfc) increases with higher stratification. Between 0 to 10% stratification, isfc is practically unaffected. Measured data showing the effect of the degree of stratification on emissions and fuel consumption for a similar prechamber configuration are presented in Fig. 15.

These experimental results are obtained for three prechamber injection quantities, namely, 0, 1.4 and 3.1 mm³ fuel per cycle, and are plotted versus engine load. The degree of stratification varied from 0 to 20%, while the total relative air-fuel ratio was constant at 1.15. As can be seen from Fig. 15, larger injection quantities increase CO emissions unacceptably at low loads, while NOx emis-

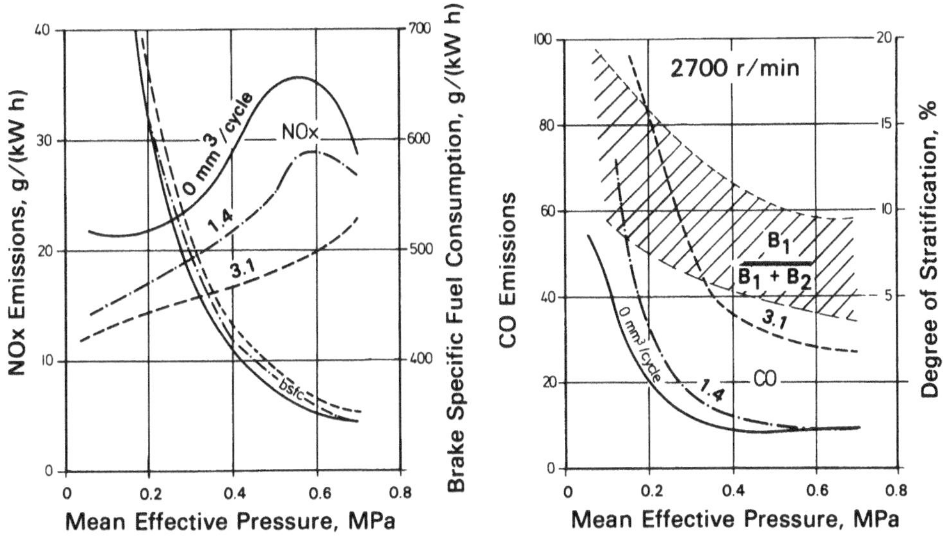

Fig. 15. Engine data showing the effect of degree of stratification on emissions and fuel economy.

References pp. 535–536.

sions are low over the entire load range. The fuel consumption increases with injection quantity. An injection quantity of 1.4 mm^3 per cycle appears to be a reasonable compromise between low emissions and good fuel economy.

The relationship between NOx emissions and fuel consumption for different stratification in a 3-valve divided-chamber engine has also been examined in a theoretical parameter study by Wall et al. [7]. Similar results were obtained.

Influence of Engine Speed at Full Load—The effect of engine speed was also investigated for unthrottled conditions. Energy release, burning duration, ignition timing, exhaust back pressure, wall temperatures, etc., were kept constant. With a prechamber volume of 21% and a connecting port size of $F_{KA}/F_K = 0.023$, the peak inflow velocity computed was 70 m/s, and three outflow peaks were noticed. (See Fig. 16.) The first maximum results from the pressure differences between the prechamber, cylinder and intake plenum shortly after the intake valve opens. The second outflow peak occurs a few degrees after ignition and has a maximum velocity of 107 m/s. Following the start of combustion in the main chamber, the mass increases in the prechamber for a short period, and then the main outflow process starts with a velocity peak of 70 m/s at approximately 30 degrees ATDC. Note that the maximum velocity in the intake port exceeds 150 m/s and is thereby higher than the maximum velocity in the connecting port. The maximum calculated outflow velocity in the exhaust port is 490 m/s.

Fig. 17 represents the cumulative mass interaction between the two combustion chambers. The outflow of mass shortly after ignition can clearly be seen. The computed peak combustion pressure of 5.0 MPa is identical with the measured average peak pressure of the four cylinders of a multicylinder engine [10].

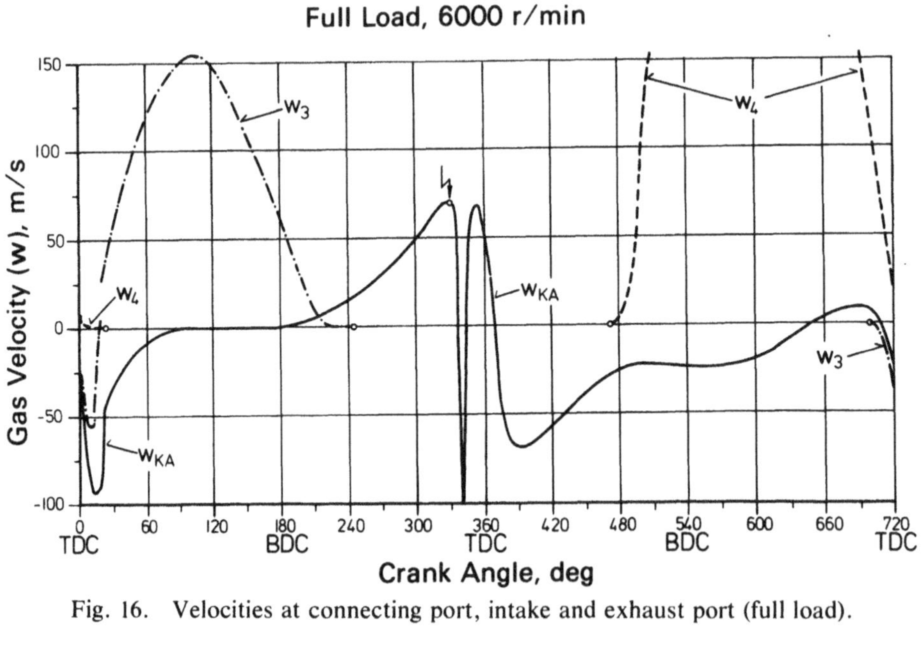

Fig. 16. Velocities at connecting port, intake and exhaust port (full load).

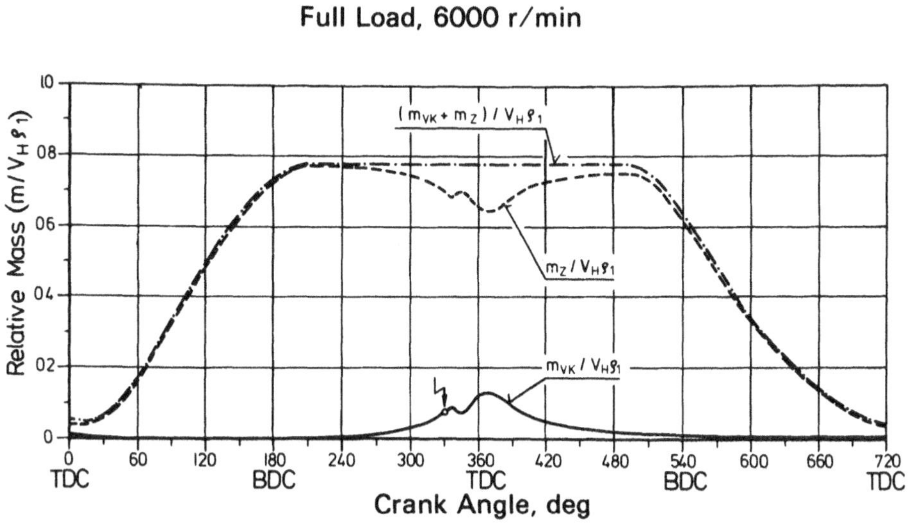

Fig. 17. Mass in prechamber and in main chamber (full load).

INVESTIGATION OF SPECIAL CASES

The simulation program has been used to study undivided combustion chambers, as well as effects of valve timing, exhaust gas recirculation, compression ratio, etc. A study of special cases is demonstrated for a connecting-port opening that varies with time.

Varying connecting-port size required a minor program expansion. A simple functional relationship was assumed. The maximum connecting port area was linearly reduced to a ring gap as the piston moved toward TDC. This could be realized by a pin on top of the piston which penetrates the prechamber near TDC. It is expected that the 2-stage combustion process could be significantly influenced by such means. Volume and injection quantities other than those studied thus far may become of interest. With the aid of this mathematical model, the most promising solutions can be optimized during the design stage, even before testing begins. Calculations were mainly made for full load and at high speed.

During several steps of the calculation it was assumed that the connecting port area was reduced by up to 95%, leaving a theoretical gap width of only 0.15 mm. In the following computations, it was concluded that the high outflow velocities would accelerate the second stage of combustion. Input modifications were made accordingly. Because of the very large change in the free cross-sectional area, large pressure differences between the two chambers result for part of the cycle. From the results obtained, it was apparent that compression ratio could be increased without exceeding allowable peak pressures.

The following combination of parameters was found to be optimum: compression ratio = 14:1, ignition = 40 degrees BTDC, area reduction of connecting port = 90%, prechamber to compression volume = 40%. Compared with the standard

References pp. 535–536.

case, the mean effective pressure increased to 1.19 MPa at lower peak pressures and favorable process temperatures. Isfc was reduced by 6%. The maximum gas velocity in the gap was less than 200 m/s.

The weakest point in these calculations is probably the assumptions made in regard of the energy-release rate.

SUMMARY AND CONCLUSIONS

From the large number of problem parameters, only the most important ones were investigated in this study. Many more quantities than those presented are available from such a calculation. Program modifications, such as the incorporation of global reaction kinetics, could easily be made in the future.

The following main conclusions can be made from this work:

(i) It is not advisable to have one universal simulation model for a large variety of problems. Because of the complexity involved, it is better to have specifically tailored programs.

(ii) The present computer program has proven helpful, particularly in assisting in data interpretation during the experimental stage. Of particular advantage is the fact that the program calculates quantities which can be measured only at great expense.

(iii) Little information is available to describe accurately the heat transfer to the prechamber wall. Because of the relatively small surface, however, overall effects on the quantities studied here are also small.

(iv) The total combustion duration that was derived from a thermodynamic analysis of measured pressure-time curves and used in many of the calculations was longer than values found in the literature. Calculations carried out with a much shorter combustion period showed no major change.

(v) In most calculated cases, a first jet of outflow after ignition was followed by flow reversal before the primary prechamber outflow occurred during the power stroke. It is recommended that the start and the rate of the second stage of combustion be coupled with this event. This has not yet been accomplished.

(vi) The calculations further indicate that during the entire exhaust valve opening period, only a negligible amount of mass leaves the prechamber. CO and unburned HC concentration found in the exhaust therefore originate mainly from incomplete combustion in the main chamber.

(vii) The residual gas fraction in the prechamber at ignition is approximately 10 to 15%. For all parameters investigated, little change was noted.

(viii) Main-chamber combustion duration is one of the most important factors influencing the quantities considered here. The effects found are similar to those known from conventional engines. Prechamber combustion duration has less effect because of the smaller mass involved.

(ix) The maximum velocity in the connecting port increases approximately linearly with prechamber volume.

(x) The mass contained in the prechamber is almost constant during 75% of the cycle time.

(xi) The pressure difference between the prechamber and main chamber is small and does not exceed 12 kPa at 2700 r/min with a connecting port diameter of 9 mm.

(xii) Similarity in the velocities developed at the connecting port could not be established for cases with identical V_{VK}/F_{KA}.

(xiii) As was also found in experiments, the calculations showed lower mean effective pressure, and therefore an increase in fuel consumption, with increasing degree of stratification.

(xiv) Following ignition, up to 60% of the fresh mixture in the prechamber is pushed out unburned and reacts in the main chamber.

(xv) At an engine speed of 6000 r/min and full load, as well as for the standard case (which also represents an optimum parameter choice from the experimental optimization), the maximum velocity at the connecting port is 100 m/s during inflow and 70 m/s during outflow.

(xvi) Calculations with a connecting port size that varies with crank position indicate that process improvements are possible. The accuracy of the assumption about the resulting combustion rate and the coupling of the first and second stage of combustion are still uncertain, however.

REFERENCES

1. *F. V. Bracco, "Modeling of Two-Phase, Two-Dimensional, Unsteady Combustion for Internal Combustion Engines," Stratified Charge Engine Conference, Proc. I. Mech. E., Paper C171/77, pp. 167–187, 1976.*
2. *H. C. Gupta, R. L. Steinberger and F. V. Bracco, "Divided-Chamber, Stratified-Charge Engine Combustion: A Comparison of Calculated and Measured Flame Propagation," SAE Paper No. 780317, 1978.*
3. *F. Pischinger, B. Kreykenbohm and W. Adams, "Experimental and Theoretical Investigations on a Stratified Charge Engine with Prechamber Injection," Stratified Charge Engine Conference, Proc. I. Mech. E., C240/76, pp. 1–12, 1976.*
4. *G. C. Davis, R. B. Krieger and R. J. Tabaczynski, "Analysis of the Flow and Combustion Processes of a Three-Valve Stratified Charge Engine with a Small Prechamber," SAE Trans., Vol. 83, Paper No. 741170, pp. 3535–3550, 1974.*
5. *S. D. Hires, A. Ekchian, J. B. Heywood, R. J. Tabaczynski and J. C. Wall, "Performance and NOx Emissions Modeling of a Jet Ignition Prechamber Stratified Charge Engine," SAE Trans., Vol. 85, Paper No. 760161, pp. 711–738, 1976.*
6. *T. Asanuma, M. K. Gajendra Babu and S. Yagi, "Simulation of Thermodynamic Cycle of a Three-Valve Stratified Charge Engine," SAE Paper No. 780319, 1978.*
7. *J. C. Wall, J. B. Heywood and W. A. Woods, "Parametric Studies of Performance and NOx Emissions of the Three-Valve Stratified Charge Engine Using a Cycle Simulation," SAE Paper No. 780320, 1978.*
8. *J. Furukawa, S. Mizumura, M. Yoshida and T. Gomi, "The Fundamental Research of Combustion in the Stratified Charge Engine with an Auxiliary Chamber," Seventeenth Congress of FISITA, Budapest, June 1978.*
9. *F. Anisits and H. Zapf, "Auswertverfahren der Druckverläufe und elektronische Berechnung des Verbrennungsverlaufs in Dieselmotoren mit unterteilten Brennräumen," MTZ, Vol. 32. Jg., December 1971.*
10. *W. R. Brandstetter, G. Decker and K. Reichel, "The Water-Cooled Volkswagen PCI-Stratified*

Charge Engine," SAE Trans., Vol. 84, Paper No. 750869, pp. 2323–2333, 1975.
11. *W. R. Brandstetter and G. Decker, "Fundamental Studies on the Volkswagen Stratified Charge Combustion Process," Combustion and Flame, Vol. 25, pp. 15–23, 1975.*
12. *J. J. Wiebe, "Brennverlauf und Kreisprozess von Verbrennungsmotoren," VEB-Verlag Technik Berlin, 1970.*
13. *R. B. Krieger and G. C. Davis, "The Influence of the Degree of Stratification on Jet-Ignition Engine Emissions and Fuel Consumption," Stratified Charge Engine Conference. Proc. I. Mech. E., C254/76, pp. 109–119, 1976.*
14. *S. Matsuoka, T. Kawakita, A. Oguri and H. Tasaka, "A New Concept 'Three Valve, Prechamber, Spark-Ignited Engine is Functioned by Three Stage Combustion Mechanism'," Seventeenth Congress of FISITA, Budapest, June 1978.*
15. *G. L. Borman, "Mathematical Simulation of Internal Combustion Engine Processes and Performance including Comparison with Experiment," Ph.D. Thesis, University of Wisconsin, 1964.*
16. *R. B. Krieger and G. L. Borman, "The Computation of Apparent Heat Release for Internal Combustion Engines," ASME Paper No. 66-WA/DGP-4, 1966.*
17. *Ralson and Wilf, "Mathematical Methods for Digital Computers," Wiley, New York, London, 1960.*
18. *A. Ekchian, J. B. Heywood and J. M. Rife, "Time Resolved Measurements of the Exhaust from a Jet Ignition Prechamber Stratified Charge Engine," SAE Trans., Vol. 86, Paper No. 770043, pp. 153–173, 1977.*
19. *S. D. Hires, R. J. Tabaczynski and J. M. Novak, "The Prediction of Ignition Delay and Combustion Intervals for a Homogeneous Charge, Spark Ignition Engine," SAE Paper No. 780232, 1978.*
20. *M. Noguchi, S. Sanda and N. Nakamura, "Development of Toyota Lean Burn Engine," SAE Trans., Vol. 85, Paper No. 760757, pp. 2358–2373, 1976.*
21. *H. A. Kuck and W. R. Brandstetter, "Investigations on a Single Cylinder Stratified Charge Engine with a Scavenged Prechamber," Combustion in Engines Conference. Proc. I. Mech. E., C92/75, pp. 105–112, 1975.*

J. H. Lienesch*

* *Mr. Lienesch of the General Motors Research Laboratories presented the paper of Dr. Brandstetter, who was unable to attend the Symposium.*

ENGINE IMPROVEMENTS THROUGH COMBUSTION MODELING

J. N. MATTAVI, E. G. GROFF, J. H. LIENESCH, F. A. MATEKUNAS and R. N. NOYES

General Motors Research Laboratories, Warren, Michigan

ABSTRACT

Previous studies have shown the rate of combustion in engines to be an important parameter affecting fuel consumption, emissions, operational smoothness, and tolerance for charge dilution. Specifically, higher fuel burning rates were shown to have many attractive attributes. Based on these findings, it was desired to increase the fuel burning rate in a single-cylinder engine having relatively long combustion duration. This paper describes how combustion modeling was used in both a diagnostic and predictive manner to identify a feasible design approach to accomplish this objective and improve engine performance.

A combustion model was developed using a quasi-dimensional propagating-flame approach. Flame motion derived from the model is compared with visual observations made using laser-schlieren cinematography in a rapid compression machine and using normal cinematography in a single-cylinder engine fitted with a transparent piston. Calculated and measured pressure histories are also compared and the validity of model assumptions examined.

Heat transfer and turbulence effects on flame speed are characterized empirically through the use of the model in a diagnostic mode in engine experiments. In this regard, a generalized correlation of flame speed with turbulent intensity and laminar flame speed in the unburned mixture is presented. Applying the diagnostic version of the model to analyze combustion in the single-cylinder engine leads to proposed chamber design modifications for improved engine performance. Integration of the predictive version into an overall engine simulation identifies the expected extent of these improvements. Finally, experimental results obtained with a single-cylinder engine incorporating the chamber design modifications are presented.

NOTATION

A	area
ATDC	after top dead center

BTDC	before top dead center
c_p	specific heat at constant pressure
c_v	specific heat at constant volume
CA	crank angle
CFR	Cooperative Fuel Research (engine)
EGR	exhaust gas recirculation
EINOx	nitric oxide emission index, g/kg
FSR	flame speed ratio, S_b/S_l
h	enthalpy
imep	indicated mean effective pressure
m	mass
MBT	minimum spark advance for best torque
NO	nitric oxide
p	pressure
Q	heat loss to surroundings
r	radius
R	gas constant
S	speed
t	time
T	temperature
TDC	top dead center
u	internal energy
u'	turbulent intensity
$\bar{u}$	average gas velocity
V	volume
α	empirical pressure coefficient

β	$1.0 - (p/R_b)\,\partial R_b/\partial p$
γ	$1.0 + (T_b/R_b)\,\partial R_b/\partial T_b$
ν	$\beta c_{v,b}/R_b + (\gamma p/R_b T_b)\,\partial u_b/\partial p$
σ	$\gamma + c_{v,b}/R_b$
ϕ	fuel-air equivalence ratio
θ	crank angle
ρ	density

Subscripts

b	burned, burning
e	expansion
f	flame
l	laminar
o	initial condition
p	propagation
u	unburned
w	wall

Superscripts

•	denotes derivative with respect to crank angle

INTRODUCTION

The use of combustion modeling in both testing and design of combustion chambers is the topic of this paper. Presented first is the formulation of a homogeneous-charge combustion model in both diagnostic and predictive modes, along with its integration into an engine simulation. Validation of the model through experiments involving visual observations of combustion in specially designed hardware is the subject of the second major section. Examples of the diagnostic application are then given. Two different combustion chambers, having open and wedge configurations, are examined to determine burning rates and flame speeds. The influence of turbulence on flame speed is then studied and a generalized correlation presented.

References pp. 578–579.

The predictive model is applied to the problem of choosing a feasible design approach to improve the performance of a homogeneous-charge single-cylinder engine have relatively long combustion duration. Experimental studies [1, 2] of combustion in spark-ignition engines have shown an increase in fuel burning rate to provide a number of favorable effects on engine performance. These include (i) more attractive tradeoffs between low nitric-oxide emissions and high thermal efficiency, (ii) lower cycle-to-cycle variations in engine output power, (iii) higher tolerance to dilution, either with excess air or with recirculated exhaust gas and (iv) higher efficiency at incipient engine knock. A combustion chamber modification intended to increase the fuel burning rate is evaluated with the model, and efficiency and nitric-oxide (NO) emissions are calculated. Results from the model prediction are then compared to experimentally determined behavior of an engine incorporating the modified chamber.

COMBUSTION MODEL

The combustion model employed in this paper can be described as a "quasi-dimensional" propagating-flame approach which considers the cylinder contents to be divided into two zones—unburned gas and burned products in thermodynamic equilibrium. The use of a two-zone thermodynamic concept in combustion modeling was introduced by Patterson and Van Wylen in 1963 [3]. Krieger and Borman [4], in 1966, extended the model to account for dissociation in the burned gases and used it to compute heat-release rates from engine pressure-time data. Their formulation represented the first publication of a two-zone combustion model in ordinary-differential-equation form.

Lancaster et al. [5], in 1976, added a geometry constraint to the thermodynamic heat-release model of Krieger-Borman and introduced the "quasi-dimensional" concept—inferring an apparent flame position from thermodynamic knowledge (burned volume) and geometric assumptions (thin spherical flame, centered at the spark plug). A limitation was that only simple combustion chamber geometries could be treated in this manner.

The model presented here extends the approach of Ref. 5 to chambers of any shape, using a separate geometry procedure. In addition to the diagnostic mode [5], in which flame position is inferred from measured pressures, our model was formulated to be used in a predictive mode to calculate pressure development and energy release from empirical flame-speed information. Data from the diagnostic results of controlled experiments are used to provide input for the predictive mode; hence both theory and experiment can be closely coupled.

Additional empiricism is introduced in the submodels used to compute heat transfer between the surrounding chamber walls and the burned and unburned zones. Care is taken, however, to ensure that the integrated values of the heat loss to the walls agree with first-law constraints. Detailed heat-release analyses of several types of engines, operated over a wide range of conditions and using two fuels, have resulted in a substantial data base for the development of turbulence and heat-transfer correlations.

Simulation of the complete engine cycle is required to characterize the thermodynamic state of the unburned gas at ignition. To accomplish this, the combustion model has been integrated into an overall engine cycle simulation. *NO* kinetics in the burned zone are computed, and a portion of the frozen *NO* is carried through to the next cycle in the residual gases, thus affecting the *NO* concentration in the unburned zone. Because the entire cycle is modeled by the simulation, such performance parameters as indicated mean effective pressure (*imep*), indicated thermal efficiency, air flow and exhaust temperature are computed.

The validity of the combustion model is tested by direct visual observations of flames and corresponding computations of heat release in both a single-stroke rapid compression machine and in a single-cylinder flame photography engine. Assumptions of flame structure, shape and position are compared directly to the combustion movies.

Model Formulation—A schematic of the two-zone flame-propagation model is shown in Fig. 1. Combustion occurs when an infinitely thin flame front of radius r_f propagates through the combustion chamber volume. At the instant shown, the flame divides the chamber into an unburned volume, V_u, and a burned volume, V_b, at average temperatures T_u and T_b, respectively, and a common pressure, p. The chamber volume is bounded at the top and bottom by cylinder-head and piston-crown surfaces of arbitrary contour.

During the combustion process, heat is transferred from each zone to its wetted boundaries. No heat is transferred across the flame front. The instantaneous heat loss in each zone is the product of an overall heat-transfer coefficient, a wetted-

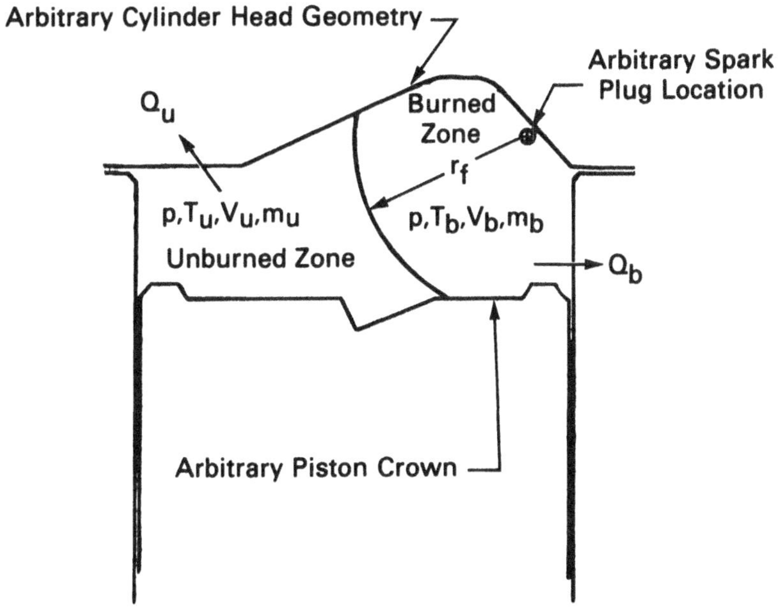

Fig. 1. Schematic of the two-zone flame-propagation model.

References pp. 578–579.

wall area, and the difference in temperature between the gas and the wall. The overall heat-transfer coefficient for each zone is computed by the method of Woschni [6].

A computer code developed at General Motors Engineering Staff was used to determine the geometric intersection of a sphere and an arbitrary combustion chamber shape at ten piston positions. These data were then placed in tables for use by the combustion model. For a given crank angle and burned volume, flame area, flame radius and wetted-wall area are obtained by bi-cubic spline interpolation from the generated tables, as indicated in Fig. 2.

All major species of the products of combustion, except *NO*, are assumed to follow a shifting equilibrium process, and are computed using the method and computer coding given in Ref. 7. *NO* concentrations are calculated from modified Zeldovich reaction kinetics, with rate constants adapted from the literature [8].

Thermodynamic properties in either the burned or the unburned zones depend

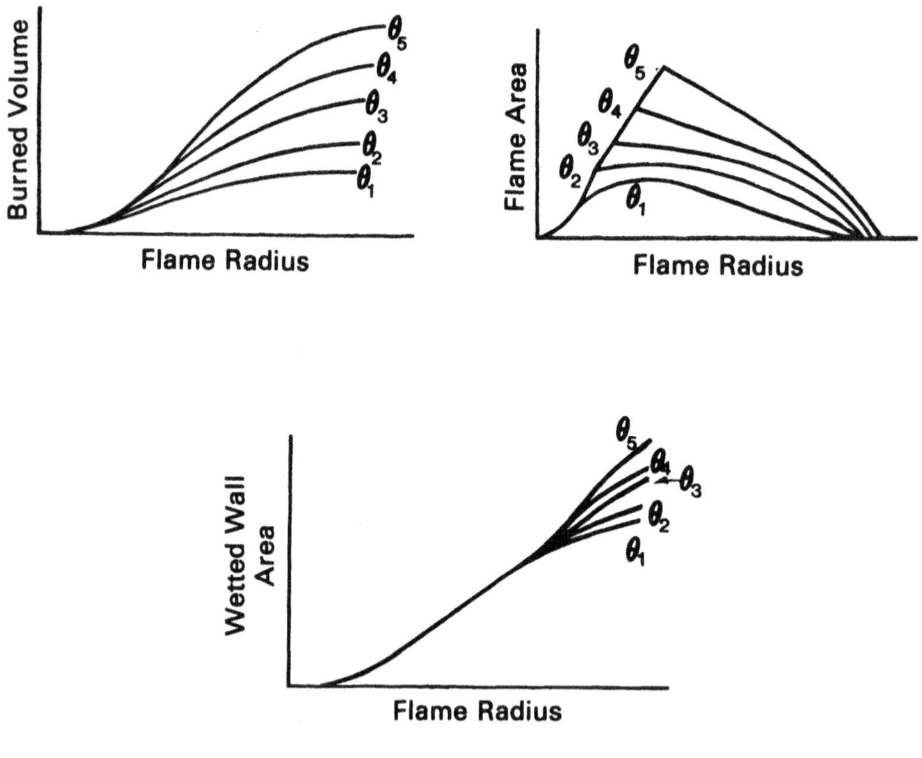

Fig. 2. Schematic of geometric sub-models used to compute flame and wetted-wall areas.

on pressure and temperature for fixed equivalence ratio and residual mass fraction. In the unburned zone, however, gas temperatures are sufficiently low so that gas composition can be considered frozen and thus independent of pressure. Employing this fact, the thermodynamic properties in either zone were also evaluated from Ref. 7. The details of laminar-flame-speed calculations are addressed in a later section.

Governing Equations—Using the notation of Fig. 1, energy equations for the burned and unburned zones can be written as follows:

$$\frac{d(mu)_u}{d\theta} = (\dot{\overline{mu}})_u = h_u \dot{m}_u - \dot{Q}_u - p\dot{V}_u \tag{1}$$

$$\frac{d(mu)_b}{d\theta} = (\dot{\overline{mu}})_b = h_u \dot{m}_b - \dot{Q}_b - p\dot{V}_b \tag{2}$$

where dots indicate derivatives with respect to crank angle. We assume that the unburned internal energy, u_u, is a function only of T_u, but that the internal energy in the burned zone, u_b, can be affected by dissociation:

$$\dot{u}_u = \frac{\partial u_u}{\partial T_u}\dot{T}_u = c_{v,u}\dot{T}_u \tag{3}$$

$$\dot{u}_b = \frac{\partial u_b}{\partial T_b}\dot{T}_b + \frac{\partial u_b}{\partial p}\dot{p} = c_{v,b}\dot{T}_b + \frac{\partial u_b}{\partial p}\dot{p} \tag{4}$$

Using Eqs. 3 and 4 and the definition of enthalpy, Eqs. 1 and 2 can be rewritten as follows:

$$m_u c_{v,u}\dot{T}_u = \frac{pV_u}{m_u}\dot{m}_u - \dot{Q}_u - p\dot{V}_u \tag{5}$$

$$m_b\left[c_{v,b}\dot{T}_b + \left(\frac{\partial u_b}{\partial p}\right)\dot{p}\right] = (h_u - u_b)\dot{m}_b - \dot{Q}_b - p\dot{V}_b \tag{6}$$

Ideal-gas equations of state can be written for each zone:

$$pV_u = m_u R_u T_u \tag{7}$$

$$pV_b = m_b R_b T_b \tag{8}$$

The assumption of no leakage from the chamber and the volume constraint results in

$$m = m_u + m_b = \text{const.} \tag{9}$$

$$V = V_u + V_b \tag{10}$$

The instantaneous mass burning rate can be related to the burning velocity, S_b, and the flame-front area, A_f:

$$\dot{m}_b = \rho_u A_f S_b \tag{11}$$

References pp. 578–579.

Finally, the geometry constraints of Fig. 2 can be expressed as follows:

$$V_b = V_b(r_f, \theta) \tag{12}$$

$$A_f = A_f(r_f, \theta) \tag{13}$$

$$A_w = A_w(r_f, \theta) \tag{14}$$

Thus Eqs. 5 through 11 and the geometric constraints shown by Eqs. 12-14 formulate our quasi-dimensional combustion model.

Diagnostic Version—At any given crank angle there are eight unknowns: $\dot{m}_u$, $\dot{m}_b$, $\dot{T}_u$, $\dot{T}_b$, $\dot{V}_u$, $\dot{V}_b$, $\dot{p}$ and S_b. In the diagnostic mode, p and $\dot{p}$ are known from measurements, allowing Eqs. 5-11 to be solved for $\dot{m}_b$ and S_b. The details of solving Eqs. 5-10 for the mass burning rate are given in Ref. 4. Burning velocities can then be calculated directly from Eq. 11, using instantaneous values of unburned density and flame area. For correlative purposes, the instantaneous computed burning velocity is divided by a laminar flame speed computed for the unburned gas. This ratio is denoted the flame speed ratio (FSR):

$$\text{FSR} = S_b/S_l \tag{15}$$

where $S_l = S_l\,(\text{fuel}, T_u, p, \phi, \text{residual})$.

Obviously the diagnostic version of the model is only as good as the data to which it is applied. Over the years, a considerable effort has been expended in the measurement and analysis of cylinder pressure data [9, 10]. Our standard data-acquisition procedure is to acquire 96 consecutive cycles of synchronous pressure data at the rate of one measurement each crank-angle degree. These data are then ensemble-averaged to generate a mean cycle for thermodynamic analysis. Along with the pressure data, measurements of air and fuel flows, engine speed and average manifold pressures are also required. Several checks are made to ensure that the pressure data are referenced properly and phased correctly. In this regard, a new phasing technique using a microwave probe in a motored engine has been developed to determine top dead center [11].

To minimize the time involved in the analysis of pressure-time data, an automated engine diagnostic software package was developed [12]. This package, consisting of six modules and 72 subroutines, processes the measured test-cell data completely to yield heat release. Its features include automatic pressure-volume phasing (using motoring data), a built-in simulation model to compute residual fraction and average wall temperatures, complete heat-release information that includes cumulative heat transfer in agreement with first-law principles as published by Samaga [13], single- and multicylinder capability, and, finally, a tabulation and storage of results in an on-line data base.

A schematic flow chart of the steps involved in using the diagnostic version of the combustion model is shown in Fig. 3. Cylinder pressure data are processed to yield the instantaneous mass burning rate, $\dot{m}_b(\theta)$, and burned volume, $V_b(\theta)$. The geometric constraints are then applied to yield the apparent flame position, $r_f(\theta)$, and flame-front area, $A_f(\theta)$. The burning velocity, $S_b(\theta)$, can then be determined from Eq. 11. The propagation velocity (velocity of the flame with

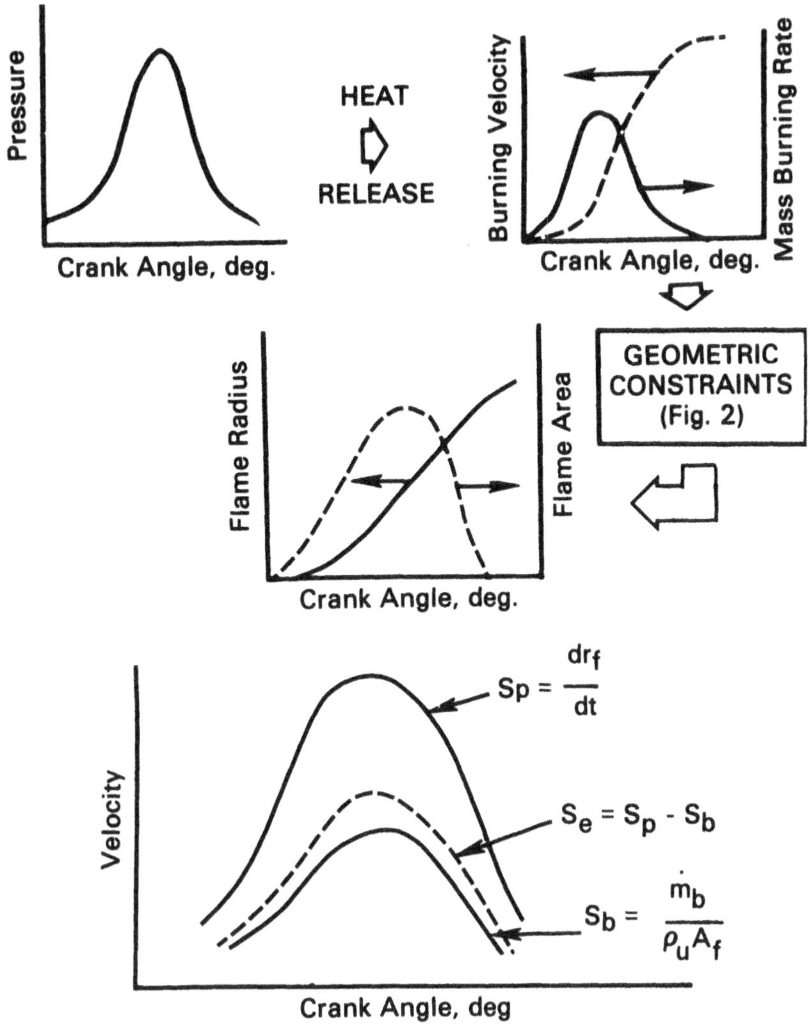

Fig. 3. Schematic flow chart showing the steps involved in using the diagnostic version of the combustion model to compute flame speeds.

respect to the chamber), $S_p(\theta)$, is determined by differentiating $r_f(\theta)$ with respect to time. Finally, the expansion velocity, $S_e(\theta)$, is simply the difference between the propagation and the burning velocities:

$$S_e = S_p - S_b \tag{16}$$

Since the thermodynamic state of the unburned gas is computed throughout the combustion event, laminar flame speed, suitably corrected for temperature, pressure, equivalence ratio and residual-gas concentration, can be determined at any given crank angle. The ratio of the diagnosed burning velocity to this laminar

References pp. 578–579.

flame speed can then be displayed as the instantaneous flame speed ratio, computed from Eq. 15.

Predictive Version—In the diagnostic mode, flame position is inferred from pressure measurement. This procedure is reversed in the predictive mode, where the motion of the flame front with respect to the unburned gas can be specified as input. This determines the instantaneous mass burning rate according to Eq. 11 and allows a reformulation of Eqs. 5–10 to compute the pressure development directly:

$$\frac{\dot{T}_u}{T_u} = \frac{\left[\gamma\left(\frac{h_u - u_b}{R_b T_b}\right) + \frac{c_{v,b}}{R_b} - \sigma\frac{V_u m_b}{V_b m_u}\right]\frac{\dot{m}_b}{m_b} - \sigma\frac{\dot{V}}{V_b} - \gamma\frac{\dot{Q}_b}{pV_b} - \left(\sigma\frac{V_u}{V_b} + \nu\right)\frac{\dot{Q}_u}{pV_u}}{\sigma\frac{c_{v,u}V_u}{R_u V_b} + \nu\frac{c_{p,u}}{R_u}} \tag{17}$$

$$\frac{\dot{p}}{p} = \frac{c_{p,u}\dot{T}_u}{R_u T_u} + \frac{\dot{Q}_u}{pV_u} \tag{18}$$

$$\frac{\dot{T}_b}{T_b} = \frac{1}{\gamma}\left[\frac{\dot{p}}{p}\left(\beta + \frac{V_u}{V_b}\right) + \frac{\dot{m}_b}{m_b}\left(\frac{m_b V_u}{m_u V_b} - 1\right) + \frac{\dot{V}}{V_b} - \frac{V_u \dot{T}_u}{V_b T_u}\right] \tag{19}$$

$$\frac{\dot{V}_b}{V_b} = \frac{\dot{m}_b}{m_b} + \gamma\frac{\dot{T}_b}{T_b} - \beta\frac{\dot{p}}{p} \tag{20}$$

where

$$\beta = \left(1 - \frac{p}{R_b}\frac{\partial R_b}{\partial p}\right)$$

$$\gamma = \left(1 + \frac{T_b}{R_b}\frac{\partial R_b}{\partial T_b}\right)$$

$$\sigma = \frac{c_{v,b}}{R_b} + \gamma$$

$$\nu = \beta\frac{c_{v,b}}{R_b} + \frac{\gamma}{R_b}\left(\frac{\partial u_b}{\partial p}\right)\frac{p}{T_b}$$

Although this predictive technique is limited to those chamber geometries and operating conditions tested, it offers an opportunity, within this limitation, to identify the physical reasons why one chamber may be superior to another.

Engine Simulation—An engine cycle simulation was formulated which includes the predictive version of the combustion model and utilizes developments in the computation of burned gas properties [7], and numerical integration procedures [14]. Models for valve flow coefficients, using data from Refs. 15 and 16, and detailed analyses of flow reversals during gas exchange were incorporated. In addition to the combustion Eqs. 5–11, sets of ordinary differential equations describing compression of the unburned charge, expansion of the post-combustion gases, and a comprehensive gas-exchange process are solved. The intake and exhaust plena are modeled as infinite reservoirs at constant pressure. Backflows of gas into the intake plenum from the cylinder and into the cylinder from the exhaust plenum are modeled explicitly with differential equations. The primary assumption during gas exchange is that perfect mixing occurs in the cylinder.

MODEL VERIFICATION

Since the calculations suggested here require rather involved computation of flame geometry, laminar flame speed and thermodynamic properties, it is desirable to test the model in experiments of increasing complexity. Two facilities used for this process of model verification are a single-stroke, rapid compression machine (RCM) and a transparent-piston flame-photography engine (FPE).

Rapid Compression Machine—The RCM, depicted in Fig. 4, provides a well controlled environment where details of the nature of the combustion process may be examined. In operation, retaining jaws are opened, allowing the compression piston to be driven from left to right by a larger drive piston at a speed determined by the pressure in the drive-air reservoir. The oil-filled snubbing chamber provides a damping force which increases near the end of the stroke. As the compression piston reaches the end of the stroke, retaining pins are activated to hold it in the position of near minimum volume. The combustion chamber is a simple cylinder. A quartz window, located in the head, allows a view of half the combustion space. A solenoid-actuated intake valve is located in the other half of the head, along with a pressure transducer. Holes in the side of the cylinder provide access for a spark plug of variable reach, a hot-wire anemometer probe and a resistance thermometer.

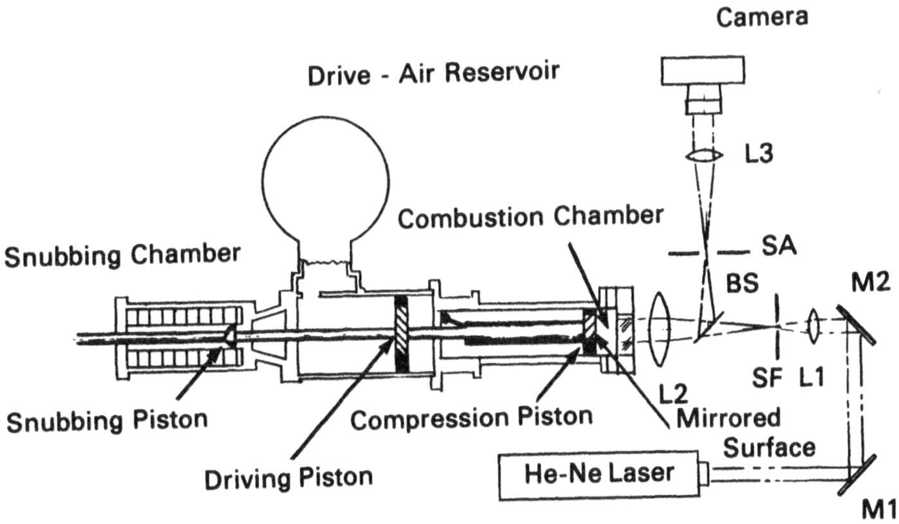

Fig. 4. The rapid compression machine test facility.

References pp. 578–579.

Rather than normal photography to film the flame process, laser-schlieren was used to provide visualization of flows as well as the flame propagation. The schlieren system consists of a beam from a 15-mW He-Ne laser focused by a 10-power microscope objective lens, L1, onto a 25-micron spatial filter, SF. The diverging beam, after passing through a beam splitter, is collimated by a 1.0-m focal length lens, L2. The parallel light then enters the chamber through the window in the head and is reflected from a first-surface mirror on the piston. The beam emerging from the chamber is focused by lens L2 onto a 2-mm aperture, SA, after reflection from the beam splitter. The light not deflected by gradients in the combustion chamber passes through the aperture and is collected by lens L3, which focuses the combustion space onto the film plane of a HyCam 16-mm camera operating at a nominal rate of 5000 frames/s.

Experimental Procedure—In operation, with the piston at its bottom position, the chamber and a manifold attached to the intake port are evacuated. The manifold is then filled with a propane-air mixture, metered by critical-flow orifices. A system of synchronized relays with adjustable delays was used to start the camera, open the intake valve, release the drive-piston retaining jaws, and fire the spark plug. Adjustment of these delays allows variation of the periods between the intake, compression and spark events, thus controlling the level of turbulence and the timing of the combustion event. Cylinder pressure is recorded concurrently with films of the flame.

Analytical Procedure—The combustion model was applied to the RCM tests to predict flame motion and pressure development during combustion. In this application, the disc geometry of the RCM's combustion chamber allowed the use of closed-form relationships to describe the geometric parameters, i.e., the frontal area, burned volume, and wetted areas, associated with the propagating flame. Laminar conditions were simulated in the model by assigning a FSR value of unity.

Near-laminar conditions were generated by imposing a long delay between the end of the intake process and the start of compression. Matekunas [17] showed that the flows generated by the piston alone had no significant influence on the flame for this geometry. Instantaneous measurements of cylinder gas temperature and pressure before ignition were used to set the initial thermodynamic state of the mixture in the model computations. The primary independent variables were the location of ignition and the mixture composition, with both central and wall ignition examined at equivalence ratios of 0.95, 1.05, and 1.30.

Results—Shown in Fig. 5 is a laser-schlieren film sequence of combustion for central ignition and an equivalence ratio of 1.05. The attainment of near-laminar conditions is demonstrated by the non-wrinkled spherical structure of the flame front. Note that the spark plug electrodes trip local perturbations, causing an increase in flame area in that region.

From this film sequence, the flame location was determined as a function of time by averaging the flame position for each frame along seven equally spaced

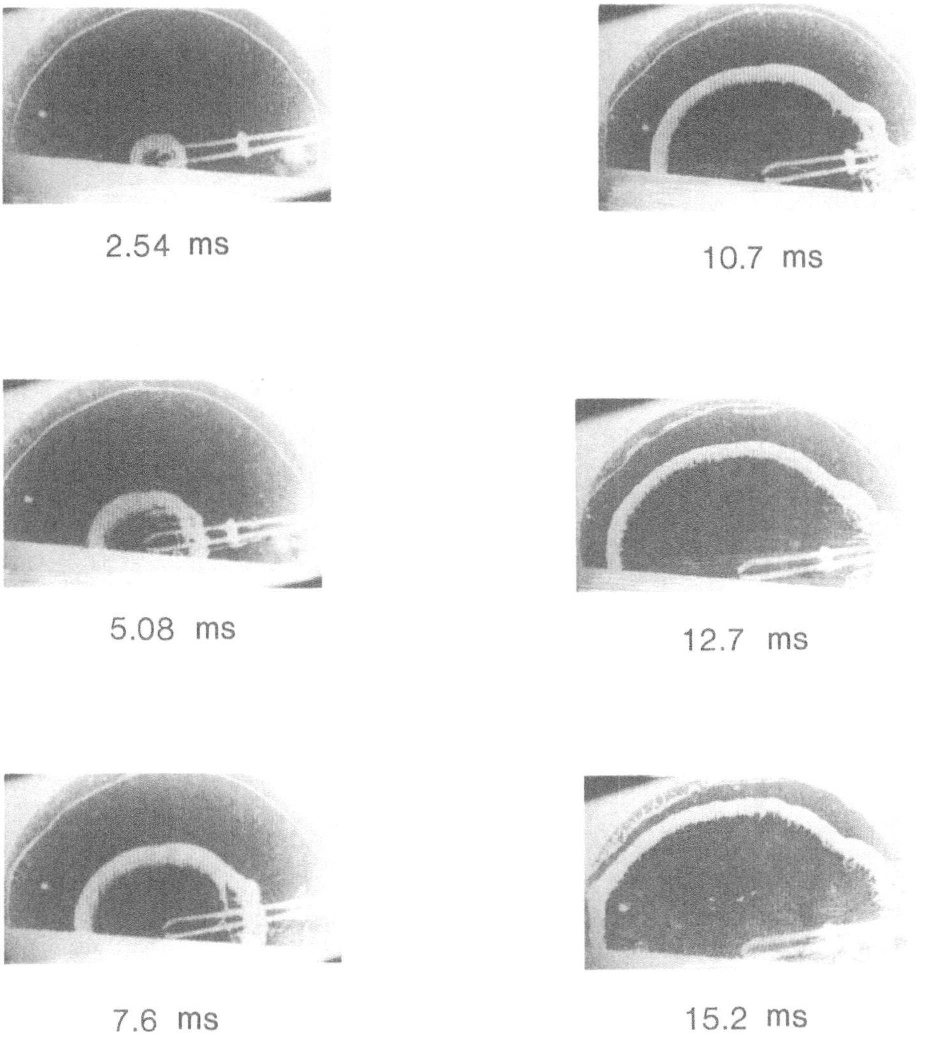

Fig. 5. RCM laser-schlieren film sequence of combustion: central ignition and an equivalence ratio of 1.05.

radii. The resulting flame radius, along with the measured pressure, is shown in Fig. 6. Overlayed on this figure are the flame radius and pressure development predicted by the model. Observed and calculated results at the alternate equivalence ratios and with the spark plug moved to the wall showed similar agreement.

The correspondence between the predicted and observed values indicates that the combustion model adequately describes the process for the laminar case. That is, the influence of the expanding burned gases on the flame speed is predictable using a two-zone thin-flame model with complete reaction in the flame. Also implied is the validity of the correlation for laminar flame speed of the mixture as a function of temperature, pressure and equivalence ratio. In the

References pp. 578–579.

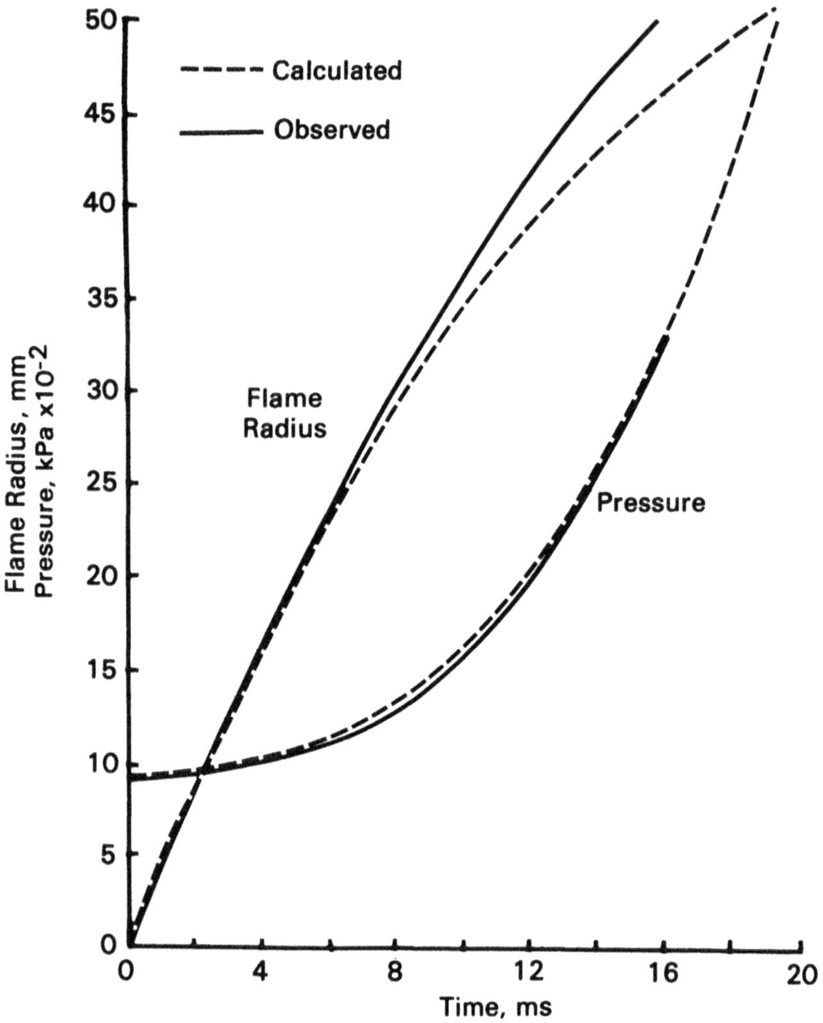

Fig. 6. Comparison of predicted combustion data with RCM experimental results.

next section, the diagnostic mode of the combustion model is tested under turbulent conditions by drawing upon experimental observations of flame motion in the flame photography engine.

Flame Photography Engine—Fig. 7 shows the arrangement of the FPE piston and head incorporating a wedge-shaped combustion chamber. The engine had a bore of 92.1 mm, a stroke of 76.2 mm, and a compression ratio of 6.85. Combustion was observed through an elongated piston fitted with a quartz window, patterned after the concept described by Bowditch [18]. A Fastax model WF-17T rotating-prism camera, equipped with a 50-mm, f/2.0 lens set at maximum aperture, filmed the combustion process at rates between 2000 and 3000 frames/s.

Experimental Procedure—Propane and air flows were metered with critical-flow orifices and mixed by a manifold in a tank located upstream of the intake. Exhaust gas recirculation (EGR) was provided by extracting exhaust gas from a tap near the cylinder head and adding it to the intake flow upstream of the mixing tank. EGR flow rate was measured by the CO_2 tracer method. Film records were obtained over a range of engine operating conditions, with maximum speed being limited by design constraints to about 2050 r/min.

Pressure-time data were recorded and digitized using previously documented procedures [9]. The recorded data were coordinated with film records by a technique whereby the ignition was deactivated for several cycles during camera acceleration. When ignition was reactivated, the first firing cycle on the film corresponded to the first firing pressure cycle following the motoring cycles.

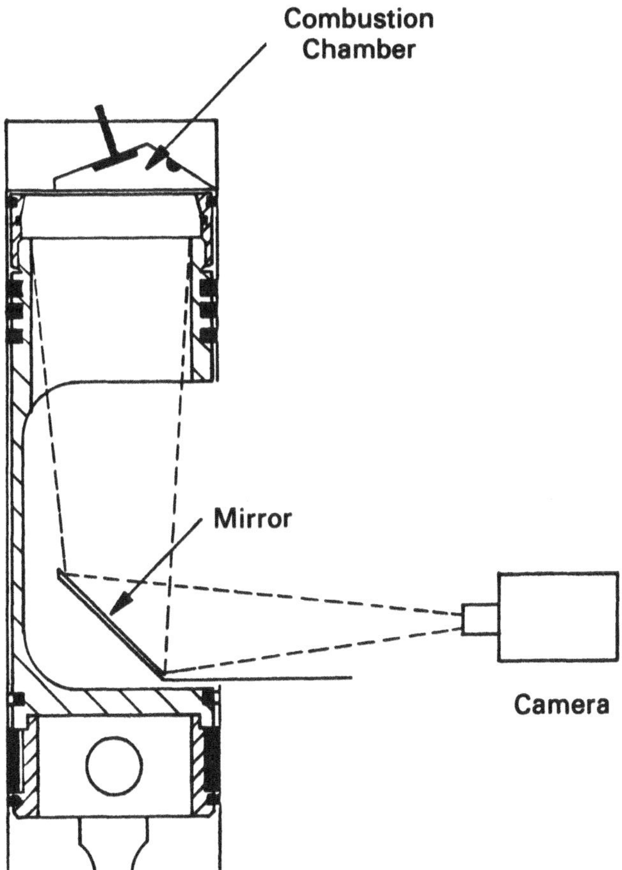

Fig. 7. Schematic of flame photography engine.

References pp. 578–579.

Event timing was provided by crank angle and spark images reflected onto the film, as well as timing marks focused onto the film by the camera timing device.

Analytical Procedure—Pressure-time data were recorded, and selected individual cycles, along with the average for 96 consecutive cycles, were analyzed. The engine diagnostic package [12] was used to process the data. The diagnostic version of the combustion model was added to the analysis procedure as an additional step to determine flame motion.

Trapped masses and combustion efficiencies for individual cycles were assumed to equal those computed for the average cycle. Individual cycles were also assumed to have a residual fraction equal to that calculated by an engine simulation for the average cycle. A first-law energy balance was obtained for each cycle by adjusting heat transfer.

Results—Fig. 8 shows consecutive frames, taken with Kodak type 7240 film at a rate of 2000 frames/s, an engine speed of 1000 r/min, an equivalence ratio of 1.0, a volumetric efficiency of 50%, a spark advance of 23 degrees, and no EGR. The field of view of the camera permits observation to a mass-burned fraction of 53% and a volume-burned fraction of 78%. A view of the remaining chamber volume is obscured by the construction of the piston, as evident in Fig. 7. The average cycle for the operating condition of Fig. 8, condition A, had an *imep* and an indicated thermal efficiency of 526.6 kPa and 29.3%, respectively. For the individual cycle shown, the respective values are 523.9 kPa and 29.2%. The image of the flywheel, showing crank-angle markers, is observed to the right of each frame. The chamber has swirl in the clockwise direction as viewed. The swirl rate decreases during compression, having an average value of about 1200 r/min.

A graphical presentation of the same film sequence, generated by digitizing sixteen points measured along the flame front, is given in the upper half of Fig. 9. Data for a second run condition, B, are given in the lower section of the figure, taken at a rate of 1955 frames/s, an engine speed of 1400 r/min, an equivalence ratio of 0.92, an EGR rate of 5%, a volumetric efficiency of 48%, and a spark advance of 32 degrees. The average cycle for condition B had an *imep* of 526.3 kPa and an indicated thermal efficiency of 32.7%. Values for the individual cycle shown were 523.7 kPa and 32.5%, again very close to the average values. Flame fronts are plotted at intervals of one frame, on the left in Fig. 9, starting at crank angles of 18.0 and 25.3 degrees BTDC in increments of 3.0 and 4.2 degrees for conditions A and B, respectively. The spark plug location and the point from which the flame is observed to propagate (labeled as "rotation point") are indicated in the figure. The intake and exhaust valves are centered along the horizontal axis to the right and to the left of the vertical axis, respectively. The outer circle represents the limits of the camera field of view, not the cylinder bore.

The average of the sixteen radii defining each flame front, measured from the rotation point, are plotted versus crank angle in Fig. 9. Smoothed curves fitted to the data are shown as solid lines. The propagation velocities given by the derivatives of these curves are also plotted as solid lines versus crank angle at

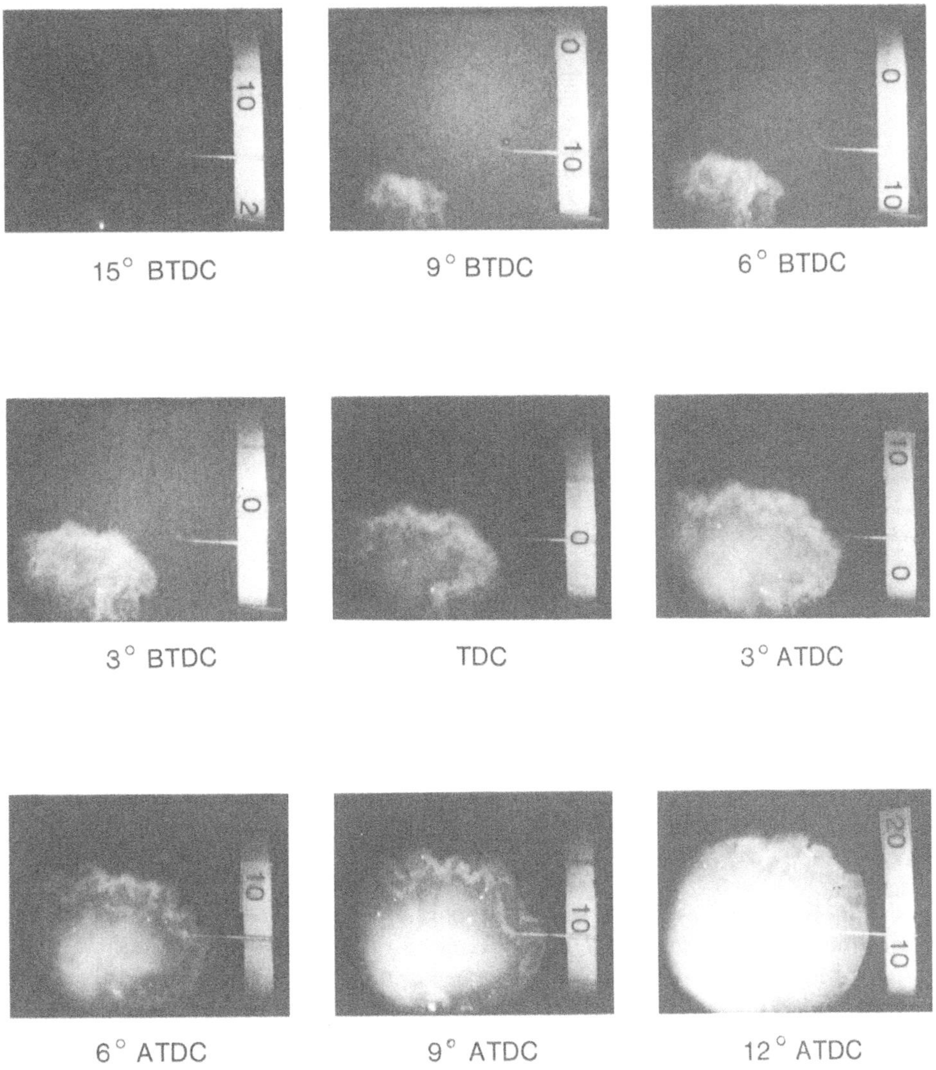

Fig. 8. Flame characteristics in flame photography engine.

the extreme right of Fig. 9. The dashed lines show the results computed from the pressure-time data by the diagnostic model, assuming the flame propagates from the rotation point. Flame positions are in close agreement at the start of combustion, but then the model results begin to lag the experimental data. At a radius of 50 mm, the difference is 5.5 mm for condition A and 9.1 mm for condition B. The computed crank angle of peak velocity is within one degree of that observed for both conditions. The magnitudes of the peak velocities from the model are 14% and 17% lower than observed values for conditions A and B respectively.

The difference between the model results and the observations can be attributed

References pp. 578–579.

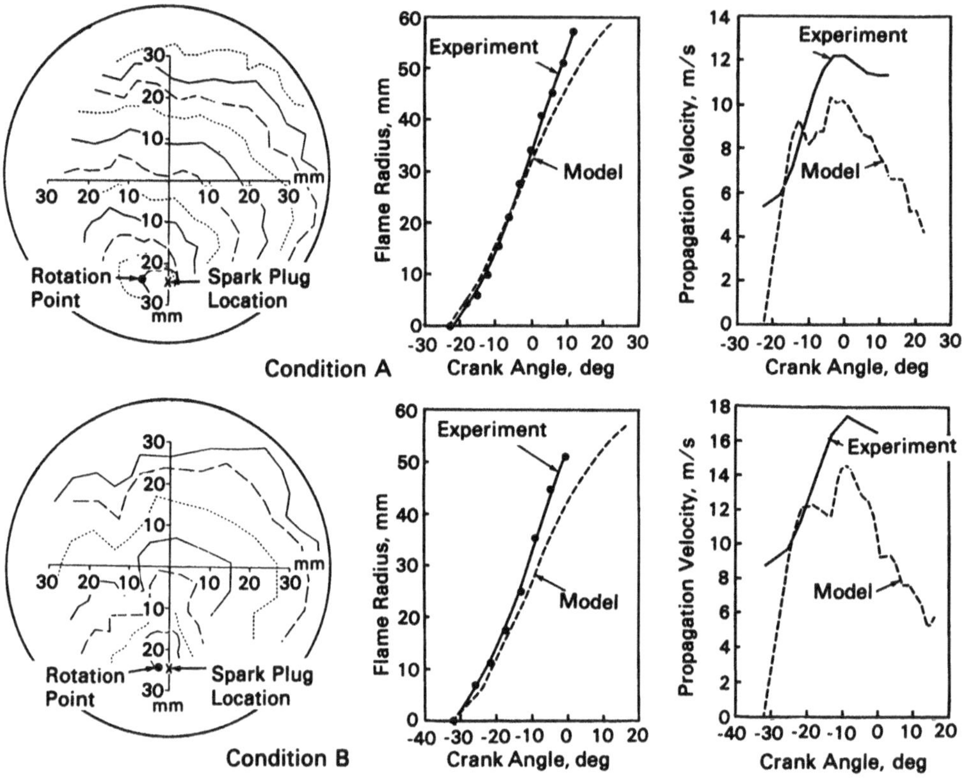

Fig. 9. Graphical presentation of FPE data compared with model results: upper, data from Fig. 8; lower, data from second operating condition.

to the model assumption of an infinitely thin flame. The film measurements record the most advanced edge of the flame as defining the front. The flames are known to be thick, on the order of 8 mm as reported in the literature [19]. Hence a measurement to the center of the flame, if it could be identified, would provide a more satisfactory comparison to the thin-flame model results. Flame thicknesses tend to increase as the flame propagates, resulting in the trends shown in Fig. 9.

One conclusion is that the diagnostic model calculates lower flame propagation velocities than those observed. The effect of model assumptions on the calculated burning velocity is less apparent, as flame areas change with flame radius. For a wedge chamber, the areas increase for less than half the flame travel and then decrease. For a given mass burning rate and unburned-gas density, the burning velocity is inversely proportional to the flame area (see Eq. 11). Therefore, errors in flame position can result in an increase or decrease in the burning velocity, depending on the flame radius.

In addition to the runs shown in Fig. 9, a total of 20–40 cycles at each of twelve other run conditions were filmed, and the trends described above were found generally to hold over a large range of engine operating conditions. Lean and high-EGR flames generally yielded larger differences between the model and observed flame positions, with larger flame thicknesses being a probable cause. These dilute flames tend to be less spherical and more affected by the bulk swirl motions. Most work was done at moderate load conditions, since low light intensity at lighter loads and with weak mixtures makes it difficult to observe the flame.

The results shown were calculated assuming the flame propagated from the apparent rotation point. Calculations were also made assuming the flame propagated from the spark plug, and no significant differences were found. This result is important for measurements in engines where the flame cannot be observed.

Both the results of the diagnostic model and the film measurements show cyclic variability in the combustion process. Fig. 10 shows six different flame-radius

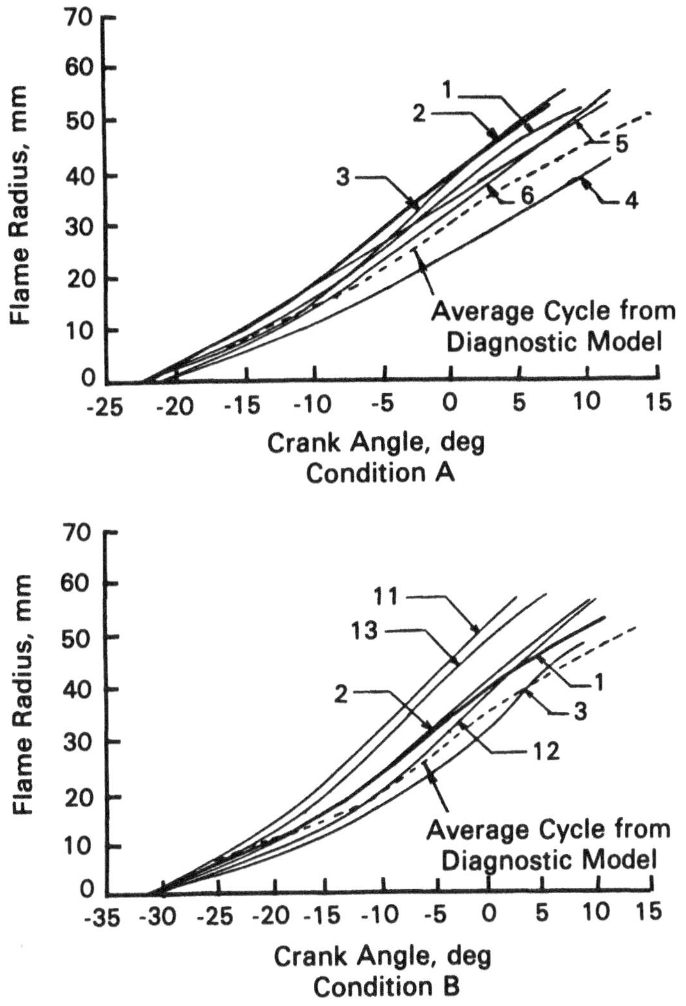

Fig. 10. Observed individual flame histories for conditions A and B, showing cyclic variation. Dashed curves show model results for average cycles.

References pp. 578–579.

histories plotted versus crank angle for run conditions A and B. The data for condition A show six consecutive cycles, while those for condition B are for three consecutive cycles at two different times. For comparison, the flame developments computed by the diagnostic model for the average cycles are shown by the dashed curves. From a statistical standpoint, the six cycles are insufficient to draw any conclusions, but a large cycle-to-cycle variability in flame development is observed. Although slow-burn cycles had significantly lower peak pressures, their *imep* values were close to the average.

Model-Verification Conclusions—The model was used in the diagnostic mode to evaluate the character of combustion and the dependence of flame speed on turbulence level. The important assumptions are that the flame propagates as a sphere, that the flame front is thin, and that the gases in the cylinder during combustion are adequately represented by two zones. A degree of tolerance is afforded the spherical-flame assumption since we are dealing with turbulent flames which are wrinkled and perturbed by the mixture motions in the chamber. The increase in surface area over that of a smooth sphere, due to wrinkling, is partly responsible for the increase in flame speed above that of the laminar flame and is accounted for in flame-speed correlations. The spherical assumption leads to errors only when large-scale motions distort the flame to the point that either the apparent origin moves far from the spark location or the actual flame front, on the average, is far from the mean sphere. One measure of the suitability of the spherical assumption is the ratio of mean gas motion relative to the flame propagation velocity, $\bar{u}/S_p$. When values of this parameter are low, the fluid motions should not affect the flame geometry significantly. High values of this parameter can be generated both when the flame is very lean, yielding low values of the expansion velocity, S_e, and early, when the flame is not yet fully developed. For a given engine configuration, the appropriateness of the spherical-flame assumption thus depends on the mean flow field and spark location. For the chambers examined here, the wedge, with its associated swirl, is most sensitive to this influence. It was observed in the FPE, however, that the influence of rotation of the apparent flame origin due to swirl did not significantly affect results.

APPLICATION OF DIAGNOSTIC VERSION OF MODEL

The diagnostic version of the model was used to analyze propagation and burning velocities throughout the combustion event from measured pressure-time data. These computed burning velocities were correlated to turbulent intensity measured under motoring conditions.

Procedure—Measurements were made in three single-cylinder engines having different combustion chamber shapes, as shown in Fig. 11. The results for a CFR engine having the disc chamber were obtained by Lancaster et al. [5] and are compared to recent results for the wedge and open chambers having irregular geometries characteristic of production engines. Both wedge- and open-chamber

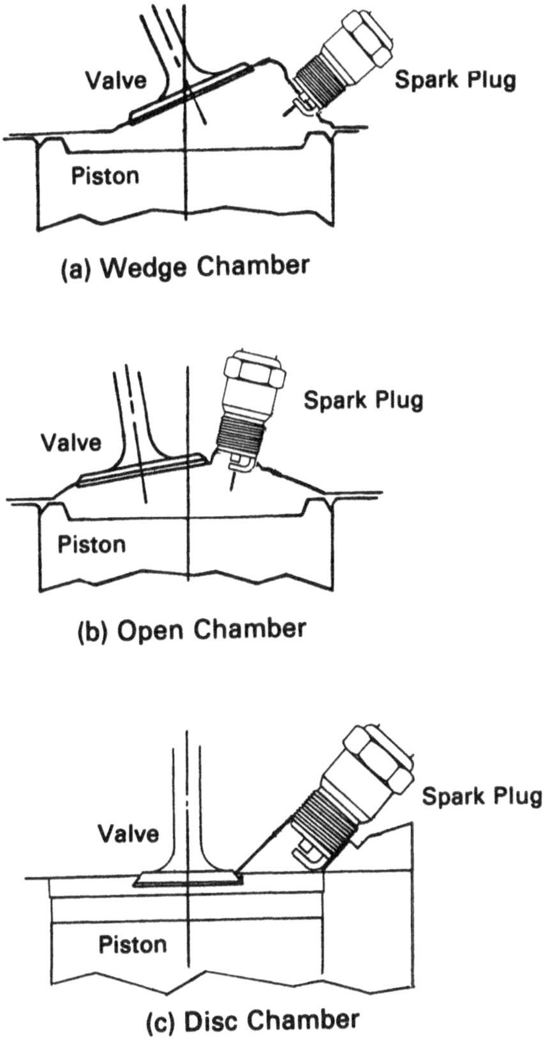

Fig. 11. Experimental combustion chamber configurations.

engines had a compression ratio of 8.4, a stroke of 88.4 mm, and a bore of 95.0 mm.

Parametric studies were conducted involving five engine operating variables: speed, load or volumetric efficiency, equivalence ratio, EGR, and spark timing. The test matrix for the disc chamber, taken from Ref. 5, is summarized in Table 1. All test conditions were run with and without a shrouded intake valve. While EGR variations were not included, effects of compression ratio were examined. The test conditions are listed in Tables 2 and 3 for the wedge and open chambers, respectively. The baseline condition for each of the three chambers is listed as the first entry in the respective table.

References pp. 578–579.

TABLE 1

Operating Conditions for CFR Combustion Tests

Run. No.	Speed (r/min)	Vol. Eff. (%)	Comp. Ratio	Equiv. Ratio	Spark Timing
1	1500	50	8.72	0.80	MBT
2	1000	50	8.72	0.80	MBT
3	2000	50	8.72	0.80	MBT
4	1500	25	8.72	0.80	MBT
5	1500	75	8.72	0.80	MBT
6	1500	50	6.84	0.80	MBT
7	1500	50	10.55	0.80	MBT
8	1500	50	8.72	1.00	MBT
9	1500	50	8.72	0.64	MBT
10	1500	50	8.72	0.80	MBT−10
11	1500	50	8.72	0.80	MBT+10

Intake Manifold Temperature = 320 K
Coolant Temperature = 363 K
Propane Fuel (Stoichiometric A/F = 15.67)
EGR = 0%

Effects of fuel type were also examined. The disc- and open-chamber tests described in Tables 1 and 3, respectively, were conducted using propane as fuel. Indolene clear was the fuel for wedge- and open-chamber tests listed in Table 2. In each chamber, turbulence was measured at two locations using a hot-wire anemometer while the engine was motored. An extensive study of turbulence in the disc chamber has been reported by Lancaster [20] over a range of engine speed, load, compression ratio, and intake geometry conditions. For a given intake geometry, turbulent intensity was found in that study to be a linear function of the volumetric flow rate, a parameter strongly affected by engine speed. In the present work this finding was used to limit the scope of subsequent turbulence measurements in the wedge and open chambers to a range of engine speeds at a fixed volumetric efficiency. Changes in load were accounted for by using a relationship from Ref. 20. The turbulence measurements were processed using the basic procedure of Ref. 20 to yield mean flow velocities and turbulent intensities during the compression stroke. For the disc chamber, the turbulence near TDC was found to be isotropic. This was not the case for the wedge chamber, as will be discussed below. The computed burning velocities were correlated to these measured turbulent intensities.

Results—The burning velocity, propagation velocity, and flame position, computed as functions of crank angle (0 degrees being TDC) using the diagnostic version of the model, are presented in Fig. 12 at the baseline run condition for the wedge and open chambers. The residual mass fractions were estimated using an engine simulation incorporating an empirical burning correlation. Model calculations were terminated after 92% of the charge was burned. Similar plots for the disc chamber are given in Ref. 5. The open chamber has a more central spark

TABLE 2

Operating Conditions for Wedge- and Open-Chamber-Engine Combustion Tests

Run No.	Speed (r/min)	Vol. Eff. (%)	Equiv. Ratio	EGR (%)	Spark Timing
			Wedge		
1	1600	39.6	0.88	0.0	MBT
2	1200	39.0	0.88	0.0	MBT
3	2000	38.8	0.88	0.0	MBT
4	1600	23.4	0.88	0.0	MBT
5	1600	54.9	0.88	0.0	MBT
6	1600	39.5	0.88	7.4	MBT
7	1600	39.6	0.88	15.2	MBT
8	1600	44.1	0.79	0.0	MBT
9	1600	39.6	0.88	0.0	MBT+6
10	1600	39.6	0.88	0.0	MBT−6
			Open		
11	1600	39.4	0.88	0.0	MBT
12	1200	39.4	0.88	0.0	MBT
13	2000	39.5	0.88	0.0	MBT
14	1600	23.5	0.88	0.0	MBT
15	1600	55.2	0.88	0.0	MBT
16	1600	39.7	0.88	7.7	MBT
17	1600	39.2	0.88	12.6	MBT
18	1600	43.5	0.79	0.0	MBT
19	1600	39.5	0.88	0.0	MBT−6
20	1600	39.4	0.88	0.0	MBT−12

Intake Manifold Temperature = 329 K
Coolant Temperature = 353 K
Indolene Clear (Fuel Stoichiometric A/F = 14.70)
Compression Ratio = 8.40

plug location yielding a shorter maximum-flame-travel distance at a given crank angle than the wedge chamber. For the given condition, the flame propagates a maximum distance of only 53 mm in the open chamber versus 74 mm in the wedge chamber. A peak burning velocity of 11.3 m/s occurs at 5 degrees ATDC in the wedge chamber, as compared to a peak value of 8.6 m/s occurring 3 degrees

TABLE 3

Operating Conditions for Open-Chamber-Engine Propane Combustion Tests

Run No.	Speed (r/min)	Vol. Eff. (%)	Equiv. Ratio	EGR (%)	Spark Timing
1	1250	40.5	1.00	0.0	MBT
2	1250	40.4	0.78	0.0	MBT
3	1650	37.5	1.00	10.0	MBT
4	1650	42.0	0.78	0.0	MBT
5	2050	36.3	0.92	15.2	MBT
6	2050	42.9	0.64	0.0	MBT

References pp. 578–579.

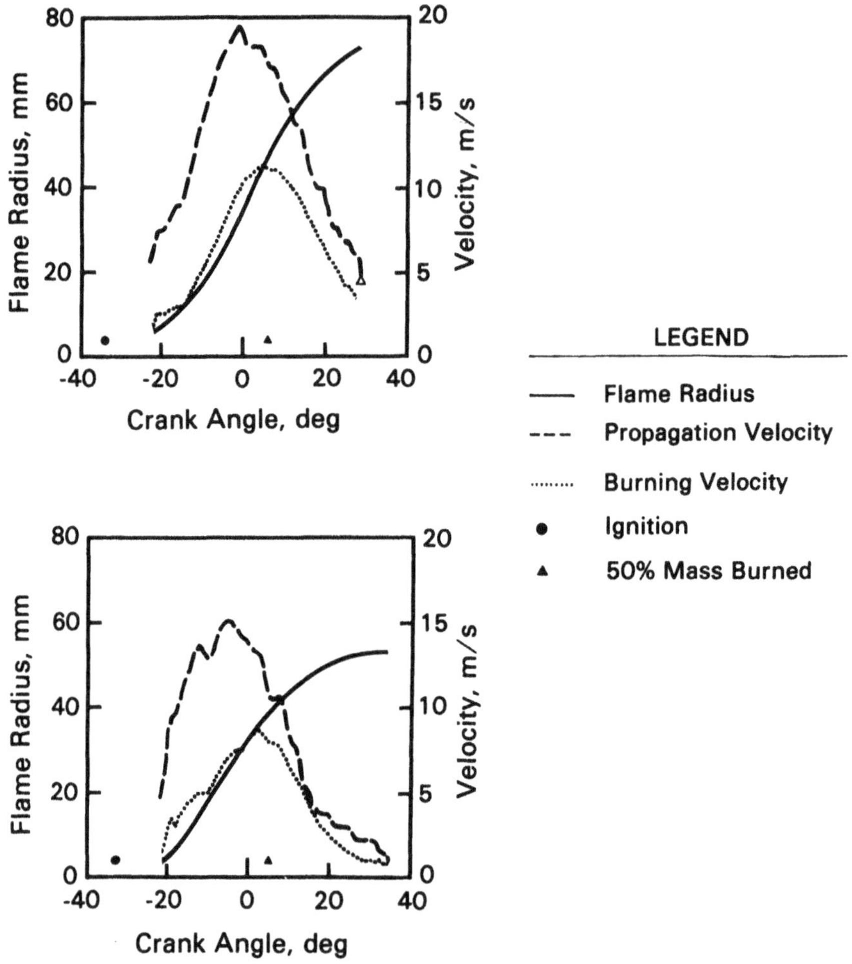

Fig. 12. Flame diagnostic results for wedge chamber (top) and open chamber (bottom).

ATDC in the open chamber. It will be shown that the higher burning velocities of the wedge chamber can be attributed to higher mixture turbulence. The propagation velocity, which is the sum of the burning velocity and the expansion velocity relative to the unburned charge, peaks in both chambers prior to the peak in the burning velocity. Similar behavior was found to occur in the disc chamber [5].

Many turbulent flame-propagation theories include a dependence of the burning velocity on laminar flame speed and turbulent intensity. Lancaster et al. [5] were successful in correlating FSR at the 50%-mass-burned point, computed by their diagnostic model, to the turbulent intensity [5] measured within a motored engine. Recent studies by Abdel-Gayed and Bradley [21] and Ballal and Lefebvre [22] addressed correlating out-of-engine steady-flow measurements of flame speed to various turbulence parameters and the laminar flame speed. These studies emphasize the importance of laminar flame speed in the interpretation of burning velocity and turbulence interactions.

Lancaster et al. [5] used a theoretical equation to correlate the atmospheric-pressure laminar-flame-speed data for propane measured by Kuehl [23]. The equation accounts for the effects of temperature, equivalence ratio, and charge dilution, details being given in Ref. 5. More recently, elevated-pressure flame speeds were measured for propane by Metghalchi [24]. Flame-speed data for Indolene clear are not available in the literature, so data for a similar fuel, isooctane, are used instead. Heimel and Weast [25] presented data for isooctane at atmospheric pressure, and Metghalchi and Keck [26] presented elevated-pressure data. The equation of Ref. 5 was found to provide a good correlation to the atmospheric-pressure data, but could not account for the observed decrease in flame speed with increasing pressure. Accordingly, an empirical term was added to account for the effects of pressure on the laminar flame speed of both fuels.*

The burning velocities computed from cylinder pressure measurements were normalized by the instantaneous laminar flame speed of the unburned mixture in an attempt, as suggested by turbulence theories, to separate chemical and turbulence effects. The resulting FSR's are plotted, along with the burning velocities, against the flame radius for the baseline conditions in Fig. 13. The FSR curves are observed to be roughly parabolic for these two chambers, reaching a maximum about 2 degrees crank angle prior to the 50%-mass-burned point indicated in the figure. The 50%-mass-burned point was chosen for correlation because: (i) it occurs in a region where geometry calculations are less sensitive to model uncertainties, (ii) with MBT spark timing, it occurs consistently at about 7 degrees ATDC for the wedge and open chambers and 12 degrees ATDC for the disc chamber and (iii) the FSR in the vicinity of this point is relatively close to its peak value.

The FSR is strongly affected by engine speed, as shown in Fig. 14. A 25% change in engine speed results in approximately a 17% and a 9% change in FSR for the wedge and open chambers, respectively. Near the end of combustion, where some model assumptions are suspect, these trends do not hold. The turbulent intensity was measured for the 45-degree crank-angle interval BTDC at the spark plug location. As shown in Fig. 15, these intensities increase with engine speed for the wedge and open chambers. Similar behavior is demonstrated in Ref. 20 for the disc chamber. Measurements of turbulence at locations away from the spark plug showed similar trends. The data were taken at a volumetric efficiency of 43%. The solid and open symbols in the figure distinguish vertical (parallel to the bore axis) and horizontal (parallel to the piston face) orientations, respectively. As shown, the flow field in the wedge chamber appears to be anisotropic, with a higher turbulent intensity measured in the bulk flow, or swirl, direction as sensed by the wire in the vertical orientation. Additional work is required to determine if the flow is truly anisotropic, or if the inability of the hot-

---

* *The equation was multiplied by* $(p/101\ kPa)^{\alpha}$. *The parameter* α *equals* $(-0.477 + 0.256\ \phi)$ *for propane and* $(-0.013 - 0.146\ \phi)$ *for isooctane.*

References pp. 578–579.

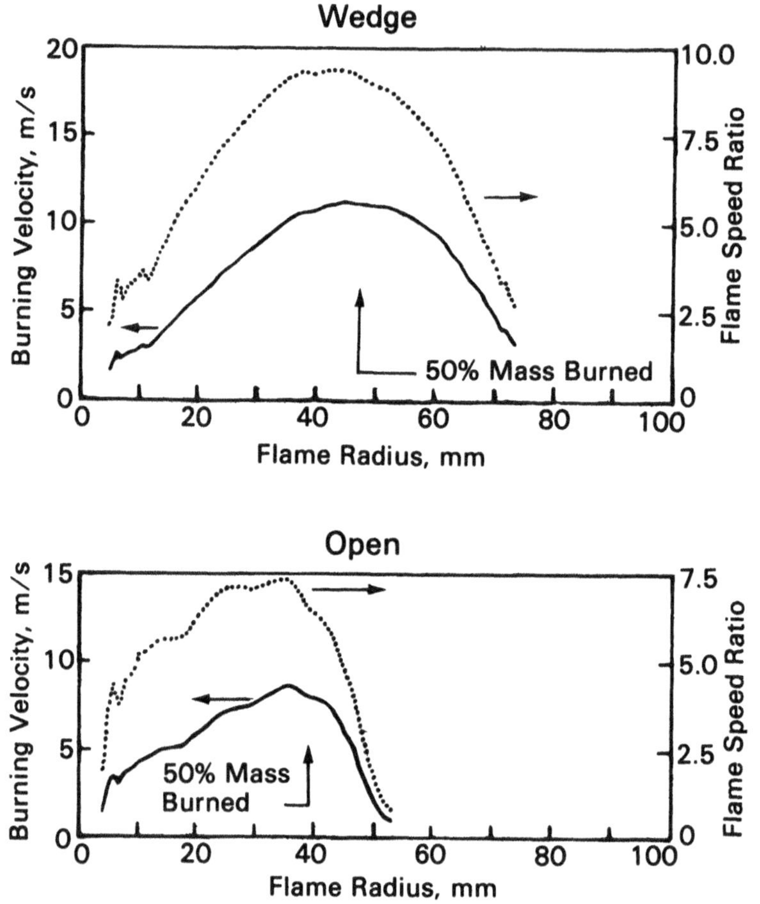

Fig. 13. FSR and burning velocities for both chambers plotted against flame radius.

wire to sense changes in flow direction when in the horizontal orientation, where no strong mean flow exists, yields an artificially low value of intensity, as discussed by Matekunas [17]. The open-chamber results are close to being isotropic, as was found for the disc chamber. In light of the above, measurements with the probe in the vertical orientation were used for correlating burning-velocity results.

The FSR, computed at the 50%-mass-burned point, was found to correlate with the ratio of the turbulent intensity to laminar flame speed, as shown in Fig. 16.

Data from tests with all three chambers at conditions listed in Tables 1 through 3 are plotted in the figure. A linear curve fit to the data forced through the theoretical intercept of unity FSR at u'/S_l equal to zero is

$$S_b/S_l = 1.0 + 4.01(u'/S_l) \tag{21}$$

Most of the data fall below a u'/S_l ratio of 3.0, with data above this value representing conditions of high EGR or lean mixtures. The works of Abdel-Gayed and Bradley [21] and Ballal and Lefebvre [22] illustrate the importance of turbulent scale in relating burning-velocity and turbulent-intensity data. Length scales are difficult to infer accurately from engine turbulence measurements, but are generally estimated in the range of 3 to 12 mm, yielding turbulent Reynolds numbers at the 50%-mass-burned point on the order of 1500 to 4000 for the run conditions of Tables 1 through 3. Our "best-fit" correlation curve to the data of Ref. 21 for this range of turbulent Reynolds numbers is also shown in Fig. 16.

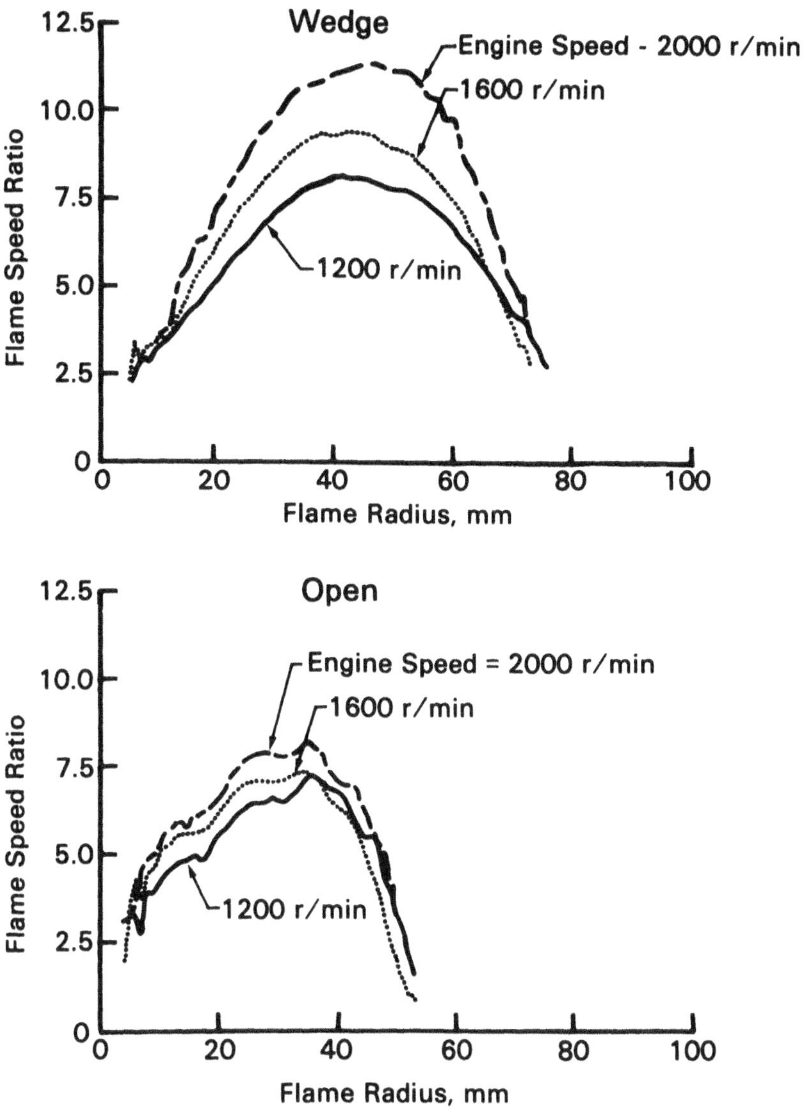

Fig. 14. Effect of engine speed on FSR for both wedge and open chambers.

References pp. 578–579.

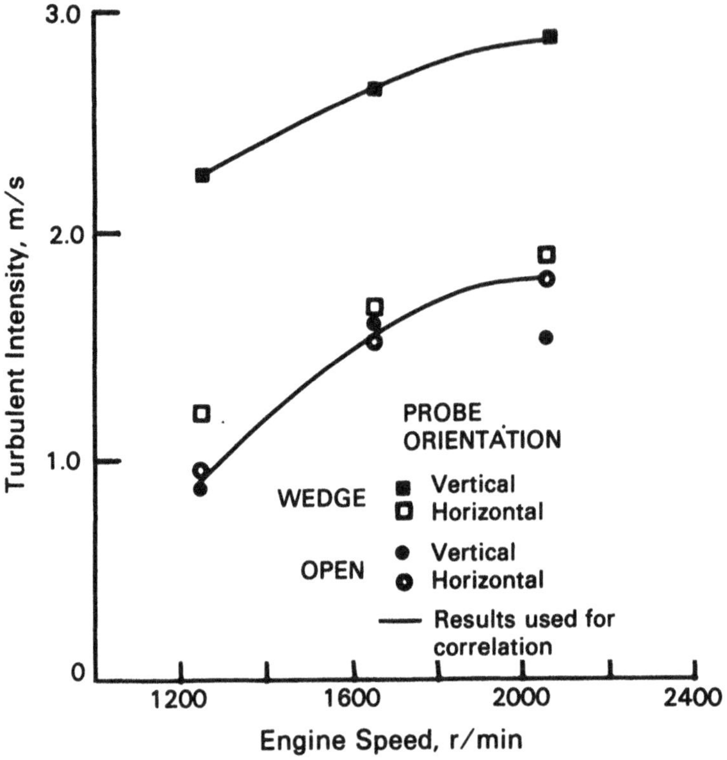

Fig. 15. Effect of engine speed on turbulent intensity for wedge and open chambers.

The data of Ballal and Lefebvre for u'/S_l less than 2.0 have reported integral scales of less than 1.0 mm, making comparisons difficult, especially since the results show a strong increase of FSR with increasing integral scale in this region. Data at the highest integral scale reported generally agree with Fig. 16. However, in the region where u'/S_l is of order 2.0, the data of Ref. 22 fall below the engine curve on Fig. 16.

An additional complication in comparing engine data to out-of-engine data is the interpretation of the turbulent-intensity value used in correlating the data. In engine experiments, the turbulent intensity used is measured approximately at the time of ignition, while the burning velocity is calculated after the flame has propagated approximately 40 to 50 mm. Calculations by Wong and Hoult [27] indicate that the initial intensity could be amplified by a factor of two as the turbulence field in the unburned charge is affected by motions resulting from the compression process. The FSR's in Fig. 16 correspond to measurements made mainly at MBT spark timing. When the spark was retarded from MBT the FSR generally fell below the points shown. Additional work is required to sort out these various effects.

The Extension of Diagnostic Results to Predictive Modeling—In applying the model in its predictive form, several additional assumptions to those made in the diagnostic application are required. These relate to the specification of the flame propagation rate in the chamber. The diagnostic application has resulted in a reasonable correlation of the FSR at the 50%-mass-burned point as a function of turbulence level. Observation of the FSR as a function of flame radius for the tested chambers indicate an acceleration from a FSR of 1.0 at initiation to the level indicated by the correlation and then a deceleration. To perform first-order calculations for a given chamber configuration, a turbulence level could be assumed along with a FSR curve. The behavior of the engine could then be examined with changes in engine speed, load and stoichiometry at MBT spark timing. However, in the application to be addressed, i.e., the calculation of the effect of a modification to the geometry of an existing tested chamber, better information is available. It will be assumed that the relationship of FSR to flame radius as measured in the unmodified engine is unchanged. Any ultimate predictive model would not require this restrictive assumption, but our present lack of understanding of the physics of the engine combustion process suggests this if

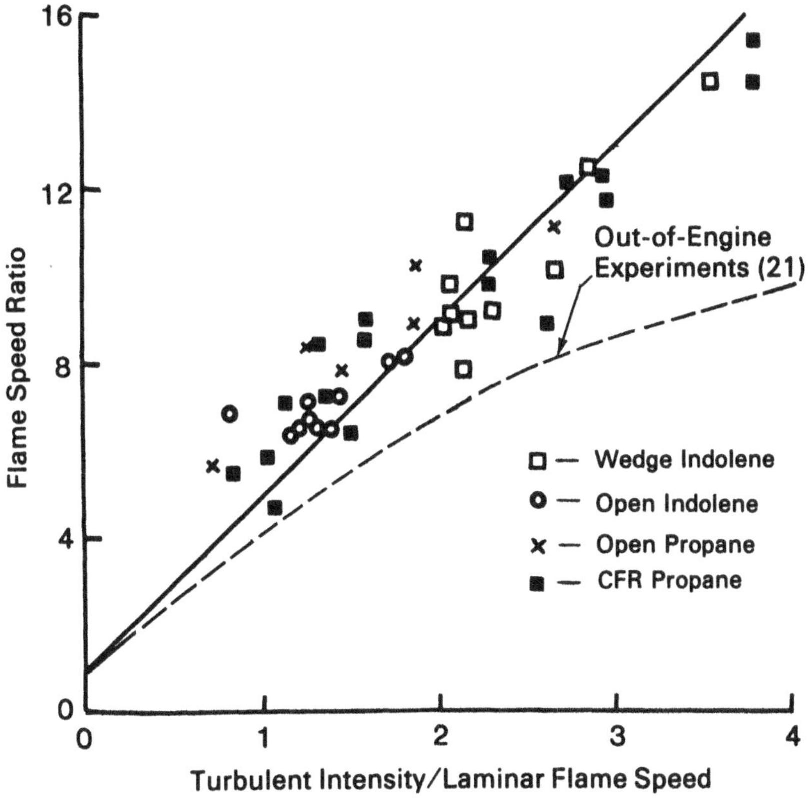

Fig. 16. Correlation of FSR at 50%-mass-burned point to u'/S_l (solid curve), along with similar data from out-of-engine experiments (dashed curve).

References pp. 578–579.

best accuracy is to be obtained. Some perspective on this assumption is obtained by reviewing other attempts to model S.I. engine combustion behavior predictively, particularly in light of the experimental results obtained here.

It has been recognized that the flames in engines accelerate for a period and then decelerate. When the flame speed relative to the unburned gases is determined from heat-release data, that determination also shows this acceleration and deceleration, as depicted in the FSR curves of Figs. 13 or 14. To replicate this behavior, several turbulent combustion models have been proposed [5, 28, 29, 30]. Lancaster et al. [5] suggested the possibility that the flame accelerates because only eddies smaller than the flame kernel can contribute to the turbulent burning velocity. As the flame grows, more of the turbulent field can contribute to the propagation. However, Matekunas [17] observed that the flame structure does not appear to change during combustion and that although the propagation velocity exhibits the acceleration-deceleration behavior, the burning velocity does not. Matekunas' experiments were performed in the regime of relatively low turbulence at constant volume.

In other work it has been suggested that flame propagation influences the state of turbulence in the unburned gases. Hires et al. [28] and Tabaczynski et al. [29] modified the basic approach of Blizard and Keck [30], where combustion was modeled as eddies which burn at the laminar flame speed and which transfer energy at a rate depending on the character of the turbulence. The early part of combustion (ignition-delay period) was calculated as the time to burn a single eddy whose size was a function of the integral scale. The acceleration-deceleration behavior was produced by adjusting the turbulence by a factor, $\left(\frac{\rho}{\rho_0}\right)^{1/3}$, which corresponds to conservation of angular momentum of an eddy as it is compressed. For typical engine operation, the factor yields a 20% increase in the turbulent intensity at peak pressure. The integral scale was assumed equal to the instantaneous clearance height. This assumption also contributes to the acceleration-deceleration. Although this formulation incorporates many mechanisms to duplicate observed heat-release behavior in engines, the descriptions of the flame structure and the behavior of the turbulent field have yet to be experimentally verified. In another approach, Wong and Hoult [27], using rapid distortion theory, calculate an approximate doubling of the turbulent intensity as a result of the flame. This concept also lacks experimental verification.

In light of these suggested mechanisms of flame propagation in engines, several explanations are possible for the departure of the engine-derived data in Fig. 16 from the correlation of Abdel-Gayed and Bradley [21] for fundamental combustion experiments conducted outside of engines. If the flame process is augmenting the turbulent intensity, the engine data of Fig. 16 would be moved to the right and closer to the out-of-engine measurements. Our flame-speed measurements with spark retard yield a reduction in the FSR at a given turbulent intensity, as measured over the 45-degree crank-angle interval BTDC. This result may be explained by reduced distortion because of lower peak pressures or, alternately, by the longer times available for the decay of turbulence before the 50%-mass-burned point. Finally, the data may be reflecting the true behavior of the engine flames. Conditions in the combustion chamber, i.e., pressures, temperatures and turbulence fields, may be so different from those in the out-of-engine experiments that complete correspondence should not be expected. The resolution of these questions requires further experiments to define the turbulence ahead of the flame in the engine environment.

One important observation from our FSR correlation is that turbulent intensity is more important than laminar flame speed in the determination of the burning velocity. A change in the laminar flame speed of a typical working mixture, in itself, should have little effect on the burning velocity near the 50%-mass-burned point. It was found, in fact, that in some tests at the same load and engine speed, lean flames produced approximately the same burning velocity as stoichiometric ones. The effect of mixture strength on combustion characteristics is primarily due to changes in the expansion component of the propagation velocity. This is reinforced by data obtained from a CFR engine by Bolt and Holkeboer [31], in which ignition delay correlated to air-fuel ratio but combustion duration did not. The fact that ignition delay appears to correlate to laminar flame speed [29, 30] may only be fortuitous since the expansion velocity of the flame varies directly with the mixture strength as well.

Based on these complexities of the combustion process, the necessity of assuming an empirical FSR versus flame radius relationship in the predictive application of the model becomes more apparent. For identical intake geometries one would not expect large variations in the state of turbulence and hence the character of the burn. To expand applications of the predictive model to broader variations in chambers, spark timing and dilution will require more information on the nature of the FSR and its dependence on these parameters. Fortunately, the diagnostic mode of the model provides a tool to provide such information.

APPLICATION OF THE MODEL TO IMPROVE ENGINE PERFORMANCE

Having reviewed the combustion model and its experimental verification, we now turn to an application of the model to guide our efforts toward increasing the mass burning rate and improving the performance of a single-cylinder engine.

Insight from Diagnostic Modeling—The motivation for increasing the fuel burning rate will become evident by examining combustion in two chambers having significantly different geometries, as shown in Fig. 11a and 11b. The open chamber, with near-central ignition, promotes higher mass burning rates. In contrast, the wedge chamber, with its long flame travel, would be expected to yield lower mass burning rates. Shown in Fig. 17 are the thermal efficiency maps as functions of EGR for these chambers. The peak efficiencies for each chamber at a given EGR correspond to MBT spark timing.

Both chambers showed an initial increase in MBT thermal efficiency with increasing EGR.* The initial increase is due primarily to the higher specific-heat

* *This does not imply that addition of EGR categorically improves thermal efficiency. When these two chambers are run at constant air-fuel ratio of 16:1, adding EGR can increase efficiency, as shown in Fig. 17. If they were run at increasingly leaner mixtures with no EGR, however, a somewhat higher maximum thermal efficiency would be anticipated.*

References pp. 578–579.

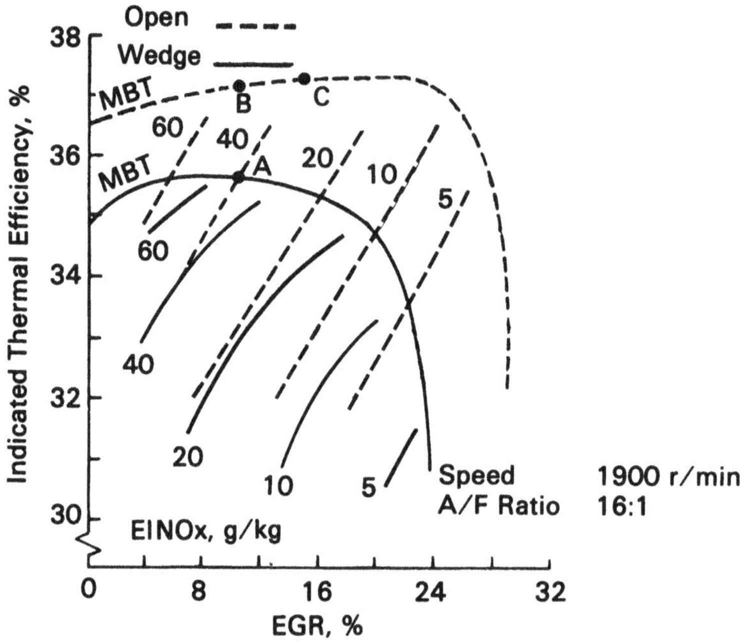

Fig. 17. Thermal efficiency maps as functions of EGR and spark timing for wedge and open chambers.

ratio associated with the lower combustion temperatures for the more dilute charge. At high EGR the reduction in flame speed with further dilution more than negates this advantage, and the efficiency drops rapidly. As shown in Fig. 17, the fast-burning open chamber has higher thermal efficiencies as well as a greater tolerance to dilution. Superimposed are lines of constant *NOx* emissions, illustrating the familiar reduction of *NOx* with spark retard and EGR. Comparing both chambers at a fixed EGR shows the higher efficiency of the open chamber occurs with a penalty of higher *NOx* emissions (see points A and B). The addition of a slight amount of EGR to the open chamber, however, reduces its *NOx* level to that of the wedge chamber while maintaining its advantage in thermal efficiency (see point C). This illustrates the favorable *NOx* versus efficiency trade-off of the open-chamber engine.

The diagnostic combustion model showed the fuel burning rate to be higher in the open chamber, and quantified the variation in flame speed and frontal area of the propagating flame throughout combustion, as shown in Fig. 18. The data of Fig. 18 shows that the burning velocity in the wedge is greater than that in the fast-burning open chamber. The lower mass burning rate in the wedge chamber is attributable to its geometry, which restricts the frontal area of the propagating flame. A potentially viable approach for increasing burning rate in the wedge chamber, therefore, is to increase its flame area.

The design selected to increase this area incorporated a cavity in the crown of the piston, as shown in Fig. 19. Note that this modification does not require a major chamber redesign, i.e., changes are confined to the piston.

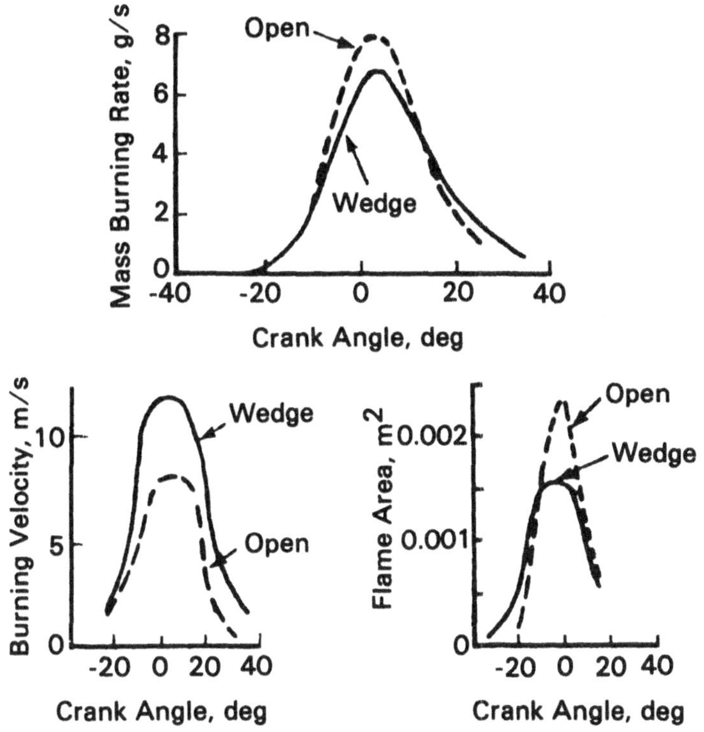

Fig. 18. Diagnostic flame-propagation results for wedge and open chambers.

Insight from Predictive Modeling—To analyze the expected improvements associated with this modified chamber, the predictive model was applied, using the engine simulation discussed earlier. The procedure was as follows. Measured engine data and diagnosed FSR data were used to simulate the unmodified wedge-chamber engine at one particular operating condition. Then, by changing only the input chamber geometry to suit the modified piston, a second simulation was run.

The operating condition selected for study was a moderate speed, moderate load point for the wedge chamber at MBT spark advance, stoichiometric mixture

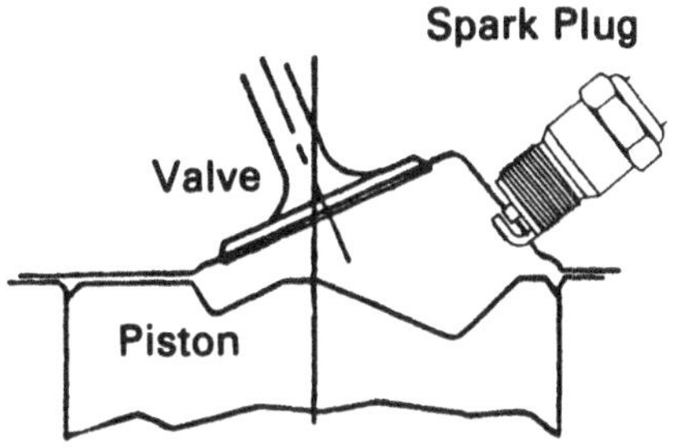

Fig. 19. Design of modified piston to increase flame area in wedge chamber.

References pp. 578–579.

ratio and 10% EGR. Experimental measurements of intake and exhaust pressures, mixture temperature and composition, and engine speed were used. An equivalent 0.04 mm lash for both valves was used to reflect compliance in the valve train. In addition, certain parameters had to be specified. These included estimated combustion chamber wall temperatures, heat-transfer coefficient multipliers, and the initial size of the burned-volume kernel. The heat-transfer information used was consistent with measured temperatures and diagnostic results from engine tests. The initial volume of the burned zone was adjusted until agreement was obtained between measured and calculated cylinder pressure at TDC. Table 4 lists the operating conditions and other assumed parameters for the wedge-chamber engine. The FSR data from the diagnostic mode and the relationship used in the predictive model are shown in Fig. 20.

Results of the simulation for the standard wedge chamber are compared to test-cell measurements and diagnostic results in Table 5 and Fig. 21. The agreement in pressure and heat release is good since empirical FSR data were used. Airflows also agree, indicating that the gas exchange and heat transfer are adequately modeled in the simulation. The important conclusion from the wedge simulation is that the torque, burning history and airflow were adequately modeled.

The simulation was rerun with the conditions of Table 4 and the modified piston geometry. As expected, an increase in maximum burning rate was projected. Table 6 and Fig. 22 show the computed performance of the modified chamber compared to that of the baseline chamber. The improvement in power and efficiency, though small, was significant. From a detailed analysis of these simulation results, it was determined that the primary mechanism contributing to the improved performance of the modified chamber at this operating condition

TABLE 4

Operating Conditions and Assumed Parameters for Wedge Chamber Simulation

Operating Conditions	
Engine speed ....1900 r/min	Spark timing ....34 BTDC
Lower heating value 4.349(10^7)J/kg	Fuel structure ....$C_8H_{15.28}$
Equivalence ratio ...0.989	Mass fraction EGR ...0.098
Intake pressure .....72.12 kPa	Exhaust pressure .....109.9 kPa
Intake pressure .....335 K	Combustion efficiency 98.0%
Assumed Parameters	
Valve lash ....0.04 mm (both valves)	
Woschni heat-transfer correlation multiplier 1.15	
Piston temperature..........450 K	
Head temperature ....430 K	
Liner temperature ....405 K	
Intake-valve temperature ....385 K	
Exhaust-valve temperature ..660 K	
Initial burned-kernel volume 0.30 ml	

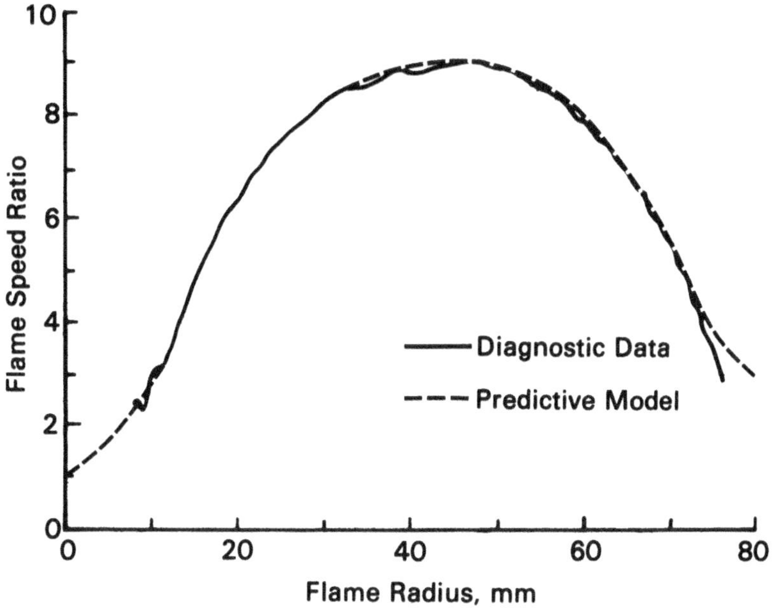

Fig. 20. FSR versus r_f; solid line, results from diagnostic tests for the wedge chamber; dashed line, input to predictive model.

was its lower in-cylinder heat transfer. Fig. 23 shows the relationship between the ratio of the burned-zone wetted area to the total wetted area as a function of volume fraction burned. The solid curve represents the standard wedge chamber while the dashed line represents the modified chamber, with both calculations made at TDC. Because of its geometry, the modified chamber enjoys a significantly lower wall area exposed to the burned gases for most of the flame travel. This advantage translates into less heat transfer during combustion and, ultimately, into higher indicated thermal efficiency.

Table 6 shows that the calculated *NO* in the exhaust from the modified chamber was 14% greater due to the higher temperatures and pressures associated with

TABLE 5

Wedge Simulation Results Compared to Diagnostic Data

Parameter	Measured or Diagnosed	Computed	Units
imep	550.3	563.2	kPa
Indicated thermal efficiency	34.62	34.61	%
Air flow	19.45	19.68	kg/h
Peak pressure	2400	2368	kPa
Location of peak pressure	15	15	CA
Peak fuel burning rate	0.6445	0.6498	mg/CA
Location of peak burning rate	9	9	CA

References pp. 578–579.

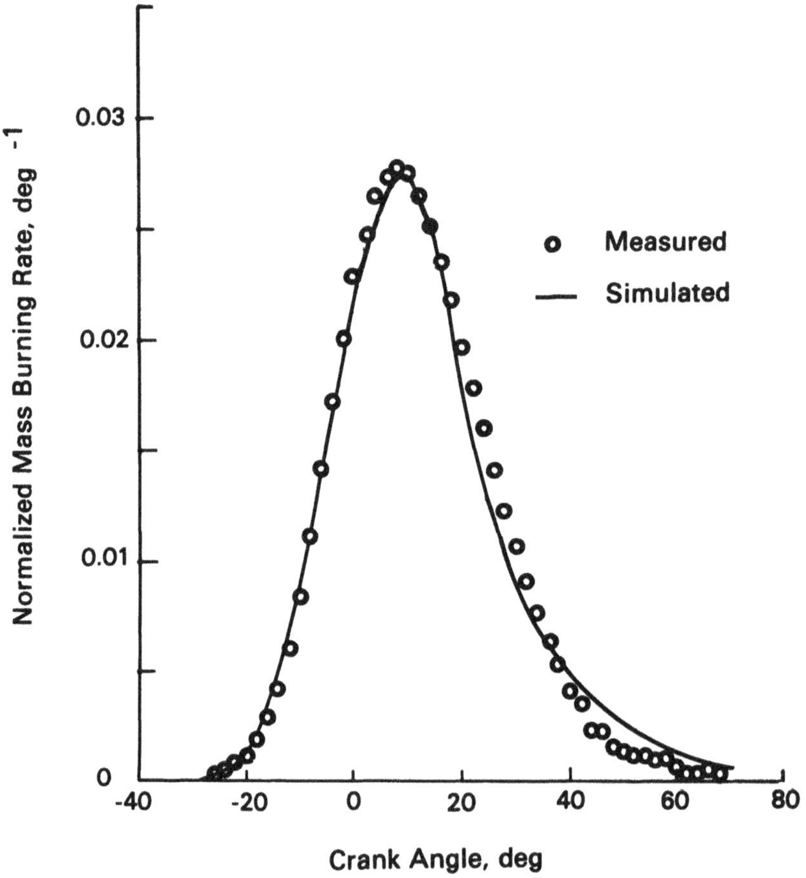

Fig. 21. Comparison of measured and simulated burning rate versus crank angle for wedge chamber.

the faster burn. Addition of 1% more EGR accompanied by a three CA degree increase in spark advance, however, reduced the computed exhaust *NO* level of the modified chamber to that of the wedge chamber at essentially the same power and economy, as shown in the last column of Table 6.

Experimental Evaluation of Modified Engine—The calculated performance improvement of the modified chamber was judged sufficient to warrant experimental evaluation. Therefore a modified piston was fabricated and installed in the single-cylinder test engine used for the wedge-chamber tests. Heat-release results for the engine operating conditions of Table 4 are shown in Fig. 24, along with the predicted data transposed from Fig. 21. As shown, the measured burning rate in the modified chamber was increased, even beyond that amount predicted. The MBT spark timing was found to be 4 CA degrees retarded from that of the unmodified chamber. The diagnosed flame speed ratio for the modified chamber is superimposed on the wedge-chamber data in Fig. 25. As seen, the peak FSR is greater than that of the wedge chamber, which translates into higher burning rates. Finding specific reasons for these changes in FSR provide a worthy subject for further research.

The measured efficiency and NOx emission levels associated with the wedge chamber are compared to data for the modified chamber in Fig. 26. The format of this figure parallels that of Fig. 17 for the wedge and open chambers. The efficiency of the modified chamber was improved beyond predictions. The tolerance to EGR was also increased. This is important in capitalizing on the superior efficiency potential of the modified chamber. For example, suppose the wedge chamber were run at MBT, 9% EGR, yielding an NOx emission index of 60 g/kg fuel (point A). Unless EGR is increased toward 14% (point B) with the modified chamber, it must accept an NOx penalty to take advantage of its efficiency capability. Because of its dilution tolerance, however, the modified chamber can be run at MBT and an even lower NOx level, e.g., 40 g/kg fuel (point C), without the efficiency penalty associated with the wedge chamber at higher EGR.

When present restrictions of 91 Research Octane Number (RON) gasoline are imposed, it is often impossible to run at MBT spark advance. Although not illustrated in Fig. 26, the modified chamber retains its efficiency advantage even at the knock limit.

Another consequence of the increased dilution tolerance of the modified chamber is a reduction in cycle-to-cycle variability in cylinder pressure and output power. Although the results are not presented here, this cyclic dispersion was less at comparable operating conditions in the modified chamber than in the

TABLE 6

Comparison of Simulation Results for Both Chambers

Parameter	Wedge Chamber	Modified Chamber	Modified Chamber (1% more EGR)	Units
imep	563.2	573.7	565.0	kPa
Indicated thermal efficiency	34.61	35.25	35.17	%
Peak fuel burning rate	0.6498	0.6991	0.6688	mg/CA
Location of peak burning rate	9	7	6	CA
Combustion duration (1%-90%)	59	55	57	CA
Exhaust NO concentration relative to wedge chamber*	1.00	1.14	.98	-
Cumulative heat transfer	175.0	169.4	167.6	J

* The computed NO concentrations are substantially lower than the measured values. This has been previously observed by others using two-zone [32], multizone adiabatic [8], and thermal boundary layer/adiabatic core [33] combustion models. In general, however, computed relative NO changes with engine design and operating conditions have been found to be representative of experimental data. For example, for the operating condition indicated above, the measured NO increase was found to be 16% versus the predicted 14%.

References pp. 578–579.

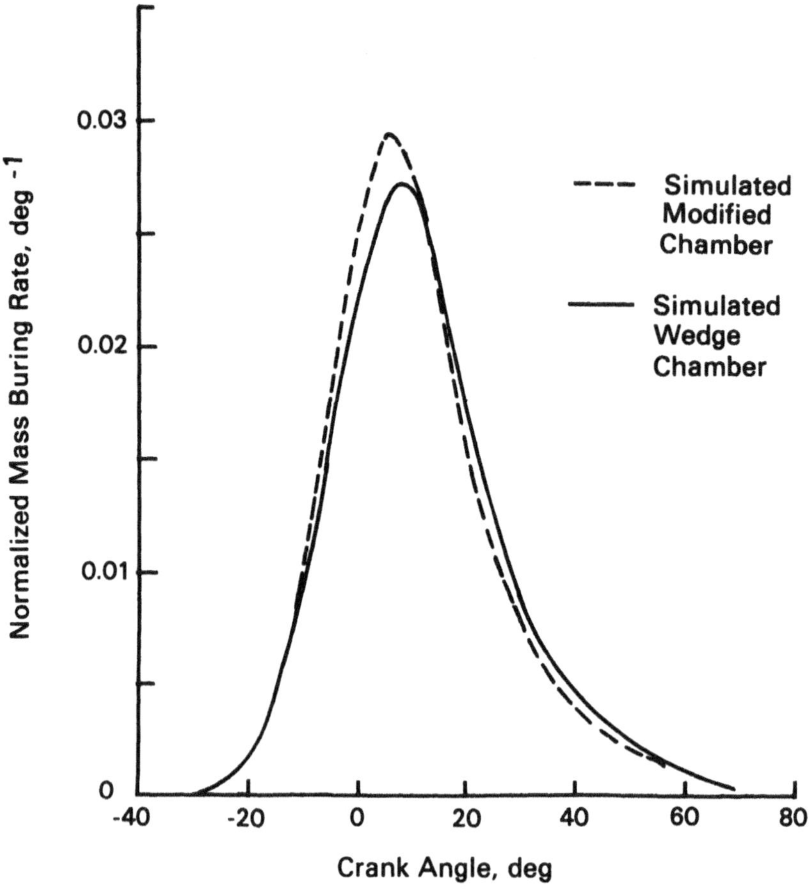

Fig. 22. Computed burning history of modified chamber compared to that of baseline wedge chamber.

wedge. Evidence suggests that this characteristic of the modified chamber may bear favorably on one aspect of vehicle driveability.

Of course there is a long road from tests in a single-cylinder engine to practical application of a chamber design in a vehicle. Manufacturing considerations are involved. Individual cylinders must be combined into a multicylinder engine that performs satisfactorily, responding quickly during transients and satisfying other driveability criteria. Not to be overlooked are the emissions of unburned hydrocarbons in the exhaust, an issue not addressed in this paper. All of these performance standards must be met over a broad range of engine speeds and loads, not just the limited number of operating points on which this research study was based. Nonetheless, the favorable results, partially demonstrated in Fig. 26, provide encouragement for further development and application of combustion modeling techniques.

SUMMARY

A quasi-dimensional propagating-flame approach was taken to develop a model of combustion in engines. Flame propagation and pressure histories derived from the model were compared with pressure measurements and visual observations recorded in specially designed research apparatus. These comparisons showed the model assumptions to be realistic.

Empirical relationships used in the model to characterize heat transfer and turbulence effects on combustion were developed by applying the model in a diagnostic mode to experimental data from a number of different engines. Resulting from this procedure was a generalized correlation of flame speed with turbulent intensity and laminar flame speed in the unburned mixture. This correlation allows the model to be used in the predictive mode to investigate the effect of chamber design and operating conditions on engine performance. The model was used to examine combustion in single-cylinder engines. Results led to

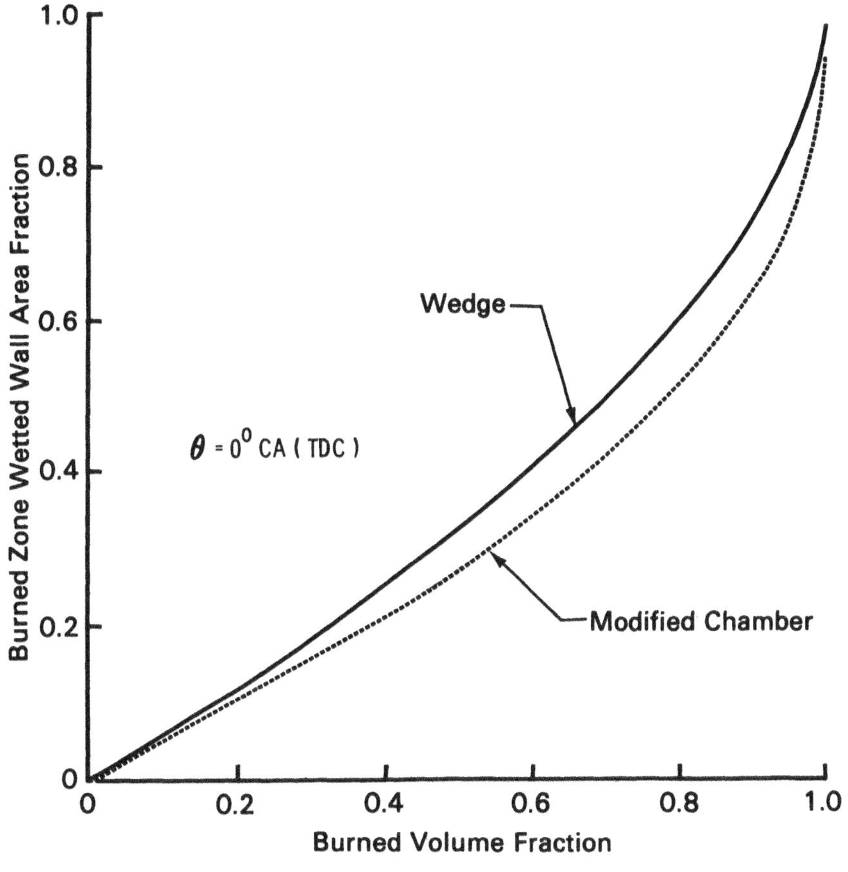

Fig. 23. Fraction of chamber surface wetted by burned volume as a function of volume fraction burned for both wedge and modified chambers at TDC.

References pp. 578–579.

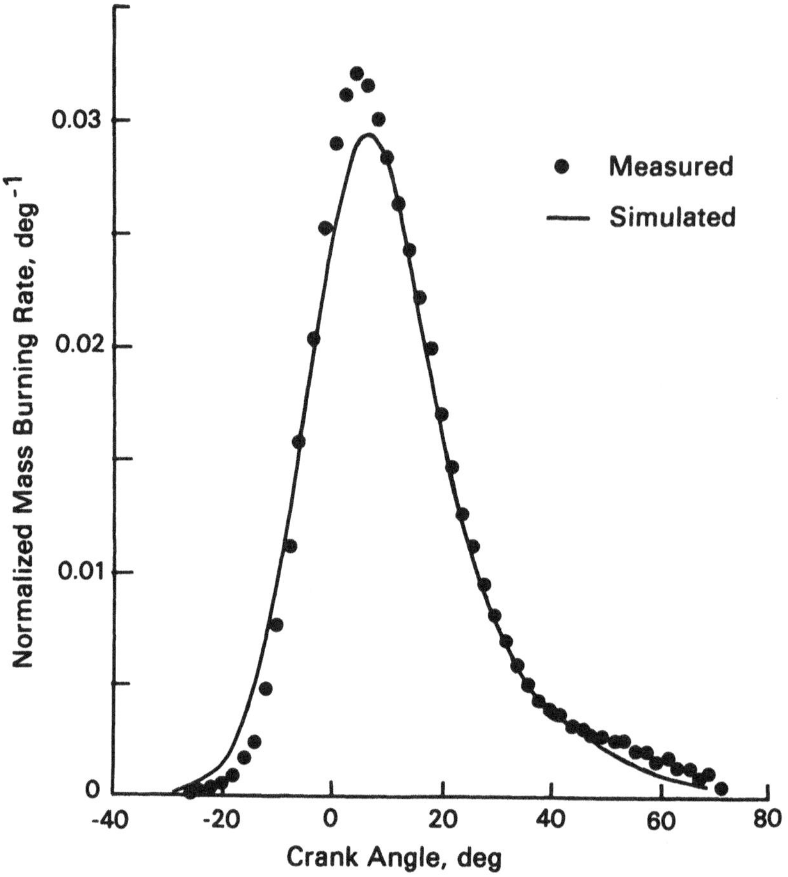

Fig. 24. Measured burning rate compared to simulation results for modified chamber.

modifications for improving the performance of a chamber having a relatively long combustion duration. Subsequent experiments with engine hardware incorporating these modifications demonstrated significant improvements. As such, the model provides direction to the engine designer when reference engine performance data are available. An important limitation, however, is the uncertainty surrounding the required *a priori* estimation of turbulence in the combustion chamber.

Because it is based on a spherical flame geometry, the model is not completely appropriate when fluid motions severely distort the flame shape. Finally, it should be pointed out that additional factors, such as transient operation, knock, total emissions tradeoffs, driveability, multicylinder performance, and manufacturing techniques, must be considered in any new engine design and development program.

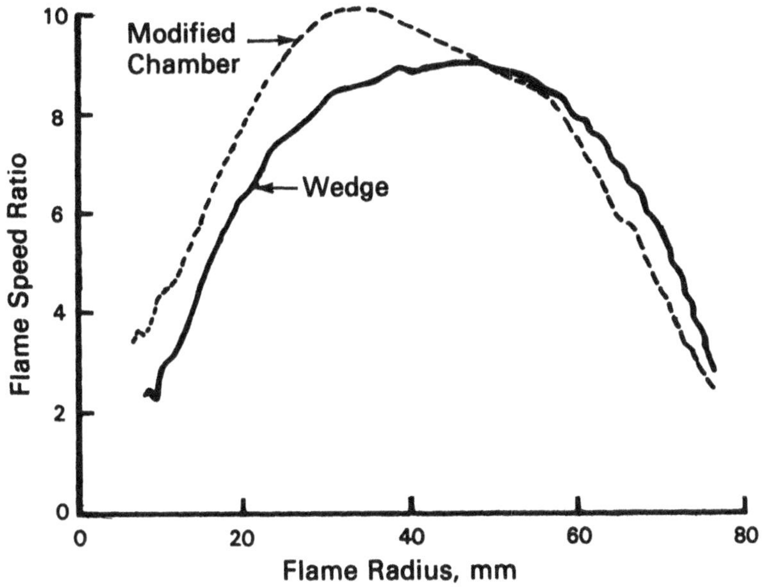

Fig. 25. Diagnosed FSR for the modified chamber compared to that for the wedge chamber.

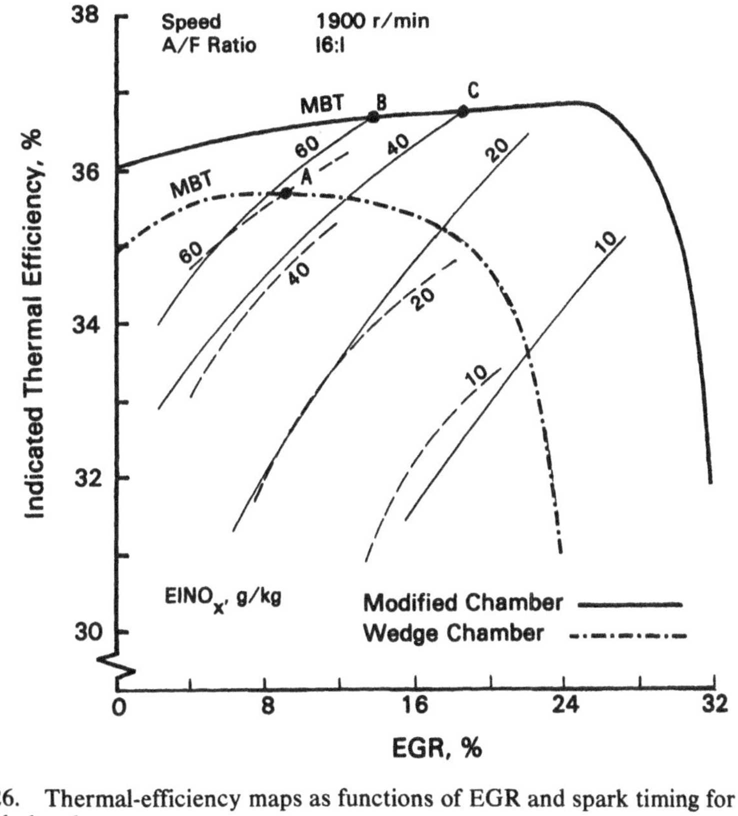

Fig. 26. Thermal-efficiency maps as functions of EGR and spark timing for wedge and modified chambers.

ACKNOWLEDGMENT

The authors would like to express their gratitude to R. R. Toepel and J. H. Tuttle, who provided engine experimental data, M. B. Young, who contributed to the engine diagnostic software development, J. C. DeSantis and R. Diepenhorst, who assisted in programming activities, and V. P. Michalak Jr. and R. E. Gutowski for their assistance during experimental tests. They are further indebted to the Engine Research Department's Design Group for its contributions in the design of the test facilities and to the Engineering Test Group for its assistance during the experimental phase of the study.

REFERENCES

1. *J. H. Tuttle and R. R. Toepel, "Increased Burning Rates Offer Improved NOx Emissions Trade-Offs in Spark-Ignition Engines," SAE Paper No. 790388, 1979.*
2. *H. Kuroda, J. Nakajima, S. Muranaka, K. Sugihara and Y. Takagi, "The Fast Burn with Heavy EGR, New Approach for Low NOx and Improved Fuel Economy," SAE Paper No. 780006, 1978.*
3. *D. J. Patterson and G. J. Van Wylen, "A Digital Computer Simulation for Spark-Ignited Engine Cycles," SAE Paper No. 633F, 1963.*
4. *R. B. Krieger and G. L. Borman, "The Computation of Apparent Heat Release for Internal Combustion Engines," ASME Paper No. 66-WA/DGP-4, 1966.*
5. *D. R. Lancaster, R. B. Krieger, S. C. Sorenson, and W. L. Hull, "Effects of Turbulence on Spark-Ignition Engine Combustion," SAE Trans., Vol. 85, Paper No. 760160, pp. 689–710, 1976.*
6. *G. Woschni, "A Universally-Applicable Equation for the Instantaneous Heat Transfer Coefficient in the Internal Combustion Engine," SAE Trans., Vol. 76, Paper No. 670931, pp. 3065–3083, 1967.*
7. *C. Olikara and G. L. Borman, "A Computer Program for Calculating Properties of Equilibrium Combustion Products with Some Applications to I.C. Engines," SAE Paper No. 750468, 1975.*
8. *G. A. Lavoie, J. B. Heywood and J. C. Keck, "Experimental and Theoretical Study of Nitric Oxide Formation in Internal Combustion Engines," Comb. Sci. and Tech., Vol. 1, pp. 313-326, 1970.*
9. *D. R. Lancaster, R. B. Krieger and J. H. Lienesch, "Measurement and Analysis of Pressure Data," SAE Trans., Vol. 84, Paper No. 750026, pp. 155–172, 1975.*
10. *R. V. Fisher and J. P. Macey, "Digital Data Acquisition with Emphasis on Measuring Pressure Synchronously with Crank Angle," SAE Paper No. 750028, 1975.*
11. *J. H. Lienesch and M. K. Krage, "Using Microwaves to Phase Cylinder Pressure to Crankshaft Position," SAE Paper No. 790103, 1979.*
12. *M. B. Young and J. H. Lienesch, "An Engine Diagnostic Package (EDPAC)—Software for Analyzing Cylinder Pressure-Time Data," SAE Paper No. 780967, 1978.*
13. *B. S. Samaga, "Assessment of a Heat Transfer Formulation for Reciprocating Combustion Engines," Indian Journal of Technology, Vol. 13, pp. 484–487, 1975.*
14. *L. F. Shampine and M. K. Gordon, "Computer Solution of Ordinary Differential Equations," W. H. Freeman and Co., New York, New York, 1975.*
15. *W. A. Woods and S. R. Kahn, "An Experimental Study of Flow Through Poppet Valves," Proc. I. Mech. E., Vol. 180, Pt. 3N, 1965-66.*
16. *E. S. Dennison, T. C. Kuchler and D. W. Smith, "Experiments on Flow of Air Through Engine Valves," Trans. ASME, Vol. 53 (OGP-53-6), pp. 79–97, 1931.*
17. *F. A. Matekunas, "A Schlieren Study of Combustion in a Rapid Compression Machine Simulating the Spark Ignition Engine," Seventeenth Symposium (International) on Combustion, The Combustion Institute, Pittsburgh, Pennsylvania, 1978.*
18. *F. W. Bowditch, "A New Tool for Combustion Research, A Quartz Piston Engine," SAE Trans., Vol. 69, Paper No. 150B, pp. 17–23, 1960.*

19. S. Ohigashi, Y. Hamamoto and A. Kizima, "Effects of Turbulence on Flame Propagation in a Closed Vessel," Bull. JSME, Vol. 14, No. 74, pp. 849-858, 1971.
20. D. R. Lancaster, "Effects of Engine Variables on Turbulence in a Spark-Ignition Engine," SAE Trans., Vol. 85, Paper No. 760159, pp. 671-688, 1976.
21. R. G. Abdel-Gayed and D. Bradley, "Dependence of Turbulent Burning Velocity on Turbulent Reynolds Number and Ratio of Laminar Burning Velocity to R.M.S. Turbulent Velocity," Symposium (International) on Combustion, The Combustion Institute, Pittsburgh, Pennsylvania, pp. 1725-1735, 1976.
22. D. R. Ballal and A. H. Lefebvre, "Turbulence Effects on Enclosed Flames," Acta Astronautica, Vol. 1, pp. 471-483, 1974.
23. D. K. Kuehl, "Laminar-Burning Velocities of Propane-Air Mixtures," Eighth Symposium (International) on Combustion, The Combustion Institute, Pittsburgh, Pennsylvania, pp. 510-521, 1962.
24. M. Metghalchi, Massachusetts Institute of Technology, Personal Communication, September 1978.
25. S. Heimel and R. C. Weast, "Effect of Initial Temperature on the Burning Velocity of Benzene-Air, n-Heptane-Air, and Isooctane-Air Mixtures," Sixth Symposium (International) on Combustion, The Combustion Institute, Pittsburgh, Pennsylvania, pp. 296-302, 1957.
26. M. Metghalchi and J. C. Keck, "Laminar Burning Velocity of Isooctane-Air, Methane-Air and Methanol-Air Mixtures at High Temperature and Pressure," Meeting of the Eastern States Section of the Combustion Institute, November 1977.
27. D. P. Hoult and V. W. Wong, "The Generation of Turbulence in an Internal Combustion Engine," General Motors Research Laboratories Symposium on Combustion Modeling in Reciprocating Engines, Warren, Michigan, November, 1978.
28. S. D. Hires, R. J. Tabaczynski and J. M. Novak, "The Prediction of Ignition Delay and Combustion Intervals for a Homogeneous Charge, Spark Ignition Engine," SAE Paper No. 780232, 1978.
29. R. J. Tabaczynski, C. R. Ferguson and K. Radhakrishnan, "A Turbulent Entrainment Model for Spark-Ignition Engine Combustion," SAE Trans., Vol. 86, Paper No. 770647, pp. 2414-2433, 1977.
30. N. C. Blizard and J. C. Keck, "Experimental and Theoretical Investigation of Turbulent Burning Model for Internal Combustion Engines," SAE Trans., Vol. 83, Paper No. 740191, pp. 846-864, 1974.
31. J. A. Bolt and D. H. Holkeboer, "Lean Fuel/Air Mixtures for High-Compression Spark-Ignited Engines," SAE Trans., Vol. 70, Paper No. 380-D, pp. 195-202, 1962.
32. K. K. Chen and R. B. Krieger, "A Statistical Analysis of the Influence of Cyclic Variation on the Formation of Nitric Oxide in Spark Ignition Engines," Comb. Sci. and Tech., Vol. 12. pp. 125-134, 1976.
33. P. N. Blumberg, G. A. Lavoie and R. J. Tabaczynski, "Phenomenological Models for Reciprocating Internal Combustion Engines," Department of Energy/Division of Power Systems Workshop on Modeling of Combustion in Practical Systems, Los Angeles, California, January 1978.

DISCUSSION

D. J. Patterson *(University of Michigan)*

I was fortunate to have the paper in advance, so I will make a couple of comments. First of all, I think the authors are to be congratulated because this is really quite a substantial effort. From my view this is kind of an optimum marriage of the theoretical and the experimental because the bottom line was to get useful results, which obviously were obtained. Now I should like to make a couple of comments beyond that. First of all, I would like to say that a two-zone model for the *NOx* would not, of course, be expected to be too good. Also, since your results depend quite a bit upon your heat-transfer calculations, I would like to know a little bit more about the assumptions that went into those calculations. Finally, the flame speed ratio seemed to be quite high—higher than I would have guessed off the top of my head. I wondered if you might have any comments on that matter.

J. N. Mattavi

Regarding, first, your comments concerning *NO*, the critical thing, as was pointed out earlier, is to get the correct heat-transfer correlation. I think we all agree that the extended Zeldovich mechanism seems to be appropriate for calculating *NO* formation, and the trick is to come up with the proper temperature in the chamber. There have been some investigations of the effect of temperature stratification in the burned region. In fact, data* by Blumberg and his co-workers suggested that it might contribute to something like a 15% difference in the calculated *NO*. We feel that the Woschni** correlation, which was applied in our model, is ample for predicting the trends. To draw an analogy I'll go back to the example given yesterday by Bill Reynolds. He mentioned that people working with an airfoil couldn't predict the exact lift on the foil, but they seemed to get

J. N. Mattavi

* *P. N. Blumberg and J. T. Kummer, "Prediction of NO Formation in Spark-Ignited Engines—An Analysis of Methods of Control," Comb. Sci. and Tech., Vol. 4, pp. 73–95, 1971.*

** *G. Woschni, "A Universally Applicable Equation for the Instantaneous Heat Transfer Coefficient in the Internal Combustion Engine," SAE Trans., Vol. 76, Paper No. 670931, pp. 3065–3083, 1968.*

a pretty good handle as to how a change in the thickness of the airfoil would affect its lift. In our case, we predicted, for example, that the *NO* would increase by about 14% with the faster burning chamber. Experimentally, it increased by 16%, so we felt that the trend was amply predicted.

The flame speed ratios were indeed comparatively high. To explain this behavior, one has to go back to the fundamental data on laminar flame speed. For example, the flame speed ratio is obviously dependent on the laminar flame speed data. When we compared the flame motion predicted by the model with the film records made for a laminar condition using propane in the rapid compression machine, the close correspondence suggested to us that we had a pretty good handle on the correlation of laminar flame speed for the propane and air mixture as a function of pressure and temperature. This correlation is based on data obtained by Metghalchi and Keck*, who found an inverse pressure dependency to occur. That is, as the pressure increased in their tests, there was a reduction in the laminar flame speed. This behavior, of course, would contribute to a larger flame speed ratio. What I am suggesting is that if one didn't recognize this more recent data and were to apply some of the older laminar flame speed data which did not have this pressure correction, one would get lower flame speed ratios than those which we reported.

D. P. Hoult *(Massachusetts Institute of Technology)*

I would like to complement the authors. I think this is a superb job and indicates how things are going to go in the future. I want to make a couple of comments on ways in which the scatter in the correlation between turbulent flame speed, laminar flame speed, and turbulent intensity could be reduced. First of all, the correlation is meant in the original sense to apply only to average conditions. Even though, as I understand it, you measured for a given cycle the turbulent flame speed, presumably one would have to average over a number of cycles to apply the correlation in the manner originally suggested in the literature. It was meant to be an average flame speed. Secondly, at least as far as how rapid distortion theory might apply to this problem, as you vary the EGR and the stoichiometry, then the amplification predicted by that theory would also vary. It's not exactly a constant factor of two. I suggest that those two effects might greatly reduce the scatter in the correlation.

J. N. Mattavi

As it turns out, the data were obtained from pressure measurements that were averaged over 96 consecutive engine cycles, so it does represent average data. It was collected over a wide range of engine speeds, loads, EGR rates and equivalence ratios. As a matter of fact, we were surprised that the correlation

---

* *M. Metghalchi and J. C. Keck, "Laminar Burning Velocity of Propane—Air Mixtures at High Temperature and Pressure," Paper No. CSS/CI-79-04, Central States Section, The Combustion Institute, Columbus, Indiana, April 1979.*

was as "tight" as it turned out to be, considering the variation in fuels and chambers. That is, we had two different types of fuels and three different types of combustion chambers.

F. V. Bracco *(Princeton University)*

I also want to complement the authors. I think it was a very good paper. I would like to relate the success of your model to what I was trying to say yesterday, one way or the other.

Your combustion chamber is open with no throats or things like that. The spherical flame assumption is pretty good. If you leave out wall heat transfer, it satisfies the condition that I was trying to point out yesterday as being necessary and sufficient for a zero- or quasi-dimensional model to be *the* best model to be used, with not much more to be gained in going to a three-dimensional unsteady model, even if it were possible. The fact that your model was successful in achieving your goal more or less substantiates what I was trying to say. Given that two conditions are satisfied (spherical flame and small flame thickness), your model is the best for the application. That's my opinion, of course.

The other side of the coin, and I don't want to leave a wrong impression here, is that yours is a very special case. Most of the action today is in engines in which the combustion chamber is not so simple. Divided chambers are used, stratified charges are of interest, two-phase flows are of interest, boundary layers on the walls are of interest. Those are all cases in which, by the same definition, zero-dimensional models would fail. Only a multidimensional model would be of some use.

J. C. Keck *(Massachusetts Institute of Technology)*

Your correlation is so impressive that I'm almost afraid to make my comments, but I'm very much afraid that it may be fortuitous. Three times, in fact, I have seen the flame speed ratio introduced, and every time I have had a little concern. It is well known that the turbulent intensity in engines correlates with engine speed and, probably more properly, with inlet flow velocity. Therefore, the first thing that concerns me is the use of laminar flame speed to make a nondimensional quantity. One thing that you ought to keep in mind is that in every one of these plots, what is plotted is the turbulent flame speed divided by the laminar flame speed as a function of turbulent intensity divided by the laminar flame speed. You could have divided both by either the inlet flow velocity or the engine speed without destroying the correlations.

Now historically people have argued, and I think with some justification, that since burning angles (combustion duration, expressed in crank angle degrees) in most engines were independent of engine speed, this was strong evidence that laminar flame speed had very little to do with combustion in engines, at least for the fast-burn phase.

The way in which I think the laminar flame speed does indeed enter into the problem is in the ignition delay period. Therefore, I would expect that during the fast burning period, you should have a correlation based on either inlet flow velocity or piston speed for that part of the phase, and then if you can distinguish an ignition delay period, you should correlate it using the laminar flame speed. In that case the correlation is going to involve a laminar flame speed. But I don't believe that a correlation involving the sum of both ignition delay plus burning time can physically be universal. For one thing, I know that the ignition delay is going to be inversely proportional to the laminar flame speed because if it is associated, for example, with a characteristic eddy burning time, it is going to appear in the denominator.

The correlation, as I say, will not be as simple as the one you suggest. I think if you were to test the model over a wider range of conditions, in particular over a wider range of spark advance, or for a variety of fuels in which the laminar flame speed varies by a significant amount, you might well find that it isn't going to work.

It's hard to make this comment without undercutting my own work on laminar flame speed, which I hate to do, because I would like to feel that laminar flame speed does have a lot to do with combustion in engines. But I think it has mostly to do with what happens during the ignition delay period, and what happens at the walls after the flame impinges on the wall—the quench layer formation—as well as with the afterburning associated with the burn-out of the flame.

J. N. Mattavi

I tend to agree with you, and I guess historically people have nondimensionalized the burning velocity with the laminar flame speed, both in engines as well as in flat-flame burners and flame tubes. Let me use this blackboard, if for no other reason, to justify its existence.

The correlation showed the burning velocity, S_b, divided by the laminar flame speed, S_l, to be equal to one plus the product of a constant, C, and the ratio of the turbulent intensity, u', to the laminar flame speed. That is

$$\frac{S_b}{S_l} = 1 - C\frac{u'}{S_l} \tag{22}$$

If one multiplies this expression by S_l, one gets

$$S_b = S_l + C\,u' \tag{23}$$

We examined combustion in a number of chambers in detail, and we found some confirmation of the remark you just made. That is, if you look at the strength of the S_l parameter versus the $C\,u'$ parameter, where C incidentally turns out to be approximately 4, S_l is about 1 to 1.25 m/s, and u' is about 2 to 4 m/s, you can see that the burning velocity indeed depends very strongly on the turbulent intensity and is somewhat independent of the laminar flame speed. This applies after the ignition process, when fully developed propagation of the flame across the chamber is occurring. So indeed we have seen some confirmation of your observation.

H. C. Watson *(University of Melbourne, Australia)*

I, too, would like to offer my congratulations for what I think is a very excellent piece of work. I'd like to comment on some experiences we've been having in attempting a similar, though less ambitious, type of program of analysis. We have not had the facility actually to measure our turbulent intensities and have therefore mainly resorted to published data in the literature. By and large, the data have come from your co-workers, so at least we have used common data. The work has been done in a disc-shaped chamber.

We performed our analysis, perhaps incorrectly, on a cycle-by-cycle basis, because our main interest was in cyclic dispersion. Our results showed a number of pictures which seemed to be very poor. First, in establishing the turbulent-to-laminar flame speed ratio we needed to handle the geometry of the combustion chamber very carefully, so my first question would be, "How did you handle this problem?" The surface area in the early growth stages of the flame kernel seems to be very important in its effect on the heat transfer and its subsequent effects. I would like to say that we eventually resorted to a numerical evaluation of this. Then, because we ended up with so much numerical data and interpolation took time in the computer program, we ended up with spline curve-fitting functions for these data to get them smooth, rather than a curve with lots of little peaks up and down in flame speed ratio.

The final conclusion of our results was that for mixtures around the stoichiometric value, we observed a two-to-one spread of flame speed ratio, which ranged from two upwards to four and a half in the 256 cycles that we analyzed. At the lean limit of 0.75 equivalence ratio, the flame speed ratio seemed to flatten out to a more constant value. Furthermore, the general trend with mass-burned fraction was for a more constant value, rather than the range of rising and falling values which your curves characteristically showed. I would be happy if you could comment perhaps a bit on this matter in your own work.

J. N. Mattavi

As far as the area determination goes, we used a subroutine that was developed at our Engineering Staff. This subroutine uses a vector approach to calculate the frontal area of a spherical flame as it propagates across *any* shaped chamber, so it's quite flexible. We're also carrying on some geometry modeling work here in the Laboratories to analyze the surface area of wetted walls, for example, to calculate heat transfer and hydrocarbon quenching. I would also like to point out that these combustion and flame motion data were obtained for three particular chambers. We would like to carry on our work and extend it to chambers, for example, where there is a large degree of swirl—that is, chambers where we have more turbulence introduced by the way of a shrouded valve or cropped valve or what have you. So I don't want you to leave with the impression that the shape of the flame speed ratio curves for these three chambers is universal.

F. F. Pischinger *(Institute of Applied Thermodynamics, Germany)*

I have a small comment concerning this flame speed problem. I think it is always very useful to try fuels with extreme values of laminar flame speed—for instance, hydrogen. We did this and I think this formula is quite good. Running with hydrogen does cause problems. Because of its high laminar flame speed, hydrogen produces very high combustion rates which, at certain air-fuel ratios, can cause real problems with harsh combustion. With higher engine speeds the increase is the same as with gasoline, and then you can get quite good combustion. So I think the use of hydrogen for comparisons is always a good idea because it is a very extreme test.

J. N. Mattavi

That is a good suggestion. The flame speeds of the fuels we used only varied by about 13% from one to the other. But if that flame speed ratio formulation is correct, then we would expect the role of laminar flame speed to be much more important in the case of hydrogen. It would be an interesting experiment to see if indeed that were the case.

F. F. Pischinger

Just a second comment, regarding the two-zone model. We tried to use a multi-zone model for the burned region and found that normally there is no big difference in the end result if one uses many or just a few zones. Of course there is a very strong stratification of the nitrogen oxide in the burned charge, but this is always well mixed in the exhaust. But we always came to nearly the same result for normal operation. So fortunately, I think one can rely on the two-zone model, but I have no explanation why.

T. Morel *(General Motors Research Laboratories)*

We have heard two opinions in support of Eqs. 22 and 23, namely, the experimental evidence of Professor Pischinger and your own data. Yet, I feel that dividing S_b, and especially u', by S_l does not seem to be physically correct. Perhaps one could reconcile this problem by rearranging Eq. 23 to obtain

$$S_b - S_l = C\,u' \tag{24}$$

P. Eyzat *(Institut Francais Du Petrole, France)*

This is a very useful tool, sometimes more useful than theory can forecast. But one must be careful in applying the formulation [Eq. 23]—particularly on the lean side as the flame propagation limit is approached. For example, as the laminar flame speed approaches zero, the formulation suggests that the burning velocity, S_b, can have a finite value, even at the lean limit of propagation, if the turbulence intensity, u', is sufficiently large. This is not possible. The limit of propagation is always given by the laminar flame speed. The influence of turbulence on flame propagation given by Eq. 23 does not apply at the limit of propagation. That is, turbulence never widens the limits of flame propagation. When S_l goes to zero, S_b must also go to zero.

P. N. Blumberg *(Ford Motor Company)*

I am not sure I have much new to add, but I would like to congratulate the authors again on a very fine piece of work. It is the first piece of work in the literature that I have seen in the conventional homogeneous spark ignited engine area that really addresses a question involving combustion chamber design. Roger [Krieger], I think when you give your very fine overview again, you'll have a lot more to say about the engine design applications side of your slide. I think this is a real contribution.

A second point. That formula has always bothered me because it indicates that the laminar flame speed, S_l, has only a very weak influence on the burning velocity, S_b, and I think we all know that there is an equivalence-ratio effect on the basic flame speed in the engine, with flame speed peaking on the rich side of stoichiometric, and becoming very slow as one goes lean. This behavior is independent of piston effects. So I think the thing to consider for incorporation into your model would be the theories* that have been proposed by Rod Tabaczynski and co-workers, whereby the effect of equivalence ratio is incorporated into the changing value of turbulent intensity, u', with the density variation in the unburned gas. The EGR effect is also incorporated in those theories. I guess this gets into what Dave Hoult was also talking about. I would just suggest that might be an avenue that could be used.

F. A. Matekunas *(General Motors Research Laboratories)*

I would like to make a comment relative to a couple of these questions. One must appreciate that a change in the laminar flame speed also changes the heat release per unit volume. This in turn changes drastically the expansion velocity (flame propagation velocity minus the burning velocity), thereby affecting the nature of the combustion process. But this is based on the burning velocity which is relative to the unburned gases. Tabaczynski's model still has to be evaluated in terms of our data, particularly with respect to its ability to predict the relative acceleration and deceleration of the flame. The difficulty is that nobody really has a good handle on the integral scales. They have assumed integral scale to be

* *R. J. Tabaczynski, C. R. Ferguson and K. Radhakrishnan, "A Turbulent Entrainment Model for Spark-Ignition Engine Combustion," SAE Trans., Vol. 86, Paper No. 770647, pp. 2414–2433, 1977.*

S. D. Hires, R. J. Tabaczynski and J. M. Novak, "The Prediction of Ignition Delay and Combustion Intervals for a Homogeneous Charge, Spark Ignition Engine," SAE Paper No. 780232, 1978.

equal to the piston clearance height. I think that is where we really need more data. The ignition delay, as pointed out in the paper, goes hand-in-hand with the air-fuel ratio and the expansion velocity of the flame. Early in the propagation of the flame, the expansion velocity is very high relative to the burning velocity because the density ratio is high. It may be that the nature of the early kernel development is more strongly related to this expansion velocity than to the burning velocity.

SYMPOSIUM SUMMARY

C. A. AMANN

General Motors Research Laboratories, Warren, Michigan

The end objective of modeling is useful application. Roger Krieger has reminded us that models may be oriented either to help us in the *design* process or to help us in *understanding* combustion. The program these past two days has been designed to improve our understanding of the modeling process, and I feel that it has helped us toward that objective.

MODEL CLASSIFICATION

Following a classification scheme attributed to Fred Bracco, John Heywood, in his overview, defined zero-dimensional, quasi-dimensional and multidimensional models. Although some objection was heard to this terminology, it does seem to meet our needs and appears to be accepted by a growing number of people.

The interest in pursuing the multidimensional model is high because after you have satisfied conservation of mass, energy, momentum and species in a model that incorporates accurate reaction kinetics, what else is left? The problem is that it is easier to write the conservation equations than it is to substitute the correct numbers in them. This prompted Gary Borman to observe that some people are trying to solve the most difficult problems and getting lost in the process, rather than settling for simpler techniques that may be only approximate but do provide insight. In this connection it was interesting to hear Fred Bracco, long an advocate of multidimensional modeling, say that he could see cases where the zero-dimensional model did as good a job as the multidimensional model. Walter Brandstetter's paper, so ably summarized by Jack Lienesch, provided an example of a constructive application of zero-dimensional modeling. The work described by Jim Mattavi represents a case of useful results coming from a quasi-dimensional model.

But Bracco also cited cases where zero-dimensional models do not suffice. Certainly one such situation would be the direct-injection stratified-charge engine modeled by Larry Cloutman and his colleagues at Los Alamos. Impressive as those results are, we must not overlook the remaining shortcomings. For example, dissatisfaction was expressed with the restriction of axial symmetry, exclusion of the intake process, the assumption of constant eddy diffusivity as a turbulence model, the single-step kinetics, inadequately treated wall heat transfer, etc., etc. Yet those shortcomings can be overcome in time. When one looks at the results Cloutman showed and then contemplates where this multidimensional approach stood, say, five years ago, one must be impressed. John Heywood reminded us that in viewing progress in modeling, one needed "patience," and he was so right. Meanwhile the utility of existing zero- and quasi-dimensional models must not be overlooked.

FLUID MOTIONS

Bill Reynolds has made several important points regarding the role played by fluid motions. First, he made the point that the turbulent field in which combustion occurs originates primarily in shear layers at the intake port. Our ability to handle the intake-generated turbulence, tracking it through the compression stroke, is not yet totally satisfactory. It is interesting that the two examples of multidimensional modeling presented in this Symposium, those shown by Gosman and by Cloutman, both began after completion of the intake stroke, with the piston initially at bottom dead center.

Reynolds also discussed the wide variations in turbulent scales. The largest scale of motion is represented by multidimensional models, as he demonstrated with Bill Ashurst's movies. These movies, incidentally, suggested asymmetric vortices that did not necessarily repeat from one cycle to the next. Such behavior could have significant implications for the typical combustion model, which admits to no cyclic variability.

As for the smallest scale of turbulence, handling that numerically is ominous because the mesh size has to be of the same order as the eddy size. As the eddies get smaller, the limit is an individual molecule, I suppose, and that is a frightening thought. I am sure we will not have to go that far, but David Gosman did warn us that if modeling is to be attractive, it must cost less than the experiments it is intended to supplant.

Reynolds also discussed boundary-layer formation time and concluded it was unlikely that fully developed boundary layers exist in engines. This is an important point for those engaged in multidimensional modeling.

David Hoult discussed application of a rapid distortion theory that suggests substantial augmentation of turbulent intensity just ahead of the flame front. This concept is a novel one that needs experimental verification. That measurement presents the experimentalist with a real challenge.

Despite all the uncertainties surrounding fluid motions, David Gosman did what any good engineer must do—the best he can with the information at hand. He demonstrated the leading edge of today's state of the art, with its limitations carefully placed alongside the insights it offers.

CHEMICAL CONSIDERATIONS

Hank Newhall noted the general success with which chemical submodels could be decoupled from other phenomena and treated separately. He observed that the Zeldovich mechanism seemed an adequate foundation for *NOx* models, but that hydrocarbon emission modeling was in a much sorrier state.

George Lavoie demonstrated the state of the hydrocarbon modeling art. The model is admirably simple but somewhat empirical, involving seven adjustable parameters. It appears to be a valuable teaching tool. Its discrimination among the roles of wall quench, ring-crevice volume, and post-flame oxidation is noteworthy. To achieve this breakdown experimentally is a difficult task. On the other hand, its failure to reproduce hydrocarbon emission trends with changes in engine speed is worrisome and indicates that further refinement is in order.

Ron Herrin presented a global kinetics expression derived from experiment that, once combined with a suitable mixing model, provides better agreement with available measured results than other known techniques. However, discussers warned of dangers inherent in applying a method derived in this manner over too broad a range of hardware configurations.

It seems to me that both the Lavoie and the Herrin papers are examples of constructing the most useful models one can at the time in order to gain understanding now, rather than awaiting the arrival of perfection. To me that is a valid and desirable approach as long as one is wise enough to exercise proper precautions in interpreting results.

Dr. Kirsch appears to have made remarkable progress in unraveling, from a workable analytical standpoint, the mysteries of engine knock. I found it interesting that most of the questions on his paper were raised by attendees from outside the United States, where our stringent emission standards and commitment to unleaded gasoline are not yet rampant.

FUEL SPRAYS

In contrast to the situation in a homogeneous charge, the state of modeling combustion of fuel sprays is both primitive and alarmingly complex. Shahed and his colleagues deserve admiration for assembling a model that has demonstrated its usefulness in engine development. The spray experiments described by Dent and Hiroyasu represent the sort of research needed to put spray-combustion modeling on a more firm footing. Professor Hiroyasu also has extended his experiments to analytical models that time restrictions unfortunately prevented him from showing us, but this work is covered in his written paper.

I would like to endorse Hank Newhall's suggestion that modelers turn some attention to soot modeling in heterogeneous combustion. With the mounting interest in diesel engines as a means for conserving petroleum resources, soot formation is certain to become a greater issue than in the past.

ROLE OF EXPERIMENTS

John Heywood said, "One cannot expect to model something for which one has no physical understanding." That physical understanding comes from experimentation. It is not surprising, therefore, that the need for more experiments was expressed over and over again during this Symposium on modeling. You may recall Hank Newhall's recipe: Add 5 modelers to 95 experimenters and mix thoroughly.

Those experiments must be well planned and well executed. They may be done in an actual engine, or they may be done in a special non-engine apparatus which is often specially constructed to answer a specific question. The "sobering complexity" of engine combustion, as Heywood put it, steers one toward the special apparatus. Indeed, Tony Oppenheim voiced a special plea for intermediate experiments done on such equipment to fill the gap between the analytical engine model and the complex real engine. But Gary Borman appropriately reminded us that only an engine can properly reproduce all the boundary conditions. I cannot help but smile when I recall a previous meeting at which one of these special pieces of apparatus—a single-stroke device, I believe—was referred to as an "engine." That reference drew the objection, "An engine is something from which one obtains useful work."

As a result of this conflict between unmanageable realism in the engine and manageable simplicity in special rigs, I cannot see that either alternative will displace the other. The important thing is that lines of communication between the experimenter and the modeler be established and kept strong. We hope that this Symposium has contributed toward that goal.

C. A. Amann

PARTICIPANTS

Adamczyk, A. A.
Ford Motor Company
Dearborn, Michigan

Agnew, W. G.
General Motors Research Laboratories
Warren, Michigan

Alkidas, A. C.
General Motors Research Laboratories
Warren, Michigan

Amann, C. A.
General Motors Research Laboratories
Warren, Michigan

Arpaci, V. S.
University of Michigan
Ann Arbor, Michigan

Ashurst, W. T.
Sandia Livermore Laboratories
Livermore, California

Asmus, T. W.
Chrysler Corporation
Detroit, Michigan

Ballal, D. R.
Purdue University
West Lafayette, Indiana

Beaman, R. T.
General Motors Research Laboratories
Warren, Michigan

Bechtel, J. H.
General Motors Research Laboratories
Warren, Michigan

Bidwell, J. B.
General Motors Research Laboratories
Warren, Michigan

Blint, R. J.
General Motors Research Laboratories
Warren, Michigan

Blumberg, P. N.
Ford Motor Company
Dearborn, Michigan

Boni, A. A.
Science Applications, Inc.
LaJolla, California

Borman, G. L.
University of Wisconsin
Madison, Wisconsin

Bowman, C. T.
Stanford University
Stanford, California

Bracco, F. V.
Princeton University
Princeton, New Jersey

Bryzik, W.
U. S. Army Tank Automotive Command
Warren, Michigan

Butler, T. D.
Los Alamos Scientific Laboratory
Los Alamos, New Mexico

Butterworth, A. V.
General Motors Research Laboratories
Warren, Michigan

Buzan, L. R.
General Motors Research Laboratories
Warren, Michigan

Caplan, J. D.
General Motors Research Laboratories
Warren, Michigan

Caton, J. A.
General Motors Research Laboratories
Warren, Michigan

Cavendish, J. C.
General Motors Research Laboratories
Warren, Michigan

Cernansky, N. P.
Drexel University
Philadelphia, Pennsylvania

Chenea, P. F.
General Motors Research Laboratories
Warren, Michigan

Chock, D. P.
General Motors Research Laboratories
Warren, Michigan

Chraplyvy, A. R.
General Motors Research Laboratories
Warren, Michigan

Cloutman, L. D.
Los Alamos Scientific Laboratory
Los Alamos, New Mexico

Coon, C. W.
Southwest Research Institute
San Antonio, Texas

Cornelius, W.
General Motors Research Laboratories
Warren, Michigan

Cornetti, G. M.
Fiat
Torino, Italy

Creighton, J. R.
Lawrence Livermore Laboratory
Livermore, California

Daily, J. W.
University of California/Berkeley
Berkeley, California

Daniel, W. A.
General Motors Engineering Staff
Warren, Michigan

DeGregoria, A. J.
Exxon Research and Engineering Company
Linden, New Jersey

DeNagel, S. F.
General Motors Research Laboratories
Warren, Michigan

Dent, J. C.
Loughborough University of Technology
Leicestershire, England

Diwakar, R.
General Motors Research Laboratories
Warren, Michigan

Dyer, T. M.
Sandia Livermore Laboratories
Livermore, California

Eberle, M. K.
Sulzer Brothers, Ltd.
Winterthur, Switzerland

Eyzat, P.
Institut Francais du Pétrole
Rueil-Malmaison, France

Fansler, T. D.
General Motors Research Laboratories
Warren, Michigan

Fendell, F. E.
TRW Systems
Redondo Beach, California

Ferguson, C. R.
Purdue University
West Lafayette, Indiana

Fleming, J. D.
General Motors Research Laboratories
Warren, Michigan

Flynn, P. F.
Cummins Engine Company
Columbus, Indiana

Frey, W. H.
General Motors Research Laboratories
Warren, Michigan

Fry, D. L.
General Motors Research Laboratories
Warren, Michigan

Gallopoulos, N. E.
General Motors Research Laboratories
Warren, Michigan

Gardels, K. D.
General Motors Research Laboratories
Warren, Michigan

Gast, R. A.
General Motors Research Laboratories
Warren, Michigan

Genslak, S. L.
General Motors Engineering Staff
Warren, Michigan

Gibeling, H. J.
Scientific Research Associates, Inc.
Glastonbury, Connecticut

Gosman, A. D.
Imperial College of Science & Technology
London, England

Gouldin, F. C.
Cornell University
Ithaca, New York

Goulish, J. N.
Buick Motor Division, GMC
Flint, Michigan

Griffiths, R. H.
General Motors Research Laboratories
Warren, Michigan

Groff, E. G.
General Motors Research Laboratories
Warren, Michigan

Gumaer, J. I.
Chrysler Corporation
Detroit, Michigan

Gupta, H. C.
Princeton University
Princeton, New Jersey

Hammond, Jr., D. C.
General Motors Research Laboratories
Warren, Michigan

Harrington, D. L.
General Motors Research Laboratories
Warren, Michigan

Harris, S. J.
General Motors Research Laboratories
Warren, Michigan

Hartley, D. L.
Sandia Livermore Laboratories
Livermore, California

Haselman, L. C.
Lawrence Livermore Laboratory
Livermore, California

Heffner, F. E.
General Motors Research Laboratories
Warren, Michigan

Henein, N. A.
Wayne State University
Detroit, Michigan

Herrin, R. J.
General Motors Research Laboratories
Warren, Michigan

Heywood, J. B.
Massachusetts Institute of Technology
Cambridge, Massachusetts

Hilden, D. L.
General Motors Research Laboratories
Warren, Michigan

Hilke, L. C.
Pontiac Motor Division, GMC
Pontiac, Michigan

Hiroyasu, H.
University of Hiroshima
Hiroshima, Japan

Hittler, D. L.
American Motors Corporation
Detroit, Michigan

Hoegberg, T. A.
AB Volvo
Gothenburg, Sweden

Hornbeck, R.
Chevrolet Motor Division, GMC
Warren. Michigan

Hoult, D. P.
Massachusetts Institute of Technology
Cambridge, Massachusetts

Hutchinson, P.
AERE Harwell
Oxfordshire, England

Jamerson,, F. E.
General Motors Research Laboratories
Warren, Michigan

Johnson, R. H.
Cadillac Motor Car Division, GMC
Detroit, Michigan

Kadlec, R. H.
University of Michigan
Ann Arbor, Michigan

Kamal, M. M.
General Motors Research Laboratories
Warren, Michigan

Karim, G. A.
University of Calgary
Calgary Alberta, Canada

Kauffman, C. W.
University of Michigan
Ann Arbor, Michigan

Keck, J. C.
Massachusetts Institute of Technology
Cambridge, Massachusetts

Kempke, Jr., E. E.
NASA-Lewis Research Center
Cleveland, Ohio

Khan, I. M.
Renault
Cedex, France

Kirsch, L. J.
Shell Research, Ltd.
London, England

Klomp, E. D.
General Motors Research Laboratories
Warren, Michigan

Kowalski, M. F.
Chevrolet Motor Division, GMC
Warren, Michigan

Krieger, R. B.
General Motors Research Laboratories
Warren, Michigan

Lancaster, D. R.
General Motors Research Laboratories
Warren, Michigan

Lavoie, G. A.
Ford Motor Company
Dearborn, Michigan

Lawson, C. K.
Chevrolet Motor Division, GMC
Warren, Michigan

Lestz, S. S.
Pennsylvania State University
University Park, Pennsylvania

Lienesch, J. H.
General Motors Research Laboratories
Warren, Michigan

LoRusso, J. A.
Ford Motor Company
Dearborn, Michigan

Lyn, W. T.
Cummins Engine Company
Columbus, Indiana

Malik, M. J.
General Motors Engineering Staff
Warren, Michigan

Mansour, N. N.
General Motors Research Laboratories
Warren, Michigan

Marks, C.
General Motors Engineering Staff
Warren, Michigan

Matekunas, F. A.
General Motors Research Laboratories
Warren, Michigan

Mattavi, J. N.
General Motors Research Laboratories
Warren, Michigan

Matthes, W. R.
General Motors Research Laboratories
Warren, Michigan

McDonald, G. C.
General Motors Research Laboratories
Warren, Michigan

McDonald, H.
Scientific Research Associates, Inc.
Glastonbury, Connecticut

McDonald, R. J.
General Motors Research Laboratories
Warren, Michigan

Mellor, A. M.
Purdue University
West Lafayette, Indiana

Monaghan, M. L.
Ricardo Consulting Engineers, Ltd.
Sussex, England

Montalenti, U.
National Research Council of Italy
Torino, Italy

Morel, T. A.
General Motors Research Laboratories
Warren, Michigan

Muench, N. L.
General Motors Research Laboratories
Warren, Michigan

Murray, J. J.
U. S. Army Research Office
Research Triangle, North Carolina

Myers, J. P.
General Motors Research Laboratories
Warren, Michigan

Nefske, D. J.
General Motors Research Laboratories
Warren, Michigan

Newhall, H. K.
Chevron Research Company
Richmond, California

Novak, J. M.
Ford Motor Company
Dearborn, Michigan

Noyes, R. N.
General Motors Research Laboratories
Warren, Michigan

Oppenheim, A. K.
University of California/Berkeley
Berkeley, California

Patterson, D. J.
University of Michigan
Ann Arbor, Michigan

Peters, B. D.
General Motors Research Laboratories
Warren, Michigan

Pischinger, F. F.
Institute of Applied Thermodynamics
Aachen, Germany

Plee, S. L.
General Motors Research Laboratories
Warren, Michigan

Pozniak, D. J.
General Motors Research Laboratories
Warren, Michigan

Pratt, D. T.
University of Michigan
Ann Arbor, Michigan

Quader, A. A.
General Motors Research Laboratories
Warren, Michigan

Ramshaw, J. D.
Los Alamos Scientific Laboratory
Los Alamos, New Mexico

Rask, R. B.
General Motors Research Laboratories
Warren, Michigan

Reuss, D. L.
General Motors Research Laboratories
Warren, Michigan

Reynolds, W. C.
Stanford University
Stanford, California

Rife, J. M.
Massachusetts Institute of Technology
Cambridge, Massachusetts

Robertson, G. F.
General Motors Research Laboratories
Warren, Michigan

Robinson, C. W.
Sandia Livermore Laboratories
Livermore, California

Rostenbach, R. E.
National Science Foundation
Washington, D.C.

Sanders, B. R.
Sandia Livermore Laboratories
Livermore, California

Schilke, N. A.
General Motors Research Laboratories
Warren, Michigan

Schoene, A. Y.
General Motors Research Laboratories
Warren, Michigan

Schwing, R. C.
General Motors Research Laboratories
Warren, Michigan

Shahed, S. M.
Cummins Engine Company
Columbus, Indiana

Sheridan, D. C.
General Motors Research Laboratories
Warren, Michigan

Shier, R. K.
Cadillac Motor Car Division, GMC
Detroit, Michigan

Siegla, D. C.
General Motors Research Laboratories
Warren, Michigan

Siewert, R. M.
General Motors Research Laboratories
Warren, Michigan

Sinnamon, J. F.
General Motors Research Laboratories
Warren, Michigan

Sirignano, W. A.
Princeton University
Princeton, New Jersey

Skellenger, G. D.
General Motors Research Laboratories
Warren, Michigan

Sloane, T. M.
General Motors Research Laboratories
Warren, Michigan

Smith, G. E.
University of Michigan
Ann Arbor, Michigan

Sorenson, S. C.
University of Illinois
Urbana, Illinois

Sovran, G.
General Motors Research Laboratories
Warren, Michigan

Speck, C. E.
General Motors Research Laboratories
Warren, Michigan

Springer, G. S.
University of Michigan
Ann Arbor, Michigan

Stebar, R. F.
General Motors Research Laboratories
Warren, Michigan

Steiner, J. C.
General Motors Research Laboratories
Warren, Michigan

Steinhilper, E. A.
General Motors Research Laboratories
Warren, Michigan

Stephens, T. G.
Cadillac Motor Car Division, GMC
Detroit, Michigan

Stivender, D. L.
General Motors Research Laboratories
Warren, Michigan

Tabaczynski, R. J.
Ford Motor Company
Dearborn, Michigan

Teets, R. E.
General Motors Research Laboratories
Warren, Michigan

Thoreson, T. R.
General Motors Research Laboratories
Warren, Michigan

Thurston, K. W.
Oldsmobile Division, GMC
Lansing, Michigan

Toepel, R. R.
General Motors Research Laboratories
Warren, Michigan

Tracy, J. C.
General Motors Research Laboratories
Warren, Michigan

Tuteja, A. D.
Detroit Diesel Allison Division, GMC
Detroit, Michigan

Tuttle, J. H.
General Motors Research Laboratories
Warren, Michigan

Vickers, P. T.
General Motors Research Laboratories
Warren, Michigan

Victoria, K. J.
Science Applications, Inc.
LaJolla, California

Wallace, T. F.
Buick Motor Division, GMC
Flint, Michigan

Wang, Neng-Ming
General Motors Research Laboratories
Warren, Michigan

Watson, H. C.
University of Melbourne
Melbourne, Australia

Wendland, D. W.
General Motors Research Laboratories
Warren, Michigan

Wentworth, J. T.
General Motors Research Laboratories
Warren, Michigan

Westbrook, C. K.
Lawrence Livermore Laboratory
Livermore, California

Wicke, B. G.
General Motors Research Laboratories
Warren, Michigan

Williams, F. A.
University of California/San Diego
LaJolla, California

Witze, P. O.
Sandia Livermore Laboratories
Livermore, California

Wong, V. W.
Cummins Engine Company
Cambridge, Massachusetts

Woods, W. A.
University of Liverpool
Liverpool, England

Woodwark, P. R.
Lawrence Livermore Laboratory
Livermore, California

Young, M. B.
General Motors Research Laboratories
Warren, Michigan

SUBJECT INDEX

GPSR Compliance

The European Union's (EU) General Product Safety Regulation (GPSR) is a set of rules that requires consumer products to be safe and our obligations to ensure this.

If you have any concerns about our products, you can contact us on ProductSafety@springernature.com

In case Publisher is established outside the EU, the EU authorized representative is:

Springer Nature Customer Service Center GmbH
Europaplatz 3
69115 Heidelberg, Germany

Batch number: 10838292

Printed by Printforce, the Netherlands